Ecosystem and Species Habitat Modeling for Conservation and Restoration

Shalini Dhyani • Dibyendu Adhikari •
Rajarshi Dasgupta • Rakesh Kadaverugu
Editors

Ecosystem and Species Habitat Modeling for Conservation and Restoration

Springer

Editors
Shalini Dhyani
Critical Zone Research Group, Water Technology and Management Division
CSIR-NEERI
Nagpur, India

Dibyendu Adhikari
Plant Ecology & Climate Change Science
CSIR-National Botanical Research Institute
Lucknow, India

Rajarshi Dasgupta
School of Public Policy
Indian Institute of Technology
Delhi, India

Rakesh Kadaverugu
Cleaner Technology and Modeling Division
CSIR-National Environmental Engineering Research Institute
Nagpur, India

ISBN 978-981-99-0130-2 ISBN 978-981-99-0131-9 (eBook)
https://doi.org/10.1007/978-981-99-0131-9

This Springer imprint is published by the registered company Springer Nature Singapore Pte Ltd.
The registered company address is: 152 Beach Road, #21-01/04 Gateway East, Singapore 189721, Singapore

Foreword

Origins

The field of distributional ecology revolves around the question of why a species is where it is, and why is the species not where it is. This question may seem simple, in the sense that ranges of species have been a central focus in biology for more than a century. Now, for many species, it is quite easy to find a range map, or occurrence data, or some source of information about the "where" question. So why should answering this question be so difficult?

These challenges of obtaining high-quality distributional information for species have been approached using myriad frameworks and tools. Early approaches centered on so-called habitat suitability modeling (e.g., Scott et al. 1996), and saw application of several multivariate statistical approaches to the question (e.g., Austin et al. 1990). A next generation of progress brought the many advantages (and disadvantages) of machine learning approaches (Stockwell and Peters 1999; Phillips et al. 2006), which had the in hindsight-dubious quality of being able to fit more complex response types (Elith et al. 2006).

Curiously, at least in my own humble opinion, a next (and key) major advance was that of bringing a more rigorous conceptual underpinning to distributional ecology. Building on the foundational work of Grinnell (Grinnell 1917a, b) and Hutchinson (Hutchinson 1957, 1978) and a few subsequent authors (Austin 1987; Austin et al. 1990; Austin et al. 1994; Pulliam 2000), a consortium of authors published a first conceptual framework for the field (Peterson et al. 2011). Based on this framework, many additional advances became possible, such as comprehending the importance of accessible areas in fitting models (Barve et al. 2011; Machado-Stredel et al. 2021), establishing appropriate approaches for evaluating models (Peterson et al. 2008), etc.

Key Elements in the Process

The typical niche modeling application is a multi-step process, which (to be honest) is not laid out in any standard methodology in any one textbook or standard reference book. Nonetheless, it is generally a process of (1) assembling point occurrence data for the species in question, (2) assembling relevant environmental data for the region in question, (3) choosing a region over which to fit models, (4) actually fitting the model, and (5) post-processing and interpretation of model outputs to respond to the questions that were originally posed. These steps have been achieved via standard statistical tools (Guisan et al. 2017), modular sets of tools (Cobos et al. 2019), or via standalone platforms that package many or all of the necessary steps (Kass et al. 2018). Beyond these basics, however, a few points remain important to emphasize, as follows.

A key emphasis is on the use of *primary* biodiversity data—i.e., data that document the presence of an individual of a species at a particular place at a particular time—as the basis for these modeling efforts. Although it is certainly tempting to appeal to easier and more readily available sources of occurrence information, such as range maps or atlas summaries, use of secondary sources of biodiversity data for model inputs introduces significant noise into the results. In essence, the primary occurrence data and the environmental data should go hand in hand in terms of spatial grain and resolution, such that neither is too general, and such that discords and mismatches are not pervasive in the modeling effort. The subjectivity introduced by secondary data sources is an additional source of uncertainty and confounding effects for the models, such that important features of the distributional ecology may be lost from the analysis.

A further emphasis should be on the use of methods that are quantitative, repeatable, scalable, and portable, at all points in the process. Although many present-day analyses simply provide general, text-based descriptions of methodological steps, program code (e.g., in R) or full-blown workflows can now be developed or implemented that make the methodology entirely portable, transferrable, and scalable. The code can be shared as part of the publication process, which makes the methodology repeatable, and ready for application in any other analysis by any investigator.

Challenges

Although this methodology has now been used in thousands of analyses and thousands of published papers, its development is not complete. That is, a number of advances remain to be explored and documented, so that the approach is as maximally informative and useful as is possible. The following are several such areas that remain under exploration and development, but that can certainly be seen as fruitful areas for future research.

Estimate the Right Sort of Object Fundamental ecological niches are likely to be relatively simple, convex objects in environmental space (Maguire 1973). Nonetheless, the methods in vogue currently in distributional ecology often estimate objects that are quite a bit more complex, with gaps, holes, and infoldings—in this sense, workers in this field are using inappropriate tools for the task. As such, an important step forward will be to develop and use tools that estimate objects that "look like" fundamental ecological niches, and are simple and convex, and that do not have bimodal environmental responses, or any other such complexities. Some initial steps have been taken toward such a methodology (Jiménez et al. 2019; Jiménez and Soberón 2022), but much work remains to be done.

Use the Right Environmental Information Workers in the field of distributional ecology have long used environmental information in the form of long-term average values to characterize species' occurrences in terms of the environments that are manifested at the site of occurrence. It is well known, however, that an average can be a poor representation of the conditions at any particular moment, and an individual or a population can be extinguished with even a short period of time spent under unsuitable conditions. As such, recent research efforts (Ingenloff and Peterson 2020) have explored the potential for representing environmental conditions associated with occurrences of species as a function of latitude, longitude, *and time*, such that conditions specific to an occurrence are identified more precisely.

Consider Dispersal Ecological niche models, if done well, present a view of the area that is suitable for a species in terms of abiotic conditions (note that the next item in this list refers to the question of suitability in biotic terms). A crucial consideration, however, is that the ability of the species to access those suitable sites is not generally considered. As a consequence, too often, conclusions in ecological niche modeling studies are based on rather simple assumptions about dispersal ability (e.g., no dispersal or universal dispersal). A few efforts have now been made to incorporate dispersal processes more powerfully into these methods (Engler and Guisan 2009; Machado-Stredel et al. 2021), but applications have been relatively few—adding this component into modeling efforts and interpretations is crucial to making this methodology and the resulting conclusions more powerful.

Incorporate Biotic Interactions A further dimension that is too often left out of ecological niche model-based studies is that of biotic dimensions—in essence, the set of biotic considerations that makes a site suitable or unsuitable for a species. As has been pointed out in several conceptual treatments, consideration of the full dynamics of the broad suite of potential biotic interactors for any given species may prove to be impossible. Nonetheless, it is feasible to incorporate at least known interactor species in two- or multi-species models (e.g., Anderson 2017; Ashraf et al. 2021), and network analysis approaches may be relevant to identifying such interactor species more rigorously (Fath et al. 2007).

This Book

This volume, entitled *Ecosystem and Species Modeling for Conservation and Restoration: Mainstreaming Modeling Approaches in Policy Planning*, comprises a set of papers that revolve around models of ecosystems and species, and their niches and distributions, in the context of guiding policy. Although I have not yet had the opportunity to read each of the contributions, the list of titles, topics, and authors is impressive—this volume will create a rich picture of the state of the field and will illustrate many of the possible applications of this methodology. As a consequence, I am so very pleased to have been invited to preface the volume with a few thoughts, ideas, and comments.

University of Kansas Biodiversity Institute A. Townsend Peterson
Lawrence, KS, USA 28 October 2022

Literature Cited

Anderson RP (2017) When and how should biotic interactions be considered in models of species niches and distributions? J Biogeogr 44:8–17

Ashraf U, Chaudhry MN, Peterson AT (2021) Ecological niche models of biotic interactions predict increasing pest risk to olive cultivars with changing climate. Ecosphere 12:e03714

Austin M (1987) Models for the analysis of species' response to environmental gradients. Vegetatio 69:35–45

Austin MP, Nicholls AO, Doherty MD, Meyers JA (1994) Determining species response functions to an environmental gradient by means of a beta-function. J Veg Sci 5:215–228

Austin MP, Nicholls AO, Margules CR (1990) Measurement of the realized qualitative niche: environmental niches of five *Eucalyptus* species. Ecol Monogr 60: 161–177

Barve N, Barve V, Jimenez-Valverde A, Lira-Noriega A, Maher SP, Peterson AT, Soberón J, Villalobos F (2011) The crucial role of the accessible area in ecological niche modeling and species distribution modeling. Ecol Model 222:1810–1819

Cobos ME, Peterson AT, Barve N, Osorio-Olvera L (2019) kuenm: an R package for detailed development of ecological niche models using Maxent. PeerJ 7:e6281

Elith J, Graham C, Anderson RP, Dudík M, Ferrier S, Guisan A, Hijmans RJ, Huettmann F, Leathwick JR, Lehmann A, Li J, Lohmann LG, Loisell BA, Manion G, Moritz C, Nakamura M, Nakazawa Y, Overton J, Peterson AT, Phillips SJ, Richardson K, Scachetti-Pereira R, Schapire E, Soberón J, Williams S, Wisz MS, Zimmerman NE (2006) Novel methods improve prediction of species' distributions from occurrence data. Ecography 29:129–151

Engler R, Guisan A (2009) MigClim: predicting plant distribution and dispersal in a changing climate. Divers Distrib 15:590–601

Fath BD, Scharler UM, Ulanowicz RE, Hannon B (2007) Ecological network analysis: network construction. Ecol Model 208:49–55

Grinnell J (1917a) Field tests of theories concerning distributional control. Am Nat 51:115–128

Grinnell J (1917b) The niche-relationships of the California Thrasher. Auk 34:427–433

Guisan A, Thuiller W, Zimmermann NE (2017) Habitat suitability and distribution models: with applications in R. Cambridge University Press, Cambridge

Hutchinson GE (1957) Concluding remarks. Cold Spring Harb Symp Quant Biol 22: 415–427

Hutchinson GE (1978) An introduction to population ecology. Yale University Press, New Haven

Ingenloff K, Peterson AT (2020) Incorporating time into the traditional correlational distributional modeling framework: a proof-of-concept using the Wood Thrush (*Hylocichla mustelina*). Methods Ecol Evol 12:311–321

Jiménez L, Soberón J (2022) Estimating the fundamental niche: accounting for the uneven availability of existing climates in the calibration area. Ecol Model 464: 109823

Jiménez L, Soberón J, Christen JA, Soto D (2019) On the problem of modeling a fundamental niche from occurrence data. Ecol Model 397:74–83

Kass JM, Vilela B, Aiello-Lammens ME, Muscarella R, Merow C, Anderson RP (2018) Wallace: a flexible platform for reproducible modeling of species niches and distributions built for community expansion. Methods Ecol Evol 9:1151–1156

Machado-Stredel F, Cobos ME, Peterson AT (2021) A simulation-based method for selecting calibration areas for ecological niche models and species distribution models. Front Biogeogr 13:e48814

Maguire B (1973) Niche response structure and the analytical potentials of its relationship to the habitat. Am Nat 107:213–246

Peterson AT, Papeş M, Soberón J (2008) Rethinking receiver operating characteristic analysis applications in ecological niche modelling. Ecol Model 213:63–72

Peterson AT, Soberón J, Pearson RG, Anderson RP, Martínez-Meyer E, Nakamura M, Araújo MB (2011) Ecological niches and geographic distributions. Princeton University Press, Princeton

Phillips SJ, Anderson RP, Schapire RE (2006) Maximum entropy modeling of species geographic distributions. Ecol Model 190:231–259

Pulliam HR (2000) On the relationship between niche and distribution. Ecol Lett 3: 349–361

Scott JM, Tear TH, Davis FW (eds) (1996) Gap analysis: a landscape approach to biodiversity planning. American Society for Photogrammetry and Remote Sensing, Bethesda, MD

Stockwell DRB, Peters DP (1999) The GARP modelling system: problems and solutions to automated spatial prediction. Int J Geogr Inf Sci 13:143–158

Acknowledgments

Without the constant assistance and contributions of numerous devoted expert authors, reviewers, copy as well as language editors, it would not have been feasible to compile a book on the somewhat futuristic subject on Ecosystem and Species Modeling that is relevant in the present context and has been mainstreamed in many international and national decisions to understand plausible future scenarios. Despite all the challenges and limitations, we have put together this important scientific contribution. We owe a debt of gratitude to the numerous subject matter experts, academicians, scientists, researchers, and practitioners who agreed to contribute, sacrificed their time away from their grueling academic obligations to write as well as many others who facilitated a thorough review of the submitted book chapters within constrained time frames. Many of our diligent and committed research academics also deserve our sincere gratitude as without their prompt assistance, our efforts to publish this book on time would not have been possible. We would also like to especially thank Mr. Manu Thomas (CSIR-NEERI) for his assistance in overseeing formatting and language correction to avoid significant errors, record keeping of all email interactions prior to the book's completion, and Ms. Radhika Sood (CSIR-NEERI) for formatting a few chapters. Finally, we would also like to express our gratitude to the Springer Nature production team, in particular Mrs. Aakanksha Tyagi and Mr. Jayesh Kalleri, for helping us with our submissions and also being kind with extending the deadline to ensure that we submit our best effort. Despite our best efforts to ensure accuracy, it is still possible that a very small number of mistakes possibly typographical or syntax-related may have crept in because of the quick pace of production and the deadlines we had to adhere to. Chapter authors and co-authors were mostly responsible for finalizing their chapters with active guidance and review comments from editors and reviewers. The editors have read through all of the chapters and personally reviewed them in accordance with international standards, including publication ethics that specifically aim to reduce plagiarism and resemblance from existing publications. If readers of this book have any helpful criticism or ideas to help us improve our future releases, we would be eternally grateful to all of them. The editors and chapter authors are jointly responsible for any inaccuracies that may be present in this book. We express our gratitude to our researchers and families, especially our spouse and children in particular, for their support, understanding, and encouragement as we continue and successfully complete this mammoth task.

We conclude by expressing our deep gratitude to Prof. Townsend Peterson, University of Kansas Biodiversity Institute Lawrence, Kansas, for authoring the foreword for this book.

Contents

Editors and Contributors

About the Editors

Shalini Dhyani is a Senior Scientist with the Critical Zone Group of Water Technology and Management Division of CSIR-NEERI, India. She is a seasoned ecologist with two decades of experience. She uses observational, empirical, and modeling approaches to investigate and understand issues related to the environment, loss of natural and urban greenspaces, and interlinkages between ecological and social systems through sustainability science approaches. She is Asia Vice Chair member of CEM (Commission on Ecosystems Management) and also Steering Committee member.

Dibyendu Adhikari is a Principal Scientist at CSIR-National Botanical Research Institute (NBRI), India. He is a seasoned researcher with over 15 years of experience in ecology and environmental science. He is skilled in terrestrial ecosystem restoration, threatened plant conservation, forest carbon assessment, ecological data analysis and modeling.

Rajarshi Dasgupta is Assistant Professor at IIT Delhi. He was previously with the Institute for Global Environmental Strategies (IGES), Kanagawa, Japan. He holds diverse research interests in the field of landscape ecology and planning, which include Ecosystem-based Disaster Risk Reduction (Eco-DRR), spatial quantification of ecosystem services, land change simulation, development of socio-ecological scenarios, participatory conservation, and social forestry.

Rakesh Kadaverugu is a Senior Scientist associated with CSIR-National Environmental Engineering Research Institute. He has more than 10 years of research experience in environmental systems modeling and his is work is focused to better understand the socio-environmental systems at multiple spatial and temporal scales using geospatial, soft-computing, and process-based modeling approaches.

Contributors

Rovshan Abbasov is the head of Khazar University's Department of Geography and Environment. He is a member of the IPBES's Multidisciplinary Expert Panel and ICOMOS National Committee and Coordinating Lead Author in the National Ecosystem Assessments.

C. S. Abhijitha is from WWF, India, with proficiency in the research areas related to biodiversity, conservation and geospatial applications.

Santoshkumar Abujam is currently working as a Research Associate under a DST-funded research project in Rajiv Gandhi University, Rono Hills, Doimukh, Arunachal Pradesh (India).

Prasannajit Acharya is a Ph.D. Scholar at the Institute of Technical Education and Research, Department of Chemistry, Siksha 'O' Anusandhan (Deemed to be University), Bhubaneswar, Odisha, India.

Dibyendu Adhikari works as a Principal Scientist in the Plant Ecology and Climate Change Science Division at CSIR-National Botanical Research Institute, Lucknow.

Peerzada Ishtiyak Ahmad is Assistant Professor cum Scientist at Faculty of Forestry, Sher-e-Kashmir University of Agricultural Sciences & Technology of Kashmir, India.

Mohammad Alizadeh-Noughani is a graduate from Ferdowsi University of Mashhad-Iran.

G. Areendran works at WWF, India, and is addressing conservation issues occurring in various landscapes, with a high focus on tiger, elephant, and rhinoceros to overseeing several institutional GIS-based projects.

A. Arunachalam is the Director of the ICAR-Central Agroforestry Research Institute, Jhansi. He is also Project Coordinator for the All India Coordinated Research Project on Agroforestry, and Task Force Coordinator for the National Mission for Sustaining Himalayan Ecosystem (NMSHE).

A. Arya is pursuing her Ph.D. in Bits Pilani G BITS, Pilani, Goa campus.

Ratul Baishya is associated with the Department of Botany, University of Delhi, as an Associate Professor.

Aritra Bandopadhyay is pursuing his Ph.D. in Climate Change and Ecology in the Department of Humanities and Social Science in BITS-Pilani, Goa Campus.

Harish Barewar is currently working as project Associate-II in CSIR-National Environmental Engineering and Research Institute (NEERI), Nagpur.

Rajesh Biniwale is a Senior Principal Scientist in CSIR National Environmental Engineering Research Institute, Nagpur.

Milben A. Bragais is a Licensed Forester and Environmental Planner with a background in hydrological modeling, watershed characterization, and vulnerability assessment.

Manish Kuntal Buragohain is currently working as a Project Associate-I in CSIR-National Environmental Engineering and Research Institute (NEERI), Nagpur.

Dhoni Bushi is currently pursuing a Ph.D. in the Department of Geography, Rajiv Gandhi University, Arunachal Pradesh.

Joseph G. Campang is an Assistant Professor at the Institute of Biological Sciences (IBS) of University of the Philippines Los Baños (UPLB) with research specialisation on lakes.

Yves Christian L. Cabillon with a bachelor's degree is an experienced researcher in water quality assessment, plankton taxonomy, environmental impact assessment, and environmental awareness.

Bipin Charles is Consultant at the Institute for Biodiversity Conservation and Training, Bangalore, India.

Sudipto Chatterjee is an Associate Professor at TERI School of Advanced Studies. He is the Principal Investigator of this study supported by UNU ProsperNet. He is the Co-Lead for the South Asia Nitrogen Hub (SANH) project on the study of impacts of pollution on biodiversity with Lichens as an indicator.

Rajiv Kumar Chaturvedi is an Assistant Professor at BITS, Pilani, Goa campus, and is an UN expert on GHG inventory in the land-use and forestry sector.

Arun Chettri is working as an Assistant Professor at Sikkim University.

Shahid Ahmad Dar is working as a postdoctoral fellow at Zoological Survey of India, Kolkata.

Mira Das is a Professor at the Institute of Technical Education and Research, Department of Chemistry, Siksha 'O' Anusandhan (Deemed to be University), Bhubaneswar, Odisha, India.

Vedika Dutta received a Master of Science in Environmental Studies and Resource Management at TERI School of Advanced Studies, New Delhi, India.

Shalini Dhyani is a Senior Scientist in CSIR National Environmental Engineering Research Institute, Nagpur.

Piyali Dias has a Master of Science in Biodiversity and Taxonomy of Plants from the University of Edinburgh, UK.

Wisdom M. D. Dlamini is an environmental and geospatial scientist at the University of Eswatini.

Jennifer D. Edrial is a postgraduate from the University of the Philippines Los Baños and is currently a member of the Sustainability Team of Ayala Land, Inc.

Azita Farashi is Associate professor at the Faculty of Natural Resources and Environment, Ferdowsi University of Mashhad, Iran.

Marlon Flores leads the global sector-centered economic valuation approach Targeted Scenario Analysis (TSA) at UNDP's Food and Agricultural Commodity Systems (FACS) Practice.

S. A. Gangoo is Dean at Faculty of Forestry, Sher-e-Kashmir University of Agricultural Sciences & Technology of Kashmir, India.

Anjaly George is a Ph.D. scholar at the Kerala University of Fisheries and Ocean Studies.

Sonali Ghosh is a Senior Indian Forest Service Officer with more than 25 years of work experience at state and center and specialization in wildlife. She has also worked at UNESCO C2C on World Natural Heritage Centre at Wildlife Institute of India.

Shailendra Goel is Professor in the Department of Botany, University of Delhi.

Kiran Hungund is a Project Assistant working on air quality changes and their visualization with the Energy, Environment and Climate Change Programme at National Institute of Advanced Studies, Bangalore.

Tanvi Hussain is a Research Scholar in the Department of Environmental Science, Gauhati University, and presently with World Agroforestry (CIFOR-ICRAF).

Nazrul Islam is pursuing his Ph.D. in Eastern Swamp Deer Ecology in Manas National Park from Gauhati University and working as biologist with Wildlife Trust of India in the Greater Manas landscape.

Jonson M. Javier is an Assistant Professor of Agricultural Economics and Environmental Science at the Western Philippines University.

Aroma Caroline John has completed an M.Sc. in Geographical Information Systems and Remote Sensing.

Karun Jose is pursuing his Ph.D. in Climate Change and Forest Phenology in the Department of Humanities and Social Science in BITS-Pilani, Goa Campus.

Shijo Joseph is an Assistant Professor and Coordinator of the Centre for Climate Resilience and Environment Management at the Kerala University of Fisheries and Ocean Studies, Kochi.

Ashok Kadaverugu is Head of Civil Engineering department in a Government Polytechnic college, Nalgonda, Telangana.

Rakesh Kadaverugu is a Senior Scientist in CSIR National Environmental Engineering Research Institute, Nagpur.

Sarbeswar Kalita is the Former Head and expert of seismology, climatology, and meteorology from the Department of Environmental Science, Gauhati University.

Sharanjeet Kaur worked as a DBT Senior Project Fellow in the Department of Botany, University of Delhi.

Siddhartha Kaushal is associated as a Ph.D. research student with the Department of Botany, University of Delhi.

B. S. P. C. Kishore is a Ph.D. scholar in the Department of Geoinformatics, Central University of Jharkhand, Ranchi, India.

Amit Kumar is working as an Assistant Professor in the Department of Geoinformatics, Central University of Jharkhand, India.

Gajendra Kumar is a Ph.D. scholar in the Department of Geoinformatics, Central University of Jharkhand, Ranchi, India.

Sudip Kumar Kundu is a Ph.D. scholar at Manipal Academy of Higher Education (MAHE), Bengaluru.

Bibhuti Prasad Lahkar is a Senior Scientist with Aaranyak, Assam. He has extensively researched grassland ecosystems in the Indian state of Assam.

Suvha Lama is currently working as a Scientist in CSIR-National Environmental Engineering and Research Institute (NEERI), Nagpur.

Felino P. Lansigan is a Professor of Statistics at the Institute of Statistics, UPLB. He specializes in stochastic modeling and crop insurance under climate change.

Jeoffrey M. Laruya is from College of Forestry and Natural Resources, University of the Philippines Los Baños, Laguna, Philippines.

Linda Loffler is an independent consultant with vast experience in botanical research in Southern Africa.

Nethanel Jireh A. Larida holds a Master of Science in Botany degree from the University of the Philippines at Baños (UPLB) and works on diversity of aquatic macrophytes and riparian diversity in freshwater lakes.

Damasa B. Magcale-Macandog is a Professor of Plant Ecology at the Institute of Biological Sciences, College of Arts and Sciences, University of the Philippines Los Baños.

Kristina S. Mago was a Student Assistant at the Ecoinformatics Laboratory of the Institute of Biological Sciences, University of the Philippines Los Baños.

Ranjit Mahato is currently pursuing a Ph.D. in the Department of Geography, Rajiv Gandhi University, Arunachal Pradesh.

Marc Bryan Manlubatan is a Research Associate in the NRCP-Funded project with a degree in Agricultural Engineering, University of the Philippines Los Baños.

Teodorico L. Marquez Jr is Student Assistant at the Ecoinformatics Laboratory of the Institute of Biological Sciences at the University of the Philippines Los Baños.

T. H. Masoodi is Registrar at Sher-e-Kashmir University of Agricultural Sciences & Technology of Kashmir, India.

Abhinav Mehta is CEO at the Geographic Information Lab (TGIS), Gujarat.

Pradipta R. Muduli is presently working as a Scientific Officer at the Wetland Research and Training Centre (WRTC), Chilika Development Authority (CDA), Dept. of Forest and Environment, Govt. of Odisha, India.

Tahir Mushtaq is Assistant Professor cum Scientist at Faculty of Forestry, Sher-e-Kashmir University of Agricultural Sciences & Technology of Kashmir, India.

Jerry Naayos is Field Guide in the research and survey experiments in Banaue.

Anukul Nath is serving at UNESCO C2C at the Wildlife Institute of India with an extensive experience of work in the Manas landscape.

Gibji Nimasow is Professor in the Department of Geography, Rajiv Gandhi University, Rono Hills, Doimukh, Arunachal Pradesh (India).

Oyi Dai Nimasow is Assistant Professor in the Department of Geography, Rajiv Gandhi University, Rono Hills, Doimukh, Arunachal Pradesh (India).

Mir Muskan Un Nisa is a M.Sc. Forestry student at Faculty of Forestry, Sher-e-Kashmir University of Agricultural Sciences & Technology of Kashmir, India

Sarath Nissanka is a Professor in Crop Science at the University of Peradeniya, Govt of Sri Lanka. Sarath implements the SANH project in Sri Lanka.

Jaderick P. Pabico is a Professor of Computer Science at the Institute of Computer Science, UPLB.

Vasundhara Pandey has a Master of Science in Environmental Studies and Resource Management at TERI School of Advanced Studies, New Delhi, India.

Ma. Grechelle Lyn D. Perez is Master's in Environmental Science at the University of the Philippines, Los Baños, Laguna, Philippines and affiliated with the University of the Philippines Rural High School handling biology, earth science, field study courses, and capstone research.

John Vincent R. Pleto is an Assistant Professor at the Institute of Biological Sciences, College of Arts and Sciences, University of the Philippines Los Banos and specializes in freshwater ecology.

Randy Porciocula is administrative staff of the NRCP-Funded project at the University of the Philippines Los Baños.

Aditya Pradhan is currently working as an Assistant Professor at SRM University Sikkim.

K. Preeti has a Master of Science in Environmental Studies and Resource Management at TERI School of Advanced Studies, New Delhi, India.

Sarena Grace L. Quiñones is a Development Communication graduate from the University of the Philippines Los Baños (UPLB).

Marc Anthony F. Rabena is an Assistant Professor at the Institute of Biological Sciences, College of Arts and Sciences, University of the Philippines Los Baños.

Krishna Raj is a Specialist in Geospatial Application and has done a Master's in Geography from Banaras Hindu University, Varanasi, India, and has been working in various major research projects of WWF-India.

Shrey Rakholia is Consultant at the Geographic Information Lab (TGIS), Ahmedabad, Gujarat.

Hemanthi Ranasinghe is a Senior Professor in Forestry and Environmental Science at the University of Jayewardenepura, Govt of Sri Lanka.

Kumar Ranjan has completed his postgraduation in GIS and Remote Sensing, and his expertise lies in varied fields of allied disciplines of geospatial analysis.

Kottapalli Sreenivasa Rao is Senior Professor and former Head, Department of Botany, University of Delhi.

Marlon A. Reblora is the Research Associate at the University of the Philippines Los Baños.

Mohan Reddy is Lead in Carbon & Sustainability Division at Nurture Agtech Private Limited, Bellandur, Bengaluru-560103, India.

Uttam Kumar Sahoo is a Professor at the Department of Forestry, School of Earth Sciences and Natural Resource Management, Mizoram University.

Purabi Saikia is working as an Assistant Professor at the Central University of Jharkhand, Ranchi.

Arnold R. Salvacion is an Associate Professor at the Department of Community and Environmental Resource Planning, College of Human Ecology, University of the Philippines Los Baños.

R. Sanil is currently working as an Associate Professor at the Department of Zoology, Government Arts College, Udhagamandalam, India.

Harini Santhanam is an Associate Professor and Head, Department of Public Policy (DPP), Manipal Academy of Higher Education (MAHE) Manipal, Bengaluru Campus, Bengaluru, India.

Thekke Thumbath Shameer is working as a scientist (APO projects) at the Centre for Conservation Education Advanced Institute for Wildlife Conservation, Tamil Nadu Forest Department.

Jahnavi Sharma is an independent researcher working on Energy and Environment policy and is based in Bangalore.

Prachi Sharma is working as a Ph.D. research student in the Department of Botany, University of Delhi.

Shilky is a Ph.D. scholar in the Department of Environmental Sciences, Central University of Jharkhand, Ranchi, India.

Alolika Sinha is Senior Wildlife Biologist with Aarnyak, Assam, with a Ph.D. in endangered cervid—the hog deer.

Anshu Siwach is pursuing her Ph.D. in the Department of Botany, University of Delhi.

P. A. Sofi is Associate Professor at the Faculty of Forestry, Sher-e-Kashmir University of Agricultural Sciences & Technology of Kashmir, India.

Radhika Sood is AcSIR PhD at CSIR-NEERI, Nagpur working on ecosystem health and mapping cultural ecosystem services.

Sundaram Suresh Ramanan is a scientist at ICAR-Central Agroforestry Research Institute, Jhansi.

Rajesh Tandon is a senior professor in the Department of Botany, University of Delhi.

Malsha Thejani is doing her BSc at the University of Jayewardenepura, Govt of Sri Lanka.

Keshia N. Tingson is an Assistant Professor at the College of Forestry and Natural Resources, University of the Philippines Los Baños.

Prem Lal Uniyal is Senior Professor in the Department of Botany, University of Delhi.

Kalidas Upadhyaya currently works at the Department of Forestry, Mizoram University. His research interests are agroforestry, soil ecology, and restoration ecology.

S. Varshini is a Project Assistant with the Energy, Environment and Climate Change Programme at National Institute of Advanced Studies (NIYAS), Bangalore.

Buddhika Weerakoon is a Ph.D. scholar at the University of Peradeniya, Govt of Sri Lanka.

Macrina T. Zafaralla is a Professor Emeritus at the University of the Philippines Los Baños. She specializes in freshwater ecology and phytoplankton.

Abbreviations

3-PG	Physiological Processes Predicting Growth
3Rs	Reduce, Reuse, Recycle
AAQMS	Ambient Air Quality Monitoring Station
ABA	Abscisic acid
ACF	Autocorrelation Function
ADB	Asian Development Bank
AEKOS	Australian Ecological Knowledge and Observation System
AHP	Analytic Hierarchy Process
AIC	Akaike Information Criterion
ANN	Artificial Neural Network
ANOVA	Analysis of Variance
AOD	Aerosol Optical Depth
AR4	Fourth Assessment Report
AR5	Fifth Assessment Report
AR6	Sixth Assessment Report
ArcGIS	Aeronautical Reconnaissance Coverage Geographic Information System
ASCII	American Standard Code for Information Interchange
ASTER	Advanced Spaceborne Thermal Emission and Reflection Radiometer
AUC	Area Under the Curve
AZN	Manat (Azerbaijani currency)
BAU	Business as Usual
BBA	Blue to Built-up Area
BCC	Basic Carrying Capacity
BCVs	Bioclimatic Variables
BFAR	Bureau of Fisheries and Aquatic Resources
BGM	Generalized Boosting Model
Bio	Bioclimatic Variables
BIOCLIM	Bioclimatic Models
BNF	Biological Nitrogen Fixation
BoB	Bay of Bengal
BOD	Biological Oxygen Demand

BP	Background Points
BRT	Boosted Regression Trees
BSIP	Birbal Sahni Institute of Palaeosciences
ca.	circa (meaning around)
CAAQMS	Continuous Air Quality Monitoring Stations
CART	Classification and Regression Tree
CASA	Carnegie-Ames-Stanford Approach
CBD	Convention on Biological Diversity
CC	Carrying Capacity
CCi	Actual Carrying Capacity Level
CCimax	Carrying Capacity Limit
CEM	Climate Envelope Model
Chl-a	Chlorophyll-a
CHL-A	Chlorophyll-a
CI	Consistency Index
CLUP	Comprehensive Land Use Plan
CMFRI	Central Marine Fisheries Research Institute
CMIP	Coupled Model Intercomparison Project
CO2	Carbon dioxide
CPCB	Central Pollution Control Board
CPT	Conditional Probability Tables
CR	Consistency Ratio
CR	Critically Endangered
CRD	Completely Randomized Design
CS	Central Sector
C-SDM	Correlative Species Distribution Model
CSi	Carrying Capacity Level
CSI	Consortium for Spatial Information
CSR	Corporate Social Responsibility
CTA	Classification Tree Analysis
CTI	Compound Topography Index
CV	Cross Validations
CWC	Canopy Water Content
CWE	Corrected Taxonomic Weighted Endemism
CWPE	Corrected Weighted Phylogenetic Endemism
DBF	Day Before Fish Kill
DEM	Digital Elevation Model
DENR AO	Department of Environment and Natural Resources
DGVM	Dynamic Global Vegetation Model
DO	Dissolved Oxygen
DOST	Department of Science and Technology
DOT	Department of Tourism
ECC	Ecological Carrying Capacity
EcoSIS	Ecological Spectral Information System
EDA	Exploratory Data Analyses

eDNA	Environmental DNA
ENFA	Environmental Niche Factor Analysis
ENM	Ecological Niche Model
ENM	Ecological Niche Modeling
ERA	European Center for Medium-Range Weather Forecasts Re-Analysis
ES	Ecosystem Services
ESD	Eastern Swamp Deer
ESM	Earth System Model
ESRI	Environmental Systems Research Institute
FAO	Food and Agriculture Organization
FAPAR	Fraction of Absorbed Photosynthetically Active Radiation
FARE	Food Analysis and Research
FARMC	Fisheries and Aquatic Resource Management Council
FDA	Flexible Discriminant Analysis
FF	Favorability Function
FGD	Focus Group Discussion
FHWAR	Fishing, Hunting, & Wildlife-Associated Recreation
FSLF	Friends of the Seven Lakes Foundation, Inc
FvCB	Farquhar, von Caemmerer, and Berry
GA	Gibberellic acid
GAM	Generalized Additive Model
GARP	Genetic Algorithm for the Rule Set Production
GBIF	Global Biodiversity Information Facility
GBM	Gradient Boosting Machine
GBR	Green Blue Ratio
GCM	Global Circulation Model
GCS WGS	Geographical Coordinate System World Geodetic System
GDEM	Global Digital Elevation Model
GDP	Gross Domestic Product
GEF	Global Environmental Facility
GEMS	Global Environment Monitoring System
GHG	Greenhouse Gas
GI	Green Infrastructure
GIOVANNI	Geospatial Interactive Online Visualization and Analysis Infrastructure
GIS	Geographic Information System
GIZ	German International Cooperation
GJAM	Generalized Joint Attribute Model
GLCF	Global Land Cover Facility
GLM	Generalized Linear Model
GLR	Generalized Linear Regression
GNF	Global Nature Fund
GOI	Government of India

GPP	Gross primary production
GPS	Global Positioning System
GSA	Global Sensitivity Analysis
GSI	Geological Survey of India
H+	Hydrogen ion
HCA	Hierarchical Cluster Analysis
HEC	Human-Elephant Conflict
HII	Human Influence Index
HL	Hidden Layers
HN	Hidden Network
HPP	Hydropower Plant
H-SDM	Hybrid Species Distribution Model
IAA	Indole Acetic Acid
IAP	Invasive Alien Plant
IBIS	Integrated Biosphere Simulator
ICAR	Indian Council of Agricultural Research
ICH	Indian Central Himalaya
IHR	Indian Himalayan Region
IL	Input layers
INCOIS	Indian National Centre for Ocean Information Services
INSAT 3D	Indian National Satellite 3D
IPBES	Intergovernmental Platform on Biodiversity and Ecosystem Services
IPCC	Intergovernmental Panel on Climate Change
IRMS	Isotope-Ratio Mass Spectrometer
ISA	Impervious Surface Area
ISRIC	International Soil Reference and Information Centre
IUCN	International Union for Conservation of Nature
JSDM	Joint Species Distribution Model
JULES	Joint U.K. Land Environment Simulator
KAP	Knowledge, Attitudes, Practices
KARB	Kura-Araz river basin
KfW	German government-owned development bank (Reconstruction Credit Institute)
KII	Key Informant Interview
$kJ\ m^{-2}\ day^{-1}$	kilo joules per square meters per day
km^2	square kilometers
KML	Key Markup Language
KMZ	Keyhole Markup Language
KRC	Knowledge Resource Centre
LAI	Leaf Area Index
LD	Land Degradation
LDN	Land Degradation Neutrality
LGU	Local Government Unit
LID	Low Impact Development

LLDA	Laguna Lake Development Authority
LPI	Largest Patch Index
LPJ-DGVM	Lund Potsdam Jena—Dynamic Global Vegetation Model
LPJ-GUESS	Lund Potsdam Jena—General Ecosystem Simulator
LPJmL	Lund-Potsdam-Jena managed Land
LPX	Land Surface Processes and Exchanges
LULC	Land Use Land Cover
m.s.a.l	meters above sea level
MARPOL	The International Convention for the Prevention of Pollution from Ships
MARS	Multivariate adaptive regression splines
MaxEnt	Maximum Entropy
MAXENT	Maximum Entropy
MDP	Master Development Plan
MDS	Multi-dimensional scale
MF	Mixed Forest
MFA	Marine Fishery Advisories
$Mg\ ha^{-1}$	Megagram per hectare
MINARS	National Programme of Monitoring of Indian National Aquatic Resources
MIR	Model Improvement Ratio
MIROC5	Model for Interdisciplinary Research on Climate Version Five
ML	Machine learning
MLP	Multi-Layer Perceptron
MNP	Manas National Park
MODIS	Moderate Resolution Imaging Spectroradiometer
MOEFCC	Ministry of Environment, Forest and Climate Change
MOSES	Modular Observation Solutions for Earth Systems
MPA	Marine Protected Area
MPCA	Minnesota Pollution Control Agency
MRF	Material Recovery Facility
M-SDM	Mechanistic Species Distribution Model
MSE	Mean of squared residuals
MT	Metric Ton
MUSIC	Model for Urban Stormwater Improvement Conceptualization
MV	Market value
MVLR	Multivariate linear regression
N2	Nitrogen Gas
NASA	National Aeronautics and Space Administration
NAT	NbS Aiding Technologies
NBM	Naive Bayesian Model
NbS	Nature-Based Solutions
NCAP	National Clean Air Program
NCP	Nature's Contributions to People
NDVI	Normalized Difference Vegetation Index

NFF	Nature Future Framework
NGO	Non-Governmental Organization
NH3-	Ammonia
NMPB	National Medicinal Plants Board
NMSHE	National Mission for Sustaining Himalayan Ecosystem
NOAA	National Oceanic and Atmospheric Administration
NOx	Nitrogen Oxide
NP	National Park
NPK	Nitrogen, Phosphorus and Potassium
NPP	Net Primary Production
NPP	Nuclear Power Plant
Nr	Nitrogen
NRCP	National Research Council of the Philippines
NS	Northern sector
NSM	Niche Suitability Models
O3	Trioxygen
OC	Outer channel
OCM	Ocean Color Monitor
OECD	Organization for Economic Cooperation and Development
OLI	Operational Land Imager
OM	Organic Matter
OOB	Out-of-bag
ORCHIDEE	Organizing Carbon and Hydrology in Dynamic Ecosystems
p	probability value
P10	10 percentile training presence
PA	Presence-Absence
PAGASA	Philippine Atmospheric Geophysical and Astronomical Services Administration
PAUC	Partial Area Under ROC curve
PB	Presence-Background
PBL	Planetary Boundary Layer
PBM	Process-Based Model
PCAMRD	Philippine Council for Aquatic and Marine Research and Development
PCC	Percent of Sites Correctly Classified
PCC	Potential Carrying Capacity
PD	Phylogenetic Diversity
PET	Potential Evapotranspiration
PFT	Plant Functional Type
PFZ	Potential Fishing Zone
pH	Potential of Hydrogen or Power of Hydrogen
PHIVOLCS	Philippine Institute of Volcanology and Seismology
PK	Pundasyon ng Kalikasan
PM	Particulate Matter
PM10	Particulate matter having size less than 10 μm and more than 2 μm

PM2.5	Particulate matter having size less than 2 μm
PNET (C.N)	Photosynthetic/Evapotranspiration model (Carbon Nitrogen)
PPM	Parts Per Million
ppmv	parts per million by volume
PRA	Participatory Rural Appraisal
pROC	partial area under the Receiver Operating Curve
PSA	Philippine Statistics Authority
p-value	Probability Value
QGIS	Quantum Geographic Information System
R&D	Research and Development
RA	Republic Act
RC	Rotation Coefficient
RCC	Real Carrying Capacity
RCI	Random Consistency Index
RCM	Regional Circulation Model
RCP	Representative Concentration Pathways
RET	Rare Endangered Threatened (Species)
RF	Random Forest
RF	Regularization Factor
RMSE	Root Mean Squared Error
RoA	Republic of Azerbaijan
ROC	Receiver Operating Characteristic
RS	Remote Sensing
RSFs	Resource selection functions
RTI	Right to Information
SAFAR	System of Air Quality and Weather Forecasting and Research
SCLWMC	Seven Crater Lakes and Watershed Management Council
SCS	State Committee of Statistics of Azerbaijan
SD	Secchi Disk Depth
SD	Standard Deviation
SDBM	Simple Diagnostic Biosphere Model
SDG	Sustainable Development Goals
SDGVM	Sheffield Dynamic Global Vegetation Model
SDM	Spatial Distribution Modeling
SDM	Species Distribution Model
SDMs	Species distribution models
SDVD	Secchi Disk Visibility Depth Index
Se	Sensitivity
SECC	Socio-economic Carrying Capacity
SEDAC	Socio-economic Data and Applications Centre
SEIB-DGVM	Spatially Explicit Individual Based-Dynamic Global Vegetation Model
SEM	Sustainable Ecosystem Management
SES	Social-ecological systems
SESAM	Spatially explicit species assemblage modeling

SFSLR	Step-wise Forward Selection Logistic Regression
SHPP	Small Hydropower Plant
SIMWAL	Simulated Walnut
SMLP	Samahang Mangingisda ng Lawa ng Pandin
SOC	Soil Organic Carbon
SOI	Survey of India
SOx	Sulfur Oxide
Sp	Specificity
SPCB	State Pollution Control Board
SpThin	Spatial Thinning
SR	Species Richness
SRE	Surface Range Envelop
SRES	Special Report on Emission Scenarios
SROCC	Special Report on the Ocean and Cryosphere in a Changing Climate
SRTM	Shuttle Radar Topography Mission
SS	Southern Sector
SSDM	Stacked-SDMs
SSE	Sum of Square Error
SSP	Shared Socio-economic Pathway
SST	Sea Surface Temperature
SVAT	Surface Vegetation Atmosphere Transfer
SVM	Support Vector Machine
SWM	Solid Waste Management
TAR	Third Assessment Report
TCC	Tourism Carrying Capacity/Total Carrying Capacity
TDS	Total Dissolved Solids
TECM	Terrestrial Ecosystem Carbon Model
TEEB	The Economics of Environment and Biodiversity
TERN	Terrestrial Ecosystem Research Network
TIRS	Thermal Infrared Sensor
TL	Total Length
TLI	Trophic Level Index
TN	Total Nitrogen
TOF	Trees Outside Forests
TP	Total Phosphorus
TPP	Thermal Power Plant
TRIFFID	Top-down Representation of Interactive Foliage and Flora Including Dynamics
TS	Total Sensitivity
TSA	Targeted Scenario Approach
TSPCB	Telangana State Pollution Control Board
TSS	Total Suspended Solid
TSS	True Skill Statistic
TSS	True Skill Statistics

TURB	Turbidity
UC ANR	University of California Agriculture and Natural Resources
UNDP	United Nations Development Program
UNEP	United Nations Environment Program
UNFCCC	United Nations Framework Convention on Climate Change
UPLB-FEWS	University of the Philippines Los Baños—Fish Kill Early Warning System
USAID	United States Agency of International Development
USD	United States Dollar
USDM	Uncertainty Analysis for Species Distribution Models
USGS	United State Geological Survey
UT	Union Territory
UTM	Universal Transverse Mercator
UYRDC	Uttarakhand Youth Rural Development Centre
VIF	Variance Inflation Factor
WB	Water Body
WE	Taxonomic Weighted Endemism
WECC	Water Ecological Carrying Capacity
WG	Western Ghats
WGS84	World Geodetic System 1984
WHO	World Health Organization
WPE	Weighted Phylogenetic Endemism
WQBCC	Water Quality and Biodiversity Carrying Capacity
WT	Water temperature
WTO	World Tourism Organization
WTTC	World Travel and Tourism Council
WWF	World Wildlife Fund
XGB	XGBoost Model

Modelling Tools and Plausible Scenarios in Science-Policy to Improve Evidence-Based Decision-Making for Human Well-Being

1

Shalini Dhyani and Radhika Sood

Abstract

In the last few decades, there has been a tremendous interest among the global researchers and intergovernmental panels on climate and biodiversity for using modelling tools in science-policy assessments and evidence-based decision-making for conservation and human well-being. There is growing relevance of mainstreaming models and scenarios in global environmental policy planning to improve ecosystem management, species conservation, and restoration of degraded landscapes. Ongoing global environmental policy discussions especially for climate and biodiversity policy continue to stress upon improving and enhancing existing modelling tools, for providing accurate and scenario projections. It is expected that both short-term and long-term conservation efforts will depend on the accuracy of the modelled outputs with lesser uncertainty and more integration of the socio-ecological concerns. The present edited book includes chapters developed from existing research knowledge and wide-ranging experience of researchers, academicians coming from diverse fields of science, policy, and practice to improve the knowledge base on effectively using modelling tools and leveraging their potentials to broadly understand climate vulnerability and the different impacts on ecosystems as well as to explore habitat suitability. Further, the book covers and provides an overview on the state-of-the-

S. Dhyani (✉) · R. Sood
Critical Zone Research Group, Water Technology and Management Division, CSIR-NEERI, Nagpur, India

Academy of Scientific and Innovative Research (AcSIR), Ghaziabad, India

International Union for the Conservation of Nature (IUCN), Commission on Ecosystem Management (CEM), Geneva, Switzerland
e-mail: s_dhyani@neeri.res.in

S. Dhyani et al. (eds.), *Ecosystem and Species Habitat Modeling for Conservation and Restoration*, https://doi.org/10.1007/978-981-99-0131-9_1

art existing modelling approaches; their use for addressing the diverse concerns related to plant, animal and habitat conservation as well as restoration; and their relevance to the ongoing policy discussions and planning. Some of the highlighted cases in this book also describe the status of threatened plants as well as wild species and restoration projects of interest to the global practitioners. Chapters spread over different sections of the book cover scientific evidence to support evidence-based policy planning by including plausible alternative scenarios to improve ecological, social, and economic benefits. The present chapter provides an overview of the growing relevance of modelling tools and key enabling and constraining concerns followed by a synopsis of the chapters.

Keywords

Modelling · Ecosystem · Species · Habitat suitability · Science-policy · Conservation · Restoration · Management

1.1 Introduction

What does the future have in store for the global ecosystems and the advantages that humans derive from them?

Ecosystems closely linked to the well-being and functionality of the entire biosphere have been impacted by the global climate change. Threats to biodiversity and ecosystems are ubiquitous and growing worse because of global warming (Dhyani et al. 2020). Globally, biodiversity is disappearing at previously undiscovered rates, undermining the basic pillars of our economies, lifestyle, employment, access to food, good health, and human well-being (IPBES 2018). Land management practices and climate change have emerged as two of the crucial mediators of future biodiversity change.

Broad alterations in productivity and species interactions have led to enhanced vulnerability from biological invasions and other emergent attributes because of adaptations by species to climate change and direct impacts of climate change on ecosystems (Weiskopf et al. 2020). Tipping points, where ecosystem thresholds are surpassed and lead to significant changes to the structure and functions of ecosystems, are of particular concern (Sintayehu 2018). When combined with other changes, climate change interacts with other stressors on ecosystems, like degradation, defaunation, as well as fragmentation (Malhi et al. 2020). Distributions of species have moved to higher altitudes at a median pace of 11.0 m and 16.9 km/decade to higher latitudes as a consequence of global warming. Furthermore, the extinction rates of 1103 species under different migration scenarios are as follows: 21–23% for unrestrained migration and 38–52% when there is no migration (Muluneh 2021). Majority of scientists across the world irrespective of their nationalities acknowledge that humanity is in the midst of a climate emergency (IPBES 2018; Pettorelli et al. 2021). For natural protection and sustainable development, it is crucial to comprehend how anthropogenic activities affect human

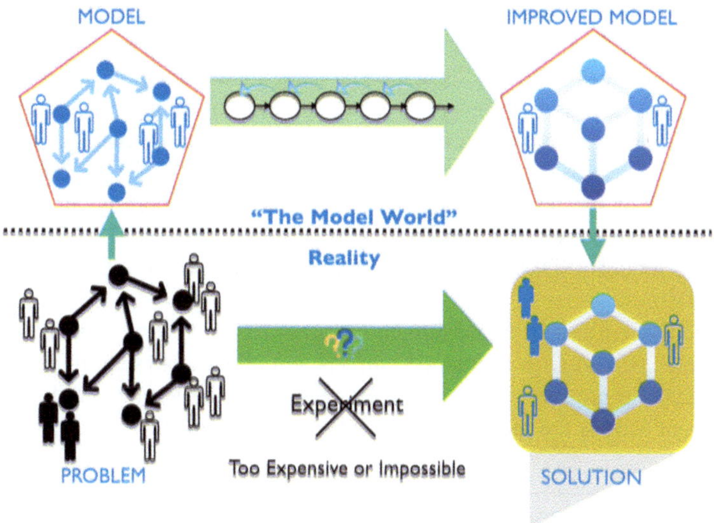

Fig. 1.1 Relevance of modelling tools in ecosystem and species scenario projections

societies and biodiversity (Kim et al. 2018). Addressing the ecological dynamics of these climatic impacts is crucial, as is identifying hotspots of fragility and resilience along with the management strategies that can boost the biosphere's ability to climate change adaptation. Ecosystems can contribute to climate change adaptation and mitigation too; hence it is essential to look into and analyse the mechanisms, opportunities, and limitations of such nature-based climate solutions (Malhi et al. 2020). In order to better understand ecosystems and guide ecosystem and species management, science has made significant advances. What does the future hold for ecosystems around the world and what advantages do people derive from them (Kok et al. 2017)? In a world threatened by both natural and human-induced change, predicting ecosystem effects is vital (Caron-Lormier et al. 2009). To comprehend and forecast ecological patterns and processes, models have emerged as helpful tools (Fig. 1.1). Models assist decision-makers in anticipating the effects of policies on ecosystems and people; for example, it is crucial to improve our capacities to depict interactions between human actions and ecological systems in order to identify methods to achieve the Sustainable Development Goals given climate, biodiversity, and restoration targets (Weiskopf et al. 2022). They can substantially aid in decision-making for conservation and restoration under current climate and biodiversity change, as well as help create suitable management methods for an uncertain future (Zurell et al. 2022).

Ecological models are being used to develop effective management methods, forecast potential scenarios under multiple scenarios of global change, and expand our understanding of how ecological systems function (Mokany et al. 2016). Scenarios and models are convincing tools for predicting the plausible future scenarios of different social-ecological development pathways that can help inform existing and future policy decisions (Lundquist et al. 2021). The creation of scenarios has been cited by the Intergovernmental Platform on Biodiversity and Ecosystem Services (IPBES) as a crucial step in assisting decision-makers in identifying the effects of various policy alternatives (Kok et al. 2017). An important step toward the co-production of knowledge for use in resource management choices is the coordinated use of multiple models (Lewis et al. 2021). Comprehensive understanding of species distributions, drivers of loss, and their dynamics is necessary for effective conservation; habitat/niche suitability models (NSM) and species distribution models (SDMs) are frequently used to forecast these trends (Leitão and Santos 2019).

An ecosystem-based management strategy called "integrated ecosystem assessments" (IEAs) aims to include humans and other ecosystem members in the decision-making process so that managers may weigh trade-offs and decide which management actions are most likely to achieve goals (Howell et al. 2021).

1.2 Modelling Tools

Significant changes in the provisions of ecosystem services as well as severe declines in biodiversity worldwide have been triggered by extensive human-induced pressures and interferences driven by the demand for agriculture intensification and forestry products. These developments are projected to endure as long as the world remains to flourish without giving much thought to how environment provisions human well-being. Acquiring the skill of developing useful models and scenarios is critical since they can be valuable tools to policymakers and decision-makers in anticipation of the effects of their decisions (Rosa et al. 2020). In recent times, such understanding of complex systems, particularly social and economic systems, has been improved by agent-based modelling (Table 1.1). The objective of modelling

Table 1.1 Examples of biological levels for modelling, including compositional, structural, and functional biodiversity variables, are chosen to represent levels of biodiversity that demand attention in environmental monitoring and assessment programmes

Level	Composition	Structure	Function
Individuals	Genes	Genetic structure	Genetic processes, metabolism
Populations	Presence, abundance, cover, biomass, density	Population structure, range, morphological variability	Demography, dispersion, phenology
Communities	Species richness, evenness and diversity, similarity	Canopy structure, habitat structure	Species interactions (herbivory, predation, competition, parasitism), decomposition
Ecosystems	Habitat richness	Spatial heterogeneity, fragmentation, connectivity	Ecosystem processes (hydrologic processes, geomorphic processes), disturbances

Source: https://ipbes.net/scenarios

complicated systems is to reduce the system to simple agents that follow straightforward rules. The model then shows these agents' emergent interactions with one another and their surroundings (Engler and Kusiak 2011). Setting and implementing strategic initiative to prevent biodiversity loss depends on biodiversity predictions with uncertainty projections under various climate, land-use, and policy scenarios. An ecologically valuable objective and issue continues to be assessing and enhancing biodiversity projections to inform policy decisions. To make more accurate prediction about biodiversity, a thorough approach to assessing and reducing the uncertainty of model outputs versus observed data and numerous models is necessary (Myers et al. 2021).

1.3 Recognition and Acceptance of Modelling Tools in Environmental Decision-Making

Since the turn of the century, the range of areas in which complexity science approaches have been applied has grown more. Applications to business strategies and public policy have multiplied together, in particular (Fig. 1.2). Agent-based modelling has made a name for itself as a powerful instrument in science. Adoption by policymakers is still in short supply, though. Given the vast number of outstanding, effective applications of complexity science in the most diverse academic fields, policy is prepared to develop into a real field of in-depth and beneficial applications (Furtado et al. 2019). Since the 1970s, long-term global scenarios have served as the foundation for analysis and research on environmental change throughout the world.

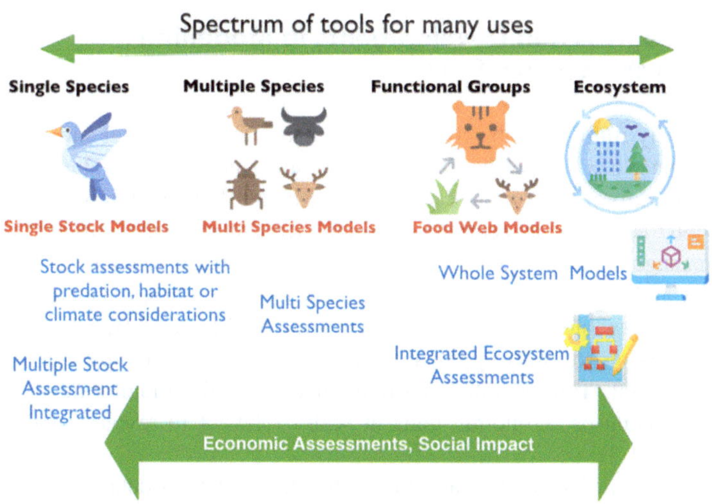

Fig. 1.2 Different mathematical model types are suitable for various analyses and recommendations. To answer a variety of scientific research issues and to provide managerial guidance, a spectrum of models is used. Adapted from NOAA Fisheries (https://www.fisheries.noaa.gov/national/ecosystems/ecosystem-modeling)

In order to allow integrated research and uniform assessment to inform policy, the community of climate change researchers has created a scenario framework during the past 10 years that combines different futures for the climate and society. The paradigm is mostly fulfilling urgent needs and has received widespread adoption throughout research communities. The development and application of this paradigm, however, must take numerous new routes due to some mixed results and a shifting policy and research environment (O'Neill et al. 2020). The focus of multilateral negotiations and conventions has transitioned to scenarios and models relating to ecosystem services and biodiversity as a result of growing global concern for biodiversity conservation. One of the initial rounds of quick assessment activities, the Methodological Assessment on Scenarios and Models of Biodiversity and Ecosystem Services, was included in the Intergovernmental Science-Policy Platform on Biodiversity and Ecosystem Services (IPBES) Work Program 2014–2018. At the fourth IPBES Plenary, the Assessment Report and accompanying Summary for Policymakers were accepted (Pan et al. 2018).

The IPBES Expert Group on Scenarios and Models compared biodiversity and ecosystem services models using harmonized scenarios to support the assessments of the IPBES (BES-SIM). The outcomes of collaborative modelling project supported the ongoing global assessment of IPBES, strengthened the connections between IPBES and the scenarios and modelling methods of the Intergovernmental Panel on Climate Change (IPCC), and offered guidance to the Convention on Biological Diversity (CBD) on the development of its post-2020 global biodiversity framework and conservation objectives that further influenced the formation of a new generation of nature-centred nature-future scenarios (Kim et al. 2018).

1.4 Progress and Developments in Modelling on the Science Front

To effectively fulfil their objectives and mandates, environmental and conservation agencies must use science-based natural resource management. Hence, there is a growing need for models in biodiversity assessments, but which models are suitable for the job? However, as ecosystems experience exceptional events that put traditional management frameworks to the test, this scientific foundation needs to be improved and developed (Spooner et al. 2021). The inherent heterogeneity and sparseness of raw biodiversity data are overcome by the use of models and remotely sensed covariates to inform predictions that are contiguous in space and time and global in extent. This essential information enables the monitoring of single or aggregate spatial or taxonomic units at scales relevant to research and decision-making (Fig. 1.3).

When combined with ancillary environmental or species data, this fundamental species population information directly underpins a range of biodiversity and ecosystem function indicators (Jetz et al. 2019).

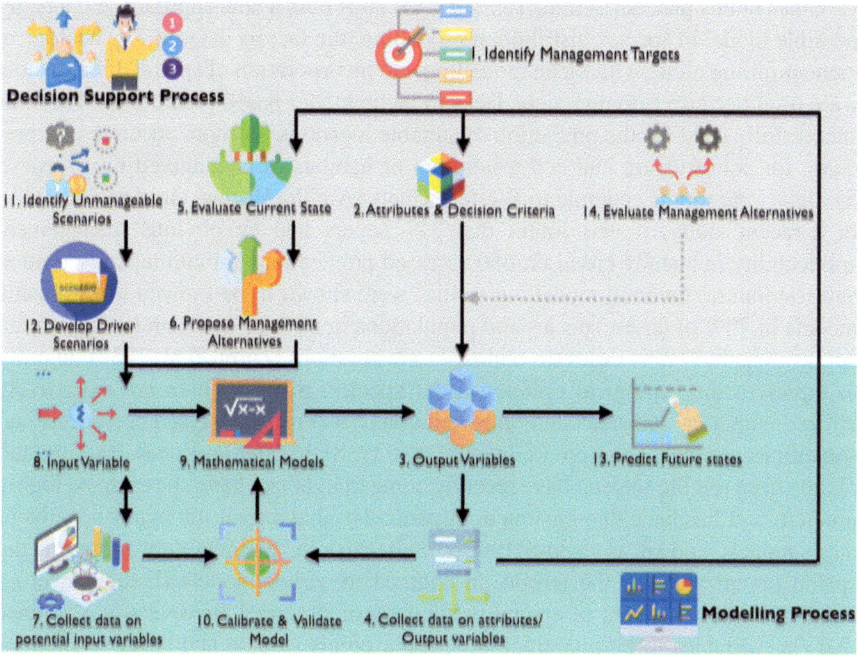

Fig. 1.3 Schematic illustration of how modelling is integrated into a management decision-support system (Source: Schuwirth et al. 2019)

1.4.1 Species Modelling

In order to design surveys for new populations, guide spatial prioritization choices for management activities, and support regulatory decision-making and compliance, species distribution models have already been utilized (Sofaer et al. 2019). The contributions of biodiversity to ecosystem functions are frequently measured by species richness; however, modelling interactions within species and/or evenness, in addition to richness, can result in a deeper understanding of diversity-driven advancements in ecosystem functions (Brophy et al. 2017). Decisions about species management and conservation heavily rely on information of where they are found; however this information is frequently vague or unreliable. Species distribution models can be used to map habitats and can generate reliable, reproducible data that can be used to support decisions. Nevertheless, because of their sensitivity to methodological and data inputs, it is crucial to evaluate the validity and applicability of model predictions (Sofaer et al. 2019). Advancements may make it easier to monitor different species through time and space, ultimately assisting in the identification of important conservation areas, determining the viability of potential habitats, and providing early indicators of shifts in species distributions. The formation of habitat suitability patterns can be significantly influenced by bioclimatic factors, terrain, and forest structure in general, indicating a major role for bioclimatic variables in this process (Amini Tehrani et al. 2021). As a new approach and a better possible model in species distribution modelling, the factors determining ecological niche profiling along with niche modelling and incorporation of spatial threat factors were used because Eltonian niche lacks in niche model-based predictions. This was successfully used for the prediction of suitable locations for more accurate interpretation for conservation and eco-restoration of habitats of endangered *Cryptocarya anamalayana* that is endemic to Western Ghats (Amitha Bachan and Devika 2022). In a recent study, it was found that 217 papers had an obvious management applicability in a study based on 650 assessed publications for animal conservation and restoration. Overall, modelling studies were shown to be skewed toward static models in 79% of cases, species and populations in 80% of cases, and conservation applications in 71% of cases (as opposed to restoration) (Zurell et al. 2022). In order to represent the ecological preferences of species, SDMs, which are extensively utilized numerical tools, depend on correlations between geo-located presences (and sometimes absences) and environmental factors. SDMs that use deep learning and photos from remote sensing have recently come to light and have shown to be highly predictive. It has been demonstrated, in particular, that one of the major benefits of these models known as deep-SDMs is their capacity to accurately represent the spatial organization of the terrain, in contrast to earlier models (Estopinan et al. 2022). The development of spatial predictions of Essential Biodiversity Variables (EBVs), variables to be quantified at specific points in time and space to monitor variations in biodiversity using SDMs, is a unique approach based on diversity metrics, such as the distribution functions of important bird habitats at a regional scale. As a spatial "species distribution" EBV (SD EBV), the suitability computed in accordance with the SDMs can be used to indicate the habitat quality, trends in land use, and climatic impacts on populations of bird species (Amini Tehrani et al. 2021).

1.4.2 Ecosystem Modelling

It takes scientific guidance to manage ecosystems sustainably because so many of them are in danger. Only a small portion of the models used to study ecological dynamics and reactions to disturbances are ever utilized to guide ecosystem management, despite the fact that many academics from all over the world use them (DeAngelis et al. 2021). Constant failures from single-species management have led to test and implement ecosystem-level management. Ecosystem management comprises understanding complex range of interacting living beings, processes, as well as interdisciplinary and multidisciplinary inputs. Considering interfaces, feedback loops, and interdependencies among ecosystem constituents is thus vital for understanding as well as managing ecosystems (Geary et al. 2020). Dynamic models have been a staple management tool for ecological and economic systems for a long time, and they were heavily utilized in the early stages of resilience research. The majority of model applications have been concerned with evaluating policies, creating the best management plans, or analysing system stability. Modelling can also be employed to promote participatory processes, explain system reactions that result from intricate interconnections between system components, and examine the effects of complicated human behavioural patterns, among many other things. Research on social-ecological systems (SESs) has a lot of potential given the variety of aims, forms, and applicability of models (Schlüter et al. 2019). Plant physiological trait-based metrics, which may be directly observed in the field, are rapidly replacing empirical parameters in ecosystem models. This has encouraged the development of new models and helps anticipate long-term terrestrial ecosystem dynamics under climate change. The ability to directly integrate observed plant ecophysiology with model processes through trait-based modelling of terrestrial ecosystems increases the potential to reduce uncertainty and enhance forecasts under unique climatic conditions. A rigorous model design, systematic intercomparisons, and benchmarking for model responses to both climatic extremes and long-term trends are however necessary due to the increased model complexity (Xu and Trugman 2021).

1.5 Enabling and Constraining

Ecosystem modelling is difficult, especially when trying to strike a balance between the need to describe all of an ecosystem's components and the data constraints and modelling goal. Therefore, the main issue is explicitly taking into account various forms of uncertainty. Existing modelling techniques typically aim to achieve one or a combination of the following: (1) define and separate ecosystem components and interactions; (2) predict future ecosystem conditions; and (3) provide guidance for decision-making by contrasting alternative strategies and highlighting significant uncertainties (Geary et al. 2020). In a world that is changing quickly, predictive models are at the core of many scientific disciplines and are essential for informing management decisions. Confidence in their forecasts is jeopardized by a poor grasp of the accuracy and precision of models translated to novel situations (i.e. their "transferability"). The combined need to expand research on the factors influencing ecological predictability, such as species attributes and data quality, and create best practices for transferring models, is driven by these technical and fundamental problems. Finding a set of transferability metrics that are broadly applicable and have the right tools for quantifying the causes and effects of prediction uncertainty in novel circumstances is of utmost importance (Yates et al. 2018). Because modelling approaches come from various disciplines, are based on various assumptions, concentrate on various levels of analysis, and employ various analytical tools, they have a wide range of goals, types, and applications that have led to a great deal of confusion. Because of this variability, it can be challenging to decide on the best strategy for dealing with a certain challenge. Modelling approaches can be categorized into "modelling for social-ecological systems research" (ModSES) along two dimensions, the degree of realism and the degree of knowledge management, in order to account for context dependence and the intertwined nature of SESs as systems of humans embedded in nature across multiple scales, as well as to acknowledge different issue framings and understandings (Schlüter et al. 2019).

Modelling work frequently focuses on certain technical aspects (e.g. refining model precision, and data processing methods, addressing data gaps) as well. From a technical standpoint, problems that might help or impact model adoption range across the entire modelling process, including the thematic focus of a model or scope as well as its assumptions, resolution, and size (spatiotemporal). It is crucial to determine whether the model is appropriate for usage in the situation where it is utilized since model relevance is frequently more context-dependent than is commonly acknowledged (Weiskopf et al. 2022). Finding the model construction with the best complexity performance is still a major challenge. Beyond the deliberate testing of a collection of different models, there are still relatively fewer methods for empirically quantifying structural uncertainty, despite the fact that there are numerous methods for parameter uncertainty (Lewis et al. 2021). The most frequently utilized model type is the correlative niche models. Dynamic models, the gene to individual level, and the community to ecosystem level are under-represented, and only 10% of the research has utilized explicit cost optimization methods (Zurell et al. 2022).

Monitoring changes in the distribution and abundance of species has an impact on the total biodiversity and is essential for the effective conservation of the integrity and functions of species populations. Acquiring precise data on biodiversity at broad spatial scales can be difficult since it is often spotty, inadequate, or even non-existent (Amini Tehrani et al. 2021). To predict a species' present and future distribution patterns and ecological niche, many researchers have used species distribution models. A study that assessed 79 publications that were published between 2010 and December 2020 found that the quantity of papers on SDM has increased significantly over time. Asia (41%), Europe (24%), and Africa (2%), in that order, made up the majority of these. The majority of the studies considered (38%) concentrated on theoretical ecology, the effects of climate change, and conservation policy and planning (22%). The majority of studies focused on transdisciplinary, ecological, or biodiversity conservation domains. In the majority of studies (81%), the level of uncertainty was not revealed. Future rare and endemic species SDMs ought to express the degree of uncertainty and projections of mistakes in the modelling procedure (Qazi et al. 2022). Utilizing the suggested standards and principles, 400 modelling studies over the last 20 years were examined and graded (Araújo et al. 2019). Overall, low model adequacy was established; however there were a clear upward trend in model creation and a downward trend in biological data and model evaluation with time. The adoption of generally accepted criteria for models used in biodiversity assessments will encourage transparency and repeatability and eventually result in models and inferences used in assessments being of a higher calibre. The extension and continued development of the SDM standards and guidelines invite broad community engagement. Even while IPBES has identified scenario generation as a crucial step in supporting decision-makers to understand the potential effects of various policy alternatives, the organization initially lacked a long-term scenario strategy. IPBES deals with a wide range of local contexts and takes into account global tele-coupling of local locales in order to capture the social-ecological dynamics of biodiversity and ecosystem services (Kok et al. 2017). The scenarios for biodiversity and ecosystem services currently under development have significant flaws and limitations that limit their applicability for reversing the dangerously falling trend of nature's contributions to people (NCP) and human contributions to nature. Most of the current scenarios and related analyses, particularly at the global and regional levels, are restricted to evaluating the effects of drivers on a small number of aspects of nature and NCP, frequently omitting to account their linkages or feedbacks over multiple spatial scales, or to take into account policy objectives pertaining to nature conservation. Additionally, they are not always able to take into account common standards, beliefs, and policy goals pertaining to the preservation of the environment and a high standard of living. Instead of identifying desirable futures for nature and people along with providing alternative pathways to reach them, current approaches frequently focus on negative trends and drivers (Lundquist et al. 2021).

1.6 Mainstreaming Modelling Tools in Policy Planning

Globally, the ranges and abundances of species are evolving fast. This emphasizes the importance of trustworthy, adequate data for directing and evaluating actions and policies intended to manage and protect the numerous functions and benefits of species (Jetz et al. 2019). The creation of solid and thorough long-term strategies can be facilitated by the use of powerful analytical frameworks and tools, most often models. Models are essential for formulating long-term policies and implementing them quickly in a manner that is consistent with a nation's aspirations for socioeconomic growth. [1] In order to integrate human well-being as a key outcome, international science organizations are moving toward transdisciplinary and inclusive research (Spooner et al. 2021). At the intersection of academia and public and commercial sector policymaking, scenario planning is becoming a significant area of study. The application, effects, and effectiveness of scenario planning in the development of public policy have received less attention in the academic literature than research methodologies, which are thoroughly covered (Volkery and Ribeiro 2009). The need for transformative, multiscale global scenarios as strategies to stop the loss of biodiversity and accomplish sustainability goals has been made repeatedly by scientists. Researchers from the IPBES scenarios and models expert group engaged in an adaptive, interactive approach that resulted in the creation of the Nature Futures Framework (NFF) as a first step toward achieving this. The NFF is a cognitive method that depicts the many, beneficial interactions between people and nature as a triangle. It can be used as a boundary component to keep bringing in more diverse viewpoints while generating acceptable nature scenarios as well as a workable framework for creating consistent nature scenarios at various scales (Pereira et al. 2020). The most comprehensive and up-to-date evaluation of how oceans and

[1] https://www.wri.org/climate/expert-perspective/role-modeling-and-scenario-development-long-term-strategies-0.

the cryosphere are changing, how they are projected to change, and the consequences of those changes, as well as a variety of response options, is believed to be using the findings of the 2019 *Special Report on the Ocean and Cryosphere in a Changing Climate* (SROCC) by the Intergovernmental Panel on Climate Change (IPCC). These discoveries are extremely important for the conservation of South Ocean ecosystems (Cavanagh et al. 2021). There are three ways to improve the calibration and validation of dynamic modelling approaches, develop best-practice guidelines for using these models, and create a toolbox with a variety of easier-to-use methods. These are the main recommendations that suggest how to increase the use of spatially explicit models for decision support. Additionally, by combining several modelling approaches to measure uncertainty and putting models at the centre of adaptive management, more robust decision-making can be obtained. To secure the best results for conservation and restoration, these initiatives must be supported by long-term funding for modelling and monitoring as well as increased communication between research and practice (Zurell et al. 2022). In the future too, it will be necessary to handle political and institutional environment issues with greater caution. It takes more than just thorough analysis to make better decisions in highly unpredictable situations. Political will, more stable institutional environments, organizational capacities to develop trust, and expertise with adaptive, flexible process formats are all necessary (Volkery and Ribeiro 2009).

1.7 Structure of the Book

The present book includes 26 chapters that are further distributed into 5 core sections. Part I of this book, "Ecosystem and Species Modelling Tools and Relevance", is a gathering of six chapters. Chapter 1 presents an overview about the growing relevance of ecosystems and species modelling, its relevance in scenario predictions, and mainstreaming modelling tools in environmental policy planning. Chapter 1 also reflects on the enabling and constraining conditions for improving effectiveness of results and scenarios generated by modelling tools. In Chap. 2 Azita Farashi provides an Introduction of species distribution modelling and discusses the relevance of species modelling in the context of global climate change. Shameer and Sanil (Chap. 3) present the case of machine learning-based predictive modelling approaches for better understanding of the evolutionary history, distribution, and niche occupancy, by sharing experiences from Western Ghats. Chapter 4 by Barewar et al. offers an overview of mapping the impact of climate change on eco-sensitive hotspots using SDMs (gaps, challenges, and future perspectives). The review concentrates on several SDMs, its application in various ecosystems and their management, the gaps in the models and modelling techniques, and the challenges in their applicability. To investigate the variables utilized for modelling the future projections of the species distribution, several SDMs are presented in this chapter. The chapter by George and Joseph (Chap. 5) presents approaches for modelling the climate change impacts on ecosystems. This chapter provides an overview on how the scenarios are being created and their evolutionary changes in the last two decades starting from the IPCC's SRES scenario to the recent SSP scenario. Further, the downscaling of physically based climate models to biosphere-based Earth system models is also presented. Two approaches to Earth system modelling, i.e. process-based dynamical global vegetation models and classical climate envelope models, are described in detail to model the response of ecosystems to climate change in this chapter. Chapter 6 by Macandog et al. outlines the importance of developing a Bayesian model of climate-induced lake overturn in Talisay, Taal Lake, a freshwater caldera lake in the province of Batangas, on the island of Luzon in the Philippines. Kadaverugu et al. (Chap. 7) bring insights from global sensitivity and uncertainty analysis of MaxEnt model and exploring the implications in species habitat projections using *banj* oak as important species dominant in moist temperate forests of Uttarakhand.

Part II of the book "Habitat Modeling for Conservation of Threatened Plants and Restoration of Habitats" includes six chapters and the section cover examples of modelling approaches used for habitat suitability for their effective conservation and habitat restoration in case of habitat loss. The opening chapter of the section by Dlamini and Loffler (Chap. 8) presents a case from Eswatini, a landlocked country from the south of Africa. The chapter highlights the tree species diversity and richness patterns that reveal high priority areas for conservation. Chapter 9 by Pradhan and Chettri presents overview on prioritizing suitable habitat for improving the current status of threatened tree (*Acer sikkimensis* Miq. syn. *Acer hookeri* Miq.) through regeneration and use of ecological niche modelling tools. Kaushal et al. (Chap. 10) provide perspectives from ecological niche modelling of the endemic Himalayan near-threatened treeline conifer *Abies spectabilis* in the Indian Central Himalaya. In Chap. 11, Bushi et al. provide scenario projections from species distribution modelling of an endangered medicinal plant *Oroxylum indicum* (L.) Kurz in Arunachal Pradesh. Peerzada et al. (Chap. 12) present a habitat suitability modelling effort for *Aconitum heterophyllum* in temperate Himalayan forest ecosystems. Habitat of *Aconitum heterophyllum* is highly vulnerable to climate shifts and anthropogenic pressure and therefore needs immediate restoration in the wild and systematic domestication in the potential areas. The last chapter of Part II (i.e. Chap. 13) by Shilky et al. outlines the relevance of SDM in conservation and restoration of forest ecosystems. Conservation management to aid in the recovery of threatened species requires an understanding of their habitat availability and preference. Part III of the book "Habitat Suitability Modeling for Protecting Animals and Their Habitat" comprises four chapters, with the opening chapter by Hussain et al. (Chap. 14) highlighting the relevance of habitat suitability analysis of Asiatic elephants (*Elephas maximus*) in the tropical moist deciduous forest of Assam using analytic hierarchy process (AHP). The study analyses the habitat suitability and also maps the corridor of Asiatic elephants in Barduar and Mayang Hill situated on both sides of Chandubi Lake in Kamrup District. Nath et al. (Chap. 15) provide an overview about the factors affecting the habitat suitability of eastern swamp deer (*Rucervus duvaucelii ranjitsinhi*; Groves 1982) in Manas National Park and implication for Terai grassland restoration. In this chapter, an attempt has been made to analyse the patterns of swamp deer occurrence as determined by habitat variables using random forest algorithm models that indicate optimal habitats of swamp deer in the large grassland patches with wet climatic conditions. These findings have significant implications for the conservation of the threatened grassland habitat and its obligate species in the Terai grasslands of the region. Fragmentation has now emerged as a major global problem, with anthropogenic activities regarded as one of the main causes, primarily for the loss of habitat suitability. Chapter 16 by Areendran evaluates the potential habitats of chital, sloth bear, and jungle cat in selected areas of Central Indian landscape. Significant overlaps of potential habitats have been observed between the species mostly within the protected areas. Mahato et al. (Chap. 17) bring instances from fisheries and present scenario projections using habitat suitability modelling of *Tor tor* (Hamilton 1822) in the Indian drainage systems using MaxEnt. Part IV of the book deals with the application of modelling tools and approaches, comprising six chapters. Kundu and Santhanam (Chap. 18) model the reduction of carbon dioxide through the use of Marine Fishery Advisories under varying climate change scenarios for the Bay of Bengal using the CMIP approach with implications for techno-policymaking. In Chap. 19, Preeti et al. bring forth the impacts of pollution on tropical montane and temperate forests of South Asia: preliminary studies by postgraduate students in India and Sri Lanka. Chapter 20 by Sharma and Santhanam highlights the selection of strategic sampling sites for river quality assessments near mined areas as a policy handle for low-impact development and biodiversity conservation using the case of river Godavari. Ramanan et al. in Chap. 21 use ecological niche modelling to predict the potential area for cultivation of *Melia dubia*: a promising tree species for agroforestry. Santhanam et al. (Chap. 22) provide an overview on proportions of change in the airborne particulate matter (PM10) concentrations across selected states in Peninsular India using study of decadal, pre-pandemic trends for planning restoration. In Chap. 23 Macandog et al. use decomposition of sunflower cuttings to study its impact on soil fertility of payoh (rice terraces) in Banaue, Ifugao, Philippines.

The last section of the book with Part V "Ecosystem and Species Modelling for Evidence-Based Decision-Making" has four chapters. The opening Chap. 24 in this section by Jose et al. discusses the relevance of forest ecosystem modelling for policy planning to address international climate targets. Chapter 24 provides review of different ecosystem modelling approaches, exploring their potential applications to understand changing forest dynamics and climate change adaptation options in forest ecosystems. It helps to get insights into the advantages and limitations of the various modelling-based approaches, providing a guideline for systematic execution of policy assessment according to a defined criterion (e.g. uncertainty management, data required, spatial and temporal dynamics, adaptation measures integration, and level of complexity). Macandog et al. (Chap. 25) provide an overview on ecological carrying capacity modelling and sustainability assessment of the seven lakes of San Pablo City, Laguna, Philippines. Chapter 26 by Abbasov assesses the contribution of freshwater ecosystem services to the sustainable development in the Kura-Aras River Basin in Azerbaijan. The study used a basic Targeted Scenario Analysis (TSA) approach. The TSA evaluates the present value of ecosystem services under "business as usual (BAU)" ecosystem management practices. To evaluate costs and possible benefits (or losses) of switching from BAU to SEM, it compares sector output indicators with potential "sustainable ecosystem management (SEM)" outputs. The last chapter (Chap. 27) provides prediction of eutrophication in aquatic ecosystems with hybrid neural networks: a case study from Chilika lagoon, India. The model evidenced an acceptable level of prediction when compared with the results of the field observations. This model's most important determinant variables were those having a high random forest (RF) model permutation relevance ranking, that reduced the networks structure and led to a more accurate process. This model that can be considered for formulation of the management and conservation action plan of other aquatic ecosystems too in other parts of the world. Through this book volume, the editors as well as authors expect decent response from wide-ranging audience and stakeholders encouraging productive reviews, insightful and progressive professional deliberations on the benefits and limitations of the modelling tools to predict species and ecosystem scenarios and also identify emerging research issues for the active execution, and greater mainstreaming of modelling results to advance the understanding and applicability of different modelling tools and approaches in addressing intractable sustainable development challenges. Finally, the edited volume focuses on wider application of these modelling tools for effective conservation and restoration planning that can help countries meet Kunming-Montreal Global Biodiversity Framework, the UN decade on restoration targets and localization of SDGs.

Acknowledgements Authors are grateful to Knowledge Resource Centre (KRC) CSIR-NEERI for extending help for using software iThenticate for avoiding any similarity and plagiarism in the text, and the support is acknowledged under the number CSIR-NEERI/KRC/2022/DEC/WTMD/1.

References

Amini Tehrani N, Naimi B, Jaboyedoff M (2021) Modeling current and future species distribution of breeding birds as regional essential biodiversity variables (SD EBVs): a bird perspective in Swiss Alps. Glob Ecol Conservat 27:e01596. https://doi.org/10.1016/j.gecco.2021.e01596

Amitha Bachan KH, Devika MA (2022) Niche profiling and niche modelling of endangered Cryptocarya anamalayana endemic to Western Ghats for conservation and restoration. In: DellaSala DA, Goldstein MI (eds) Imperiled: the encyclopedia of conservation. Elsevier, Oxford, pp 751–764

Araújo MB, Anderson RP, Márcia Barbosa A et al (2019) Standards for distribution models in biodiversity assessments. Sci Adv 5:eaat4858. https://doi.org/10.1126/sciadv.aat4858

Brophy C, Dooley Á, Kirwan L et al (2017) Biodiversity and ecosystem function: making sense of numerous species interactions in multi-species communities. Ecology 98:1771–1778. https://doi.org/10.1002/ecy.1872

Caron-Lormier G, Bohan DA, Hawes C et al (2009) How might we model an ecosystem? Ecol Model 220:1935–1949. https://doi.org/10.1016/j.ecolmodel.2009.04.021

Cavanagh RD, Trathan PN, Hill SL et al (2021) Utilising IPCC assessments to support the ecosystem approach to fisheries management within a warming Southern Ocean. Mar Policy 131:104589. https://doi.org/10.1016/j.marpol.2021.104589

DeAngelis DL, Franco D, Hastings A et al (2021) Towards building a sustainable future: positioning ecological modelling for impact in ecosystems management. Bull Math Biol 83:107. https://doi.org/10.1007/s11538-021-00927-y

Dhyani S, Kadaverugu R, Pujari P (2020) Predicting impacts of climate variability on Banj oak (*Quercus leucotrichophora* A. Camus) forests: understanding future implications for Central Himalayas. Reg Environ Change 20:113. https://doi.org/10.1007/s10113-020-01696-5

Engler J, Kusiak A (2011) Modeling an innovation ecosystem with adaptive agents. Int J Innov Sci 3:55–68. https://doi.org/10.1260/1757-2223.3.2.55

Estopinan J, Servajean M, Bonnet P et al (2022) Deep species distribution modeling from sentinel-2 image time-series: a global scale analysis on the orchid family. Front Plant Sci 13:839327

Furtado BA, Fuentes MA, Tessone CJ (2019) Policy modeling and applications: state-of-the-art and perspectives. Complexity 2019:e5041681. https://doi.org/10.1155/2019/5041681

Geary WL, Bode M, Doherty TS et al (2020) A guide to ecosystem models and their environmental applications. Nat Ecol Evol 4:1459–1471. https://doi.org/10.1038/s41559-020-01298-8

Howell D, Schueller AM, Bentley JW et al (2021) Combining ecosystem and single-species modeling to provide ecosystem-based fisheries management advice within current management systems. Front Mar Sci 7:607831

IPBES (2018) Summary for policymakers of the regional assessment report on biodiversity and ecosystem services for Asia and the Pacific of the Intergovernmental Science-Policy Platform on Biodiversity and Ecosystem Services. IPBES Secretariat, Bonn

Jetz W, McGeoch MA, Guralnick R et al (2019) Essential biodiversity variables for mapping and monitoring species populations. Nat Ecol Evol 3:539–551. https://doi.org/10.1038/s41559-019-0826-1

Kim H, Rosa IMD, Alkemade R et al (2018) A protocol for an intercomparison of biodiversity and ecosystem services models using harmonized land-use and climate scenarios. Geosci Model Dev 11:4537–4562. https://doi.org/10.5194/gmd-11-4537-2018

Kok MTJ, Kok K, Peterson GD et al (2017) Biodiversity and ecosystem services require IPBES to take novel approach to scenarios. Sustain Sci 12:177–181. https://doi.org/10.1007/s11625-016-0354-8

Leitão PJ, Santos MJ (2019) Improving models of species ecological niches: a remote sensing overview. Front Ecol Evol 7:9

Lewis KA, Rose KA, de Mutsert K et al (2021) Using multiple ecological models to inform environmental decision-making. Front Mar Sci 8:625790

Lundquist C, Hashimoto S, Denboba MA et al (2021) Operationalizing the Nature Futures Framework to catalyze the development of nature-future scenarios. Sustain Sci 16:1773–1775. https://doi.org/10.1007/s11625-021-01014-w

Malhi Y, Franklin J, Seddon N et al (2020) Climate change and ecosystems: threats, opportunities and solutions. Philos Trans R Soc Lond B Biol Sci 375:20190104. https://doi.org/10.1098/rstb.2019.0104

Mokany K, Ferrier S, Connolly SR et al (2016) Integrating modelling of biodiversity composition and ecosystem function. Oikos 125:10–19. https://doi.org/10.1111/oik.02792

Muluneh MG (2021) Impact of climate change on biodiversity and food security: a global perspective—a review article. Agric Food Secur 10:36. https://doi.org/10.1186/s40066-021-00318-5

Myers BJE, Weiskopf SR, Shiklomanov AN et al (2021) A new approach to evaluate and reduce uncertainty of model-based biodiversity projections for conservation policy formulation. Bioscience 71:1261–1273. https://doi.org/10.1093/biosci/biab094

O'Neill BC, Carter TR, Ebi K et al (2020) Achievements and needs for the climate change scenario framework. Nat Clim Chang 10:1074–1084. https://doi.org/10.1038/s41558-020-00952-0

Pan Y, Tian Y, Xu J et al (2018) Methodological assessment on scenarios and models of biodiversity and ecosystem services and impacts on China within the IPBES framework. Biodivers Sci 26:89. https://doi.org/10.17520/biods.2017228

Pereira LM, Davies KK, den Belder E et al (2020) Developing multiscale and integrative nature–people scenarios using the Nature Futures Framework. People Nat 2:1172–1195. https://doi.org/10.1002/pan3.10146

Pettorelli N, Graham NAJ, Seddon N et al (2021) Time to integrate global climate change and biodiversity science-policy agendas. J Appl Ecol 58:2384–2393. https://doi.org/10.1111/1365-2664.13985

Qazi AW, Saqib Z, Zaman-ul-Haq M (2022) Trends in species distribution modelling in context of rare and endemic plants: a systematic review. Ecol Process 11:40. https://doi.org/10.1186/s13717-022-00384-y

Rosa IMD, Lundquist CJ, Ferrier S et al (2020) Increasing capacity to produce scenarios and models for biodiversity and ecosystem services. Biota Neotrop 20. https://doi.org/10.1590/1676-0611-BN-2020-1101

Schlüter M, Müller B, Frank K (2019) The potential of models and modeling for social-ecological systems research: the reference frame ModSES. Ecol Soc 24. https://doi.org/10.5751/ES-10716-240131

Schuwirth N, Borgwardt F, Domisch S et al (2019) How to make ecological models useful for environmental management. Ecol Model 411:108784. https://doi.org/10.1016/j.ecolmodel.2019.108784

Sintayehu DW (2018) Impact of climate change on biodiversity and associated key ecosystem services in Africa: a systematic review. Ecosyst Health Sustain 4:225–239. https://doi.org/10.1080/20964129.2018.1530054

Sofaer HR, Jarnevich CS, Pearse IS, Smyth RL, Auer S, Cook GL, Edwards TC Jr, Guala GF, Howard TG, Morisette JT, Hamilton H (2019) Development and delivery of species distribution models to inform decision-making. Bioscience 69(7):544–557. https://doi.org/10.1093/biosci/biz045

Spooner E, Karnauskas M, Harvey CJ et al (2021) Using integrated ecosystem assessments to build resilient ecosystems, communities, and economies. Coast Manag 49:26–45. https://doi.org/10.1080/08920753.2021.1846152

Volkery A, Ribeiro T (2009) Scenario planning in public policy: understanding use, impacts and the role of institutional context factors. Technol Forecast Soc Chang 76:1198–1207. https://doi.org/10.1016/j.techfore.2009.07.009

Weiskopf SR, Rubenstein MA, Crozier LG et al (2020) Climate change effects on biodiversity, ecosystems, ecosystem services, and natural resource management in the United States. Sci Total Environ 733:137782. https://doi.org/10.1016/j.scitotenv.2020.137782

Weiskopf SR, Harmáčková ZV, Johnson CG et al (2022) Increasing the uptake of ecological model results in policy decisions to improve biodiversity outcomes. Environ Model Softw 149:105318. https://doi.org/10.1016/j.envsoft.2022.105318

Xu X, Trugman AT (2021) Trait-based modeling of terrestrial ecosystems: advances and challenges under global change. Curr Clim Change Rep 7:1–13. https://doi.org/10.1007/s40641-020-00168-6

Yates KL, Bouchet PJ, Caley MJ et al (2018) Outstanding challenges in the transferability of ecological models. Trends Ecol Evol 33:790–802. https://doi.org/10.1016/j.tree.2018.08.001

Zurell D, König C, Malchow A-K et al (2022) Spatially explicit models for decision-making in animal conservation and restoration. Ecography 2022. https://doi.org/10.1111/ecog.05787

Part I

Ecosystem and Species Modelling Tools and Relevance

Basic Introduction to Species Distribution Modelling

2

Azita Farashi and Mohammad Alizadeh-Noughani

Abstract

Species distribution models (SDMs) have become the most widely used method for wildlife management and have been applied in the fields of ecology, biogeography, and conservation.

Species distribution modelling commonly requires two categories of data: (1) species data and (2) environmental data. Species data can be nominal (presence/absence records), ordinal (ranked abundances), or ratio (abundance and richness). Environmental data refers to both biotic and abiotic conditions. The most common types of environmental data in species distribution modelling are climatic and topographical variables since these two sets of variables represent, respectively, the large-scale conditions relevant to species' physiology and small-scale conditions which affect solar energy input and availability of moisture. There are various techniques for species distribution modelling. The choice of modelling technique is affected by the availability of data and in turn affects modelling outcomes. The accuracy of SDMs can be measured with respect to two characteristics: discrimination capacity and reliability; generally, discrimination capacity has been seen as a more crucial metric of model performance. Accuracy is an important challenge faced by SDMs. Several factors affect the accuracy of SDMs such as environmental data, species data, the ecology of the species, available computational resources, the model being utilized, and spatial resolution.

A. Farashi (✉) · M. Alizadeh-Noughani
Department of Environment, Faculty of Natural Resources and Environment, Ferdowsi University of Mashhad, Mashhad, Iran
e-mail: farashi@um.ac.ir

S. Dhyani et al. (eds.), *Ecosystem and Species Habitat Modeling for Conservation and Restoration*, https://doi.org/10.1007/978-981-99-0131-9_2

Keywords

Distribution · Species · Habitat · Prediction · Modelling

2.1 Introduction

Species distribution models (SDMs) have become the most widely used method for predicting the impacts of global change on species and have been applied in the fields of ecology, biogeography, and conservation. SDMs have also found applications in evolutionary biology and ecology, management of invasive species, design of protected areas, and predicting the impacts of climate change (Guillera-Arroita et al. 2015). The increased interest and attention to these approaches is a result of increased availability of digital data, user-friendly software, and accessibility of guides and educational material (Zurell et al. 2020). Current SDMs are the result of integrating ideas from natural history and ecology with modern innovations in statistics and information technology. The history of this approach in ecology can be traced to past studies which linked biological patterns with environmental variations such as geographical gradients (e.g., Grinnell 1904). Furthermore, works which highlighted the unique response of individual species, rather than communities, to environmental variables motivated the development of approaches to model individual species (Miller 2010). Correlative SDMs infer species-environment relationships and use them to predict species distributions. That is, an SDM infers relationships between the distributions of species (as occurrence or abundance records) and environmental variables at those locations to provide a picture of potential distributions at the landscape level. In the literature, these models have also been referred to as resource selection functions (RSFs), bioclimatic models, range maps, ecological niche models (ENMs), habitat models, climate envelopes, and correlative models or spatial models. In this chapter, we review studies of SDMs and aim to systematically review the available knowledge on species distribution modelling to offer ecologically relevant insights.

2.2 The Modelling Process

The process and input of modelling are largely determined by its goals. However, any SDM regardless of its specific goals and methodology includes three main elements: (1) data on species, (2) environmental covariates, and (3) a modelling algorithm. In the following sections, we review these three essential steps through discussing (1) data preparation, (2) variable selection, (3) model construction, and (4) model evaluation (Fig. 2.1).

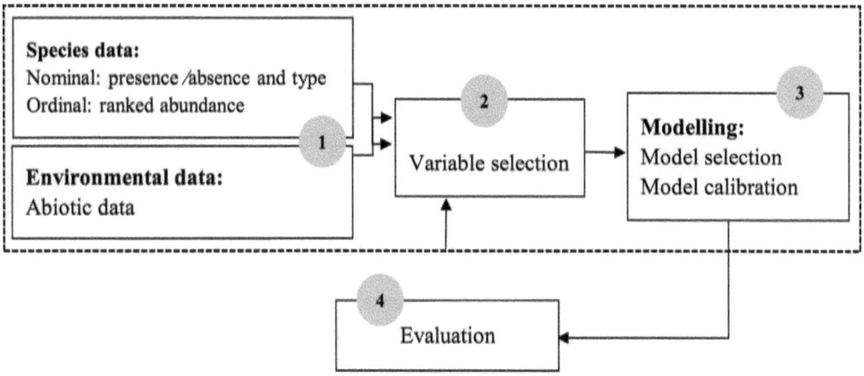

Fig. 2.1 The four major steps involved in SDMs

2.3 Data Types

Species distribution modelling commonly requires two categories of data: (1) species data and (2) environmental data.

2.3.1 Species Data

Species data can be nominal (presence/absence records), ordinal (ranked abundances), or ratio (abundance and richness). The type of available data partially determines the appropriate modelling approach and therefore affects modelling output (e.g., whether the model generates suitability values or expected abundances) (Miller 2010). In addition to the level of measurement, the ratio between observations and predictor variables also informs the modelling process. Vaughan and Ormerod (2003) recommend a minimum ratio of ten observations for each predictor, while Franklin (2010) recommends a 20:1 ratio in general and a 40:1 ratio when stepwise correlation is utilized. Miller et al. (2007) caution that these ratios can be affected by spatial autocorrelation.

Collection of species data is often one of the expensive steps in the process of creating SDMs. Species data often comes from past biodiversity surveys/counts and records at natural history museums. Data collected through these methods often represent only part of the ecological reality due to spatial, temporal, and taxonomic biases (Rocchini et al. 2011). For instance, available data on most Neotropical species is usually limited, and the recorded locations are inaccurate and the metadata inappropriate. Using such biased date to construct, calibrate, or test SDMs leads to increased error, particularly in the case of species for which few points have been recorded (Rocchini et al. 2011; Kamino et al. 2012).

Two general solutions can reduce and counter the biases and errors in species data. First, the reliability of records should be evaluated using automatic workflows and robust and regular revisions to omit unreliable data while preserving outliers. Taking into consideration the route taken by expeditions and collection dates can assist with deducing range shifts over time as well differentiating errors from outliers. During the preparation of species data, the expertise of the data collector in identifying species and locations, and fieldwork in general, should also be considered in resolving suspicious cases. The other approach to solving bias and inaccuracy in species data is to undertake additional surveys to supplement data from the available repositories. This solution is especially suitable for rare or under-studied species (Kamino et al. 2012).

Another limitation of many species occurrence records is the lack of absence data, rendering such data unfit as input for many modelling algorithms and evaluation approaches. One solution proposed to overcome the unavailability of absence data is to generate "pseudo-absences" where species have not been observed. Several studies have focused on the different methods to generate pseudo-absences and the effects of including absence data on model performance (e.g., Lobo and Tognelli 2011; Cerasoli et al. 2017). Other issues arising from uncertainty can also impact species data. For instance, it is possible that a species is regarded as absent despite being present at a certain location due to low detectability. This issue is more frequent when the organism of interest is highly mobile or hard to detect. Although detectability is often associated with animal species, plants can also go undetected despite being present in the seed bank or as their seasonal cryptic form (e.g., as underground organs in winter) (Franklin 2010). Absences might also be recorded in suitable habitat due to biotic interactions, dispersal limitations, and disturbances, among other factors.

There is also the implicit assumption in SDMs that the locations of occurrence records for a certain species are independent although this is not necessarily the case. As a result, the geographical distance between occurrence points heavily affects the disparity in the value of variables. Spatial autocorrelation (i.e., similarity between places in close proximity) is a common feature in most spatial data (Segurado et al. 2006) and can influence geospatial analyses (Anselin et al. 2004). Spatial autocorrelation increases the rate of false positives (type I errors) and can affect parameter estimation and therefore model selection (Lennon 2000). Spatial Analysis in Macroecology (SAM; Rangel et al. 2006) calculates and graphs Moran's correlograms to visualize spatial autocorrelation for occurrence records according to the distance between points. To evaluate spatial autocorrelation in occurrence data, Farashi and Alizadeh-Noughani (2021) evaluated the significance of Moran's I using a randomization test with 9999 Monte Carlo permutations, adjusted for multiple testing. When spatial autocorrelation was present in the data, the testing and training datasets were restricted according to the following steps: first, a distance threshold was imposed on the data based on the distance lags which showed spatial autocorrelation (10–25 km in their study). Next, points which were closer to each other than the threshold values were aggregated and considered to occupy the same partition (Parolo et al. 2008).

2.3.2 Environmental Data

Environmental data refers to both biotic and abiotic conditions. The most common types of environmental data in species distribution modelling are climatic and topographical variables since these two sets of variables represent, respectively, the large-scale conditions relevant to species' physiology (by representing temperature and moisture) and small-scale conditions which affect solar energy input and availability of moisture. Geological and edaphic variables can also be used as environmental input for SDMs, but these variables are usually available at relatively low resolutions due to their categorical nature. Distance from natural or man-made features such as sources of water and roads is sometimes used to construct models to represent distance to relevant landscape features such resources or disturbances (Miller 2010). Moreover, the distribution of other species can be included to represent biotic interactions such as predation or competition, or to stratify sampling schemes to enhance the detection of rare species (Heikkinen et al. 2007).

Remote sensing products and metrics from landscape ecology can offer data on habitat structure, biophysical conditions, landscape patterns, and heterogeneity (Lausch et al. 2015; Dorph et al. 2021). Satellite imagery offers a wide array of data on variables such as land cover, soil water content, vegetation, fraction of photosynthetically active radiation, and leaf area indices. Although the Internet has made remotely sensed data widely available, using these products should be done with due consideration for their resolution, accuracy, and interpretability. Remote sensing products should only be used in species distribution modelling if the variable and the resolution of the product can competently capture the ecological phenomenon of interest (Kamino et al. 2012).

The hypothesis to be addressed by modelling determines which environmental variables should be used in modelling since hypothesis entails inherent assumptions about the variables and mechanisms affecting species distributions. Therefore, it is essential to pay close attention to species' ecology and the history of the study area in order to select environmental variables which capture the salient variables for the modelled species (Kamino et al. 2012).

Modellers should also be cautious about collinearity among variables since it can confound the identification of the most important predictors, especially with small sample sizes. Sadly, the exact impact of collinearity on model predictions is yet unknown since the effects of multicollinearity have rarely been assessed in the literature. As a solution, some studies have suggested performing principal component analysis prior to modelling and using the results with the greatest explanatory power as the environmental predictors (e.g., Townsend Peterson et al. 2007; Raney and Leopold 2018). Other studies have used correlation coefficients (such as Pearson's r) to screen modelling inputs (e.g., Hosseini et al. 2019; Moghadam et al. 2021; Farashi and Karimian 2021). To do so, pairwise correlation is performed between all variables, and those with $|r| > 0.7$ or 0.8 are considered to be strongly correlated and are therefore eliminated as environmental predictors.

SDMs suffer from some limitations with respect to the possible inputs because the algorithms are not capable of understanding all factors affecting distribution as some factors cannot be easily represented as formulae. For instance, the static nature of SDMs means they cannot account for population dynamics or spatial dynamics such as connectivity and fragmentation. However, SDMs can still provide some utility even in such cases. For instance, an SDM can first predict the potential distribution of a species, and its output can then be filtered based on the effect of connectivity or other habitat variables at different time steps.

2.4 Modelling Techniques

2.4.1 Single Algorithm Techniques

Table 2.1 presents a list of SDM algorithms. The choice of modelling technique is affected by the availability of data and in turn affects modelling outcomes (Jiménez-Valverde et al. 2008). This highlights the need for evaluating models after modelling is performed. For instance, Farashi and Alizadeh-Noughani (2021) compared different SDMS with respect to their performance using environmental variables that were known to affect the distribution of species and found MaxEnt and general additive models (GAMs) to have the best performance. However, such evaluations of performance should not be the sole criteria for model selection since other considerations such as model complexity and number of parameters, incorporation of ecological mechanisms, input type, and generalizability should also be considered.

Models are by nature incomplete representations of real conditions, but the degree to which a model attempts to capture reality depends on whether realism, generalizability, or precision is the priority; better results in any one of these three dimensions often come at the expense of the other two. Overall, modellers should not try to choose models which best fit the data. Rather, the focus should be on choosing a technique that can address the research question most accurately. In many cases, a simpler model with sound theoretical foundations is preferable to a more complex model that overfits the data (Kamino et al. 2012).

2.4.2 Ensemble Techniques

The availability of numerous modelling algorithms, the unique strengths and weaknesses of each algorithm, and the fact that the choice of model can affect predictions have led to the development of ensemble modelling techniques. Ensemble modelling is founded on the notion that each model reveals some true "signal" about real-world relationships and some noise generated due to the data and the model's shortcomings. Thus, ensemble modelling combines different models to better distinguish the signal from the noise (Dormann et al. 2018). Ensemble modelling has been applied in fields other than ecology when systems with high degrees of complexity have to be modelled, including in meteorology, economics, and Gregory et al. (2001). This approach is also used in machine learning to combine

Table 2.1 Species distribution models (*P* presence, *A* absence, *B* background)

Type	Name	Description	Type of variable	Reference
Regression models: A classic method to evaluate the relationship between species data (presence/absence or counts) and environmental data	Generalized linear models (GLMs)	The independent variables in GLMs can include interaction and polynomial terms. GLMs are preferred for simple nonlinear species-environment relationships	P/A	Guisan et al. (2002)
	Generalized additive models (GAMs)	A nonparametric extension of GLMs, thus potentially better able to fit some data than GLMs. GAMs use data-defined smoothing functions to fit nonlinear species-environment relationships. GAMs are preferred when species-environment relationships are more complex and not adequately captured by GLMs	P/A	Guisan et al. (2002)
	Multivariate adaptive regression splines (MARS)	An extension of linear models that automatically models nonlinearities and interactions. These models are especially useful for datasets with many predictors and lower-order interaction effects	P/A	Friedman (1991)
Classification model: Compared with regression models, classification models are much more robust against outliers	Flexible discriminant analysis (FDA)	An extension of linear discriminant analysis. Linear discriminant analysis is closely related to general linear models. FDA is based on mixture models	P/A	Hastie et al. (1994)
	Classification and regression tree (CART)	These models recursively partition the data into smaller homogenous parts. CART models are able to detect complex interactions between explanatory variables, which might be	P/A	Vayssières et al. (2000)

(continued)

Table 2.1 (continued)

Type	Name	Description	Type of variable	Reference
		overlooked by other multivariate techniques		
Complex models: Frameworks that contain simple models. Complex models can detect the hidden features of the data. These models are suitable for datasets with strongly correlated variables or extensive correlation structures. However, complex models might reproduce minor details of training data, leading to over fitting and lowering generalizability	Random forest (RF)	These models are based on decision trees, implemented using Breiman's random forest algorithm for classification and regression. RF models are robust against multicollinearity, missing data, and unbalanced datasets. They can also be used to detect variable interactions	P/A	Breiman (2001)
	The genetic algorithm for rule set production (GARP)	An SDM that develops rule sets for determining species distributions. The rule set includes mathematical rules to limit environmental (e.g., species range) and species-environment interactions (e.g., regression patterns)	P/B	Stockwell and Peters (1999)
	The maximum entropy (MaxEnt) method	A general-purpose machine learning algorithm that estimates target probabilities by finding the distribution with maximum entropy (i.e., closest to uniform) under the constraints that the expected value of each environmental variable be equal to its empirical average (average value of the variable at a sample of points from species distribution)	P/B	Phillips et al. (2006)
	Artificial neural network (ANN)	Nonlinear models with a very large number of parameters, making them flexible enough to represent any smooth function. ANNs can function as multiple	P/A	Lek and Guégan (1999)

(continued)

Table 2.1 (continued)

Type	Name	Description	Type of variable	Reference
		regression for continuous variables or classification for categorical variables. The accuracy of these algorithms is largely determined by weight decay of the links and number of hidden neurons. ANNs can update model parameters based on new observations		

simple modelling units into complex classifiers. Ensemble models have shown superior performance than individual models in many cases (Seni and Elder 2010). Ensemble modelling approaches utilize a number of methods to construct models and have been fundamental in the development of some SDM approaches (e.g., Guo et al. 2015; Kindt 2018; Hao et al. 2020; Kaky et al. 2020).

The literature includes a variety of methods for combining models. The simplest, and the most intuitive, is calculating the mean or median across predictions regardless of data type. More complex schemes assign weights to predictions by different models based on some measure of predictive performance. The weights are often obtained after model validation on some of the test data. Weighting improves predictive performance of the ensemble but takes more effort to implement since individual models need to be validated before being combined into the ensemble (Araújo and New 2007).

2.5 Model Evaluation

The accuracy of SDMs can be measured with respect to two characteristics: discrimination capacity and reliability (Pearce and Ferrier 2000); generally, discrimination capacity has been seen as a more crucial metric of model performance (Ash and Shwartz 1999). This metric captures the ability of the model to distinguish presence sites from absence sites. Reliability captures the degree of consistency between predicted and observed presence sites (Pearce and Ferrier 2000). For models which output continuous results, both dimensions can be evaluated; however, only discrimination capacity can be evaluated for models with binary output. Several metrics have been proposed to assess discrimination capacity and reliability, some of which are applicable exclusively to binary results or continuous results converted to binary using a threshold value (threshold-dependent measures). In contrast, metrics which can be directly applied to continuous predictions are called threshold-independent measures. By applying systematic variations to threshold values, the optimal value of a threshold-dependent metric can be obtained according to a pre-defined definition of what constitutes optimal. Since this process utilizes several threshold values, the obtained optimal value of the threshold-dependent measure can be treated in same manner as threshold-independent measures (Liu et al. 2011).

2.5.1 Threshold-Dependent Measures

Threshold-dependent measures take advantage of a confusion matrix (Table 2.2). Table 2.3 presents some of the threshold-dependent measures used to evaluate SDMs. Sensitivity (Se), specificity (Sp), true skill statistic (TSS), and kappa are based on conditional probability and are some of the most common metrics for model performance. The Se represents the probability that presence is accurately predicted at a site, and Sp represents the likelihood that a species is accurately predicted as absent at a site. TSS and Youden's index (J) are equivalent and are widely used in dichotomous diagnostic tests in medicine. In the context of SDMs, these measures are defined as mean net prediction success rate for presence and absence sites. TSS and Youden's index have gained popularity for evaluating the performance of SDMs (e.g., Farashi and Shariati 2017; Farashi and Erfani 2018; Makori et al. 2022). Kappa measures the extent to which agreement between predicted and observed values is greater than what is expected from chance alone. This metric has been developed to overcome the problem of overestimating accuracy and has been employed for meteorological applications since the nineteenth century (Murphy 1996), where it is referred to as Heidke skill score (Stephenson 2000).

2.5.2 Threshold-Independent Measures

Table 2.3 presents threshold-independent metrics used in model evaluation. Area under the receiver operating characteristic curve (AUC) is one of the most commonly used metrics of model accuracy, including SDMs (e.g., Sobek-Swant et al. 2012; Wei et al. 2018; Soilhi et al. 2022). In species distribution modelling, AUC is equal to the probability that a randomly selected presence site will be ranked higher than a randomly selected absence site (Pearce and Ferrier 2000). AUC can also be

Table 2.2 The confusion matrix used to represent the accuracy of classification for binary data. *a* and *d* represent correct predictions (true positives and negatives, respectively); *b* and *c* represent incorrect predictions (false positives and negatives, respectively)

		Observed	
		Present	Absent
Predicted	Present	*a*	*b*
	Absent	*c*	*d*

Table 2.3 Metrics used to evaluate the accuracy of SDMs

Type	Metric	Reference
Threshold-dependent	Overall accuracy	Finley (1884)
	True skill statistic	Peirce (1884)
	Kappa	Cohen (1960)
	Normalized mutual information	Finn (1993)
	Sensitivity	Fielding and Bell (1997)
	Specificity	Fielding and Bell (1997)
	Positive predictive value	Fielding and Bell (1997)
	Negative predictive value	Fielding and Bell (1997)
	Positive likelihood ratio	Glas et al. (2003)
	Negative likelihood ratio	Glas et al. (2003)
	Odds ratio	Glas et al. (2003)
	F measure	Daskalaki et al. (2006)
	Yule's Y	Kraemer (2006)
	Yule's Q	Kraemer (2006)
	Phi coefficient	Kraemer (2006)
	Extreme dependency score	Stephenson et al. (2008)
Threshold-independent accuracy index	Mean square error	Brier (1950)
	Point biserial correlation coefficient	Tate (1954)
	Rank biserial correlation coefficient	Glass (1966)
	Proportion of explained deviance	Mittlböck and Schemper (1996)
	Maximum kappa	Guisan et al. (1998)
	Coefficient of determination	Ash and Shwartz (1999)
	Maximum vertical distance	Lek and Guégan (1999)
	Adjusted proportion of explained deviance	Guisan and Zimmermann (2000)
	Gini index	Hand (2001)
	Maximum overall accuracy	Stockwell and Peterson (2002)
	Area under ROC curve (AUC)	Mason and Graham (2002)
	Mean absolute prediction error	Schemper (2003)
	Root mean square error	Caruana and Niculescu-Mizil (2004)
	Mean cross entropy	Caruana and Niculescu-Mizil (2004)
	Partial area under ROC curve (PAUC)	He and Escobar (2008)

calculated as the average value of Se over all possible values of Sp, or vice versa (Jiang et al. 1996). AUC has drawn some criticism due to utilizing parts of the prediction range which do not have practical applications, leading to an inaccurate evaluation of model performance (Lobo et al. 2008). In response, McClish (1989) developed the partial AUC (PAUC).

2.6 Application of SDMs in Paleobiogeography

SDMs have found a number of paleobiogeographical applications (see Svenning et al. 2011), including studying the effects of climate change on the distribution of species over time (McGuire and Davis 2013), nature and causes of extinctions (e.g., Meseguer et al. 2018), locations of glacial refugia (e.g., Carnaval and Moritz 2008; Schmickl et al. 2010; Gavin et al. 2014), preservation of traits which are tied to an organism's niche over time (e.g., McDonald and Bryson 2010), and the impact of historical changes in climate on modern genetics structures (Alexandrino et al. 2007). SDMs have improved our knowledge of factors affecting species' distribution and evolution, their response to environmental change, the impacts of extreme climate events, and the role of glacial refugia in determining the current distribution of species.

However, the application of SDMs in paleobiogeography faces some challenges. For instance, similar to the challenges associated with present species data, the data on the occurrence of species might not fully reflect the environmental conditions that are suitable for species. Such underestimations of species' range could lead to predictions that underestimate historic distributions. However, this problem can be diminished if the model utilizes the entirety of a species' temporal and geographic range and does not use absence data as input. In the absence of data on a species' climatic niche, more conservative approaches and interpretations are advised. Regardless, the application of SDMs for the past periods still faces three main issues: the difficulty of obtaining species data (presence/absence), the difficulty of making extended predictions, and the difficulty of validating projections (Varela et al. 2011).

2.7 Niche Theory in SDM

The idea of ecological niche forms the theoretical foundations of SDMs. Although niche is a fundamental idea in ecology, it has been interpreted in different ways both in general ecological literature and specifically for SDMs (Araujo and Guisan 2006; Hirzel and Le Lay 2008; Franklin 2010). Hutchinson's definition of niche is one of the most widely used definitions of this concept. According to him, the niche of a species is an "n-dimensional hypervolume" which can indefinitely support the organism (Hutchinson 1957). This concept is further divided into the fundamental niche (the range of potentially habitable conditions) and the realized niche (the portion of the fundamental niche inhabited as a function of biotic interactions). However, some challenges have been posed to Hutchinson's definition of ecological niche. Araujo and Guisan (2006) state that Hutchinson limited biotic interactions to negative interactions, while positive interactions would compromise the hypervolume definition of niche. Such interactions often occur at spatial and temporal scales that are too small to be captured by SDMs. Also, biotic processes such as dispersal are similarly fundamental to the distribution of species, yet a temporally static concept of niche does not allow for such variations over time (Araujo and Guisan 2006; Soberón 2007).

It is crucial that the phenomena being modelled or mapped are clearly defined for the application of SDMs; the assumptions and limitations of the data and methodology should also be clarified in advance. Although the concept of ecological niche is fundamental to SDMs, what SDMs are actually modelling in most cases are species' habitats. To delineate these two concepts, Kearney (2006) places the two concepts in a hierarchy, with niche models being the outcome of mechanistic analyses considering morphology, physiology, behavior, and species-environment interactions and habitat models referring to the outcome of descriptive/correlative analysis often using environmental variables. Environmental variables can be considered as the dimensions of n-dimensional hypervolume proposed by Hutchinson and species' responses to those variables as the determinants of their distributions. Based on this interpretation, the exact properties of a species' response curve for an environmental covariate reflect some measure of species presence (abundance or occurrence) under changes in that environmental variable. Response curves often display the minima and maxima where a species is expected to occur (tolerances) and a mode representing optimal conditions with respect to that environmental variable.

Another issue associated with Hutchinson's definition of niche arises from the distinction between fundamental and realized niche. It follows from Hutchinson's definition that two species which utilize, and are limited by, a common resource cannot co-occur at the same place since the superior competitor is expected to drive out the other species. In other words, we expect not to observe intersecting realized niches. In the context of SDMs, the significance and applicability of this concept is heavily affected by the interaction between the scale of the analysis (both in terms of resolution and extent) and the type of organism being studied. While Hutchinson mostly focused on species with small ranges and formulated his ideas at the scale of communities, SDMs are commonly applied at much larger scales (regions of even continents). Also, the resolution of input data in SDMs can be quite coarse (1–50 km), meaning that competing species can occur at the same location as far as the model is concerned but move to avoid competitors in the real world. In effect, this means that species whose realized niches overlap can be present at the same location. Even if fine-grained data is used, such species could still reach local equilibrium. For instance, weakly competitive species could randomly establish at unoccupied sites (Hutchinson's "fugitive" species) or temporally partition their resource use (e.g., nocturnal and diurnal species using the same resource). Thus, species whose realized niches overlap can coexist in space and time (Araujo and Guisan 2006).

2.8 Challenges in SDM

SDMs have become an important tool in theoretical and applied research in biogeography. However, important conceptual ambiguities must be addressed before models can generate more reliable outputs (Rodríguez-Rey et al. 2013). Among the major issues faced by SDMs is the implicit assumption that species inhabit their environment at equilibrium states (Franklin 2010). However, a state of equilibrium is relatively uncommon in nature (Gaston 2009). In fact, the need for modelling a species distribution is more strongly felt when there is no equilibrium or the equilibrium has been disrupted, such as the first stages of invasion (Peterson 2003; Srivastava et al. 2019).

Accuracy is another challenge faced by SDMs. Several factors affect the accuracy of SDMs (Allouche et al. 2008) such as environmental data (e.g., type and variance of data; Aguirre-Gutiérrez et al. 2013), species data (e.g., geographical accuracy, sample size, field survey limitations, or autocorrelation; Huettmann and Diamond 2006), the ecology of the species (e.g., distribution, abundance, niche; Beale et al. 2008; Saupe et al. 2012), available computational resources (very fine resolutions can be computationally demanding), the model being utilized (e.g., presence only, presence-absence), and spatial resolution (changes in spatial resolution can affect spatial patterns) (Graham and Hijmans 2006; Farashi and Alizadeh-Noughani 2021). Many researchers have evaluated the effects of these factors on the performance of models. For example, Stockwell and Peterson (2002) reported the effects of spatial resolution on the performance of a genetic algorithm and a logistic regression algorithm for predicting the distribution of birds. Similarly, Hernandez et al. (2006) studied the effect of spatial resolutions on 17 vertebrates and 1 insect in California. Using 400,000 records of 328 species over 200 years in the Netherlands, Aguirre-Gutiérrez et al. (2013) evaluated the effects of a number of factors on model performance including models, environmental variables, spatial distribution, and spatial resolution. Farashi and Alizadeh-Noughani (2021) evaluated how the outputs of different models are affected by the spatial resolution of input data. However, no study has comprehensively investigated the effects of input parameters on the performance of SDMs.

Finally, the concept of realized ecological niche also poses a challenge to SDMs. According to Hutchinson (1957), the spatial distribution of a species is shaped by its environmental tolerances (fundamental niche) and biotic interactions (realized niche). However, decoupling the influence of these two aspects is not easy. Often only species-environment (climate and topography) interactions are explicitly accounted for in modelling (Austin 2002), while biotic interactions are only implicitly considered (Dormann et al. 2012) despite their significant influence on species' distributions (Wisz et al. 2013; Pollock et al. 2014). Recent techniques such as joint species distribution models (JSDMs) have attempted to take the realized ecological niche of species into account. Pollock et al. (2014) offer a broad description of the JSDM used for frogs and eucalypt trees in Victoria, Australia.

References

Aguirre-Gutiérrez J, Carvalheiro LG, Polce C, van Loon EE, Raes N, Reemer M, Biesmeijer JC (2013) Fit-for-purpose: species distribution model performance depends on evaluation criteria–Dutch hoverflies as a case study. PLoS One 8(5):e63708

Alexandrino J, Teixeira J, Arntzen JW, Ferrand N (2007) Historical biogeography and conservation of the golden-striped salamander (*Chioglossa lusitanica*) in northwestern Iberia: integrating ecological phenotypic and phylogeographic data. In: Phylogeography of southern European refugia. Springer, Dordrecht, pp 189–205. https://doi.org/10.1007/1-4020-4904-8_6

Allouche O, Steinitz O, Rotem D, Rosenfeld A, Kadmon R (2008) Incorporating distance constraints into species distribution models. J Appl Ecol 45(2):599–609. https://doi.org/10.1111/j.1365-2664.2007.01445.x

Anselin L, Bongiovanni R, Lowenberg-DeBoer J (2004) A spatial econometric approach to the economics of site-specific nitrogen management in corn production. Am J Agric Econ 86(3):675–687. https://doi.org/10.1111/j.0002-9092.2004.00610.x

Araujo MB, Guisan A (2006) Five (or so) challenges for species distribution modelling. J Biogeogr 33(10):1677–1688. https://doi.org/10.1111/j.1365-2699.2006.01584.x

Araújo MB, New M (2007) Ensemble forecasting of species distributions. Trends Ecol Evol 22(1):42–47. https://doi.org/10.1016/j.tree.2006.09.010

Ash A, Shwartz M (1999) R2: a useful measure of model performance when predicting a dichotomous outcome. Stat Med 18(4):375–384. https://doi.org/10.1002/(SICI)1097-0258(19990228)18:4<375::AID-SIM20>3.0.CO;2-J

Austin MP (2002) Spatial prediction of species distribution: an interface between ecological theory and statistical modelling. Ecol Model 157(2–3):101–118. https://doi.org/10.1016/S0304-3800(02)00205-3

Beale CM, Lennon JJ, Gimona A (2008) Opening the climate envelope reveals no macroscale associations with climate in European birds. Proc Natl Acad Sci 105(39):14908–14912

Breiman L (2001) Random forests. Mach Learn 45(1):5–32. https://doi.org/10.1023/A:1010933404324

Brier GW (1950) Verification of forecasts expressed in terms of probability. Mon Weather Rev 78(1):1–3

Carnaval AC, Moritz C (2008) Historical climate modelling predicts patterns of current biodiversity in the Brazilian Atlantic Forest. J Biogeogr 35(7):1187–1201. https://doi.org/10.1111/j.1365-2699.2007.01870.x

Caruana R, Niculescu-Mizil A (2004) Data mining in metric space: an empirical analysis of supervised learning performance criteria. In: Proceedings of the tenth ACM SIGKDD international conference on knowledge discovery and data mining, pp 69–78. https://doi.org/10.1145/1014052.1014063

Cerasoli F, Iannella M, D'Alessandro P, Biondi M (2017) Comparing pseudo-absences generation techniques in Boosted Regression Trees models for conservation purposes: a case study on amphibians in a protected area. PLoS One 12(11):e0187589. https://doi.org/10.1371/journal.pone.0187589

Cohen J (1960) A coefficient of agreement for nominal scales. Educ Psychol Meas 20(1):37–46. https://doi.org/10.1177/001316446002000104

Daskalaki S, Kopanas I, Avouris N (2006) Evaluation of classifiers for an uneven class distribution problem. Appl Artif Intell 20(5):381–417. https://doi.org/10.1080/08839510500313653

Dormann CF, Schymanski SJ, Cabral J, Chuine I, Graham C et al (2012) Correlation and process in species distribution models: bridging a dichotomy. J Biogeogr 39(12):2119–2131. https://doi.org/10.1111/j.1365-2699.2011.02659.x

Dormann CF, Calabrese JM, Guillera-Arroita G, Matechou E, Bahn V et al (2018) Model averaging in ecology: a review of Bayesian information-theoretic and tactical approaches for predictive inference. Ecol Monogr 88(4):485–504. https://doi.org/10.1002/ecm.1309

Dorph A, Swan M, Di Stefano J, Penman TD (2021) Relating mammal species richness to landscape patterns across multiple spatial scales. Landsc Ecol 36(4):1003–1022. https://doi.org/10.1007/s10980-021-01208-8

Farashi A, Alizadeh-Noughani M (2021) Predicting the invasion risk of non-native reptiles as pets in the Middle East. Glob Ecol Conserv 31:e01818. https://doi.org/10.1016/j.gecco.2021.e01818

Farashi A, Erfani M (2018) Modeling of habitat suitability of Asiatic black bear (Ursus thibetanus gedrosianus): in Iran in future. Acta Ecol Sin 38(1):9–14. https://doi.org/10.1016/j.chnaes.2017.07.003

Farashi A, Karimian Z (2021) Assessing climate change risks to the geographical distribution of grass species. Plant Signal Behav 16(7):1913311. https://doi.org/10.1080/15592324.2021.1913311

Farashi A, Shariati M (2017) Biodiversity hotspots and conservation gaps in Iran. J Nat Conserv 39:37–57. https://doi.org/10.1016/j.jnc.2017.06.003

Fielding AH, Bell JF (1997) A review of methods for the assessment of prediction errors in conservation presence/absence models. Environ Conserv 24(1):38–49. https://doi.org/10.1017/S0376892997000088

Finley JP (1884) Tornado predictions. Am Meteorol J 1(3):85

Finn JT (1993) Use of the average mutual information index in evaluating classification error and consistency. Int J Geogr Inf Sci 7(4):349–366. https://doi.org/10.1080/02693799308901966

Franklin J (2010) Mapping species distributions: spatial inference and prediction. Cambridge University Press

Friedman JH (1991) Multivariate adaptive regression splines. Ann Stat 19(1):1–67. https://doi.org/10.1214/aos/1176347963

Gaston KJ (2009) Geographic range limits: achieving synthesis. Proc R Soc B Biol Sci 276(1661):1395–1406. https://doi.org/10.1098/rspb.2008.1480

Gavin DG, Fitzpatrick MC, Gugger PF, Heath KD, Rodríguez-Sánchez F, Dobrowski SZ et al (2014) Climate refugia: joint inference from fossil records species distribution models and phylogeography. New Phytol 204(1):37–54. https://doi.org/10.1111/nph.12929

Glas AS, Lijmer JG, Prins MH, Bonsel GJ, Bossuyt PM (2003) The diagnostic odds ratio: a single indicator of test performance. J Clin Epidemiol 56(11):1129–1135. https://doi.org/10.1016/S0895-4356(03)00177-X

Glass GV (1966) Note on rank biserial correlation. Educ Psychol Meas 26(3):623–631

Graham CH, Hijmans RJ (2006) A comparison of methods for mapping species ranges and species richness. Glob Ecol Biogeogr 15(6):578–587. https://doi.org/10.1111/j.1466-8238.2006.00257.x

Gregory AW, Smith GW, Yetman J (2001) Testing for forecast consensus. J Bus Econ Stat 19(1):34–43. https://doi.org/10.1198/07350010152472599

Grinnell J (1904) The origin and distribution of the chest-nut-backed chickadee. Auk 21(3):364–382

Guillera-Arroita G, Lahoz-Monfort JJ, Elith J, Gordon A, Kujala H, Lentini PE et al (2015) Is my species distribution model fit for purpose? Matching data and models to applications. Glob Ecol Biogeogr 24(3):276–292. https://doi.org/10.1111/geb.12268

Guisan A, Zimmermann NE (2000) Predictive habitat distribution models in ecology. Ecol Model 135(2–3):147–186. https://doi.org/10.1016/S0304-3800(00)00354-9

Guisan A, Theurillat JP, Kienast F (1998) Predicting the potential distribution of plant species in an alpine environment. J Veg Sci 9(1):65–74. https://doi.org/10.2307/3237224

Guisan A, Edwards TC Jr, Hastie T (2002) Generalized linear and generalized additive models in studies of species distributions: setting the scene. Ecol Model 157(2–3):89–100. https://doi.org/10.1016/S0304-3800(02)00204-1

Guo C, Lek S, Ye S, Li W, Liu J, Li Z (2015) Uncertainty in ensemble modelling of large-scale species distribution: effects from species characteristics and model techniques. Ecol Model 306:67–75. https://doi.org/10.1016/j.ecolmodel.2014.08.002

Hand DJ (2001) Measuring diagnostic accuracy of statistical prediction rules. Statistica Neerlandica 55(1):3–16. https://doi.org/10.1111/1467-9574.00153

Hao T, Elith J, Lahoz-Monfort JJ, Guillera-Arroita G (2020) Testing whether ensemble modelling is advantageous for maximising predictive performance of species distribution models. Ecography 43(4):549–558. https://doi.org/10.1111/ecog.04890

Hastie T, Tibshirani R, Buja A (1994) Flexible discriminant analysis by optimal scoring. J Am Stat Assoc 89(428):1255–1270. https://doi.org/10.1080/01621459.1994.10476866

He Y, Escobar M (2008) Nonparametric statistical inference method for partial areas under receiver operating characteristic curves with application to genomic studies. Stat Med 27(25): 5291–5308. https://doi.org/10.1002/sim.3335

Heikkinen RK, Luoto M, Virkkala R, Pearson RG, Körber JH (2007) Biotic interactions improve prediction of boreal bird distributions at macro-scales. Glob Ecol Biogeogr 16(6):754–763. https://doi.org/10.1111/j.1466-8238.2007.00345.x

Hernandez PA, Graham CH, Master LL, Albert DL (2006) The effect of sample size and species characteristics on performance of different species distribution modeling methods. Ecography 29(5):773–785. https://doi.org/10.1111/j.0906-7590.2006.04700.x

Hirzel AH, Le Lay G (2008) Habitat suitability modelling and niche theory. J Appl Ecol 45(5): 1372–1381. https://doi.org/10.1111/j.1365-2664.2008.01524.x

Hosseini M, Farashi A, Khani A, Farhadinia MS (2019) Landscape connectivity for mammalian megafauna along the Iran-Turkmenistan-Afghanistan borderland. J Nat Conserv 52:125735. https://doi.org/10.1016/j.jnc.2019.125735

Huettmann F, Diamond AW (2006) Large-scale effects on the spatial distribution of seabirds in the Northwest Atlantic. Landsc Ecol 21(7):1089–1108. https://doi.org/10.1007/s10980-006-7246-8

Hutchinson GE (1957) Concluding remarks. In: Cold spring harbor symposia on quantitative biology, vol 22, pp 415–427

Jiang Y, Metz CE, Nishikawa RM (1996) A receiver operating characteristic partial area index for highly sensitive diagnostic tests. Radiology 201(3):745–750. https://doi.org/10.1148/radiology.201.3.8939225

Jiménez-Valverde A, Lobo JM, Hortal J (2008) Not as good as they seem: the importance of concepts in species distribution modelling. Divers Distrib 14(6):885–890. https://doi.org/10.1111/j.1472-4642.2008.00496.x

Kaky E, Nolan V, Alatawi A, Gilbert F (2020) A comparison between ensemble and MaxEnt species distribution modelling approaches for conservation: a case study with Egyptian medicinal plants. Eco Inform 60:101150. https://doi.org/10.1016/j.ecoinf.2020.101150

Kamino LH, Stehmann JR, Amaral S, De Marco P Jr, Rangel TF, de Siqueira MF et al (2012) Challenges and perspectives for species distribution modelling in the neotropics. Biol Lett 8: 324–326. https://doi.org/10.1098/rsbl.2011.0942

Kearney M (2006) Habitat environment and niche: what are we modelling? Oikos 115(1):186–191. https://doi.org/10.1111/j.2006.0030-1299.14908.x

Kindt R (2018) Ensemble species distribution modelling with transformed suitability values. Environ Modell Softw 100:136–145. https://doi.org/10.1016/j.envsoft.2017.11.009

Kraemer HC (2006) Correlation coefficients in medical research: from product moment correlation to the odds ratio. Stat Methods Med Res 15(6):525–545. https://doi.org/10.1177/0962280206070650

Lausch A, Blaschke T, Haase D, Herzog F, Syrbe RU, Tischendorf L, Walz U (2015) Understanding and quantifying landscape structure–a review on relevant process characteristics data models and landscape metrics. Ecol Model 295:31–41. https://doi.org/10.1016/j.ecolmodel.2014.08.018

Lek S, Guégan JF (1999) Artificial neural networks as a tool in ecological modelling an introduction. Ecol Model 120(2–3):65–73. https://doi.org/10.1016/S0304-3800(99)00092-7

Lennon JJ (2000) Red-shifts and red herrings in geographical ecology. Ecography 23(1):101–113. https://doi.org/10.1111/j.1600-0587.2000.tb00265.x

Liu C, White M, Newell G (2011) Measuring and comparing the accuracy of species distribution models with presence–absence data. Ecography 34(2):232–243. https://doi.org/10.1111/j.1600-0587.2010.06354.x

Lobo JM, Tognelli MF (2011) Exploring the effects of quantity and location of pseudo-absences and sampling biases on the performance of distribution models with limited point occurrence data. J Nat Conserv 19(1):1–7. https://doi.org/10.1016/j.jnc.2010.03.002

Lobo JM, Jiménez-Valverde A, Real R (2008) AUC: a misleading measure of the performance of predictive distribution models. Glob Ecol Biogeogr 17(2):145–151. https://doi.org/10.1111/j.1466-8238.2007.00358.x

Makori DM, Abdel-Rahman EM, Ndungu N, Odindi J, Mutanga O, Landmann T et al (2022) The use of multisource spatial data for determining the proliferation of stingless bees in Kenya. GIScі Remote Sens 59(1):648–669. https://doi.org/10.1080/15481603.2022.2049536

Mason SJ, Graham NE (2002) Areas beneath the relative operating characteristics (ROC): and relative operating levels (ROL): curves: statistical significance and interpretation. Quart J R Meteorol Soc 128(584):2145–2166. https://doi.org/10.1256/003590002320603584

McClish DK (1989) Analyzing a portion of the ROC curve. Med Decis Mak 9(3):190–195. https://doi.org/10.1177/0272989X8900900307

McDonald HG, Bryson RA (2010) Modeling Pleistocene local climatic parameters using macrophysical climate modeling and the paleoecology of Pleistocene megafauna. Quat Int 217(1–2):131–137. https://doi.org/10.1016/j.quaint.2009.10.010

McGuire JL, Davis EB (2013) Using the palaeontological record of Microtus to test species distribution models and reveal responses to climate change. J Biogeogr 40(8):1490–1500. https://doi.org/10.1111/jbi.12106

Meseguer AS, Lobo JM, Cornuault J, Beerling D, Ruhfel BR, Davis CC et al (2018) Reconstructing deep-time palaeoclimate legacies in the clusioid Malpighiales unveils their role in the evolution and extinction of the boreotropical flora. Glob Ecol Biogeogr 27(5):616–628. https://doi.org/10.1111/geb.12724

Miller J (2010) Species distribution modeling. Geography. Compass 4(6):490–509. https://doi.org/10.1111/j.1749-8198.2010.00351.x

Miller J, Franklin J, Aspinall R (2007) Incorporating spatial dependence in predictive vegetation models. Ecol Model 202(3–4):225–242. https://doi.org/10.1016/j.ecolmodel.2006.12.012

Mittlböck M, Schemper M (1996) Explained variation for logistic regression. Stat Med 15(19):1987–1997. https://doi.org/10.1002/(SICI)1097-0258(19961015)15:19<1987::AID-SIM318>3.0.CO;2-9

Moghadam ZR, Farashi A, Rashki A (2021) Development of a framework to predict the effects of climate change on birds. Ecol Complex 47:100952. https://doi.org/10.1016/j.ecocom.2021.100952

Murphy AH (1996) The Finley affair: a signal event in the history of forecast verification. Weather Forecast 11(1):3–20. https://doi.org/10.1175/1520-0434(1996)011<0003:TFAASE>2.0.CO;2

Parolo G, Rossi G, Ferrarini A (2008) Toward improved species niche modelling: Arnica montana in the Alps as a case study. J Appl Ecol 45(5):1410–1418. https://doi.org/10.1111/j.1365-2664.2008.01516.x

Pearce J, Ferrier S (2000) Evaluating the predictive performance of habitat models developed using logistic regression. Ecol Model 133(3):225–245. https://doi.org/10.1016/S0304-3800(00)00322-7

Peirce CS (1884) The numerical measure of the success of predictions. Science 93:453–454. https://doi.org/10.1126/science.ns-4.93.453.b

Peterson AT (2003) Predicting the geography of species' invasions via ecological niche modeling. Q Rev Biol 78(4):419–433. https://doi.org/10.1086/378926

Phillips SJ, Anderson RP, Schapire RE (2006) Maximum entropy modeling of species geographic distributions. Ecol Model 190(3–4):231–259. https://doi.org/10.1016/j.ecolmodel.2005.03.026

Pollock LJ, Tingley R, Morris WK, Golding N, O'Hara RB et al (2014) Understanding co-occurrence by modelling species simultaneously with a Joint Species Distribution Model (JSDM). Methods Ecol Evol 5(5):397–406. https://doi.org/10.1111/2041-210X.12180

Raney PA, Leopold DJ (2018) Fantastic wetlands and where to find them: modeling rich fen distribution in New York state with Maxent. Wetlands 38(1):81–93. https://doi.org/10.1007/s13157-017-0958-5

Rangel TFL, Diniz-Filho JAF, Bini LM (2006) Towards an integrated computational tool for spatial analysis in macroecology and biogeography. Glob Ecol Biogeogr 15(4):321–327

Rocchini D, Hortal J, Lengyel S, Lobo JM, Jimenez-Valverde A, Ricotta C et al (2011) Accounting for uncertainty when mapping species distributions: the need for maps of ignorance. Prog Phys Geogr 35(2):211–226. https://doi.org/10.1177/0309133311399491

Rodríguez-Rey M, Jiménez-Valverde A, Acevedo P (2013) Species distribution models predict range expansion better than chance but not better than a simple dispersal model. Ecol Model 256:1–5. https://doi.org/10.1016/j.ecolmodel.2013.01.024

Saupe EE, Barve V, Myers CE, Soberón J, Barve N, Hensz CM et al (2012) Variation in niche and distribution model performance: the need for a priori assessment of key causal factors. Ecol Model 237:11–22. https://doi.org/10.1016/j.ecolmodel.2012.04.001

Schemper M (2003) Predictive accuracy and explained variation. Stat Med 22(14):2299–2308. https://doi.org/10.1002/sim.1486

Schmickl R, Jørgensen MH, Brysting AK, Koch MA (2010) The evolutionary history of the Arabidopsis lyrata complex: a hybrid in the amphi-Beringian area closes a large distribution gap and builds up a genetic barrier. BMC Evol Biol 10(1):1–18. https://doi.org/10.1186/1471-2148-10-98

Segurado P, Araujo MB, Kunin WE (2006) Consequences of spatial autocorrelation for niche-based models. J Appl Ecol 43(3):433–444. https://doi.org/10.1111/j.1365-2664.2006.01162.x

Seni G, Elder JF (2010) Ensemble methods in data mining: improving accuracy through combining predictions. In: Synthesis lectures on data mining and knowledge discovery, vol 2, no, 1, pp 1–126. https://doi.org/10.2200/S00240ED1V01Y200912DMK002

Sobek-Swant S, Kluza DA, Cuddington K, Lyons DB (2012) Potential distribution of emerald ash borer: what can we learn from ecological niche models using Maxent and GARP? For Ecol Manag 281:23–31. https://doi.org/10.1016/j.foreco.2012.06.017

Soberón J (2007) Grinnellian and Eltonian niches and geographic distributions of species. Ecol Lett 10(12):1115–1123

Soilhi Z, Sayari N, Benalouache N, Mekki M (2022) Predicting current and future distributions of Mentha pulegium L. in Tunisia under climate change conditions using the MaxEnt model. Eco Inform 68:101533. https://doi.org/10.1016/j.ecoinf.2021.101533

Srivastava V, Lafond V, Griess VC (2019) Species distribution models (SDM): applications benefits and challenges in invasive species management. CAB Rev 14:1–13. https://doi.org/10.1079/PAVSNNR201914020

Stephenson DB (2000) Use of the "odds ratio" for diagnosing forecast skill. Weather Forecast 15(2):221–232. https://doi.org/10.1175/1520-0434(2000)015<0221:UOTORF>2.0.CO;2

Stephenson SL, Schnittler M, Novozhilov YK (2008) Myxomycete diversity and distribution from the fossil record to the present. Biodivers Conserv 17:285–301

Stockwell DRB, Peters D (1999) The GARP Modeling System: problems and solutions to automated spatial prediction. Int J Geograph Inform Sci 13:143–158. https://doi.org/10.1080/136588199241391

Stockwell DR, Peterson AT (2002) Effects of sample size on accuracy of species distribution models. Ecol Model 148(1):1–13. https://doi.org/10.1016/S0304-3800(01)00388-X

Svenning JC, Fløjgaard C, Marske KA, Nógues-Bravo D, Normand S (2011) Applications of species distribution modeling to paleobiology. Quat Sci Rev 30(21–22):2930–2947

Tate RF (1954) Correlation between a discrete and a continuous variable. Point-biserial correlation. Ann Math Stat 25(3):603–607

Townsend Peterson A, Papeş M, Eaton M (2007) Transferability and model evaluation in ecological niche modeling: a comparison of GARP and Maxent. Ecography 30(4):550–560. https://doi.org/10.1111/j.0906-7590.2007.05102.x

Varela S, Lobo JM, Hortal J (2011) Using species distribution models in paleobiogeography: a matter of data predictors and concepts. Palaeogeogr Palaeoclimatol Palaeoecol 310(3–4):451–463. https://doi.org/10.1016/j.palaeo.2011.07.021

Vaughan IP, Ormerod SJ (2003) Improving the quality of distribution models for conservation by addressing shortcomings in the field collection of training data. Conserv Biol 17(6):1601–1611. https://doi.org/10.1111/j.1523-1739.2003.00359.x

Vayssières MP, Plant RE, Allen-Diaz BH (2000) Classification trees: an alternative non-parametric approach for predicting species distributions. J Veg Sci 11(5):679–694. https://doi.org/10.2307/3236575

Wei B, Wang R, Hou K, Wang X, Wu W (2018) Predicting the current and future cultivation regions of Carthamus tinctorius L. using MaxEnt model under climate change in China. Glob Ecol Conserv 16:e00477. https://doi.org/10.1016/j.gecco.2018.e00477

Wisz MS, Pottier J, Kissling WD, Pellissier L, Lenoir J, Damgaard CF et al (2013) The role of biotic interactions in shaping distributions and realised assemblages of species: implications for species distribution modelling. Biol Rev 88(1):15–30. https://doi.org/10.1111/j.1469-185X.2012.00235.x

Zurell D, Franklin J, König C, Bouchet PJ, Dormann CF, Elith J et al (2020) A standard protocol for reporting species distribution models. Dent Echo 43(9):1261–1277. https://doi.org/10.1111/ecog.04960

3 Machine Learning-Based Predictive Modelling Approaches for Effective Understanding of Evolutionary History, Distribution, and Niche Occupancy: Western Ghats as a Model

Thekke Thumbath Shameer and Raveendranathanpillai Sanil

Abstract

Machine learning enables computers to learn similarly to humans. In recent years, the use of machine learning in ecology has skyrocketed. Advances in computer science have allowed us to better combine the ever-increasing volumes of data we acquire with our knowledge of how natural systems function. These enhancements in process comprehension are essential for accurate ecological predictions. The MaxEnt software implements the maximum-entropy approach, a spatial distribution modelling (SDM) tool for biological entities. MaxEnt uses machine learning to produce accurate predictions about whether a particular species will be found in a particular place based solely on the locations in which it has previously been observed. These locations along with various environmental variables preferable by the targeted species can produce predictive distribution models. Variables like climate, topography, land use, vegetation, human impacts, etc. are commonly used. These variables are available as open sources and can be downloaded from multiple open-access database online. These future climate variables for various Representative Concentration Pathways (RCP) are available and are popularly used in these models for climate change predictions. The present research discusses the output of various studies conducted in Western Ghats (WG) using these models and points out benefits of using ML-based models for lesser-known species conservation.

T. T. Shameer
Advanced Institute for Wildlife Conservation, Tamilnadu Forest Department, Chennai, Tamil Nadu, India

R. Sanil (✉)
Molecular Biodiversity Lab, Department of Zoology and Wildlife Biology, Government Arts College, Udhagamandalam, Tamil Nadu, India

S. Dhyani et al. (eds.), *Ecosystem and Species Habitat Modeling for Conservation and Restoration*, https://doi.org/10.1007/978-981-99-0131-9_3

Keywords

Machine learning · MaxEnt · Species distribution model · Climate change · Western Ghats

3.1 Introduction

Machine learning is a fascinating computer science technology that allows computers to work without being explicitly programmed (Mitchell 2006). Machine learning, as the name suggests, allows computers to learn in the same way that humans do, and it is widely used in all aspects of life. This field of artificial intelligence learns like a person and improves its accuracy over time using data and algorithms. The three categories of machine learning methodologies are supervised, semi-supervised, and unsupervised (Grira et al. 2004). Models are used in machine learning technologies to create precise predictions (Liakos et al. 2018). In supervised machine learning, as input data is entered into the model, the weights are adjusted until the model is adequately fitted (Choi et al. 2018). This is done as part of the cross-validation procedure to avoid over fitting or under fitting the model. In supervised learning, neural networks, naive Bayes, linear regression, logistic regression, and random forest are just a few of the approaches used. The purpose of supervised learning, also known as supervised machine learning, is to efficiently classify data or predict outcomes. As input data is entered into the model, the weights are adjusted until the model is adequately fitted. This is done as part of the cross-validation procedure to avoid over fitting or under fitting the model.

Machine learning is used to assess and cluster unlabelled data sets using a method known as unsupervised machine learning (Chegini et al. 2019). These algorithms can detect patterns or groups of data without requiring human input. For camera trap data and conflict analysis, such techniques are employed in picture and pattern recognition. Unsupervised learning techniques include neural networks, k-means clustering, and probabilistic clustering. Semi-supervised learning falls between supervised and unsupervised learning. During training, a smaller, labelled data set guides classification and feature extraction from an unlabelled data set. Semi-supervised learning can be used to overcome the lack of labelled data (or the inability to afford to label enough data) (Chapelle et al. 2009).

It is now easier than ever to produce precise and unbiased predictions regarding the state of the environment. Three events occurred simultaneously to cause this: To begin with, previously lacking information about ecosystems is now easily accessible. Ecology is swiftly transitioning from a period of scarce data to one awash in information as a result of the advent of big data. There has been a major cultural shift in the scientific community in recent years toward making ecological data accessible to the public (Shameer et al. 2021a). These recent methodological discoveries have also enabled us to better combine the ever-increasing volumes of data we are collecting with our understanding of how natural systems work. These improvements in process understanding are critical for good ecological forecasting, as we face a future with no analogue conditions. Finally, the increasing availability of high-performance infrastructure for scientific computing and an increase in processing capacity in general serve as the technological foundation for both of the aforementioned tendencies. The quick uptake of machine learning in ecology can be attributed to these three novelties. While machine learning methods were not widely used (Olden et al. 2008) until recently, the popularity has skyrocketed in the past several years. Given its limited application thus far, deep learning in ecology has only been put to a few select uses.

Soberón (2010) argues that accurate distribution mapping in a sustainable habitat can be achieved by the integration of environmental data with species occurrence data. SDM, also known as ecological niche modelling (ENM) (Peterson 2006), helps in the conservation of less well-known species by resolving range assessment and optimal habitat prediction (Whittaker et al. 2005; Warren and Seifert 2011; Fourcade et al. 2014). With the help of presence data, MaxEnt (Phillips et al. 2006) can mimic the ecological niche of a wide variety of taxa, including flora and fauna (Raman et al. 2020a, b). With insufficient information, the MaxEnt machine language has proven capable of estimating the range, preferred environment, and niche suitability of species (Phillips et al. 2006; Elith et al. 2011). Distribution models will give us detailed information about the habitat, which will help us learn more about the needs and less important parts of the ecosystem that affect a species with a high management plan score. This article aims to provide an overview of prediction models, providing numerous examples of how they are used to anticipate the habitat of lesser-known species and how climate change affects species distribution.

3.2 MaxEnt Modelling

The MaxEnt software implements the maximum-entropy method, a technique for modelling the spatial distribution of biological organisms (Phillips et al. 2006). MaxEnt uses machine learning to make accurate predictions about whether a certain species will be found in a certain area based only on where that species has been seen before. This method of prediction is gaining popularity since it outperforms similar algorithms in terms of precision. Using environmental factors as backdrop points, the software may be used to determine the greatest entropy in a geographical data set of target species. This is comparable to the concept of increasing the log likelihood of species presence data and removing it from the penalty term, which is akin to the concept of AIC. Each of the used environment variables is assigned a weight based on the amount of complexity it adds. In addition, an empirically derived regularization parameter will be integrated into the weighting. The total of these weights defines how a penalty should be given to the probability to prevent over fitting. The MaxEnt's regularization parameters are derived from a study conducted by Phillips and Dudik (2008); however, users have the ability to adjust this value, which is advised in the default circumstance.

The best model is the one with the highest entropy under particular conditions. MaxEnt is the model of choice when it comes to extrapolating species distributions with remarkable exactness (Bosso et al. 2018; Soucy et al. 2018; Zhang et al. 2018). Assuming a uniform distribution, the software begins to operate and runs continuous iterations, increasing the chances of finding an appropriate spot (Merow et al. 2013). Logistic output is often used. This is the probability of a species' binary argument given the environmental variables (Merow et al. 2013). Using logical output, we can discern between appropriateness of different sites. The settings of the regularization multiplier (rm) can be modified to change the models. To change the model's complexity, several feature types can be utilized, such as linear (L), product (P), quadratic (Q), and hinge (H). A bias grid can also be created by computing the Gaussian kernel density of sample localities while taking into account the possibility of bias in the data. Using a subsampling technique with a number of repeats, a "N" number of iterations can be used to train the models, and a "100-N" number of iterations may be used to test them. The jackknife method can be used to assess the significance of all environmental variables.

3.3 The Climatic Variables

Any properly assigned variable can be used in MaxEnt habitat suitability modelling. Bioclim variables are a collection of 19 climatic variables from data sets offered by the WorldClim database (worldclim.org). Combining monthly temperature and precipitation values yielded these bioclimatic variables. Bioclimatic factors are frequently employed in species distribution prediction models. The bioclimatic variables are given annual temperature, annual precipitation, or climatic extremes. Table 3.1 provides information about bioclimatic factors. The variables 1 to 4 (BIO1 to BIO4) reflect the annual temperature, while the variable 12 (BIO12) represents the annual precipitation. The climatic variables 5 to 11 (BIO5 to BIO11) display the varying quarterly or monthly temperature extremes. Quarters are a group of 3 months in a year, with four quarters such as cold, warm, wet, and dry. BIO13 and BIO14 provide the precipitation data for 2 severe months, while BIO15 is the coefficient of variance for this data (precipitation seasonality). Precipitation data for the four quarters is represented by the numbers 16 to 19 (BIO16, BIO17, BIO18, and BIO19).

3.4 Climate Change and Habitat Suitability

Numerous websites provide future climate statistics based on the three Representative Concentration Pathways (RCP). WorldClim, CHELSA, CliMond, ecoClimate, ENVIREM, and MERRAclim are the most important data-supply databases. RCP describes the possible future climate based on the Intergovernmental Panel on Climate Change's (IPCC) greenhouse gas emission scenarios. For the prediction and analysis of climate change, four basic paths are frequently used. They had RCP values of 2.6, 4.5, 6, and 8.5. In RCP 2.6, it is assumed that greenhouse gas emissions will begin to decline by 2020, with carbon dioxide emissions reaching

Table 3.1 Bioclimatic variables used for modelling (See Bioclimatic variables—WorldClim 1 documentation for more details)

BIO1	Annual mean temperature
BIO2	Mean diurnal range (mean of monthly (max temp—min temp))
BIO3	Isothermality (BIO2/BIO7) (×100)
BIO4	Temperature seasonality (standard deviation ×100)
BIO5	Max temperature of warmest month
BIO6	Min temperature of coldest month
BIO7	Temperature annual range (BIO5–BIO6)
BIO8	Mean temperature of wettest quarter
BIO9	Mean temperature of driest quarter
BIO10	Mean temperature of warmest quarter
BIO11	Mean temperature of coldest quarter
BIO12	Annual precipitation
BIO13	Precipitation of wettest month
BIO14	Precipitation of driest month
BIO15	Precipitation seasonality (coefficient of variation)
BIO16	Precipitation of wettest quarter
BIO17	Precipitation of driest quarter
BIO18	Precipitation of warmest quarter
BIO19	Precipitation of coldest quarter

zero by the same year. In RCP 4.5, greenhouse gas emissions peak in 2040 and then fall, whereas in RCP 6, emissions peak in 2080 and then decline. Under RCP 8.5, greenhouse gas emissions will continue throughout the twenty-first century (Sharma et al. 2017). Using these notions, it is possible to model the habitat appropriateness of different species under different RCP scenarios. This will aid comprehension of the alternating distribution ranges of several species. The aforementioned databases can be used to find bioclimatic factors that can be used to make predictions about the distribution of species.

3.5 Model Appraisal

Area under the receiver operating characteristic curve (AUC) and actual skill statistics are two metrics that can be used to assess models (TSS). AUC is a threshold-independent metric used to evaluate model performance by measuring the model's ability to distinguish between random and background data. Not all models with a high AUC score have great predictive value (Phillips et al. 2006), and evaluations based only on the AUC score are not accurate. The TSS formula is sensitivity plus specificity equals one, where sensitivity and specificity are evaluated relative to the probability threshold at which they are greatest (Allouche et al. 2006).

3.6 The Western Ghats and Climate Change

From Gujarat to Goa, Kerala, through Karnataka and Tamil Nadu in India, there is a 1600-kilometer range of mountain chains known as the Western Ghats (WG). The faulted ridges of an elevated plateau make up the WG, which is not a mountain in the traditional sense (Bhat 2017). During the continental drift, when the Indian subcontinent moved close to Reunion Island 120–130 million years ago, the mountain chain was built by volcanic eruptions. Volcanic eruptions contributed to the extinction of many reptiles, including the dinosaurs, during this period. The Southern WG's 2000-million-year-old rocks provide evidence of domal uplift, which raised the WG. The fauna and geography of Peninsular India were altered by the Eocene alterations (40–45 million years ago) in the region (Karanth 2006). The high rates of uplift resulted in high heights, slopes, and gorges, which served as the cradle of speciation, resulting in the current amount of endemism. This hilly, rolling area with a wide range of landscapes and plants has a big effect on the climate of Peninsular India (Gunnell 1997).

The WG is a delicately diverse environment that hosts a wide range of rare and endangered species, making it a biodiversity hotspot (Cincotta et al. 2000; Myers et al. 2000; Shameer et al. 2019). The mountain ranges on the west coast of Peninsular India are unique due to a variety of topography, varied altitudes, different climates, and a variety of habitats. There are humid tropical conditions at lower elevations and a temperate environment with an annual average temperature of 150 ° C at higher elevations. Many instances of parapatric and allopatric species have been found in high altitudes where frigid climates are prevalent (Vijayakumar et al. 2016). Deforestation, forest encroachment, infrastructural developments, agricultural expansion, hydroelectric projects, mining, timber logging, and the extraction of forest products are some of the human-induced stresses that the WG faces today (Menon and Bawa 1997; Priti et al. 2016; Sen et al. 2016; Raman et al. 2020a, b; Shameer et al. 2021a). Tropical montane ecosystems, like WG, are undergoing fast change, but the exact rate and pattern of this change remain a mystery. Variations in the pattern of land use and land cover have a significant impact on the fragile ecosystem's biodiversity (Sukumar et al. 1995; Menon and Bawa 1997). Many species have already gone extinct due to habitat fragmentation, tourism schemes that are not based on facts, and the expansion of exotic/invasive species. Changing the landscape by removing shola-grasslands and replacing them with exotics has unpredictable consequences.

Animal metabolism and development are directly influenced by changes in CO_2 concentration, temperature, or precipitation (Hughes 2000). Due to temperature changes, the reproductive requirements, habitat selection, and feeding strategies of species may also have differential effects, which may represent an extra risk for their survival. According to the Intergovernmental Panel on Climate Change (IPCC), if global temperatures rise by 2–3 degrees Celsius, 20–30% of species will become extinct (Stocker et al. 2013; Warren et al. 2013). Climate change is causing WG's delicate biological equilibrium to be upended, resulting in an increase in dependent fauna and changes in floral composition (Shukla et al. 2003). According to the vulnerability index (Gopalakrishnan et al. 2011), the WG is more vulnerable to climate change than the northeastern forests. Local variety may suffer as a result of climate change's negative influence on water supplies (Wagner and Weitzman 2015). Changes in the trophic structure have been observed in locations that are particularly vulnerable to climate change. Because of the altered climate, non-native and invasive organisms have an advantage over their native counterparts (Hellmann et al. 2008). Because of the rising impact of humans and the introduction of invasive species, tropical montane ecosystems like WG host many threatened taxa with a restricted distribution that are vulnerable to local extinction (Arasumani et al. 2019). Invasive species (flora and fauna) can spread rapidly in a changing environment because they are able to take advantage of new niches. The successful invaders are projected to be species that are phenologically flexible and occupy the temporal niche of the indigenous species (Moran and Alexander 2014). An alien flora has the same features as native flora and disperses in the same manner. External variables like climate change play a significant role in reshaping the trophic system. Sukumar et al. (1995) found that fragile mountain ecosystems are especially at risk from climate change because of their complicated topography and biogeographic history.

3.7 Habitat Suitability Model of an Endemic Mammal

We were able to model the ideal habitat for the Western Ghats' endemic brown palm civet using the MaxEnt (Shameer et al. 2021b). Prediction models are important for reviewing or creating data for a less-known species since they understand the target species' core niche. The brown palm civet, an endemic species, is difficult to monitor because of its nocturnal habits and elusive nature (Mudappa 2006; Patou et al. 2010). A thorough understanding of a species' natural habitat helps researchers and conservationists plan suitable actions and undertake extensive monitoring. It has been suggested that the brown palm civet lives at elevations ranging from 500 to 1300 m above sea level (Rajamani et al. 2002). The Western Ghats' brown palm civet has only been studied in terms of its occurrence, diet, pelage variation, and taxonomy (Pocock 1933; Hutton 1949; Schreiber 1989; Ramachandran 1990; Ashraf et al. 1993; Ganesh 1997; Rajamani et al. 2002; Mudappa et al. 2010). It is not enough to know about a species' natural history and biology to devise an effective conservation plan. An in-depth understanding of the species' range and ideal habitat is even more important (Papeş and Gaubert 2007). Please refer to Fig. 3.1 (adapted from Shameer et al. 2021b) for a visual representation of the predicted habitat areas. According to our research, the brown palm civet was previously widespread in the Western Ghats but is now confined to just four isolated blocks. The brown palm civet's habitat was broken up by the destruction of dense rainforest, which was caused by a lot of human activity.

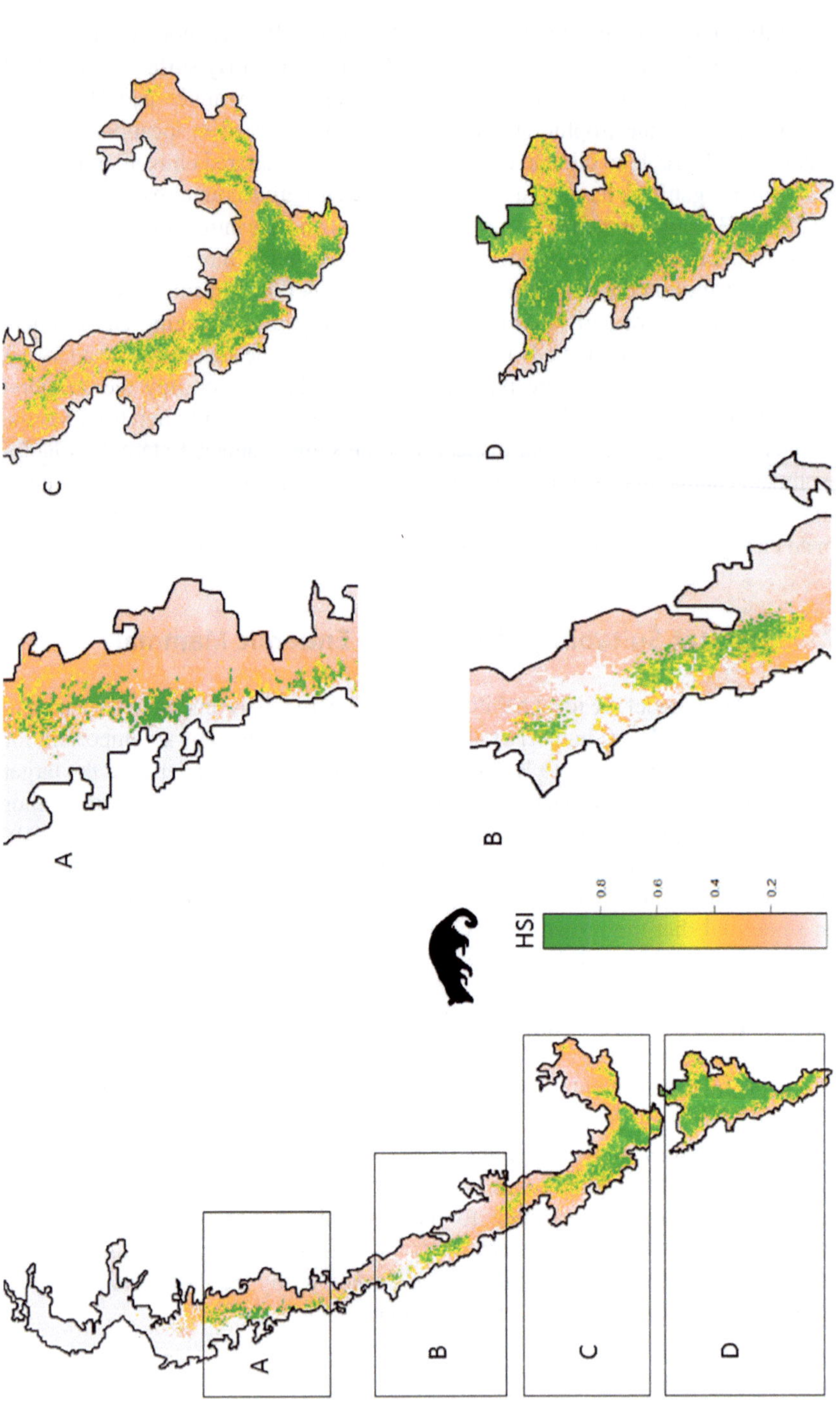

Fig. 3.1 Predicted SDM of endemic brown palm civet in Western Ghats. **(a)** The Nilgiri region; **(b)** Anamalai region; **(c)** Periyar and adjoined reserves; **(d)** Agasthyamalai region

3.8 Consequences of Climate Change on Endemic Animals

Many species' distributions, abundances, and life cycles are directly impacted by climate change as a result of global warming (Thuiller et al. 2006). In order to protect biodiversity for the future, planners and politicians must pay direct attention to the impact of climate change around the globe (Pacifici et al. 2017). Climate change is expected to have a considerable impact on species' geographic ranges, resulting in a decrease in their abundance (Warren et al. 2013). An ecological process in which climatic variables influence species niches at the spatiotemporal scale is of interest in a long-term study (MacFadyen et al. 2018). As a result, climate variables that influence species abundance and distribution can be predicted based on their response to current climatic conditions. Climate change has been implicated in numerous studies around the world, which have found that species' geographic ranges are shrinking as a result (Walther et al. 2002; Hickling et al. 2006; Priti et al. 2016; Bhattacharyya et al. 2019). Because of their diverse ecological patterns and processes, high-altitude ecosystems, also known as "sky islands," are particularly vulnerable to climate change (Raman et al. 2020b).

An essential role in the study of lesser-known species and the geographical simulation of the prospective effects of future environmental circumstances on various species has been played by ecological niche modelling (ENM) (Guisan and Zimmermann 2000). It is a very climatic-dependent species with a special geographic affinity for sky islands; it is a data-deficient high-altitude species. Two endangered species, the brown mongoose and Salim Ali's fruit bat, were modelled to examine the effects of various levels of greenhouse gas emissions. Brown mongoose and other related species would experience considerable shifts in range due to climate change. The brown mongoose's estimated range map is provided in Fig. 3.2 (adopted from Raman et al. 2020b), as is Salim Ali's fruit bat's expected range map in Fig. 3.3 (adopted from Raman et al. 2020a). The brown mongoose's range will be significantly affected by climate change in the changing climatic circumstances, according to the findings. The expected shift in Salim Ali's fruit bat's trophic composition indicates that the WG's floral composition is shifting, and this shift is reflected in the change predicted for this species. Because of this, we expect the floral and faunal composition of WG may change as a result of the shifting climatic condition.

3.9 Paleoclimatic Model and Allopatric Speciation

The shifts in endemic species' geographic distribution that have led to the current patterns can be traced back to the quaternary climate change (Hewitt 2000; Hewitt and Griggs 2004; Bose 2016; Ray et al. 2018). Climate change has resulted in a shrinking of existing ranges, culminating in the creation of new species from isolated meta-populations (Hewitt and Griggs 2004; Provan and Bennett 2008; Stewart et al. 2010; Bose 2016). During the Eocene, the Indian plate was migrating, and this is when the *Dravidogecko* evolved and became a distinct species (Chaitanya et al.

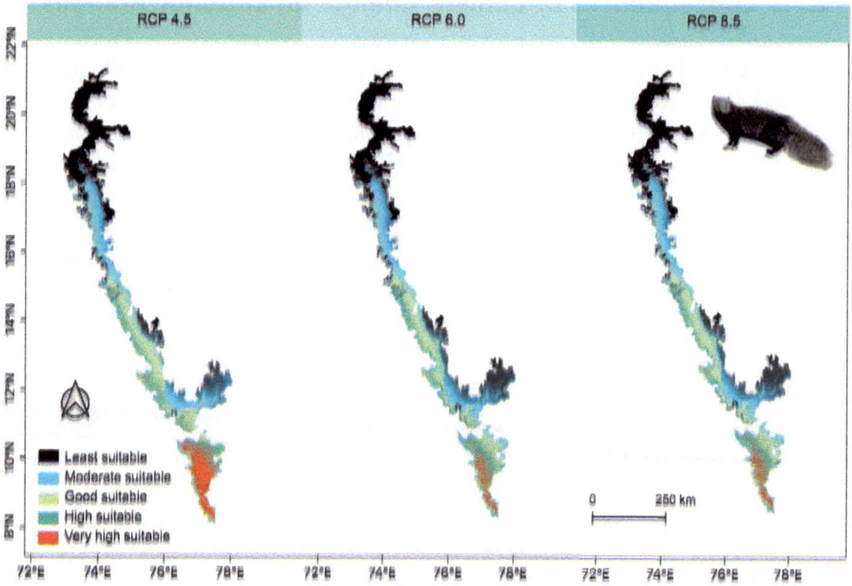

Fig. 3.2 Predicted habitat change of endemic brown mongoose in Western Ghats

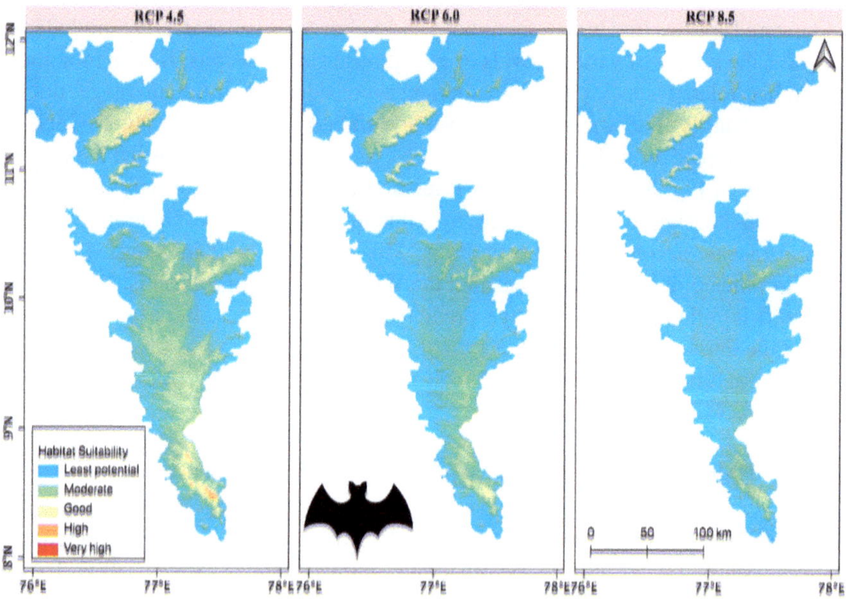

Fig. 3.3 Predicted habitat change of endemic Salim Ali's fruit bat in Western Ghats

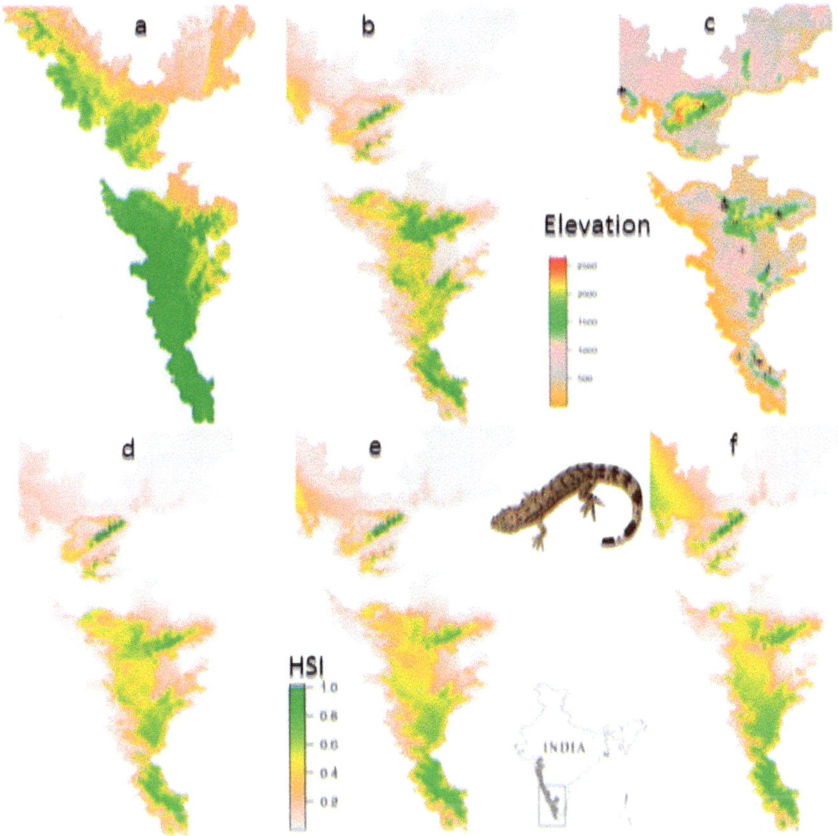

Fig. 3.4 Species distribution model of *Dravidogecko* in Western Ghats. (**a**) SDM under Pleistocene climate, (**b**) SDM under current climate, (**c**) occurrence points on the elevation map of the WG, (**d**) SDM under RCP 4.5, (**e**) SDM under RCP 6, and (**f**) SDM under RCP 8.5

2018). Many species arrived in the WG in the late Miocene to early Quaternary period when the paleoclimate was suitable, according to Gupta (2010). Insights gained from phylogeography and paleoniche modelling studies (Robin et al. 2010; Ray et al. 2018) shed light on life during and after the ice age, and the WG's constant precipitation made these locations ideal for human settlement and expansion. Based on paleoclimate theory, the dry glacial epoch may have led to species diversification. WG species diversification and its causes have been hypothesized using data collected from mountain ranges (Robin et al. 2015). During the glacial and interglacial periods, temperature changes affected forests and grasslands in the high mountains. Species may grow during the ice age and shrink during the interglacial period, with the former being more likely. By simulating historical, present-day, and future climate scenarios on the distribution patterns of old endemic reptile genera like *Dravidogecko* (Fig. 3.4), we were able to test this notion. The Nilgiris (Western Ghats) were the focus of our 1-year survey, which covered 58 diverse sites. For the

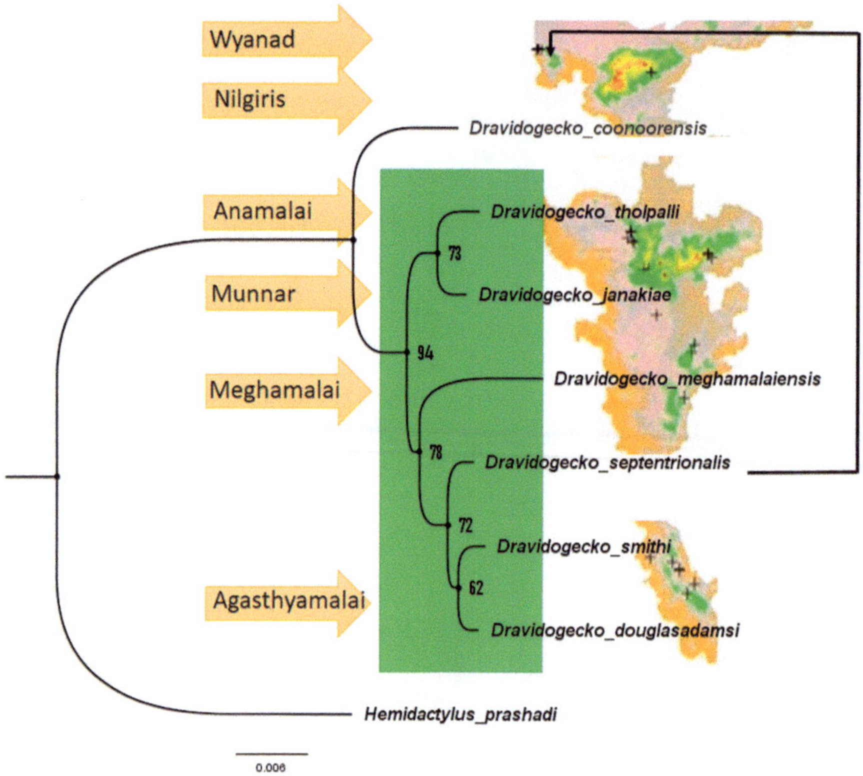

Fig. 3.5 Connecting speciation and SDM and phylogeny

distribution model, we employed environmental variables such as diurnal range, isothermality, and altitude. Modelling of past climate suggests that species currently found in the Southern Western Ghats will have existed throughout the WG during Pleistocene times. Foreseeing a new species from the Western Ghats, we combined our findings with DNA analysis (Fig. 3.5).

3.10 Limitations of Research

Species distribution models are superior machine learning techniques for predicting and mapping species' possible habitats in space and time. Consequently, these methods are now acknowledged as sustainable biodiversity management instruments (Qazi et al. 2022). However, they are not always appropriate and, if their limitations are not acknowledged by decision-makers, can lead to ineffective and costly mitigation and compensation (Carneiro et al. 2016). Robust models that account for species detectability, such as occupancy (MacKenzie et al. 2006) models, require recurrent presence and absence records. MaxEnt solely employs presence data, which has been criticized as a significant restriction. This can be circumvented by giving accurate sample data. In much of the research, the modelling relies on incorrect secondary data, posing the greatest problem and leading to inaccurate predictions. Obtaining presence-absence records for lesser-known, uncommon, or elusive species is frequently difficult for researchers. If this is the case, MaxEnt outperforms occupancy models and generates a valid species distribution map using only presence data. It is important to sample correctly if you want to make accurate predictions about where less-known species live.

3.11 Future Prospects

Modelling the spread of different species can be greatly assisted by the recent and future breakthroughs in machine learning and artificial intelligence. Species distribution modelling could benefit from including recent developments in ecological theory. Using algorithms that take into account how prey and predators interact, how competition works, and how niches change over time makes species modelling more effective.

3.12 Conclusion

The addition of environmental data to the occurrence data of species aids in the precise mapping of their distribution in a plausible habitat. Ecological niche modelling (ENM) is a technique that aids in the conservation of lesser-known species by resolving range assessment and preferred habitat prediction. Based on the presence data, MaxEnt predicts the ecological niche of a variety of species, including both plants and animals. MaxEnt's machine language has demonstrated the ability to estimate the geographic distribution, preferred habitat, and niche compatibility of species with minimal data. Hence this method can be used to identify the potential habitats of lesser-known species and develop long-term conservation plans for these species.

References

Allouche O, Tsoar A, Kadmon R (2006) Assessing the accuracy of species distribution models: prevalence, kappa and the true skill statistic (TSS). J Appl Ecol 43(6):1223–1232

Arasumani M, Khan D, Vishnudas CK, Muthukumar M, Bunyan M, Robin VV (2019) Invasion compounds an ecosystem-wide loss to afforestation in the tropical grasslands of the Shola Sky Islands. Biol Conserv 230:141. https://doi.org/10.1016/j.biocon.2018.12.019

Ashraf NVK, Kumar A, Johnsingh AJT (1993) Two endemic viverrids of the Western Ghats, India. Oryx 27:109. https://doi.org/10.1017/S0030605300020640

Bhat HR (2017) Forest guardians in the Western Ghats. Sahyadri E-News 2017(57):2

Bhattacharyya S, Mungi NA, Kawamichi T, Rawat GS, Adhikari BS, Wilkening JL (2019) Insights from present distribution of an alpine mammal Royle's pika (Ochotona roylei) to predict future climate change impacts in the Himalaya. Reg Environ Chang 19:2423. https://doi.org/10.1007/s10113-019-01556-x

Bose PS (2016) Vulnerabilities and displacements: adaptation and mitigation to climate change as a new development mantra. Area 48:168–175. https://doi.org/10.1111/area.12178

Bosso L, Smeraldo S, Rapuzzi R, Sama G, Garonna AP, Russo D (2018) Nature protection areas of Europe are insufficient to preserve the threatened beetle Rosalia alpina (Coleoptera: Cerambycidae): evidence from species distribution models and conservation gap analysis. Ecol Entomol 43(2):192. https://doi.org/10.1111/een.12485

Carneiro LR d A, Lima AP, Machado RB, Magnusson WE (2016) Limitations to the use of species-distribution models for environmental-impact assessments in the Amazon. PLoS One 11(1): e0146543. https://doi.org/10.1371/journal.pone.0146543

Chaitanya R, Lajmi A, Giri VB (2018) A new cryptic, rupicolous species of Hemidactylus Oken, 1817 (Squamata: Gekkonidae) from Meghamalai, Tamil Nadu, India. Zootaxa 4374:49–70

Chapelle O, Scholkopf B, Zien A (2009) Semi-supervised learning (Chapelle, O. et al. eds.; 2006) [book reviews]. IEEE Trans Neural Netw 20:542

Chegini M, Bernard J, Berger P, Sourin A, Andrews K, Schreck T (2019) Interactive labelling of a multivariate dataset for supervised machine learning using linked visualisations, clustering, and active learning. Vis Inform 3:9

Choi C, Kim J, Kim J, Kim D, Bae Y, Kim HS (2018) Development of heavy rain damage prediction model using machine learning based on big data. Adv Meteorol 2018:5024930

Cincotta RP, Wisnewski J, Engelman R (2000) Human population in the biodiversity hotspots. Nature 404:990–992. https://doi.org/10.1038/35010105

Elith J, Phillips SJ, Hastie T, Dudík M, Chee YE, Yates CJ (2011) A statistical explanation of MaxEnt for ecologists. Divers Distrib 17:43. https://doi.org/10.1111/j.1472-4642.2010.00725.x

Fourcade Y, Engler JO, Rödder D, Secondi J (2014) Mapping species distributions with MAXENT using a geographically biased sample of presence data: a performance assessment of methods for correcting sampling bias. PLoS One 9(5):e97122. https://doi.org/10.1371/journal.pone.0097122

Ganesh T (1997) Occurrence of the brown palm civet in the wet forest of Kalakad Mundanthurai tiger reserve, Tamil Nadu. J Bombay Nat Hist Soc 94:556. https://www.biodiversitylibrary.org/page/48601934#page/608/mode/1up

Gopalakrishnan R, Jayaraman M, Bala G, Ravindranath NH (2011) Climate change and Indian forests. Curr Sci 101:348

Grira N, Crucianu M, Boujemaa N, (2004) Unsupervised and semi-supervised clustering: a brief survey. In: A review of machine learning techniques for processing multimedia content, vol 1, p 9

Guisan A, Zimmermann NE (2000) Predictive habitat distribution models in ecology. Ecol Model 135(2–3):147. https://doi.org/10.1016/S0304-3800(00)00354-9

Gunnell Y (1997) Relief and climate in South Asia: the influence of the western ghats on the current climate pattern of peninsular India. Int J Climatol 17:1169. https://doi.org/10.1002/(SICI)1097-0088(199709)17:11<1169::AID-JOC189>3.0.CO;2-W

Gupta AK (2010) Evolution of the Indian monsoon since late Miocene intensification—marine and land proxy records. J Palaeontol Soc 55:1

Hellmann JJ, Byers JE, Bierwagen BG, Dukes JS (2008) Five potential consequences of climate change for invasive species. Conserv Biol 22:534

Hewitt G (2000) The genetic legacy of the quaternary ice ages. Nature 405(6789):907. https://doi.org/10.1038/35016000

Hewitt CD, Griggs DJ (2004) Ensembles-based predictions of climate changes and their impacts. Eos 85:566. https://doi.org/10.1029/2004EO520005

Hickling R, Roy DB, Hill JK, Fox R, Thomas CD (2006) The distributions of a wide range of taxonomic groups are expanding polewards. Glob Chang Biol 12:450. https://doi.org/10.1111/j.1365-2486.2006.01116.x

Hughes L (2000) Biological consequences of global warming: is the signal already apparent? Trends Ecol Evol 15(2):56. https://doi.org/10.1016/S0169-5347(99)01764-4
Hutton AF (1949) Notes on the snakes and mammals of the high wavy mountains, Madura District, South India Part II-Mammals. J Bombay Nat Hist Soc 48:681
Karanth KP (2006) Out-of-India Gondwanan origin of some tropical Asian biota. Curr Sci 90(6): 789–792. http://www.jstor.org/stable/24089190
Liakos KG, Busato P, Moshou D, Pearson S, Bochtis D (2018) Machine learning in agriculture: a review. Sensors 18:2674
MacFadyen S, Zambatis N, Van Teeffelen AJ, Hui C (2018) Long-term rainfall regression surfaces for the Kruger National Park, South Africa: a spatio-temporal review of patterns from 1981 to 2015. Int J Climatol 38:2506
MacKenzie D, Nichols J, Royle J et al (2006) Occupancy estimation and modelling: inferring patterns and dynamics of species occurrence. Academic Press, Burlington, MA, p 324
Menon S, Bawa KS (1997) Applications of geographic information systems, remote-sensing, and a landscape ecology approach to biodiversity conservation in the Western Ghats. Curr Sci 73:134
Merow C, Smith MJ, Silander JA (2013) A practical guide to MaxEnt for modeling species' distributions: what it does, and why inputs and settings matter. Ecography 36:1058. https://doi.org/10.1111/j.1600-0587.2013.07872.x
Mitchell TM (2006) The discipline of machine learning, vol 9. Carnegie Mellon University, School of Computer Science, Machine Learning Department, Pittsburgh
Moran EV, Alexander JM (2014) Evolutionary responses to global change: lessons from invasive species. Ecol Lett 17:637. https://doi.org/10.1111/ele.12262
Mudappa D (2006) Day-bed choice by the brown palm civet (Paradoxurus jerdoni) in the Western Ghats, India. Mamm Biol 71:238. https://doi.org/10.1016/j.mambio.2006.01.003
Mudappa D, Kumar A, Chellam R (2010) Diet and fruit choice of the brown palm civet Paradoxurus jerdoni, a viverrid endemic to the Western Ghats rainforest, India. Trop Conserv Sci 3:282. https://doi.org/10.1177/194008291000300304
Myers N, Mittermeler RA, Mittermeler CG, Da Fonseca GAB, Kent J (2000) Biodiversity hotspots for conservation priorities. Nature 403:853. https://doi.org/10.1038/35002501
Olden JD, Lawler JJ, Poff NL (2008) Machine learning methods without tears: a primer for ecologists. Q Rev Biol 83:171. https://doi.org/10.1086/587826
Pacifici M, Visconti P, Butchart SHM, Watson JEM, Cassola FM, Rondinini C (2017) Species' traits influenced their response to recent climate change. Nat Clim Chang 7(3):205–208. https://doi.org/10.1038/nclimate3223
Papeş M, Gaubert P (2007) Modelling ecological niches from low numbers of occurrences: assessment of the conservation status of poorly known viverrids (Mammalia, Carnivora) across two continents. Divers Distrib 13(6):890–902. https://doi.org/10.1111/j.1472-4642.2007.00392.x
Patou ML, Wilting A, Gaubert P, Esselstyn JA, Cruaud C, Jennings AP, Fickel J, Veron G (2010) Evolutionary history of the Paradoxurus palm civets - a new model for Asian biogeography. J Biogeogr 37:2077. https://doi.org/10.1111/j.1365-2699.2010.02364.x
Peterson AT (2006) Uses and requirements of ecological niche models and related distributional models. Biodivers Inform. https://doi.org/10.17161/bi.v3i0.29
Phillips SJ, Dudík M (2008) Modeling of species distributions with Maxent: new extensions and a comprehensive evaluation. Ecography 31:161. https://doi.org/10.1111/j.0906-7590.2008.5203.x
Phillips SJ, Anderson RP, Schapire RE (2006) Maximum entropy modeling of species geographic distributions. Ecol Model 190(3):231–259. https://doi.org/10.1016/j.ecolmodel.2005.03.026
Pocock RI (1933) The Palm Civets or 'Toddy Cats' of the genera Paradoxurus and Paguma inhabiting British India. J Bombay Nat Hist Soc 36:856
Priti H, Aravind NA, Uma Shaanker R, Ravikanth G (2016) Modeling impacts of future climate on the distribution of Myristicaceae species in the Western Ghats, India. Ecol Eng 89:14–23
Provan J, Bennett K (2008) Phylogeographic insights into cryptic glacial refugia. Trends Ecol Evol 23:564. https://doi.org/10.1016/j.tree.2008.06.010

Qazi AW, Saqib Z, Zaman-ul-Haq M (2022) Trends in species distribution modelling in context of rare and endemic plants: a systematic review. Ecol Process 11:40. https://doi.org/10.1186/s13717-022-00384-y

Rajamani N, Mudappa D, Van Rompaey H (2002) Distribution and status of the Brown Palm Civet in the Western Ghats, South India. Small Carniv Conserv 27:6. http://nebula.wsimg.com/d42c49f79b262a71ec8426b2f9629b49?AccessKeyId=35E369A09ED705622D78&disposition=0&alloworigin=1

Ramachandran KK (1990) Recent evidence of the Brown Palm Civet, Paradoxurus jerdoni, from Silent Valley National Park, India. Mustel Viverr Conserv 3:15

Raman S, Shameer TT, Charles B, Sanil R (2020a) Habitat suitability model of endangered Latidens salimalii and the probable consequences of global warming. Trop Ecol 61:570. https://doi.org/10.1007/s42965-020-00114-5

Raman S, Shameer TT, Sanil R, Usha P, Kumar S (2020b) Protrusive influence of climate change on the ecological niche of endemic brown mongoose (Herpestes fuscus fuscus): a MaxEnt approach from Western Ghats, India. Model Earth Syst Environ 6:1795. https://doi.org/10.1007/s40808-020-00790-1

Ray PA, Bonzanigo L, Wi S, Yang YCE, Karki P, García LE, Rodriguez DJ, Brown CM (2018) Multidimensional stress test for hydropower investments facing climate, geophysical and financial uncertainty. Glob Environ Chang 48:168–181. https://doi.org/10.1016/j.gloenvcha.2017.11.013

Robin VV, Sinha A, Ramakrishnan U (2010) Ancient geographical gaps and paleo-climate shape the phylogeography of an endemic bird in the sky islands of Southern India. PLoS One 5(10): e13321. https://doi.org/10.1371/journal.pone.0013321

Robin VV, Vishnudas CK, Gupta P, Ramakrishnan U (2015) Deep and wide valleys drive nested phylogeographic patterns across a montane bird community. Proc R Soc B Biol Sci 282(1810): 20150861. https://doi.org/10.1098/rspb.2015.0861

Schreiber A (1989) Weasels, civets, mongooses, and their relatives: an action plan for the conservation of mustelids and viverrids. IUCN

Sen S, Gode A, Ramanujam S, Ravikanth G, Aravind NA (2016) Modeling the impact of climate change on wild Piper nigrum (Black Pepper) in Western Ghats, India using ecological niche models. J Plant Res 129(6):1033–1040. https://doi.org/10.1007/s10265-016-0859-3

Shameer TT, Ramesh B, Easa PS (2019) Recent records of rusty spotted cat from southern Western Ghats, India. CAT News 70:12

Shameer TT, Nittu G, Mohan G, Backer SJ, Khedkar GD, Sanil R (2021a) Consequences of climate change in allopatric speciation and endemism: modeling the biogeography of Dravidogecko. Model Earth Syst Environ. https://doi.org/10.1007/s40808-021-01284-4

Shameer TT, Backer SJ, Yogesh J, Mujawar AN, Ali SZ, Raman S, Kaushal KK, Reddy SR, Sanil R (2021b) Phenotypic variations, habitat suitability, and diel activity of the endemic brown palm civets. Geol Ecol Landscap 13:1

Sharma J, Upgupta S, Jayaraman M, Chaturvedi RK, Bala G, Ravindranath NH (2017) Vulnerability of forests in India: a national scale assessment. Environ Manag 60(3):544–553. https://doi.org/10.1007/s00267-017-0894-4

Shukla PR, Sharma SK, Ravindranath NH, Garg A, Bhattacharya S (2003) Climate change and India: vulnerability assessment and adaptation. Universities Press, Hyderabad

Soberón JM (2010) Niche and area of distribution modeling: a population ecology perspective. Ecography 33:159. https://doi.org/10.1111/j.1600-0587.2009.06074.x

Soucy JPR, Slatculescu AM, Nyiraneza C, Ogden NH, Leighton PA, Kerr JT, Kulkarni MA (2018) Vector Borne Zoonotic Dis 18:235. https://doi.org/10.1089/vbz.2017.2234

Stewart BT, Venkat AN, Rawlings JB, Wright SJ, Pannocchia G (2010) Cooperative distributed model predictive control. Syst Cont Lett 59:460. https://doi.org/10.1016/j.sysconle.2010.06.005

Stocker TF, Qin D, Plattner G-K et al (2013) Technical summary. In: Stocker TF, Qin D, Plattner G-K et al (eds) Climate change 2013: the physical science basis Contribution of Working Group

I to the Fifth Assessment Report of the Intergovernmental Panel on climate change. Cambridge University Press, Cambridge, p 33

Sukumar R, Suresh HS, Ramesh R (1995) Climate change and its impact on tropical montane ecosystems in southern India. J Biogeogr 22:533. https://doi.org/10.2307/2845951

Thuiller W, Lavorel S, Sykes MT, Araújo MB (2006) Using niche-based modelling to assess the impact of climate change on tree functional diversity in Europe. Divers Distrib 12:49. https://doi.org/10.1111/j.1366-9516.2006.00216.x

Vijayakumar SP, Menezes RC, Jayarajan A, Shanker K (2016) Glaciations gradients and geography: multiple drivers of diversification of bush frogs in the Western Ghats Escarpment. Proc R Soc B Biol Sci 283:20161011. https://doi.org/10.1098/rspb.2016.1011

Wagner G, Weitzman ML (2015) Climate shock: the economic consequences of a hotter planet. Princeton University Press, Princeton

Walther GR, Post E, Convey P, Menzel A, Parmesan C, Beebee TJC, Fromentin JM, Hoegh-Guldberg O, Bairlein F (2002) Ecological responses to recent climate change. Nature 416:389. https://doi.org/10.1038/416389a

Warren DL, Seifert SN (2011) Ecological niche modeling in Maxent: the importance of model complexity and the performance of model selection criteria. Ecol Appl 21:335. https://doi.org/10.1890/10-1171.1

Warren R, Vanderwal J, Price J, Welbergen JA, Atkinson I, Ramirez-Villegas J, Osborn TJ, Jarvis A, Shoo LP, Williams SE, Lowe J (2013) Quantifying the benefit of early climate change mitigation in avoiding biodiversity loss. Nat Clim Chang 3:678. https://doi.org/10.1038/nclimate1887

Whittaker RJ, Araújo MB, Jepson P, Ladle RJ, Watson JEM, Willis KJ (2005) Conservation biogeography: assessment and prospect. Divers Distrib 11:3. https://doi.org/10.1111/j.1366-9516.2005.00143.x

Zhang K, Yao L, Meng J, Tao J (2018) Maxent modeling for predicting the potential geographical distribution of two peony species under climate change. Sci Total Environ 634:1326. https://doi.org/10.1016/j.scitotenv.2018.04.112

4 Mapping the Impact of Climate Change on Eco-sensitive Hotspots Using Species Distribution Modelling (SDM): Gaps, Challenges, and Future Perspectives

Harish Barewar, Manish Kuntal Buragohain, and Suvha Lama

Abstract

Climate change's impact on biodiversity is expected to be significant in the twenty-first century. Climate change will influence ecologically sensitive areas, and managing these changes will be critical. This chapter focuses on the utilization of species distribution models (SDMs) in assessing climate change impacts and its associated variables on species distribution, leading to population shift, migration, and species vulnerability. The review concentrates on several species distribution models (SDMs), its application in various ecosystems and their management, the gaps in the models and modelling techniques, and the challenges in their applicability. To investigate the variables utilized for modelling the future projections of the species distribution, several SDMs were explored.

Additionally, the most commonly used SDM parameters are assessed in relation to their data inputs. However, the applicability of this metric is also evaluated for various ecosystems. Further, different SDMs were contrasted regarding how their algorithms utilized the input variables. A conventional review was conducted to examine the applicability of various SDMs in relation to climate change. The assessment concentrates on (1) climate change impacts on biodiversity and related ecologically sensitive hotspots, (2) various SDMs employed for biodiversity management, (4) SDM variables used to account for climate change, (5) the parameters and factors that influence the outcomes of SDMs, (6) how SDMs are applied in different ecosystems, and (7) a comparative of different SDMs currently used with the algorithms and variables they employ.

H. Barewar · M. K. Buragohain · S. Lama (✉)
CSIR-National Environmental Engineering Research Institute, Nagpur, Maharashtra, India
e-mail: s.lama@neeri.res.in

S. Dhyani et al. (eds.), *Ecosystem and Species Habitat Modeling for Conservation and Restoration*, https://doi.org/10.1007/978-981-99-0131-9_4

Our research includes the discussion of gaps and challenges with the use of different SDM models, such as the lack of appropriate data and the noninclusion of biotic factors. But it also discusses the future perspectives and direction of research that needs to be conducted. Given our analysis, the use of SDMs will be critical in comprehending the future effect of climate change on species dispersal and distribution in the future; however there is a need to improve the robustness of these models so accurate assessments and predictions can be made.

Keywords

Climate change · Eco-sensitive hotspots · Biodiversity · Spatial distribution · Species distribution model

4.1 Introduction

Since the advent of the Industrial Revolution, greenhouse gases (GHGs) concentrations in the atmosphere have risen to extreme levels (The Royal Society 2022). As per the Global Monitoring Laboratory, Hawaii, the CO_2 concentration has risen to 416.45 ppm (September 10, 2022), the highest level seen in the last 800,000 years. According to NOAA's 63-year record, the increase of 2.58 ppm for 2021 is the fifth highest annual increase in CO_2 levels (GML-NOAA 2013; Ahmed et al. 2022; NOAA 2022). The US Environmental Protection Agency states that this unnatural increase in CO_2 levels since the Industrial Revolution is due to anthropogenic activities (US EPA 2016), which is seconded by IPCC's AR6 report. As these emissions rise, they will warm the atmosphere, causing numerous changes in the planet's atmosphere, land, and oceans (US EPA 2016). With climate change leading to frequent extreme weather events, increasing ocean levels, melting of mountain glaciers, and warmer oceans, it poses difficult challenges for the continued survival of flora and fauna. These challenges can lead to habitat loss and food security issues (WWF 2022). The IPCC Assessment Report (AR4) concludes that many aspects of biological variables are impacted by climate change, which may have a consequence on ecosystems, the species that make up those ecosystems, the genetic diversity of such species, and ecological interactions. As a significant threat to the world's biodiversity, anthropogenic climate change can potentially wipe out thousands of species over the next century. Given that it may be challenging to preserve different species even within forest/wildlife reserves, climate change is seen as a dangerous hazard and threat (IPBES 2019).

Furthermore, there may be significant interactions between climate change and other anthropogenic effects such as addition of CO_2 and other greenhouse gases (Thomas et al. 2004). Future effects of climate change have been extensively debated in research since it has already significantly influenced species in various ways, including range shifts in a wide range of taxa. Range borders may not be as vulnerable to climate change as species abundances, which undergo a binary presence/absence, shift. However, the effects of climate change that have already taken place on species abundances are far less known. Numerous environmental factors, including habitat loss and degradation, pollution, invasive species, and exploitation, impact population abundance and occurrence (WWF 2016; Bowler et al. 2017). However, due to its effects like drought, floods, and wind, as well as indirectly due to changes in the patterns of wildfires, insects, and disease outbreaks, climate change can directly alter the distribution of species. Changes in growth, reproduction, and death impact species distributions, and there is a growing possibility that these changes will become more pronounced in the following decades (Iverson and McKenzie 2014).

Eventually, a wide range of flora and fauna population sizes and species richness are anticipated to decrease as the climate warms, and changes in species distribution occur. Numerous studies have found that the effects of climate change are already being felt by a variety of species, leading to population changes, species extinctions, phonological changes, and geographic range expansion and contraction (Parmesan 1996, 2006; Rabaiotti and Woodroffe 2019). The United Nations Framework Convention on Climate Change (UNFCCC) has long advocated that alterations as far as how nature is managed might aid in addressing the climate issue. One of the most important scientific and political recommendations for tackling the risks posed by climate and biodiversity conservation is the increased preservation of seascapes and landscapes. Geographic areas, on the other hand, that are exceptionally rich in species and ecologically distinct and/or have a high endemism species (species that occur only in that defined geographic area and nowhere else) are widely recognized as eco-sensitive hotspots and prioritized for conservation. Species in biodiversity hotspots are already transforming as a result of climate change (Kiessling et al. 2022). There are 36 hotspots around the world. Their intact habitats cover only 2.5% of the Earth's land surface, but they are available to more than half of the world's plant species.

Efficiency and effectiveness are more important than ever for area-based conservation due to rising climate change, increased human land use, and unfulfilled conservation goals (Hoffmann 2022). Although it can be challenging to predict how biodiversity will react to various sources of change, current research employs models and findings to inform risk assessment, management, and conservation efforts. To analyse biodiversity, several researchers have employed species distribution models (SDMs), which they have used for habitat restoration, species translocation, and projecting the impact of climate change on biodiversity (Araújo et al. 2019). The effects of climate change on many components of biodiversity and ecosystems are shown in Fig. 4.1.

As a crucial tool for preserving biodiversity, species distribution models have vital applications such as spatially prioritizing conservation efforts and illuminating the connections between environmental predictors and species responses. These models are most effective for conservation managers when they contain easily manipulable elements (Swan et al. 2021). Therefore, the effects of climate change on ecologically sensitive hotspots must be examined to identify components that may be utilized for management or included in various models to assess the change in species distribution. However, research has demonstrated that predictions from

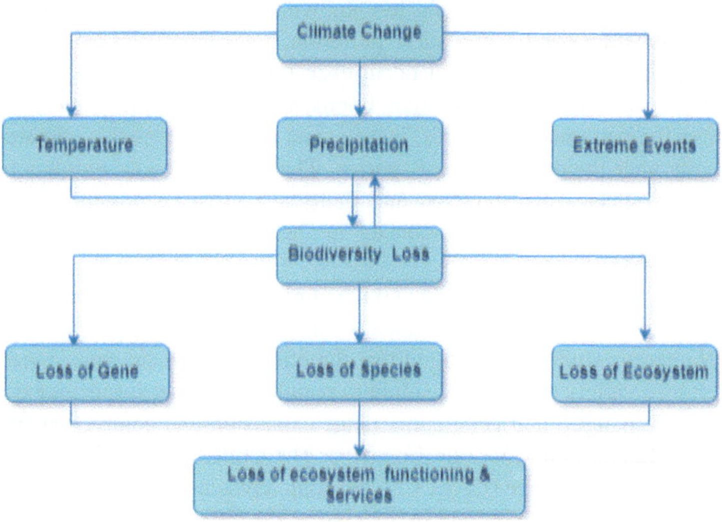

Fig. 4.1 Impacts of climate change on biodiversity (Adapted from Sintayehu 2018)

correlative and mechanistic modelling techniques should be employed in tandem rather than in opposition. A combination of correlative and mechanistic SDMs aids in guiding conservation actions in the context of climate change, as well as detecting data gaps and focusing data gathering activities. However, when compared to the RCP 8.5 scenario, the projection result revealed the rate of climate change and its influence on species distribution (Rougier et al. 2015). Furthermore, SDMs' forecasting has become a strong tool for conservation practitioners and resource managers in the face of changing climates and dwindling habitats for many species. SDMs can forecast changes in a species' geographic range under various climate change scenarios. Representative concentration pathways (RCPs) are the developments of scenario sets combining emissions, concentrations, and land use trajectories that depict these climate change scenarios. RCPs anticipate a hypothetical future situation and allow SDMs to capture alterations in a species' appropriate habitat. This is an excellent tool for proactively monitoring and planning conservation activities for specialized species at risk of extinction and dwindling habitat due to climate change (Guisan et al. 2013; Driver et al. 2020). The following section will summarize the effects of changing climate on eco-sensitive hotspot and the use of various species distribution models to understand these effects.

4.2 Impact of Climate Change on Eco-sensitive Hotspots

Humanity's biggest concern is the continual reduction of biodiversity, which undermines ecosystems' capacity to acclimatize to changing environmental conditions and hinders the provision of ecosystem services (Sala et al. 2000). The importance of forests in ecosystem conservation and management is highlighted by the sensitivity of forest biodiversity and the dependency of the majority of terrestrial species on forest ecosystems (Parrotta et al. 2012; Bellard et al. 2012).

Nearly 25% of terrestrial biodiversity hotspot regions have undergone drying 5.4% or wetting 19.3%, according to a study on global biodiversity hotspots, which are both biodiversity reservoirs and severely vulnerable (Aukema et al. 2017). The Himalaya Hotspot, Indian Western Ghats and moist forest regions in South Western Ghats, Western Himalaya temperate forest region, African savanna region and Horn of Africa Hotspot, Sri Lanka Hotspot, Coastal Forest Hotspot in the East Africa, and moist forest of Sri Lanka are the priority areas for upkeep and management as these eco-sensitive areas show the greatest change in precipitation and population (>60%) (Aukema et al. 2017).

There is much disagreement over the scope, persistence, and implications of climatic variability for the emergence of species that are less resilient to environmental change. Compared to flora, fauna is poorly suited to endure future climatic shifts due to climate stability throughout evolutionary periods (Malcolm et al. 2006). Traits like restricted climatic endurance, high habitat specificity, low dispersion capability, weak dormancy potential, small population densities, and/or low genetic variation and diversity make them more sensitive to climate change impact (Harrison and Noss 2017).

In the tropics, species typically inhabit substantially smaller temperature regimes than their temperate equivalents. Elevational range size often declines with latitude. In addition to those experiencing current stability, biodiversity hotspots that have not had exceptionally high levels of climatic fluctuation across centuries are expected to be particularly vulnerable to climate change (Trew and Maclean 2021). Climate change, in particular temperature and rainfall variability, will critically impact wildlife resources. The increase in climate-related extreme events could have an impact on species range shift, migration, or even extinction directly or indirectly. Extreme weather events can further combine with other anthropogenic stresses to affect changes in the distribution and availability of animal resources leading to changes in the species distribution (IPCC 2018).

These consequences of climate variability render biodiversity vulnerable, necessitating risk assessment and management. By predicting habitat appropriateness in regions with few or no occurrence data, the species distribution model may be used to fill informational gaps. These models may also be used to predict how environmental changes will affect species distribution. Given the severe threat of invasive species, land use change, and climate change posed to the functions of ecosystems in general, it is critical to understand the impacts of future GHG emission scenarios and implement conservation and management strategies. Here SDMs may play a very critical role, especially in areas where physical verification is not possible, in helping policy-makers understand where the interventions are required. Therefore, the next section discusses the various species distribution models currently in use.

4.3 Types of Species Distribution Models Currently Used

Understanding species distributions is crucial for environmental management (Robinson et al. 2017). Species distribution modelling (SDM) is based on fundamental ecological and biogeographical ideas concerning the interaction between species distributions and the physical environment. SDMs are quantitative, empirical models of the connections between species and their environments that are often created utilizing information on species distribution and related environmental variables. SDMs may be applied to any taxon, including marine, terrestrial, and freshwater species, and at any granularity and extent, provided that the necessary data is available (Elith and Franklin 2017). Because models sometimes lack a mechanical foundation and rely on unworkable assumptions under climate change, the accuracy of SDM projections has been questioned (Kearney et al. 2010). The descriptions of several SDM types currently in use are listed below. The different type of SDMs, their data requirements, evaluation and validation requirement and final outcomes is shown in Fig. 4.2.

4.3.1 Empirical Model

Empirical models use data from the entire census or a representative sample. It investigates how variables, recorded at various resolutions, may be used to simulate spatial patterns (Buckland and Elston 1993). Identifying a species' presence and/or future habitat relies on empirical models. The empirical models, which project variations in distribution, are also referred to as niche or habitat suitability models, or bioclimatic envelope models. To determine the appropriateness of a habitat, empirical models analyse correlations between species distributions and abiotic variables (Franklin 2010).

The predictor variables are frequently chosen using an understanding of the species' physiology. Alternatively, they could be picked using empirical best fit that has no direct connection or reference to physiology. As a result, the non-incorporation of the mechanistic depiction of abiotic and biotic interactions has been questioned in empirical models (Estes et al. 2013). According to research, SDMs using environmental change scenarios alone are insufficient to evaluate the danger of extinction for most species (Akçakaya et al. 2006). It has been argued that including more realistic assumptions about the dispersion or species migration into the empirical SDMs might help improve the estimate of climate change's impact on species distributions. A simple method has been to assume a constant migratory rate for the species examined (Thuiller et al. 2008).

4.3.2 Correlative Model

Correlative species distribution models (C-SDMs) establish mathematical correlations between environmental variables and observed species occurrence

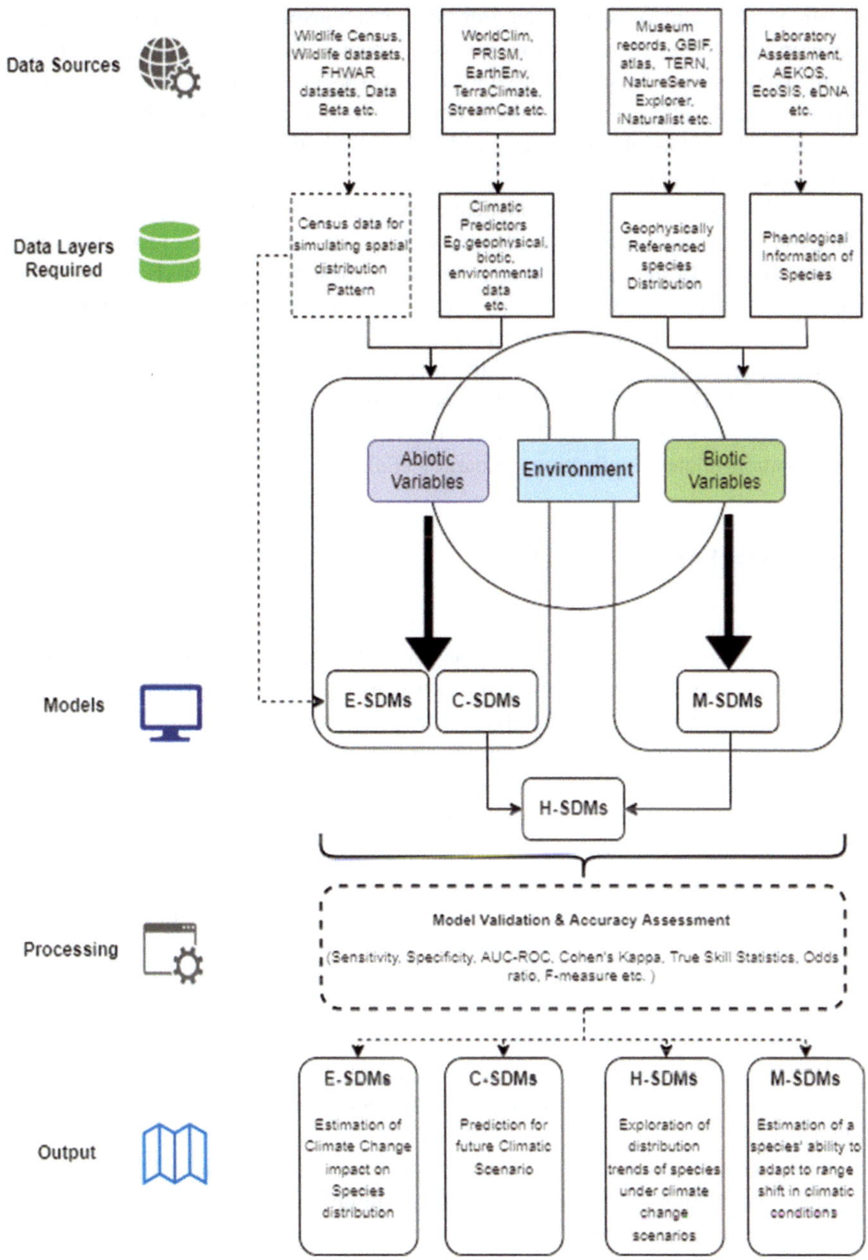

Fig. 4.2 Framework illustrating the different species distribution models (SDMs), their data requirements, validation, and final outcomes

locations. Correlative species-environment relationships are frequently used to develop hypotheses regarding the drivers of species distributions instead of testing them. They are also frequently used to forecast species occurrence or its favourable environmental settings and climatic conditions in locations that have not been sampled. These models are widely utilized in ecological forecasting with climate change scenarios (Jarnevich et al. 2015).

Correlative models use the statistical correlation between regional environmental data and occurrence records to identify mechanisms that impose restrictions on a species' range. In the past 20 years, the use of these SDM methods has increased rapidly (Elith and Leathwick 2009). Because of their flexibility and simplicity of their data needs, relative ease of access and application within open-source packages, and variety of relationships that can be modelled (biotic and abiotic), correlative SDMs have practical benefits over more mechanistic modelling approaches. Correlative SDMs have therefore been extensively employed in conservation-related applications (Kearney et al. 2010).

4.3.3 Mechanistic Model

The mechanistic species distribution models (M-SDMs) consider how the surrounding environment influences the physiological performance of the species in question. Then, a process of elimination is used to anticipate future distribution, excluding any regions that interfere with physiological function from the final distribution to the point that the ability to survive, develop, or reproduce is damaged (Kearney and Porter 2009). An increasing body of research supports the advantages of utilizing models with mechanistic variables to relate predicted climate change to mechanisms that influence species distributions (Kearney et al. 2008).

Given their capacity to extrapolate beyond known circumstances and extract variables that impact biogeography, mechanistic models have also been claimed to be the best method to project environmental impact and plan management strategies (Cuddington et al. 2013). Furthermore, mechanisms that restrict distributions are explicitly included in mechanistic SDMs (Kearney and Porter 2009). Environmental restrictions with physiological roots often influence species abundance and their spatiotemporal distribution. These physiological processes are closely linked to energy fluxes and mass as individuals of a species engage with their surrounding environment.

The effects of climate change on biological diversity are spread to upper echelons of the organization, such as populations, communities, and ecological systems, through such mechanisms. To build mechanistic models of range boundaries, which are unconstrained to species' current range, the study of biophysical ecology offers a platform for estimating the physiological repercussions of various environmental and bioclimatic variables on a spatial scale as a function of climate, topography, and vegetation. Physiologically based SDMs can more accurately forecast the effects of climate change since they explicitly integrate recognized processes in the model (Kearney and Porter 2009; Kearney et al. 2010).

4.3.4 Hybrid Model

Hybrid species distribution models (H-SDMs) are often termed as niche population models (Fordham et al. 2013a). In addition to interspecific interactions, individual variability and local adaptation, dispersion or transport, or demography, H-SDMs build on C-SDMs, which represent filtering by the abiotic environment. These hybrid model-based studies indicated changes in range predictions with better accuracy than those models solely based on abiotic environmental data because they parameterized biotic processes using extra ecological knowledge (Swab et al. 2015; Singer et al. 2018).

Regarding model complexity and data needs, the H-SDM technique can be categorized as midway between correlative and mechanistic species distribution models. It consists of two components: the relationship between observations of a species and its abiotic environment and biological phenomena and processes essential for the species dispersion, which include population dynamics, prey-predator relationships, and dispersal dynamics. Incorporating all of these factors, instead of only accounting for abiotic environmental factors, can enhance predictions of species distributions and possible range changes (Singer et al. 2018). These hybrid models incorporate spatially explicit mechanisms that function on both finer and coarser spatial scales and can be used to first identify the presence and absence of a species by taking environmental variables into consideration first, followed by biological and ecological aspects (Singer et al. 2016; Barber-O'Malley et al. 2022).

4.4 Parameters Influencing Species Distribution Models

Over the past 20 years, over 6000 peer-reviewed scientific articles have used SDMs for biodiversity assessment; half of these articles applied their findings to at least 1 type of biodiversity assessment, such as predicting the effects of climate change on biodiversity, choosing locations for protected areas, habitat restoring, species migration, etc. (Araújo et al. 2005), using several parameters and datasets such as species distribution data, meteorological data, elevation, cold-air drainage, topography, solar irradiation, soil moisture, etc. (Morán-Ordóñez et al. 2017). These are known to be influencing the outcomes of any SDMs. However, several factors, such as the choice of modelling tools, assumption inadequacy, lack of biotic components, issues with spatial and temporal scales, and inherent characteristics of the species being modelled, pose additional challenges to the predictive potential of SDMs (Fernandes et al. 2019; Luan et al. 2021). Apart from these, the resolution of variables and the data representing them, the availability and quality of presence data, the applied algorithms and their configurations, and other factors may influence the final model (La Marca et al. 2019).

4.4.1 Species Distribution Data

One of the essential datasets required for species distribution modelling is species distribution data, often given as coordinates of locations or areas where the target species occurs. Such data can be labelled into *presence-background* (PB) *data*, *presence-absence* (PA) *data*, and *occupancy detection data* (DET) (Guillera-Arroita et al. 2015), and different approaches of SDMs can be used with different species data. The presence-background (PB) data comprises the location of areas where individuals of the target species have been spotted but often lack information regarding absences or places where a species has not been detected (Wang and Stone 2019). However, it is impossible to reliably forecast the actual abundance and geographical distribution of a species for models based purely on PB data, despite repeated attempts (Koshkina et al. 2017).

Contrarily, presence-absence (PA) statistics reveal whether a species was found at a group of sample sites or not. PA data offer more reliable results when the presence-only data is sparse or ambiguous. This is so that over-prediction and extension into unknown territory can be minimized by absence and/or pseudo-absence points (Senay et al. 2013). Accordingly, the occupancy detection data (DET) model shows the likelihood that a species is present but undetected at a specific location with respect to the sampling effort that has been undertaken to account for poor detection while calculating the likelihood of a species' occupancy (Beery et al. 2021).

4.4.2 Bioclimatic or Environmental Data

Another parameter that influences any species distribution model and its predictability is bioclimatic or environmental data. Some of the most commonly used bioclimatic predictors in SDMs include annual mean temperature, precipitation seasonality, precipitation of the driest period, temperature seasonality, isothermality, total annual precipitation, temperature annual range, etc. (Gardner et al. 2019). These data points are available from different sources such as WorldClim (Fick and Hijmans 2017), TerraClimate (Abatzoglou et al. 2018), EarthEnv (Domisch et al. 2015), StreamCat (Hill et al. 2016), etc.

However, the choice of environmental variables to employ as predictors presents a recurrent challenge in SDM. Although techniques for assisting in predictor selection have been established, there is still no agreement on the predictors that should be used in SDMs (Bucklin et al. 2015). The impact of various climate datasets on operational metrics, interpretation, and spatial accuracy of SDMs remains unknown notwithstanding wider accessibility and availability of climate and environmental datasets (Abdulwahab et al. 2022).

Selection of climatic datasets from particular sources also influences the outcome of any SDM. Although some sources have comparable predictor variables, the extent of spatial coverage and resolution (WorldClim, PRISM), temporal resolutions, weather station data used in generating the coverages and ranges, and the insinuation methods utilized in producing spatially consistent meshes of the variables can vary between datasets from these sources (Abdulwahab et al. 2022). Additionally, while some environmental datasets tend to cover the entire landscape with continuous grids of consistent size (WorldClim, TerraClimate) (Fick and Hijmans 2017; Abatzoglou et al. 2018), others generally encompass only a fraction of it, and the coverage's cell size might vary significantly (StreamCat, EarthEnv) (Domisch et al. 2015; Hill et al. 2016).

Often, the use of multiple and multilevel datasets can result in the increase of uncertainty in SDMs, which can be either model uncertainty or measurement uncertainty (Beale and Lennon 2012), with the former resulting from model constraints, generalizations, or assumptions when modelling extremely complicated processes (Thibaud et al. 2014) while the latter resulting from incorporating inaccurate geo-location for a species' sightings (Fernandes et al. 2019), or climatic and environmental datasets that were inconsistently compiled from a multitude of meteorological stations and time frames and interpolated during the mapping procedure (Shabani et al. 2018), which in turn can influence the accuracy of any SDM.

4.4.3 Indicators of Good Fit (Accuracy)

SDM projection can be used to assess the viability of a habitat, the effects of climate change, the influence of land use management, and the selection of areas for species rehabilitation. However, all of these exercises' success relies on how accurate the models are (Liu et al. 2009). Hence, the determination of the accuracy of any SDM model is crucial before being employed. Discrimination capacity and reliability are the two key components used in evaluating model accuracy, with the former being the model's power to distinguish between species presence sites and absence sites while the latter determining the connection between the proportions of observed species presences and the projected probabilities of species presence (Shabani et al. 2018).

Several indicators, including sensitivity and specificity (Liu et al. 2009), Cohen's kappa (Cohen 1960), true skill statistics (TSS) (Allouche et al. 2006), and the area under the receiver operating characteristic curve (AUC-ROC) (Fourcade et al. 2018), among others, have been developed to evaluate SDM's degree of accuracy. The receiver operating characteristic's (ROC) area under the curve (AUC), which compares true positive rate (sensitivity) versus false positive rate (1—specificity; commission error), is presently the most often used statistic for evaluating accuracy. Alternative methods have also been suggested, primarily because of the well-established drawbacks of AUC (reliance on the calibration area, disregarding geographic allocation of errors, and relying on the hierarchy of specificity/sensitivity along all threshold values) (Fernandes et al. 2019). Cohen's kappa and the true skill statistics (TSS) are the most popular substitutes. Kappa corrects the aggregate accuracy of the model's predicted results by adjusting it for the accuracy that is assumed to occur by coincidence, while TSS rectifies kappa's reliance on prevalence (Xu et al. 2021).

Research indicates that, among other factors, sample size, habitat heterogeneity, species body size, and range extent have an impact on the accuracy of current SDM forecasts (M. McPherson and Jetz 2007; Pöyry et al. 2008; Kharouba et al. 2009; Morán-Ordóñez et al. 2012). Some of these parameters may also affect the precision of future forecasts since they affect how well the fitted model performs.

4.5 Application of Species Distribution Models in Various Ecosystems

The majority of SDM applications are found within the broad domains of ecology and biodiversity conservation, primarily in connection with species range and habitat shifts and climate change analysis (Srivastava et al. 2019). Over time, SDM has gained prominence for a wide range of applications, including monitoring climatic change impact, detecting biological diversification through spatiotemporal patterns, regulating invasive and exotic species, tracing the distribution of vector-borne diseases, and selecting protected zones for preservation and species rehabilitation in almost all ecosystems (Chapman et al. 2016; West et al. 2017; Rahman et al. 2019). Some of its applications are discussed below.

4.5.1 Application in Urban Ecosystem

Urbanization is frequently linked to excessive anthropogenic impact on the urban ecosystem because of overpopulation, anthropogenic pollution, and forest degradation. Such effects can be experienced locally, regionally, and even globally. Given that cities currently accommodate well over 50 per cent of the world's population, it is crucial to assess the effects of climate change on urban ecosystems (Kang et al. 2020).

Urban SDMs have thus facilitated comparisons of the implications of socio-ecological variables brought on by anthropogenic activities to those resulting from natural environmental fluctuations (Liu et al. 2019). They also facilitated investigative research on the consequences of environmental factors on urban biological diversity (Fröhlich and Ciach 2019) to assess the impacts of the said variables on the urban ecosystem.

Another impact of climate change on urban ecology is the development of urban pests, which threatens the already limited urban vegetation. In contemporary urban contexts, trees offer a range of ecosystem services. However, pest pressure on city trees is often higher than that on trees in nearby natural settings, which puts them under more stress (Parsons and Frank 2019). Therefore, SDMs are applied in predicting the potential transmission and distribution of pests in urban ecosystems by performing a climate-based pest risk assessment. This results in the detection and identification of plausible pest species. This can direct surveillance programmes to regions most likely to be plagued, establish a foundation for sharing containment expenses, and promote the systematic elimination of pests between jurisdictions (Elith and Franklin 2017).

4.5.2 Application in Forest Ecosystem

One of the major applications of SDMs is in the conservation of species in the forest ecosystem. Many forest species are already endangered by climate change as well as due to anthropogenic activities. It has become crucial to identify such species and their distribution in any forest ecosystem to employ conservation measures. To conserve an endangered and threatened species, SDMs are applied to identify and distinguish corridors and passageways between protected forest lands, allowing the transfer of a species across temperature gradients. It can be used to evaluate functional redundancy in the establishment of conservation areas, connect major demographic indicators with global paradigm change models and prioritize critical habitats, and plan and prepare for an increase in the frequency of extreme weather events (Barlow et al. 2021; Qazi et al. 2022).

In a forest ecosystem, the growth and survival of a species depend on the site's bioclimatic, physicochemical, and ecological attributes. However, significant changes in forest area conditions triggered by climate change increase the risk of diminished growth species. Hence, planning and forest management are crucial to minimizing this risk (Falk and Mellert 2011). In this context, SDMs are applied in forest management and planning by conducting risk assessments for species to future climate change scenarios to determine whether a species can survive in the predicted climatic conditions. Besides SDMs characterises ecological aspects as grounds for species selection, to substitute a species having a lesser probability of adapting to future climatic conditions with a species having a higher probability of survival in the long run (Booth 2018; Pecchi et al. 2019).

Some other applications of SDMs include enhancing sampling techniques and strategies for rare, threatened, and endangered species, designating priority regions and establishing networks of conservation areas, investigating the implications of land use/land cover changes and anthropogenic footprints on the distribution and dispersion of species in forest fringes, and directing contingency plan for reintroduction of target species in the desired forest ecosystem and ecological restoration (Angelieri et al. 2016; Srivastava et al. 2019).

4.5.3 Application in Marine Ecosystem

The application of a multitude of SDMs in the marine ecosystem has been seen in recent years, with a focus on climate change's impact on marine organisms and the marine environment. The majority of SDM applications have been seen concentrating on developing initiatives for conservation measures, analysing climate impacts on marine flora and fauna, tracking the spread of invading exotic species, and comprehending the interactions between marine species and their physical and chemical environment (Gormley et al. 2015; Cheung et al. 2016). As a strategy for choosing priority conservation sites for marine species threatened by climate change, SDMs have also been applied with other notable techniques, including connectivity analysis (Robinson et al. 2017).

One prominent application of SDM in marine ecology is the management of marine fisheries and marine protected areas (MPAs). SDMs are employed to identify environmental and ecological factors that potentially affect spatiotemporal patterns of species concentrations and assemblage characteristics. Upon identification, SDMs are used to assist in designing precise spatial conservation strategies and measures for MPAs. These include incorporating crucial habitats in proposed MPAs and estimating the overall population of protected marine species. This can be used to quantify larval production rate and understand how protected populations of marine species contribute to population restocking inside and outside of MPAs (Botsford et al. 2014; Young and Carr 2015a, b).

4.6 Comparative of Different SDMs for Mapping of Impact of Climate Change

Models of species distribution use relationships between environmental factors and known species' records of occurrence to pinpoint environmental factors that might be conducive to the occurrence of populations. Using a geographic information system (GIS), base data layers map occurrences using data on temperature, vegetation, rainfall, and other meteorological parameters. There are several data sources that offer information on global climate, occurrence, and abundance. One of the most widely used sources of climatic data is WorldClim, which contains 19 bioclimatic variables, including annual mean temperature, mean diurnal range (mean of monthly maximum and minimum temperatures), isothermality, seasonality of temperature, maximum and minimum temperatures for each month, temperature annual range, mean temperature for the wettest and driest quarters, mean temperature for the warmest and coldest quarters, and more. The most common data sources include annual precipitation, precipitation during the wettest and driest months, precipitation seasonality (coefficient of variation), precipitation during the wettest and driest quarters, precipitation during the warmest and coldest quarters, and elevation.

Given a specific climate change scenario, this data may then be used to estimate where a species could relocate to or be able to persist. Early initiatives that were important in the field's development, such as BIOCLIM and DOMAIN, are examples of commonly used SDMs. MaxEnt (maximum entropy modelling), ENFA (environmental niche factor analysis), and BRT are modern systems that are regarded as more reliable. However, a package like BIOMOD includes ten algorithms, such as the artificial neural network (ANN), classification tree analysis (CTA), flexible discriminant analysis (FDA), generalized additive model (GAM), generalized boosting model (GBM), generalized linear model (GLM), multiple adaptive regression splines (MARS), maximum entropy (MaxEnt), random forest (RF), and surface range envelope (SRE), which are state-of-the-art modelling techniques to describe and model the relationships between species and climatic data.

Table 4.1 shows a variety of models and modelling algorithms used with different predictor variables to achieve particular results for the future prediction of species distribution. These models are also used to examine species distribution while maintaining climate change's impact on biodiversity.

4.7 Gaps and Challenges Associated with Mapping the Impact of Climate Change

During the last century, the biosphere has seen drastic environmental changes, leading to the reduction of the resilience of the ecosystem. In this scenario, it becomes crucial to assess the shift and predict the changes in the distribution of species due to climate change (Dutra Silva et al. 2019). These change scenarios are frequently assessed using species distribution models (SDMs). SDMs are specifically used when knowledge of species physiology is lacking; incorporating relevant environmental variables counterbalances this deficiency (Moullec et al. 2022). These empirical statistical models are prevalent in forecasting the distribution and dispersal of species diversity in the past, present, and future under diverse climate scenarios. However, when it comes to predicting future distributions under new environmental and climatic conditions, these models have certain limitations (Elith and Leathwick 2009; Moullec et al. 2022).

One of the significant limitations faced by SDMs is that they do not consider critical ecological processes like (1) species interactions, (2) adaptation, (3) population dynamics, (4) dispersal capacity, and (5) species migration (Morin and Lechowicz 2008; Elith and Leathwick 2009; Moullec et al. 2022). Generally, SDMs rely on the absence-presence of data rather than the abundance, which could lead to the under- or overestimation of species distribution at a given condition (location, time, and environment). The inclusion of species interactions has been shown to impact the performance and uncertainty of SDMs; however, how these inclusions can alter the prediction of future distributions is yet to be investigated in detail (Moullec et al. 2022). Further, these biotic interactions are generally considered to be static and have not been verified as of now.

Empirical or correlative SDMs don't factor in biological parameters; on the other hand, mechanistic SDMs (M-SDMs) or process-based models look to incorporate these dynamics into the model (Barber-O'Malley et al. 2022). However, M-SDMs require high computational capacities and are very data-intensive. The data required to compute these models are not available on a large spatial scale, leading to high uncertainty in the models in question. Current SDMs do not factor in critical biological interactions like dispersal capacities and species migration. Harrison (1991) and Hanski (1998) state that dispersal is a crucial aspect that drives the range shift of organisms to climate change especially seen in aquatic ecosystems (HARRISON 1991; Hanski 1998). Stressed populations may return to the original spawning location (homing) and shift to a non-natal location (straying). However,

Table 4.1 Comparative of various SDMs, their modelling algorithm, variables used

Sl. no.	Model	Modelling algorithm/ packages used	Climatic predictor/ variables	Outcome of the modelling	Reference
1.	SDM for marine fishes	BIOMOD	Ocean depth, distance to shore, mean sea surface temperature, salinity, and current velocity	Predicted the suitability of its habitat for both the current climate and several climate change scenarios	Zhang et al. (2019)
2.	SDM for *Cynops orientalis*	BIOMOD	19 bioclimatic variables	The shift in *C. orientalis* optimal environment was predicted	Guo et al. (2021)
3.	Acoustic species distribution models (aSDMs)	Regression models and boundary models	Acoustic data is used for calling behaviour and climatic data such as precipitation, temperature, etc.	The acoustic SDM estimated geographical and phonological variations under climate change scenarios and assessed the environmental appropriateness for calling behaviour	Desjonquères et al. (2022)
4.	Single species distribution models (SSDMs)	Generalized joint attribute model (GJAM)	Abundance data for over 250 fish species, depth, temperature (bottom and surface), and salinity (bottom and surface)	Implemented to fit distribution models for each targeted species separately to abiotic environmental factors combination to comprehend and forecast the distribution and abundance of marine fishes	Roberts et al. (2022)
5.	Ecological Envelope Model	MaxEnt	Abundance data and 19 bioclimatic variables	Planning future conservation and afforestation efforts may be aided by having helpful knowledge of	Ksiksi et al. (2019)

(continued)

Table 4.1 (continued)

Sl. no.	Model	Modelling algorithm/ packages used	Climatic predictor/ variables	Outcome of the modelling	Reference
				species, their suitable habitats, and distribution. The model is effective in forecasting species distribution under various climate change scenarios	
6.	SDM for 108 tree species	BIOMOD and an ensemble prediction based on these 10 algorithms	Precipitation, temperature, elevation, soil, other physiographic variables, species presence/ absence, abundance, and basal area	Research concludes that SDMs may have limitations in their capacity to forecast species distributions outside of the environmental variables utilized for model fitting	Charney et al. (2021)
7.	Species distribution model (SDM) for *Portunus trituberculatus*	GAM, GLM, GBM, RF, classification tree analysis (CTA), ANN, surface range envelope (SRE), flexible discriminant analysis (FDA), and MaxEnt	Species occurrence point, occurrence records, sea surface temperature, surface salinity, current velocity, and offshore distance	To anticipate potential season-appropriate habitats for the target species under the present and future climatic circumstances, the study built an ensemble SDM for swimming crabs	Liu et al. (2022b)
8.	SDM for *benthic species*	Random forest modelling	Occurrence data, water depth, wave exposure, salinity, temperature, and sediments	Climate change predictions were developed to forecast the species distribution	Torn et al. (2020)
9.	SDM for *Aloe vera*	MaxEnt	Occurrence data,	Future predictions	

(continued)

Table 4.1 (continued)

Sl. no.	Model	Modelling algorithm/ packages used	Climatic predictor/ variables	Outcome of the modelling	Reference
			19 bioclimatic variables	indicate the appropriate distribution, area and dispersion	Hussein and Workeneh (2021)
10.	SDM for Indo-Pacific humpback dolphins	BIOMOD	Ocean depth, distance to shore, mean sea surface temperature, salinity, and current velocity	Predicts the suitability of its habitat for both the current climate and several climate change scenarios	Zhang et al. (2019)
11.	SDM for *L. polyphyllum*	MaxEnt model	Elevation above mean sea level, aspect and slope, 19 bioclimatic variables	With the use of future scenario predictions, the study assists in the identification of potential conservation areas and the provision of climate change protection	Dhyani et al. (2021)

this aspect is not considered in the current niche-based modelling and dispersal is assumed to be unlimited or zero. This could be due to methodological or data-related limitations (Barber-O'Malley et al. 2022).

Another critical factor to be included in SDMs would be the aspect of 'migration'. Generally, a static SDM would consider that the distribution of species is in equilibrium with climate, and the species will react at the local level to climate change. However, the inclusion of the migration capacity of various species could be used to improve the estimations of species' resilience to changing climate (Pecchi et al. 2019).

Statistical SDMs are generally based on free-air, synoptic, or ambient temperature conditions. These observations are of coarser resolution and fail to capture the impact of local temperature (microclimate) on urban ecosystems (Lembrechts et al. 2019). Not incorporating the impact of micro-refugia could lead to overestimation in the prediction of future distribution of species, especially on the urban scale (Lenoir et al. 2013). These micro-refugia could help improve climate resilience; however, in the large-scale spatial data (macroclimate) used by current SDMs, this buffering capacity remains undetected (Lenoir et al. 2013, 2017).

Another critical challenge that needs to be addressed while working with SDMs is the inherent bias in prediction incorporated due to the different methodologies used in modelling the future species distribution (Warren et al. 2021). Warren et al. (2021) conducted a Monte Carlo analysis on different methodologies. They concluded that input parameters like emission scenarios, climate models used, study area, and SDMs could introduce bias on the effects of climate change on habitat suitability and species distribution.

4.8 Future Perspectives

Though SDMs have been proven to be critical in understanding the future distribution of species facing the impact of climate change, they have certain limitations that might introduce bias, uncertainty, or estimation errors in the prediction and modelling process. This is especially true when handling projections at the local scale when a static or stand-alone SDM is used (Moullec et al. 2022). Stand-alone SDMs sometimes produce overtly optimistic climate change-induced projections relating to species richness. This aspect was documented by Moullec et al. (2022) in the Mediterranean Sea, where gains were overestimated and losses were underestimated.

However, various studies have been undertaken to improve upon these shortcomings. These studies can be categorized into two aspects: (1) focus on species abundance rather than its presence or absence and (2) integration of empirical/statistical SDMs with mechanistic SDMs. Waldock et al. (2022) emphasize using empirical-statistical-correlative abundance models to understand and quantify the changes in spatial patterns in species abundance due to climate change. The use of abundance trends also can provide an early warning of population depletion and collapse, which cannot be adequately predicted using occurrence data (Waldock et al. 2022).

Many researchers have pointed out the importance of integrated or ensemble models providing more accurate and robust predictions compared to stand-alone (Liu et al. 2022a, b). Ensemble models have been included in the Biomod2 software, which can be used in R. Further, the use of integrated models can be termed as an intermediate between C-SDMs and M-SDMs in terms of model complexity and data requirement (Singer et al. 2016). Integrated models can be categorized into two components: (1) correlation of species observation with environmental parameters and (2) biological processes related to species distributions (Singer et al. 2018; Barber-O'Malley et al. 2022).

The incorporation of biological processes like population dynamics, dispersal, migration, etc. has shown to improve the estimation of species distribution and shift of ranges in comparison to stand-alone or static SDMs (Fordham et al. 2013b; Singer et al. 2018). Madzokere et al. (2020) also talks about integrated models called joint SDMs (JSDMs), which utilize latent parameters, generalized linear regression (GLR), and neural network processes to integrate biotic interactions and abiotic environmental parameters. These integrated models, or JSDMs, have been shown to improve the understanding of underlying processes and reduce the bias associated with the prediction of species distribution (Madzokere et al. 2020).

This study also underlines three methods for integrating C-SDMs and M-SDMs. The first is comparing both the outputs from C-SDMs and M-SDMs for the same species. The second method uses M-SDMs to generate highly proximal geographical outputs for the basis of correlative/statistical modelling. The final method states the linking of C-SDMs and M-SDMs using the M-SDMs to define the geographical scope of the C-SDMs. Integrating multiple models uses climate, and land use data reduces bias and uncertainty, makes the model more scalable, and provides robust and realistic species distribution projections. Further, to improve the accuracy of SDMs, especially at the urban level, a combination of high-resolution spatial data, in situ measurements, and long-term historical data can be used (Lembrechts et al. 2019).

Incorporating new data like species abundance and using multilevel monitoring like high-resolution spatial monitoring along with on-site/in situ monitoring can improve the accuracy and reduce bias and uncertainty of the model being used to predict species distribution. However, integrating different frameworks like C-SDMs with M-SDMs can utilize the best of both worlds. Although rarely, an integrated modelling framework is being in utilization to comprehend how climate change affects the distribution of flora and fauna. However, these integrated models provide very robust, realistic, and reliable predictions of the habitat changes and the understanding of the underlying biotic and abiotic processes affecting the distributions of the species. These outcomes can then be utilized to improve the decision and policymaking related to the management of various ecosystems like forests, marine ecosystems, etc.

SDMs are already being used in policymaking, allowing stakeholders to identify any species impacted by climate change as well as any area affected by human activity, such as deforestation and habitat destruction, that requires conservation priorities and directs a large percentage of resources toward conservation of such areas and the target species (Rahman et al. 2019). Although the incorporation of SDMs in policymaking at national and international level is yet to be fully realized as most of the SDMs are normally used for research purpose only, it can be used as an additional tool for decision-making and designing policies that mostly deal with activities such as deforestation, afforestation, decarbonization, etc.

However, a few examples of how SDMs are currently being used in policymaking at the international level include the use of SDMs in forecasting and tracing climate risk and threats as well as population dynamics of Bornean orangutan (*Pongo pygmaeus*) in Kalimantan, Indonesia, and in developing policies and improved design management techniques to support and strengthen the Indonesian government's conservation initiatives to stabilize orangutan populations (Abram et al. 2015); the use of SDMs in assessing the impacts of deforestation, habitat degradation, and fragmentation on the chimpanzee population of western Tanzania and in the decision and policymaking for sustainable, precise, and cost-effective surveillance and monitoring of the species and their conservation (Dickson et al. 2020); etc.

In the Indian subcontinent, the integration of SDMs in decision and policymaking is yet to be fully realized; however, for a country with a vast species and forest resource, it can be a very useful tool in species and forest management and conservation. A few examples include the use of SDMs in identifying the suitable habitat as well as new occurrence of the critically endangered species *Gymnocladus assamicus*, which is endemic to North-East India, and in policymaking to strengthen the ongoing measures for in situ conservation of the endangered plant (Menon et al. 2010); the use of SDMs in policymaking for restoration and conservation planning of a species of medicinal and therapeutic plant found in the lower Himalayan foothills, *Justicia adhatoda* (Yang et al. 2013); etc.

Most national and international forest policies put forth to date are often focused on conservation and preservation, lowering stress on forests, and providing biomass to the significant population that depends on forests for its fuel and fodder needs, thereby limiting the use of SDMs in any decision-making process. However, with the growing issues of deforestation and requirement for reforestation or afforestation, SDMs have become necessary in policymaking for the mentioned issues (Murthy and Kumar 2019). When it comes to phenomena such as deforestation, afforestation, SDMs can be used to identify frequently logged tree species and their distribution areas, the frequency of logging leading to deforestation. Therefore SDMs can also be utilized to assist in policymaking for reforestation or afforestation of the degraded landscape by means of providing information on potential alternate species for restoration, alternative to slash-and-burn agriculture, adoption of sustainable agroforestry, sustainable logging, agro-pastoral production systems, etc.

Author Contributions
SL contributed to the study conception and design. All authors contributed equally to the drafting of the chapter: *HB* and *MKB* wrote the abstract. *HB* wrote the sections containing the introduction, the impact of climate change on eco-sensitive hotspots, the types of SDMs, and the comparative of different SDMs used for mapping climate change impacts. *MKB* wrote the sections containing the parameters and factors that influence SDMs and their outcomes and the application of SDMs in various ecosystems. *SL* wrote the sections containing the gaps and challenges of climate change impact mapping and the future perspectives of SDMs. The figures and tables were conceptualized and designed by *HB*, *MKB*, and *SL*. *SL* reviewed the first draft. *MKB* and *HB* contributed to the revision of the first draft. All authors have read and approved the final manuscript.

Acknowledgement The authors extend their gratitude to the Director of CSIR-National Environmental Engineering Research Institute (CSIR-NEERI) for providing all means of support in carrying out this work. The Knowledge Resource Centre (KRC) of CSIR-NEERI has been very helpful in providing manuscript processing services having the reference no. CSIR-NEERI/KRC/2022/SEP/EMD/3.

References

Abatzoglou JT, Dobrowski SZ, Parks SA, Hegewisch KC (2018) TerraClimate, a high-resolution global dataset of monthly climate and climatic water balance from 1958–2015. Sci Data 5: 170191. https://doi.org/10.1038/sdata.2017.191

Abdulwahab UA, Hammill E, Hawkins CP (2022) Choice of climate data affects the performance and interpretation of species distribution models. Ecol Model 471:110042. https://doi.org/10.1016/j.ecolmodel.2022.110042

Abram NK, Meijaard E, Wells JA et al (2015) Mapping perceptions of species' threats and population trends to inform conservation efforts: the Bornean orangutan case study. Divers Distrib 21:487–499. https://doi.org/10.1111/ddi.12286

Ahmed F, Ali I, Kousar S, Ahmed S (2022) The environmental impact of industrialization and foreign direct investment: empirical evidence from Asia-Pacific region. Environ Sci Pollut Res 29:29778–29792. https://doi.org/10.1007/s11356-021-17560-w

Akçakaya HR, Butchart SHM, Mace GM et al (2006) Use and misuse of the IUCN Red List Criteria in projecting climate change impacts on biodiversity. Glob Change Biol 12:2037–2043. https://doi.org/10.1111/j.1365-2486.2006.01253.x

Allouche O, Tsoar A, Kadmon R (2006) Assessing the accuracy of species distribution models: prevalence, kappa and the true skill statistic (TSS). J Appl Ecol 43:1223–1232. https://doi.org/10.1111/j.1365-2664.2006.01214.x

Angelieri CCS, Adams-Hosking C, de Barros Ferraz M et al (2016) Using species distribution models to predict potential landscape restoration effects on puma conservation. PLoS One 11: e0145232. https://doi.org/10.1371/journal.pone.0145232

Araújo MB, Pearson RG, Thuiller W, Erhard M (2005) Validation of species–climate impact models under climate change. Glob Change Biol 11:1504–1513. https://doi.org/10.1111/j.1365-2486.2005.01000.x

Araújo MB, Anderson RP, Márcia Barbosa A et al (2019) Standards for distribution models in biodiversity assessments. Sci Adv 5:eaat4858. https://doi.org/10.1126/sciadv.aat4858

Aukema JE, Pricope NG, Husak GJ, Lopez-Carr D (2017) Biodiversity areas under threat: overlap of climate change and population pressures on the world's biodiversity priorities. PLoS One 12: e0170615. https://doi.org/10.1371/journal.pone.0170615

Barber-O'Malley B, Lassalle G, Chust G et al (2022) HyDiaD: a hybrid species distribution model combining dispersal, multi-habitat suitability, and population dynamics for diadromous species under climate change scenarios. Ecol Model 470:109997. https://doi.org/10.1016/j.ecolmodel.2022.109997

Barlow MM, Johnson CN, McDowell MC et al (2021) Species distribution models for conservation: identifying translocation sites for eastern quolls under climate change. Glob Ecol Conserv 29:e01735. https://doi.org/10.1016/j.gecco.2021.e01735

Beale CM, Lennon JJ (2012) Incorporating uncertainty in predictive species distribution modelling. Philos Trans R Soc B Biol Sci 367:247–258. https://doi.org/10.1098/rstb.2011.0178

Beery S, Cole E, Parker J et al (2021) Species distribution modeling for machine learning practitioners: a review. In: ACM SIGCAS conference on computing and sustainable societies (COMPASS). ACM, Virtual Event Australia, pp 329–348

Bellard C, Bertelsmeier C, Leadley P et al (2012) Impacts of climate change on the future of biodiversity. Ecol Lett 15:365–377. https://doi.org/10.1111/j.1461-0248.2011.01736.x

Booth TH (2018) Species distribution modelling tools and databases to assist managing forests under climate change. For Ecol Manag 430:196–203. https://doi.org/10.1016/j.foreco.2018.08.019

Botsford LW, White JW, Carr MH, Caselle JE (2014) Marine protected area networks in California, USA. Adv Mar Biol 69:205–251. https://doi.org/10.1016/B978-0-12-800214-8.00006-2

Bowler DE, Hof C, Haase P et al (2017) Cross-realm assessment of climate change impacts on species' abundance trends. Nat Ecol Evol 1:1–7. https://doi.org/10.1038/s41559-016-0067

Buckland ST, Elston DA (1993) Empirical models for the spatial distribution of wildlife. J Appl Ecol 30:478–495. https://doi.org/10.2307/2404188

Bucklin DN, Basille M, Benscoter AM et al (2015) Comparing species distribution models constructed with different subsets of environmental predictors. Divers Distrib 21:23–35. https://doi.org/10.1111/ddi.12247

Chapman DS, Makra L, Albertini R et al (2016) Modelling the introduction and spread of non-native species: international trade and climate change drive ragweed invasion. Glob Change Biol 22:3067–3079. https://doi.org/10.1111/gcb.13220

Charney ND, Record S, Gerstner BE et al (2021) A test of species distribution model transferability across environmental and geographic space for 108 western North American Tree Species. Front Ecol Evol 9:689295

Cheung WWL, Jones MC, Reygondeau G et al (2016) Structural uncertainty in projecting global fisheries catches under climate change. Ecol Model 325:57–66

Cohen J (1960) A coefficient of agreement for nominal scales. Educ Psychol Meas 20:37–46. https://doi.org/10.1177/001316446002000104

Cuddington K, Fortin M-J, Gerber LR et al (2013) Process-based models are required to manage ecological systems in a changing world. Ecosphere 4:art20. https://doi.org/10.1890/ES12-00178.1

Desjonquères C, Villén-Pérez S, De Marco P et al (2022) Acoustic species distribution models (aSDMs): a framework to forecast shifts in calling behaviour under climate change. Methods Ecol Evol 13(10):2275–2288. https://doi.org/10.1111/2041-210X.13923

Dhyani A, Kadaverugu R, Nautiyal BP, Nautiyal MC (2021) Predicting the potential distribution of a critically endangered medicinal plant Lilium polyphyllum in Indian Western Himalayan region. Reg Environ Chang 21:30. https://doi.org/10.1007/s10113-021-01763-5

Dickson R, Baker M, Bonnin N et al (2020) Combining deforestation and species distribution models to improve measures of chimpanzee conservation impacts of REDD: a case study from Ntakata Mountains, Western Tanzania. Forests 11:1195. https://doi.org/10.3390/f11111195

Domisch S, Amatulli G, Jetz W (2015) Near-global freshwater-specific environmental variables for biodiversity analyses in 1 km resolution. Sci Data 2:150073. https://doi.org/10.1038/sdata.2015.73

Driver S, Chris MS, Unger D, Kulhavy D (2020) Species distribution modeling for arid adapted habitat specialists in Zion National Park. Electronic Theses and Dissertations

Dutra Silva L, Brito de Azevedo E, Vieira Reis F et al (2019) Limitations of species distribution models based on available climate change data: a case study in the Azorean forest. Forests 10:575. https://doi.org/10.3390/f10070575

Elith J, Franklin J (2017) Species distribution modeling. In: Reference module in life sciences. Elsevier, p B9780128096338024000

Elith J, Leathwick JR (2009) Species distribution models: ecological explanation and prediction across space and time. Annu Rev Ecol Evol Syst 40:677–697. https://doi.org/10.1146/annurev.ecolsys.110308.120159

Estes LD, Bradley BA, Beukes H et al (2013) Comparing mechanistic and empirical model projections of crop suitability and productivity: implications for ecological forecasting. Glob Ecol Biogeogr 22:1007–1018. https://doi.org/10.1111/geb.12034

Falk W, Mellert KH (2011) Species distribution models as a tool for forest management planning under climate change: risk evaluation of Abies alba in Bavaria. J Veg Sci 22:621–634. https://doi.org/10.1111/j.1654-1103.2011.01294.x

Fernandes RF, Scherrer D, Guisan A (2019) Effects of simulated observation errors on the performance of species distribution models. Divers Distrib 25:400–413. https://doi.org/10.1111/ddi.12868

Fick SE, Hijmans RJ (2017) WorldClim 2: new 1-km spatial resolution climate surfaces for global land areas. Int J Climatol 37:4302–4315. https://doi.org/10.1002/joc.5086

Fordham DA, Akçakaya HR, Araújo MB et al (2013a) Tools for integrating range change, extinction risk and climate change information into conservation management. Ecography 36: 956–964. https://doi.org/10.1111/j.1600-0587.2013.00147.x
Fordham DA, Mellin C, Russell BD et al (2013b) Population dynamics can be more important than physiological limits for determining range shifts under climate change. Glob Change Biol 19: 3224–3237. https://doi.org/10.1111/gcb.12289
Fourcade Y, Besnard AG, Secondi J (2018) Paintings predict the distribution of species, or the challenge of selecting environmental predictors and evaluation statistics. Glob Ecol Biogeogr 27:245–256. https://doi.org/10.1111/geb.12684
Franklin J (2010) Moving beyond static species distribution models in support of conservation biogeography. Divers Distrib 16:321–330. https://doi.org/10.1111/j.1472-4642.2010.00641.x
Fröhlich A, Ciach M (2019) Nocturnal noise and habitat homogeneity limit species richness of owls in an urban environment. Environ Sci Pollut Res 26:17284–17291. https://doi.org/10.1007/s11356-019-05063-8
Gardner AS, Maclean IMD, Gaston KJ (2019) Climatic predictors of species distributions neglect biophysiologically meaningful variables. Divers Distrib 25:1318–1333. https://doi.org/10.1111/ddi.12939
GML-NOAA (2013) CO_2 at NOAA's Mauna Loa Observatory reaches new milestone: tops 400 ppm. https://gml.noaa.gov/news/7074.html. Accessed 9 Feb 2022
Gormley KSG, Hull AD, Porter JS et al (2015) Adaptive management, international co-operation and planning for marine conservation hotspots in a changing climate. Mar Policy 53:54–66. https://doi.org/10.1016/j.marpol.2014.11.017
Guillera-Arroita G, Lahoz-Monfort JJ, Elith J et al (2015) Is my species distribution model fit for purpose? Matching data and models to applications. Glob Ecol Biogeogr 24:276–292. https://doi.org/10.1111/geb.12268
Guisan A, Tingley R, Baumgartner JB et al (2013) Predicting species distributions for conservation decisions. Ecol Lett 16:1424–1435. https://doi.org/10.1111/ele.12189
Guo K, Yuan S, Wang H et al (2021) Species distribution models for predicting the habitat suitability of Chinese fire-bellied newt Cynops orientalis under climate change. Ecol Evol 11: 10147–10154. https://doi.org/10.1002/ece3.7822
Hanski I (1998) Metapopulation dynamics. Nature 396:41–49. https://doi.org/10.1038/23876
Harrison S (1991) Local extinction in a metapopulation context: an empirical evaluation. Biol J Linn Soc 42:73–88. https://doi.org/10.1111/j.1095-8312.1991.tb00552.x
Harrison S, Noss R (2017) Endemism hotspots are linked to stable climatic refugia. Ann Bot 119: 207–214. https://doi.org/10.1093/aob/mcw248
Hill RA, Weber MH, Leibowitz SG et al (2016) The stream-catchment (StreamCat) dataset: a database of watershed metrics for the conterminous United States. JAWRA J Am Water Resour Assoc 52:120–128. https://doi.org/10.1111/1752-1688.12372
Hoffmann S (2022) Challenges and opportunities of area-based conservation in reaching biodiversity and sustainability goals. Biodivers Conserv 31:325–352. https://doi.org/10.1007/s10531-021-02340-2
Hussein A, Workeneh S (2021) Modeling the impacts of climate changes on the distribution of aloe vera species in Ethiopia. J Earth Sci Clim Chang 12:567
IPBES (2019) Global assessment report on biodiversity and ecosystem services of the Intergovernmental Science-Policy Platform on Biodiversity and Ecosystem Services. IPBES secretariat, Bonn
IPCC (2018) Summary for policymakers. In: Global warming of 1.5°C. An IPCC Special Report on the impacts of global warming of 1.5°C above pre-industrial levels and related global greenhouse gas emission pathways, in the context of strengthening the global response to the threat of climate change, sustainable development, and efforts to eradicate poverty. IPCC, Cambridge, New York, NY
Iverson L, McKenzie D (2014) Species distribution and climate change. U.S. Department of Agriculture, Forest Service, Climate Change Resource Center

Jarnevich CS, Stohlgren TJ, Kumar S et al (2015) Caveats for correlative species distribution modeling. Ecol Inform 29:6–15. https://doi.org/10.1016/j.ecoinf.2015.06.007
Kang W, Chon J, Kim G (2020) Urban ecosystem services: a review of the knowledge components and evolution in the 2010s. Sustainability 12:9839. https://doi.org/10.3390/su12239839
Kearney M, Porter W (2009) Mechanistic niche modelling: combining physiological and spatial data to predict species' ranges. Ecol Lett 12:334–350. https://doi.org/10.1111/j.1461-0248.2008.01277.x
Kearney M, Phillips BL, Tracy CR et al (2008) Modelling species distributions without using species distributions: the cane toad in Australia under current and future climates. Ecography 31:423–434. https://doi.org/10.1111/j.0906-7590.2008.05457.x
Kearney M, Wintle B, Porter W (2010) Correlative and mechanistic models of species distribution provide congruent forecasts under climate change. Conserv Lett 3:203–213. https://doi.org/10.1111/j.1755-263X.2010.00097.x
Kharouba HM, Algar AC, Kerr JT (2009) Historically calibrated predictions of butterfly species' range shift using global change as a pseudo-experiment. Ecology 90:2213–2222. https://doi.org/10.1890/08-1304.1
Kiessling W, Maharaj S, Price J, Talukdar GH (2022) Cross-chapter paper 1: biodiversity hotspots. In: Climate change 2022: impacts, adaptation and vulnerability. Contribution of working group II to the sixth assessment report of the intergovernmental panel on climate change. IPCC
Koshkina V, Wang Y, Gordon A et al (2017) Integrated species distribution models: combining presence-background data and site-occupancy data with imperfect detection. Methods Ecol Evol 8:420–430. https://doi.org/10.1111/2041-210X.12738
Ksiksi TS, Remya K, Mousa MT et al (2019) Climate change-induced species distribution modeling in hyper-arid ecosystems. F1000Research 8:978
La Marca W, Elith J, Firth RSC et al (2019) The influence of data source and species distribution modelling method on spatial conservation priorities. Divers Distrib 25:1060–1073. https://doi.org/10.1111/ddi.12924
Lembrechts JJ, Nijs I, Lenoir J (2019) Incorporating microclimate into species distribution models. Ecography 42:1267–1279. https://doi.org/10.1111/ecog.03947
Lenoir J, Graae BJ, Aarrestad PA et al (2013) Local temperatures inferred from plant communities suggest strong spatial buffering of climate warming across Northern Europe. Glob Change Biol 19:1470–1481. https://doi.org/10.1111/gcb.12129
Lenoir J, Hattab T, Pierre G (2017) Climatic microrefugia under anthropogenic climate change: implications for species redistribution. Ecography 40:253–266. https://doi.org/10.1111/ecog.02788
Liu C, White M, Newell G (2009) Assessing the accuracy of species distribution models more thoroughly. 18th World IMACS MODSIM Congr Cairns Aust 13–17 July 2009 7
Liu J, Bai H, Ma H, Feng G (2019) Bird diversity in Chinese urban parks was more associated with natural factors than anthropogenic factors. Urban For Urban Green 43:126358. https://doi.org/10.1016/j.ufug.2019.06.001
Liu D, Lei X, Gao W et al (2022a) Mapping the potential distribution suitability of 16 tree species under climate change in northeastern China using Maxent modelling. J For Res 33:1739–1750. https://doi.org/10.1007/s11676-022-01459-4
Liu X, Han X, Han Z (2022b) Effects of climate change on the potential habitat distribution of swimming crab Portunus trituberculatus under the species distribution model. J Oceanol Limnol 40:1556–1565. https://doi.org/10.1007/s00343-021-1082-1
Luan J, Zhang C, Ji Y et al (2021) Matching data types to the objectives of species distribution modeling: an evaluation with marine fish species. Front Mar Sci 8:771071. https://doi.org/10.3389/fmars.2021.771071
Madzokere ET, Hallgren W, Sahin O et al (2020) Integrating statistical and mechanistic approaches with biotic and environmental variables improves model predictions of the impact of climate

and land-use changes on future mosquito-vector abundance, diversity and distributions in Australia. Parasit Vectors 13:484. https://doi.org/10.1186/s13071-020-04360-3

Malcolm JR, Liu C, Neilson RP et al (2006) Global warming and extinctions of endemic species from biodiversity hotspots. Conserv Biol 20:538–548. https://doi.org/10.1111/j.1523-1739.2006.00364.x

McPherson JM, Jetz W (2007) Effects of species' ecology on the accuracy of distribution models. Ecography 30:135–151. https://doi.org/10.1111/j.0906-7590.2007.04823.x

Menon S, Choudhury B, Khan M, Townsend Peterson A (2010) Ecological niche modeling and local knowledge predict new populations of Gymnocladus assamicus a critically endangered tree species. Endanger Species Res 11:175–181. https://doi.org/10.3354/esr00275

Morán-Ordóñez A, Suárez-Seoane S, Elith J et al (2012) Satellite surface reflectance improves habitat distribution mapping: a case study on heath and shrub formations in the Cantabrian Mountains (NW Spain). Divers Distrib 18:588–602. https://doi.org/10.1111/j.1472-4642.2011.00855.x

Morán-Ordóñez A, Lahoz-Monfort JJ, Elith J, Wintle BA (2017) Evaluating 318 continental-scale species distribution models over a 60-year prediction horizon: what factors influence the reliability of predictions? Glob Ecol Biogeogr 26:371–384. https://doi.org/10.1111/geb.12545

Morin X, Lechowicz MJ (2008) Contemporary perspectives on the niche that can improve models of species range shifts under climate change. Biol Lett 4:573–576. https://doi.org/10.1098/rsbl.2008.0181

Moullec F, Barrier N, Drira S et al (2022) Using species distribution models only may underestimate climate change impacts on future marine biodiversity. Ecol Model 464:109826. https://doi.org/10.1016/j.ecolmodel.2021.109826

Murthy IK, Kumar P (2019) Forests policies and programmes in India: implications for climate change adaptation. Open J For 9:226–240. https://doi.org/10.4236/ojf.2019.93012

NOAA (2022) Climate change: atmospheric carbon dioxide. In: Climate.Gov. https://www.climate.gov/news-features/understanding-climate/climate-change-atmospheric-carbon-dioxide. Accessed 9 Feb 2022

Parmesan C (1996) Climate and species' range. Nature 382:765–766. https://doi.org/10.1038/382765a0

Parmesan C (2006) Ecological and evolutionary responses to recent climate change. Annu Rev Ecol Evol Syst 37:637–669. https://doi.org/10.1146/annurev.ecolsys.37.091305.110100

Parrotta JA, Wildburger C, Mansourian S (eds) (2012) Understanding relationships between biodiversity, carbon, forests and people: the key to achieving REDD+ objectives. A global assessment report prepared by the Global Forest Expert Panel on Biodiversity, Forest Management, and REDD+. U.S. Department of Agriculture, Forest Service, Vienna

Parsons SE, Frank SD (2019) Urban tree pests and natural enemies respond to habitat at different spatial scales. J Urban Ecol 5:juz010. https://doi.org/10.1093/jue/juz010

Pecchi M, Marchi M, Burton V et al (2019) Species distribution modelling to support forest management. A literature review. Ecol Model 411:108817. https://doi.org/10.1016/j.ecolmodel.2019.108817

Pöyry J, Luoto M, Heikkinen RK, Saarinen K (2008) Species traits are associated with the quality of bioclimatic models. Glob Ecol Biogeogr 17:403–414. https://doi.org/10.1111/j.1466-8238.2007.00373.x

Qazi AW, Saqib Z, Zaman-ul-Haq M (2022) Trends in species distribution modelling in context of rare and endemic plants: a systematic review. Ecol Process 11:40. https://doi.org/10.1186/s13717-022-00384-y

Rabaiotti D, Woodroffe R (2019) Coping with climate change: limited behavioral responses to hot weather in a tropical carnivore. Oecologia 189:587. https://doi.org/10.1007/s00442-018-04329-1

Rahman AA, Mohamed M, Tokiman L, Mohd Sanget M-S (2019) Species distribution modelling to assist biodiversity and conservation management in Malaysia. IOP Conf Ser Earth Environ Sci 269:012041. https://doi.org/10.1088/1755-1315/269/1/012041

Roberts SM, Halpin PN, Clark JS (2022) Jointly modeling marine species to inform the effects of environmental change on an ecological community in the Northwest Atlantic. Sci Rep 12:132. https://doi.org/10.1038/s41598-021-04110-0
Robinson NM, Nelson WA, Costello MJ et al (2017) A systematic review of marine-based species distribution models (SDMs) with recommendations for best practice. Front Mar Sci 4:421. https://doi.org/10.3389/fmars.2017.00421
Rougier T, Lassalle G, Drouineau H et al (2015) The combined use of correlative and mechanistic species distribution models benefits low conservation status species. PLoS One 10:e0139194. https://doi.org/10.1371/journal.pone.0139194
Sala OE, Stuart Chapin F III et al (2000) Global biodiversity scenarios for the year 2100. Science 287:1770–1774. https://doi.org/10.1126/science.287.5459.1770
Senay SD, Worner SP, Ikeda T (2013) Novel three-step pseudo-absence selection technique for improved species distribution modelling. PLoS One 8:e71218. https://doi.org/10.1371/journal.pone.0071218
Shabani F, Kumar L, Ahmadi M (2018) Assessing accuracy methods of species distribution models: AUC, specificity, sensitivity and the true skill statistic. Glob J Hum Soc Sci B 18:7–18
Singer A, Johst K, Banitz T et al (2016) Community dynamics under environmental change: how can next generation mechanistic models improve projections of species distributions? Ecol Model 326:63–74. https://doi.org/10.1016/j.ecolmodel.2015.11.007
Singer A, Schweiger O, Kühn I, Johst K (2018) Constructing a hybrid species distribution model from standard large-scale distribution data. Ecol Model 373:39–52. https://doi.org/10.1016/j.ecolmodel.2018.02.002
Sintayehu DW (2018) Impact of climate change on biodiversity and associated key ecosystem services in Africa: a systematic review. Ecosyst Health Sustain 4:225–239. https://doi.org/10.1080/20964129.2018.1530054
Srivastava V, Lafond V, Griess V (2019) Species distribution models (SDM): applications, benefits and challenges in invasive species management. CAB Rev Perspect Agric Vet Sci Nutr Nat Resour 14:1–13. https://doi.org/10.1079/PAVSNNR201914020
Swab RM, Regan HM, Matthies D et al (2015) The role of demography, intra-species variation, and species distribution models in species' projections under climate change. Ecography 38:221–230. https://doi.org/10.1111/ecog.00585
Swan M, Le Pla M, Di Stefano J et al (2021) Species distribution models for conservation planning in fire-prone landscapes. Biodivers Conserv 30:1119–1136. https://doi.org/10.1007/s10531-021-02136-4
The Royal Society (2022) The basics of climate change. https://royalsociety.org/topics-policy/projects/climate-change-evidence-causes/basics-of-climate-change/. Accessed 9 Nov 2022
Thibaud E, Petitpierre B, Broennimann O et al (2014) Measuring the relative effect of factors affecting species distribution model predictions. Methods Ecol Evol 5:947–955. https://doi.org/10.1111/2041-210X.12203
Thomas CD, Cameron A, Green RE et al (2004) Extinction risk from climate change. Nature 427:145–148. https://doi.org/10.1038/nature02121
Thuiller W, Albert C, Araújo MB et al (2008) Predicting global change impacts on plant species' distributions: future challenges. Perspect Plant Ecol Evol Syst 9:137–152. https://doi.org/10.1016/j.ppees.2007.09.004
Torn K, Herkül K, Peterson A, Suursaar Ü (2020) Predicting potential effects of climate change on benthic species: current and future distribution of native and non-native charophytes and amphipods. pp 85–95
Trew BT, Maclean IMD (2021) Vulnerability of global biodiversity hotspots to climate change. Glob Ecol Biogeogr 30:768–783. https://doi.org/10.1111/geb.13272
US EPA (2016) Climate change indicators: weather and climate. https://www.epa.gov/climate-indicators/weather-climate. Accessed 16 Jan 2022
Waldock C, Stuart-Smith RD, Albouy C et al (2022) A quantitative review of abundance-based species distribution models. Ecography 2022. https://doi.org/10.1111/ecog.05694

Wang Y, Stone L (2019) Understanding the connections between species distribution models for presence-background data. Theor Ecol 12:73–88. https://doi.org/10.1007/s12080-018-0389-9

Warren DL, Dornburg A, Zapfe K, Iglesias TL (2021) The effects of climate change on Australia's only endemic Pokémon: measuring bias in species distribution models. Methods Ecol Evol 12: 985–995. https://doi.org/10.1111/2041-210X.13591

West AM, Evangelista PH, Jarnevich CS et al (2017) Using multi-date satellite imagery to monitor invasive grass species distribution in post-wildfire landscapes: an iterative, adaptable approach that employs open-source data and software. Int J Appl Earth Obs Geoinformation 59:135–146. https://doi.org/10.1016/j.jag.2017.03.009

WWF (2016) Living planet report 2016. Risk and resilience in a new era. WWW International, Gland

WWF (2022) Effects of climate change. In: World wildlife fund. https://www.worldwildlife.org/threats/effects-of-climate-change. Accessed 9 Feb 2022

Xu Y, Huang Y, Zhao H et al (2021) Modelling the effects of climate change on the distribution of endangered Cypripedium japonicum in China. Forests 12:429. https://doi.org/10.3390/f12040429

Yang X-Q, Kushwaha SPS, Saran S et al (2013) Maxent modeling for predicting the potential distribution of medicinal plant, Justicia adhatoda L. in lesser Himalayan foothills. Ecol Eng 51: 83–87. https://doi.org/10.1016/j.ecoleng.2012.12.004

Young M, Carr M (2015a) Assessment of habitat representation across a network of marine protected areas with implications for the spatial design of monitoring. PLoS One 10: e0116200. https://doi.org/10.1371/journal.pone.0116200

Young M, Carr MH (2015b) Application of species distribution models to explain and predict the distribution, abundance and assemblage structure of nearshore temperate reef fishes. Divers Distrib 21:1428–1440. https://doi.org/10.1111/ddi.12378

Zhang Z, Xu S, Capinha C et al (2019) Using species distribution model to predict the impact of climate change on the potential distribution of Japanese whiting Sillago japonica. Ecol Indic 104:333–340. https://doi.org/10.1016/j.ecolind.2019.05.023

Approaches for Modelling the Climate Change Impacts on Ecosystems

5

Anjaly George and Shijo Joseph

Abstract

Computer-based models have become important tools for examining the responses to changes in any system. A climate model simulates every aspect of the planet's climate including how climate has changed in the past and how it may change in the future. The climate models are built on scenarios, which offer a method for assessing how plausible futures might develop. Recent global and regional assessments that project future environments based on shifting driving forces were built on the foundation of these modelling and scenario tools. This chapter addresses how the scenarios are being created and their evolutionary changes in the last two decades starting from the IPCC's SRES scenario to the recent SSP scenario. Further, the downscaling of physically based climate models to biosphere-based Earth system models is also discussed. Two approaches to Earth system modelling, i.e. process-based dynamical global vegetation models and classical climate envelope models, are described in detail to model the response of ecosystems to climate change. The chapter contributes to our understanding on various approaches to model the impact of climate change on ecosystems, parameterization of models and responses of ecosystems to changing environmental conditions.

Keywords

Climate change · Ecosystem modelling · Process-based models · DGVM · Climate envelope models

A. George · S. Joseph (✉)
Centre for Climate Resilience and Environment Management, Kerala University of Fisheries and Ocean Studies (KUFOS), Puduveypu, Kochi, Kerala, India
e-mail: shijo@kufos.ac.in

S. Dhyani et al. (eds.), *Ecosystem and Species Habitat Modeling for Conservation and Restoration*, https://doi.org/10.1007/978-981-99-0131-9_5

5.1 Introduction

The ecosystems experience exceptional negative disturbances in recent times attributed to global warming, climate change, extreme weather events and anthropogenic pressures (Krishnaswamy et al. 2014; Cheng et al. 2018; Pan et al. 2020; Ren et al. 2022). Unchecked population growth, modernization, industrialization and urbanization have resulted in a sharp increase in the atmospheric concentration of greenhouse gases (GHGs) (Mandal et al. 2021). This in turn affects the temperature profile of the atmosphere by trapping more heat within the atmospheric blanket. The temperature is highly linked to other physical variables of the Earth system, and the resultant manifestations even impact biological systems. It is by now unambiguous that climate change stresses every ecosystem throughout the world (IPCC 2021). Climate change affects the distribution of ecosystems, species ranges, structure and functions, seasonality, changes in productivity, forest fire regimes and many more (Melillo 1999; IPCC 2001; Koca et al. 2006; Joseph et al. 2013; Grimm et al. 2013; Krishnaswamy et al. 2014; George et al. 2019). Accurate projections of the effects on ecosystems in the coming decades are urgently needed in order to design effective mitigation and adaptation measures to maintain ecosystem services and function in the face of evidence of accelerated climate change (MEA 2005; Sutherland 2006; Morin and Thuiller 2009; Joseph et al. 2013; de Sassi et al. 2015; Bos et al. 2017).

The development of tools and methods for reliable climate change impact projections of different ecosystems has been a research priority in recent times. It is impossible to predict how the climate will change over the next century and beyond in a deterministic, conclusive manner. The only possibility is to create scenarios which are likely to occur in future, which include time series of emissions and concentrations of GHGs and aerosols and chemically active gases in the atmosphere (IPCC 1996). This in turn aligned with the forcing and underlying agents of climate change such as energy use, land use patterns, economic activity, technology and climate policy. The scenarios are essentially the result of creative collaboration between experts in emission inventories, climate modelling, terrestrial ecosystem modelling and integrated assessment modelling.

5.2 IPCC Modelling Scenarios

In order to track the changing status of Earth's climate system, to expand our understanding and to develop new tools and approaches for preserving and repairing resilient biological and social systems, a range of likely possibilities known as emission scenarios is described in the Assessment Reports (AR) of the Intergovernmental Panel on Climate Change (IPCC).

5.2.1 SRES Scenarios

The SRES scenarios were based on the Special Report on Emissions Scenarios published by IPCC in 2000. The SRES scenarios were used in the Third Assessment Report (TAR) published in 2001 and Fourth Assessment Report (AR4) published in 2007. This report used the 'storyline approach' in which the future storylines were defined based on possible socio-economic changes in the future (Pedersen et al. 2022). They included economic, demographic, social, technological and environmental factors. It also describes economic and environmental and global and regional connections. The implications for future greenhouse gas emissions for each story would be estimated and emission scenarios were defined. The SRES were divided into four scenario families designated with the letters A1, A2, B1 and B2 standing for economic (A) or environmental (B) concerns and global (1) or regional (2) development patterns, respectively.

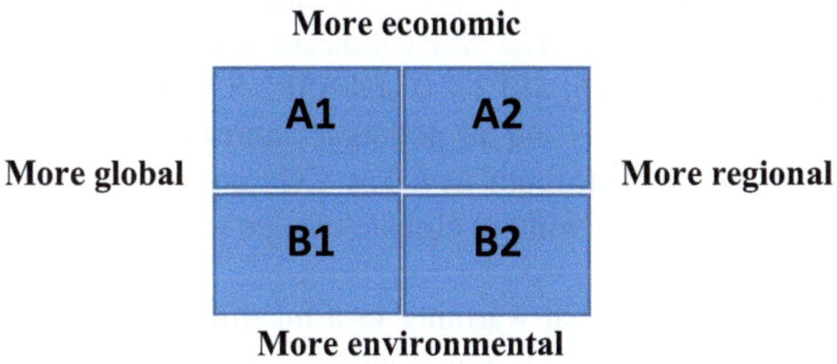

Families A1 and A2 tend towards more economic growth and B1 and B2 tend towards more environmental protection. The two on the left A1 and B1 describe areas that are more globally interconnected, and the two on the right A2 and B2 describe the areas where regional connections are more important. Thus, SRES framework was often used as a reference document for modelling the diverse dimensions of the impact of climate change (Gaffin et al. 2004).

5.2.2 RCP Scenarios

In 2007, the IPCC developed a new set of scenarios as an updation and expansion in scope to the existing SRES scenarios, known as Representative Concentration Pathways (RCP), which formed the basis of the Fifth Assessment Report (AR5) (Van Vuuren et al. 2011). In this approach, a condition is defined for Earth by the amount of extra energy that might be added to the climate system by 2100, compared to the pre-industrial era. After describing an endpoint condition of radiative forcing for the year 2100, the representative way to get there is defined which is called as the Representative Concentration Pathways. The concentration part relates to the concentrations of GHGs in the atmosphere along the way. The names of the RCP are based on the target amount of radiative forcing for 2100. The RCP include a stringent mitigation scenario (RCP2.6 where the radiative forcing will lead to 2.6 W/m^2 by 2100 with an atmospheric CO_2 concentration of 490 ppm), two intermediate scenarios (RCP4.5 (i.e. 4.5 W/m^2 and 650 ppm radiative forcing and atmospheric concentration of CO_2, respectively, by 2100) and RCP6.0 (approximately 850 ppm CO_2 eq)) and one scenario with very high GHG emissions (RCP8.5 (approximately 1370 ppm CO_2 eq)).

5.2.3 SSP Scenarios

After the RCP scenarios, a range of new 'pathways' collectively known as the 'Shared Socioeconomic Pathways' (SSPs) were developed as part of the Sixth Assessment Report (AR6) to examine how global society and economics or the socio-economic factors might change over the next century. SSPs are projections of anticipated worldwide socio-economic trends through the year 2100. These include elements like population, economic expansion, education, urbanization and the pace of technological development. The SSPs indicate five illustrative SSP scenarios, SSP1-1.9, SSP1-2.6, SSP2-4.5, SSP3-7.0 and SSP5-8.5. These scenarios analyse a wider series of greenhouse gas and air pollutant futures than those assessed in past publications, and they include both new low-CO_2 emission pathways as well as high-CO_2 emission pathways without climate change mitigation.

- SSP1-1.9: By keeping global warming to a maximum of 1.5 °C, this scenario envisions a world in which CO_2 emissions are reduced to zero globally by 2050.
- SSP1-2.6: With assumed net zero emissions in the second half of the century, global warming stays below 2 °C.
- SSP2-4.5: CO_2 emissions hover around current levels and do not reach net zero by 2100. Development and income are expanding unevenly, and sustainability progress is modest. In this case, the century's conclusion will see a 2.7 °C increase in temperature.
- SSP3-7.0: Temperatures and emissions both increase steadily and by 2100, CO_2 emissions will have roughly doubled from current levels. The century's conclusion will see an increase in average temperatures of 3.6 °C.
- SSP5-8.5: Current CO_2 emissions levels roughly double by 2050, and temperature will reach 4.4 °C by 2100.

These estimates of emissions are then fed into Global Circulation Models (GCMs) and Regional Circulation Models (RCMs) to know how these scenarios affect global climate change. The GCM or RCM output is then downscaled to the finer spatial resolution that is more meaningful for analysing the impact of climate change on the ecosystems.

5.3 Model Downscaling

Studies of the effects of climate change on ecosystems often use climate scenarios produced by GCMs with resolutions of a few hundred kilometres, which are coarse compared to the scales of relevance in regional studies (Mearns et al. 2001). This is where downscaling is required. Downscaling is a process of generating higher-resolution data from relatively coarse-resolution GCMs. The RCM is the down-scaled version of the GCM and has a much higher grid resolution than GCM. There are two established methods for downscaling the GCMs to RCMs: the dynamical and the statistical or empirical method. These techniques are complementary and both have strengths and weaknesses. Statistical downscaling compares GCM output for a particular period in the past with observations during the same time by comparing model projections in actual climate data observations. A statistical relationship is established between global and regional climate patterns. This statistical relationship is then applied to predict future climate projections. Dynamical downscaling is a method where the GCM output is simulated to smaller scales using another high-resolution dynamical regional model. The regional model for the area of interest is chosen such that it is large enough to capture important weather processes in the region of interest while ensuring that this region of interest is far enough from the boundaries of the model. Regional climate models provide more geographic resolution and incorporate regional physiography in an effort to address GCM's drawbacks (McGregor 1997; Mearns et al. 2001; Koca et al. 2006). RCMs are in the order of a few kilometres ranging from 5 kms to 25 kms, and even 50 km. However, such scales are again coarser to model the impact at the species, habitat and ecosystem levels, and hence further downscaling is important considering other biophysical variables. With this, Hijmans et al. (2005) and Fick and Hijmans (2017) made unique contributions to downscale the climate models to 1 km resolution keeping account of weather station data with covariates including elevation, distance to the coast and satellite-derived covariates of land surface temperature as well as cloud cover, obtained with the MODIS (Moderate Resolution Imaging Spectroradiometer) satellite platform using a thin-plate spline algorithm.

5.4 Earth System Models (ESMs)

The integration of biosphere interaction to the physical climate models is important in the assessment of climate change impacts, and this is normally achieved through the Earth system models (ESMs). ESM acts as an interface between traditional climate science and the synergy of other sciences including life science and social science and hence is applied to a wide range of issues related to the mitigation and adaptation of climate change. Two approaches of ESMs include either simulating the physiological processes of a system to externalities of climate change or placing the system to the envelope of climate, and studying the shift in patterns. The former is a process-based model, for example, a Dynamic Global Vegetation Model (DGVM), and the latter is known as a climate envelope model, e.g. BIOCLIM model.

5.4.1 Process-Based Models (PBMs)

Process-based modelling is a technique that derives a system's behaviour from a collection of functional components and their interactions with one another and the surroundings of the system through gradual physical and mechanical processes (Mäkelä et al. 2000). They are primarily based on theoretical understanding of germane ecological processes or functions and provide a useful framework to integrate precise responses to disturbed environmental conditions.

Dynamic Global Vegetation Models (DGVMs) are process-based models that simulate the changes in the vegetation in response to varying environmental conditions at a particular location. It uses sub-models for photosynthesis, plant carbon balance and other factors (Sitch et al. 2003). DGVMs have historically been used to model assemblages of species with related forms and functions in ecosystems. They do not seek to focus on specific species distributions, despite recent developments that have led to the introduction of hybrid DGVM-individual-based models that concentrate on specific well-known dominant species, such as LPJ-GUESS and HYBRID (Koca et al. 2006; Morin and Thuiller 2009). Table 5.1 explains the commonly used DGVMs applied in ecosystem studies to measure the responses to climate change.

TRIFFID (Top-down Representation of Interactive Foliage and Flora Including Dynamics) is a Dynamic Global Vegetation Model that simulates the distribution of plants and soil carbon based on climate-sensitive CO_2 fluxes (Cox et al. 2000; Hughes et al. 2006). The plant distribution with respect to five plant function types (PFTs), i.e. broadleaf tree, needleleaf tree, C_3 grass, C_4 grass and shrub, is addressed in each grid box. The carbon fluxes for each PFT are computed every 30 min as a function of the climate and atmospheric CO_2 concentration, and every 10 days, the cumulative fluxes are added to update the soil and vegetation carbon. Local litterfall or widespread disturbance that causes carbon loss from the vegetation is deposited into the soil carbon pool, where it is broken down by microorganisms that release CO_2 back into the atmosphere. For every 10 K of warming, it is predicted that the rate of soil respiration will double. This rate also depends on the moisture level of the soil. The land-atmosphere exchange of fluxes is also established through a feedback loop mechanism.

LPJ (Lund-Potsdam-Jena)-DGVM simulates vegetation biogeography and biogeochemistry in a modular framework. It dynamically computes the composition of transient vegetation in terms of plant functional groups, together with their related carbon and water budgets, using climatic, soil and atmospheric data as input, while LPJ-GUESS (General Ecosystem Simulator) was built for the purpose of modelling the growth of an entity, typically a tree, on a number of replicate patches, corresponding in size approximately to the maximum area of influence of one large adult tree on its neighbours. The Land surface Processes and eXchanges (LPX) model works on the wildfire regimes that describe how terrestrial biogeochemical processes interact with the environment to govern wildfire disturbance and the changes in vegetation (Prentice et al. 2011). Although LPX works well globally, it has trouble simulating fire patterns and vegetation composition in savanna (Kelley

Table 5.1 The Dynamic Global Vegetation Models that are applied to measure the response of ecosystems to climate change

Model	Properties	References
TRIFFID (Top-down Representation of Interactive Foliage and Flora Including Dynamics)	Updates the plant distribution and soil carbon based on climate-sensitive CO_2 fluxes at the land-atmosphere interface	Cox (2001), Sitch et al. (2008) and Huang et al. (2021)
LPJ-DGVM (Lund-Potsdam-Jena Dynamic Global Vegetation Model)	An area-based model that simulates the dynamics of terrestrial vegetation by representation of biogeochemical processes, with different properties prescribed for PFTs rather than individual plants	Sitch et al. (2003)
LPJ-GUESS (General Ecosystem Simulator)	An individual-based model in which individuals compete for light and soil water within the same patch	Smith et al. (2001, 2014)
SDBM (Simple Diagnostic Biosphere Model)	A simple light-use efficiency and water-balance model driven by observed precipitation, temperature and remotely sensed observations of FAPAR (fraction of absorbed photosynthetically active radiation)	Kelley et al. (2013)
LPX (Land surface Processes and eXchanges) model	Fire intensity, spread, residence time and carbon flux estimations based on fuel moisture content, seasonality and climate data on a daily time step	Prentice et al. (2011) and Kelley et al. (2013)
HYBRID	Simulates individual trees and grass layers, competing for light, moisture and nitrogen on a grid box	Friend et al. (1997) and Cramer et al. (2001)
IBIS (Integrated Biosphere Simulator) model	PFTs compete for light and moisture within each canopy of trees and grasses. Generates net carbon exchange and runoff	Foley et al. (1996), Hughes et al. (2006) and Jinxun et al. (2022)
SEIB-DGVM (Spatially Explicit Individual-Based Dynamic Global Vegetation Model)	Simulates the local interactions among individual trees where they compete for light and space. This method offers the benefit of simulating a spatially detailed distribution of vegetation	Sato et al. (2007)
SDGVM (Sheffield Dynamic Global Vegetation Model)	By taking into account the biogeochemical distribution of the planet's main PFTs, this model simulates daily carbon, water and nutrient cycles at sizes ranging from forest stands to the entire planet	Woodward and Lomas (2004) and Walker et al. (2017)

(continued)

Table 5.1 (continued)

Model	Properties	References
ORCHIDEE (Organising Carbon and Hydrology In Dynamic Ecosystems)	Specifically simulates the phenomenon of terrestrial carbon cycle related to vegetation and soil microbial activities	Krinner et al. (2005)

et al. 2013). The HYBRID model takes into account the daily cycling of carbon, nitrogen and water, both within the biosphere and between the biosphere and the atmosphere to determine the net exchanges in the land-atmosphere interface. The Integrated Biosphere Simulator (IBIS) is intended to be a comprehensive representation of the Earth's biosphere that captures a variety of processes, such as carbon and nitrogen cycle, soil surface physics, canopy physiology, plant morphology, dynamics and competition. The model generates terrestrial carbon balance, surface water balance and vegetation structure. ORCHIDEE (Organising Carbon and Hydrology In Dynamic Ecosystems) combines dynamic biogeography to a surface-vegetation-atmosphere transfer (SVAT) scheme and models the terrestrial carbon cycle associated with vegetation and soil decomposition processes, as well as changes in vegetation distributions with respect to the changes in climate systems.

5.4.2 Climate Envelope Models (CEMs)

Climate envelope models (CEMs) are widely used to develop adaptation strategies for the species vulnerable to climate change by forecasting the effect of climate change on species distributions. The main strategy is to define the spectrum of climate conditions that the species currently experiences (the climate envelope) and to predict the future spatial distribution of the climate envelope based on projections of various scenarios (Franklin 2009; Watling et al. 2012). 'Bioclimate' variables, which are mostly obtained from seasonal connections between precipitation and temperature, or 'monthly climate' variables are used to create CEMs (Watling et al. 2012). CEMs describe areas where climate is apt for the species or anticipated to become suitable as it includes only climate variables. There are many factors other than climate, such as habitat availability and fragmentation, and competition with other species that may limit the species distributions that CEMs do not take into account. Climate envelope modelling recognizes vital links between predictor variables and relevant responses, thus relating species and environment responses to climate change. Several CEMs have been developed around the world including BIOCLIM, DOMAIN, GAM and MaxEnt (Table 5.2).

Bioclimatic envelope models, also known as 'ecological niche models', 'species distribution models' or 'habitat suitability models', are associated with climate and the probability of occurrence of species in their area of interest. Species distribution models are used extensively in the field of conservation biogeography for supporting conservation planning and evaluating the probable impacts of climate change (Booth

Table 5.2 The climate envelope models used for measuring the impact of climate change on species and ecosystem distributions

Model name	Properties	References
BIOCLIM (percentile distributions)	This model relates species distributions to climatic conditions as multiple one-tailed percentile distributions. BIOCLIM model maps the climatically fitting sites of a species by determining the environmental settings required by them when information on the occurrence of the species is given	Nix (1986), Busby (1991) and Hijmans and Graham (2006)
DOMAIN (distance metric)	In the DOMAIN model, the Gower distance is calculated to measure similarity in environmental variables between the species occurrence points	Carpenter et al. (1993) and Hijmans and Graham (2006)
GAM (general additive modelling)	GAMs model nonlinear trends between predictor variables and response variables using nonparametric functions	Lehmann et al. (2002) and Hijmans and Graham (2006)
MaxEnt (maximum entropy)	MaxEnt models the likelihood of a species being present from the distribution of maximum entropy subject to the constraint that the projected value of each environmental variable under this estimated distribution matches its empirical average	Phillips et al. (2006) and Hijmans and Graham (2006)

et al. 2014). The first software tool to link spatially detailed species occurrence data with maps of environmental variables was called BIOCLIM. It compares the values of the environmental factors at a site to the percentile distribution of the values from other locations in order to estimate the likelihood of a species occurring there. Species are likely to maintain viable populations under a set of parameters that the BIOCLIM model defines (Araújo and Peterson 2012). Species distribution models use rules or mathematical functions to describe associations between species occurrence and environmental conditions. The variables used in species distribution models may include climate, land cover, topography or any other variable relevant to the species being modelled (Watling et al. 2013).

MaxEnt modelling evaluates the possibility for a species' occurrence on a site based on environmental constraints (Phillips et al. 2006). MaxEnt modelling uses an algorithm that calculates the probability of occurrence of a species relative to background conditions in the area of interest as a function of environmental or climatic conditions where the species occurs. It is used for modelling species geographic distributions with presence-only data. It also avoids over fitting, a serious concern for other niche modelling methods (Phillips and Dudík 2008). Likewise, GAM is a model which allows the linear model to learn nonlinear relationship trends between dependent (species presence or absence) and independent (environment) variables. The DOMAIN model calculates the potential distribution based on a standardized point-to-point similarity metric and provides a simple and reliable method for modelling the potential distribution of plant and animal species.

5.5 Conclusion

Climate change is projected to impact all the ecosystems around the world in one way or the other. Since its inception in 1988, the IPCC has published an array of extensive assessment reports on our climate, the potential impacts of changing climate and options for response strategies. The emission scenarios developed by IPCC are used as important inputs for modelling the climate change impacts on ecosystems. The emission scenarios are integrated with the Earth system models by way of process-based models or climate envelope models. The parameterization and mechanisms of these models vary, and the choice of the model depends on the type of output variables required. This chapter collated the various models that are in practice now and elaborated further on the input variables and the mechanism adopted in various models. Some approaches for modelling are better fitted than others to address the problem of prediction under a set of criteria. The development of high-quality climate predictions has become the need of the hour for understanding the impacts of different greenhouse gas emission scenarios and for mitigating and adapting to the resulting climate changes and thus to meet the climate goals of the Paris Agreement.

Acknowledgements The authors would like to thank Hon'ble Vice Chancellor, Director of Research, Dean of Faculty of Ocean Sciences and Technology and Head of the Department of Climate Variability and Aquatic Ecosystems of Kerala University of Fisheries and Ocean Studies (KUFOS) for providing facilities to conduct the study.

References

Araújo MB, Peterson AT (2012) Uses and misuses of bioclimatic envelope modelling. Ecology 93(7):1527–1539

Booth TH, Nix HA, Busby JR, Hutchinson MF (2014) BIOCLIM: the first species distribution modelling package, its early applications and relevance to most current MAXENT studies. Divers Distrib 20(1):1–9

Bos AB, Duchelle AE, Angelsen A, Avitabile V, Sy VD, Herold M, Joseph S, Sassi CD, Sills EO, Sunderlin WD, Wunder S (2017) Comparing methods for assessing the effectiveness of subnational REDD+ initiatives. Environ Res Lett 12:074007

Busby JR (1991) BIOCLIM—a bioclimate analysis and prediction system. In: Margules CR, Austin MP (eds) Nature conservation: cost effective biological surveys and data analysis. CSIRO, Melbourne, pp 64–68

Carpenter G, Gillison AN, Winter J (1993) DOMAIN: a flexible modelling procedure for mapping potential distributions of plants and animals. Biodivers Conserv 2(6):667–680

Cheng X, Chen L, Sun R, Kong P (2018) Land use changes and socio-economic development strongly deteriorate river ecosystem health in one of the largest basins in China. Sci Total Environ 616–617:376–385. https://doi.org/10.1016/j.scitotenv.2017.10.316

Cox PM (2001) Description of the "TRIFFID" dynamic global vegetation model. Hadley Centre Technical Note 24, Met Office

Cox PM, Betts RA, Jones CD, Spall SA, Totterdell IJ (2000) Acceleration of global warming due to carbon-cycle feedbacks in a coupled climate model. Nature 408(6809):184–187

Cramer W, Bondeau A, Woodward FI, Prentice IC, Betts RA, Brovkin V, Young-Molling C (2001) Global response of terrestrial ecosystem structure and function to CO_2 and climate change: results from six dynamic global vegetation models. Glob Chang Biol 7(4):357–373
de Sassi C, Joseph S, Bos AB, Duchelle AE, Ravikumar A, Herold M (2015) Towards integrated monitoring of REDD+. Curr Opin Environ Sustain 14:93–100
Fick SE, Hijmans RJ (2017) WorldClim 2: new 1-km spatial resolution climate surfaces for global land areas. Int J Climatol 37:4302–4315
Foley JA, Prentice IC, Ramankutty N, Levis S, Pollard D, Sitch S, Haxeltine A (1996) An integrated biosphere model of land surface processes, terrestrial carbon balance, and vegetation dynamics. Glob Biogeochem Cycles 10:603–628
Franklin J (2009) Mapping species distributions: spatial inference and prediction. Cambridge University Press, New York
Friend AD, Stevens AK, Knox RG, Cannell MGR (1997) A process-based, terrestrial biosphere model of ecosystem dynamics (Hybrid v3.0). Ecol Model 95:249–287
Gaffin SR, Rosenzweig C, Xing X, Yetman G (2004) Downscaling and geo-spatial gridding of socio-economic projections from the IPCC Special Report on Emissions Scenarios (SRES). Glob Environ Chang 14(2):105–123
George A, Joseph S, Sebastian A, Sajeev TV (2019) Impact of past climate change and socio-economic drivers on different crops in agroforestry systems of Wayanad, India. In: 4th World Congress on agroforestry—book of abstracts, p 62
Grimm NB, Chapin FS III, Bierwagen B, Gonzalez P, Groffman PM, Luo Y, Williamson CE (2013) The impacts of climate change on ecosystem structure and function. Front Ecol Environ 11(9): 474–482
Hijmans RJ, Graham CH (2006) The ability of climate envelope models to predict the effect of climate change on species distributions. Glob Chang Biol 12(12):2272–2281. https://doi.org/10.1111/j.1365-2486.2006.01256.x
Hijmans RJ, Cameron SE, Parra JL, Jones P, Jarvis A (2005) Very high resolution interpolated climate surfaces for global land areas. Int J Climatol 25:1965–1978
Huang H, Xue Y, Liu Y, Li F, Okin GS (2021) Modeling the short-term fire effects on vegetation dynamics and surface energy in southern Africa using the improved SSiB4/TRIFFID-Fire model. Geosci Model Dev 14:7639–7657
Hughes JK, Valdes PJ, Betts R (2006) Dynamics of a global-scale vegetation model. Ecol Model 198(3–4):452–462
IPCC (1996) Report of the twelfth session of the IPCC. Mexico City, 11–13 September 1996
IPCC (2001) In: McCarthy JJ, Canziani OF, Leary NA, Dokken DJ, White KS (eds) Intergovernmental Panel on Climate Change. Climate Change 2001: impacts, adaptation and vulnerability. Cambridge University Press, Cambridge
IPCC (2007) The physical science basis. Contribution of working group I to the fourth assessment report of the Intergovernmental Panel on Climate Change, vol 996. Cambridge University Press, Cambridge and New York, NY, pp 113–119
IPCC (2021) Climate change 2021: the physical science basis. Contribution of working group I to the sixth assessment report of the Intergovernmental Panel on Climate Change. Cambridge University Press, Cambridge
Jinxun L, Xuehe L, Qiuan Z, Wenping Y, Quanzhi Y, Zhen Z, Qingxi G, Carol D (2022) Terrestrial ecosystem modeling with IBIS: progress and future vision. J Resour Ecol 13:2–16
Joseph S, Anitha K, Murthy M (2009) Forest fire in India: a review of the knowledge base. J For Res 14:127–134
Joseph S, Herold M, Sunderlin WD, Verchot LV (2013) REDD+ readiness: early insights on monitoring, reporting and verification systems of project developers. Environ Res Lett 8:034038
Kelley DI, Prentice IC, Harrison SP, Wang H, Simard M, Fisher JB, Willis KO (2013) A comprehensive benchmarking system for evaluating global vegetation models. Biogeosciences 10:3313–3340

Koca D, Smith B, Sykes MT (2006) Modelling regional climate change effects on potential natural ecosystems in Sweden. Clim Chang 78(2):381–406

Krinner G, Viovy N, de Noblet-Ducoudré N, Ogée J, Polcher J, Friedlingstein P, Prentice IC (2005) A dynamic global vegetation model for studies of the coupled atmosphere-biosphere system. Glob Biogeochem Cycles 19(1):GB1015. https://doi.org/10.1029/2003GB002199

Krishnaswamy J, John R, Joseph S (2014) Consistent response of vegetation dynamics to recent climate change in tropical mountain regions. Glob Chang Biol 20:203–215

Lehmann A, Overton JMC, Leathwick JR (2002) GRASP: generalized regression analysis and spatial predictions. Ecol Model 157:189–207

Mäkelä A, Landsberg J, Ek AR, Burk TE, Ter-Mikaelian M, Ågren GI, Puttonen P (2000) Process-based models for forest ecosystem management: current state of the art and challenges for practical implementation. Tree Physiol 20(5–6):289–298

Mandal S, Islam MS, Biswas MHA, Akter S (2021) Modeling the optimal mitigation of potential impact of climate change on coastal ecosystems. Heliyon 7(7):e07401

McGregor JL (1997) Regional climate modelling. Meteorog Atmos Phys 63(1):105–117

MEA (2005) Ecosystems and human well-being: biodiversity synthesis. World Resources Institute, Washington, DC

Mearns LO, Easterling W, Hays C, Marx D (2001) Comparison of agricultural impacts of climate change calculated from high and low resolution climate change scenarios: part I. The uncertainty due to spatial scale. Clim Chang 51(2):131–172

Melillo JM (1999) Warm, warm on the range. Science 283:183–184

Morin X, Thuiller W (2009) Comparing niche-and process-based models to reduce prediction uncertainty in species range shifts under climate change. Ecology 90(5):1301–1313

Nix HA (1986) A biogeographic analysis of Australian elapid snakes. In: Longmore R (ed) Atlas of elapid snakes of Australia: Australian flora and fauna series 7. Bureau of Flora and Fauna, Canberra, pp 4–15

Pan Z, He J, Liu D, Wang J (2020) Predicting the joint effects of future climate and land use change on ecosystem health in the Middle Reaches of the Yangtze River economic belt. China Appl Geogr 124:102293. https://doi.org/10.1016/j.apgeog.2020.102293

Pedersen JTS, van Vuuren D, Gupta J, Santos FD, Edmonds J, Swart R (2022) IPCC emission scenarios: how did critiques affect their quality and relevance 1990–2022? Glob Environ Chang 75:102538

Phillips SJ, Dudík M (2008) Modeling of species distributions with Maxent: new extensions and a comprehensive evaluation. Ecography 31(2):161–175

Phillips SJ, Anderson RP, Schapire RE (2006) Maximum entropy modelling of species geographic distributions. Ecol Model 190:231–259

Prentice IC, Kelley DI, Foster PN, Friedlingstein P, Harrison SP, Bartlein PJ (2011) Modeling fire and the terrestrial carbon balance. Glob Biogeochem Cycles 25:GB3005

Ren Y, Zhang F, Li J, Zhao C, Jiang Q, Cheng Z (2022) Ecosystem health assessment based on AHP-DPSR model and impacts of climate change and human disturbances: a case study of Liaohe River Basin in Jilin Province, China. Ecological Indicators 142:109171

Sato H, Itoh A, Kohyama T (2007) SEIB–DGVM: a new dynamic global vegetation model using a spatially explicit individual-based approach. Ecol Model 200(3–4):279–307

Sitch S, Smith B, Prentice IC, Arneth A, Bondeau A, Cramer W, Kaplan JO, Levis S, Lucht W, Sykes MT, Thonicke K, Venevsky S (2003) Evaluation of ecosystem dynamics, plant geography and terrestrial carbon cycling in the LPJ dynamic global vegetation model. Glob Chang Biol 9:161–185

Sitch S, Huntingford C, Gedney N, Levy PE, Lomas M, Piao SL, Woodward FI (2008) Evaluation of the terrestrial carbon cycle, future plant geography and climate-carbon cycle feedbacks using five Dynamic Global Vegetation Models (DGVMs). Glob Chang Biol 14(9):2015–2039

Smith B, Prentice IC, Sykes MT (2001) Representation of vegetation dynamics in the modelling of terrestrial ecosystems: comparing two contrasting approaches within European climate space. Glob Ecol Biogeogr 10:621–637

Smith B, Wårlind D, Arneth A, Hickler T, Leadley P, Siltberg J, Zaehle S (2014) Implications of incorporating N cycling and N limitations on primary production in an individual based dynamic vegetation model. Biogeosciences 11:2027–2054. https://doi.org/10.5194/bg-11-2027-2014

Sutherland WJ (2006) Predicting the ecological consequences of environmental change: a review of the methods. J Appl Ecol 43(4):599–616

Van Vuuren DP, Edmonds JA, Kainuma M, Riahi K, Weyant J (2011) A special issue on the RCPs. Clim Chang 109(1):1–4

Walker AP, Quaife T, Van Bodegom PM, De Kauwe MG, Keenan TF, Joiner J, Woodward FI (2017) The impact of alternative trait-scaling hypotheses for the maximum photosynthetic carboxylation rate (Vcmax) on global gross primary production. New Phytol 215(4):1370–1386

Watling JI, Romanach SS, Bucklin DN, Speroterra C, Brandt LA, Pearlstine LG, Mazzotti FJ (2012) Do bioclimate variables improve performance of climate envelope models? Ecol Model 246:79–85

Watling JI, Brandt LA, Mazzotti FJ, Romañach SS (2013) Use and interpretation of climate envelope models: a practical guide. University of Florida

Woodward FI, Lomas MR (2004) Vegetation dynamics–simulating responses to climatic change. Biol Rev 79(3):643–670

Developing a Bayesian Model of Climate-Induced Lake Overturn in Talisay, Taal Lake

6

Damasa B. Magcale-Macandog, Arnold R. Salvacion, Jaderick P. Pabico, Keshia N. Tingson, Marlon A. Reblora, Jennifer D. Edrial, Felino P. Lansigan, and Macrina T. Zafaralla

Abstract

Significant fish kills in Taal Lake in Talisay, Batangas Province, which created major economic setbacks in the area, were mainly attributed to lake overturn. A predictive model of climate-induced lake overturn for the lake was developed using a combination of data exploratory analysis, correlation, autocorrelation, logistic regression, and Bayesian causality modeling. Climatic data and reports of fish kill events in the lake were analyzed for trends and patterns. Statistical tests of possible relationships were done via logistic regression modeling, while stepwise logistic regression method was applied to identify climatic variables that significantly affected lake overturn.

D. B. Magcale-Macandog (✉) · M. A. Reblora · J. D. Edrial · M. T. Zafaralla
Institute of Biological Sciences, College of Arts and Sciences, University of the Philippines Los Baños, College, Los Baños, Laguna, Philippines
e-mail: dmmacandog@up.edu.ph

A. R. Salvacion
Institute for Governance and Rural Development, College of Public Affairs, and School of Environmental Science and Management, University of the Philippines Los Baños, College, Los Baños, Laguna, Philippines

J. P. Pabico
Institute of Computer Science, College of Arts and Sciences, University of the Philippines Los Baños, College, Los Baños, Laguna, Philippines

K. N. Tingson
Institute of Biological Sciences, College of Arts and Sciences, and School of Environmental Science and Management, University of the Philippines Los Baños, Los Baños, Laguna, Philippines

F. P. Lansigan
Institute of Statistics, College of Arts and Sciences, and School of Environmental Science and Management, University of the Philippines Los Baños, College, Los Baños, Laguna, Philippines

S. Dhyani et al. (eds.), *Ecosystem and Species Habitat Modeling for Conservation and Restoration*, https://doi.org/10.1007/978-981-99-0131-9_6

Wind speed and minimum air temperature were the two significant climatic variables found to have induced lake overturn. The results of the logistic probability analyses were transformed to develop the conditional probability tables (CPT) required in the development of the model.

Keywords

Lake overturn · Bayesian model · Fish kill

6.1 Introduction

Volcanic Taal Lake, the third largest lake in the Philippines, is uniquely located inside the Taal Caldera. It lies in the heart of Batangas Province and occupies a total area of 260 km^2. It has an average depth of 65 m and a maximum depth of 180 m. Feeding into it are 37 river tributaries, with 1 river outlet, the Pansipit River, flowing into Balayan Bay (Schiemer et al. 2001; ADB 2004; You et al. 2013).

Lake Taal Lake has multiple uses and benefits such as for open water fisheries, commercial aquaculture, recreational activities, navigation routes, and water source. Of particular interest are the immense aquaculture activities in the lake that started in the 1980s by which tilapia (*Oreochromis niloticus*) and milkfish (*Chanos chanos*) culture was introduced (Papa and Mamaril 2011). The proliferation of fish pens and cages has affected the water quality of the lake. It is estimated that 64% of the nitrogen and 81% of the phosphorus contents of fish feed are released into the lake environment (Edwards 1993). Yambot (2000) calculated that for every 1.5 tons of fish feed given, 16 kg of phosphorus is released into Taal Lake waters. Further, the excess fish feeds and fish feces contribute to the increased organic material that settles at the bottom of the lake. Decomposition of these organic matter releases hydrogen sulfide (H_2S) and other toxic gases (White et al. 2007).

Significant fish kill occurrences in Taal Lake have created major economic setbacks in the area. One noteworthy incident was the 2011 massive fish kill that disrupted the socio-economic activities in the lake, with recorded losses of approximately PHP 140 million. The event was attributed to an interplay of factors such as lake overturn, water pollution, change in season (i.e., summer going to rainy season), changes in wind stress, and intermittent rainfall (BFAR 2011).

The participants of the participatory activities conducted with the local communities in Taal Lake expressed their local ecological knowledge on the occurrence of fish kill (Magcale-Macandog et al. 2014). They perceive that a combination of climatic, volcanic, and anthropogenic factors causes localized fish kill episodes in the lake. These factors include oxygen depletion, volcanic activities at the bottom of the lake, strong winds, hydrothermal vents, polluted water, and aquaculture activities.

The dynamic thermal changes of lake water and above water surface create a condition for possible lake overturn. With cool winds coupled with heavy rainfall as a function of the annual seasonal shift, the surface water becomes cooler and thus denser relative to the water column below the thermocline. The surface water subsequently sinks down, displacing the warmer hypoxic bottom layer which in return pushes up into the surface (Rosana 2011).

Many fish kills recorded in Taal Lake are caused by lake overturn. Increase in wind turbulence and low atmospheric temperature cools the lake water surface layer (epilimnion) and erodes the thermal stratification of the water column (ADB 2004; Balistrieri et al. 2006; Caliro et al. 2008; Marti-Cardona et al. 2008). In combination with the pressure of strong winds, mixing of water occurs. This transports the low dissolved oxygen and reduced chemical substances such as H_2S, nitrite (NO_2), and ammonia (NH_3) from the lake bottom to the water surface, as well as mixing them in localized portions of the lake. The lake then goes into a state of hypoxia characterized by low dissolved oxygen, that is, below 2 mg/L. This undesirable water quality subsequent to lake overturn triggers fish kills in Taal Lake.

In the seven crater lakes of San Pablo City, Philippines, sporadic fish kill events have been recorded during the cool months from December to February due to lake overturn resulting to the upwelling of anoxic water (Brillo 2015, 2016a, b, c, d, e; Diana 2009; Paller et al. 2021). Another possible reason for these fish kill events is the deterioration of water quality due to aquaculture activities characterized by excessive use of commercial fish feeds (Orosa 2014). Domestic waste from increasing human settlements also contributes to lake water quality deterioration.

Similar phenomena were observed in stratified lakes around the world. For instance, lake overturn caused the fish kill in Lake Averno, Italy, last 2005. After the fish kill, researchers discovered that the lake's water was unstratified chemically and isotopically. Increasing H_2S and methane (CH_4) concentrations and decreasing sulfate (SO_4) levels in the lake with depth were observed. The presence of CH_4 and sulfide (SO_2) in the surface water, as well as the anoxic condition along the water column, proved the occurrence of lake overturn. Moreover, white water was noted and found to contain high H_2S level (10–20 mg/L). These data supported the hypothesis that lake overturn and high H_2S levels were the main causes of fish kill in Lake Averno (Caliro et al. 2008).

By the same token, Lake Valencia in Venezuela experienced a massive fish and zooplankton mortality in 1977. It was found that the lake overturned due to a shift in physical conditions such as minimum air temperature and maximum wind strength during the months of December to March. Moreover, the presence of H_2S dominated in the lake as evidenced by its strong odor. Discolorations in lake water approximately 1 km or more in diameter were also observed. Fish and zooplankton mortality was attributed to anoxic lake water (i.e., between 20 m and 30 m depth) and toxic level of H_2S in epilimnion (de Infante et al. 1979).

A Bayesian network consists of a graphical structure and a probabilistic description of the relationships among variables in a system. The graphical structure explicitly represents cause-and-effect assumptions that allow a complex causal chain linking actions to outcomes to be factored into an articulated series of conditional relationships (Borsuk et al. 2004). Bayesian network modeling is believed to provide the most feasible method of estimating parameters in complex systems, such as biogeochemical cycles involving nitrate, ammonium, dissolved organic nitrogen, phytoplankton, zooplankton, and bacteria, while including random processes and variables to model uncertainty (Borsuk et al. 2001). Bayesian networks provide a methodology for combining expert knowledge of causal structure and aggregate ecosystem response with condensed models that are identifiable from available data.

Logistic regression modeling has been widely used to model events or phenomena with dichotomous response (i.e., yes or no) (Salem et al. 2004). The same method was used by Can et al. (2005) to assess the susceptibility of shallow earthflows triggered by heavy rainfall in three catchments in Turkey. Van Den Eeckhaut et al. (2006) produced the landslide susceptibility map of Flemish Ardennes in Belgium applying the same method. In Taiwan, Chang et al. (2007) modeled landslide events based on the occurrence of rainfall and earthquakes. Hu and Lo (2007) modeled urban growth in Atlanta, Georgia, through the use of logistic regression. Yang et al. (2006) used logistic regression with geographic information system (GIS) to map the distribution of matsutake mushrooms in Yunnan, southwest China. The same methodology was employed by Ozdemir (2011) to map the groundwater spring potential in the Sultan Mountains, Turkey. Chen developed a model to predict financial distress by integrating logistic regression with decision tree classification technique. Piwczynski et al. (2012) used logistic regression to determine factors affecting lamb mortality.

This study aimed to model the occurrence of overturn in Taal Lake using logistic regression techniques and Bayesian network modeling using weather variables (i.e., rainfall, maximum temperature, minimum temperature, wind speed, and wind direction) and their duration. The probabilistic predictions of the model can give stakeholders and decision-makers a realistic appraisal of the chances of fish kills in the future, which is critical to the decision process. The corresponding recommendations may help advance the stakeholders' ability to forecast fish kill events and consequently allow the implementation of preventative measures to reduce the frequency and magnitude of fish kills in Taal Lake and other lakes in the country.

6.2 Materials and Methods

6.2.1 Data Collection

Secondary data on the water biophysicochemical properties and geological and climatic attributes of Taal Lake were gathered from various agencies including the Philippine Atmospheric, Geophysical and Astronomical Services Administration (PAGASA), Philippine Institute of Volcanology and Seismology (PHIVOLCS), Bureau of Fisheries and Aquatic Resources (BFAR), fish cage operators, and UPLB-FEWS (Fish Kill Early Warning System) Program Research Team 1. Data on occurrences of lake overturn and fish kill were lifted from reports and announcements disseminated by BFAR to the municipalities around Taal Lake for the period 1998–2012. Thirty-two years (1980–2012) of weather data from the nearest agroclimatic station (Ambulong, Tanauan, Batangas) were acquired from PAGASA.

6.2.2 Exploratory Data Analysis

The collected secondary data were subjected to exploratory data analyses (EDA) to discover trends, patterns, correlations, and relationships. The availability of long-term weather data (32 years) enabled the conduct of EDA on the weather variables (i.e., rainfall, temperature, wind direction, and wind speed). EDA on the weather variables included construction of various graphs such as scatter plots to detect linear relationships, conditional density plot to determine probability distribution, and time series plots to visualize trends. Discernible patterns of duration of sustained wind speed and minimum temperature were explored graphically. Data on fish kill events were superimposed over the graphs of the weather variables to visualize trends and possible relationships between fish kill events and weather variables.

Possible relationships among weather variables prior to and during lake overturn were analyzed graphically. Correlations of lag periods of each weather variable were also analyzed to determine if prior weather conditions (e.g., 7 days before) had some relationships with the occurrence of lake overturn.

6.2.3 Statistical Test

After exploring possible relationships between lake overturn and weather variables graphically, statistical test of the possible relationship was done via logistic regression modeling. Stepwise forward selection logistic regression (SFSLR) was applied to identify which among the six weather variables (wind speed, wind direction, minimum air temperature, maximum air temperature, mean air temperature, and rainfall) had significant effect on lake overturn.

6.2.4 Development of Bayesian Model of Lake Overturn

R Console was used to determine the respective conditional probability relationships between the occurrences of lake overturn and each of the weather variables found significant by SFSLR above. The results of these logistic probability analyses were transformed to develop the conditional probability tables (CPT) for each of the weather variables and their combination. The CPT were used in the development of the Bayesian network of models for lake overturn with the aid of Netica.

6.3 Results and Discussion

6.3.1 Patterns of Wind Velocity and Wind Direction

The average weekly trends and patterns of wind velocity in Taal Lake (Fig. 6.1) are generally higher (>1.5 mps) from the 4th week of October to the 3rd week of April. This period coincides with the cool months. The prevailing wind direction during this period is from the NE (Fig. 6.2). From May to June, the wind direction gradually shifts from NE to SW (Fig. 6.2). From July to September, the prevailing wind is from the SW direction, coinciding with the monsoon or rainy season in the country. On the 4th week of April, the change in wind direction from NE to SW coincides with a decline in wind velocity (<1.5 mps) (Fig. 6.1). Likewise, on the 3rd week of October, the change in wind direction from SW to NE coincides with increase in wind velocity (>1.5 mps).

6.3.2 Fish Kill Events and Wind Patterns

Data on occurrence of fish kill events were superimposed over the graphs of wind speed and direction to assess possible relationships (Fig. 6.3). Lake overturn, sulfur upwelling, oxygen depletion, and pollution are the common causes of fish kills in Taal Lake. Fish kill events due to lake overturn are reported to occur mostly during the times when the wind direction shifts from NE to SW in May and June (weeks 18–25). The average wind velocity during this period is 1.25 mps.

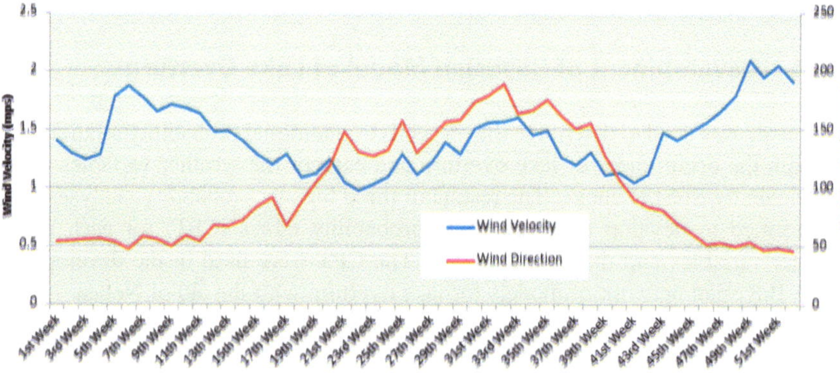

Fig. 6.1 Average weekly trends and patterns of wind velocity and wind direction for 30 years (1980–2009) in Ambulong Station, Tanauan, Batangas

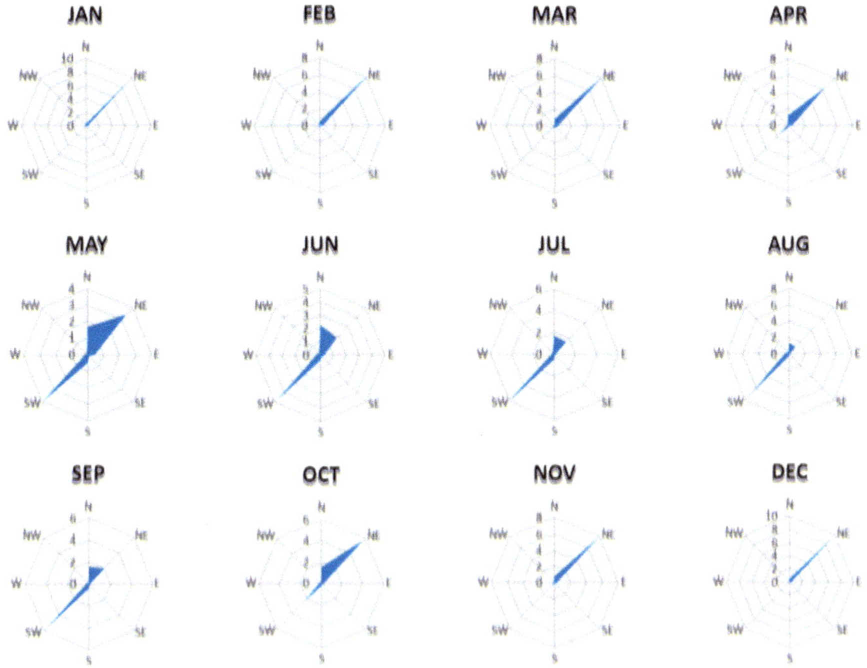

Fig. 6.2 Monthly wind direction in Ambulong Station, Tanauan, Batangas, based on historical weather data from 1980 to 2009

6.3.3 Rainfall

The average weekly pattern of the 32-year rainfall data shows that the rainy season in the Taal Lake area occurs from June to October. The average weekly rainfall during this season varies from 6 cm to 12 cm (Fig. 6.4). In the dry months of December to April, the average weekly rainfall is only 1 cm.

Most of the recorded fish kill events occurred at the onset of the rainy season particularly during intermittent light to moderate rains in the area (Fig. 6.4). In fact, stakeholders reported some fish kills happening 7 days following consecutive rainfall events, with a total cumulative rainfall of at least 260 mm after several months of no rainfall event. Note the absence of rainfall event during the summer months of March to before mid-May, with the middle of May as the onset of rainy season.

6.3.4 Spatial Occurrence of Fish Kill in Taal Lake

Reported fish kill events from 1998 to 2011 were mapped to visualize possible clustering of the events with respect to their location and reported cause. Lake overturn was the major cause of fish kill in the Laurel and San Nicolas

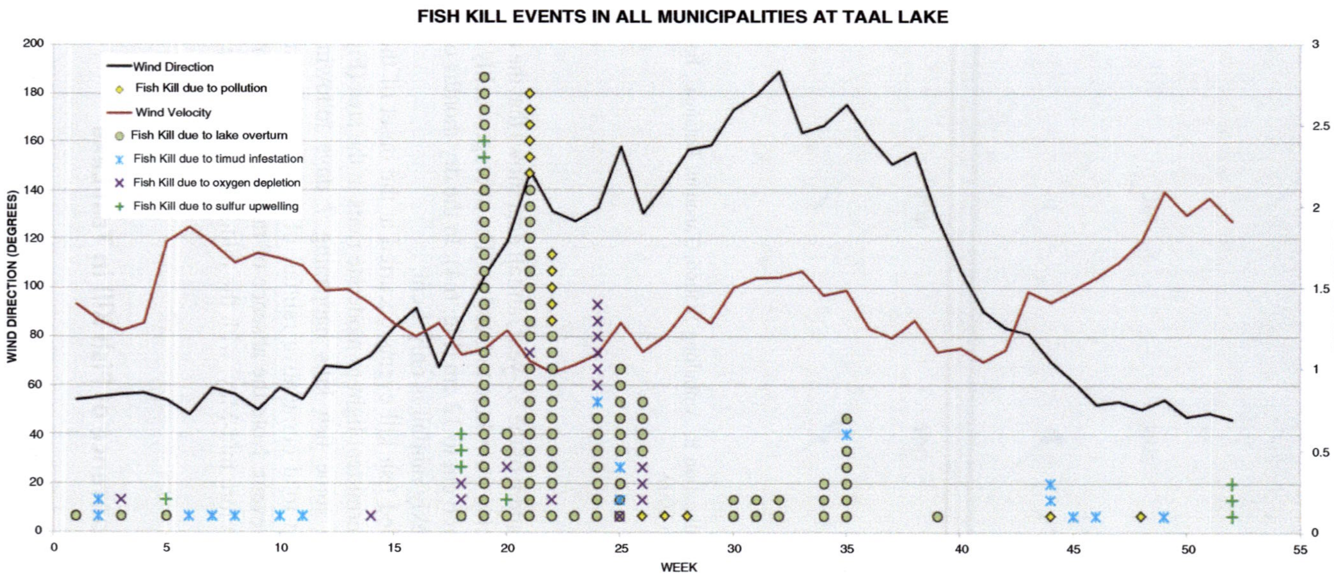

Fig. 6.3 Occurrence of fish kills in Taal Lake due to various factors including lake overturn, oxygen depletion, sulfur upwelling, and isopod infestation based on BFAR announcements and reports from 1998 to 2011

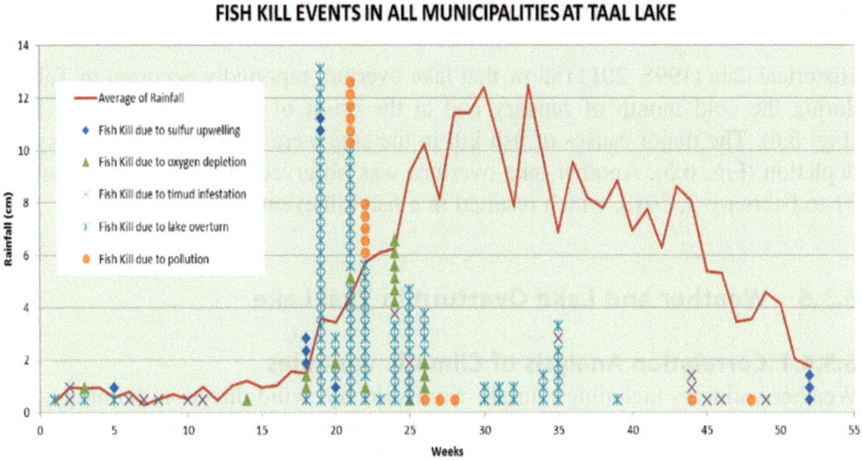

Fig. 6.4 Average weekly rainfall pattern (1980–2011), fish kill events (1998–2011), and corresponding causes of fish kill events in Taal Lake

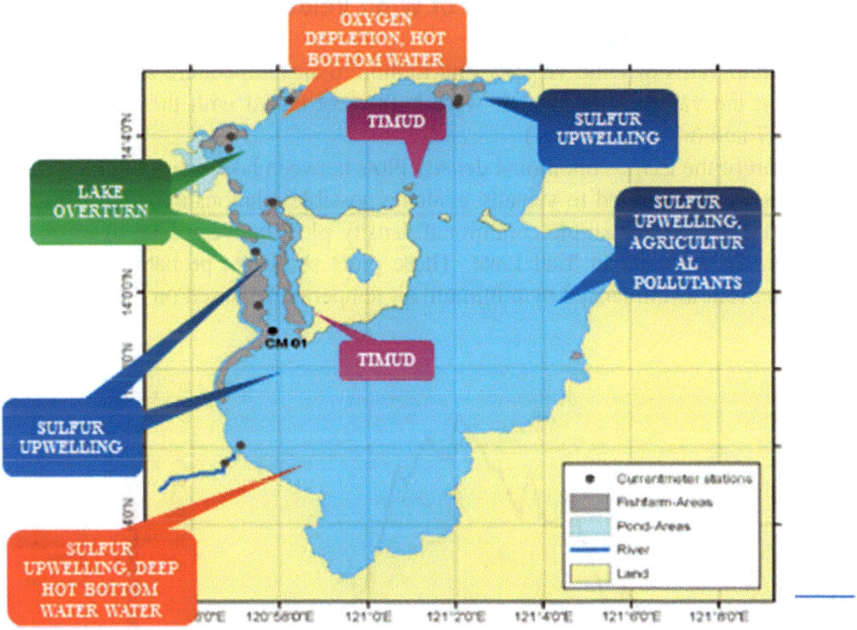

Fig. 6.5 Major causes of fish kills in various fishing and aquaculture areas in Taal Lake. (Base map source: White et al. 2007)

municipalities, while it was sulfur upwelling in Talisay and San Nicolas municipalities (Fig. 6.5). This suggests that causes of fish kills vary spatially across the lake.

6.3.5 Fish Kill Due to Lake Overturn in Talisay

Historical data (1998–2011) show that lake overturn reportedly occurred in Talisay during the cold month of January and at the onset of the rainy season in May (Fig. 6.6). The major causes of fish kill in the area were lake overturn and oxygen depletion (Fig. 6.6). Another lake overturn was observed in Talisay from January 31 to February 2, 2013, which resulted in a fish kill event.

6.3.6 Weather and Lake Overturn in Taal Lake

6.3.6.1 Correlation Analysis of Climatic Variables

Weather variables including rainfall, wind velocity, wind direction, humidity, and minimum air temperature and the occurrence of lake overturn/fish kill in the lake were subjected to correlation analysis to identify highly correlated variables. Results revealed that wind velocity and minimum air temperature significantly affected lake overturn in Taal Lake.

6.3.6.2 Autocorrelation Analysis of Wind Velocity and Minimum Air Temperature

Autocorrelograms of wind velocity and minimum air temperature showed that for variables, the values 1 day before were highly correlated with the current values (Figs. 6.7 and 6.8, respectively).

As part of the EDA, conditional density plots between lake overturn and weather variables were generated to visually evaluate possible relationships between them. Figure 6.9 shows the sample conditional density plots between lake overturn and minimum temperature in Taal Lake. These plots show the probability of a lake overturn event as influenced by minimum air temperature. Based on the conditional

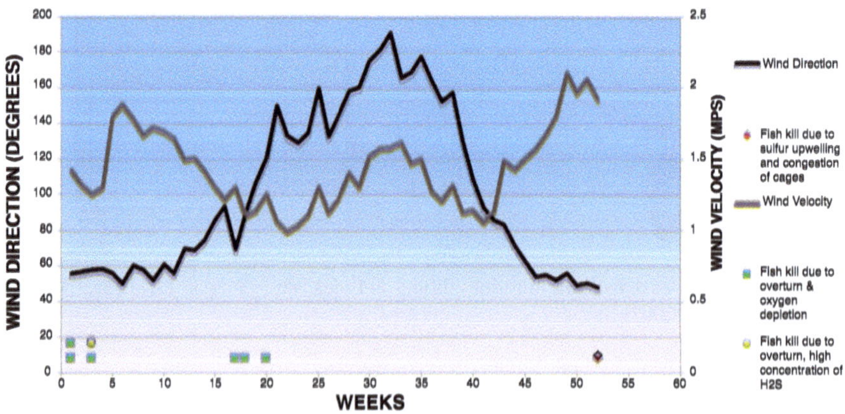

Fig. 6.6 Occurrence of fish kill in Talisay due to various factors, 1998–2011

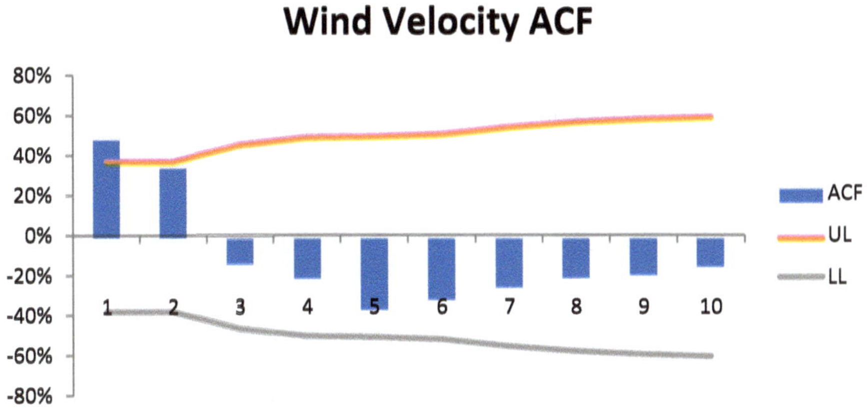

Fig. 6.7 Autocorrelation of wind speed during and up to 7 days before the fish kill

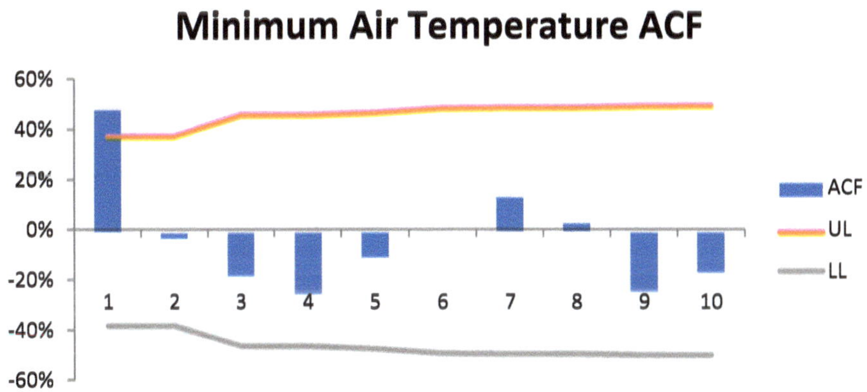

Fig. 6.8 Autocorrelation of minimum temperature data during and up to 7 days before the fish kill

density plots (g), the probability of a lake overturn is almost 100% when the minimum temperature is below 21 °C and 0% when the minimum temperature is above 25 °C, or 6 days prior to the event. Such information is valuable in developing a predictive model for lake overturn event given the minimum air temperature. The same procedure was also applied to the other weather variables (i.e., rainfall, wind direction, and wind speed).

6.3.6.3 Logistic Regression of Minimum Air Temperature and Lake Overturn

Results of logistic regression show that among the weather variables, minimum air temperature showed significant relationship with lake overturn. Minimum air temperature 6 days and 4 days prior to lake overturn showed significant relationships with lake overturn with p-values of 0.031 and 0.059, respectively (Table 6.1). However, the two lag periods showed contrasting effects on lake overturn. The

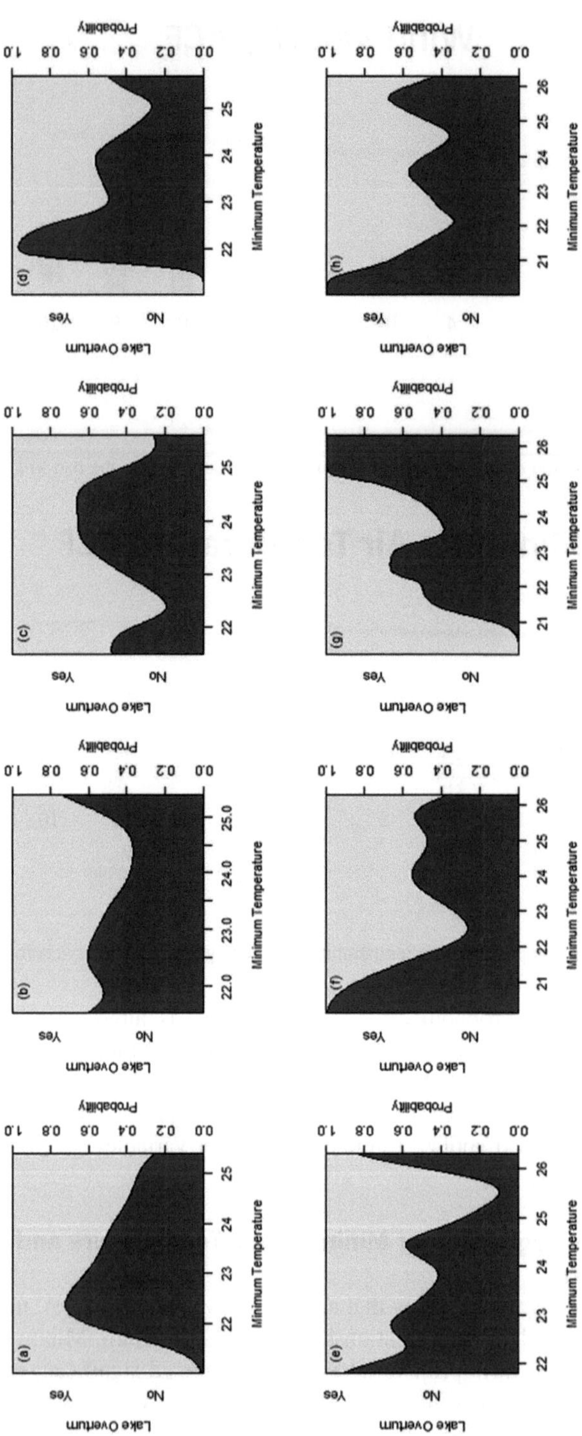

Fig. 6.9 Conditional density plots between lake overturn events and minimum air temperature: (**a**) during; (**b**) 1 day before; (**c**) 2 days before; (**d**) 3 days before; (**e**) 4 days before; (**f**) 5 days before; (**g**) 6 days before; and (**h**) 7 days before the lake overturn

Table 6.1 Logistic regression coefficient of minimum temperature lags with lake overturn event as dependent variable

Variable	Coefficient	*p*-value
(Intercept)	−0.6605	0.9570
Minimum temp 4 days prior	1.2095	0.0592*
Minimum temp 4 days prior	−1.1910	0.0312*

*Significant at $\alpha = 0.1$

probability of lake overturn increases with lower minimum temperature 6 days prior to the event but decreases with lower minimum temperature 4 days prior to the event. Colder air temperature (1 °C decrease in minimum temperature) 6 days before increases the chances of lake overturn by 69.6%. On the other hand, warmer air temperature 4 days before increases the odds of lake overturn by 235%. Table 6.1 summarizes the results of the stepwise logistic regression analysis on lake overturn and weather variables.

Cooler minimum air temperature 6 days before followed by warmer minimum air temperature 4 days before tended to increase the probability of lake overturn in Taal Lake. This result suggests that changes in minimum temperature, from colder to warmer temperature, can significantly affect vertical changes in the density of the lake water, leading to overturn (Caliro et al. 2008)

The lag periods (i.e., 6 days and 4 days) of the independent variable (i.e., minimum temperature) suggest that occurrence of overturn in Taal Lake can be predicted 4 days to 6 days in advance provided the transition or shift in air temperature can be determined. Such information is very useful in the development of Bayesian networks of model for predicting fish kill event.

6.3.7 Lake Overturn Due to Combined Wind Speed and Minimum Temperature

Results showed that among the six weather variables analyzed, two significantly affected lake overturn, i.e., wind speed and minimum air temperature.

Lake overturn may be due to the combined effects of wind speed and cool air. It is hypothesized that the increase in wind velocity of NE winds blowing from the shores of Talisay toward the Taal Volcano Island during the cool months of January and February may induce lake overturn. This is because the heavy cool surface water is being pushed and sinks downward, eventually breaking the thermal stratification of the water column that leads to overturn (ADB 2004; Balistrieri et al. 2006; Caliro et al. 2008; Marti-Cardona et al. 2008). Logistic regression of the combined data on wind speed and minimum temperature with the occurrence of lake overturn predicted the probability of lake overturn and fish kill (Table 6.2).

To find out if there were discernible wind speed and minimum temperature patterns several days before a fish kill event from among the 16 observed and recorded fish kill events from 1998 to 2011, the daily wind speed and daily minimum

Table 6.2 Partial results of logistic regression showing predicted probability of lake overturn due to the combined effects of wind speed and minimum air temperature

Chance		% Probability	Reset	Close
Combine...	Combine...	Combine...	Yes	No
Yes	Yes	Yes	100	0
Yes	Yes	Yes	40	60
Yes	Yes	Yes	50	50
Yes	Yes	Yes	20	80
Yes	Yes	No	60	40
Yes	Yes	No	0	100
Yes	Yes	No	0	100
Yes	Yes	No	0	100
Yes	No	Yes	70	30
Yes	No	Yes	10	90
Yes	No	Yes	10	90
Yes	No	Yes	0	100
Yes	No	No	10	90

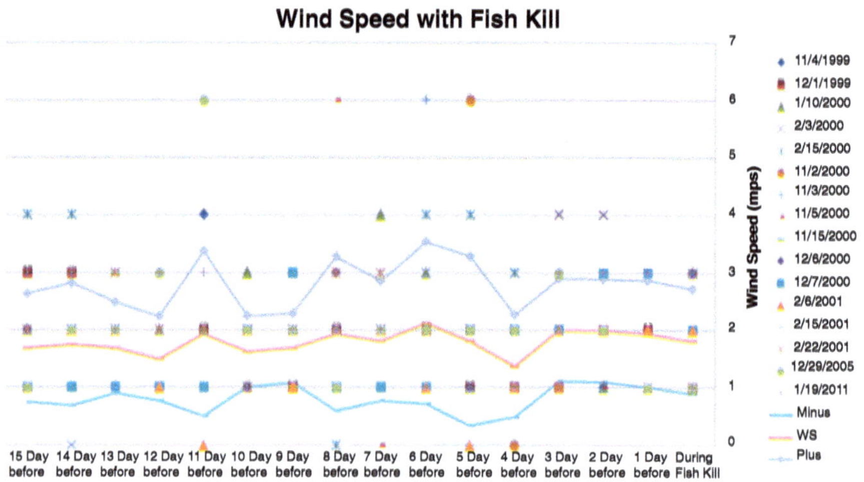

Fig. 6.10 Scatter plot and spread of the deviation (represented as lines) of observed daily wind speed starting from 15 days prior to and up to the day of the fish kill event (the deviation spread was computed as the one standard deviation from the mean)

temperature from 15 days before the event up to the day of the event were considered and plotted, respectively, in Figs. 6.10 and 6.11.

The lines in both figures represented the spread of the scatter plot with twice the standard deviation of the daily data. The line at the center represented the mean daily observation.

From Fig. 6.10, a discernible pattern of constant wind speed ranging from 1 mps to 3 mps was observed 3 days before the fish kill event. There was no discernible wind speed pattern observed prior to these days.

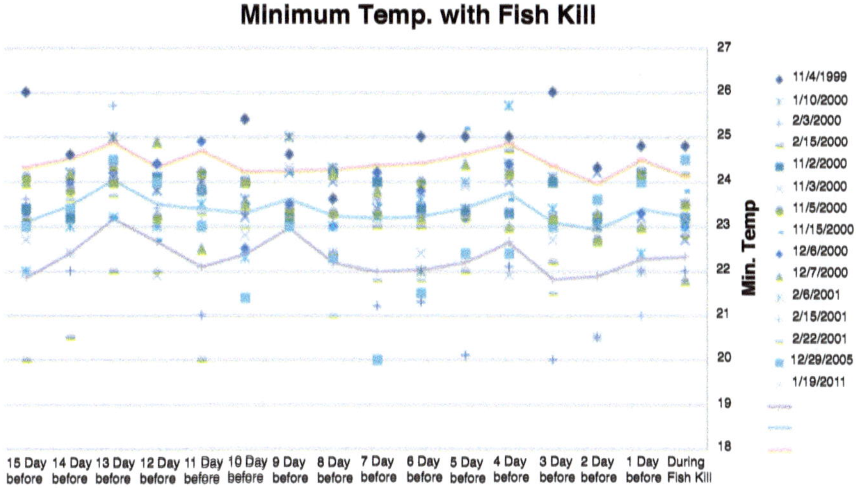

Fig. 6.11 Scatter plot and spread of the deviation (represented as lines) of observed daily minimum temperature starting from 15 days prior to and up to the day of the fish kill event. The deviation spread was computed as the one standard deviation from the mean

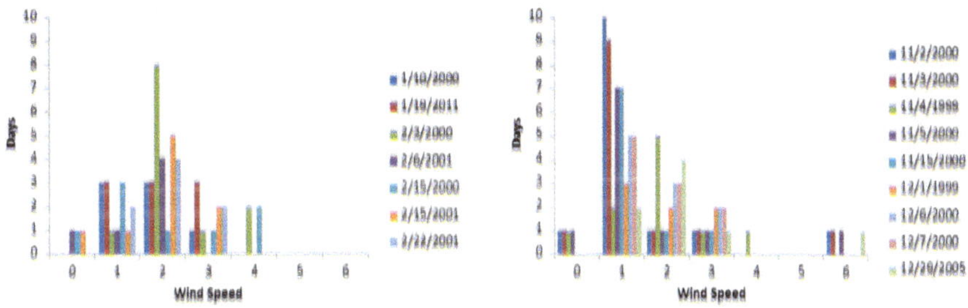

(a) (b)

Fig. 6.12 Pattern of sustained wind speed (in days) within 15 days before fish kill events in January through February (**a**) and November through December (**b**)

From Fig. 6.11, a discernible pattern of minimum temperature dynamics was observed as far back as 1 week before the fish kill event, i.e., a non-varying minimum temperature from 7 days to 6 days before the fish kill event ranging from 22 °C to 25.5 °C. The pattern showed a slow increase that peaked until 4 DBF, with the minimum temperature ranging from 22.8 °C to 25 °C. The minimum temperature dropped considerably 3 DBF (from 24.5 °C to 21.9 °C) but slowly peaked again at 1 DBF (from 22.3 °C to 24.5 °C). A sudden drop in the pattern (from 24.1 °C to 22.3 °C) occurred during the day of the fish kill event.

To find out whether there was a pattern on the duration (in days) of sustained wind speed, the contiguous days were plotted when a given wind speed was observed during the respective 15 days prior to the fish kill events. Figure 6.12a, b shows these patterns for wind speeds observed in January to February and November to December, respectively. There was a sustained wind speed of 2 m/s ranging from 3 consecutive days up to 8 consecutive days from January through February. From November through December, however, a sustained wind speed of 1 m/s was observed ranging from 3 consecutive days up to 10 consecutive days.

Cold air temperature (20–23 °C) coupled with strong wind velocity (3–4 mps) for a duration of 5 days (December 21–25, 1999), followed by a combination of cold air temperature (22–23 °C) and strong wind velocity (3–4 mps) on January 2–3, 2000, preceded the occurrence of lake overturn and fish kill on January 10, 2000.

Continuous strong wind velocity ranging from 3 mps to 4 mps for 7 days (January 14–20, 2013) combined with cold air temperature (19–22 °C) for 12 days (January 13–25, 2013) and followed by strong wind velocity (3 mps) on January 31–February 1, 2013, preceded the occurrence of lake overturn and fish kill on February 2, 2013. During this period, the surface water temperature cooled from 27 °C on January 8, 2013, to 25.5 °C on January 29–30, 2013. Both events show that fish kill due to lake overturn can be predicted 7 days before the event given a combination of continuous cold air temperature of 19–23 °C and strong wind velocity of 304 mps for 5–7 days.

6.3.8 Bayesian Model of Lake Overturn Due to Wind Speed and Minimum Air Temperature

The Bayesian model of lake overturn was developed using Netica. Results of the logistic probability analyses were transformed to develop the conditional probability tables (CPT) for wind speed and minimum air temperature. The CPT were inputted in the Bayesian models for fish kill due to lake overturn (Fig. 6.13).

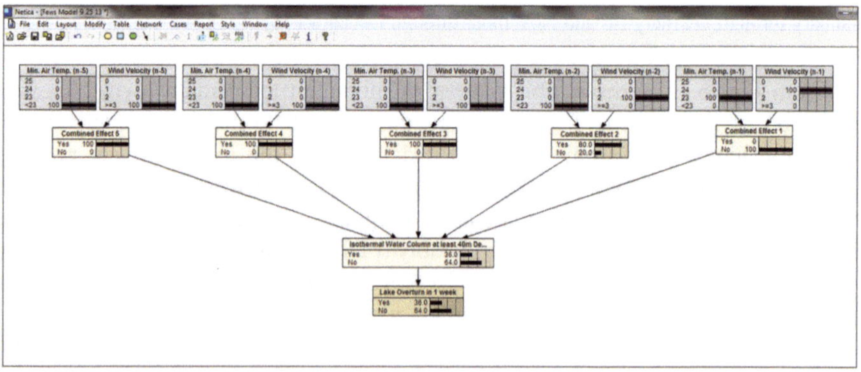

Fig. 6.13 Bayesian model of fish kill in Taal Lake due to lake overturn using Netica

6.4 Conclusion

This study demonstrates the use of a combination of analytical tools (exploratory data analysis, correlation, autocorrelation, logistic regression, scatter plots, conditional probability) in developing a predictive model of lake overturn in Taal Lake, Philippines. Results of this study can provide a good starting point in the development of a predictive model of fish kill due to overturn for Taal Lake. The logistic regression method used in this study can be applied to other factors hypothesized to influence fish kill event in the lake. Also, using logistic regression method minimizes the number of factors or variables that have statistical significance on the occurrence of lake overturn and fish kill. The analysis also provides a minimum dataset that should be monitored to obtain early warning on lake overturn or fish kill.

Acknowledgments The project team acknowledges the support and fund provided by the Department of Science and Technology (DOST) through the Philippine Council for Aquatic and Marine Research and Development (PCAMRD) for the Fishkill Early Warning System (FEWS) program. Our specific project is entitled Development of a Predictive Model for the Occurrence of a Fish Kill in Volcanic Taal Lake. We would also like to extend our gratitude to the officials and Municipal Agriculture Officer of Talisay, Batangas, and the participation of fisher folks and locals from San Nicolas, Talisay, Laurel, and Agoncillo municipalities.

References

Asian Development Bank (ADB) (2004) Tilapia cage farming in Lake Taal, Batangas, Philippines. http://www.adb.org/sites/default/files/tilapia-cage-farming-phi.pdf. Accessed 3 May 2013

Balistrieri LS, Tempel RN, Stillings LL, Shevenell LA (2006) Modeling spatial and temporal variations in temperature and salinity during stratification and overturn in Dexter Pit Lake, Tuscarora, Nevada, USA. Appl Geochem 21:1184–1203

Borsuk ME, Higdon D, Stow CA, Reckhow KH (2001) A Bayesian hierarchical model to predict benthic oxygen demand from organic matter loading in estuaries and coastal zones. Ecol Model 143:165–181

Borsuk ME, Stow CA, Reckhow KH (2004) A Bayesian network of eutrophication models for synthesis, prediction, and uncertainty analysis. Ecol Model 173:219–239

Brillo BBC (2015) Development issues regarding Bunot Lake: the lesser lake among the seven lakes of San Pablo City, Philippines. Lakes Reserv Res Manag 20:155–165. https://doi.org/10.1111/lre.12096

Brillo BBC (2016a) The case of Yambo lake of San Pablo City, Nagcarlan and Rizal, Laguna, Philippines. Soc Sci 11:5693–5702. https://doi.org/10.3923/sscience.2016.5693.5702

Brillo BBC (2016b) An assessment of development of a transboundary small lake: Calibato Lake, San Pablo City and Rizal, Laguna, the Philippines. Asian J Water Environ Pollut 13:55–67. https://doi.org/10.3233/AJW-160017

Brillo BBC (2016c) Developing a small lake: the case of Palakpakin Lake, San Pablo City, Philippines. Water Resour 43:611–620. https://doi.org/10.1134/S0097807816040035

Brillo BBC (2016d) Urban lake governance and development in the Philippines: the case of Sampaloc Lake, San Pablo City. Taiwan Water Conserv 64:66–81

Brillo BBC (2016e) Developing Mohicap Lake, San Pablo City, Philippines. Soc Sci 11:283–290

Bureau of Fisheries and Aquatic Resources (BFAR) (2011) NDRRMC Update Sitrep No. 2 on Taal Fishkill. http://www.ndrrmc.gov.ph/attachments/article/219/NDRRMC%20Update%20%20re%20Taal%20Lake%20Fish%20Kill%2002%20june%202011.pdf. Accessed 7 May 2013

Caliro S, Chiodini G, Izzo G, Minopoli C, Signorini A, Avino R, Granieri D (2008) Geochemical and biochemical evidence of lake overturn and fish kill at Lake Averno, Italy. J Volcanol Geotherm Res 178:305–316

Can T, Nefeslioglu HA, Gokceoglu C, Sonmez H, Duman TY (2005) Susceptibility assessments of shallow earthflows triggered by heavy rainfall at three catchments by logistic regression analyses. Geomorphology 72:250–271

Chang K, Chiang S, Hsu M (2007) Modeling typhoon- and earthquake-induced landslides in a mountainous watershed using logistic regression. Geomorphology 89:335–347

De Infante A, Infante O, Marquez M, Lewis W, Weibezahn F (1979) Conditions leading to mass mortality of fish and zooplankton in Lake Valencia Venezuela. Acta Cient Venez 30:67–73

Diana JS (2009) Aquaculture production and biodiversity conservation. BioScience 59:27–38. https://doi.org/10.1525/bio.2009.59.1.7

Edwards P (1993) Environmental issues in integrated agriculture-aquaculture and wastewater-fed fish culture systems. In: Pullin RSV, Rosenthal H, Maclean JL (eds) Environment and aquaculture in developing countries, ICLARM conference proceedings 31, Manila, pp 193–170

Hu Z, Lo CP (2007) Modeling urban growth in Atlanta using logistic regression. Comput Environ Urban 31:667–688

Magcale-Macandog DB, de la Cruz CPP, Edrial JD, Reblora MA, Pabico JP, Salvacion AR, Marquez TL Jr, Macandog PBM, Perez DKB (2014) Eliciting local ecological knowledge and community perception on fishkill in Taal Lake through participatory approaches. J Environ Sci Mag 2(17):1–16

Marti-Cardona B, Steissberg TE, Schladow SG, Hook SJ (2008) Relating fish kills and wind patterns in the Salton Sea. Hydrobiologia 604:85–95

Ozdemir A (2011) Using a binary logistic regression method and GIS for evaluating and mapping the groundwater spring potential in the Sultan Mountains (Aksehir, Turkey). J Hydrol 405:123–136

Paller VG, Macandog D, de Chavez ER, Paraso MG, Tsuchiya MC, Campang J, Pleto JV, Bandal M, Cabillon YC, Elepano A, Macaraig JR, Mendoza S (2021) The seven lakes of San Pablo: assessment and monitoring toward a sustainable lake ecosystem. Philipp Sci Lett 14(1): 157–178. https://scienggj.org/2021/PSL%202021-vol14-no01-p158-179-Paller%20et%20al.pdf

Papa RD, Mamaril A (2011) History of the biodiversity and limno-ecological studies on Lake Taal with notes on the current state of Philippine limnology. Philipp Sci Lett 4(1):1–10

Piwczynski D, Sitkowska B, Wisniewska E (2012) Application of classification trees and logistic regression to determine factors responsible for lamb mortality. Small Rumin Res 103:225–231

Rosana MR (2011) Chronic and massive fish kills in Taal Lake, Southern Luzon (May-June 2011): the roles of changing weather condition, overturn and elevated nutrients levels. Paper presented to Bureau of Fisheries and Aquatic Resources (BFAR) Region 4A Quezon City on Sept 14, 2011. 27 p

Salem AM, Rekab K, Whittaker JA (2004) Prediction of software failures through logistic regression. Inform Software Tech 46:781–789

Schiemer F, Amarasinghe US, Frouzova J, Sricharoendham B, Silva EIL (2001) Ecosystem structure and dynamics—a management basis for Asian reservoirs and lakes. Australian Centre for International Agricultural Research (ACIAR) Proceedings, Canberra, ACT, pp 215–226

Van Den Eeckhaut M, Vanwalleghem T, Poesen J, Govers G, Verstraeten G, Vandekerckhove L (2006) Prediction of landslide susceptibility using rare events logistic regression: a case-study in the Flemish Ardennes (Belgium). Geomorphology 76:392–410

White P, Christensen GN, Palerud R, Legovic T, Rosario WR, Lopez N, Regpala RR, Gecek S, Hernandez J (2007) Environmental monitoring and modelling of aquaculture in risk areas of the Philippines. National Integrated Fisheries Technology Development Center-Bureau of Fisheries and Aquatic Resources (NIFTDC-BFAR). 39 p

Yambot AV (2000) Problems and issues of Nile tilapia cage farming in Taal Lake, Philippines. In: Liao IC, Lin CK (eds) Proceedings on cage aquaculture in Asia. AFS; WAS-SC, Manila

Yang X, Skidmore AK, Melick DR, Zhoua Z, Xu J (2006) Mapping non-wood forest product (matsutake mushrooms) using logistic regression and a GIS expert system. Ecol Model 198: 208–218

You SH, Gung Y, Lin CH, Konstantinou KI, Chang TM, Chang ETY, Solidum R (2013) A preliminary seismic study of Taal Volcano, Luzon Island Philippines. J Asian Earth Sci 65: 100–106

Global Sensitivity and Uncertainty Analysis of MaxEnt Model: Implications in Species Habitat Projections

7

Rakesh Kadaverugu, Shalini Dhyani, Ashok Kadaverugu, and Rajesh Biniwale

Abstract

MaxEnt is a widely used species distribution model (SDM) that works on the principle of maximizing entropy. Despite large body of species habitat research carried out using MaxEnt, till now there is no standardized accepted modeling procedure for obtaining reproducible research outcomes. There is a need to understand the nuances in the selection of model parameters and resulting uncertainties in the outcomes. We studied the global sensitivity and uncertainty in habitat projections of the species *Quercus leucotrichophora* (Banj oak) over Uttarakhand State of India in the Central Himalayas by varying the model parameters—regularization factor (RF), background points (BP), and k-fold cross-validations (CVs). The Sobol variance decomposition sensitivity analysis on the model outcomes indicates that high probable habitats and potential habitats are sensitive to RF and BP, while prediction of less probable habitats is relatively sensitive to the number of k-fold CVs. Accuracy of the model is also highly correlated with the RF ($r = -0.75, p < 0.001$), which has influenced the extent of

R. Kadaverugu (✉) · R. Biniwale
Cleaner Technology and Modeling Division, CSIR-National Environmental Engineering Research Institute, Nagpur, India
e-mail: r_kadaverugu@neeri.res.in

S. Dhyani
Critical Zone Research Group, Water Technology and Management Division, CSIR-NEERI, Nagpur, India

Academy of Scientific and Innovative Research (AcSIR), Ghaziabad, India

International Union for the Conservation of Nature (IUCN), Commission on Ecosystem Management (CEM), Geneva, Switzerland

A. Kadaverugu
Department of Civil Engineering, Government Polytechnic, Nalgonda, Telangana, India

S. Dhyani et al. (eds.), *Ecosystem and Species Habitat Modeling for Conservation and Restoration*, https://doi.org/10.1007/978-981-99-0131-9_7

potential and high probable habitat projections. We conclude that SDMs should be supplemented with the information on sensitive model parameters and the uncertainty associated with the model parameters for improved objectivity and reproducibility of research findings related to species conservation planning.

Keywords

Sobol sensitivity · MaxEnt model · Species distribution modeling · Uncertainty analysis · Potential habitats · Conservation planning

7.1 Introduction

Understanding the behavior of complex ecological systems is important to anticipate and avoid any impending potential tipping points. Climate change is one of the mega drivers that is accelerating the impact on sensitive ecosystems across the globe. Dynamic simulation models that account for the behavior of the socio-ecological system can simulate the shift in ecological states, hence providing an idea about the system's travel path (Pickett et al. 2005; Liu et al. 2007). The choice between complex deterministic models or simple statistical correlative models has always perplexed the ecosystem researchers and decision-makers. Also, the uncertainty and sensitivity of these models play a decisive role in projecting the future states of the ecosystem under study. This concern has led to the development of new techniques to study model uncertainty and for its effect on model projections (Saltelli et al. 2008).

Uncertainty in a wide range of ecosystem models stems mainly from (a) errors in input data representing the systems, (b) errors in the model parameters, and (c) incorrect structural formulation of the models with dynamism and feedbacks (Perz et al. 2013). Less complex models are known to have high uncertainty in the output owing to structural limitations, whereas highly complex models tend to have reduced uncertainty in the predictions (Snowling and Kramer 2001). As the applications of simplified ecosystem models such as habitat suitability models are widely used in the decision-making and in conservation planning, there is a need to study such model's uncertainty in habitat predictions to the variations in the parameters.

Ecological niche models (ENMs) or species distribution models (SDMs) are interchangeably used to represent a set of machine learning methods that fit a point process relationship between the occurrences of a species and surrounding environmental data, and these trained models estimate the geographical extent of potential habitats of the species (Dhyani et al. 2018, 2020). The role of ENMs in ecological studies is ever-increasing and is incorporated into conservation planning (Feng et al. 2019). There are several machine learning algorithms, viz., MaxEnt, artificial neural network (ANN), random forest, support vector machines, generalized additive model (GAM), generalized linear model (GLM), multivariate adaptive regression splines (MARS), flexible discriminant analysis (FDA), surface range envelope (SRE), and classification tree analysis (CTA), which have been applied to study the association between species occurrence and environmental data (Hallgren et al. 2019). These algorithms require either presence-only or presence and absence/pseudo-absence species locations for training the SDMs. The validity in the assumptions about occurrence locations is difficult to ascertain, which leads to bias in the modeling. But the MaxEnt model (Elith et al. 2006), which is based on the principle of maximum entropy (Jaynes 1957), relies maximally noncommittal to what is unknown, and is based on the presence-only type of occurrence data, and is hence prone to less bias. This is why the MaxEnt model is being widely used among the ecological modeling community for a variety of applications not limited to conservation planning, domestication studies, reintroduction of wild species, invasion of alien species, and climate-sensitive restoration planning (Dhyani et al. 2020). Several global assessments like IUCN, UNEP-World Conservation Monitoring Centre, European Union, Convention on Biological Diversity, IPBES (Intergovernmental Science-Policy Platform on Biodiversity and Ecosystem Services), and World Wildlife Fund have also applied MaxEnt for studying the multi-scale habitat dynamics. Due to a user-friendly interface, the model has gained popularity among the ecological modeling community. Further, with the availability of GBIF (Global Biodiversity Information Facility; https://www.gbif.org) occurrence data on a variety of plants, mammals, and reptiles, the applications of SDMs have tremendously increased. Apart from habitat modeling, MaxEnt has also been applied in other multidisciplinary areas such as spatial prediction of groundwater contamination, drinking water sources, and soil erosion-susceptible zones.

The MaxEnt species habitat modeling follows a typical structure with the sequence of steps as (1) collection of species occurrence locations from the area of interest, (2) preparation of spatial environmental layers, usually consisting of bioclimatic data and topographic information, (3) training the MaxEnt model with around 80% of the occurrence data and corresponding environmental data, (4) verification of the model accuracy with AUC (area under the curve metric) or with other metrics occasionally, (5) calculation of the model's threshold based on the sensitivity and specificity analysis with the remaining 20% of the occurrence data, and (6) projection of logistic probability of occurrence of the species on each pixel of the study area using the full-scale environmental input layers. The model-projected output raster layer consists of logistic probability values ranging between 0 and 1 indicating the least and strong chances of the suitability of the species habitats, respectively. The pixels with a probability greater than the threshold value are statistically confident to be treated as potential habitats. The potential habitats can be further classified into high probable and less probable classes for better interpretation of the habitat projections (Dhyani et al. 2018, 2020). We have observed that the majority of MaxEnt studies have a recurring pattern in methodology and interpretation of results. The input environmental data are bioclimatic raster layers obtained from the WorldClim portal (www.worldclim.org/current) and topographic data (containing elevation, slope, and aspect). The typical outcome of any study shows an alarming decline in the potential habitats or range contractions of the species due to future climate variability (e.g., Dhyani et al. 2018, 2020; Abdelaal et al. 2019;

Raman et al. 2020; Purohit and Rawat 2021), while very few studies have predicted an increase in the species habitats in the future (Yi et al. 2016).

More than 50% of the ENM studies have not provided sufficient reference to the choice of model parameters and settings (Feng et al. 2019). To this effect, several studies have highlighted the inadequacies in overall documentation and understanding about the SDM algorithms and found that the justification about the settings and model configurations are largely missing in the SDM works (Convertino et al. 2014; Hallgren et al. 2019; Feng et al. 2019). Lack of documentation and justification about the model parameters prevents the flow of understanding and reproducibility of the research. Further, there are a very few studies on in-depth analysis of SDM sensitivity and uncertainty (Perz et al. 2013; Convertino et al. 2014; Hallgren et al. 2019). Although Feng et al. (2019) have highlighted the need to provide bare minimum details about the ENM studies for increasing reproducibility and objectivity in the habitat projection studies, there is also a need to include the information on model sensitivity and uncertainty in the results. Out of 302 research papers retrieved from the Web of Science database that are having MaxEnt in title, only 5 (1.6%) have reported the model sensitivity. Studies devoid of such vital information make it difficult for reproducible and objective research. The model parameter settings, such as weighted response weights, maximum number of interactions, threshold hinge features, and regularization multiplier, have been studied earlier by Hallgren et al. (2019) by using non-variance decomposition methods, in which the parameters were varied at fixed levels.

Out of eight SDM algorithms studied by Hallgren et al. (2019), MARS, FDA, GAM, SRE, and CTA are found to be more sensitive to the model parameter settings and have a significant influence on habitat projections, while MaxEnt and GLM have shown less sensitivity. Unlike other methods, the MaxEnt and GLM have fewer settings that influence the model complexity, and these models are more based on the ecological theory which provides greater control to the modeler with sound ecological understanding (Hallgren et al. 2019). However, some studies have mentioned that the regularization factor (Cao et al. 2013) and the number of background points (Lobo and Tognelli 2011; Barbet-Massin et al. 2012; Merow et al. 2013) significantly influence the model over-fitting or under-fitting, which is a major concern and can lead to inadequacy in decision-making on species conservation. Anderson and Gonzalez (2011) have found a strong influence of the regularization parameter on MaxEnt outcome. Alsamadisi et al. (2020) and Convertino et al. (2014) have studied the MaxEnt model's sensitivity and uncertainty concerning the input data and have also quantified the variations in the future habitats of a particular species, but these studies have not mentioned the uncertainty and sensitivity due to the model parameters. The size and spatial biases in the species occurrence data also play an important role in habitat predictions along with the threshold values and algorithms (Bean et al. 2012). Several earlier studies have shown that when the MaxEnt model is applied with default settings on new areas, the model under-predicted the species habitat ranges, due to over-fitting (Townsend Peterson et al. 2007; Anderson and Gonzalez 2011). Merow et al. (2013) have emphasized exploring the MaxEnt model behavior toward various parameter choices. The MaxEnt model applies by default a regularization scheme called "L1" to control over-fitting; however, it can be modified by user-specific input (Phillips et al. 2006). Hence in this study, we have quantified the model sensitivity and uncertainty due to the parameters, viz., regularization factor, k-folds in cross-validation, and the number of background locations on the model-projected potential habitats and less and high probable habitats, model accuracy (AUC), and threshold, using a more robust method—Sobol variance decomposition method.

We strongly advocate that the studies on SDM should discuss the uncertainty in the habitat projections arriving due the model settings and also the habitat projections should be expressed along with the confidence intervals for an improved understanding and decision-making. Sensitivity analysis (SA) is a systemic computational experiment performed on a model to identify the output behavior corresponding to the variations in the input variables or model parameters or factors (Saltelli et al. 2010; Girard et al. 2016). The traditional SA method one-at-a-time (OAT), which measures output variation by varying a single variable at a time, doesn't adequately capture the nonadditive responses among the variables (Saltelli and Annoni 2010; Girard et al. 2016; Jaxa-Rozen and Kwakkel 2018), whereas global sensitive analysis (GSA) that depends on variance decomposition technique helps in identifying the influential factors to which the model is most sensitive or the factor that causes maximum variation in the output. See Razavi and Gupta (2016) for a detailed review of the SA. Despite several other SA methods like Morris method (elementary effect test), regional SA, regression-based SA, FAST (Fourier amplitude sensitivity test), and extended FAST, the variance-based Sobol sensitivity analysis remains robust to account for all main and interaction effects based on statistical theory on variance decomposition (Koo et al. 2020). To date, Sobol and Morris approaches are the most rigorous GSA methods that are based on the law of total variance for the decomposition of output variance (Razavi and Gupta 2016). These methods have become the standard in studying the propagation of uncertainty in the model due to various complex interactions of the factors, and the SA analysis has become a key step in understanding the complex environmental models (Saltelli and Annoni 2010; Nossent et al. 2011; Pianosi and Wagener 2015). The Morris method has a drawback in accounting for the relative importance among the multiple input factors (Brockmann and Morgenroth 2007).

Although Sobol SA is one of the most computationally demanding methods, it delivers precise estimates of model sensitivity and factor interactions (Girard et al. 2016). The Sobol SA provides first-order (main effects) and total effect (interactions) indices, which quantify the fraction of model output variance contributed by an individual factor and by the sum of individual and higher-order interactions of the factors (Jaxa-Rozen and Kwakkel 2018), respectively. The Sobol SA has been applied on models from various disciplines such as atmospheric dispersion modeling (Girard et al. 2016), groundwater contamination (Kumar et al. 2020), hydrological studies (Song et al. 2013), and pharmacological studies (Zhang et al. 2015), but very few have studied in the realm of species habitat modeling.

Through this study, we attempted to conclude the sensitivity of the MaxEnt model output (potential habitats, less and high probable habitats) to the model parameter settings on the number of background points (BP) selection, k-fold cross-validation (CV) runs, and regularization factor (RF). To achieve this we have extended our earlier habitat modeling work on the species *Quercus leucotrichophora* A. Camus in the Central Himalayan region of Uttarakhand State of India (Dhyani et al. 2020) to study the model sensitivity and uncertainty. The remaining paper is organized into three sections. Section 7.2 discusses in more detail about the study area and describes species occurrence, MaxEnt model, and Sobol sensitivity and uncertainty analysis. Section 7.3 provides results drawn from the study and discusses the significance of the outcome in comparison with the other reported works in the field of habitat projections. Conclusion and broad implications in conservation planning are provided in the last Sect. 7.4.

7.2 Materials and Methods

7.2.1 Study Area and Occurrence Data

The study was carried out in *Quercus leucotrichophora* (Banj oak) forests in the Uttarakhand State of India (Fig. 7.1) in the Central Himalayas. The state extends from 77.56 to 81.02 E and 28.71 to 31.47 N, covering an area of 58,483 km^2. Moist temperate mixed broad-leaved forests are dominant in the state, which forms climatic climax from 1000 to 3500 m above mean sea level (MSL) exhibiting high

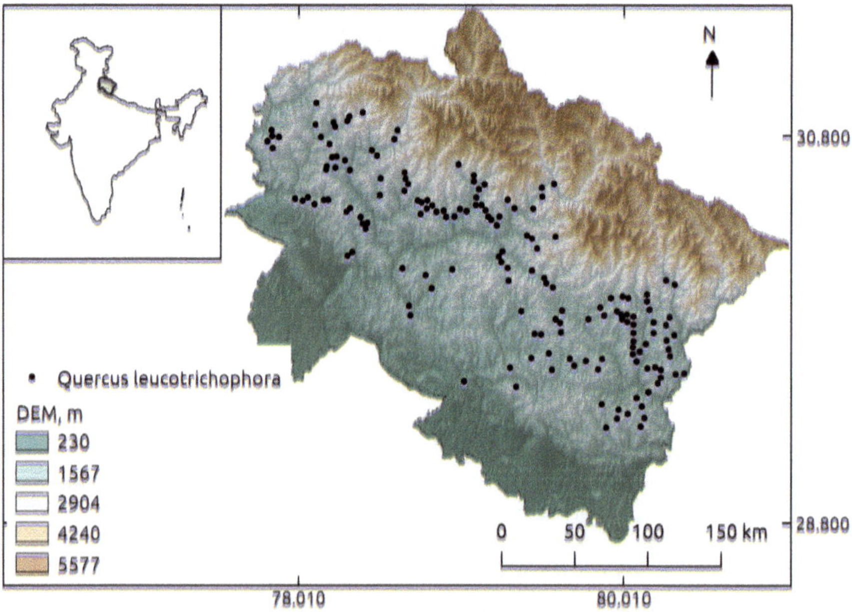

Fig. 7.1 The geogrpahical locations of *Quercus leucotrichophora* used in the MaxEnt model, in the Uttarakhand State of India

floral and faunal diversity with *Q. leucotrichophora* being a keystone species (Champion and Seth 1968). *Rhododendron arboreum*, *Lyonia ovalifolia*, *Pyrus pashia*, *Alnus nepalensis*, *Cinnamomum tamala*, etc. are common associates in these forests with rich herb and shrub diversity and significant presence of lianas and epiphytes (Dhyani et al. 2020). *Q. leucotrichophora* grows luxuriantly in moist temperate broad-leaved forests of 11 hill districts of the state and covers 5.24% of the geographical area which is equivalent to 1284 km^2 (Singh et al. 2016; Verma and Garkoti 2019). The region has an average annual temperature varying between 13 and 23 °C and an annual average rainfall of 1550 mm and experiences snowfall during winter months from November to February. Local inhabitants are dependent on the rich and diverse oak forests for their subsistence requirements especially for fuelwood, fodder, leaf litter, timber, and crop support (Dhyani et al. 2020).

Quercus leucotrichophora has a dominant presence in Garhwal province in comparison with Kumaon province of the state (Dhyani et al. 2020). A mixed approach was used to collect a maximum number of occurrences of the species for the analysis. We reviewed research papers published during 1984–2019 and collected 120 occurrence points from species growing locations, out of which we have used only 90 points to avoid any overlapping and repetition (See ESM-I of Dhyani et al. 2020). Another 30 occurrence points were also added from the primary survey in the study area (Dhyani et al. 2020) and 7 from GBIF portal. In total 127 occurrence locations were used in the study (Fig. 7.1). Handheld GPS (Global Positioning System) (make Trimble) was used in recording the exact occurrence points, and the points were later validated from the images of Google Earth. See Dhyani et al. (2020) for more information on methodology.

7.2.2 MaxEnt Model

The programmable version of the MaxEnt model v3.3.3 (Phillips et al. 2006) was used through the *dismo* library of R (R Core Team 2017). Environmental raster layers of bioclimatic data for the present time period simulated by CCSM4 (Community Climate System Model) model were downloaded from the WorldClim portal (http://www.worldclim.org/current) having 30 arc second spatial resolution. Out of 19 bioclimatic variables, 6 were identified which have the least mutual correlation (see Dhyani et al. 2018, 2020, 2021). The occurrence data were randomly divided in the ratio of 80:20 for training and testing of the model, respectively. The model training accuracy was determined by the area under the curve (AUC) metric. The outputs of k-fold cross-validation runs were averaged to produce an ensemble model. The threshold for the ensemble model was evaluated based on the theory of maximizing sensitivity and specificity (MSS), which is widely used in combination with presence-only kind of occurrence data (e.g., Dhyani et al. 2018, 2020). The model output is classified into three categories, in terms of potential habitats (area of pixels with probability>threshold), less probable habitat (threshold < probability < 0.5), and high probable habitat (0.5 < probability < 1). The threshold probability was calculated according to MSS method based on the ensemble model output layer and testing dataset.

7.2.3 Sobol Sensitivity Analysis

The variation in the model output range is studied by completely varying the input range (Sobol 1993; Saltelli et al. 1999). According to the law of total variance (also known as Eve's rule), the variance in the output Y, Var(Y), can be attributed to the sum of variance that each variable contributes (Eq. 7.1). The decomposition of the model variance into main effects and interaction effects is quite intuitive to understand the role of individual variables from Eve's rule (see Nossent et al. 2011). In general terms, Eq. (7.1) is understood as the sum of explained variance (first-order/main effect) and unexplained variance. Here, Y is the MaxEnt model output scalar representing any one of the output variables, the area of potential habitats, high probable habitats, and less probable habitats, in each case, and X is a vector representing the factors RF, BL, and CV as X_i. See Saltelli and Annoni (2010), Song et al. (2013), and Girard et al. (2016) for more details on derivations of the S_i (first-order sensitivity index or main effect; Eqs. (7.2)–(7.3)) and ST_i (total sensitivity index; Eq. (7.4)). S_i represents the main effect of the factor itself or the fraction by which the model output variance Var(Y) reduces by fixing a variable within its range, while the total effect (ST_i) of a factor consists of its main effect and all its interactions with the rest of the factors (Vanuytrecht et al. 2014). The number of evaluations (N) or rows of the matrix X equals $N = 2n(k + 1)$, where n = sample size and k = number of factors studied (here k = 3 and n = 300). The sequence of N evaluations follows a quasi-random sequence to equally cover the entire input space of the X (Saltelli 2002). The *R* library *sensobol* was used to generate the quasi-random run sequences and calculates the first-order and total Sobol's indices with bootstrapped confidence intervals. The number of bootstrap resampling was fixed at $R = 100$ in the analysis.

$$\mathrm{Var}(Y) = E[\mathrm{Var}[Y|X_i]] + Var[E[Y|X_i]] \tag{7.1}$$

$$S_i = \frac{V_i}{V} \tag{7.2}$$

$$V_i = \mathrm{Var}[E(Y|X_i)] \tag{7.3}$$

$$\mathrm{ST}_i = 1 - \frac{V_{\sim \mathrm{i}}}{V} \tag{7.4}$$

The MaxEnt model was tested against the variable regularization factor (RF), background locations (BL), and the number of cross-validations (CVs) using quasi-random sampling and bootstrap method. The ranges of these variables have been identified based on the reported literature on MaxEnt model. The values of these parameters are required to be explicitly stated for reproducibility of the species habitat modeling results. The beta multiplier to the regularization factor (RF) was allowed to vary between 0 and 1, as earlier studies reported that beta value >1 was only occasionally needed for optimal MaxEnt performance (Anderson and Gonzalez 2011; Cao et al. 2013). However, Merow et al. (2013) suggested exploring a range of regularization coefficients for optimal model performance. Theoretically MaxEnt model builds the probability distribution of the species based on the contrast between environmental variables observed at presence locations and background points (Phillips et al. 2006; Merow et al. 2013; Dhyani et al. 2018, 2020). The spread of background points hence plays a significant role in building the premise of the model, which demands sound theoretical knowledge about the species and their spatial spread for gaining confidence in the MaxEnt results. The allowable limits for each parameter were decided based on the physical acceptability. In this study, the background points (BP) were varied between 2000 and 20,000, while the number of k-fold cross-validations was varied between 2 and 20. Care should be exercised in deciding the background and cross-validation numbers as they depend on the percentage of the occurrence locations used in the model. The values of the parameters were assigned based on the uniform distribution between the limits of the respective variable. The Sobol sensitivity analysis calls the MaxEnt model with the parameter values provided as per the quasi-random sequence. A total of 2400 runs (number of rows in the parameter matrix) each row with a different combination of the parameter values were performed on the MaxEnt model. Sobol's sensitivity indices S_i and ST_i were calculated according to the Saltelli et al. (2010) method using bootstrap sampling.

The model outputs, viz., potential habitat, less probable habitats, and high probable habitats, were considered for the Sobol sensitivity and uncertainty analysis. The relation between the model outputs and the model accuracy metric (AUC) and probability threshold values were also studied. The mutual relationships among the model input parameters and the outcomes were described using matrix scatter plots and Pearson's correlations measured at a confidence level of 95%, and the distribution of the model outcomes is tested against the normality and plotted its density.

7.3 Results and Discussion

7.3.1 Model Uncertainty

The AUC of 2400 MaxEnt model runs (N = number of total runs) varied between 0.80 and 0.95 with a mean of 0.91 ± 0.03 and is negatively skewed. This is an indication that the model training accuracy has greatly varied due to the variations in the model parameters. Models with AUC between 0.7 and 0.9 are considered moderately accurate, while those with >0.9 are treated as highly accurate (Elith et al. 2006). The AUC is greater than 0.8, which indicates that all the model runs in the study were trained accurately enough and they are capable of making valid projections about the species habitats. The dependency of the AUC on model parameters and the effect of AUC on the habitat projections are discussed in the

Table 7.1 Descriptive statistical summary of the model outcomes, model accuracy metric (AUC, area under the curve), and model probability threshold values

Statistic	Model AUC	Threshold	Potential area (km^2)	Less probable area (km^2)	High probable area (km^2)
Minimum	0.80	0.16	4543	0	1447
Maximum	0.95	0.51	24549	18165	11352
Mean	0.91	0.35	12309	7830	4479
Median	0.91	0.35	11529	7662	4171
Standard deviation	0.03	0.06	3776	3236	2008
Kurtosis	3.26	2.33	3.06	2.49	2.29
Skewness	−0.69	0.09	0.76	0.21	0.49

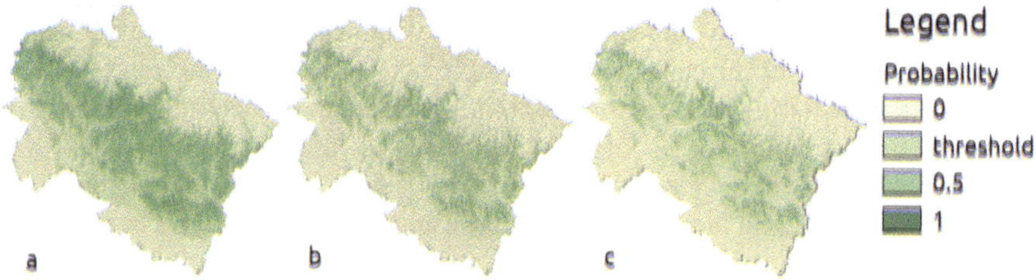

Fig. 7.2 Potential habitat at (**a**) maximum, (**b**) mean, and (**c**) minimum levels having threshold values 0.21, 0.28, and 0.46, respectively

Sect. 7.3.2. Similarly, the threshold values for the 2400 model runs ranged between 0.16 and 0.51 with a mean of 0.35 ± 0.06. Threshold value infers the model's confidence in classifying a pixel as a potential habitat only when the predicted logistic probability over the pixel is greater than the threshold value (Dhyani et al. 2020). The variation in the threshold value significantly influences the model habitat projections (see Sect. 7.3.2).

The potential habitat projections of the species varied between 4543 and 24,549 km^2 with a mean of 12,309 ± 3776 km^2 and follow a normal distribution with positive skewness. Similarly, the less probable and high probable habitats vary between 0–18,165 and 1447–11,352 km^2, respectively, and have positive skewness and are platykurtic (Table 7.1). The results suggest that carrying out a single MaxEnt run with specified model parameters will not provide a complete picture of the nature of the variability of the potential and high probable habitats. As there are no definite rule sets for the selection of a particular combination of model parameters, we reiterate the need to evaluate the uncertainty imparted from the choice of model parameters. Future studies on the SDM have to make an extra effort in providing the confidence intervals for the habitat projections based on the uncertainty analysis. The variations in the habitat suitability predictions resulted from the model uncertainty are visualized in Fig. 7.2 at maximum, mean, and minimum values of potential habitats. The spatial variations in the predicted habitats due to the variations in the model parameters and bias in occurrence data are not studied here.

7.3.2 Sensitivity Analysis

The scatter plots, histograms, and the density plots (Fig. 7.3) of the model outcomes and input parameters infer that the model parameters obviously follow a uniform distribution, whereas the outcome variables follow a normal distribution with various degrees of skewness and kurtosis (Table 7.1). Strong Pearson's correlation is observed between the pairs—RF and AUC ($r = -0.75$ at $p < 0.001$) and RF and high probable habitat ($r = 0.94$ at $p < 0.001$); moderate level of correlation is

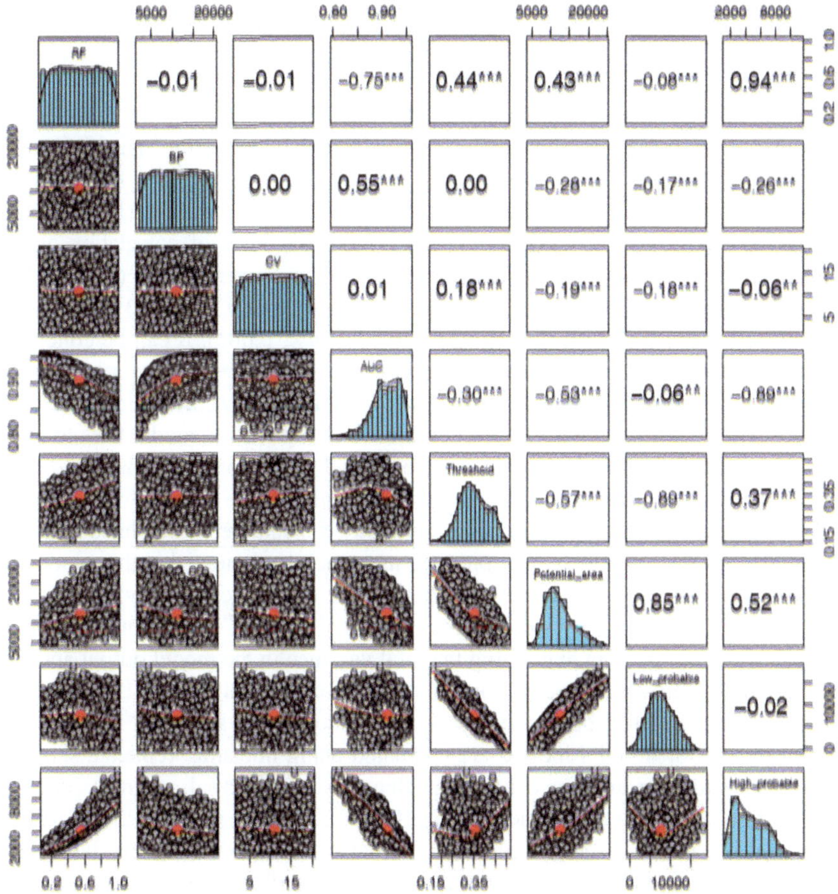

Fig. 7.3 Density, scatter plots, and correlations between the model parameters (*BP* background points, *CVs* k-fold cross-validations, *RF* regularization factor), model outcomes (potential habitats, less and high probable habitats), model accuracy (AUC), and probability threshold value

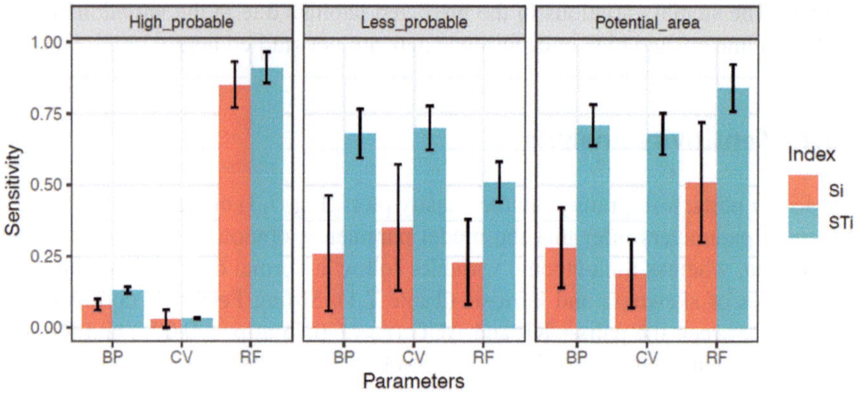

Fig. 7.4 Sobol's first-order (main effect) and total effect indices along with standard error. *BP* background points, *CV* k-fold cross-validations, *RF* regularization factor

observed between BP and AUC ($r = 0.55$ at $p < 0.001$). Among the model outcomes, a high degree of correlation is observed between the pairs AUC and high probable habitat ($r = -0.89$ at $p < 0.001$) and threshold and less probable habitat ($r = -0.89$ at $p < 0.001$) and between potential habitat and less probable habitat ($r = 0.85$ at $p < 0.001$), whereas a moderate level of correlation among model outcomes is observed between the pairs AUC and potential area ($r = -0.53$ at $p < 0.001$) and threshold and potential area ($r = -0.57$ at $p < 0.001$) and between potential area and high probable area ($r = 0.52$ at $p < 0.001$). It can be concluded that the AUC has a significant influence in determining the model projections on high probable habitats. Besides, the AUC is in turn negatively correlated with the RF and positively correlated with BP. When the RF is higher, the model tries to avoid the over-fitting; hence the AUC of the model is negatively correlated with RF. Similarly, low accuracy models tend to overestimate the high probable habitats and potential habitats (Fig. 7.3). The threshold is negatively correlated with the potential and less probable habitat predictions, which is obvious that the window of the probability of species occurrence will be higher with a lower threshold.

The first-order sensitivity indices rank the contribution of specific input variables according to their share in explaining the total variance in the model output and also help to rank the input variables according to the level of significance. The first-order and total effect Sobol's indices for the model outcomes are shown in Fig. 7.4. The standard errors of Sobol's indices are estimated using bootstrap replicate samples. Sensitivity values greater than 0.1 are considered to be influential, whereas those less than 0.01 are treated as insignificant (Saltelli et al. 2008; Zhan et al. 2013). Results show that the model-projected potential habitat of the species is sensitive to RF, followed by BP and CV, having main effect Sobol's indices (± standard error) 0.51 (±0.21), 0.28 (±0.14), and 0.19 (±0.12), respectively. The potential area represents the suitable habitat of the species at geographical locations having probability values greater than the threshold probability. The number of k-fold CVs used in the model ensemble output probability layer is relatively affecting the less probable habitats (having pixel probability between threshold and 0.5). The factors affecting the less probable habitats should be interpreted with caution as the standard error interval of the main effects is wider compared with other model parameters. Categorically, results show that the model-predicted high probable habitat area is highly sensitive to RF, followed by BP, having S_i first-order indices 0.88 (±0.08) and 0.11 (±0.02), respectively. The simulation runs having high RF have lower model accuracy (AUC) but greater than 0.8, and the lower model accuracy runs have overpredicted the potential and high probable habitats (reflected in Fig. 7.3). Hence, significant care must be exercised in choosing the values of RF; the criteria for such selection should be explained and justified in the habitat modeling studies.

Results show that first-order effect indices are lesser than total effect for all input parameters and model output variables. This is in conformity with the statistical theory that the first-order effect is already included in the total effect, and similar conformity has been observed in other studies (Nossent et al. 2011; Kumar et al. 2020).

Although Hallgren et al. (2019) mentioned that the MaxEnt model is less complicated to run and goes well with default parameter settings, our results conclude that RF and BP significantly influence the habitat predictions. Specifically, over-fitting and under-fitting of the model are governed by the RF, which inevitably influences the AUC and modifies the high probable and potential habitat predictions. The geographical spread of BP and their spatial biases play a crucial role, while training the model, which is hard to account for, nevertheless, by varying the number of background points, a priori in assuming the species occurrence locations can be adjusted, and hence its effect can be captured on the model outcomes. In the present study, the results show that BP is one of the main factors that influence the model accuracy.

Model sensitivity toward sampling biases is largely unknown due to lack of computational justification of the bias, as many species that are endangered or threatened are justified to be sampled from a narrower spatial location than from widespread areas. Hence proper ecological knowledge is warranted for justifying a priori in sampling biases. Selection of background points and weight assignment to the occurrence locations will be required for more generalized habitat predictions in such specific scenarios. Alternatively, methods based on target group sampling are applied (TGS; see Phillips et al. 2009; Merow et al. 2013). For projecting future habitat scenarios, care must be taken for model transferability to the future scenarios, especially the model assumptions like background (pseudo-absence locations), sampling biases, and regularization factors, along with more importantly the biological reality (response curves of the species with the environmental variables), which have to be biologically verified and ecologically accepted to be sound.

7.3.3 Implication on Conservation Management

Although SDM has gained respect and popularity in the last few years as an important tool for predicting species response to climate change, still, there are many limitations and assumptions that need to be considered. Most of the climate models in recent times have been predicting terrible changes in the global environment with large and unforeseen impacts (Voosen 2019). Selection of occurrence data, environmental layers, and choice of model parameters significantly introduce bias into the model projections, and the estimated predictive performance can be overoptimistic in comparison to the actual predictive performance. There are pieces of evidence that projections by alternative models are so different that they lose their importance in guiding conservation policy decisions (Araujo and New 2007). Conservation planning and strategies are constrained due to the considerable amount of uncertainty that is inherent in the climate projections because of the magnitude, rate, and ecological impacts of projected climate variability (Kujala et al. 2013). Growing concern about the uncertainties associated with selected factors is either not appropriately incorporated or not even considered in many SDM studies (Porfirio et al. 2014). Over or under projections of the climate impacts on the species cannot only significantly alter the conclusions, but largely affect the conservation planning efforts and fund allocation in the long run. This will further result in under/over consideration and mismatched financial support for species conservation, its sustainable use, and restoration in natural conditions. Delay in receiving appropriate conservation support may result in complete loss or extinction of species too, whereas over financial and conservation support to a species may ignore many species that are in dire need of conservation support. Hence, it is vital to appropriately and accurately predict the species projections in the future.

7.4 Conclusion

The prediction of species niches and their bioclimatic requirements are vital in planning and implementing conservation efforts. Ecological community is increasingly utilizing the niche models, especially the MaxEnt model for species habitat predictions, which relies on the presence-only kind of species occurrence data. But the model is highly sensitive to certain key parameters such as regularization factor (RF), the number of background points (BP), and k-fold cross-validation (CV) replicates, which are either ignored or often not sufficiently discussed in the modeling studies. Due to the lack of transparent modeling guidelines in species distribution modeling as noted by several earlier studies, the results are not only reproducible but are prone to high uncertainty. The present study quantifies the MaxEnt model's global sensitivity and uncertainty using the Sobol variance decomposition method, based on the 2400 model runs by varying the sensitive parameters under consideration. The variation in the MaxEnt model-projected habitats of the species *Quercus leucotrichophora* (Banj oak) in Uttarakhand State of India in Central Himalayan region was analyzed.

The results indicate that the MaxEnt model-predicted high probable and potential habitats are highly sensitive to RF followed by the BP. In turn, the model accuracy (area under the curve, AUC) is also highly sensitive to RF, and they are negatively correlated with the AUC, whereas the potential habitats and less probable habitats are sensitive toward the RF and number of CV, which are also correlated with the model threshold probability. The overall model-predicted potential habitats followed a normal distribution due to any random combination of the input parameters considered in the present study. Hence, we recommend to perform at least a few combinations of the input parameters as a sample in order to assess the variance and mean of the global distribution of potential habitats in consideration. This approach will certainly hint about the uncertainty of the habitat projections. Further studies are required to analyze the spatial variation in the habitat suitability predictions due to the changes in model parameters. We conclude that the choice of model parameters and its justification should be sufficiently discussed in SDM studies, as these results have potential ramifications in planning the strategic global conservation policies post-2020, and also affect the global biodiversity strategies in achieving the targets of UN decade on restoration (2021–2030) at local, regional, and global levels.

Acknowledgments The authors sincerely thank the support extended by the Knowledge Resource Center of CSIR-National Environmental Engineering Research Institute in processing the manuscript having the reference number CSIR-NEERI/KRC/2021/JAN/CTMD-WTMD/1.

References

Abdelaal M, Fois M, Fenu G, Bacchetta G (2019) Using MaxEnt modeling to predict the potential distribution of the endemic plant Rosa arabica Crép. in Egypt. Eco Inform 50:68–75. https://doi.org/10.1016/j.ecoinf.2019.01.003

Alsamadisi AG, Tran LT, Papeş M (2020) Employing inferences across scales: integrating spatial data with different resolutions to enhance Maxent models. Ecol Model 415:108857. https://doi.org/10.1016/j.ecolmodel.2019.108857

Anderson RP, Gonzalez I (2011) Species-specific tuning increases robustness to sampling bias in models of species distributions: an implementation with Maxent. Ecol Model 222:2796–2811. https://doi.org/10.1016/j.ecolmodel.2011.04.011

Araujo M, New M (2007) Ensemble forecasting of species distributions. Trends Ecol Evol 22:42–47. https://doi.org/10.1016/j.tree.2006.09.010

Barbet-Massin M, Jiguet F, Albert CH, Thuiller W (2012) Selecting pseudo-absences for species distribution models: how, where and how many?: how to use pseudo-absences in niche modelling? Methods Ecol Evol 3:327–338. https://doi.org/10.1111/j.2041-210X.2011.00172.x

Bean WT, Stafford R, Brashares JS (2012) The effects of small sample size and sample bias on threshold selection and accuracy assessment of species distribution models. Ecography 35:250–258. https://doi.org/10.1111/j.1600-0587.2011.06545.x

Brockmann D, Morgenroth E (2007) Comparing global sensitivity analysis for a biofilm model for two-step nitrification using the qualitative screening method of Morris or the quantitative variance-based Fourier Amplitude Sensitivity Test (FAST). Water Sci Technol 56:85–93. https://doi.org/10.2166/wst.2007.600

Cao Y, DeWalt RE, Robinson JL et al (2013) Using Maxent to model the historic distributions of stonefly species in Illinois streams: the effects of regularization and threshold selections. Ecol Model 259:30–39. https://doi.org/10.1016/j.ecolmodel.2013.03.012

Champion HG, Seth SK (1968) A revised survey of the forest types of India. Manager of Publications, Delhi

Convertino M, Muñoz-Carpena R, Chu-Agor ML et al (2014) Untangling drivers of species distributions: global sensitivity and uncertainty analyses of MaxEnt. Environ Model Softw 51:296–309. https://doi.org/10.1016/j.envsoft.2013.10.001

Dhyani S, Kadaverugu R, Dhyani D et al (2018) Predicting impacts of climate variability on habitats of Hippophae salicifolia (D. Don) (Seabuckthorn) in Central Himalayas: future challenges. Eco Inform 48:135–146. https://doi.org/10.1016/j.ecoinf.2018.09.003

Dhyani S, Kadaverugu R, Pujari P (2020) Predicting impacts of climate variability on Banj oak (Quercus leucotrichophora A. Camus) forests: understanding future implications for Central Himalayas. Reg Environ Chang 20:113. https://doi.org/10.1007/s10113-020-01696-5

Dhyani A, Kadaverugu R, Nautiyal BP, Nautiyal MC (2021) Predicting the potential distribution of a critically endangered medicinal plant Lilium polyphyllum in Indian Western Himalayan Region. Reg Environ Chang 21:30. https://doi.org/10.1007/s10113-021-01763-5

Elith J, Graham CH, Anderson R et al (2006) Novel methods improve prediction of species' distributions from occurrence data. Ecography 29:129–151. https://doi.org/10.1111/j.2006.0906-7590.04596.x

Feng X, Park DS, Walker C et al (2019) A checklist for maximizing reproducibility of ecological niche models. Nat Ecol Evol 3:1382–1395. https://doi.org/10.1038/s41559-019-0972-5

Girard S, Mallet V, Korsakissok I, Mathieu A (2016) Emulation and Sobol' sensitivity analysis of an atmospheric dispersion model applied to the Fukushima nuclear accident. J Geophys Res Atmos 121:3484–3496. https://doi.org/10.1002/2015JD023993

Hallgren W, Santana F, Low-Choy S et al (2019) Species distribution models can be highly sensitive to algorithm configuration. Ecol Model 408:108719. https://doi.org/10.1016/j.ecolmodel.2019.108719

Jaxa-Rozen M, Kwakkel J (2018) Tree-based ensemble methods for sensitivity analysis of environmental models: a performance comparison with Sobol and Morris techniques. Environ Model Softw 107:245–266. https://doi.org/10.1016/j.envsoft.2018.06.011

Jaynes ET (1957) Information theory and statistical mechanics. Phys Rev 106:620

Koo H, Iwanaga T, Croke BFW et al (2020) Position paper: sensitivity analysis of spatially distributed environmental models- a pragmatic framework for the exploration of uncertainty sources. Environ Model Softw 134:104857. https://doi.org/10.1016/j.envsoft.2020.104857

Kujala H, Moilanen A, Araújo MB, Cabeza M (2013) Conservation planning with uncertain climate change projections. PLoS One 8:e53315. https://doi.org/10.1371/journal.pone.0053315

Kumar D, Singh A, Kumar P et al (2020) Sobol sensitivity analysis for risk assessment of uranium in groundwater. Environ Geochem Health 42:1789–1801. https://doi.org/10.1007/s10653-020-00522-5

Liu J, Dietz T, Carpenter SR et al (2007) Complexity of coupled human and natural systems. Science 317:1513–1516. https://doi.org/10.1126/science.1144004

Lobo JM, Tognelli MF (2011) Exploring the effects of quantity and location of pseudo-absences and sampling biases on the performance of distribution models with limited point occurrence data. J Nat Conserv 19:1–7. https://doi.org/10.1016/j.jnc.2010.03.002

Merow C, Smith MJ, Silander JA (2013) A practical guide to MaxEnt for modeling species' distributions: what it does, and why inputs and settings matter. Ecography 36:1058–1069. https://doi.org/10.1111/j.1600-0587.2013.07872.x

Nossent J, Elsen P, Bauwens W (2011) Sobol' sensitivity analysis of a complex environmental model. Environ Model Softw 26:1515–1525. https://doi.org/10.1016/j.envsoft.2011.08.010

Perz SG, Muñoz-Carpena R, Kiker G, Holt RD (2013) Evaluating ecological resilience with global sensitivity and uncertainty analysis. Ecol Model 263:174–186. https://doi.org/10.1016/j.ecolmodel.2013.04.024

Phillips SJ, Anderson RP, Schapire RE (2006) Maximum entropy modeling of species geographic distributions. Ecol Model 190:231–259. https://doi.org/10.1016/j.ecolmodel.2005.03.026

Phillips SJ, Dudík M, Elith J et al (2009) Sample selection bias and presence-only distribution models: implications for background and pseudo-absence data. Ecol Appl 19:181–197. https://doi.org/10.1890/07-2153.1

Pianosi F, Wagener T (2015) A simple and efficient method for global sensitivity analysis based on cumulative distribution functions. Environ Model Softw 67:1–11. https://doi.org/10.1016/j.envsoft.2015.01.004

Pickett STA, Cadenasso ML, Grove JM (2005) Biocomplexity in coupled natural–human systems: a multidimensional framework. Ecosystems 8:225–232. https://doi.org/10.1007/s10021-004-0098-7

Porfirio LL, Harris RMB, Lefroy EC et al (2014) Improving the use of species distribution models in conservation planning and management under climate change. PLoS One 9:e113749. https://doi.org/10.1371/journal.pone.0113749

Purohit S, Rawat N (2021) MaxEnt modeling to predict the current and future distribution of Clerodendrum infortunatum L. under climate change scenarios in Dehradun district, India. Model Earth Syst Environ 8:2051–2063. https://doi.org/10.1007/s40808-021-01205-5

R Core Team (2017) R: a language and environment for statistical computing. R Foundation for Statistical Computing, Vienna

Raman S, Shameer TT, Sanil R et al (2020) Protrusive influence of climate change on the ecological niche of endemic brown mongoose (Herpestes fuscus fuscus): a MaxEnt approach from Western Ghats, India. Model Earth Syst Environ 6:1795–1806. https://doi.org/10.1007/s40808-020-00790-1

Razavi S, Gupta HV (2016) A new framework for comprehensive, robust, and efficient global sensitivity analysis: 1. Theory. Water Resour Res 52:423–439. https://doi.org/10.1002/2015WR017558

Saltelli A (2002) Making best use of model evaluations to compute sensitivity indices. Comput Phys Commun 145:280–297. https://doi.org/10.1016/S0010-4655(02)00280-1

Saltelli A, Annoni P (2010) How to avoid a perfunctory sensitivity analysis. Environ Model Softw 25:1508–1517. https://doi.org/10.1016/j.envsoft.2010.04.012

Saltelli A, Tarantola S, Chan KP-S (1999) A quantitative model-independent method for global sensitivity analysis of model output. Technometrics 41:39–56. https://doi.org/10.1080/00401706.1999.10485594

Saltelli A, Ratto M, Andres T et al (2008) Global sensitivity analysis: the primer. Wiley

Saltelli A, Annoni P, Azzini I et al (2010) Variance based sensitivity analysis of model output. Design and estimator for the total sensitivity index. Comput Phys Commun 181:259–270. https://doi.org/10.1016/j.cpc.2009.09.018

Singh G, Padalia H, Rai I et al (2016) Spatial extent and conservation status of Banj oak (Quercus leucotrichophora A. Camus) forests in Uttarakhand, Western Himalaya. Trop Ecol 57:255–262

Snowling SD, Kramer JR (2001) Evaluating modelling uncertainty for model selection. Ecol Model 138:17–30. https://doi.org/10.1016/S0304-3800(00)00390-2

Sobol IM (1993) Sensitivity estimates for nonlinear mathematical models. Mathemat Modell Comput Exper 1:407–414

Song W, Kim E, Lee D et al (2013) The sensitivity of species distribution modeling to scale differences. Ecol Model 248:113–118. https://doi.org/10.1016/j.ecolmodel.2012.09.012

Townsend Peterson A, Papeş M, Eaton M (2007) Transferability and model evaluation in ecological niche modeling: a comparison of GARP and Maxent. Ecography 30:550–560. https://doi.org/10.1111/j.0906-7590.2007.05102.x

Vanuytrecht E, Raes D, Willems P (2014) Global sensitivity analysis of yield output from the water productivity model. Environ Model Softw 51:323–332. https://doi.org/10.1016/j.envsoft.2013.10.017

Verma AK, Garkoti SC (2019) Population structure, soil characteristics and carbon stock of the regenerating banj oak forests in Almora, Central Himalaya. Forest Science and Technology. https://www.tandfonline.com/doi/abs/10.1080/21580103.2019.1620135

Voosen P (2019) New climate models predict a warming surge. Science. https://doi.org/10.1126/science.aax7217

Yi Y, Cheng X, Yang Z-F, Zhang S-H (2016) Maxent modeling for predicting the potential distribution of endangered medicinal plant (H. riparia Lour) in Yunnan, China. Ecol Eng 92:260–269. https://doi.org/10.1016/j.ecoleng.2016.04.010

Zhan C, Song X, Xia J, Tong C (2013) An efficient integrated approach for global sensitivity analysis of hydrological model parameters. Environ Model Softw 41:39–52. https://doi.org/10.1016/j.envsoft.2012.10.009

Zhang X, Trame M, Lesko L, Schmidt S (2015) Sobol sensitivity analysis: a tool to guide the development and evaluation of systems pharmacology models. CPT Pharmacometrics Syst Pharmacol 4:69–79. https://doi.org/10.1002/psp4.6

Part II

Habitat Modeling for Conservation of Threatened Plants and Restoration of Habitats

Tree Species Diversity and Richness Patterns Reveal High Priority Areas for Conservation in Eswatini

8

Wisdom M. D. Dlamini and Linda Loffler

Abstract

The Kingdom of Eswatini (formerly Swaziland) is characterized by high plant species richness and endemism. In this study, stacked species distribution models derived from maximum entropy and random forest models are applied on tree species distribution data to estimate and map taxonomic and phylogenetic diversity and endemism using six indices: species richness (SR), taxonomic weighted endemism (WE), corrected taxonomic weighted endemism (CWE), phylogenetic diversity (PD), weighted phylogenetic endemism (WPE) and corrected weighted phylogenetic endemism (CWPE). In addition, hotspots were identified by mapping the 95% percentile of the values from each index. Although weakly correlated, the hotspots overlap particularly in mountainous areas mainly in the north-western, eastern and mid- to south-central parts of the country. A combined hotspot measuring 1642 km^2 or 9.42% of the total land area was also mapped, showing the priority areas for conservation. Between 69% and 85% of the identified hotspots are not protected. Conservation gaps were also mapped and quantified by overlaying protected areas with the identified hotspots. The combined hotspot of all indices indicates an overall conservation gap of 82.03% indicating that only 14.8% is covered by existing protected areas and another 3.17% within ungazetted conservation areas. Areas of priority conservation are highlighted.

W. M. D. Dlamini (✉)
Department of Geography, Environmental Science and Planning, Faculty of Science and Engineering, University of Eswatini, Kwaluseni, Eswatini
e-mail: wdlamini@uniswa.sz

L. Loffler
Independent Consultant and Field Botanist, Mbabane, Swaziland
e-mail: lindad@realnet.co.sz

S. Dhyani et al. (eds.), *Ecosystem and Species Habitat Modeling for Conservation and Restoration*, https://doi.org/10.1007/978-981-99-0131-9_8

Keywords

Conservation · Diversity · Endemism · Eswatini · Hotspot · Trees

8.1 Introduction

The survival of human beings relies on ecosystem services such as food production, clean water, clean air, nutrient cycling, plant pollination, carbon sequestration, medicinal plants, climate stability and recreation, among many others that are provided by a well-functioning ecosystem (Yessoufou and Davies 2016). The conservation of biodiversity is, therefore, imperative and intricately linked to human well-being. However, effective conservation of biodiversity requires, among other things, knowledge on the geographic distribution and diversity of species (Guisan et al. 2013). Spatially explicit information on species distribution patterns is required to determine areas that have high concentrations of species and have high endemism and those which have a high number of endangered species (Mcshea 2014). It has also been observed that records of species occurrences characteristically provide information on only a subset of areas occupied by a species (Rondinini et al. 2006). Such records also do not provide information on areas not surveyed (Guisan et al. 2013). In addition, evolutionary diversity and the evolution of species have not been adequately considered in identifying important biodiversity areas due to the biased focus long-term on the evolution of different species (Scherson et al. 2017). Nevertheless, the interest in species' geographic distribution continues to grow due to the need to understand the impact of environmental change, including climate change, and other anthropogenic pressures on ecosystems.

Despite this interest, the analysis of species' geographic distribution patterns is important in macroecology and conservation biology. The geographic patterns of species are determined by both past and current biophysical factors and biogeographical regionalization (Kreft and Jetz 2010; Morrone 2018). In studying both basic and applied questions on the distribution of species, these spatial patterns provide a useful background against which to explain biogeographical research findings. The increasing availability of large datasets on species distributions is accompanied by the need for robust techniques for analysing biogeographical patterns. Since existing collection localities do not necessarily represent random samples from the existing environmental gradients, statistical procedures that consider this are critical for obtaining accurate biodiversity maps. Species distribution models (SDMs) are now predominantly used to determine distribution and diversity patterns at large geographical scales through combining environmental predictors with species occurrence information (Elith and Leathwick 2009; Guisan et al. 2013). SDM-based maps have also been shown to correctly identify conservation priority areas and gaps (Di Febbraro et al. 2018; Moradi et al. 2019). Subsequently, robust analytical methods have been developed and applied to conservation planning using several taxa across various biogeographic regions (Guisan et al. 2013).

Information on species distribution that accounts for multiple dimensions of biodiversity, including phylogeny, taxonomy and traits, is required for spatial conservation planning (González-Orozco et al. 2016; Pollock et al. 2017). However, some studies have shown a high correlation among these diversity indices, especially those indices affected by species richness (Morris et al. 2014; del Valle and Astorkiza 2018). Nonetheless, the relationships among these biodiversity dimensions are least understood and vary depending on the diversity index used and the analysis scale (Mazel et al. 2017; Santo-Silva et al. 2018). Conservation policy requires knowledge of how these dimensions relate to each other and how they can be applied in conservation planning. Recently, the biodiversity hotspot identification has been generalized to include other aspects such as the number of species, number of threatened species and evolutionary history, in addition to endemic species and habitat loss (Brum et al. 2017; Rosauer et al. 2017; Daru et al. 2019). Even though taxonomic diversity was the focus of earlier studies, recent studies have also focused on evolutionary processes (Tucker and Cadotte 2013; González-Orozco et al. 2014; Laity et al. 2015; Marchese 2015; Xu et al. 2019), hence the increasing interest on phylogenies (evolutionary trees) for the identification of areas of biodiversity importance. Phylogenetic diversity, such as is measured through phylogenetic branch length, is sometimes preferred because it shows relationships between current species and provides more information about long-term evolutionary processes (Faith 1992). As a result, there is an increasing recognition of the need to link phylogeny with distribution data in spatial conservation prioritization (Laity et al. 2015; Pollock et al. 2015; González-Orozco et al. 2016; Daru et al. 2019). This is also based on the premise that taxonomic diversity alone does not show complementarity and can result in the prioritization of areas with similar species assemblages, at the cost of protecting unique assemblages (Brown et al. 2015; Brum et al. 2017).

In Eswatini (formerly Swaziland), studies have been undertaken to look at the broader geographic distribution of species, some of which have focused on plants. Although the floristic composition of the country has been studied since the early twentieth century (e.g. Compton 1976; Galpin et al. 2002; Pott 1920), there are currently no high-resolution plant diversity maps for the country despite the robust software and hardware that are accessible today. The existing herbarium collection datasets generally provide the key occurrence data for such analyses. However, the validity of such data may be questionable considering the changes that may have occurred over the years since the specimen collections begun over a century ago. The tree atlas by Loffler and Loffler (2005) is the most comprehensive geo-referenced species occurrence dataset available for Eswatini and was the first to provide the broader tree species richness patterns albeit at a coarse resolution for the size of the country. Hence, such information could not provide detailed or landscape-scale distribution of taxonomic and phylogenetic diversity. In this study we use the tree atlas data to develop stacked species distribution models and phylogenetic information in mapping the country's phytogeographical hotspots and their protection status.

8.2 Methodological Approach

8.2.1 Study Area

Eswatini, a country located in the southern African region, covers an area of 17,365 km^2 and is characterized by a highly divergent topography largely determined by the underlying geology (Fig. 8.1). The altitudinal range from the low-lying east to the western highlands, only 130 km from east to west, is approximately 1800 m. To the west of the country lies an escarpment area which extends from the Drakensberg, and the Lubombo mountain range in the east separates the country's lowlands from the Mozambique coastal plains. This topographic heterogeneity results in steep environmental gradients including diverse climate systems and ecosystems. The country's subtropical climate is typified by distinct dry cold winters and wet warm summers. The eastern part of the country is inherently dry with rainfall averaging 600 mm, whilst the western highlands are humid averaging at about 1300 mm (van Waveren and Nhlengetfwa 1992). In contrast, mean annual temperature for the 2000–2016 period varies from an average of 17 °C in the north-west uplands increasing to 22 °C in the southeast lowlands albeit with varying microclimates driven by localized topographic features.

The divergent landscapes and climatic systems give rise to diverse ecosystems (Sweet and Khumalo 1994). The western part of the country is a grassy Highveld dominated by rocky outcrops separated by narrow river valleys. Patches of

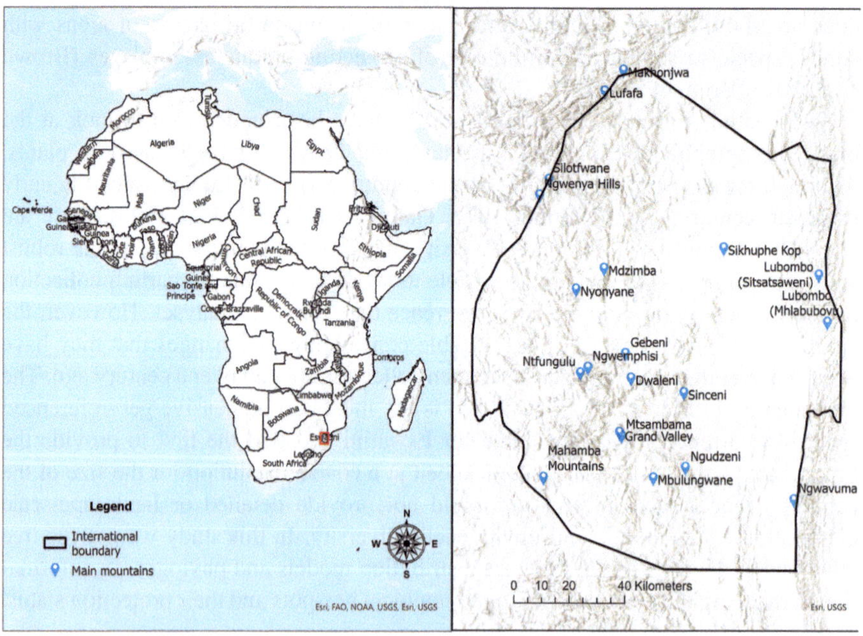

Fig. 8.1 Location of Eswatini. The main map also shows the country's topography

Afromontane forest are also found along the mountain ranges, particularly in areas above the mist belt. The central part of the country, called the Middleveld, is characterized by tall grasses interspersed within forests and thickets surrounded by rocky outcrops. The eastern lowlands, the Lowveld, are dominated by broad-leaved savanna which merges acacia woodlands towards the eastern flatter plains. The steep Lubombo escarpment is adjacent to the flat Lowveld and is dissected with steep gorges that support patches of scarp and *Androstachys* forest. The escarpment also harbours a *Combretum* Bushveld that gradually changes to a grassy plateau, bounded by rocky outcrops and cliff faces (Loffler and Loffler 2005).

8.2.2 Species Data

The Eswatini tree atlas data (which also includes selected shrubs) from Loffler and Loffler (2005) was extracted from a database provided through the Eswatini National Trust Commission flora database (http://eswatininaturereserves.com/flora/index.asp, accessed September 2021). The data used in this study was collected by Loffler and Loffler (2005) from sample plots derived from 2 km transects over a 6-year period beginning in early 1999 during which efforts were made to cover flowering, fruiting and growing seasons. Notably, the transects used to collect the field data had been randomly located within systematic 11 × 11 km grid squares with additional data being collected from extensive field surveys during environmental impacts assessments and other projects. In addition, the sample sites were revisited by Loffler and Loffler (2005) to reduce omission, and such areas included places affected by disturbances such as floods or bush clearing. Additional sample plots were also added whenever a different vegetation type was encountered within a transect.

Furthermore, the tree atlas database has continually been updated with more field data over the years (L. Loffler and K. Braun, pers. comm.). The data used in this study had been updated up to 2014. We then cleaned the dataset to ensure every data point was geo-referenced, resulting in 26,802 geo-referenced presence points from 630 sample sites (Fig. 8.2). The data contained 659 species ranging from rare species with a single record to common species with 264 records. All the species which had less than five records were excluded from further analyses.

8.2.3 Environmental Predictors

We initially considered a total of 38 environmental predictors for developing the species distribution models. These included 19 bioclimatic predictors from the WorldClim 2.0 dataset (Fick and Hijmans 2017), which is averaged over the period 1970–2000, covering Eswatini at 30 arc-second (~880 m) resolution. Additionally, 19 other variables representing topographic, anthropogenic and geomorphological factors were used. All the datasets were resampled to a standard 30 arc-second resolution resulting in the whole country being covered by 39,744 grid cells. Categorical and continuous variables were resampled using the nearest neighbour

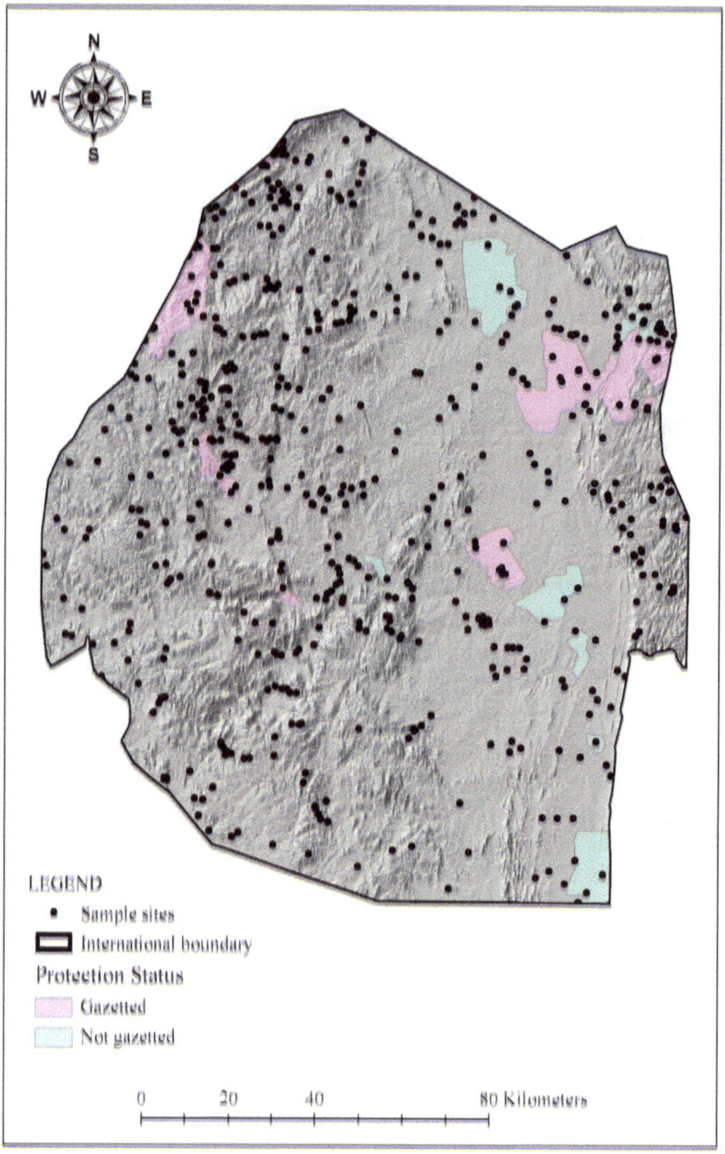

Fig. 8.2 Location of sampling plots for the tree atlas data overlaid over protected areas in Eswatini

and bilinear interpolation techniques, respectively. The former technique performs a nearest neighbour assignment, whilst the latter involves a bilinear interpolation of the grid cell values and determines the new value of a grid cell based on a weighted distance average of the four nearest input grid cell centres.

Multicollinearity, however, arises when highly correlated independent variables are included and must be estimated. Dormann et al. (2013) argue that multicollinearity is certain when the correlation coefficient among variables is at least 0.7 and the VIF values are above 10 (Dormann et al. 2013). The correlation coefficient and variance inflation factor (VIF; Naimi et al. 2014) were used to verify multicollinearity. The final set of variables included in the models was selected using a stepwise procedure aimed at excluding highly correlated variables, using the VIF as a measure of collinearity. Variable bias was also reviewed by visually evaluating the outputs and finding overly dominant variables (Ng et al. 2018). Using the R package 'usdm' (Naimi 2017), the VIF was calculated for all variables and in each step excluding the variable with the highest VIF (>10). This procedure was repeated until there were no variables with a VIF value greater than 10. In total 22 predictors were retained and used to develop the species distribution models (Table 8.1).

8.2.4 Species Distribution Modelling and Bias Correction

We implemented stacked SDMs (SSDMs) (D'Amen et al. 2015a) to model the species distributions using the R package 'ssdm' (Schmitt et al. 2017). Ensembles of the 'maxent' (ver. 3.4.1k; Phillips et al. 2019) and 'randomforest' algorithms (Breiman 2001) were used to develop the SSDMs for each species. Maxent uses a maximum entropy approach to model species distributions with presence-only data (Phillips et al. 2006) and is a consistent best-performing algorithm, especially when there are a limited number of records (Wisz et al. 2008). It is also least affected by geographical errors in the species occurrences (Graham et al. 2008). The random forest algorithm is an efficient and robust ecological modelling algorithm that provides a better model fit even for under-sampled areas (Mi et al. 2017).

The SSDMs were then calibrated using bootstrap resampling during which 75% of each species' occurrence data was used as training data. Random resampling was performed 100 times using the remaining 25% of the dataset as a test data to evaluate model accuracy. The accuracy was evaluated using the area under the curve (AUC) of the receiver operating characteristic (ROC) plot. We used a bias-corrected null model (Syfert et al. 2013; Fithian et al. 2015) to evaluate the AUC value for each SSDM developed with all occurrence records against the AUC values expected by chance. A sampling bias grid was produced by summing the number of occurrences found within the 30 arc-second grid cells which were aligned to the environmental grid cell resolution. Using the approach of Calabrese et al. (2014) and Zellmer et al. (2019), the outputs from both the random forest and Maxent models were combined to reduce individual modelling algorithm biases.

Notably, Loffler and Loffler (2005) also minimized data collection bias through the following:

- Using fine-scale (11 km grid square) stratified random sampling (accounting for vegetation and land use types, i.e. focusing on tree-covered areas)
- Focusing on areas previously under-collected

Table 8.1 List of variables used in the species distribution models

Variable	Type	Source	Variance inflation factor (VIF)
Slope aspect (cardinal directions)	Categorical	Derived from digital elevation model (Jarvis et al. 2008)	1.023059
Land cover fragmentation index	Continuous	Derived from land cover data	1.068082
Land tenure	Categorical	Modified from Remmelzwaal and Vilakati (1994)	1.211217
Protection status	Categorical	Derived from cadastral data and Roques (2002)	1.295225
Lithology	Categorical	Vegter (1995)	1.296875
Soil type	Categorical	Murdoch (1968)	1.332099
Distance to water (rivers/ reservoirs)	Continuous	Derived from land cover data	1.484975
Land cover	Categorical	Muyambi (2016)	1.58556
Precipitation seasonality	Continuous	Fick and Hijmans (2017)	1.615047
Topographic position index	Continuous	Derived from digital elevation model (Jarvis et al. 2008)	1.676945
Distance to human-disturbed land	Continuous	Derived from land cover data	1.767758
Solar radiation duration	Continuous	Zomer et al. (2008)	1.999564
Slope	Continuous	Derived from digital elevation model (Jarvis et al. 2008)	2.262022
Human population density	Continuous	(Central Statistical Office 2018)	2.805749
Human settlement density	Continuous	Calculated using data from Facebook Connectivity Lab and Center for International Earth Science Information Network (CIESIN), Columbia University (2016)	2.879977
Surface form	Continuous	Derived from digital elevation model	2.910769
Isothermality		Fick and Hijmans (2017)	2.968866
Precipitation of driest month	Continuous	Fick and Hijmans (2017)	3.440109
Mean diurnal range	Continuous	Fick and Hijmans (2017)	4.408422
Solar radiation total/annum	Continuous	Zomer et al. (2008)	6.200382

(continued)

Table 8.1 (continued)

Variable	Type	Source	Variance inflation factor (VIF)
Mean annual number of frost days	Continuous	Schulze et al. (2008)	7.272504
Minimum temperature of coldest month	Continuous	Fick and Hijmans (2017)	8.365488

- Multiple visits to sampling sites
- Multi-seasonal sampling (to account for flowering, fruiting and growing seasons)
- Extensive use of regional expertise in species identification
- Use of multiple herbaria including South African herbaria
- Linking field visits to herbarium information on occurrences

8.2.5 Species Richness and Diversity Patterns

The species richness maps were computed using the probability ranking rule (D'Amen et al. 2015a, b). This method estimates community composition by ranking the species in decreasing order of their predicted probability up to the species richness prediction. This rule assumes that species with the highest habitat suitability are competitively superior (Schmitt et al. 2017). The analysis made use of the spatially explicit species assemblage modelling (SESAM) framework (Guisan and Rahbek 2011) which uses macroecological models to set a limit to the number of species predicted by the stacked distribution models.

The diversity indices calculated in this study were taxonomic species richness (SR, the number of species in an area), taxonomic weighted endemism (WE) and corrected taxonomic weighted endemism (CWE) (Crisp and Linder 2001). The endemism indices indicate that a species is unique to a defined geographical locality (Crisp and Linder 2001; Moraes Mónica et al. 2014) and avoid using a threshold by applying a continuous weighting function by assigning high weights to species with small occurrence ranges and progressively smaller weights to those with larger ranges (Schmitt et al. 2017).

$$\mathrm{WE} = \sum_{t \in T} \frac{r_t}{R_t}$$

where t is a label (taxon) in the set of labels (taxa) T in neighbour set 1, r_t is the local (Eswatini) geographic range (the number of elements containing label t within neighbour sets 1 and 2), and R_t is the global range of label t across the dataset.

$$\mathrm{CWE} = \sum_{t \in T} \frac{1}{r_t} R_t$$

Since the aim was to identify local (Eswatini-specific) conservation hotspots, the geographic range was estimated as the number of grid cells in which a species was found within Eswatini. Phylogeny-based diversity measurement methods have been rapidly receiving interest (Scherson et al. 2017; Millar et al. 2017) and are being used in plant conservation (Escudero et al. 2003; Millar et al. 2017). Hence, we also measured phylogenetic diversity (PD) which is the proportion of the total tree length present in a grid cell (Faith 1992), weighted phylogenetic endemism (WPE) and corrected weighted phylogenetic endemism (CWPE) (Rosauer et al. 2009) using the Biodiverse v2.0 software (Laffan et al. 2010). The phylogenetic indices were derived using a time-calibrated species-level supertree from TimeTree (Kumar et al. 2017) and the Open Tree Of Life (Hinchliff et al. 2015). The outputs from the stacked species distribution models were used as geographic range inputs in the estimation of PD, WPE and CWPE.

$$\mathrm{PD} = \sum_{c \in C} L_c$$

where L_c is the length of branch c and C is the set of branches in the minimum spanning path connecting the species.

$$\mathrm{WPE} = \sum_{\lambda \in \Lambda} L_\lambda \frac{r_\lambda}{R_\lambda} / L$$

where Λ is the set of branches found across neighbour sets 1 and 2, L_λ is the length of branch λ, r_λ is the local range of branch λ (the number of groups in neighbour sets 1 and 2 containing it), R_λ is the global range of branch λ (the number of groups across the entire dataset containing it), and L is the sum of all branch lengths in the trimmed phylogenetic tree.

$$\mathrm{CWPE} = \frac{\mathrm{WPE}}{\mathrm{PD}}$$

PD evaluates the evolutionary diversity within each grid cell using phylogenetic branch lengths, whilst WPE is a range-weighted index which assesses the branches of the phylogenetic tree restricted or endemic to a given geographical location (Rosauer et al. 2009). Hence, WPE represents the sum of weighted branch lengths, where each branch is weighted by the fraction of its geographic range represented by the given area. In this study, PD, WPE and CWPE were scaled to represent the proportion of variation within the tree represented by the taxa and the total length of the tree divided between branches according to their relative lengths. The statistical significance of each diversity index was then evaluated using a randomization test with the standard null model (Mishler et al. 2014).

To assess the effect of each diversity index on the estimate of protection, we overlaid each diversity map with the protection status map. We then conducted the analyses using three protection levels: gazetted (by law), protected (but not gazetted) and unprotected. This enabled us to evaluate the protection status and diversity value of each grid cell, which we then used when calculating the percentages of hotspots under each protection level. Even though we used the 95th percentiles as hotspot thresholds, we also calculated protection status between the 0 and 99th percentiles in order to explore how protection status varies with the choice of threshold.

8.3 Results

8.3.1 Diversity and Endemism Patterns

The AUC values of the SSDMs for all the species included and studied ranged from a minimum of 0.88 to a highest value of 0.98 (mean = 0.907) indicating high predictive performance. The derived geographic patterns of species richness (SR) and phylogenetic diversity (PD) are not uniform but geographically similar (Fig. 8.3). The SR and PD maps indicate that eastern and north-western Eswatini have the highest species richness. Mountain ranges and their valleys around the country are especially high in diversity, including most of the Lubombo Mountains

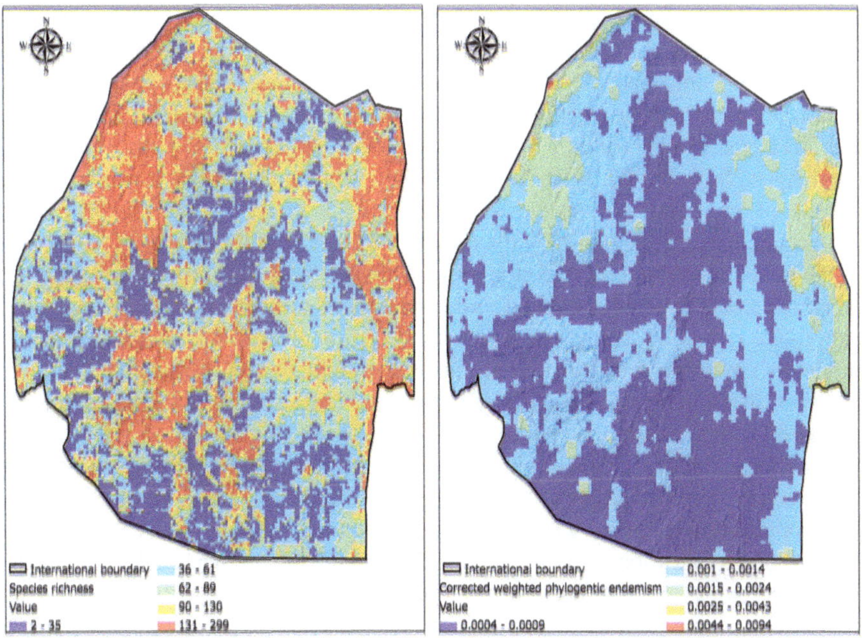

Fig. 8.3 Maps of estimated species diversity and endemism in Eswatini

(particularly Jilobi forest, Muti-muti, Mhlumeni, Mahhuku, Manzimnyama, Usuthu Gorge and Shewula), Lufafa-Makhonjwa Mountains, Mdzimba Mountains, Mahamba Mountains, Sinceni Mountains, Mkhondvo, Ntfungulu Mountains, Grand Valley, Mtsambama Mountains and Gebeni-Dwaleni Mountains.

Notably, there is a strong similarity in the geographic patterns of the WE and WPE ($r^2 = 0.99$, $p < 0.005$) and between CWE and CWPE ($r^2 = 0.98$) (Fig. 8.4). There is, however, a relatively weak but significant correlation of SR with both CWE and CWPE ($r^2 = 0.36$, $p < 0.05$ and $r^2 = 0.41$, $p < 0.05$, respectively). Similarly, PD was weakly correlated to both CWE ($r^2 = 0.37$, $p < 0.05$) and CWPE ($r^2 = 0.41$, $p < 0.05$) which suggests these indices may be showing different features of geographic diversity (Fig. 8.4).

8.3.2 Geographic Patterns of Conservation Gaps

The spatial distribution of the derived hotspots is similar, with small differences in both size and location (Fig. 8.5). The sizes of the total combined hotspot areas derived from the six indices range in size from 727 km^2 for the WE-based hotspot to 755 km^2 for the CWPE-based hotspot, representing an average of 4.26% of the total land area. The geographic distribution of the current protected area network and the derived hotspots in Eswatini shows clear protection gaps (Fig. 8.5 and Table 8.2). When considering native tree species richness (SR), the area of overlap between protected areas and hotspots is only 12.88%. Therefore, 83.70% of the taxonomic richness-based hotspot is not covered by protected areas, whilst 3.42% is covered by conservation areas not legally proclaimed. When using WE and CWE, 23.66% and 24.40% of the hotspots are, respectively, covered by protected areas, whilst another 6.05% and 5.60% are, respectively, covered by the ungazetted protected areas. When PD, WPE and CWPE are used to derived hotspots, the areas under strict protection are, respectively, 10.72%, 23.67% and 25.30%. This indicates that a relatively low number of range-restricted species are covered by the protected area network. These protected areas include the Mlawula and Malolotja Nature Reserves.

There is considerable spatial overlap between hotspots identified using the weighted and corrected weighted endemism (both taxonomic and phylogenetic) and those based on taxonomic richness. Significant portions of the Lubombo mountain range in the east and the north-western mountain ranges are such notable hotspots. In addition, small patches of fragmented remnants of the species-rich areas as earlier identified also remain important when considering the WE and CWE. In all the analyses, the coincidence with areas with no form of protection was largest, followed by gazetted areas (Table 8.2). Protected areas such as Mlawula and Malolotja Nature Reserves, as well as ungazetted private conservation areas, provide notable protection especially when considering endemism. When all the hotspots from each index are used, the combined hotspot area increases to 1642 km^2 which

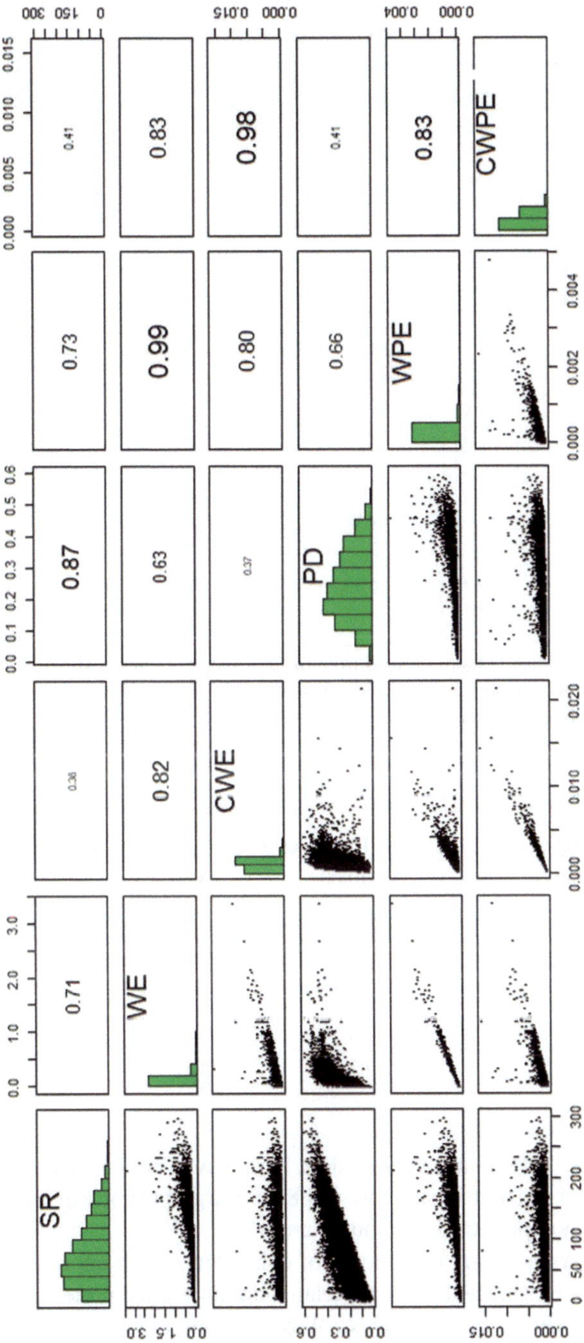

Fig. 8.4 Pairwise correlation between the diversity and endemism indices

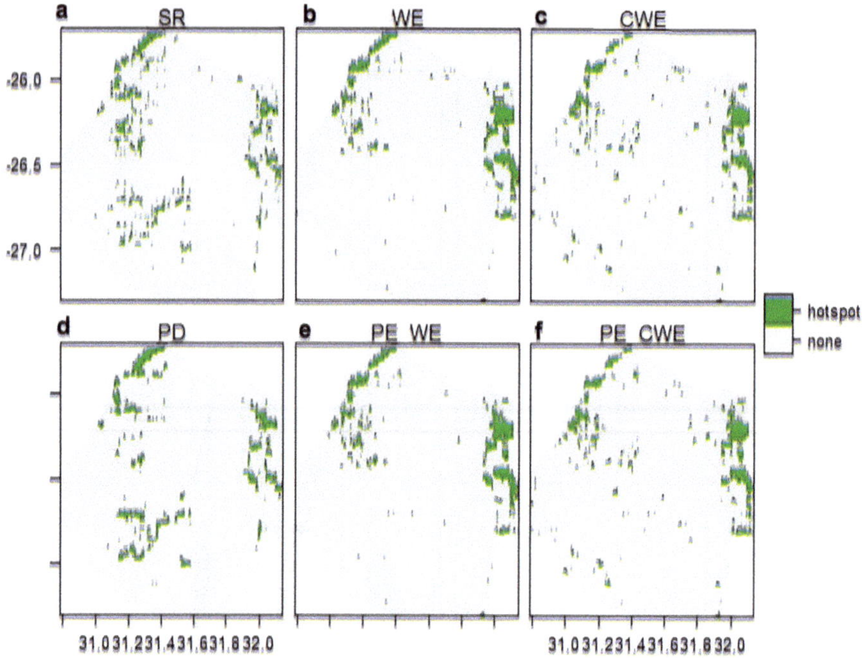

Fig. 8.5 Eswatini's hotspot maps derived from (**a**) tree species richness, (**b**) weighted endemism, (**c**) corrected weighted endemism, (**d**) phylogenetic diversity, (**e**) weighted phylogenetic endemism and (**f**) corrected phylogenetic endemism

Table 8.2 Percentage of derived hotspots covered under each protection category

Index	Hotspot size (km^2)	Unprotected	Protected-not gazetted	Gazetted
CWE	750	70.00	5.60	24.40
WE	727	70.29	6.05	23.66
SR	730	83.70	3.42	12.88
CWPE	755	69.01	5.70	25.30
WPE	731	70.31	6.02	23.67
PD	746	85.25	4.02	10.72
Combined	1642	82.03	3.17	14.80

equates to 9.42% of the total land area (Fig. 8.6). However, the area under protection for the combined hotspot is still low (14.8%). Large portions of the hotspots are predominantly in unprotected communal areas.

For each diversity index, the amount of coincidence with unprotected lands was largest, followed by gazetted areas. Overlap with ungazetted protected areas that are managed for biodiversity was the smallest. Within the same protection status, however, our estimates of biodiversity hotspot protection varied in astonishing

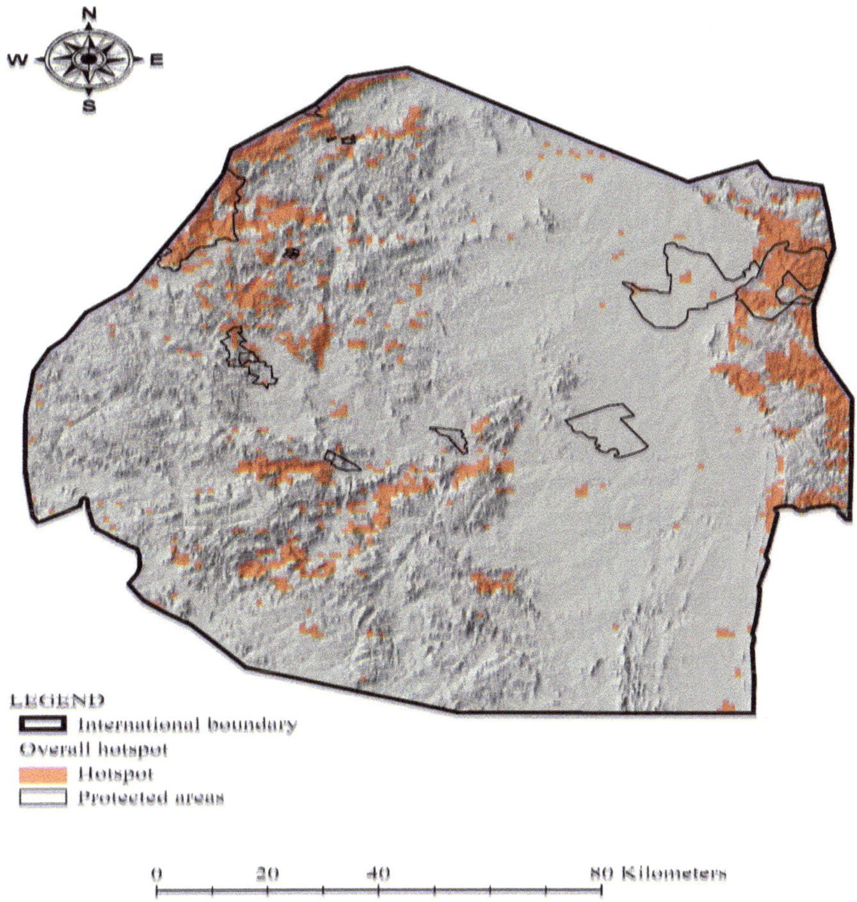

Fig. 8.6 The combined hotspot derived from all the species diversity and endemism indices

ways. Both the diversity index and the percentile threshold used to estimate hotspots influenced the estimates of protection, and their relationships were not consistent among the studied levels of protection (Fig. 8.7). The overlap between the protection levels and the hotspots identified by the different indices generally increases as the percentile threshold decreases from the 99th to the 1st percentile. The percentages of hotspot grid cells that were in unprotected areas were high for all percentiles dropping sharply up to the 50th percentile for all levels of protection (Fig. 8.7). When using a 95th percentile threshold, the estimates of protection were most sensitive to the protection status, more so when using WPE. Other indices show larger hotspots at the 50th percentile.

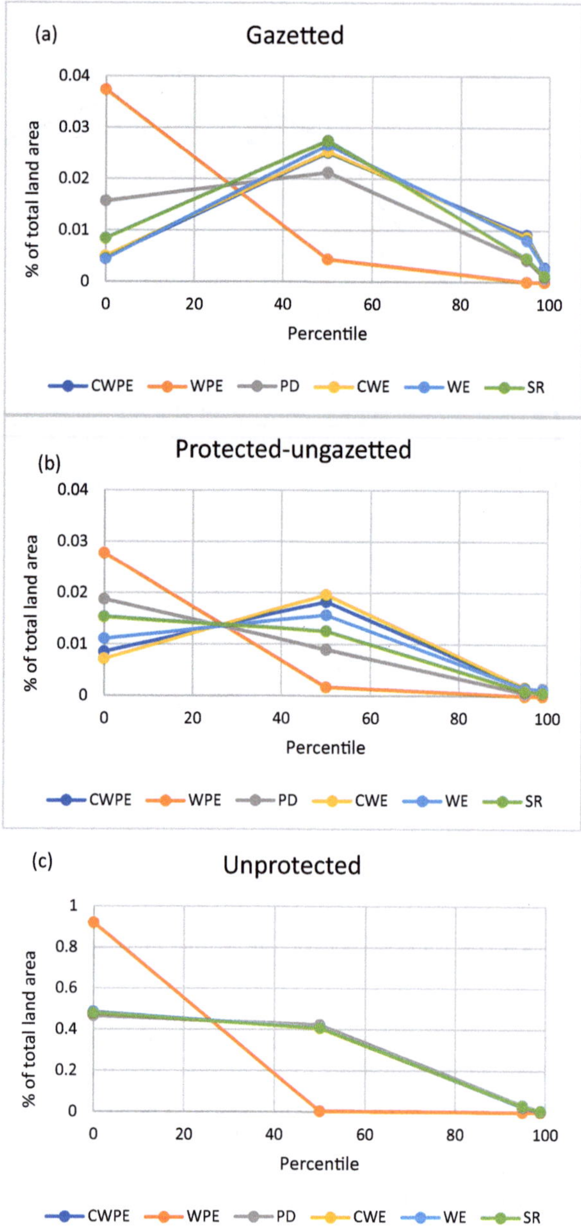

Fig. 8.7 Relationship between the percentiles to be used as the threshold for mapping hotspots and the hotspot size within each protection status

8.4 Discussion

8.4.1 Tree Species Richness and Diversity Patterns

Despite its small size, Eswatini is characterized by high species diversity resulting from high topographic and bioclimatic diversity. Mapping the geographic patterns of plant species diversity is critical in the conservation of biodiversity in the country. The models used in this study produced high accuracies, although these may still contain errors that can contribute to some errors in species diversity maps (Guisan et al. 2013). Geographical collection bias remains one of the most common issues associated with disparate collection effort which may lead to incomplete information on actual species ranges (Robertson and Barker 2006; Sardà-Palomera et al. 2012).

It should be noted, though, that even high-resolution maps may include errors of commission because both the accuracy and precision of SDMs are limited by both the availability and resolutions of the predictor variables used (Liang et al. 2013; McKerrow et al. 2018). The efforts to reduce omission errors from previous coarse-resolution maps reduced the underestimation in species diversity maps and revealed diversity hotspots in locations that the coarser maps did not identify. The resulting phylogenetic diversity and endemism maps are new in the country. Despite varying levels of correlations, the results show notable spatial overlaps in the distribution patterns of Eswatini's overall taxonomic and phylogenetic diversity and endemism.

Eswatini's native tree species and phylogenetic diversity are weakly correlated, and the tree diversity hotspots obtained from the analysis also have a high geographic congruence (Table 8.2, Fig. 8.5). This observed high diversity congruence is important since these indices are most often used for planning the protection of species and may be helpful in guiding biodiversity conservation strategies. For instance, when the hotspots derived from each of the indices are relatively concentrated in the same areas, species of concern can be protected to the greatest extent possible using less resources.

The country's tree species diversity and endemism are neither uniformly nor randomly distributed, and the indices used exhibited an overall trend of high values within mountain ranges, particularly the eastern and north-western mountain ranges, and fewer species being distributed in the low-lying areas of the country. The highest numbers of species and phylogenetic diversity are found in the Lubombo Mountains in the east, the Mdzimba Mountain extending to the Makhonjwa-Lufafa-Bulembu mountain range in the north-west and the south-central mountain complexes including Ngwempisi and Sinceni mountains. Loffler and Loffler (2005) observed that some coastal species generally restricted to the Lubombo Mountains also appear on Sinceni Mountain in the interior of the country, resulting in similarities and connections in the geological history of the coastal, dune, Lubombo and inland forests. Loffler and Loffler (2005) also note that some southern African coastal species such as *Strychnos gerrardii*, *Ficus burtt-davyi*, *Deinbollia oblongifolia*, *Pavetta gerstneri* and *Dovyalis longispina* are also found in the inland mountains on the eastern and interior mountain ranges in the country.

The identified species diversity and endemism hotspots are notably within known regional phytochoria that are globally significant centres of plant diversity and endemism. The Lubombo Mountains in the eastern part of the country, for instance, lie within the Maputaland Centre of Plant Endemism (Mittermeier et al. 2015). The interior and north-western hotspots are part of the Barberton Centre of Plant Endemism which is an extension of the Drakensberg Afromontane Regional System (Margules and Pressey 2000). Both these hotpots support high concentrations of endemic taxa and high tree species diversity (Mittermeier et al. 2015). It is worth highlighting, therefore, that this study recovers these areas of regional endemism and diversity albeit with localized details such as the mountain 'islands' of diversity within the Afromontane archipelago originally described by White (1981) and subsequently by Grimshaw (2001). The high taxonomic and phylogenetic diversity of the Lubombo escarpment result from the diversity of habitats emanating from both geological and evolutionary history of the area (van Wyk and Smith 2001). The strong relationship between the country's vegetation communities and elevation and geology has been observed before (Dlamini 2011a).

Overall, the patterns of endemism have a high similarity with those of species diversity largely due to habitats in areas with high endemism being fragile and mountainous. In addition, most endemic species have a restricted range and are susceptible to outside interference as well as being vulnerable to becoming endangered (Myers et al. 2000; Zhao et al. 2016). As a result, the geographic patterns of the hotspots derived from the six indices geographically overlap. However, the endemism-based indices show localized and smaller hotspots. This may be due to the fact that, aside from the association with species distribution itself, the distribution patterns of endemic species are closely correlated with human disturbance or threat factors (Zhao et al. 2016). Additionally, the results not only highlight current tree diversity hotspots but also provide insights into the biogeographical distribution and evolutionary history of Eswatini's flora as similarly observed by Daru et al. (2015) for southern Africa.

Researchers have previously highlighted sampling biases in plant collection data including taxonomic/phylogenetic, spatial and temporal biases (Meyer et al. 2016; Daru et al. 2018). Although not explicitly and quantitatively assessed in this study, techniques were employed during both the data collection and analyses stages to reduce spatial bias. This was evidenced by the significant number of new records of species, including relic species, that were recorded since the atlassing commenced two decades ago. Nevertheless, there is a need for continuous surveys within the hotspots especially those that have not been adequately studied before.

8.4.2 Diversity and Endemism Patterns in Relation to Protection

The study of biodiversity-rich areas and areas of endemism is important in the assessment of protection strategies. The intersections between the derived hotspots are very important for biodiversity protection. Using the tree data for terrestrial biodiversity conservation targeting is primarily because trees are well studied in the country and, in the absence of detailed data on other plant forms, such data provides the available baseline for conservation planning. In addition, vascular plant diversity patterns may be used as bioindicators for other taxa due to plants being the main primary producers in terrestrial ecosystems (Brunbjerg et al. 2018). Andersen et al. (2013) attest that plant bioindication may be a cost-effective method to estimate general habitat quality.

The tree diversity hotspots are important because different indices highlight different aspects of biodiversity, such as species richness, geographic range and phylogeny. Hence, targeting only the species-rich hotspots would protect the greatest number of species per unit area. However, if the goal is to conserve areas with ancient flora, phylogenetic diversity might be an appropriate consideration (Xu et al. 2017). Focusing conservation efforts on areas with high taxonomic and phylogenetic endemism stresses the protection of range-restricted species where species composition similarity may be low (Xu et al. 2017). Rosauer and Jetz (2015) observed that protecting areas with high phylogenetic endemism may address conservation concerns on lineages of plants that are both evolutionarily distinct and geographically restricted.

Focusing on all the identified hotspots would be a cost-efficient approach to biodiversity conservation by concentrating efforts at those areas with the most species of concern. There is notable spatial overlap among the hotspots defined using the various diversity indices in Eswatini, making it relatively less intricate to identify the boundaries of priority conservation areas. However, the size of hotspots is reduced when using the endemism indices compared to using taxonomic richness and phylogenetic diversity. Therefore, whilst there is need to define a clear and agreed index when prioritizing areas for conservation, where there are significant spatial overlaps, combining the hotspots will likely result in a greater conservation outcome with limited resources. This is important because Eswatini has established a few protected areas whose efficiency is not ideal because of structural weaknesses and limited financing (Dormann et al. 2013).

The findings indicate that most of the tree diversity hotspots for Eswatini are poorly covered by the current protected area network. The existing protected areas cover approximately 4.2% of the country's total land area, but the area covered is geographically biased, especially in relation to the identified hotspots. Thus, there are still considerable conservation gaps for the country's flora since hotspot protection is low (<25%) when considering both taxonomic and phylogenetic diversity. The large spatial overlap between the hotspots derived from different indices could also be an indicator of similarities between drivers of taxonomic richness, phylogenetic diversity and biodiversity protection. It is important, though, to highlight that the geographic patterns of tree species richness are determined by both the biogeographic evolution of species and land use practices, whilst the location of Eswatini's protected areas has also evolved over the years in response to historical circumstances and land tenure (Hackel and Carruthers 1993). Most of the protected areas were historically disease-infested areas and marginal lands that were considered unsuitable for productive use. These characteristics helped reduce human presence and impact in these areas; hence those areas appear to be better protected and preserved. However, as most (endemic) species are found in the eastern escarpment and north-western mountain ranges, very few portions of those hotspots are within the protected area network.

8.4.3 Priority Areas for Enhanced Conservation

This study reveals a spatial overlap in the priority areas for the taxonomic and phylogenetic dimensions of tree species diversity in Eswatini. This is in contrast to the observations from Daru et al. (2015) who observed spatial incongruence between tree diversity indices in hotspots and currently protected areas, SR and PD. Their study was, however, notable at a higher scale and lower resolution compared to this study (50 km vs 1 km grid size). The lack of geographic overlap among diversity hotspots could, therefore, be indicative of the small size of hotspot areas (such as revealed by this study) compared to their grid size. Nevertheless, the geographic distribution of the hotspots is in general agreement with those of Daru et al. (2015). There are notable differences between the hotspot maps when considering the level of protection, suggesting that effective conservation planning should be based on both taxonomic and phylogenetic diversity.

The results of this study corroborate other studies that showed the percentile threshold and diversity index used influence the estimates of protection and the design of protected area networks (McKerrow et al. 2018). The variations in the hotspot size as a function of the percentile can be attributed to the geographic disparities in terms of both protection levels and the species distribution. We also demonstrate that the diversity index used influences the hotspot maps and their protection, which underscores the need to account for both taxonomic and phylogenetic diversity when designing systems of protected areas. Based on this study, we suggest that the combined hotspot of approximately 1642 km^2 (Fig. 8.6) be prioritized for protection of the country's biodiversity and hence be considered in protected area expansion strategies within the short to medium term. The identified hotspot constitutes 9.46% of the country's total land area and its conservation would significantly to the achievement of the country's target of protecting at least 10% of the land area (Swaziland Environment Authority 2016). We suggest that the effectiveness of the existing protected areas, in their various forms and governance regimes, be enhanced through revisiting areas previously demarcated and identified as protection-worthy by Roques (2002) and Deall et al. (2000). We also propose that for those protected parts of the hotspots which still harbour inherently threatened species, such as those within Mlawula and Malolotja Nature Reserves in the east and north-west, the efforts should focus on improving in situ conservation and expansion of range size. Areas undergoing severe spatial fragmentation due to deforestation (as identified by Dlamini 2017), especially those in the north-west of the country such as the north-western mountain range and the central and southern parts of the country such as the Mdzimba Mountains and the Ngwempisi-Sinceni mountain complex stretching to Mahamba, Ngudzeni and Mkhondvo, have a high risk of local extinction. Reducing habitat disturbance within these identified hotspots is the preferred conservation strategy. For the highly isolated hotspots with high species endemism, both in situ and ex situ strategies must be employed because numerous narrow-range and threatened species are found in these areas. Most importantly, the country needs to take active steps to protect most of the mountain ecosystems which are evidently hotspots and are a key habitat for most of the tree species studied.

The urgency to protect the mountain ecosystems is necessitated by the increasing expansion of human settlements and illicit cultivation of cannabis in many of the remote mountain valleys, riparian zones and indigenous forests (pers. obs.). Rising human population and expanding demand of land for agriculture, human settlements and other infrastructure have meant that large areas of forests and woodlands are lost annually (Dlamini 2016, 2017). Numerous species are threatened with extinction due to habitat loss in unprotected and highly disturbed low-lying and flatter areas. It should also be noted that most areas within the hotspots are predicted to be undergoing notable change and spatially shifted bioclimatic conditions (Dlamini 2011b). Hence, a climate-adaptive protected area network and conservation programmes could buffer against the potential impacts of climate change. Tree diversity has the potential to enhance ecosystem functionality, including biomass production and carbon sequestration (Hulvey et al. 2013; Ratcliffe et al. 2016; Mori 2018).

The high-resolution maps of tree diversity, therefore, indicate areas with potential co-benefits for both biodiversity conservation and climate change mitigation at local levels (Soto-Navarro et al. 2020). Hence, the extent of various categories of protected areas should also be periodically reviewed to consider not only the taxonomic but also the phylogenetic diversity. Most areas with high diversity and endemism are within unprotected communal areas and privately owned land parcels. It is, therefore, necessary that conservation strategies consider this through the adoption of innovative community-public-private conservation partnerships to ensure maximum protection. The identified hotspots could be protected as individual units, particularly where there is high ecosystem fragmentation, or as a continuum linked to the existing protected area network.

Notwithstanding the high predictive performance of the models across the species studied, it is important to highlight that SDMs are without their limitations. These have been studied and discussed widely by other researchers (e.g. Jarnevich et al. 2015) and can be implied to apply in this study too.

8.5 Conclusion

This study produced high-resolution tree species distribution maps for the entire country, revealing subtle patterns of species diversity and endemism in Eswatini. The use of disparate diversity indices revealed comparable distribution patterns of native tree richness and endemism. Three broad hotspots were identified by taxonomic and phylogenetic diversity. In addition, a few areas in the central to southern part of the country are also observed to be important tree conservation sites. All these hotspots are predominantly in the mountainous areas not covered by the current protected area system. The findings indicate that extensive species-rich and phylogenetically diverse areas are unprotected. The study reveals that for effective conservation of Eswatini's taxonomic and phylogenetic tree diversity and endemism, the country's mountain ecosystems need to be protected. More importantly, the areas in each of the identified hotspots should be considered with special priority for effective conservation. In-depth surveys are required in areas with high endemism and phylogenetic diversity including areas that have not been previously surveyed.

Acknowledgements We sincerely appreciate the efforts of Kate Braun in curating the data used in this study.

References

Andersen DK, Nygaard B, Fredshavn JR, Ejrnæs R (2013) Cost-effective assessment of conservation status of fens. Appl Veg Sci 16:491–501. https://doi.org/10.1111/avsc.12020

Breiman L (2001) Random forests. Mach Learn 45:5–32. https://doi.org/10.1023/A:1010933404324

Brown CJ, Bode M, Venter O et al (2015) Effective conservation requires clear objectives and prioritizing actions, not places or species. Proc Natl Acad Sci U S A 112:E4342

Brum FT, Graham CH, Costa GC et al (2017) Global priorities for conservation across multiple dimensions of mammalian diversity. Proc Natl Acad Sci U S A 114:7641–7646. https://doi.org/10.1073/pnas.1706461114

Brunbjerg AK, Bruun HH, Dalby L et al (2018) Vascular plant species richness and bioindication predict multi-taxon species richness. Methods Ecol Evol 9:2372–2382

Calabrese JM, Certain G, Kraan C, Dormann CF (2014) Stacking species distribution models and adjusting bias by linking them to macroecological models. Glob Ecol Biogeogr 23:99–112. https://doi.org/10.1111/geb.12102

Central Statistical Office (2018) The 2017 population and housing census: preliminary results. Mbabane

Compton RH (1976) The flora of Swaziland. J South African Bot/Suppl 11:684

Crisp L, Linder M (2001) Endemism in the Australian flora. J Biogeogr 28:183–198. https://doi.org/10.1046/j.1365-2699.2001.00524.x

D'Amen M, Dubuis A, Fernandes RF et al (2015a) Using species richness and functional traits predictions to constrain assemblage predictions from stacked species distribution models. J Biogeogr 42:1255–1266. https://doi.org/10.1111/jbi.12485

D'Amen M, Pradervand JN, Guisan A (2015b) Predicting richness and composition in mountain insect communities at high resolution: a new test of the SESAM framework. Glob Ecol Biogeogr 24:1443–1453. https://doi.org/10.1111/geb.12357

Daru BH, van der Bank M, Davies TJ (2015) Spatial incongruence among hotspots and complementary areas of tree diversity in southern Africa. Divers Distrib 21:769–780. https://doi.org/10.1111/ddi.12290

Daru BH, Park DS, Primack RB et al (2018) Widespread sampling biases in herbaria revealed from large-scale digitization. New Phytol 217:939–955. https://doi.org/10.1111/nph.14855

Daru BH, le Roux PC, Gopalraj J et al (2019) Spatial overlaps between the global protected areas network and terrestrial hotspots of evolutionary diversity. Glob Ecol Biogeogr 28:757–766. https://doi.org/10.1111/geb.12888

Deall GB, Dobson L, Masson PH, et al (2000) Assessment of the protection value of remaining indigenous forests and woodlands in Swaziland. Mbabane del Valle I, Astorkiza K (2018) Exploring cross correlation among diversity indices. Fish Res 204: 103–115. https://doi.org/10.1016/j.fishres.2018.02.008

Di Febbraro M, Sallustio L, Vizzarri M et al (2018) Expert-based and correlative models to map habitat quality: which gives better support to conservation planning? Glob Ecol Conserv 16: e00513. https://doi.org/10.1016/j.gecco.2018.e00513

Dlamini W (2011a) Probabilistic spatio-temporal assessment of vegetation vulnerability to climate change in Swaziland. Glob Chang Biol 17:1425–1441. https://doi.org/10.1111/j.1365-2486.2010.02317.x

Dlamini WM (2011b) A data mining approach to predictive vegetation mapping using probabilistic graphical models. Eco Inform 6:111–124. https://doi.org/10.1016/j.ecoinf.2010.12.005

Dlamini WM (2016) Analysis of deforestation patterns and drivers in Swaziland using efficient Bayesian multivariate classifiers. Model Earth Syst Environ 2:1–14. https://doi.org/10.1007/s40808-016-0231-6

Dlamini WM (2017) Mapping forest and woodland loss in Swaziland: 1990–2015. Remote Sens Appl Soc Environ 5:45–53. https://doi.org/10.1016/j.rsase.2017.01.004

Dormann CF, Elith J, Bacher S et al (2013) Collinearity: a review of methods to deal with it and a simulation study evaluating their performance. Ecography 36:027–046. https://doi.org/10.1111/j.1600-0587.2012.07348.x

Elith J, Leathwick JR (2009) Species distribution models: ecological explanation and prediction across space and time. Annu Rev Ecol Evol Syst 40:677–697. https://doi.org/10.1146/annurev.ecolsys.110308.120159

Escudero A, Iriondo JM, Torres ME (2003) Spatial analysis of genetic diversity as a tool for plant conservation. Biol Conserv 113:351–365. https://doi.org/10.1016/S0006-3207(03)00122-8

Facebook Connectivity Lab and Center for International Earth Science Information Network - CIESIN - Columbia University (2016) High Resolution Settlement Layer (HRSL). Source imagery for HRSL © 2016 DigitalGlobe. Accessed DAY MONTH YEAR. In: Columbia University. https://ciesin.columbia.edu/data/hrsl/#acknowledgements

Faith DP (1992) Conservation evaluation and phylogenetic diversity. Biol Conserv 61:1–10. https://doi.org/10.1016/0006-3207(92)91201-3

Fick SE, Hijmans RJ (2017) WorldClim 2: new 1-km spatial resolution climate surfaces for global land areas. Int J Climatol 37:4302–4315. https://doi.org/10.1002/joc.5086

Financing S (2013) Towards sustainable financing of protected areas: a brief overview of pertinent issues. Int J Biodiver Conserv 5:436–445. https://doi.org/10.5897/IJBC11.238

Fithian W, Elith J, Hastie T, Keith DA (2015) Bias correction in species distribution models: pooling survey and collection data for multiple species. Methods Ecol Evol 6:424–438. https://doi.org/10.1111/2041-210X.12242

Galpin E, Bolus H, Wood M et al (2002) A first check-list of flowering plants and ferns of the Transvaal and Swaziland. Ann Transv Mus 3:1–30

González-Orozco CE, Ebach MC, Laffan S et al (2014) Quantifying phytogeographical regions of Australia using geospatial turnover in species composition. PLoS One 9:e92558. https://doi.org/10.1371/journal.pone.0092558

González-Orozco CE, Pollock LJ, Thornhill AH et al (2016) Phylogenetic approaches reveal biodiversity threats under climate change. Nat Clim Chang 6:1110–1114. https://doi.org/10.1038/nclimate3126

Graham CH, Elith J, Hijmans RJ et al (2008) The influence of spatial errors in species occurrence data used in distribution models. J Appl Ecol 45:239–247. https://doi.org/10.1111/j.1365-2664.2007.01408.x

Grimshaw JM (2001) What do we really know about the Afromontane Archipelago? In: Systematics and geography of plants, pp 949–957

Guisan A, Rahbek C (2011) SESAM - a new framework integrating macroecological and species distribution models for predicting spatio-temporal patterns of species assemblages. J Biogeogr 38:1433–1444

Guisan A, Tingley R, Baumgartner JB et al (2013) Predicting species distributions for conservation decisions. Ecol Lett 16:1424–1435. https://doi.org/10.1111/ele.12189

Hackel JD, Carruthers EJ (1993) Swaziland's twentieth century wildlife preservation efforts: the present as a continuation of the past. Environ Hist Rev 17:61–84. https://doi.org/10.2307/3984605

Hinchliff CE, Smith SA, Allman JF et al (2015) Synthesis of phylogeny and taxonomy into a comprehensive tree of life. Proc Natl Acad Sci U S A 112:12764–12769. https://doi.org/10.1073/pnas.1423041112

Hulvey KB, Hobbs RJ, Standish RJ et al (2013) Benefits of tree mixes in carbon plantings. Nat Clim Chang 3:869–874. https://doi.org/10.1038/nclimate1862

Jarnevich CS, Stohlgren TJ, Kumar S et al (2015) Caveats for correlative species distribution modeling. Eco Inform 29:6–15

Jarvis A, Reuter HI, Nelson A, Guevara E (2008) Hole-filled SRTM for the globe Version 4, available from the CGIAR-CSI SRTM 90m Database

Kreft H, Jetz W (2010) A framework for delineating biogeographical regions based on species distributions. J Biogeogr 37:2029–2053. https://doi.org/10.1111/j.1365-2699.2010.02375.x

Kumar S, Stecher G, Suleski M, Hedges SB (2017) TimeTree: a resource for timelines, timetrees, and divergence times. Mol Biol Evol 34:1812–1819. https://doi.org/10.1093/molbev/msx116

Laffan SW, Lubarsky E, Rosauer DF (2010) Biodiverse, a tool for the spatial analysis of biological and related diversity. Ecography 33:643–647. https://doi.org/10.1111/j.1600-0587.2010.06237.x

Laity T, Laffan SW, González-Orozco CE et al (2015) Phylodiversity to inform conservation policy: an Australian example. Sci Total Environ 534:131–143. https://doi.org/10.1016/j.scitotenv.2015.04.113

Liang Y, He HS, Fraser JS, Wu ZW (2013) Thematic and spatial resolutions affect model-based predictions of tree species distribution. PLoS One 8:e67889. https://doi.org/10.1371/journal.pone.0067889

Loffler L, Loffler P (2005) Swaziland Tree Atlas—including selected shrubs and climbers. Southern African Botanical Diversity Network (SABONET), Pretoria

Marchese C (2015) Biodiversity hotspots: a shortcut for a more complicated concept. Glob Ecol Conserv 3:297–309. https://doi.org/10.1016/j.gecco.2014.12.008

Margules CR, Pressey RL (2000) Systematic conservation planning. Nature 405:243–253. https://doi.org/10.1038/35012251

Mazel F, Mooers AO, Riva GVD, Pennell MW (2017) Conserving phylogenetic diversity can be a poor strategy for conserving functional diversity. Syst Biol 66:1019–1027. https://doi.org/10.1093/sysbio/syx054

McKerrow AJ, Tarr NM, Rubino MJ, Williams SG (2018) Patterns of species richness hotspots and estimates of their protection are sensitive to spatial resolution. Divers Distrib 24:1464–1477. https://doi.org/10.1111/ddi.12779

Mcshea WJ (2014) What are the roles of species distribution models in conservation planning? Environ Conserv 41:93–96. https://doi.org/10.1017/S0376892913000581

Meyer C, Weigelt P, Kreft H (2016) Multidimensional biases, gaps and uncertainties in global plant occurrence information. Ecol Lett 19:992–1006

Mi C, Huettmann F, Guo Y et al (2017) Why choose Random Forest to predict rare species distribution with few samples in large undersampled areas? Three Asian crane species models provide supporting evidence. PeerJ 2017:e2849. https://doi.org/10.7717/peerj.2849

Millar TR, Heenan PB, Wilton AD et al (2017) Spatial distribution of species, genus and phylogenetic endemism in the vascular flora of New Zealand, and implications for conservation. Aust Syst Bot 30:134–147. https://doi.org/10.1071/SB16015

Mishler BD, Knerr N, González-Orozco CE et al (2014) Phylogenetic measures of biodiversity and neo-and paleo-endemism in Australian acacia. Nat Commun 5:4473. https://doi.org/10.1038/ncomms5473

Mittermeier RA, van Dijk PP, Rhodin AGJ, Nash SD (2015) Turtle hotspots: an analysis of the occurrence of tortoises and freshwater turtles in biodiversity hotspots, high-biodiversity wilderness areas, and turtle priority areas. CEMEX, Mexico City
Moradi S, Sheykhi Ilanloo S, Kafash A, Yousefi M (2019) Identifying high-priority conservation areas for avian biodiversity using species distribution modeling. Ecol Indic 97:159–164. https://doi.org/10.1016/j.ecolind.2018.10.003
Moraes Mónica R, Ríos-Uzeda B, Moreno LR et al (2014) Using potential distribution models for patterns of species richness, endemism, and phytogeography of palm species in Bolivia. Trop Conserv Sci 7:45–60. https://doi.org/10.1177/194008291400700109
Mori AS (2018) Environmental controls on the causes and functional consequences of tree species diversity. J Ecol 106:113–125. https://doi.org/10.1111/1365-2745.12851
Morris EK, Caruso T, Buscot F et al (2014) Choosing and using diversity indices: insights for ecological applications from the German Biodiversity Exploratories. Ecol Evol 4:3514–3524. https://doi.org/10.1002/ece3.1155
Morrone JJ (2018) The spectre of biogeographical regionalization. J Biogeogr 45:282–288. https://doi.org/10.1111/jbi.13135
Murdoch G (1968) Soils and land capability in Swaziland. Mbabane
Muyambi F (2016) Swaziland land cover, land cover change analysis and vegetation types for 1990, 2000, 2010 and 2015. Lobamba
Myers N, Mittermeler RA, Mittermeler CG et al (2000) Biodiversity hotspots for conservation priorities. Nature 403:853–858. https://doi.org/10.1038/35002501
Naimi B (2017) Package "usdm". Uncertainty analysis for species distribution models. R-Cran 18
Naimi B, Hamm NAS, Groen TA et al (2014) Where is positional uncertainty a problem for species distribution modelling? Ecography 37:191–203. https://doi.org/10.1111/j.1600-0587.2013.00205.x
Ng WT, Cândido de Oliveira Silva A, Rima P et al (2018) Ensemble approach for potential habitat mapping of invasive Prosopis spp. in Turkana, Kenya. Ecol Evol 8:11921–11931. https://doi.org/10.1002/ece3.4649
Phillips SB, Aneja VP, Kang D, Arya SP (2006) Modelling and analysis of the atmospheric nitrogen deposition in North Carolina. Int J Glob Environ Issues 6:231–252. https://doi.org/10.1016/j.ecolmodel.2005.03.026
Phillips SJ, Dudík M, Schapire RE (2019) Maxent software for modeling species niches and distributions (Version 3.4.1). Available from url: http://biodiversityinformatics.amnh.org/open_source/maxent/
Pollock LJ, Rosauer DF, Thornhill AH et al (2015) Phylogenetic diversity meets conservation policy: small areas are key to preserving eucalypt lineages. Philos Trans R Soc B Biol Sci 370: 1–10. https://doi.org/10.1098/rstb.2014.0007
Pollock LJ, Thuiller W, Jetz W (2017) Large conservation gains possible for global biodiversity facets. Nature 546:141–144. https://doi.org/10.1038/nature22368
Pott R (1920) Addendum to the first check-list of the flowering plants and ferns of the Transvaal and Swaziland. Ann Transv Mus 6:119–135
Ratcliffe S, Liebergesell M, Ruiz-Benito P et al (2016) Modes of functional biodiversity control on tree productivity across the European continent. Glob Ecol Biogeogr 25:251–262. https://doi.org/10.1111/geb.12406
Remmelzwaal A, Vilakati JD (1994) Land Tenure Map of Swaziland, Scale 1:250,000. Mbabane
Robertson MP, Barker NP (2006) A technique for evaluating species richness maps generated from collections data. S Afr J Sci 102:77–84
Rondinini C, Wilson KA, Boitani L et al (2006) Tradeoffs of different types of species occurrence data for use in systematic conservation planning. Ecol Lett 9:1136–1145. https://doi.org/10.1111/j.1461-0248.2006.00970.x
Roques KG (2002) A preliminary field assessment of protection worthy areas of Swaziland. Mbabane

Rosauer DF, Jetz W (2015) Phylogenetic endemism in terrestrial mammals. Glob Ecol Biogeogr 24: 168–179. https://doi.org/10.1111/geb.12237

Rosauer D, Laffan SW, Crisp MD et al (2009) Phylogenetic endemism: a new approach for identifying geographical concentrations of evolutionary history. Mol Ecol 18:4061–4072. https://doi.org/10.1111/j.1365-294X.2009.04311.x

Rosauer DF, Pollock LJ, Linke S, Jetz W (2017) Phylogenetically informed spatial planning is required to conserve the mammalian tree of life. Proc R Soc B Biol Sci 284:20170627. https://doi.org/10.1098/rspb.2017.0627

Santo-Silva EE, Santos BA, Arroyo-Rodríguez V et al (2018) Phylogenetic dimension of tree communities reveals high conservation value of disturbed tropical rain forests. Divers Distrib 24:776–790. https://doi.org/10.1111/ddi.12732

Sardà-Palomera F, Brotons L, Villero D et al (2012) Mapping from heterogeneous biodiversity monitoring data sources. Biodivers Conserv 21:2927–2948. https://doi.org/10.1007/s10531-012-0347-6

Scherson RA, Thornhill AH, Urbina-Casanova R et al (2017) Spatial phylogenetics of the vascular flora of Chile. Mol Phylogenet Evol 112:88–95. https://doi.org/10.1016/j.ympev.2017.04.021

Schmitt S, Pouteau R, Justeau D et al (2017) ssdm: an r package to predict distribution of species richness and composition based on stacked species distribution models. Methods Ecol Evol 8: 1795–1803. https://doi.org/10.1111/2041-210X.12841

Schulze RE, Maharaj M, Warburton ML et al (2008) South African atlas of climatology and agrohydrology. Water Research Commission, Pretoria, RSA WRC Report

Soto-Navarro C, Ravilious C, Arnell A et al (2020) Mapping co-benefits for carbon storage and biodiversity to inform conservation policy and action. Philos Trans R Soc B Biol Sci 375: 20190128. https://doi.org/10.1098/rstb.2019.0128

Swaziland Environment Authority (2016) Swaziland's second national biodiversity strategy and action plan. Mbabane

Sweet RJ, Khumalo S (1994) Range resources and grazing potentials in Swaziland. FAO report, Ministry of Agriculture and Cooperatives

Syfert MM, Smith MJ, Coomes DA (2013) The effects of sampling bias and model complexity on the predictive performance of MaxEnt species distribution models. PLoS One 8:e55158. https://doi.org/10.1371/journal.pone.0055158

Tucker CM, Cadotte MW (2013) Unifying measures of biodiversity: understanding when richness and phylogenetic diversity should be congruent. Divers Distrib 19:845–854. https://doi.org/10.1111/ddi.12087

van Waveren EJ, Nhlengetfwa JV (1992) Agro-climatic characterization of Swaziland. Food and Agriculture Organization/United Nations Development Programme/Ministry of Agriculture and Co-operatives, Mbabane

van Wyk AE, Smith GF (2001) Regions of floristic endemism in Southern Africa: a review with emphasis on succulents. Umdaus Press, Hatfield

Vegter JR (1995) Geology map of South Africa with simplified lithostratigraphy for geohydrological use. Pretoria

White F (1981) The history of the Afromontane archipelago and the scientific need for its conservation. Afr J Ecol 19:33–54. https://doi.org/10.1111/j.1365-2028.1981.tb00651.x

Wisz MS, Hijmans RJ, Li J et al (2008) Effects of sample size on the performance of species distribution models. Divers Distrib 14:763–773. https://doi.org/10.1111/j.1472-4642.2008.00482.x

Xu Y, Shen Z, Ying L et al (2017) Hotspot analyses indicate significant conservation gaps for evergreen broadleaved woody plants in China. Sci Rep 7:1859. https://doi.org/10.1038/s41598-017-02098-0

Xu Y, Huang J, Lu X et al (2019) Priorities and conservation gaps across three biodiversity dimensions of rare and endangered plant species in China. Biol Conserv 229:30–37. https://doi.org/10.1016/j.biocon.2018.11.010

Yessoufou K, Davies TJ (2016) Reconsidering the loss of evolutionary history: how does non-random extinction prune the tree-of-life? In: Pellens R, Grandcolas P (eds) Biodiversity conservation and phylogenetic systematics. Springer, Cham, pp 57–80

Zellmer AJ, Claisse JT, Williams CM et al (2019) Predicting optimal sites for ecosystem restoration using stacked-species distribution modeling. Front Mar Sci 6:3. https://doi.org/10.3389/fmars.2019.00003

Zhao L, Li J, Liu H, Qin H (2016) Distribution, congruence, and hotspots of higher plants in China. Sci Rep 6:19080. https://doi.org/10.1038/srep19080

Zomer RJ, Trabucco A, Bossio DA, Verchot LV (2008) Climate change mitigation: a spatial analysis of global land suitability for clean development mechanism afforestation and reforestation. Agric Ecosyst Environ 126:67–80. https://doi.org/10.1016/j.agee.2008.01.014

Improving the Conservation Status of a Threatened Tree (*Acer sikkimensis* Miq. syn. *Acer hookeri* Miq.) Through Standardization of Seed Germination Protocol and Using Ecological Niche Modeling

9

Aditya Pradhan and Arun Chettri

Abstract

Threatened plant conservation is fraught with numerous intrinsic and extrinsic challenges, which vary across species. For example, the absence of a standardized regeneration protocol significantly hinders the mass multiplication of seedlings, while insufficient distribution records make it difficult to model their potential distribution area for reintroduction. We standardized the germination protocol in *Acer sikkimensis*, a threatened tree of northeast India, and modeled the potential distribution area for its reintroduction in Sikkim, northeast India. The seeds collected for the germination experiment were given cold treatment (5 °C) for 3 months to break dormancy. Subsequently, the seeds were soaked in different concentrations of gibberellic acid (GA_3), abscisic acid (ABA), indoleacetic acid (IAA), and kinetin to identify the treatment that enhances germination. Control was maintained by soaking the seeds in deionized water. The potential distribution area of the species was modeled using maximum entropy distribution modeling (MaxEnt) software and averaged monthly normalized difference vegetation index (NDVI) data for the study area. The treatment with GA_3 and kinetin improved seed germination significantly compared to ABA and IAA. The MaxEnt model performed well with less number of occurrence records. The model predicted that only 38 km^2 area in Sikkim was highly suitable for the species where the species can be reintroduced. This study's seed germination protocol is less expensive than the existing micro-propagation techniques. The

A. Pradhan
Department of Botany, School of Basic Sciences, SRM University Sikkim, Gangtok, Sikkim, India

A. Chettri (✉)
Taxonomy and Biodiversity Laboratory, Department of Botany, Sikkim University, Gangtok, Sikkim, India
e-mail: achettri01@cus.ac.in

S. Dhyani et al. (eds.), *Ecosystem and Species Habitat Modeling for Conservation and Restoration*, https://doi.org/10.1007/978-981-99-0131-9_9

proposed germination protocol and the potential distribution area map can be helpful for conservationists, scientists, and local nongovernmental organizations to conserve *A. sikkimensis* and improve its conservation status.

Keywords

Acer · Conservation · Darjeeling · ENM · Regeneration · Sikkim Himalaya

9.1 Introduction

Conservation of threatened plants has numerous challenges which range from species-specific regeneration issues to the availability of suitable habitats. Specifically, the nonavailability of a standardized regeneration protocol hinders the mass multiplication of seedlings, while an inadequate number of distribution records make it difficult to summarize the species' environmental requirements or niche, making it difficult to model their potential distribution area/habitats. However, overcoming such challenges through effective methods can help successfully conserve such species.

Multiplication of plants through various asexual (e.g., cuttings, layering, grafting, budding, and other nonconventional means such as tissue culture) and sexual means of propagation (i.e., through seeds) is important to reinforce the dwindling population of threatened species. Plants propagated through seed germination are expected to be more resilient to abiotic and biotic stresses and also maintain the genetic diversity of the species in the long term. However, the germination response of seeds on the forest floor is governed by the surrounding environmental factors. For example, seeds often fail to break their dormancy because of the nonavailability of suitable environmental conditions. Therefore, it is important to understand the optimal conditions required and also to overcome the mechanical barriers which prevent the germination of seeds.

The availability of suitable areas is important for the success of species conservation programs. Such areas can be identified through field inspection, environmental characterization, and matching at a local scale, while predictive modeling tools can be used in the case of larger areas at a landscape or regional scale. In this respect, ecological niche modeling (ENM) has emerged as an effective tool to identify such suitable areas (Elith and Leathwick 2009). Mapping potentially suitable habitat for threatened and endangered species is critical for monitoring and restoring their declining native populations (Guisan and Zimmermann 2000). ENM aids in the identification of areas for species reserves, reintroduction, and the development of effective species conservation measures (Elith et al. 2006; Peterson 2006). Several ENM methods are available to predict potentially suitable habitats for a species (Guisan and Thuiller 2005). However, they are sensitive to sample size and may fail to predict threatened species distribution accurately if the sample size is small (Wisz et al. 2008). Moreover, there are far fewer examples of predictive models being used for rare and endangered plant species in India (Kumar and Stohlgren 2009; Ray et al.

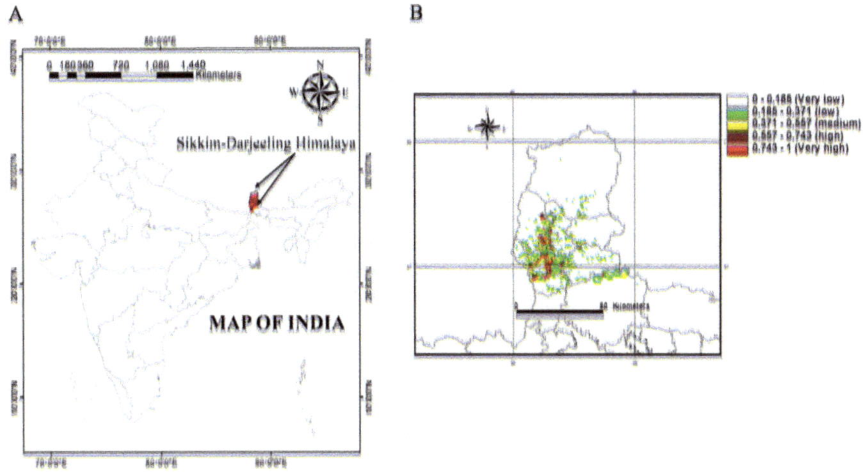

Fig. 9.1 (**a**) Study site, (**b**) predicted distribution map for *A. sikkimensis*

2011; Adhikari et al. 2012; Babar et al. 2012; Jaryan et al. 2013; Thriveni et al. 2015; Sreekumar et al. 2016; Pradhan et al. 2020). However, the maximum entropy (MaxEnt) model has shown a high success rate with sample sizes as small as five (Pearson et al. 2007; Thorn et al. 2009).

Maples belonging to the genus *Acer* are a crucial element of temperate forests (Tanai 1978). However, the maple species in the Himalayan region are under threat because of anthropogenic activities and climate change (Rana et al. 2011). Therefore, appropriate measures are needed for the conservation of this genus. There are approximately 114 species in the genus, distributed mainly in temperate climates. Thirteen *Acer* species have been described from the Darjeeling and Sikkim Himalayan region (Lama et al. 2015). *A. hookeri* Miq.—endemic to the Sikkim Himalaya—is listed as endangered in the Red Data Book of Indian Plants (Nayar and Sastry 1990; Adhikari et al. 2018). Since *A. hookeri* is currently considered a synonym of *A. sikkimensis*, the correct name *A. sikkimensis* will be used in place of *A. hookeri* hereafter in this study. The species is endemic to the state of Sikkim and the Darjeeling district of West Bengal, India (Fig. 9.1a).

The present study had the following objectives: (1) to standardize the seed germination protocol in *A. sikkimensis* and (2) to predict the distribution of suitable habitats of *A. sikkimensis* in its native range.

9.2 Materials and Method

9.2.1 Field Survey and Seed Collection

Field survey was conducted during the flowering (May–June) and fruiting (August–September) months in the state of Sikkim and the Darjeeling district of West Bengal (Fig. 9.1a). Seeds were collected from two mature trees, and associated ecological parameters such as habitat status, associated species, and phenological characteristics were noted. In addition, the fruit type, color, dispersal agent, and various threats the species faced were also recorded. Geocoordinates of the species were collected using a Global Positioning System (GPS) from six locations in the Sikkim Himalayas with an accuracy of <10 m, which were used to model the distribution of potential habitats.

9.2.2 Seed Moisture Content

The moisture content of fresh seeds was determined by the low constant temperature oven drying method (ISTA 1996). Seeds of *A. sikkimensis* were placed in an oven maintaining a temperature of 103 °C for 17 h. The moisture content as a percentage by weight (fresh weight basis) was calculated using the formula:

$$\%\text{seed moisture content} = \frac{M2 - M3}{M2 - M1} \times 100$$

where:

$M1$ = weight of the container with cover in gm
$M2$ = weight of the container with cover and seeds before drying
$M3$ = weight of the container with cover and seeds after drying

9.2.3 Seed Germination

Seeds were collected in October after they had reached maturity. The seeds were kept in a refrigerator at 5 °C for 3 months to break the dormancy (Yilmaz 2007). Before the germination experiment, the seeds were soaked in distilled water for 72 h until fully imbibed to remove the pericarp and thin papery testa (Phartyal et al. 2003). Germination experiments were performed in Petri dishes (9 cm diameter) lined with two filter papers. Seeds were soaked in different concentrations of gibberellic acid (GA_3) (100 μM, 200 μM, and 500 μM), abscisic acid (ABA) (100 μM, 200 μM, and 500 μM), indoleacetic acid (IAA) (100 μM, 200 μM, and 500 μM), and kinetin (100 μM, 200 μM, and 500 μM). Seeds soaked in deionized (2.5–3 mL) water were used as a control. The germination experiment was carried out at 25–30 °C. For each of these treatments, 3 replicates of 100 seeds were

Fig. 9.2 (**a**) Natural habitat of *Acer sikkimensis*, (**b**) flowering twig, (**c**) stages of seed germination, (**d**) germinating winged seed, (**e**) germinated seed transferred, (**f**) sapling

maintained. Seeds with a 1-mm-long visible radicle were considered germinated and counted three times a week. The germinated seeds were then placed in a germinating tray before being moved to the greenhouse (Fig. 9.2).

9.2.4 Predictor Variables

The model was developed using the average normalized difference vegetation index (NDVI) raster data for 12 months, i.e., January to December, obtained from GLCF (Global Land Cover Facility) (University of Maryland, USA) and Shuttle Radar Topography Mission (SRTM) elevation data obtained from worldclim.org. The NDVI and elevation data had a resolution of 30 arc seconds. The 12 NDVI variables were first subjected to correlated tests ($r > 0.9$) using ENM Tools 1.3 software (Warren et al. 2010). Thus out of 12 NDVI variables, 11 were used to model the distribution of *A. sikkimensis* in Sikkim Himalaya along with altitude.

9.2.5 Ecological Niche Modeling

The model was created using maximum entropy modeling (Phillips et al. 2006). We executed five bootstrap runs for the species to derive an optimized model. All other parameters were left at their default values as the program is already calibrated on various species datasets (Phillips and Dudik 2008). The average, maximum, minimum, median, and standard deviation were calculated for the replicated runs.

9.2.6 Model Evaluation

Model quality was evaluated based on area under the curve (AUC) value (Thuller et al. 2005), and the model was graded as poor ($\mathrm{AUC} < 0.8$), fair ($0.8 < \mathrm{AUC} < 0.9$), good ($0.9 < \mathrm{AUC} < 0.95$), and very good ($0.95 < \mathrm{AUC} < 1.0$).

9.2.7 Identification of Potential Habitats

The predicted ENM distribution map was exported in KMZ format using DIVA-GIS ver. 7.5 (www.diva-gis.org). The KMZ files were overlaid on Google Earth to identify the habitat of *A. sikkimensis* in Sikkim Himalaya. Subsequently, field surveys were undertaken in the identified areas to determine the status of the habitats in terms of anthropogenic disturbances.

9.3 Result

The species inhabited temperate mixed deciduous forest dominated by other tree species, viz., *Exbucklandia populnea*, *Edgeworthia gardneri*, *Castanopsis indica*, *Castanopsis tribuloides*, *Acer campbellii*, *Engelhardia spicata*, *Cryptomeria japonica*, and *Symplocos* spp.; shrubs, viz., *Viburnum erubescens*, *Gaultheria* spp., and *Rubus ellipticus*; and herbaceous species, viz., *Anaphalis contorta*, *Anaphalis*

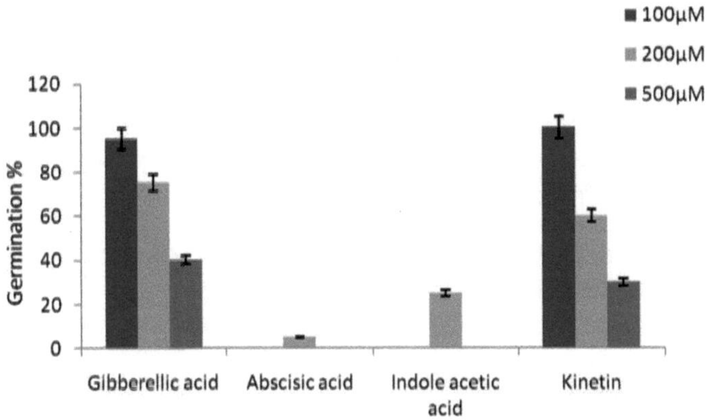

Fig. 9.3 Germination percentage of seeds at different concentrations (i.e., 100 μM, 200 μM, 500 μM) of GA_3, ABA, IAA, and kinetin

margaritacea, *Impatiens glandulifera*, *Impatiens stenantha*, and *Swertia bimaculata*.

Only six populations of *A. sikkimensis* were recorded in and around Senchal Wildlife Sanctuary, Darjeeling, at an altitude ranging from 2000 to 2500 m asl and at Rinchenpong, West Sikkim, at an altitude ranging from 1700 to 2200 m asl. Each population consisted of 10–20 individuals, and only a few were in the flowering and fruiting stages. The flowering occurs from May to June and the fruiting occurs between August and September. The fruit attains reddish-brown color upon maturity. However, no seedlings were observed in any of the survey locations. The natural population of the species is under severe abiotic anthropogenic pressure due to tree felling.

The mean seed weight of the species was 11.9 g with a standard deviation of 0.2. The seed moisture content in percent was 2.692 ± 0.19 (without wings) and 3.68 ± 0.06 (with wings). Seed germination in the species is epigeal. Increased GA_3 and kinetin concentrations resulted in a lower percentage of seeds germinating, while the effects of ABA and IAA on seed germination were insignificant. Treatment with GA_3 and kinetin was very effective in improving seed germination, with germination rates of 95% and 100% of seeds treated with GA_3 and kinetin at 100 μM concentrations, respectively. The germination rate, however, decreased as the concentrations of GA_3 and kinetin increased. Only 5% and 25% of seeds germinated after being treated with ABA and IAA at 200 μM concentrations, respectively. IAA and ABA at concentrations of 100 μM and 500 μM, respectively, inhibited seed germination (Figs. 9.2 and 9.3). The seeds with wings had a low germination rate after being treated with different hormones.

The predicted distribution map for *A. sikkimensis* is given in Fig. 9.1. The AUC_{train} and AUC_{test} values for *A. sikkimensis* were both satisfactory (AUC_{train} = 0.9923 ± 0.0042 and AUC_{test} = 0.9484 ± 0.0925). NDVI for the month of April and July was the most influential of the input environmental variables, contributing 35% and 26.9% to the MaxEnt model, respectively. The

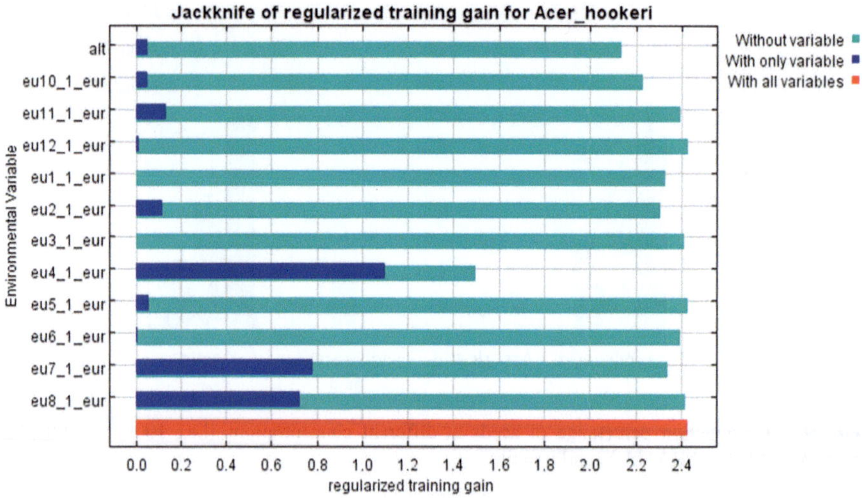

Fig. 9.4 Results of jackknife of regularized training gain for *A. sikkimensis*

Table 9.1 Relative contribution and permutation importance of the predictor variables to the MaxEnt model

Predictor variables (NDVI codes)	Months	Percent contribution	Permutation importance
eu4_1_eur	April	35	40.9
eu7_1_eur	July	26.9	12.4
altitude	–	11.1	23.4
eu10_1_eur	October	9.5	10.5
eu8_1_eur	August	6	5.2
eul_1_eur	January	5.1	0.1
eu2_1_eur	February	3.8	6.5
eu6_1_eur	June	1.5	0.3
eu11_1_eur	November	0.8	0.1
eu3_1_eur	March	0.2	0.5
eu5_1_eur	May	0.1	0
eu12_1_eur	December	0	0

remaining layers collectively contributed 38.1% to the species' habitat model (Fig. 9.4 and Table 9.1). In terms of permutation importance, NDVI for the month of April had the greatest impact on the habitat model, accounting for 40.9% of the total, while the rest contributed 59.1% (Table 9.1). In the Sikkim Himalaya, a total potential area of 1738 km^2 was predicted to be suitable for *A. sikkimensis*.

The majority of the area is classified as low suitability and covers approximately 1120 km^2. The area with the highest suitability was limited to about 38 km^2. High and medium suitability areas were limited to 162 km^2 and 418 km^2, respectively (Table 9.2). The NDVI for the months of April and July was used to determine the

Table 9.2 Habitat suitability classes of *A. sikkimensis* in Sikkim derived from the MaxEnt model

Habitat suitability classes	Area (km^2)	Area (%)
Low	1120	64.44
Medium	418	24.05
High	162	9.32
Very high	38	2.18

distribution of potential habitat for *A. sikkimensis* in its native range in this study. Interestingly, the NDVI for the months of April and July, which contributed the most to the habitat model, corresponds to the species' flowering and fruiting months, demonstrating the importance of phenology in determining the species' distribution.

9.4 Discussion

The study revealed that storing seeds in cold (5 °C) before germination considerably improved *A. sikkimensis* performance by elevating seed germination rate. Cold treatment was necessary because the *Acer* seed remains dormant (Kanazashi et al. 2014). Furthermore, due to the presence of hard seed cover in the form of wings, seed germination was poor. As a result, removing the hard covering improved seed germination. The seeds treated with different concentrations (i.e., 100 μM, 200 μM, 500 μM) of GA_3, ABA, IAA, and kinetin showed different levels of germination. Increased concentrations of GA_3 and kinetin showed a decrease in germination percentage. The germination percentage was almost 100% at 100 μM concentration of GA_3 and kinetin. Seeds did not germinate when treated to ABA and IAA at concentrations of 100 μM and 500 μM, respectively. However, seed germination was negligible when treated with 200 μM of ABA and IAA. Thus it can be concluded that in *A. sikkimensis*, seeds treated with GA_3 and kinetin accelerated seed germination, while seeds treated with ABA inhibited seed germination and development. The same has been reported for *A. pseudoplatanus* L. and *A. platanoides* (Pinfield and Stobart 1972; Tillberg and Pinfield 1981; Pinfield and Gwarazimba 1992).

In predicting the potentially suitable habitat for *A. sikkimensis*, the NDVI for April and July was critical. Since different environmental factors such as geology, soil, and climate have a credible impact on vegetation indices, the role of such environmental factors in determining the species' habitat suitability could be explained using NDVI layers (Soleimani et al. 2008). In the present study, NDVI for April and July contributed the most which correspond to leafing and flowering phase of *A. sikkimensis*. After a prolonged winter that begins in December, new leaves begin to appear in March and April, which is the reason April month is contributing more. As a result, NDVI is an effective alternative variable for reflecting the net results of multiple environmental conditions that affect the probability distribution of *A. sikkimensis*.

In the Sikkim Himalaya region, just 38 km^2 (or 2.18%) of the total 1738 km^2 predicted to be suitable falls into the very high suitable category (Table 9.2). The majority of the area in the West Bengal district of Darjeeling predicted to be suitable is covered by the Senchal Wildlife Sanctuary, whereas the majority of the area in the Sikkim state is covered by the West district. Regions like Kaluk, Rishi, Lingzo, Dalep, and Yuksom in the West district of Sikkim fall under very high suitability class and could serve as sites for in situ conservation. Therefore, it should be a primary concern to attempt to reintroduce seedlings in the areas predicted by ENM.

9.5 Conclusion

The findings presented on seed germination would aid conservationists, scientists, and local nongovernmental organizations in the mass multiplication of seedlings and reintroduction of species in their natural habitat. When compared to other micro-propagation techniques, which are both expensive and time-consuming, the process of seed germination presented here is more cost-effective. We also demonstrated how the MaxEnt model can be successfully used to predict the suitable habitat of threatened and endangered species, and we successfully superimposed the predicted habitat suitability map and identified sites for reintroduction using Google Earth. The study presented here is very promising for conservation planners and biologists working on the conservation of this species. The potential habitat distribution map for *A. sikkimensis* can aid in natural habitat planning and restoration for more effective conservation. Such forest areas predicted to be suitable would serve as in situ conservation sites for the species' reintroduction and recovery.

Acknowledgments The financial support received from DBT, Govt. of India, under a project entitled "Preventing extinction and improving conservation status of threatened plants through application of biotechnological tools" (Sanctioned letter No. BT/Env/BC/01/2010; 23rd March, 2012) is gratefully acknowledged. We also thank the Head, Department of Botany, Sikkim University, for providing the necessary laboratory facilities.

References

Adhikari D, Barik SK, Upadhaya K (2012) Habitat distribution modelling for reintroduction of Ilex khasiana Purk., a critically endangered tree species of northeastern India. Ecol Eng 40:37–43. https://doi.org/10.1016/j.ecoleng.2011.12.004

Adhikari D, Reshi Z, Datta BK, Samant SS, Chettri A, Upadhaya K, Shah MA, Singh PP, Tiwary R, Majumdar K, Pradhan A (2018) Inventory and characterization of new populations through ecological niche modelling improve threat assessment. Curr Sci 114(03):519–531. https://doi.org/10.18520/cs/v114/i03/519-531

Babar S, Amarnath G, Reddy CS, Jentsch A, Sudhakar S (2012) Species distribution models: ecological explanation and prediction of an endemic and endangered plant species (Pterocarpus santalinus L. f.). Curr Sci 102:1157–1165

Elith J, Leathwick JR (2009) Species distribution models: ecological explanation and prediction across space and time. Annu Rev Ecol Evol Syst 40:677–697. https://doi.org/10.1146/annurev.ecolsys.110308.120159

Elith J et al (2006) Novel methods improve prediction of species' distributions from occurrence data. Ecography (Cop) 2:129–151. https://doi.org/10.1111/j.2006.0906-7590.04596.x

Guisan A, Thuiller W (2005) Predicting species distribution: offering more than simple habitat models. Ecol Lett 8:993–1009. https://doi.org/10.1111/j.1461-0248.2005.00792.x
Guisan A, Zimmermann NE (2000) Predictive habitat distribution models in ecology. Ecol Model 135:147–186. https://doi.org/10.1016/S0304-3800(00)00354-9
ISTA (1996) International rules for seed testing 1996. The International Seed Testing Association, Zurich
Jaryan V, Datta A, Uniyal SK, Kumar A, Gupta RC (2013) Modelling potential distribution of Sapium sebiferum—an invasive tree species in western Himalaya. Curr Sci 105:1282–1288
Kanazashi A, Nagamitsu T, Suzuki W (2014) Seed dormancy and germination characteristics in relation to the regeneration of Acer pycnanthum, a vulnerable tree species in Japan. J For Res 20:160–166. https://doi.org/10.1007/s10310-014-0451-4
Kumar S, Stohlgren TJ (2009) Maxent modeling for predicting suitable habitat for threatened and endangered tree Canacomyrica monticola in New Caledonia. J Ecol Nat Environ 1:94–98
Lama D, Moktan S, Das AP (2015) Diversity and distribution of Acer Linnaeus (Sapindaceae) in Darjiling and Sikkim Himalayas. Pleione 9:61–73
Nayar MP, Sastry ARK (1990) Red data book of Indian plants, vol 3. Botanical Survey of India, Kolkata
Pearson RG, Raxworthy CJ, Nakamura M, Peterson AT (2007) Predicting species distributions from small numbers of occurrence records: a test case using cryptic geckos in Madagascar. J Biogeogr 34:102–117. https://doi.org/10.1111/j.1365-2699.2006.01594.x
Peterson AT (2006) Uses and requirements of Ecological Niche Models and related distributional models. Biodivers Inform 3:59–72. https://doi.org/10.17161/bi.v3i0.29
Phartyal S, Thapliyal RC, Nayal JS, Joshi G (2003) Seed dormancy in Himalayan maple (Acer caesium) I: effect of stratification and phyto-hormones. Seed Sci Technol 31:1–11. https://doi.org/10.15258/sst.2003.31.1.01
Phillips JS, Dudik M (2008) Modeling of species distributions with Maxent: new extensions and a comprehensive evaluation. Ecography (Cop) 31:161–175. https://doi.org/10.1111/j.0906-7590.2008.5203.x
Phillips SJ, Anderson RP, Schapire RE (2006) Maximum entropy modeling of species geographic distributions. Ecol Eng 190:231–259. https://doi.org/10.1016/j.ecolmodel.2005.03.026
Pinfield NJ, Gwarazimba VEE (1992) Seed dormancy in Acer: the role of abscisic acid in the regulation of seed development in Acer platanoides L. Plant Growth Regul 11:293–299. https://doi.org/10.1007/BF00024568
Pinfield NJ, Stobart AK (1972) Hormonal regulation of germination and early seedling development in Acer pseudoplatanus (L.). Planta 104:134–145. https://doi.org/10.1007/BF00386990
Pradhan A, Adhikari D, Chettri A (2020) Predicting the distribution of suitable habitats for *Pandanus Unguifer* Hook.F.—a Dwarf endemic species from Sikkim Himalayas, through Ecological Niche Modeling. Int J Conserv Sci 11(1):145–152
Rana MS, Lal M, Samant SS (2011) Status and regeneration of Himalayan maple in the Himachal Pradesh: honing red list of plants. J Sustain For 30(8):775–789
Ray R, Gururaja KV, Ramchandra TV (2011) Predictive distribution modeling for rare Himalayan medicinal plant Berberis aristata DC. J Environ Biol 32(6):725–730
Soleimani K, Kordsavadkooh T, Muosavi SR (2008) The effect of environmental factors on vegetation changes using GIS (case study: Cherat Catchment, Iran). World Appl Sci J 3:95–100
Sreekumar VB, Sakthivel RS, Sreejith KA (2016) Distribution mapping and conservation of Rhopaloblaste augusta (Kurz) H. E. Moore in Nicobar Islands, India. Trop Ecol 57:271–277
Tanai T (1978) Taxonomical investigation of the living species of the genus Acer L., based on vein architecture. J Fac Sci 18:243–282
Thorn JS, Nijman V, Smith D, Nekaris KAI (2009) Ecological niche modelling as a technique for assessing threats and setting conservation priorities for Asian slow lorises (Primates: Nycticebus). Divers Distrib 15:289–298. https://doi.org/10.1111/j.1472-4642.2008.00535.x

Thriveni HN, Gunaga SV, Babu HNR, Vasudeva R (2015) Ecological niche modeling, population status and regeneration of Coscinium fenestratum colebr. (Menispermaceae): a medicinally important liana of the central Western Ghats. Trop Ecol 56:101–110

Thuller W, Richardson DM, Pysek P, Midley GF, Hughes GO, Rouget M (2005) Niche-based modelling as a tool for predicting the risk of alien plant invasions at a global scale. Glob Chang Biol 11:2234–2250. https://doi.org/10.1111/j.1365-2486.2005.01018.x

Tillberg E, Pinfield NJ (1981) The dynamics of Indole-3-acetic-acid in Acer platanoides seeds during stratification and germination. Physiol Plant 53:34–38. https://doi.org/10.1111/j.1399-3054.1981.tb05041.x

Warren D, Richard G, Turelli M (2010) ENMTools: a toolbox for comparative studies of environmental niche models. Ecography 33:607–611. https://doi.org/10.1111/j.1600-0587.2009.06142.x

Wisz MS, Hijmans RJ, Li J, Peterson AT, Graham CH, Guisan A (2008) Effects of sample size on the performance of species distribution models. Divers Distrib 14:763–773. https://doi.org/10.1111/j.1472-4642.2008.00482.x

Yilmaz M (2007) Depth of dormancy and desiccation tolerance in Acer trautvetteri Medv. seeds. Turk J Agric For 31:201–205

Ecological Niche Modeling of the Endemic Himalayan Near-Threatened Treeline Conifer *Abies spectabilis* (D.Don) Mirb. in the Indian Central Himalaya

10

Siddhartha Kaushal, Sharanjeet Kaur, Anshu Siwach, Prachi Sharma, Prem Lal Uniyal, Rajesh Tandon, Shailendra Goel, K. S. Rao, and Ratul Baishya

Abstract

Abies spectabilis (D.Don) Mirb. is an endemic Himalayan near-threatened coniferous species and the predominant treeline-forming species in the Indian Central Himalaya (Uttarakhand). The impact of climate change and anthropogenic activities is perilous to its habitat distribution. Accurate species habitat distribution is a prerequisite for efficient conservation planning. This vital information is still missing in the Indian Himalayan region for *A. spectabilis*. This study models the habitat suitability of *A. spectabilis* in Uttarakhand and discusses the management implications. Species occurrence records from primary and secondary sources were used with environment variables for habitat suitability modeling using the MaxEnt approach. Environment variables included bioclimatic (BCVs), topographic, edaphic, and anthropogenic variables. The BCVs from two global bioclimatic databases CHELSA (model 1) and WorldClim (model 2) were used. Models were validated using threshold-independent measures and further used to build habitat suitability maps. Both models performed optimally; however, model 2 had higher average values for the area under the curve (AUC) (0.980), partial AUC (0.977), and the AUC ratio (1.954) indicating higher predictive power. The precipitation of the driest month (BIO14) was the most important predictor variable under both models. The habitat suitability area for model 1 (16,324.57 km^2) was five times greater than model 2 (3323.86 km^2). Only 826.86 km^2 area (model 2) was highly suitable for *A. spectabilis*. The habitat

Supplementary Information The online version contains supplementary material available at https://doi.org/10.1007/978-981-99-0131-9_10.

S. Kaushal · S. Kaur · A. Siwach · P. Sharma · P. L. Uniyal · R. Tandon · S. Goel · K. S. Rao · R. Baishya (✉)
Department of Botany, University of Delhi, Delhi, India
e-mail: rbaishya@botany.du.ac.in

S. Dhyani et al. (eds.), *Ecosystem and Species Habitat Modeling for Conservation and Restoration*, https://doi.org/10.1007/978-981-99-0131-9_10

suitability is concentrated predominantly in a tight group in the northern region of Uttarakhand. Chamoli (8.43%), Rudraprayag (6.20%), and Uttarkashi (0.53%) were the top 3 districts with the highest percentage of high suitability regions under model 2. The results from the habitat suitability distribution of *A. spectabilis* suggest in situ conservation of present occurrences and monitoring of the regeneration process. Stringent monitoring in the protected areas and the highly suitable habitat regions modeled must be used for assisted regeneration and plantation-based habitat enhancement of this endangered species.

Keywords

CHELSA · WorldClim · MaxEnt · *Abies spectabilis* · Alpine treeline · Uttarakhand

10.1 Introduction

Globally mountains cover ca. 25% of the land surface and act as a habitat for around 12% of the global population. Mountain ecosystems provide a range of services and are important from socioeconomic and cultural perspectives. They are the predominant source of freshwater and the head source of prominent rivers around the globe (Grabherr and Messerli 2011). Various climatic zones resulting from latitudinal zonation are condensed within the mountains, making them unique ecosystems supporting various climate zones and associated biodiversity. The steep slope and elevational gradients create unique biodiversity hotspots for various life forms. Moreover, the transition zones along the elevational gradient create ecotones, increasing the number of endemic species in the mountains (Diaz et al. 2003). Half of the global biodiversity hotspots are in the mountainous regions, representing the enormity of mountain biodiversity (Rodríguez-Rodríguez and Bomhard 2011).

Globally the impact of climate change is not uniform; mountain ecosystems especially high-elevation mountain ecosystems are acknowledged as the pioneers in expressing climate change impacts (Becker and Bugmann 2001). In contrast with the pre-industrialized temperature, the 1.3 °C to 1.6 °C increase in surface air temperature over land is one of the most recognized impacts of climate change. The elevation-dependent climatic zones on mountains undergo rapid transitions due to steep slopes and elevational gain; moreover, the high-elevation regions are relatively undisturbed from strong anthropogenic influence, making the mountain ecosystems vulnerable to climate change-induced warming impacts (Beniston 2003). In this regard, mountain ecosystems are even debated as the sentinels of climate change, especially the high-elevation treeline ecotone regions and the alpine vegetation, though there are several biotic, abiotic, and spatial complications in this regard (Malanson et al. 2019). Upward movement of treeline species and alpine flora or even changes in the composition of alpine flora is a recognized impact of climate change on mountain ecosystems; the upward movement of the species is to remain associated with their bioclimatic preference (Zisenis and Price 2011; IPCC 2019).

For the decade 2006–2015, in context with the pre-industrialization period, the global mean surface (land and ocean) temperature increase was around 0.87 °C (Allen et al. 2019). From 1901 to 2009, the annual mean temperature in India increased by 0.56 °C (Attri and Tyagi 2010), the year 2016 in India was recorded as the warmest year with a temperature increase of 0.71 °C, and the year 2020 was the eighth warmest in record since the year 1901 (Attri and Chug 2021). The Himalayan mountain ranges are experiencing a more significant impact and are warming faster than other mountain ranges globally. The Himalayas showed a significant increase in annual temperature along with a reduced number of cold days and nights than the global trend, an increase in decadal temperature rise with a positive association with elevation, temporal variation in warming, rainfall deficit, pre-monsoon drought, early snowmelt, and extreme rainfall events (Schickhoff et al. 2016). The northwestern region of the Himalayas has warmed by almost 1.1 °C in the past century (Bhutiyani 2015). Additionally, the rapid population boom leading to widespread modern construction is causing increased regional warming in the Himalayas (Pandit 2013).

The high-elevation Himalayan treeline ecotone is represented by 58 tree species (Singh et al. 2020). The treeline taxa having a specialist niche might face extirpation due to competition from the upward migration of lowland species, alpine shrubs, and krummholz species (Schickhoff et al. 2015). In the Indian Himalayan region (IHR), limited studies have investigated the impact of real-time climate change on treeline dynamics using tree rings with chronologies. Yadava et al. (2017) reported an average upward shift of 11–54 m decade^{-1} of the Himalayan pine (*Pinus wallichiana* A.B. Jacks.) treeline in the western Himalaya; the winter and early spring warming due to climate change resulted in the increased radial growth. Similarly, Singh et al. (2018) showed elevated monthly temperature during November and February to positively correlate with increased radial growth in Himalayan silver fir, i.e., *Abies spectabilis* (D.Don) Mirb., during the past century. Interestingly, the studies independently also attributed anthropogenic pressure as a regulatory factor in treeline dynamics.

Conservation planning of Himalayan treeline species, therefore, becomes crucial. Ecological niche modeling (ENM) is a requisite tool for the conservation planning approach, not only for current distribution but also for future distribution under various climate change scenarios (McShea 2014). Globally several studies have incorporated the use of ENM for plant species conservation planning (Zhang et al. 2012; Nakao et al. 2013; Fajardo et al. 2014; Spiers et al. 2018). From the IHR perspective, ENM has been used for the conservation planning of endangered plant species (Adhikari et al. 2019; Dhyani et al. 2021), medicinal plants (Yang et al. 2013; Tariq et al. 2021), endemic species (Chitale and Behera 2019; Manish and Pandit 2019), major tree species (Chakraborty et al. 2016), prominent shrub species (Dhyani et al. 2018), invasive plant species (Srivastava et al. 2018), and several others. These studies and others (Upgupta et al. 2015; Manish et al. 2016; Dhyani et al. 2020) have also focused on the potential distribution under future climate scenarios to provide appropriate management implications concerning climate change. However, in terms of treeline ecotone, there are very few studies on the IHR (Singh et al. 2012, 2021a). Those present focus primarily on broadleaf species, viz., *Betula utilis* D.Don (Singh et al. 2013, 2021b; Hamid et al. 2019) and *Quercus semecarpifolia* Sm. (Singh et al. 2021c).

The dominant timberline species of the Indian Central Himalaya (ICH) which is represented by the Uttarakhand state (Nandy et al. 2009; Negi 2022) are *Q. semecarpifolia*, *B. utilis*, *Abies pindrow* (Royle ex D.Don) Royle, and *A. spectabilis* (Negi et al. 2018; Rawal et al. 2018; Sharma et al. 2018; Tiwari et al. 2018). Among these species, *A. spectabilis* is rated as near threatened as per the IUCN Red List Criteria (Zhang et al. 2011). Though both *A. spectabilis* and *A. pindrow* are treeline species, only *A. spectabilis* generally forms the treeline in ICH, while *A. pindrow* remains a few hundred meters below (Singh et al. 2020). This makes *A. spectabilis* the only coniferous timberline – treeline-dominant forest-forming tree species in ICH. The treeline of *A. spectabilis* in the Tungnath region which lies in the ICH recorded no upward shift for the past four decades (Singh et al. 2018). Kaushal et al. (2021) showed unimodal girth class distribution of *A. spectabilis* in the Tungnath region, indicating long-term poor regeneration in the region; such distribution is perilous for the future security of this endemic Himalayan species. Poor natural regeneration of *A. spectabilis* in ICH was noted by several others; the primary reasons are anthropogenic pressure, land-use change, and winter warming trend (Rai et al. 2012a; Singh et al. 2018, 2019).

The facts mentioned above led us to choose ICH and *A. spectabilis* for the present study. For efficient management and planning of conservation strategies, it is imperative to precisely model the habitat distribution of *A. spectabilis*, which is still eluded in ICH. Our study, therefore, addresses the following broad objectives: (1) determination of an appropriate global bioclimatic database, (2) identifying the most influential climatic and non-climatic predictor variables, (3) determining the currently suitable area of *A. spectabilis* in ICH, and (4) considering necessary management implications for conservation planning.

10.2 Materials and Methods

10.2.1 Study Area and Tree Species of Interest

The study area is the Uttarakhand state of India, which constitutes the central portion of the Indian Himalayan region (Fig. 10.1). Uttarakhand, also known as Devbhumi or the land of gods/goddesses, lies between 28°43′ N to 31°27′ N latitude and 77°34′ E to 81°02′ E longitude and has a wide elevational span from 187 m.a.s.l. to 7816 m. a.s.l. The state is bordered internationally in the north by China and in the east by Nepal. On the western and southern sides, it is bordered by interstate boundaries of Himachal Pradesh and Uttar Pradesh, respectively. The state covers an area of 53,483 km^2 of which 71.05% (38,000 km^2) is recorded as forest area. The average rainfall of the state for the year 2019 was 1644 mm, while the minimum and the

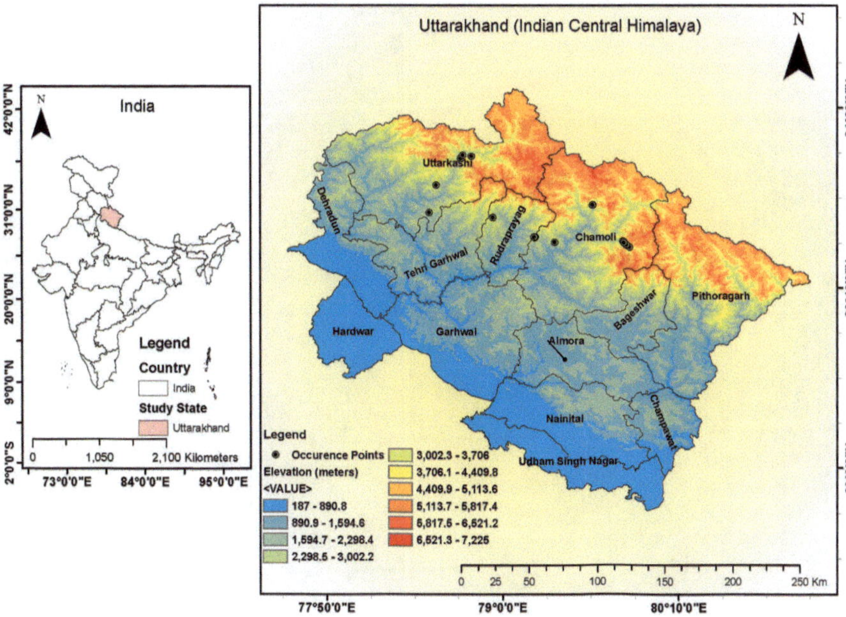

Fig. 10.1 Study area map representing the location of Uttarakhand in India and an elevational gradient map showing the distribution of *Abies spectabilis* occurrence records

maximum temperatures recorded were −2.9 °C and 40.7 °C, respectively (FSI 2019; Uttarakhand at a Glance 2020).

This study focuses on an endemic Himalayan near-threatened tree species *Abies spectabilis* (D.Don) Mirb. (Family: Pinaceae) (Fig. 10.2). The English vernacular name of the species is Himalayan silver fir, while locally in Central Himalaya it is known as Morinda or Raga. It is an evergreen tree species that attain heights of up to 45 m. The wood is of commercial value in construction and carpentry, while the decoction of the leaves and bark is antipyretic and is used to treat respiratory problems. The elevational range limit of this species is from 2400 m to 4000 m and is predominantly found on the northern to northwestern slopes between 3000 m and 4000 m. The native habitat distribution of the species ranges from Afghanistan to the Karakoram range, Jammu and Kashmir, Himachal Pradesh, Uttarakhand, Tibet, and Nepal (Gaur 1999; Zhang et al. 2011).

10.2.2 Species Occurrence Data

Gathering the species occurrence data for *A. spectabilis* is challenging since it is a treeline coniferous species. It occurs near ridge tops or high-elevation regions with undulating terrain, extreme climatic events, and poor accessibility. Even in the Global Biodiversity Information Facility database (GBIF), there were only

Fig. 10.2 *Abies spectabilis* habitat. (**a**) *A. spectabilis* treeline at Tungnath, Uttarakhand. (**b**) Mature *A. spectabilis* tree; notice the thick *Rhododendron campanulatum* D.Don krummholz in the understory. (**c**) A twig-bearing pollen cones. (**d**) Mature female cones of *A. spectabilis*

13 records of *A. spectabilis* from India with geo-coordinates. Only one record pertained to the area of interest in this study (GBIF 2022). We conducted our field survey in the Tungnath region, which lies in the core zone of the Kedarnath Wildlife Sanctuary in the Rudraprayag district of Uttarakhand. We recorded the geo-coordinate data for *A. spectabilis* occurrence using a hand-held Global Positioning System (GPS) meter (Garmin® GPS72™). Additionally, we enhanced our occurrence records with 15 occurrence points from published literature (Supplementary Table 10.1).

A total of 26 occurrence records were collected using primary (11 occurrence points) as well as secondary sources (15 occurrence points) (Supplementary Table 10.1, Fig. 10.1). The occurrence points were converted to a shapefile and projected to WGS84 projection using SDMtoolbox (version 2.5) (Brown et al. 2017) in ArcGIS (version 10.5). There were spatial clusters in occurrence points (Fig. 10.3); such clusters lead to spatial autocorrelation and biasedness in model predictions causing inflated values for model accuracy (Veloz 2009). Occurrence points were therefore spatially rarefied using a 1-km resolution using *spatially rarefy occurrence data* tool under SDMtoolbox. A 1-km resolution was selected due to small occurrence records. The distribution range of *A. spectabilis* is narrow, and the

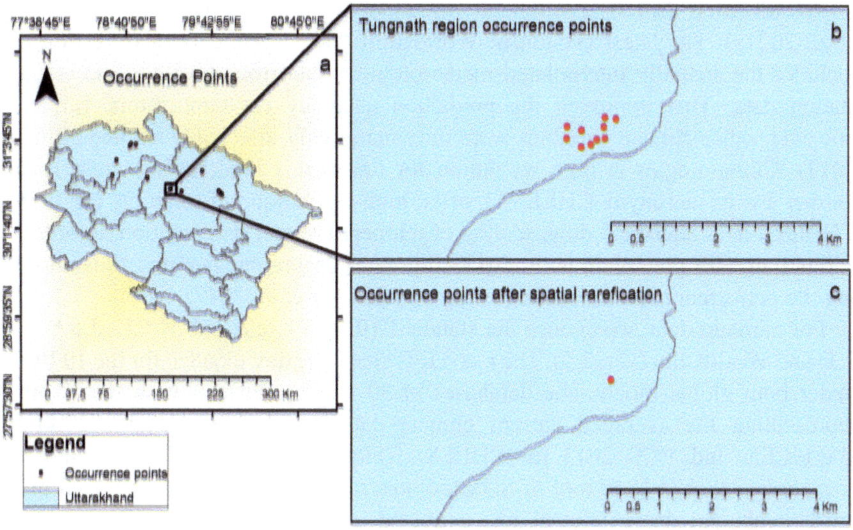

Fig. 10.3 Spatial rarefication of occurrence records. (**a**) Occurrence records of *A. spectabilis* in Uttarakhand. (**b**) One example of spatial clustering of occurrence records in the Tungnath region. (**c**) Occurrence records after spatial rarefication (resolution 1 km)

environmental layers we selected were of 30 arc-second resolution (ca. 1 km × 1 km resolution).

10.2.3 Environmental Data for Habitat Distribution Modeling

A total of 44 environmental variables were used including climatic (19 WorldClim bioclimatic variables +19 CHELSA bioclimatic variables +1 solar radiation variable), topographic (3), edaphic (1), and anthropogenic (1) variables. For climatic variables, we choose the bioclimatic variables (BCVs), which are the derivatives of monthly temperature and precipitation values that determine habitat distribution and its abundance and interactions (Noce et al. 2020). For mountainous regions having sharp climate gradients due to their topography, capturing the environmental variation precisely requires high spatial resolution (Fick and Hijmans 2017). Furthermore, different global bioclimatic databases use different methodologies, leading to the poor congruence between them. Thus, using a single database could lead to biased and unreliable predictions, especially in mountainous regions with stronger climatic variances (Morales-Barbero and Vega-Álvarez 2019). There are several global bioclimatic databases, among which only two, i.e., WorldClim ver. 2.1 (Fick and Hijmans 2017) and CHELSA ver. 2.1 (Karger et al. 2017), have high spatial resolution (30 arc-seconds) and recent averages for current BCVs. Therefore, we choose both WorldClim and CHELSA for predicting the habitat distribution of *A. spectabilis*.

WorldClim is the most prominently used global bioclimatic database (Bobrowski et al. 2021a). The latest version of WorldClim (ver. 2.1) is a refined version that includes the spatially interpolated meteorological data from satellites and weather station data. This improved the prediction accuracy for temperature variables; however, other climate variables were only marginally affected (Fick and Hijmans 2017). Climatologies at high resolution for the earth's land surface areas (better known by its acronym CHELSA) offer high-resolution information on BCVs. CHELSA uses statistical downscaling of temperature and precipitation algorithms; additionally, for precipitation, it also incorporates orographic features, thus improving the performance in mountainous regions (Karger et al. 2017).

For climatic data, we choose the standard 19 BCVs released by CHELSA (ver. 2.1) and WorldClim (ver. 2.1). The GeoTIFF files were downloaded for the 19 BCVs under both global bioclimatic databases in 30 arc-second resolution (ca. 1 km^2), representing the averaged current climate data for the period 1970–2000 for WorldClim and 1979–2013 for CHELSA (Fick and Hijmans 2017). Using the *Extract by Mask (folder)* tool of SDMtoolbox, the input rasters were clipped using the Uttarakhand shapefile mask and extracted in ASCII format. The ASCII layers were projected to WGS84 projection using the *Define Projection as WGS84 (folder)* tool under SDMtoolbox in ArcGIS. The bioclimatic variables with $r < \pm 0.90$ were retained, and highly correlating variables were omitted using the *Remove Highly Correlated Variables* tool of the SDMtoolbox to avoid any autocorrelations leading to unintentional biasedness. BCVs BIO1, BIO2, BIO3, BIO4, BIO7, BIO12, BIO14, BIO15, BIO17, and BIO18 were the variables for WorldClim after removing correlation, while for CHELSA, the BCVs were BIO1, BIO2, BIO3, BIO7, BIO12, BIO13, and BIO14. Since both the databases had a different set of BCVs so, to prevent any biasedness, we selected the common set of BCVs. Thus, for both databases we used six BCVs (Table 10.1) for habitat distribution modeling, viz., BIO1, BIO2, BIO3, BIO7, BIO12, and BIO14. In addition to the bioclimatic variables, we also used solar radiation (kJ m^{-2} day^{-1}) as an additional climatic variable. The data was obtained from WorldClim ver. 2.1 (Fick and Hijmans 2017) at 30 arc-second resolution. The monthly GeoTIFF files were averaged using *Cell Statistics* tool (Spatial Analyst toolbox).

The species of interest occurs in mountainous regions, and the BCVs also use digital elevation models (DEM); therefore, we also used three topographical variables, i.e., elevation, aspect, and slope. The elevation data was obtained from WorldClim ver. 2.1 (Fick and Hijmans 2017) at 30 arc-second resolution. The GeoTIFF elevation raster was projected in WGS84 projection and clipped using Uttarakhand shapefile. The resulting raster was used to compute aspect and slope under the Spatial Analyst toolbox in ArcGIS. A z-factor corresponding to 30 degrees latitude was chosen to calculate the slope raster in degrees.

One edaphic variable, i.e., Global Soil Organic Carbon Map V1.5 released by the Food and Agriculture Organization (FAO), was selected. Soil organic carbon (SOC) being the predominant component of soil organic matter reflects soil quality. SOC is associated with nutrient availability, water retention capacity of the soil, and even structural stability (FAO and ITPS 2018). The GeoTIFF raster was downloaded in

Table 10.1 Environmental variable used for *A. spectabilis* ENM

Model 1	
BCV codes (units)	Interpretation (scaling factor and offset)
CHELSA BIO1 (°C)	Mean annual air temperature (0.1 and − 273.15)
CHELSA BIO2 (°C)	Mean diurnal air temperature range (0.1 and 0)
CHELSA BIO3	Isothermality, i.e., BIO2/BIO7 (0.1 and 0)
CHELSA BIO7 (°C)	Annual range of air temperature (0.1 and 0)
CHELSA BIO12 (kg m^{-2})	Annual precipitation amount (0.1 and 0)
CHELSA BIO14 (kg m^{-2})	Precipitation amount of the driest month (0.1 and 0)
Model 2	
WorldClim BIO1 (°C)	Annual mean temperature
WorldClim BIO2 (°C)	Mean diurnal range
WorldClim BIO3 (%)	Isothermality, i.e., (BIO2/BIO7) × (100)
WorldClim BIO7 (°C)	Temperature annual range
WorldClim BIO12 (mm)	Annual precipitation
WorldClim BIO14 (mm)	Precipitation of the driest month
Variables common to both model 1 and model 2	
Solar radiation (kJ m^{-2} day^{-1})	
Elevation (meters)	
Aspect (degrees)	
Slope (degrees)	
SOC (Mg ha^{-1})	
HII	

All the environmental variables are at 30 arc-second resolution. For each model there are 12 sets of environmental variables each

30 arc-second resolution. It represents SOC for 0–30 cm depth on Mg ha^{-1} basis (accessed on 16 April 2022 https://storage.googleapis.com/fao-maps-catalog-data/geonetwork/gsoc/GSOCmap/GSOC map 1.5.0.tif).

For the anthropogenic variable, we choose Global Human Influence Index (HII) version 2 (1995–2004) dataset. The raster data was downloaded at 30 arc-second resolution and obtained from the Socioeconomic Data and Applications Centre (SEDAC) (WCS 2005). HII uses nine datasets broadly grouped into four categories, population density, land transformation, accessibility, and electrical power infrastructure, which describe the human footprint. The index ranges from 0 to 72, with higher scores indicating a higher anthropogenic footprint (Sanderson et al. 2002).

All the environmental layers were masked to the Uttarakhand shapefile, projected in WGS84 projection, and converted to ASCII format using ArcGIS. So, there were a set of 12 environmental variables under each of the two sets, i.e., model 1 and model 2, as indicated in Table 10.1.

10.2.4 Ecological Niche Modeling

ENM is an empirical approach that couples the species occurrences with the environmental predictor variables to model specific environmental constraints relative to the species (species realized niche), which aids in the spatial and temporal mapping of the species habitat distribution (Elith and Franklin 2013). Several species distribution modeling algorithms are based on different principles, and different methodologies exist to use the models. The models could either be standalone or in a combination approach (ensemble modeling) which is gaining trend since combining the response from various models improves the prediction accuracy (Kaky et al. 2020). We used a single-model approach and deployed the use of maximum entropy algorithm-based species distribution modeling software MaxEnt (version 3.4.4) (Phillips et al. 2004, 2006, 2017). MaxEnt has been a primary choice for the majority of ENM studies on account of its advantages, viz., it is a presence-only model, works well with both continuous and categorical environmental variables, its regularization parameters avoid over-fitting of the model, flexibility in the choice of threshold selection for binary output, and easy-to-understand distribution results as well as interpretation of environmental variables with habitat suitability (Phillips et al. 2006). Furthermore, the ensemble method requires complex computation, and with limited occurrence records, MaxEnt has proved to be equally robust and accurate in predicting habitat suitability (Kaky et al. 2020); thus, we used MaxEnt for the ENM.

Two sets of MaxEnt models were run, i.e., model 1 and model 2. Model 1 included CHELSA BCVs along with other environmental variables, and model 2 used WorldClim-based BCVs and other environmental variables (Table 10.1). MaxEnt model was executed using auto-feature mode so that MaxEnt automatically decides the best features suited to our dataset. The random test percentage was set to 30%; therefore, 70% of the occurrence records were reserved for training and 30% for testing. We used the entire spatially rarefied occurrence dataset for the maximum use of occurrence records for model building (Phillips et al. 2006). Other settings included checking the check box for creating response curves and Jackknife for testing variable importance (this allows to assign the permutation importance and contribution of each predictor variable), output format was set to logistic threshold for obtaining a binary output, and output file format was chosen as ASCII. Under the Basic Settings tab, the default random seed was selected; regularization multiplier β value was set to 1 (helps to prevent model overcomplexity); background points to default 10,000; and 50 replicates with bootstrap run the method (bootstrap method was chosen due to limited occurrence records). Under the Advance tab, Write Plot Data was selected, maximum iterations were 500, convergence threshold was 0.00001, threshold rule was set to 10-percentile training presence, and under the Experimental tab, Write Background Predictions were selected.

10.2.5 Model Performance and Model Output

We evaluated the performance of both the SDM models, i.e., model 1 (CHELSA BCVs and other variables) and model 2 (WorldClim BCVs and other variables) using threshold-independent techniques since they avoid any issues regarding the selection and influence of threshold values (Fielding and Bell 1997). We used two methods, viz., the area under the receiver operating curve (AUC) and the partial area under the receiver operating curve (pROC), as threshold-independent methods. The AUC score ranges from 0 to 1, with 1 (perfect discrimination) being the greatest predictive ability of the model and 0 showing no predictive ability. Generally, an AUC score from 0.9 to 1.0 is considered excellent, 0.8–0.9 as good, 0.7–0.8 as fair, 0.6–0.7 as poor, and 0.5–0.6 as very poor. At a 0.5 score for presence/background models, no discrimination exists between true and false proportions (random performance) (Swets 1988; Kaky et al. 2020). The AUC scores were obtained from MaxEnt average replicate runs. The pROC considers only the ROC region associated with the data and not the entire AUC range. We also computed the AUC ratio, i.e., the pROC value compared to the null AUC value expected at AUC = 0.5. The AUC ratio value ranges from 0 to 2, with the value of 1 indicating random performance (Peterson et al. 2008; Chaitanya and Meiri 2021). The pROC was calculated using NicheToolBox, an online platform to perform processing steps involved in ecological niche modeling (Osorio-Olvera et al. 2020). For pROC, we used the average ASCII model output from MaxEnt, the proportion of omission was set to 0.05, the random point percentage was set to 50%, and 100 bootstrap iterations were used. We also performed an independent sample t-test on the 100 bootstrap values of the AUC ratio between models 1 and 2 to evaluate the significant difference between the models. Furthermore, the average values of the 12 predictor variables for each of the 50 bootstrap runs were subjected to an independent sample t-test under both models 1 and 2 to determine any significant difference ($p < 0.05$) between the variables.

For both the SDM models, i.e., model 1 and model 2, the average ASCII MaxEnt output file (the result of 50 bootstrap replicates) was converted into raster format using ArcGIS. The average raster suitability files are binary (since a logistic output method was used), with habitat suitability ranging from 0 (unsuitable) to 1 (highly suitable); the value of 1 is generally not achieved. The binary outputs were reclassified into four habitat suitability classes, viz., unsuitable, low, medium, and high suitability. The average 10-percentile training presence (P10) logistic threshold was used to set the lowest limit of habitat suitability above which the suitability classes were classified. The classification was as follows, unsuitable habitat (0–P10 logistic threshold value), low suitability (P10 logistic threshold value–0.4), medium suitability (0.4–0.6), and high suitability (0.6–maximum logistic value). The P10 threshold is a conservative approach, and it considers the habitat suitability of regions lower than the lowest 10% of the training locations to be unsuitable (Di Pasquale et al. 2020). We calculated the total area under each habitat suitability class for both models using *zonal geometry as table* tool under spatial analyst tools in ArcGIS. Furthermore, to determine the percentage distribution of habitat suitability classes, the final habitat suitability raster file was extracted by mask using each district shapefile for both the models and analyzed for area under each class in each district using *zonal geometry as table* tool in ArcGIS.

10.3 Results

10.3.1 Spatial Autocorrelation of Occurrence Records and Model Performance

There were 26 total occurrence records (Fig. 10.1), of which 14 records showed spatial autocorrelation at >1-km resolution (Fig. 10.3). Therefore, after spatial thinning net 12 occurrence records of *A. spectabilis* were retained, having the least resolution of 1 km so that a grid of environmental variables has at most one occurrence record.

Both the models performed well as per threshold-independent model evaluation parameters AUC and pROC. Model 2 outperformed model 1 in terms of AUC score (Fig. 10.4). Model 1 had a good AUC score; however, model 2 had an excellent score with a very low standard deviation. The mean pROC results also showed model 2 to be the most predictive (Table 10.2). The AUC ratio for model 2 was significantly higher than model 1 [t (198) = 24.53, p < 0.0001] (Table 10.2, Fig. 10.5). The difference between the means of AUC for model prediction (AUC partial) and AUC at random was significantly different with p < 0.001 for both model 1 and model 2, indicating that both models performed well.

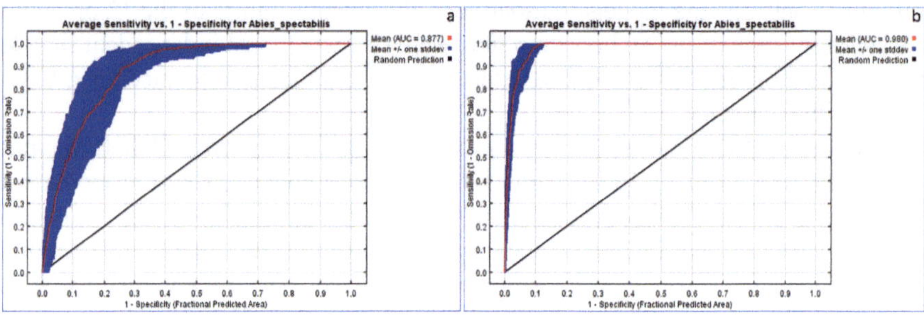

Fig. 10.4 Average area under the receiver operating curves for the two models for *A. spectabilis* with 50 bootstrap runs. (**a**) AUC for model 1 (notice the greater standard deviation) and (**b**) AUC for model 2

Table 10.2 *A. spectabilis* model evaluation parameters indicating AUC (MaxEnt, 50 replicate runs), AUC ratio (NicheToolBox, 100 replicate runs), and mean pROC (NicheToolBox, 100 replicate runs). The values indicate the mean with a standard deviation

Model	AUC	AUC ratio	Mean pROC
Model 1	0.877 ± 0.040	1.726 ± 0.09	0.863 ± 0.04
Model 2	0.980 ± 0.008	1.954 ± 0.02	0.977 ± 0.01

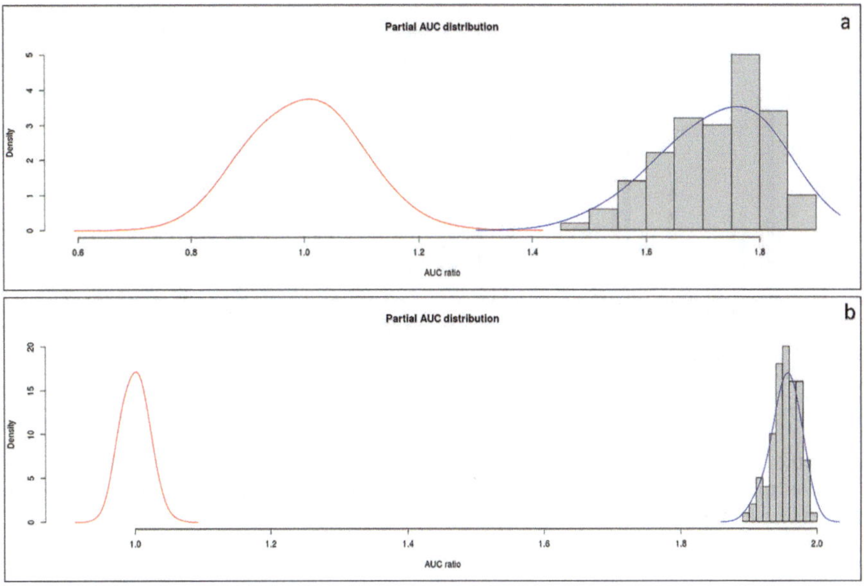

Fig. 10.5 The pROC distribution of *A. spectabilis* under two separate models. (**a**) pROC distribution under model 1 and (**b**) pROC distribution under model 2. The shaded bars indicate the frequency distribution of the AUC ratio, and the red-colored bell curve (curve at the left side) indicates the AUC ratios for random models

10.3.2 Influence of Predictor Environmental Variables

The 12 environmental variables chosen under model 1 and model 2 (Table 10.1) had BIO14 as the primary environment variable with the maximum percentage of permutation importance (Fig. 10.6). On a broader overview, the CHELSA BCVs contributed 71.91% of the total permutation importance in model 1, while the WorldClim BCVs contributed 89.43% of the total permutation importance in model 2. For model 1, the top 3 predictor variables based on permutation importance were BIO14 > BIO01 > Soil organic carbon content. For model 2, the trend was BIO14 > Solar radiation > BIO03. In terms of percentage contribution (Fig. 10.6), also BIO14 had the highest contribution among all the environmental variables in both the models.

The Jackknife test for determining variable importance (Fig. 10.7) for models 1 and 2 indicated BIO14 to contain the most useful information and show the highest gain when used in isolation. BIO14 also decreased the maximum gain when omitted, thus indicating that it has information that is not shared by other variables. The result was similar for all three Jackknife tests, i.e., for regularized training gain, test gain, and the AUC.

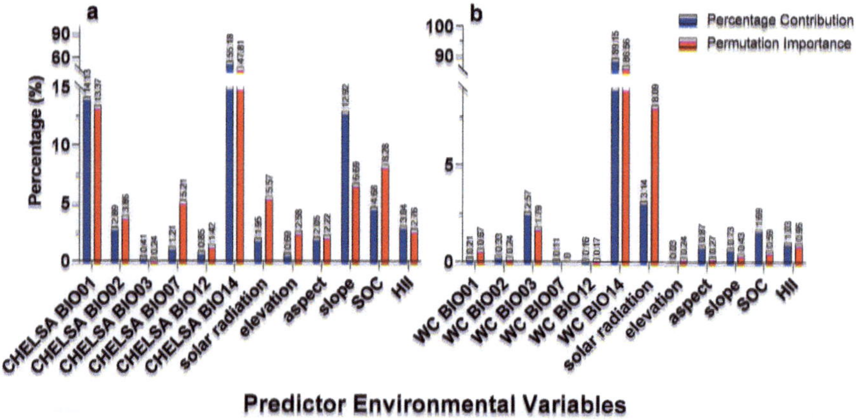

Fig. 10.6 Percentage contribution and permutation importance of the predictor environmental variables for *A. spectabilis*. (**a**) Predictor variable importance for model 1 and (**b**) predictor variable importance for model 2

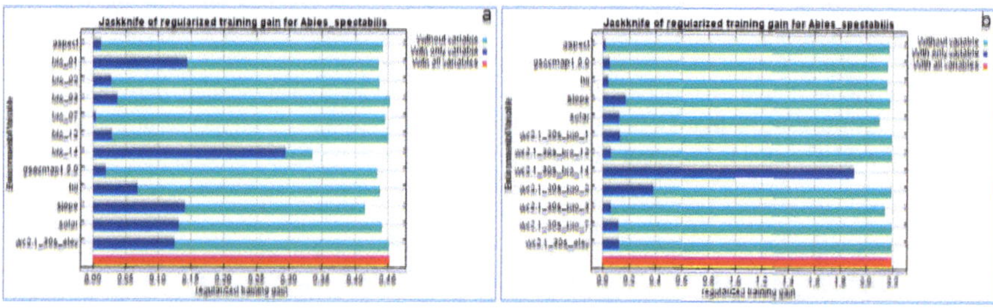

Fig. 10.7 Jackknife test for the regularized training gain for *A. spectabilis*. (**a**) Model 1 and (**b**) model 2

10.3.3 Interpretation of the Response Curves

We studied the response curves generated using the particular environmental variable alone to avoid any undue disproportionate impact of variable correlations. Furthermore, we analyzed only the top 3 variables in permutation importance for both models. For model 1, BIO14 and SOC positively influence the habitat distribution of *A. spectabilis*, showing a sigmoid curve, while BIO01 shows a negative influence with an inverse J-shaped curve (Fig. 10.8). On the other hand, in model 2, only BIO14 showed a positive influence on *A. spectabilis* habitat suitability (sigmoid curve), while solar radiation and BIO03 showed a negative trend (Inverse J curve). Model 1 BIO14 ranged from around 5 to 68 kg m^{-2} with a habitat suitability of >0.5 at around 29 kg m^{-2}. BIO01, on the other hand, declined with a suitability of >0.5 at ca. 4.35 °C. SOC showed a positive trend with values ranging from ca. 17 to 67 t ha^{-1} and a suitability of >0.5 at ca. 42.5 t ha^{-1}. Under model 2, BIO14 ranged from ca. 10 to 38 mm with a suitability of >0.5 at ca. 31 mm. Solar

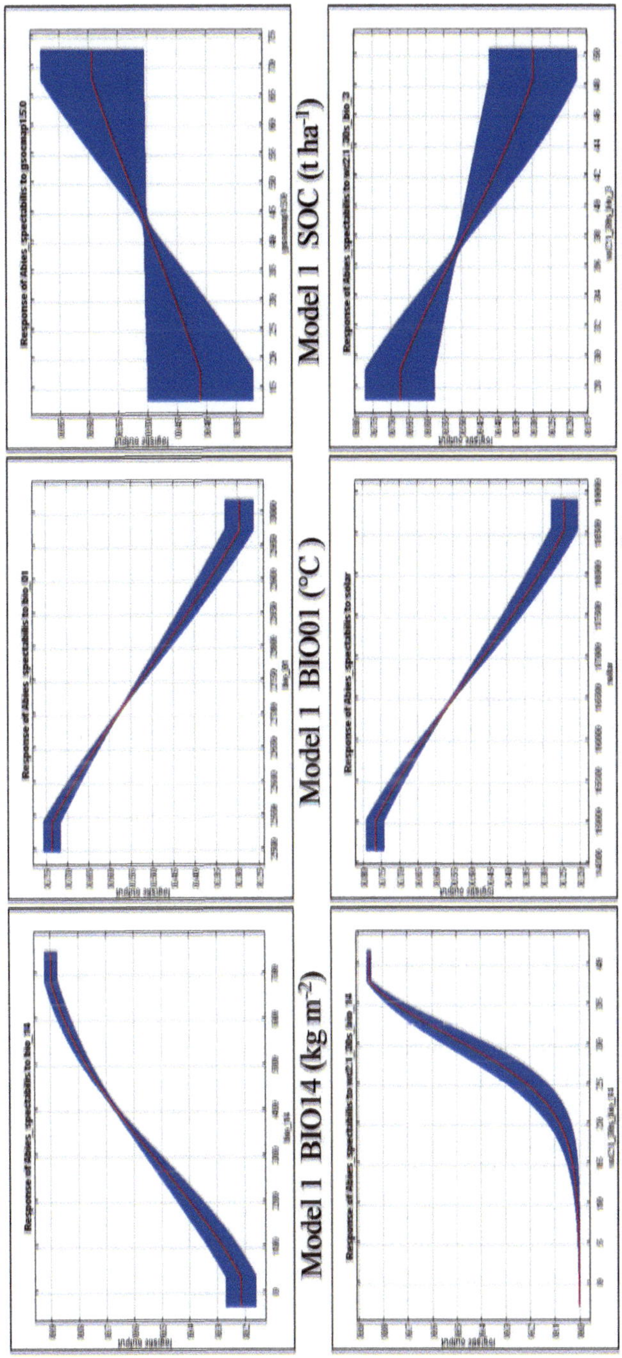

Fig. 10.8 Response curves of *Abies spectabilis* to environmental variables. X-axis represents the value of the environment variable, while the Y-axis represents the logistic value of habitat suitability. The red line represents the mean of MaxEnt runs, while the blue region represents the standard deviation

Table 10.3 Average values of 50 replicate runs of each predictor variable for the occurrence records

Environmental variable	Model 1[a]	Model 2
BIO1 (°C)	3.64 ± 0.88	6.54 ± 0.77
BIO2 (°C)	10.25 ± 0.28	9.55 ± 0.18
BIO3 (%)	36.36 ± 0.86	37.68 ± 0.65
BIO7 (°C)	28.20 ± 0.51	25.38 ± 0.51
BIO12 (mm)	2770.57 ± 165.49	1699.51 ± 60.84
Bio14 (mm)	30.56 ± 2.58	31.84 ± 1.28
Solar radiation (kJ m^{-2} day^{-1})	16758.33 ± 44.72	16782.96 ± 57.64
Elevation (meters)	3236.10 ± 89.30	3217.14 ± 108.73
Aspect (degrees)	170.23 ± 32.16	165.12 ± 34.68
Slope (degrees)	17.80 ± 2.75	18.55 ± 2.84
SOC (t ha^{-1})	43.16 ± 1.99	44.20 ± 2.72
HII	15.62 ± 1.57	16.49 ± 1.69

[a]The values for CHELSA BCVs (model 1) are corrected for scale and offset

radiation declined the habitat suitability with a suitability of >0.5 at ca. 16800 kJ m^{-2} day^{-1}. BIO03 also showed a similar trend as solar radiation with a suitability of >0.5 at ca. 38% (Fig. 10.8).

The values of the 12 environmental predictor variables over the species occurrence points were obtained from MaxEnt sample average files for each of the 50 replicates under both models 1 and 2 (Table 10.3). Only elevation, slope, and aspect did not have any statistically significant difference between the means ($p > 0.05$) as determined using an independent sample t-test, rest all the pairs between models 1 and 2 showed statistically significant difference ($p < 0.05$).

10.3.4 Current Habitat Suitability of *A. spectabilis*

The average P10 training presence logistic threshold values were 0.3784 and 0.2724 for models 1 and 2. These values were regarded as the threshold for suitability cutoff, and based on this, the current habitat suitability map was drafted (Fig. 10.9). Model 1 had the highest percentage distribution for all three suitability classes compared to model 2 (Table 10.4). Both the models showed a statistically significant strong positive correlation in their percentage distribution of habitat suitability classes indicating similar predictions trend ($r = 0.966$, $n = 4$, $p = 0.034$). *A. spectabilis* habitat suitability area in the Indian Central Himalaya as per model 1 was 16,324.57 km^2, while for model 2, it was 3323.86 km^2.

The spread of habitat suitability of *A. spectabilis* is toward the northern region of Uttarakhand, having more prominence of higher elevation (Fig. 10.1). The majority of the unsuitable and low suitable habitat suitability regions were situated in the southern part of the state (Fig. 10.9). The niche overlap map (Fig. 10.11) shows model 1 to predict higher suitable regions than model 2. Moreover, the overlap also

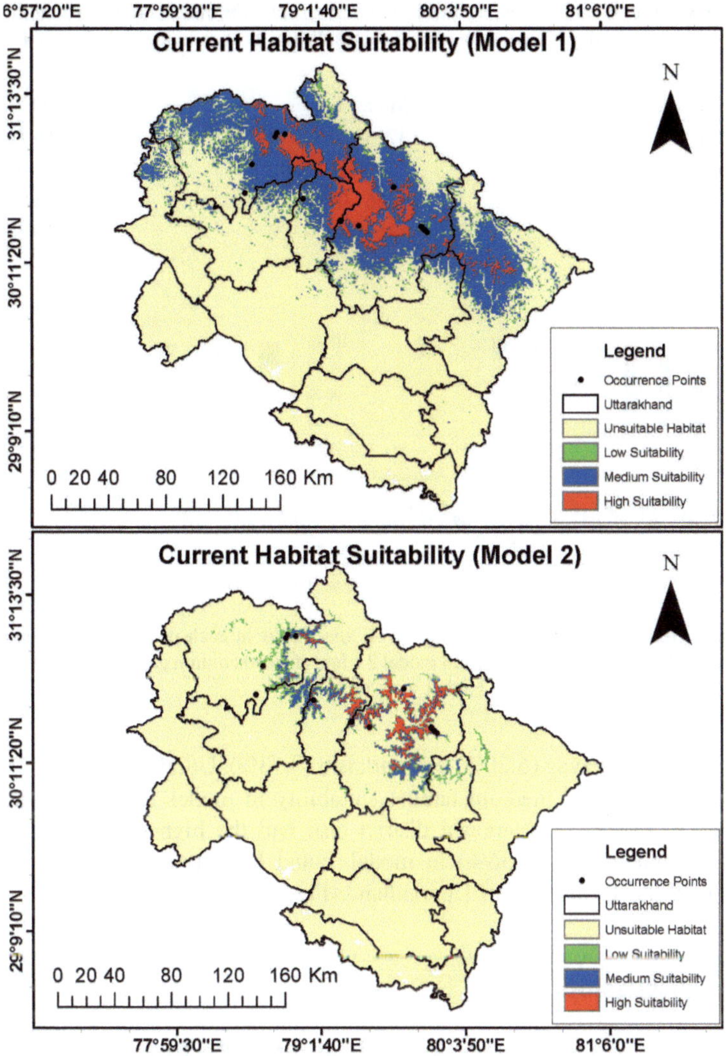

Fig. 10.9 Current habitat suitability map of *A. spectabilis* in the Indian Central Himalaya under models 1 (top) and 2 (bottom). The unsuitable class has the 10-percentile training presence average logistic threshold value as cutoff

shows that the region predicted as suitable by only model 2, excluding the overlap areas, was smaller than model 1. Haridwar and Udham Singh Nagar were the only two districts showing complete *A. spectabilis* habitat unsuitability as per model 1; however, model 2 predicted Almora, Champawat, Dehradun, and Nainital to be unsuitable in addition to Haridwar and Udham Singh Nagar. For model 1, the districts with the highest suitability region were Rudraprayag (19.55%) > Chamoli (16.72%) > Uttarkashi (8.25%), while for model 2 the trend was Chamoli

Table 10.4 Current habitat suitable area (km^2) of *A. spectabilis* under different suitability classes in both models 1 and 2. Values in parenthesis indicate the percentage of the total land area (53,483 km^2) of Uttarakhand

Habitat suitability class	Model 1	Model 2
Unsuitable	37158.43 (69.48)	50159.14 (93.79)
Low suitability	2127.15 (3.98)	1222.03 (2.28)
Medium sitability	11561.03 (21.62)	1274.97 (2.38)
High suitability	2636.39 (4.93)	826.86 (1.55)

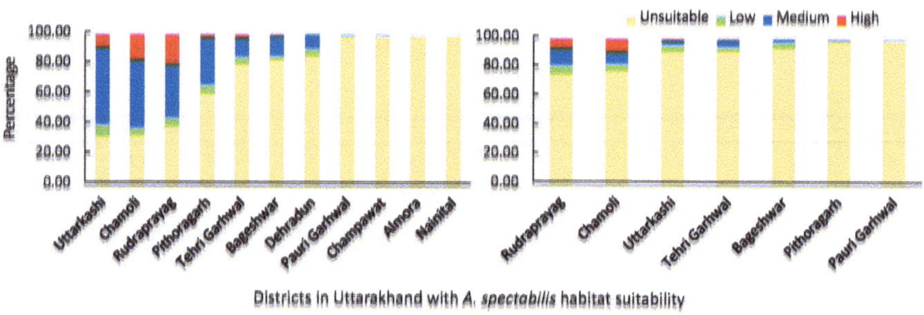

Fig. 10.10 Habitat suitability percentage of *A. spectabilis* in each district of Uttarakhand state under model 1 (left stacked column) and model 2 (right stacked column graph). Districts with 100% region as unsuitable *A. spectabilis* habitat were omitted

(8.43%) > Rudraprayag (6.20%) > Uttarkashi (0.53%). Uttarkashi (50.86%) had the highest percentage of medium habitat suitability in model 1, while Rudraprayag (11.35%) in model 2. Uttarkashi district also had the highest proportion of low habitat suitable region (8.36%) in model 1 and Rudraprayag (6.32%) in model 2 among all the 13 districts of Uttarakhand (Fig. 10.10).

10.4 Discussion

10.4.1 Species Occurrence Records and the Performance of Models

Occurrence records following spatial thinning (Fig. 10.3) for this study were relatively low (12 points). This is because *A. spectabilis* occurs in high-elevation zones in the Himalayas (Fig. 10.1) prominently in rugged and inaccessible terrains, thus making field surveys difficult. Low occurrence points are not unusual in literature; for instance, Kumar and Stohlgren (2009) used 11 occurrence records (the only known occurrence) to predict the habitat distribution of a threatened and endangered tree species *Canacomyrica monticola* Guillaumin. *A. spectabilis* is a treeline coniferous species (Fig. 10.2) that is endemic to the Himalayas and occurs in a narrow elevation zone generally spanning from 3000 to 4000 m.a.s.l. (Zhang et al. 2011). The vegetation pattern in the Indian Central Himalayas along the elevational gradient shows fairly consistent sub-alpine vegetation with selected species that can withstand harsh environmental conditions (Singh and Singh 1987; Sharma et al. 2018; Tiwari et al. 2018). Such species, including *A. spectabilis*, have a specialized niche; modeling such species even with lower occurrence records is much easier since the species with broader geographical distribution require a more significant number of occurrence records for higher model accuracy. Hernandez et al. (2006) experimentally showed that models run using occurrence points as low as 10 generated almost similar results as those run with twice the size predominantly for the species with the specialized niche. They also supported that the MaxEnt model performs the strongest even with low occurrence records on its regularization parameters. Our occurrence records (26 records) were, however, lower as compared with the records used for habitat suitability prediction of *A. spectabilis* in Nepal (94 records) (Chhetri et al. 2018). We also recommend including true absence points in the modeling of the habitat distribution of treeline species since certain environmental factors limit the spread of these species beyond the treeline. Encoding these factors in the modeling parameters will yield greater accuracy in habitat suitability prediction, especially for future habitat modeling studies under different climate change scenarios.

The predictive performance of the models was tested using three parameters, viz., AUC, pROC, and AUC ratio (Table 10.2). This method is advantageous since error and bias associated with selecting the threshold value are omitted; furthermore, different models would have different threshold cutoffs making model comparison problematic (Phillips et al. 2006). Model 2 outperformed model 1 (Table 10.2, Figs. 10.4 and 10.5) in all three model testing parameters; however, model 1 was only marginally lower than model 2, suggesting the successful habitat prediction from both models. AUC was not used as the only model evaluation statistic as the error components are weighted equally. Moreover, comparison with AUC becomes problematic since the commission errors are calculated along with the entire range (0–1) even though the predicted occurrence may not span that range; therefore, we used pROC and AUC ratio (Table 10.2, Fig. 10.5) which helps overcome these issues (Peterson et al. 2008).

Chhetri et al. (2018) reported a high AUC value (AUC = 0.89) for *A. spectabilis* habitat distribution in Nepalese Himalaya. High mean AUC values were also reported for *Picrorhiza kurroa* Royle ex Benth. (AUC = 0.915) (Rawat et al. 2022) and *Dactylorhiza hatagirea* (D.Don) Soó (AUC = 0.868) (Chandra et al. 2021); the two endangered alpine medicinal plants in Uttarakhand having similar to higher elevational range as *A. spectabilis*. *Q. semecarpifolia* (AUC = 0.982), a broadleaf treeline oak species in the Indian Central Himalaya, also showed excellent AUC value (Chakraborty et al. 2016). We do not intend to compare our AUC values with these studies but only show a near comparable range for the model prediction of these species, which occur in the same niche space as *A. spectabilis*. The comparison of AUC between different species, across different regions and using varying modeling parameters, is not valid due to the differences in the potential distribution area. However, AUC could be used to compare two environmental datasets (Fig. 10.4), provided the species and area of interest remain the same (Peterson et al. 2011).

10.4.2 Use of Two Global Bioclimatic Databases

There are only two predominant high-resolution bioclimatic databases, i.e., WorldClim and CHELSA, which differ in their BCV prediction algorithm. WorldClim uses a spatial interpolation of climate data, and this technique is considered less ideal than the statistical downscaling method used by CHELSA. The spatial interpolation methods usually perform less optimally in regions with uneven topography, like mountains. Furthermore, in extreme environments like treeline ecotone or alpine regions, the environmental features have a predominant say in shaping the habitat distribution of the species (Morales-Barbero and Vega-Álvarez 2019). All the bioclimatic variables are based upon two primary datasets temperature and precipitation. While both CHELSA and WorldClim showed congruence for temperature predictors, the prediction accuracy for CHELA precipitation pattern is more sensitive in the mountainous regions than WorldClim due to the incorporation of orographic features (Karger et al. 2017).

Furthermore, Bobrowski and Schickhoff (2017) evaluated the efficacy of global climate datasets in modeling the habitat distribution of a principal Himalayan treeline broadleaf species, *B. utilis*. They concluded CHELSA to show greater accuracy with predictions closer to actual field observations. They also found WorldClim to over-predicting the habitat distribution and thus cautioned against the use of WorldClim BCVs alone in the Himalayan landscape without scrutinizing. Therefore, we used both WorldClim- and CHELSA-based BCVs for the ENM of *A. spectabilis*. In a recent critical review on ENM, Bobrowski et al. (2021b) reported that there are only three studies in the Himalayas that used both WorldClim- and CHELSA-based BCVs to compare the modeling performance. Therefore, our study adds to the performance of the two global bioclimatic databases and their limited knowledge of the Himalayan landscape.

Furthermore, our study is probably the first to use a comparative approach to the ENM of the endemic Himalayan treeline conifer *A. spectabilis*. With model 2 having outperformed model 1 on model validation parameters, we infer that in our study, WorldClim-based BCVs (model 2) performed much better than CHELSA (model 1). The performance of CHELSA was only marginally lower, and thus we do not discriminate against its use. However, we recommend further improving the model by adding more occurrence points and including true species absence points for greater validation and accuracy.

10.4.3 Predictor Environmental Variables and their Response Curves

We focused on the permutation importance rather than percentage contribution in deciding the importance of environmental variables since the percentage contribution depends upon the path MaxEnt chosen to build the optimal model. The contribution would vary with each run due to the change in the modeling algorithm. Furthermore, we also used the Jackknife test, wherein the model was run multiple times. Each variable was first modeled alone, excluding it and including the remaining variables to interpret the degree of gain or loss for that parameter (Phillips 2017). Interestingly, models 1 and 2 had the precipitation of the driest month (BIO14) as the most influential predictor environmental variable having the highest permutation importance and percentage contribution (Fig. 10.6). Even from the Jackknife test (Fig. 10.7), BIO14 had the highest gain when used in isolation and caused the most loss in gain when omitted under both models 1 and 2, thus confirming it to be the most influential predictor variable of *A. spectabilis* in Uttarakhand. In Uttarakhand Himalaya, BIO14 was also deemed to be the most significant predictor variable for the habitat distribution of Himalayan endangered medicinal herb *P. kurroa* (Rawat et al. 2022), critically endangered medicinal plant *Lilium polyphyllum* D.Don (Dhyani et al. 2021), and multipurpose shrub species *Hippophae salicifolia* D.Don (Dhyani et al. 2018). In other parts of the IHR and similar elevational range, species like *Rheum webbianum* Royle, a vulnerable medicinal herb in northwest Himalaya, had BIO14 as the most important predictor variable (Wani et al. 2021). All the aforementioned plant species have a similar to higher elevational range as *A. spectabilis*; moreover, *P. kurroa*, *H. salicifolia*, and *R. webbianum* are also found in Tungnath region having *A. spectabilis* treeline ecotone (Rai et al. 2012b). The trend, however, is not universal, and *D. hatagirea*, a threatened medicinal orchid that also occurs in the Tungnath region along the *A. spectabilis* treeline (Rai et al. 2012b), showed a mean diurnal range (BIO2) as the most significant predictor variable with the highest permutation importance for its ENM in Uttarakhand (Chandra et al. 2021). Thus, the species occurring in similar habitats do not always share similar niche requirements. Chhetri et al. (2018) critically modeled the habitat distribution of prominent treeline species in the Nepalese Himalaya and found elevation as the strongest predictor variable for *A. spectabilis* followed by isothermality (BIO3) in their model using 94 occurrence records for *A. spectabilis* alone.

Our results indicate that *A. spectabilis* is sensitive to dry seasons and constantly requires a minimum moist climate for habitat suitability. Moreover, BIO14 for both model 1 and model 2 showed a sigmoidal response curve (Fig. 10.8), indicating a positive relationship between precipitations in the driest month with predicted suitability. We also observed that the average value for BIO14 under the occurrence records (Table 10.3) for both models was well within the response curve range for BIO14 (Fig. 10.8) and since the average current value is around 30 mm for both the models thus as per the response curve higher values will lead to more habitat suitability for *A. spectabilis*. The value for the precipitation of the driest month reflects the bare minimum amount of precipitation required for the habitat suitability in Uttarakhand region, since November is the month that usually receives the least annual rainfall (Climate-Data.org 2022). Therefore, the precipitation during this month predominantly characterizes the *A. spectabilis* habitat suitability. Singh and Negi (2018), in their phenological studies of Central Himalayan treeline species, showed October–November months as the seed maturation duration for *A. spectabilis*. Therefore, this period is critical since precipitation ensures the establishment of seed set and further seed germination. Singh et al. (2018) indicated the importance of winter temperatures, especially November and February months, in regulating the growth (radial growth) of *A. spectabilis* in the Tungnath region of Uttarakhand. Additionally, they predicted the winter warming trend to be a probable cause for the loss of regeneration for *A. spectabilis*. Our results confirm this observation since BIO1 (annual mean temperature) was the second most important predictor variable under model 1 (Fig. 10.6), which showed an inverse J trend (Fig. 10.8), inferring habitat loss with increasing warming. Solar radiation was the second most important predictor variable in model 2 which also showed a negative trend, thus indicating increasing levels of solar radiation detrimental to habitat suitability. Globally the levels of solar radiation have been studied either as a key factor (Bader et al. 2007) or as one of the factors (Barbeito et al. 2012) responsible for the alpine treeline formation, regeneration, and mortality. The third environmental variable having the highest permutation importance (Fig. 10.6) for both models 1 (SOC) and 2 (BIO3) had a high standard deviation, and the permutation importance was less than 5%. An inverse J relationship of isothermality (BIO3) with habitat suitability (Fig. 10.8) indicates that *A. spectabilis* has a narrow range to tolerate the temperature oscillations between monthly temperature range and annual temperature range, thus indicating its narrow niche requirements for its habitat suitability. BIO3 was found to be the most important BCV for three prominent alpine treeline species *A. spectabilis*, *B. utilis*, and *P. wallichiana* in the Nepal Himalaya (Chhetri et al. 2018). The importance of SOC as a predictor variable indicates the contribution of edaphic factors in regulating habitat suitability. Higher SOC values are associated with increased nutrient availability in soil on account of increased microbial and enzymatic activity in soil (Siwach et al. 2021). Moreover, the sigmoidal response curve obtained (Fig. 10.8) infers the positive association of increasing soil carbon stock with habitat suitability.

However, one significant concern is regarding the selection of the environmental variables. Since each study has either difference in study species or study region, and even if these components are similar, there would be a difference in occurrence records. This would cause differential autocorrelation in the BCVs, thus resulting in a heterogeneous selection of BCVs. Furthermore, the variables omitted to reduce spatial autocorrelation might show significant importance for the same species in different regions. Therefore, the selected predictor variables must be used cautiously. Similar concerns were also reflected by others (Kumar 2012; Chhetri et al. 2018; Yoon and Lee 2021).

10.4.4 Current Habitat Suitability of *A. spectabilis*

The strong correlation found between the habitat suitability classes for models 1 and 2 indicates similar trend prediction by both models. The suitability area under medium suitability (0.4–0.6) is much higher than low suitability (P10 threshold–0.4) since the threshold value is quite large (larger than 0.2), thereby making the size of the low suitability class smaller. The suitable area predicted by model 1 was around five times greater than model 2 (Table 10.4). Surprisingly the average values for the occurrence locations for all the environmental predictor variables

(Table 10.3) were only marginally different; still, the output predictions turned out to be substantially diverse. We attribute this difference solely to the two global bioclimatic databases used since the other environmental variables were similar under both models. In this study, BIO14 was the most important predictor variable under both models, and this variable is derived from annual precipitation values. Since precipitation-related CHELSA BCVs have been deemed more accurate due to the edge of statistical downscaling approach over spatial interpolation of WorldClim (Karger et al. 2017) therefore, CHELSA-based model 1 might have predicted a larger area. Bobrowski and Schickhoff (2017) highlighted a critical note about the distribution of weather stations in the Himalayas; not only their number is scarce, but also they are heterogeneously distributed along the Himalayan range. Moreover, those present have an almost negligible presence in the treeline elevations due to inaccessible terrain, poor connectivity, and harsh weather conditions. This makes the BCV data for higher elevations more susceptible to errors. Furthermore, due to complex topographical terrain such as slope degrees, slope aspect, windward and leeward side, etc., local site-specific weather patterns develop that modulate the habitat suitability of vegetation on a local scale. In our opinion, the global bioclimatic database that captures the maximum amount of this fine variability would yield the most accurate prediction. Though both the models performed equally well, model 2 with WorldClim BCVs performed slightly better. Therefore, we believe that the current habitat distribution from model 2 has greater accuracy.

Interestingly, the districts with high habitat suitability class (0.6–maximum value) for models 1 and 2 were the same (Fig. 10.9). This infers that *A. spectabilis* is chiefly distributed in Rudraprayag, Chamoli, and Uttarkashi districts. This is also evident from the overlay map, wherein the maximum overlay is primarily situated in these three districts (Fig. 10.11). The primary region is confined to the eastern part of Uttarkashi, the northern region of Tehri Garhwal, the northern part of Rudraprayag, central to the northern fringes of Chamoli, and the western part of Pithoragarh and northern region of Bageshwar district (Fig. 10.11). These regions have the appropriate cold sub-alpine climate and appropriate mesic conditions. Our predicted regions for *A. spectabilis* were similar to the model predictions for *B. utilis* (Singh et al. 2013) and *D. hatagirea* in Uttarakhand (Chandra et al. 2021). These species share their elevational range with *A. spectabilis*. Overall, the highly suitable regions of the species are distributed in a tight group toward the northern region of Uttarakhand. The absence in the western portion of Uttarkashi district could be due to reduced annual precipitation in the district from east to west direction; furthermore lower suitability in the east (eastern portion of Pithoragarh) is due to reduced early winter season precipitation (November) (Singh and Singh 1987) which is critical for *A. spectabilis* as per our model predictions.

10.4.5 Implications for Conservation and Future Prospects

Being a near-threatened species, the conservation of *A. spectabilis* demands priority. From our current habitat suitability map (Fig. 10.9) and the area under different

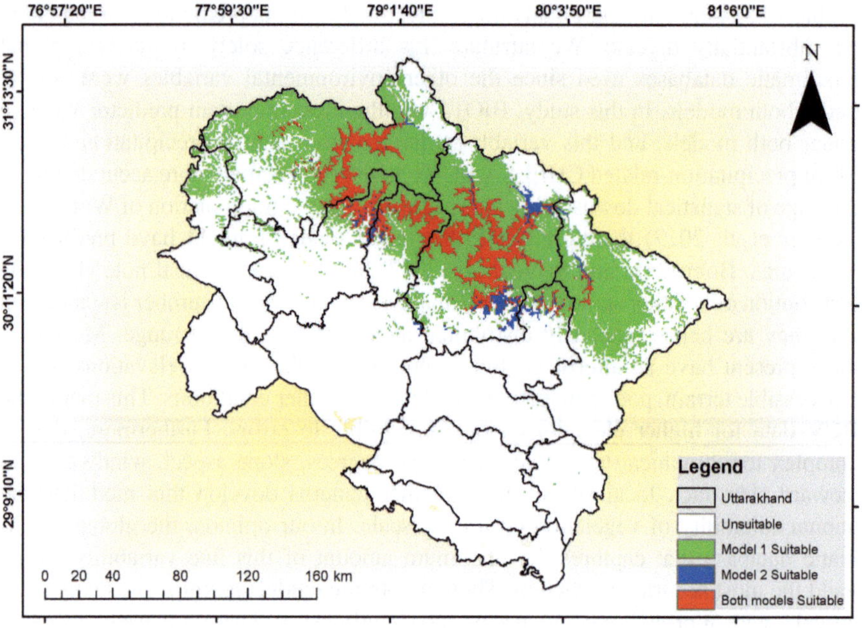

Fig. 10.11 An overlay of model 1 and model 2 habitat suitability regions for current habitat distribution of *A. spectabilis* in Uttarakhand. The poinsettia red color indicates the regions suitable under models 1 and 2

habitat suitability classes (Table 10.4), these ranges are higher than the actual occurrence of the species. This is because the MaxEnt-based habitat suitability model indicates the regions with environmental conditions suitable for *A. spectabilis*, i.e., the fundamental niche. However, the true distribution of the species is restricted by several factors such as geographic barriers, anthropogenic disturbances, and even biotic pressure (Phillips et al. 2006). These factors shape the actual distribution of the species, thus marking its realized niche. Therefore, policy planners must be cautious in using habitat suitability predictions. The regions we suggest as highly suitable or suitable represent the regions with environmental conditions conducive to *A. spectabilis*. They may or may not show their actual occurrence in those locations. However, being suitable, these regions could be selected for raising assisted regeneration and plantations of *A. spectabilis* for conservation. Furthermore, planners must have *A. spectabilis* as the primary choice for plantations in these regions.

Uttarkashi, Rudraprayag, and Chamoli are the predominant districts with high to medium habitat suitability (Fig. 10.10). Most of the highly suitable habitat regions in Uttarkashi district and the regions common to models 1 and 2 from the overlay map (Fig. 10.11) lie in the Gangotri National Park region. For Rudraprayag also, the suitable habitat falls under Kedarnath Wildlife Sanctuary. However, for Chamoli district, the eastern region is in the protected area status under Kedarnath Wildlife

Sanctuary, while the western region falls under Nanda Devi biosphere reserve; however, the central portion does not lie in any protected area category. Therefore, focusing on this region is critical since, under models 1 and 2, this region has a prominent presence of highly suitable habitat for *A. spectabilis*. Apart from these districts, the upper fringe of Tehri Garhwal, Bageshwar, and the northwestern portion of Pithoragarh also have noticeable regions with high suitability for *A. spectabilis*. Such areas should be physically monitored to assess the species' presence and its regeneration status which must be enhanced for future habitat security.

Those regions which are under the protected area must also not be considered entirely conserved. For instance, in our earlier work (Kaushal et al. 2021) in the *A. spectabilis*-dominant forest at Tungnath, Uttarakhand, we found the forest to show unimodal girth class distribution with negligible occurrence in the 10–20-cm diameter at breast height (DBH) range indicating poor regeneration. Furthermore, the region has immense grazing pressure, especially from nomadic grazers, and intense fuelwood demand from the local community, especially during the pilgrim and tourist season (Rai et al. 2012a). Stringent monitoring is essential, especially in the protected areas, though understandably, the livelihood of the local communities is dependent upon the surrounding natural resources, and the growing tourist pressure over the years has led to extensive utilization beyond the sustainable capacity. Specified regions should be demarcated for grazing, and using species with efficient regeneration for fuelwood could help reduce the anthropogenic pressure on this endemic Himalayan conifer. Our observations show that many threatened medicinal plants have similar niche requirements as *A. spectabilis*. Therefore, it is a win-win strategy where the conservation of one species would also ensure the conservation of several associated species. Thus, preserving the present habitat and using our predictions for suitable regions for expanding the habitat using *A. spectabilis*-assisted regeneration and plantation are the optimum management requirements.

In terms of future perspectives, we suggest an extensive hyper-accurate survey for delineating the occurrence records of *A. spectabilis* throughout the state. Being a treeline species determining absence points near the treeline would also help fine-tune the models for future climate predictions to accurately monitor the species shift under changing climate scenarios. Presently, we omitted majority of the BCVs on account of autocorrelation, and determination of actual predictor variables associated with the species in the future would help increase the accuracy of predictions. We avoided future climate modeling that can be taken after such issues are addressed to determine the geographical regions with the most significant habitat loss.

10.5 Conclusions

The accuracy of the habitat suitability model is as good as the input parameters. The results indicate model 2 to have greater predictive power; however, the performance of model 1 was also not poor. This methodological dilemma can be avoided by selecting the highly suitable regions under both models since CHELSA-based BCVs are more accurate with precipitation-based BCVs, and for our study the most important predictor variable was the precipitation amount in the driest month (BIO14). The information from the most important predictor environment variables must be associated with conservation planning approaches especially in plantations wherein the conditions can be modulated for *A. spectabilis* suitability. The species has mere 1.55% area as highly suitable (model 2) in Uttarakhand, and this region is further tightly grouped into five northern districts, i.e., Uttarkashi, Tehri Garhwal, Rudraprayag, Chamoli, and Pithoragarh. Though a few of these districts have protected areas that encompass the high suitability regions, our opinion is that for efficient management, the high-elevation sub-alpine to alpine regions for all these northern districts must be under a common protected area status. This would help conserve *A. spectabilis* along with other endangered species that have a similar ecological niche. The regions with current occurrence reflect the realized niche space of *A. spectabilis* and require proactive monitoring. The conservation in these regions would only be possible with the complete support of the local stakeholders and their inclusion in the conservation planning processes. This is critical since the habitat suitability of *A. spectabilis* is at high-elevation regions where regular monitoring can become problematic; in such situation making the local people aware of the conservation necessity would help the situation. The present condition of *A. spectabilis* is perilous, and without any conservation response, the situation can become much worse since this species is one of the dominant forest-forming species in high-elevation alpine timberline – treeline ecotone. Therefore, assisted regeneration, expanding habitat by plantations in the predicted suitable regions, and conservation of the current population are prudent.

Acknowledgments The authors acknowledge the Principal Chief Conservator of Forest (Chief Wildlife Warden), Uttarakhand, and DFO, Kedarnath Wildlife Sanctuary, for their necessary research permission and logistic support during the research. The corresponding author acknowledges the financial support received from the Science and Engineering Research Board (SERB), Government of India, through research project no. EEQ/2016/000164, and Institution of Eminence, University of Delhi (IoE/2021/12/FRP), Faculty Research Programme (FRP), sanctioned during 2021–2022 and 2022–2023. The first author thanks the University Grants Commission (UGC), New Delhi, for providing Senior Research Fellowship. Anshu Siwach thanks the Council of Scientific and Industrial Research, New Delhi, for providing financial assistance as Junior Research Fellowship. Prachi Sharma thanks the UGC and the University of Delhi for the non-NET fellowship.

References

Adhikari D, Tiwary R, Singh PP, Upadhaya K, Singh B, Haridasan KE, Bhatt BB, Chettri A, Barik SK (2019) Ecological niche modeling as a cumulative environmental impact assessment tool for biodiversity assessment and conservation planning: a case study of critically endangered plant *Lagerstroemia minuticarpa* in the Indian Eastern Himalaya. J Environ Manag 243:299–307. https://doi.org/10.1016/j.jenvman.2019.05.036

Allen MR, Dube OP, Solecki WA, Aragón-Durand F, Cramer W, Humphreys S, Kainuma M, Kala J, Mahowald N, Mulugetta Y, Perez R, Wairiu M, Zickfeld K (2019) Framing and context. In: Masson-Delmotte V, Zhai P, Portner H-O, Roberts D, Skea J, Shukla PR, Pirani A, Moufouma-Okia W, Péan C, Pidcock R, Connors S, Matthews JBR, Chen Y, Zhou X, Gomis MI, Lonnoy E, Maycock T, Tignor M, Waterfield T (eds) Global warming of 1.5°C. An IPCC special report on the impacts of global warming of 1.5°C above pre-industrial levels and related global greenhouse gas emission pathways, in the context of strengthening the global response to the threat of climate change. Intergovernmental Panel on Climate Change (IPCC), Geneva, p 84

Attri SD, Chug S (2021) Annual Report 2020. Information Science & Knowledge Resource Development Division, Indian Meteorology Department. New Delhi

Attri SD, Tyagi A (2010) Climate profile of India. Environment Monitoring and Research Center, India Meteorology Department, New Delhi

Bader MY, van Geloof I, Rietkerk M (2007) High solar radiation hinders tree regeneration above the alpine treeline in northern Ecuador. Plant Ecol 191(1):33–45. https://doi.org/10.1007/s11258-006-9212-6

Barbeito I, Dawes MA, Rixen C, Senn J, Bebi P (2012) Factors driving mortality and growth at treeline: a 30-year experiment of 92 000 conifers. Ecology 93(2):389–401. https://doi.org/10.1890/11-0384.1

Becker A, Bugmann H (eds) (2001) Global change and mountain regions. The mountain research initiative. IGBP Report 49, Stockholm

Beniston M (2003) Climatic change in mountain regions: a review of possible impacts. Climate Change 59:5–31. https://doi.org/10.1023/A:1024458411589

Bhutiyani MR (2015) Climate change in the Northwestern Himalayas. In: Joshi R et al (eds) Dynamics of climate change and water resources of Northwestern Himalaya. Springer, Cham, pp 85–96

Bobrowski M, Schickhoff U (2017) Why input matters: selection of climate data sets for modelling the potential distribution of a treeline species in the Himalayan region. Ecol Model 359:92–102. https://doi.org/10.1016/j.ecolmodel.2017.05.021

Bobrowski M, Weidinger J, Schickhoff U (2021a) Is new always better? Frontiers in global climate datasets for modeling treeline species in the Himalayas. Atmos 12(5):543. https://doi.org/10.3390/atmos12050543

Bobrowski M, Weidinger J, Schwab N, Schickhoff U (2021b) Searching for ecology in species distribution models in the Himalayas. Ecol Model 458:109693. https://doi.org/10.1016/j.ecolmodel.2021.109693

Brown JL, Bennett JR, French CM (2017) SDMtoolbox 2.0: the next generation Python-based GIS toolkit for landscape genetic, biogeographic and species distribution model analyses. PeerJ 5:e4095. https://doi.org/10.7717/peerj.4095

Chaitanya R, Meiri S (2021) Can't see the wood for the trees? Canopy physiognomy influences the distribution of peninsular Indian Flying lizards. J Biogeogr 49(1):1–13. https://doi.org/10.1111/jbi.14298

Chakraborty A, Joshi PK, Sachdeva K (2016) Predicting distribution of major forest tree species to potential impacts of climate change in the central Himalayan region. Ecol Eng 97:593–609. https://doi.org/10.1016/j.ecoleng.2016.10.006

Chandra N, Singh G, Lingwal S, Jalal JS, Bisht MS, Pal V, Bisht MPS, Rawat B, Tiwari LM (2021) Ecological niche modeling and status of threatened alpine medicinal plant *Dactylorhiza hatagirea* D. Don in Western Himalaya. J Sustain For:1–17. https://doi.org/10.1080/10549811.2021.1923530

Chhetri PK, Gaddis KD, Cairns DM (2018) Predicting the suitable habitat of treeline species in the Nepalese Himalayas under climate change. Mt Res Dev 38(2):153–163. https://doi.org/10.1659/MRD-JOURNAL-D-17-00071.1

Chitale V, Behera MD (2019) How will forest fires impact the distribution of endemic plants in the Himalayan biodiversity hotspot? Biodivers Conserv 28:2259–2273. https://doi.org/10.1007/s10531-019-01733-8

Climate-Data.org (2022) Climate data for cities worldwide. https://en.climate-data.org/asia/india/uttarakhand-763/. Accessed 20 Apr 2022

Dhyani S, Kadaverugu R, Dhyani D, Verma P, Pujari P (2018) Predicting impacts of climate variability on habitats of *Hippophae salicifolia* (D. Don) (Seabuckthorn) in Central Himalayas: future challenges. Eco Inform 48:135–146. https://doi.org/10.1016/j.ecoinf.2018.09.003

Dhyani S, Kadaverugu R, Pujari P (2020) Predicting impacts of climate variability on Banj oak (*Quercus leucotrichophora* A. Camus) forests: understanding future implications for Central Himalayas. Reg Environ Chang 20(4):1–13. https://doi.org/10.1007/s10113-020-01696-5

Dhyani A, Kadaverugu R, Nautiyal BP, Nautiyal MC (2021) Predicting the potential distribution of a critically endangered medicinal plant *Lilium polyphyllum* in Indian Western Himalayan region. Reg Environ Chang 21(2):1–11. https://doi.org/10.1007/s10113-021-01763-5

Di Pasquale G, Saracino A, Bosso L, Russo D, Moroni A, Bonanomi G, Allevato E (2020) Coastal pine-oak glacial refugia in the Mediterranean basin: a biogeographic approach based on charcoal analysis and spatial modelling. Forests 11(6):673. https://doi.org/10.3390/f11060673

Diaz HF, Grosjean M, Graumlich L (2003) Climate variability and change in high elevation regions: past, present and future. Clim Chang 59(1):1–4. https://doi.org/10.1023/A:1024416227887

Elith J, Franklin J (2013) Species distribution modeling. In: Encyclopedia of biodiversity, 2nd edn. Elsevier, pp 692–705

Fajardo J, Lessmann J, Bonaccorso E, Devenish C, Muñoz J (2014) Combined use of systematic conservation planning, species distribution modelling, and connectivity analysis reveals severe conservation gaps in a megadiverse country (Peru). PLoS One 9(12):e114367. https://doi.org/10.1371/journal.pone.0122159

FAO, ITPS (2018) Global soil organic carbon map (GSOCmap) Technical Report, Rome, p 162

Fick SE, Hijmans RJ (2017) WorldClim 2: new 1km spatial resolution climate surfaces for global land areas. Int J Climatol 37(12):4302–4315. https://doi.org/10.1002/joc.5086

Fielding AH, Bell JF (1997) A review of methods for the assessment of prediction errors in conservation presence/absence models. Environ Conserv 24(1):38–49. https://doi.org/10.1017/S0376892997000088

FSI (2019) India State of Forest Report 2019. Forest Survey of India, Ministry of Environment Forest and Climate Change, Government of India. Dehradun, India

Gaur RD (1999) Flora of the District Garhwal. North West Himalaya. Transmedia, Srinagar

GBIF (2022) GBIF occurrence download. https://doi.org/10.15468/dl.rfwpc8. Accessed 16 Apr 2022

Grabherr G, Messerli B (2011) An overview of the world mountain environments. In: Austrian MAB committee, pp 8–14

Hamid M, Khuroo AA, Charles B, Ahmad R, Singh CP, Araving NA (2019) Impact of climate change on the distribution range and niche dynamics of Himalayan birch, a typical treeline species in Himalayas. Biodivers Conserv 28:2345–2370. https://doi.org/10.1007/s10531-018-1641-8

Hernandez PA, Graham CH, Master LL, Albert DL (2006) The effect of sample size and species characteristics on performance of different species distribution modeling methods. Ecography 29:773–785. https://doi.org/10.1111/j.0906-7590.2006.04700.x

IPCC (2019) Summary for policymakers. In: Shukla PR, Skea J, Buendia EC, Masson-Delmotte V, Pörtner H-O, Roberts DC, Zhai P, Slade R, Connors S, van Diemen R, Ferrat M, Haughey E, Luz S, Neogi S, Pathak M, Petzold J, Pereira JP, Vyas P, Huntley E, Kissick K, Belkacemi M, Malley J (eds) Climate change and land: an IPCC special report on climate change, desertification, land degradation, sustainable land management, food security, and greenhouse gas fluxes in terrestrial ecosystems. In press

Kaky E, Nolan V, Alatawi A, Gilbert F (2020) A comparison between Ensemble and MaxEnt species distribution modelling approaches for conservation: a case study with Egyptian medicinal plants. Eco Inform 60:101150. https://doi.org/10.1016/j.ecoinf.2020.101150

Karger D, Conrad O, Böhner J, Kawohl T, Kreft H, Soria-Auza RW, Zimmermann NE, Linder HP, Kessler M (2017) Climatologies at high resolution for the earth's land surface areas. Sci Data 4: 170122. https://doi.org/10.1038/sdata.2017.122

Kaushal S, Siwach A, Baishya R (2021) Diversity, regeneration, and anthropogenic disturbance in major Indian Central Himalayan forest types: implications for conservation. Biodivers Conserv 30(8):2451–2480. https://doi.org/10.1007/s10531-021-02203-w

Kumar P (2012) Assessment of impact of climate change on Rhododendrons in Sikkim Himalayas using Maxent modelling: limitations and challenges. Biodivers Conserv 21(5):1251–1266. https://doi.org/10.1007/s10531-012-0279-1

Kumar S, Stohlgren TJ (2009) Maxent modeling for predicting suitable habitat for threatened and endangered tree *Canacomyrica monticola* in New Caledonia. Journal of Ecology and the Natural Environment 1(4):094–098

Malanson GP, Resler LM, Butler DR, Fagre DB (2019) Mountain plant communities: uncertain sentinels? Prog Phys Geograp Earth Environ 43(4):521–543. https://doi.org/10.1177/0309133319843873

Manish K, Pandit MK (2019) Identifying conservation priorities for plant species in the Himalaya in current and future climates: a case study from Sikkim Himalaya, India. Biol Conserv 233:176–184. https://doi.org/10.1016/j.biocon.2019.02.036

Manish K, Telwala Y, Nautiyal DC, Pandit MK (2016) Modelling the impacts of future climate change on plant communities in the Himalaya: a case study from Eastern Himalaya, India. Model Earth Syst Environ 2(92). https://doi.org/10.1007/s40808-016-0163-1

McShea WJ (2014) What are the roles of species distribution models in conservation planning? Environ Conserv 41(2):93–96

Morales-Barbero J, Vega-Álvarez J (2019) Input matters matter: bioclimatic consistency to map more reliable species distribution models. Methods Ecol Evol 10(2):212–224. https://doi.org/10.1111/2041-210X.13124

Nakao K, Higa M, Tsuyama I, Matsui T, Horikawa M, Tanaka N (2013) Spatial conservation planning under climate change: using species distribution modeling to assess priority for adaptive management of *Fagus crenata* in Japan. J Nat Conserv 21(6):406–413. https://doi.org/10.1016/j.jnc.2013.06.003

Nandy SN, Dhyani PP, Samal PK (2009) Resource information database of the Indian Himalaya. ENVIS Monograph 3:108

Negi GCS (2022) Trees, forests and people: the Central Himalayan case of forest ecosystem services. Trees For People 8:100222. https://doi.org/10.1016/j.tfp.2022.100222

Negi VS, Giri L, Sekar KC (2018) Floristic diversity, community composition and structure in Nanda Devi National Park after prohibition of human activities, Western Himalaya, India. Curr Sci 115(6):1056–1064

Noce S, Caporaso L, Santini M (2020) A new global dataset of bioclimatic indicators. Sci Data 7(1): 1–12. https://doi.org/10.1038/s41597-020-00726-5

Osorio-Olvera L, Lira-Noriega A, Soberón J, Townsend Peterson A, Falconi M, Contreras-Díaz RG, Martínez-Meyer E, Barve V, Barve N (2020) ntbox: an R package with graphical user interface for modeling and evaluating multidimensional ecological niches. Methods Ecol Evol 11:1199–1206. https://doi.org/10.1111/2041-210X.13452

Pandit MK (2013) The Himalayas must be protected. Nature 501(7467):283–283. https://doi.org/10.1038/501283a

Peterson AT, Papeş M, Soberón J (2008) Rethinking receiver operating characteristic analysis applications in ecological niche modeling. Ecol Model 213(1):63–72. https://doi.org/10.1016/j.ecolmodel.2007.11.008

Peterson AT, Soberón J, Pearson RG, Anderson RP, Martínez-Meyer E, Nakamura M, Araújo MB (2011) Ecological niches and geographic distributions (MPB-49). Princeton University Press

Phillips SJ (2017) A brief tutorial on maxent. Available from url: http://biodiversityinformatics.amnh.org/open_source/maxent/. Accessed 16 Apr 2022

Phillips SJ, Duduk M, Schapire RE (2004) A maximum entropy approach to species distribution modeling. In: Proceedings of the twenty-first international conference on machine learning. ACM Press, pp 472–486. https://doi.org/10.1145/1015330.1015412

Phillips SJ, Anderson RP, Schapire RE (2006) Maximum entropy modeling of species geographic distributions. Ecol Model 190(3-4):231–259. https://doi.org/10.1016/j.ecolmodel.2005.03.026

Phillips SJ, Anderson RP, Dudík M, Schapire RE, Blair ME (2017) Opening the black box: an open-source release of Maxent. Ecography 40(7):887–893. https://doi.org/10.1111/ecog.03049

Rai ID, Adhikari BS, Rawat GS, Bargali K (2012a) Community structure along timberline ecotone in relation to micro-topography and disturbances in Western Himalaya. Not Sci Biol 4(2):41–52

Rai ID, Adhikari BS, Rawat GS (2012b) Floral diversity along sub-alpine and alpine ecosystems in Tungnath area of Kedarnath wildlife sanctuary, Uttarakhand. Indian Forester 138(10):927–940

Rawal RS, Rawal R, Rawat B, Negi VS, Pathak R (2018) Plant species diversity and rarity patterns along altitude range covering treeline ecotone in Uttarakhand: conservation implications. Trop Ecol 59(2):225–239

Rawat N, Purohit S, Painuly V, Negi GS, Bisht MPS (2022) Habitat distribution modeling of endangered medicinal plant *Picrorhiza kurroa* (Royle ex Benth) under climate change scenarios in Uttarakhand Himalaya, India. Ecol Inform 68:101550. https://doi.org/10.1016/j.ecoinf.2021.101550

Rodríguez-Rodríguez D, Bomhard B (2011) Towards effective conservation in mountains: protected areas and biosphere reserves. In: Austrian MAB Committee (ed) Biosphere reserves in the mountains of the world excellence in the clouds? Austrian Sciences Academy

Sanderson EW, Jaiteh M, Levy MA, Redford KH, Wannebo AV, Woolmer G (2002) The human footprint and the last of the wild: the human footprint is a global map of human influence on the land surface, which suggests that human beings are stewards of nature, whether we like it or not. Bioscience 52(10):891–904. https://doi.org/10.1641/0006-3568(2002)052[0891:THFATL]2.0.CO;2

Schickhoff U, Bobrowski M, Böhner J, Bürzle B, Chaudhary RP, Gerlitz L, Heyken JL, Muller M, Scholten T, Schwab N, Wedegärtner R (2015) Do Himalayan treelines respond to recent climate change? An evaluation of sensitivity indicators. Earth Syst Dynam 6(1):245–265. https://doi.org/10.5194/esd-6-245-2015

Schickhoff U, Singh RB, Mal S (2016) Climate change and dynamics of glaciers and vegetation in the himalaya: an overview. In: Singh R, Schickhoff U, Mal S (eds) Climate change, glacier response, and vegetation dynamics in the Himalaya. Springer, Cham. https://doi.org/10.1007/978-3-319-28977-9_1

Sharma CM, Tiwari OP, Rana YS, Krishan R, Mishra AK (2018) Elevational behaviour on dominance–diversity, regeneration, biomass and carbon storage in ridge forests of Garhwal Himalaya, India. For Ecol Manag 424:105–120. https://doi.org/10.1016/j.foreco.2018.04.038

Singh P, Negi GCS (2018) Treeline species phenology: shoot growth, leaf characteristics and nutrient dynamics. Trop Ecol 59(2):297–311

Singh JS, Singh SP (1987) Forest vegetation of the Himalaya. Bot Rev 53(1):80–192. https://doi.org/10.1007/BF02858183

Singh CP, Panigrahy S, Thapliyal A, Kimothi MM, Soni P, Parihar JS (2012) Monitoring the alpine treeline shift in parts of the Indian Himalayas using remote sensing. Curr Sci 102(4):559–562

Singh CP, Panigrahy S, Parihar JS, Dharaiya N (2013) Modeling environmental niche of Himalayan birch and remote sensing based vicarious validation. Trop Ecol 54(3):319–327

Singh U, Phulara M, David B, Ranhotra PS, Shekhar M, Bhattacharyya A, Dhyani R, Joshi R, Pal AK (2018) Static tree line of Himalayan silver fir since last several decades at Tungnath, western Himalaya. Trop Ecol 59:351–363

Singh N, Tewari A, Shah S (2019) Tree regeneration pattern and size class distribution in anthropogenically disturbed sub-alpine treeline areas of Indian Western Himalaya. Int J Sci Technol Res 8(08):537–546

Singh SP, Gumber S, Singh RD, Singh G (2020) How many tree species are in the Himalayan treelines and how are they distributed? Trop Ecol 61:317–327. https://doi.org/10.1007/s42965-020-00093-7

Singh CP, Mohapatra J, Mathew JR, Khuroo AA, Hamid M, Malik AH (2021a) Long-term observation and modelling on the distribution and patterns of alpine treeline ecotone in Indian Himalaya. J Geom 15(1)

Singh A, Samant SS, Naithani S (2021b) Population ecology and habitat suitability modelling of *Betula utilis* D. Don in the sub-alpine ecosystem of Great Himalayan National Park, North-Western Indian Himalaya: a UNESCO World Heritage site. Proc Indian Natl Sci Acad 87:640–656. https://doi.org/10.1007/s43538-021-00055-0

Singh A, Samant SS, Naithani S (2021c) Population ecology and habitat suitability modelling of *Quercus semecarpifolia* Sm. in the sub-alpine ecosystem of Great Himalayan National Park, northwestern Himalaya, India. S Afr J Bot 141:158–170

Siwach A, Kaushal S, Baishya R (2021) Terricolous mosses impact soil microbial biomass carbon and enzymatic activity under temperate forest types of the Garhwal Himalayas. Environ Monit Assess 193(8):1–14. https://doi.org/10.1007/s10661-021-09295-5

Spiers JA, Oatham MP, Rostant LV, Farrell AD (2018) Applying species distribution modelling to improving conservation based decisions: a gap analysis of Trinidad and Tobago's endemic vascular plants. Biodivers Conserv 27(11):2931–2949. https://doi.org/10.1007/s10531-018-1578-y

Srivastava V, Griess VC, Padalia H (2018) Mapping invasion potential using ensemble modelling. A case study on *Yushania maling* in the Darjeeling Himalayas. Ecol Model 385:35–44. https://doi.org/10.1016/j.ecolmodel.2018.07.001

Swets JA (1988) Measuring the accuracy of diagnostic systems. Science 240(4857):1285–1293. https://doi.org/10.1126/science.3287615

Tariq M, Nandi SK, Bhatt ID, Bhavsar D, Roy A, Pande V (2021) Phytosociological and niche distribution study of *Paris polyphylla* smith, an important medicinal herb of Indian Himalayan region. Trop Ecol 62(2):163–173. https://doi.org/10.1007/s42965-020-00125-2

Tiwari OP, Rana YS, Krishan R, Sharma CM, Bhandari BS (2018) Regeneration dynamics, population structure, and forest composition in some ridge forests of the Western Himalaya, India. For Sci Technol 14(2):66–75. https://doi.org/10.1080/21580103.2018.1447517

Upgupta S, Sharma J, Jayaraman M, Kumar V, Ravindranath NH (2015) Climate change impact and vulnerability assessment of forests in the Indian Western Himalayan region: a case study of Himachal Pradesh, India. Clim Risk Manag 10:63–76. https://doi.org/10.1016/j.crm.2015.08.002

Uttarakhand at a Glance (2020) Uttarakhand at a Glance 2018-19. Directorate of Economics and Statistics, Department of Planning, Government of Uttarakhand, Dehradun

Veloz SD (2009) Spatially autocorrelated sampling falsely inflates measures of accuracy for presence-only niche models. J Biogeogr 36(12):2290–2299. https://doi.org/10.1111/j.1365-2699.2009.02174.x

Wani IA, Verma S, Kumari P, Charles B, Hashim MJ, El-Serehy HA (2021) Ecological assessment and environmental niche modelling of Himalayan rhubarb (*Rheum webbianum* Royle) in northwest Himalaya. PLoS One 16(11):e0259345. https://doi.org/10.1371/journal.pone.0259345

WCS—Wildlife Conservation Society, and Center for International Earth Science Information Network—CIESIN—Columbia University (2005) Last of the wild project, version 2, 2005 (LWP-2): Global Human Influence Index (HII) Dataset (Geographic). NASA Socioeconomic Data and Applications Center (SEDAC), Palisades, New York. https://doi.org/10.7927/H4BP00QC. Accessed 16 Apr 2022

Yadava AK, Sharma YK, Dubey B, Singh J, Singh V, Bhutiyani MR, Yadav RR, Misra KG (2017) Altitudinal treeline dynamics of Himalayan pine in western Himalaya, India. Quat Int 444:44–52. https://doi.org/10.1016/j.quaint.2016.07.032

Yang XQ, Kushwaha SPS, Saran S, Xu J, Roy PS (2013) Maxent modeling for predicting the potential distribution of medicinal plant, *Justicia adhatoda* L. in Lesser Himalayan foothills. Ecol Eng 51:83–87

Yoon S, Lee WH (2021) Methodological analysis of bioclimatic variable selection in species distribution modeling with application to agricultural pests (*Metcalfa pruinosa* and *Spodoptera litura*). Comput Electron Agric 190:106430. https://doi.org/10.1016/j.compag.2021.106430

Zhang D, Rushforth K, Katsuki T (2011) *Abies spectabilis*. The IUCN red list of threatened species 2011: e.T42300A10686224. https://doi.org/10.2305/IUCN.UK.2011-2.RLTS.T42300A10686224.en

Zhang MG, Zhou ZK, Chen WY, Slik JF, Cannon CH, Raes N (2012) Using species distribution modeling to improve conservation and land use planning of Yunnan, China. Biol Conserv 153: 257–264. https://doi.org/10.1016/j.biocon.2012.04.023

Zisenis M, Price MF (2011) Europe's mountain biodiversity: status and threats. In: Austrian MAB Committee (ed) Biosphere reserves in the mountains of the world excellence in the clouds? Austrian Sciences Academy

11 Modelling the Distribution of a Medicinal Plant *Oroxylum indicum* (L.) Kurz for Its Conservation in Arunachal Pradesh

Dhoni Bushi, Oyi Dai Nimasow, and Gibji Nimasow

Abstract

The natural populations of *Oroxylum indicum*—an important medicinal plant—is declining due to habitat loss and overexploitation. Modelling its potential distribution area can help in its conservation. We used maximum entropy (MaxEnt) model to predict the potential distribution of the species and identify the factors determining its niche. The model performance was consistent and good with area under curve (AUC) value of 0.827 and true skill statistic (TSS) of 0.658. The major environmental variables determining the niche of the species were precipitation of the warmest quarter (49.7%) and annual mean temperature (19.6%). The model predicted 6.01% as highly suitable, 23.29% as moderately suitable and 70.70% as least suitable. The model predictions could be is useful in identifying the critical habitats for conservation and restoration of important plant resources.

Keywords

Modelling · Habitat · Maximum entropy · Medicinal plant · Arunachal Pradesh

D. Bushi · G. Nimasow (✉)
Department of Geography, Rajiv Gandhi University, Rono Hills, Doimukh, Arunachal Pradesh, India
e-mail: dhoni.bushi@rgu.ac.in; gibji.nimasow@rgu.ac.in

O. D. Nimasow
Department of Botany, Rajiv Gandhi University, Rono Hills, Doimukh, Arunachal Pradesh, India
e-mail: oyidai.nimasow@rgu.ac.in

S. Dhyani et al. (eds.), *Ecosystem and Species Habitat Modeling for Conservation and Restoration*, https://doi.org/10.1007/978-981-99-0131-9_11

11.1 Introduction

The spatial and temporal distribution of a species is an important information in ecological studies. An understanding of the constraints in geographical distribution based on environmental factors and available occurrence records of species is vital for many purposes such as successful conservation of a species (Graham and Hijmans 2006; Glor and Warren 2011). Species distribution modelling (SDM) techniques both statistical and machine learning are gaining popularity in recent days. SDMs are also known as climate envelope modelling, habitat modelling and environmental or ecological niche modelling (Elith and Leathwick 2009; Guisan et al. 2013, 2017; Hamann and Wang 2006; Jeschke and Strayer 2008). It is a numerical tool that combines species occurrence or abundance with the environmental estimates. SDMs are used to understand the ecological and evolutionary history and to predict distributions across landscapes, sometimes requiring extrapolation in space and time. SDMs make predictions of where species may be present but unrecorded or where they might be found if anthropogenic activities had not removed them (Anderson et al. 2009) and predictions of where species may be in the future due to changes in distribution (Parmesan and Yohe 2003). Such predictions are useful in identifying spatial priorities for conservation (Vaughan and Ormerod 2003; Kremen et al. 2008) and to assess risks from climate change to particular species (Julliard et al. 2004; Huntley et al. 2008).

According to Skov (2000), the application of modelling techniques in the distribution of plants are emerging as a powerful tool that combines locations from herbarium specimens, desktop modelling softwares and geographic information system (GIS). SDMs correlate the presence/absence species records with climatic variables to predict current distribution and future potential distribution under climate change scenarios (Trisurat et al. 2011). Out of several modelling algorithms, maximum entropy (MaxEnt) is a widely used and globally accepted technique for species distribution modelling (Graham and Hijmans 2006; Baldwin 2009; Ramírez-Villegas and Bueno Cabrera 2009). Additionally, it requires presence-only data of the species, uses concise mathematical algorithm and quantitative assessment of variable contribution and generates spatially explicit habitat suitability maps (Phillips et al. 2006; Elith et al. 2010; Yi et al. 2016).

Oroxylum indicum (L.) Kurz (family: Bignoniaceae) is an ornamental tree which is 8–15 m tall and branched at top. The bark is light-brown and often with numerous corky lenticels (Deka et al. 2013). The species is mainly distributed in Cambodia, Fujian, Guangdong, Guangxi, Guizhou, Sichuan, India, Indonesia (Java, Sumatra), Laos, Malaysia, Myanmar, Nepal, the Philippines, Taiwan, Thailand, Vietnam and Yunnan (Lawania et al. 2010). In the Indian subcontinent, it is mostly found in the foothills of the Himalayas, Bhutan, southern China and Sri Lanka (Theobald et al. 1981).

The seeds are purgative and taken orally to treat throat infections and hypertension (Singh et al. 2002); fruits are stomachic, anthelmintic, effective in throat and heart diseases, piles, bronchitis and useful in leucoderma (Chopra et al. 2002; Drury 2006; Nadkarni 1982; Khare 2007); and the leaves are prescribed for snake bite

(Nadkarni 1982; Khare 2007), enlarged spleen, headaches, ulcers, analgesic and antimicrobial (Prakash 2005; Drury 2006). The indigeneous tribal communities of Arunachal Pradesh like Adi, Galo, Nyishi, Monpa, Tagin and Wancho use it in the treatment of liver diseases, stomachache, cancer, itching, inflammation, tuberculosis, diarrhoea, rheumatism, jaundice and heart diseases (Khongsai et al. 2015; Murtem and Chaudhry 2016; Tangjang et al. 2011; Tripathi and Limasenla 2017). The plant materials are also used as wood, tannins and dyestuffs. Thus, *O. indicum* is both economically and medicinally important plant. The natural population of the species is under threat due to indiscriminate harvesting and habitat degradation. Therefore, in this study we attempted a species distribution modelling of *O. indicum* in Arunachal Pradesh, India, based on the occurence records and environmental variables to assess the potential suitable areas of this important medicinal plant. The results would be helpful in evolving appropriate strategies for conservation of the plant.

11.2 Materials and Methods

11.2.1 Study Area

The study was conducted in the state of Arunachal Pradesh (26.28° N–29.30° N latitudes and 91.20° E–97.30° E longitudes) in northeast India. It shares international border with Bhutan in the west, China in the north, Myanmar in east and state boundaries with Assam and Nagaland (Fig. 11.1). The state is characterized by a series of high ridges and low valleys. The major rivers are Kameng, Dibang, Siang, Subansiri, Lohit and Tirap. The state receives an average annual rainfall of 300–500 cm and an average temperature of 15–21 °C (winter) and 22–30 °C (summer). With a wide range of floral and faunal species, including rare and endemic species, this region is one of the biodiversity hotspots of the world.

11.2.2 Occurrence Data

A total of 217 occurrence records of *O. indicum* have been collected through field survey, online database and published literature. Out of which, 186 occurrence points were collected using a handheld Garmin GPS-based field survey (January 2022–July 2022), 12 occurrence records from various published literature and only 1 occurrence point from the Global Biodiversity Information Facility (GBIF 2022). We used a spatial thinning method (spThin package) in R (Hijmans et al. 1999) to reduce spatial sampling bias, following which 76 geo-referenced points were selected for the model building.

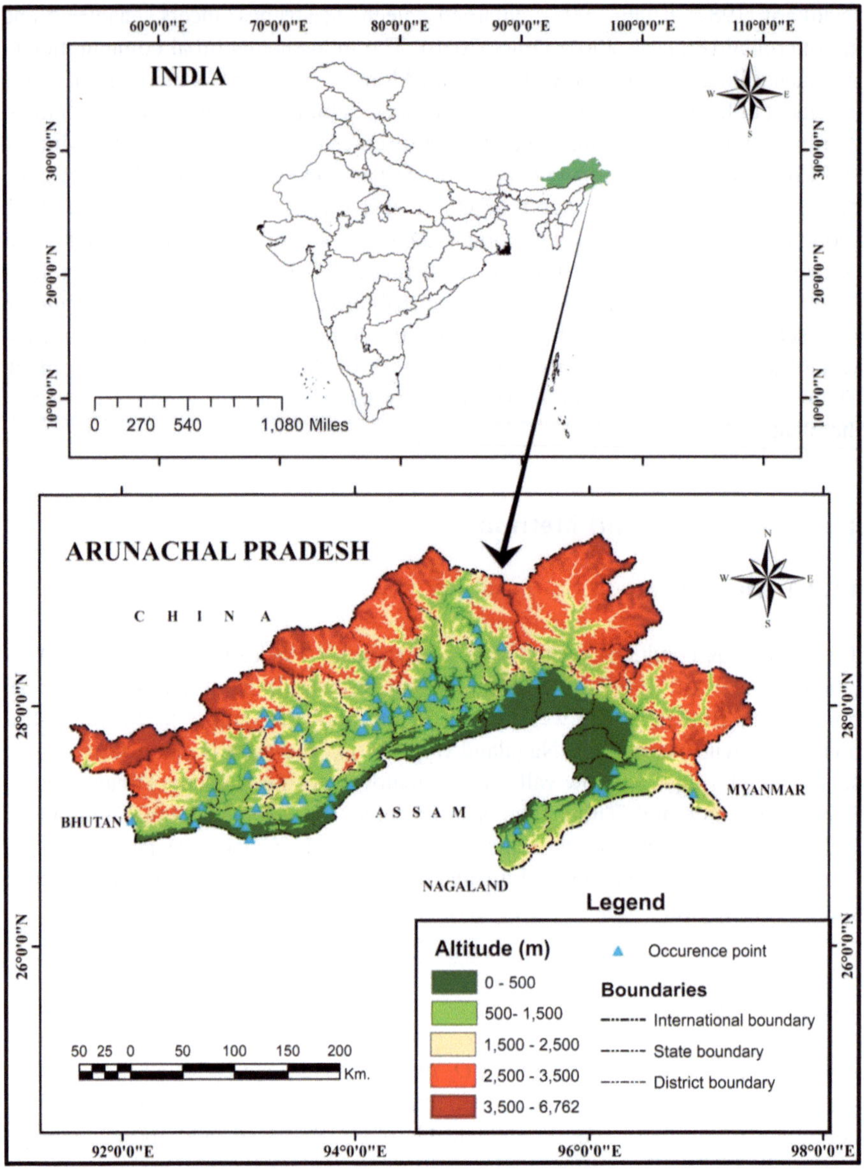

Fig. 11.1 Location map of the study area

11.2.3 Environmental Layers

To generate the distribution modelling of *O. indicum*, a set of environmental, topographical, land use/land cover and soil properties with 30 m resolution were acquired from various sources like WorldClim, ESRI land cover and World Soil

Table 11.1 Environmental variables used in the final model

S. no.	Environmental variables	Code	Unit
1.	Annual mean temperature	BIO1	°C
2.	Precipitation of the warmest quarter	BIO18	mm
3.	Soil pH	pH	g/kg
4.	Clay soil	Clay	g/kg
5.	Silt soil	Silt	g/kg
6.	Slope	Clip_slope	mm
7	Aspect	Clip_aspect	m
8.	Proximity to drainage	Proximity_drainage	–
9.	Land use/land cover	LULC	–

Information (ISRIC). The parameters include 19 bioclimatic variables, 4 soil variables and 1 land use/land cover. Additionally, the topographical variables like altitude, slope, aspect and drainage proximity were derived from the Advanced Spaceborne Thermal Emission and Reflection Radiometer Global Digital Elevation Model (ASTER GDEM) database with 30 m spatial resolution using ArcGIS 10.3. The highly correlated variables were removed as such variables negatively affect model performance and result in inaccurate predictions (Parolo et al. 2008; Merow et al. 2013; Dormann et al. 2013; Manzoor et al. 2018). Ideally, the correlation between predictor variables should be $|r| > 0.7$ as reported by various studies (Dormann et al. 2013; Manzoor et al. 2018; Sony et al. 2018; Farrell et al. 2019; Feng et al. 2019). After performing multicollinearity test using *usdm* package in *R* (Naimi and Araújo 2016), nine predictor variables, namely, annual mean temperature (BIO1), precipitation of the warmest quarter (BIO18), land use/land cover, soil pH, clay soil, silt soil, slope, aspect and proximity to drainage, were included in the final model (Table 11.1).

11.2.4 Model Settings and Evaluation

MaxEnt (version 3.4.1) was used for the potential niche modelling of *O. indicum*. MaxEnt technique is more reliable than other methods as it provides accurate prediction for small sample sizes (Hernandez et al. 2006; Wisz et al. 2008; Baldwin 2009) and resilient to spatial errors and sampling-biased occurrence data (Baldwin 2009; Graham et al. 2008). MaxEnt performs better than other methods of niche modelling (Phillips et al. 2006; Peterson et al. 2008; Phillips 2008; Radosavljevic and Anderson 2014; Nimasow et al. 2016). The model used 500 iterations, 0.00001 convergence threshold, 0.5 prevalence, 10,000 background points, 10 percentile training presence and 5 replicate model runs with cross-validation technique to ensure and assess the model reliability (Pearson et al. 2004). The individual contributions and permutation importance were analysed using variable contribution table and jackknife test (Baldwin 2009; Phillips et al. 2017a, b; Thapa et al. 2018). The threshold-independent, area under curve (AUC), receiver operating

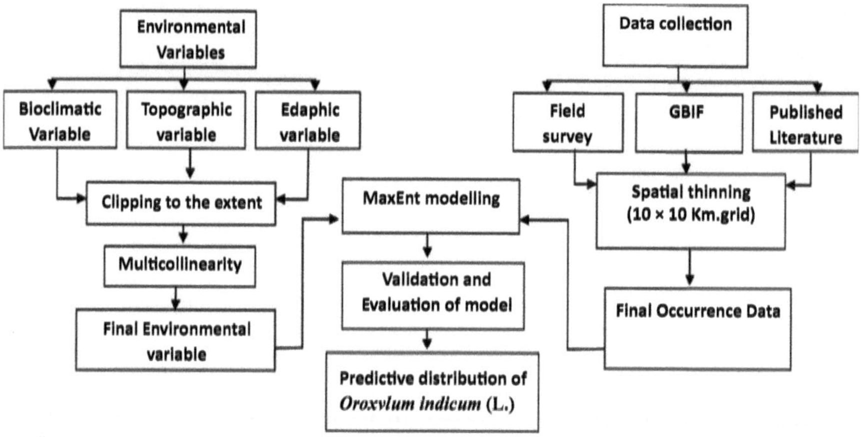

Fig. 11.2 Methodology followed in the study

characteristic (ROC) curve (Hanley and McNeil 1982; Manel et al. 2001) and true skill statistics (TSS) were used to evaluate model performance. Logistic response option was used to represent the probability of the presence of species within a range of 0–1 showing a range of suitability scale (Phillips 2008). The average logistic outputs were imported into ArcMap 10.3, and the suitability ranges were determined by employing 10 percentile training presence threshold rule showing suitable and unsuitable areas of *O. indicum*. The complete methodology applied in this study is shown in Fig. 11.2.

11.3 Results

11.3.1 Model Performance

The model used the best environmental variables which play a significant role in the growth, development and distribution of *O. indicum*. The averaged model showed high training AUC values (>0.861) throughout the models and also slightly greater than the test AUC values. The test AUC values, which exhibit the actual model predictive power, were higher than 0.827. The final model is reasonably good and consistent with mean AUC of 0.827 and TSS value of 0.658. The jackknife test revealed that the distribution of *O. indicum* was mostly constrained by the precipitation of the warmest quarter (BIO18) which accounts for 49.7% of the explained variation. Annual mean temperature (BIO01) with 19.6% and LULC 9.3% accounted for the next highest variation. The mean contribution of slope, soil pH, clay soil and proximity to drainage was greater than 3%. Other variables such as silt soil and aspect explained minor variance with less than 2% (Fig. 11.3).

The response curves show a positive relationship of *O. indicum* with precipitation of the warmest quarter (maximal at 2000–2700 mm), annual mean temperature

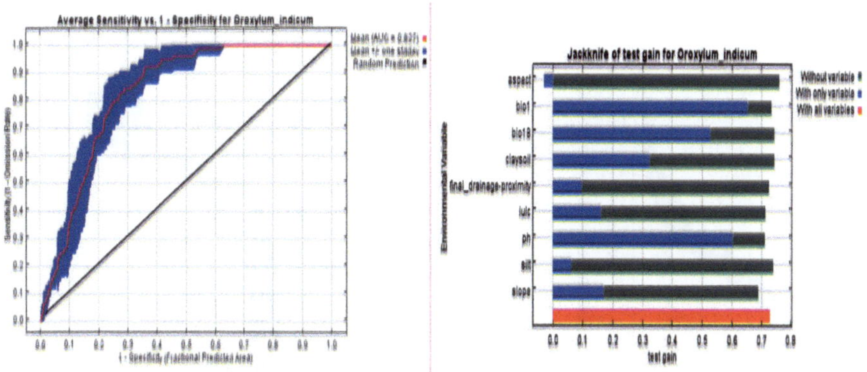

Fig. 11.3 Area under curve and jackknife test of environmental variable importance

(maximal at 19–27 °C), silt soil (maximal at 450–500 g/kg) and clay soil (maximal at 250–480 g/kg). On the other hand, there was a negative association with F (minimal from 2% to 12%), soil pH (falling from 5 to 8), proximity to drainage (2000–12,000 m) and slope (peaked towards lower values of 10–70°) as shown in Fig. 11.4.

11.3.2 Species Distribution Modelling of *O. indicum*

The suitability model predicted the distribution of *O. indicum* in Arunachal Pradesh in the range of 0–0.82 (Table 11.2 and Fig. 11.5) which was classified into three categories, namely, least suitable (0–0.30), moderately suitable (0.30–0.60) and highly suitable (>0.60). The results show an area of 59,205.3 km^2 (70.70%) as least suitable, followed by 19,504.1 km^2 (23.23%) as moderately suitable and only 5033.61 km^2 (6.29%) under highly suitable category. An examination of the final model reveals majority of northern part of the state as unsuitable for *O. indicum* due to high altitude, steep slope, low temperature and rainfall. On the other hand, the foothill areas characterized by lower slope, high temperature and rainfall showed the high suitable areas of *O. indicum*. The suitable areas have been mostly predicted the hot and humid foothill regions of Papum Pare, Lower Dibang Valley, East Siang, Pakke Kessang and Changlang district of Arunachal Pradesh. The river banks of Siang, Subansiri, Kameng and their major tributaries were also predicted under highly suitable category. In fact, the model predictions follow the current known distribution pattern of *O. indicum* in the state. The predicted area under moderately suitable (23.23%) forms the potential habitat of the species in the future which can be explored for regeneration and restoration of the species. The model predictions of only 6% under highly suitable category confirm the limited area of occupancy and endangered status of the plant.

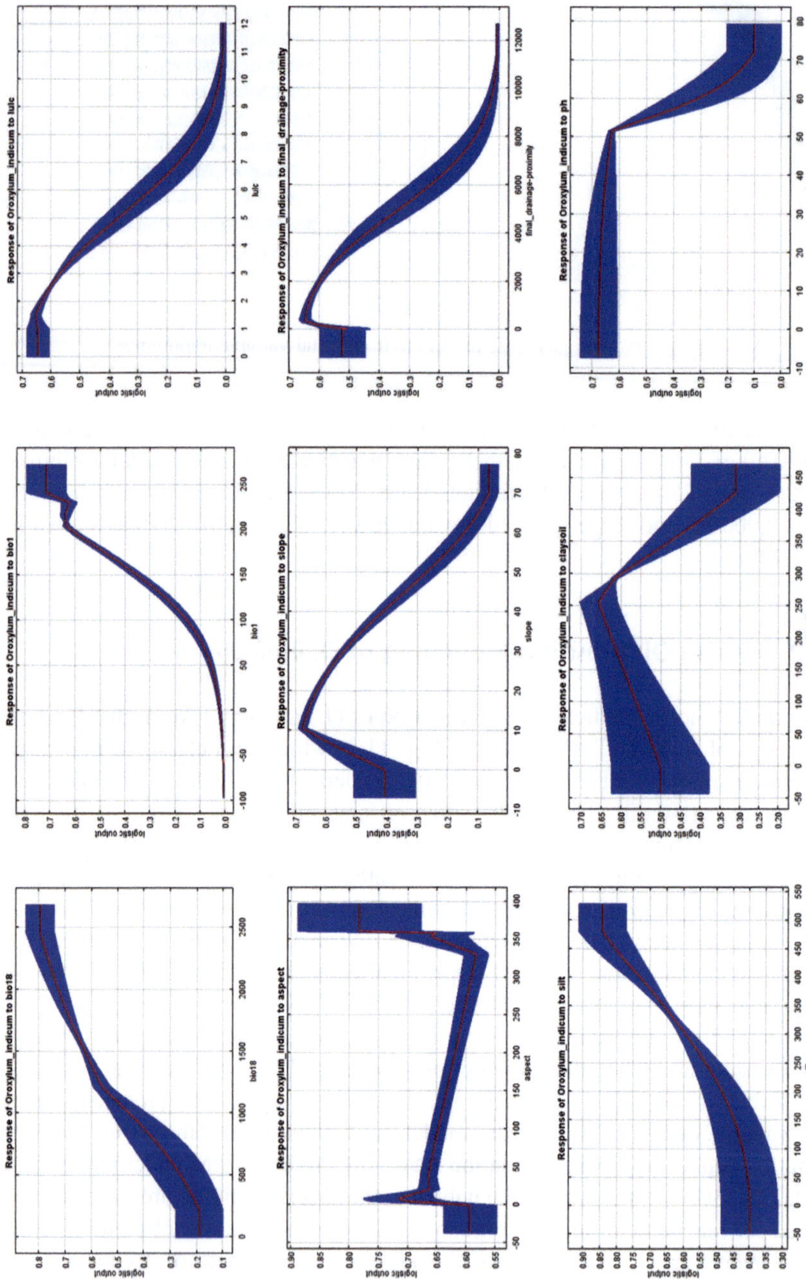

Fig. 11.4 Response curves of the selected environmental predictors to the probability of species suitability

Table 11.2 Suitable categories of *O. indicum* in Arunachal Pradesh

Suitability class	Area (km^2)	Percentage
Least suitable	59,205.3	70.70
Moderately suitable	19,504.1	23.29
Highly suitable	5033.61	6.01
Total	83,743.00	100.00

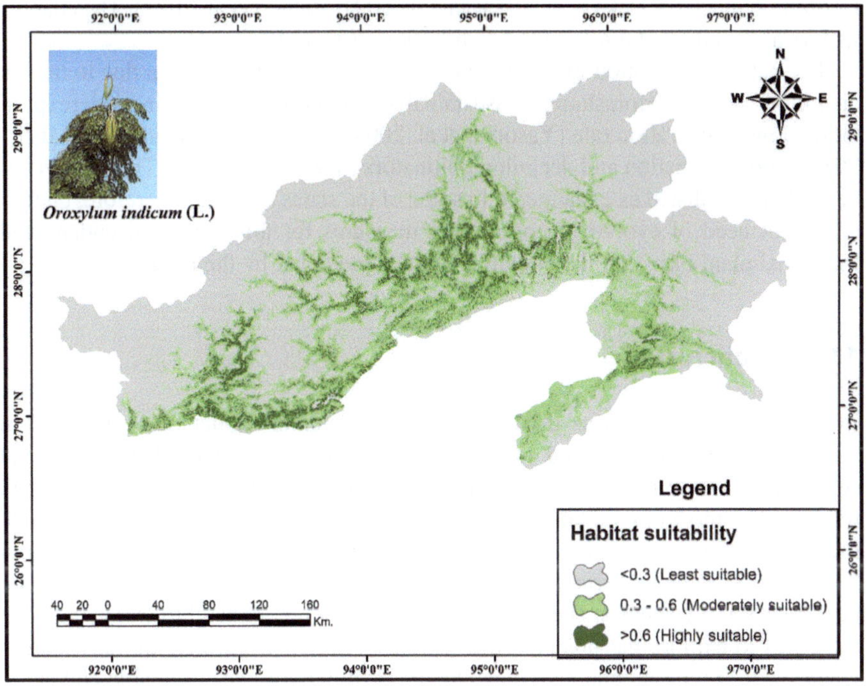

Fig. 11.5 Habitat suitabity map of *Oroxylum indicum* (L.) in Arunachal Pradesh

11.4 Discussion

The modelling results reveal that the distribution of *O. indicum* is highly influenced by variations in the precipitation of the warmest quarter (Bio18), annual mean temperature (Bio1) and land use/land cover (LULC). The slope, soil pH, clayey texture and proximity to drainage also moderately affect the distribution of *O. indicum* in the study area. The results indicate that the occurrence of *O. indicum* is highly influenced by precipitation of the warmest quarter with optimal precipitation requirement of ~1300 mm. The findings are in agreement with the reported precipitation requirement of 850–1300 mm (Behera et al. 2019) and ~1200 mm (Kumar et al. 2021) for the growth and development of *O. indicum*. We also found a positive association of *O. indicum* with moderate to high annual mean temperature in conformity with the reported optimal requirement of 22–35 °C (Kumar et al. 2021). Thus, the species has been found to grow mostly in moist and humid areas that receives moderate to high temperature (Bennet et al. 1992; Chauhan 1999; Bumrungsri et al. 2008). The model also shows a negative relationship of *O. indicum* with slope and proximity to drainage species by restricting the predictions to the foothills and river banks in agreement with the previous studies (Bennet et al. 1992; Chauhan 1999; Gu et al. 2006; Jayaram and Prasad 2008; Lawania et al. 2010; Deka et al. 2013; Jagetia 2021; Rathod et al. 2022). The edaphic factors also play a vital role in the distribution of *O. indicum*.

The natural population of *O. indicum* is on the verge of extinction due to habitat alteration such as deforestation, unsustainable extraction for medicine, overexploitation, low regeneration rate (Yasodha et al. 2004; Mishra and Kotwal 2010; Mishra 2011), poor pollination and declining pollinators (Vikas et al. 2009). The species has been already enlisted as endangered in some of the states of India. Therefore, there is an urgent need of evolving conservation measures for this important endangered medicinal plant based on the potential habitats predicted by the model.

11.5 Conclusion

In this study, MaxEnt was used to model the suitable habitat of *O. indicum* in Arunachal Pradesh. We used topographic, bioclimatic, land use/land cover and soil variables to run the model. The model performance was reasonably good and reliable. The model predicted only 6.01% of the total area as highly suitable which indicates its restricted area of occupancy. However, the model also predicted about 23% of the state as moderately suitable that needs to be explored for confirming the occurrence of the species. Based on the findings, we recommend to evolve suitable strategies to explore the potential habitats of *O. indicum*. Further, we also recommend for lessening human activities like overexploitation, unsustainable collection and deforestation over the suitable habitats of *O. indicum*.

Acknowledgements The authors are thankful to the Soil and Limnological Laboratory, Department of Geography, Rajiv Gandhi University, Rono Hills, Doimukh, Arunachal Pradesh (India), for facilitating necessary infrastructural and laboratory facilities to accomplish this work. The authors also acknowledge WorldClim, ESRI land cover and World Soil Information (ISRIC) and GBIF for making available various data sources used in this study.

References

Anderson BJ, Arroyo BE, Collingham YC, Etheridge B, Fernandez-De-Simon J, Gillings S, Gregory RD, Leckie FM, Sim IMW, Thomas CD, Travis J, Redpath SM (2009) Using distribution models to test alternative hypotheses about a species' environmental limits and recovery prospects. Biol Conserv 142(3):488–499. https://doi.org/10.1016/j.biocon.2008.10.036

Baldwin R (2009) Use of maximum entropy modeling in wildlife research. Entropy 11:854–866. https://doi.org/10.3390/e11040854

Behera LK, Gunaga RP, Thakur NS, Mehta AA, Sukhadiya M, Dholariya CA (2019) *Oroxylum indicum*: detail on a medicinally important and rare tree species of India. MFP News 29(3):9–12

Bennet SSR, Gupta PC, Rao RV (1992) Venerated plants. Indian Council of Forestry Research & Education (ICFRE), Dehradun, pp 147–149

Bumrungsri S, Harbit A, Benzie C, Carmouche K, Sridith K, Racey P (2008) The pollination ecology of two species of Parkia (Mimosaceae) in southern Thailand. J Trop Ecol 24(5): 467–475. https://doi.org/10.1017/S026646740800521X

Chauhan NS (1999) Medicinal and aromatic plants of Himachal Pradesh. Indus Publishing, New Delhi, pp 96–298

Chopra RN, Nayar SL, Chopra IC (2002) Glossary of Indian medicinal plants. National Institute of Science Communication and Information Resources, New Delhi, p 182

Deka DC, Kumar V, Prasad C, Kumar K, Gogoi BJ, Singh L, Srivastava RB (2013) *Oroxylum indicum*—a medicinal plant of North East India: an overview of its nutritional, remedial, and prophylactic properties. J Appl Pharmaceut Sci 3(Suppl. 1):S104–S112. https://doi.org/10.7324/JAPS.2013.34.S19

Dormann CF, Elith J, Bacher S, Buchmann C, Carl G, Carré G, Marquéz RG, Gruber B, Lafourcade B, Leitão PJ, Münkemüller T, McClean C, Osborne PE, Reineking B, Schröder B, Skidmore AK, Zurell D, Lautenbach S (2013) Collinearity: a review of methods to deal with it and a simulation study evaluating their performance. Ecography 36(1):27–46. https://doi.org/10.1111/j.1600-0587.2012.07348.x

Drury CH (2006) Ayurvedic useful plants of India. Asiatic Publishing House, Delhi, p 360

Elith J, Leathwick JR (2009) Species distribution models: ecological explanation and prediction across space and time. Annu Rev Ecol Evol Syst 40:677–697. https://doi.org/10.1146/annurev.ecolsys.110308.120159

Elith J, Kearney M, Phillips S (2010) The art of modelling range-shifting species. Methods Ecol Evol 1:330–342. https://doi.org/10.1111/j.2041-210X.2010.00036.x

Farrell A, Wang G, Rush SA, Martin JA, Belant JL, Butler AB, Godwin D (2019) Machine learning of large-scale spatial distributions of wild turkeys with high-dimensional environmental data. Ecol Evol 9(10):5938–5949. https://doi.org/10.1002/ece3.5177

Feng X, Park DS, Liang Y, Pandey R, Papeş M (2019) Collinearity in ecological niche modeling: confusions and challenges. Ecol Evol 9(18):10365–10376. https://doi.org/10.1002/ece3.5555

GBIF, The Global Biodiversity Information Facility (2022) What is GBIF? https://www.gbif.org/what-is-gbif. Accessed 5 Aug 2022

Glor RE, Warren D (2011) Testing ecological explanations for biogeographic boundaries. Evolution 65:673–683. https://doi.org/10.1111/j.1558-5646.2010.01177.x

Graham CH, Hijmans RJ (2006) A comparison of methods for mapping species ranges and species richness. Glob Ecol Biogeogr 15(6):578–587. https://doi.org/10.1111/j.1466-8238.2006.00257.x

Graham CH, Elith J, Hijmans RJ, Guisan A, Townsend Peterson A, Loiselle BA, NCEAS Predicting Species Distributions Working Group (2008) The influence of spatial errors in species occurrence data used in distribution models. J Appl Ecol 45(1):239–247. https://doi.org/10.1111/j.1365-2664.2007.01408.x

Gu Q, Luo H, Zheng W, Liu Z, Huang Y (2006) *Pseudonocardia oroxyli* sp. nov., a novel actinomycete isolated from surface-sterilized *Oroxylum indicum* root. Int J Syst Evol Microbiol 56(9):2193–2197. https://doi.org/10.1099/ijs.0.64385-0

Guisan A, Tingley R, Baumgartner JB, Naujokaitis-Lewis I, Sutcliffe PR, Tulloch AIT, Regan TJ, Brotons L, Mcdonald-Madden E, Mantyka-Pringle C, Martin TG, Rhodes JR, Maggini R, Setterfield SA, Elith J, Schwartz MW, Wintle BA, Broennimann O, Austin M, Ferrier S, Kearney MR, Possingham HP, Buckley YM (2013) Predicting species distributions for conservation decisions. Ecol Lett 16:1424–1435. https://doi.org/10.1111/ele.12189

Guisan A, Thuiller W, Zimmermann NE (2017) Habitat suitability and distribution models with applications in R (Ecology, Biodiversity and Conservation, p. I). Cambridge University Press, Cambridge. https://doi.org/10.1017/9781139028271

Hamann A, Wang T (2006) Potential effects of climate change on ecosystem and tree species distribution in British Columbia. Ecology 87(11):2773–2786

Hanley JA, McNeil BJ (1982) The meaning and use of the area under a receiver operating characteristic (ROC) curve. Radiology 143(1):29–36. https://doi.org/10.1148/radiology.143.1.7063747

Hernandez PA, Graham CH, Master LL, Albert DL (2006) The effect of sample size and species characteristics on performance of different species distribution modeling methods. Ecography 29(5):773–785. https://doi.org/10.1111/j.0906-7590.2006.04700.x

Hijmans RJ, Schreuder M, De la Cruz J, Guarino L (1999) Using GIS to check coordinates of germplasm accessions. Genet Resour Crop Evol 46:291–296. https://doi.org/10.1023/A:1008628005016

Huntley B, Collingham YC, Willis SG, Green RE (2008) Potential impacts of climatic change on European breeding birds. PLoS One 3(1):e1439. https://doi.org/10.1371/journal.pone.0001439

Jagetia GC (2021) A review on the medicinal and pharmacological properties of traditional ethnomedicinal plant sonapatha, *Oroxylum indicum*. Sinusitis 5(1):71–89. https://doi.org/10.3390/sinusitis5010009

Jayaram K, Prasad MNV (2008) Genetic diversity in *Oroxylum indicum* (L.) vent. (Bignoniaceae), a vulnerable medicinal plant by random amplified polymorphic DNA marker. Afr J Biotechnol 7(3):254–262. https://doi.org/10.5897/AJB07.096

Jeschke JM, Strayer DL (2008) Usefulness of bioclimatic models for studying climate change and invasive species. Ann N Y Acad Sci 1134:1–24. https://doi.org/10.1196/annals.1439.002

Julliard R, Jiguet F, Couvet D (2004) Common birds facing global changes: what makes a species at risk? Glob Chang Biol 10(1):148–154. https://doi.org/10.1111/j.1365-2486.2003.00723.x

Khare CP (2007) Indian medicinal plants. Springer Science Business Media, LLC, Berlin, p 453

Khongsai M, Kayang H, Saikia SP (2015) Ethnomedicinal plants used by different tribes of Arunachal Pradesh. Indian J Tradit Knowl 10(3):541–546

Kremen C, Cameron A, Moilanen A, Phillips SJ, Thomas CD, Beentje H, Dransfield J, Fisher BL, Glaw F, Good TC, Harper GJ, Hijmans RJ, Lees DC, Louis E Jr, Nussbaum RA, Raxworthy CJ, Razafimpahanana A, Schatz GE, Vences M, Vieites DR, Wright PC, Zjhra ML (2008) Aligning conservation priorities across taxa in Madagascar with high-resolution planning tools. Science 320(5873):222–226. https://doi.org/10.1126/science.1155193

Kumar D, Rawat S, Joshi R (2021) Predicting the current and future suitable habitat distribution of the medicinal tree Oroxylum indicum (L.) Kurz in India. J Appl Res Med Aromat Plants 23:100309. https://doi.org/10.1016/j.jarmap.2021.100309

Lawania RD, Mishra A, Gupta R (2010) *Oroxylum indicum*: a review. Pharm J 2(9):304–310. https://doi.org/10.5530/pj.2011.24.10

Manel S, Williams HC, Ormerod SJ (2001) Evaluating presence–absence models in ecology: the need to account for prevalence. J Appl Ecol 38(5):921–931. https://doi.org/10.1046/j.1365-2664.2001.00647.x

Manzoor SA, Griffiths G, Lukac M (2018) Species distribution model transferability and model grain size–finer may not always be better. Sci Rep 8(1):1–9. https://doi.org/10.1038/s41598-018-25437-1

Merow C, Smith MJ, Silander JA Jr (2013) A practical guide to MaxEnt for modeling species' distributions: what it does, and why inputs and settings matter. Ecography 36(10):1058–1069. https://doi.org/10.1111/j.1600-0587.2013.07872.x

Mishra M (2011) Conservation of biodiversity in the natural forests of Central India: a case of critically endangered medicinal species Safed musli in Bhopal Forest (MP) India. Biosci Discov 2:299–308

Mishra M, Kotwal PC (2010) Unripe collection of musli (*Chlorophytum* spp.) and its impact on raw material quality-a case of Dhamtari forest Chhattisgarh, India. Life Sci Leafl 3:79–89

Murtem G, Chaudhry P (2016) An ethnobotanical note on wild edible plants of Upper Eastern Himalaya, India. Braz J Biol Sci 3(5):63–81

Nadkarni AK (1982) Indian materia medica. Popular Prakashan Pvt. Ltd., Bombay

Naimi B, Araújo MB (2016) Sdm: a reproducible and extensible R platform for species distribution modelling. Ecography 39(4):368–375. https://doi.org/10.1111/ecog.01881

Nimasow G, Nimasow OD, Rawat JS, Tsering G, Litin T (2016) Remote sensing and GIS-based suitability modeling of medicinal plant (*Taxus baccata* Linn.) in Tawang district, Arunachal Pradesh, India. Curr Sci 110:219–227. https://doi.org/10.18520/cs/v110/i2/219-227

Parmesan C, Yohe G (2003) A globally coherent fingerprint of climate change impacts across natural systems. Nature 421(6918):37–42. https://doi.org/10.1038/nature01286

Parolo G, Rossi G, Ferrarini A (2008) Toward improved species niche modelling: Arnica montana in the Alps as a case study. J Appl Ecol 45(5):1410–1418. https://doi.org/10.1111/j.1365-2664.2008.01516.x

Pearson RG, Dawson TP, Liu C (2004) Modelling species distributions in Britain: a hierarchical integration of climate and land-cover data. Ecography 27:285–298. https://doi.org/10.1111/j.0906-7590.2004.03740.x

Peterson AT, Papeş M, Soberón J (2008) Rethinking receiver operating characteristic analysis applications in ecological niche modeling. Ecol Model 213(1):63–72. https://doi.org/10.1016/j.ecolmodel.2007.11.008

Phillips SJ (2008) Transferability, sample selection bias and background data in presence-only modelling: a response to Peterson et al. (2007). Ecography 31(2):272–278. https://doi.org/10.1111/j.0906-7590.2008.5378.x

Phillips SJ, Anderson RP, Schapire RE (2006) Maximum entropy modeling of species geographic distributions. Ecol Model 190(3–4):231–259. https://doi.org/10.1016/j.ecolmodel.2005.03.026

Phillips SJ, Anderson RP, Dudík M, Schapire RE, Blair ME (2017a) Opening the black box: an open-source release of Maxent. Ecography 40(7):887–893. https://doi.org/10.1111/ecog.03049

Phillips SJ, Dudík M, Schapire RE (2017b) Maxent software for modeling species niches and distributions (Version 3.4.1). http://biodiversityinformatics.amnh.org/open_source/maxent/. Accessed 20 Aug 2022

Prakash P (2005) Indian medicinal plants. Chaukhamba Sanskrit Pratishthan, Delhi

Radosavljevic A, Anderson RP (2014) Making better Maxent models of species distributions: complexity, overfitting and evaluation. J Biogeogr 41(4):629–643. https://doi.org/10.1111/jbi.12227

Ramírez-Villegas J, Bueno Cabrera A (2009) Working with climate data and niche modeling: I. creation of bioclimatic variables. International Center for Tropical Agriculture (CIAT), Cali

Rathod K, Ram Mayur L, Jaliya R, Dhaka R, Jha S, Desai BS (2022) *Oroxylum indicum*: ethnobotany, phytochemistry and therapeutic uses. Pharma Innov J SP-11(1):200–220

Singh HB, Prasad P, Rai LK (2002) Folk medicinal plants in the Sikkim Himalayas of India. Asian Folk Stud 61:295–310

Skov F (2000) Potential plant distribution mapping based on climatic similarity. Taxon 49:503–515. https://doi.org/10.2307/1224346

Sony RK, Sen S, Kumar S, Sen M, Jayahari KM (2018) Niche models inform the effects of climate change on the endangered Nilgiri Tahr (*Nilgiritragus hylocrius*) populations in the southern Western Ghats, India. Ecol Eng 120:355–363. https://doi.org/10.1016/j.ecoleng.2018.06.017

Tangjang S, Namsa ND, Aran C, Litin A (2011) An ethnobotanical survey of medicinal plants in the Eastern Himalayan zone of Arunachal P, Shrestha BB (2018) understanding the dynamics in distribution of invasive alien plant species under predicted climate change in Western Himalaya. PLoS One 13(4):e0195752. https://doi.org/10.1371/journal.pone.0195752

Thapa S, Chitale V, Rijal SJ, Bisht N, Shrestha BB (2018) Understanding the dynamics in distribution of invasive alien plant species under predicted climate change in Western Himalaya. PLoS One 13(4):e0195752. https://doi.org/10.1371/journal.pone.0195752

Theobald WL, Dassanayake MD, Fosberg MR (1981) A revised handbook to the flora of Ceylon. Amerind Publishing Co. Pvt. Ltd., New Delhi

Tripathi AK, Limasenla RS (2017) Ethno-medicinal plants used by Nyishi tribe of Arunachal Pradesh, India. World J Pharm Pharm Sci 6(5):1246–1253. https://doi.org/10.20959/wjpps20175-9213

Trisurat Y, Shrestha RP, Kjelgren R (2011) Plant species vulnerability to climate change in Peninsular Thailand. Appl Geogr 31(3):1106–1114. https://doi.org/10.1016/j.apgeog.2011.02.007

Vaughan IP, Ormerod SJ (2003) Improving the quality of distribution models for conservation by addressing shortcomings in the field collection of training data. Conserv Biol 17(6):1601–1611. https://doi.org/10.1111/j.1523-1739.2003.00359.x

Vikas GM, Tandon R, Ram HYM (2009) Pollination ecology and breeding system of *Oroxylum indicum* (Bignoniaceae) in the foothills of the Western Himalaya. Appl Geogr 25:93–96. https://doi.org/10.1017/S0266467408005634

Wisz MS, Hijmans RJ, Li J, Peterson AT, Graham CH, Guisan A, NCEAS Predicting Species Distributions Working Group (2008) Effects of sample size on the performance of species distribution models. Divers Distrib 14(5):763–773. https://doi.org/10.1111/j.1472-4642.2008.00482.x

Yasodha R, Ghosh M, Santan B, Gurumurthi K (2004) Importance of biotechnological research in tree species of Dashmula. Indian Forester 130:79–88. https://doi.org/10.36808/if/2004/v130i1/2221

Yi YJ, Cheng X, Yang ZF, Zhang SH (2016) MaxEnt modeling for predicting the potential distribution of endangered medicinal plant (*H. riparia* lour) in Yunnan, China. Ecol Eng 92: 260–269. https://doi.org/10.1016/j.ecoleng.2016.04.010

Habitat Suitability and Niche Modelling for Conservation and Restoration of *Aconitum heterophyllum* Wall. in Temperate Himalayan Forest Ecosystem

12

Peerzada Ishtiyak Ahmad, T. H. Masoodi, S. A. Gangoo, P. A. Sofi, Tahir Mushtaq, Mir Muskan Un Nisa, Mohan Reddy, Abhinav Mehta, Shrey Rakholia, and Bipin Charles

Abstract

Plant-based medicine has played a significant role in maintaining the human health since the dawn of civilization. Identification of bioactive compounds from medicinal plants has led to the discovery of a number of valuable drugs that are being extensively used in modern and traditional practices of medicine. The escalating commercial demand of medicinal plants has led to the overexploitation of many species from the wild, which has resulted in the loss of their natural populations. Consequently, several valuable medicinal plants including *Aconitum heterophyllum* have been put under different categories of threat due to their overexploitation and habitat degradation. Moreover, the abysmal consequences of climate change on habitat range shifts at the ecosystem and species level have threatening impacts on ecosystem functional traits. The Rare, Endangered, Threatened (RET) and endemic species, which have narrow geographic distribution, undersized population and low reproductive capacity, are extremely susceptible to such variations and are at higher risk of extinction. Hence, the habitat suitability modelling of these plant species is highly significant for monitoring, rehabilitation and conservation of their diminishing populations and natural

P. I. Ahmad (✉) · T. H. Masoodi · S. A. Gangoo · P. A. Sofi · T. Mushtaq · M. M. U. Nisa
Faculty of Forestry, Sher-e-Kashmir University of Agricultural Sciences and Technology of Kashmir, Srinagar, Jammu and Kashmir, India
e-mail: drishtiyak@skuastkashmir.ac.in

M. Reddy
Carbon and Sustainability Division at Nurture Agtech Private Limited, Bellandur, Bengaluru, India

A. Mehta · S. Rakholia
The Geographic Information Lab (TGIS), Ahmedabad, Gujarat, India

B. Charles
Institute for Biodiversity Conservation and Training, Bangalore, India

S. Dhyani et al. (eds.), *Ecosystem and Species Habitat Modeling for Conservation and Restoration*, https://doi.org/10.1007/978-981-99-0131-9_12

habitats. Therefore, for predicting the distribution range and niche area of *Aconitum heterophyllum*, the field collected data was synchronized with environmental and bioclimatic variables using MaxEnt machine learning programme. We attempted to study the role of different variables on the habitat of *Aconitum heterophyllum*, modelling of the current distribution range across the Kashmir Himalaya and evaluating the niche area using models to formulate management strategies for its conservation and restoration. It is concluded that the habitat of *Aconitum heterophyllum* is highly vulnerable to climate shifts and anthropogenic pressure; therefore it needs immediate restoration in the wild and systematic domestication in the potential areas.

Keywords

Aconite · Habitat · Conservation · MaxEnt niche modelling · Kashmir Himalaya

12.1 Introduction

Plant-based medicine has played a significant role in maintaining the human health since the dawn of civilization. Herbal medicines were used mostly in their crude form as infusion, tincture and decoction or applied externally as balm. However, from the late nineteenth and early twentieth century onwards, scientists started isolation, purification and identification of bioactive compounds from medicinal plants, which led to the discovery of a number of valuable drugs that are being extensively used in the modern system of medicine. The World Health Organization (WHO) reports that 60% of the world's population and around 80% of the developing countries' population depend entirely on herbal medicine for their primary healthcare needs (WHO 2002).

The global herbal medicine market size is estimated at US$71.19 billion (Hexa-Research 2017). With an annual growth rate of 15–25%, the World Health Organization (WHO) has estimated the demand for medicinal plants to increase more than US$5 trillion by 2050 (Muir and Simona 2018). In India, about 1178 medicinal plant species are in trade, out of which 242 species have a demand of more than 100 metric tonnes/year (Goraya and Ved 2017). The consumption of herbal raw drug in India has been estimated at 5, 12,000 MT (during the years 2014–2015) with a corresponding trade value of Rs. 5500 crore. The export value has shown a record increase from Rs. 345.80 crore in 2005–2006 to Rs. 3211 crore in 2014–2015, registering a ninefold increase during a decade (Goraya and Ved 2017).

In spite of such a glorious history and ever-increasing market demand, herbal medicine could not integrate fully into modern healthcare system due to increasing concerns about the quality and sustainability of the resources. This sector has also suffered due to lack of proper R & D on the habitat suitability, regenerating capacity and efficacy of premature medicinal plants supplied through shady marketing channels. Nevertheless, the prospect of advancing the research in herbal medicine has renewed the interest in the medicinal plant sector (Peerzada et al. 2021).

Therefore, the medicinal plant sector offers an exceptional investment opportunity for rapid and sustainable growth (Tahir et al. 2019; Peerzada et al. 2022). However, there are certain issues with regulation, standardization and quality assurance in the manufacturing of herbal medicine, due to complexity of their diverse secondary metabolites, which principally depend on genetic factors, age and geographical location of the plant species. The variability in phytochemical contents as well as occurrences of adulteration has an abysmal impact on uniform standards of herbal medicines.

The escalating commercial demand of medicinal plants for their therapeutic uses has led to the overexploitation of many species from the wild, which has resulted in the loss of their natural populations (Pimm et al. 1995). Consequently, several valuable medicinal plants have been put under different categories of threat due to their overexploitation and habitat degradation. Prominent medicinal plants from Kashmir, which are under certain categories of threat, are *Aconitum heterophyllum* Wallich., *Arnebia benthamii* Wallich., *Saussurea costus* (Falc.) Lipsch. and *Trillium govanianum* Wall. ex D.Don. Among which *Aconitum heterophyllum*, one of the significant medicinal plants, with known antipyretic, analgesic, anti-venom and anti-inflammatory properties, is being heavily exploited to cater the exponential demand at local, national and international markets (Srivastava et al. 2010; Peerzada et al. 2021). Therefore, there is an immense need to rehabilitate these species in the wild by identifying the suitable habitats. This will have an impeccable impact on the modern medical sciences and shall help in informed decision-making in the herbal medicine system.

12.1.1 Target Species

Aconitum is a large genus in the Ranunculaceae family with over 250 species, among which around 33 species are found in the Himalayas, from Afghanistan in the west to Myanmar (Burma) in the east. Aconites are also found in Europe, where they are used in traditional systems of medicine. However, *Aconitum heterophyllum*, the only non-poisonous member of the Ranunculaceae family (Wani et al. 2021b), is the most exploited and traded aconite species in the Northwestern Himalayas (Peerzada et al. 2021).

Aconitum heterophyllum Wall. ex Royle is commonly called as *aconite*, *ativisha*, *atis*, *patrees* and *monkshood*. In India, the plant is generally found in Jammu and Kashmir, Himachal Pradesh and Uttarakhand (Nagarajan et al. 2015; Balaramnavar et al. 2021). In Jammu and Kashmir, the scattered populations of sparsely distributed individuals of this species are confined to sub-alpine and alpine habitats between 2200 m and 4200 m altitudes (Beigh et al. 2008; Peerzada and Sofi 2017; Peerzada et al. 2018). This species is adapted to the inflexible locations having severe site factors and extreme climatic conditions (Beigh et al. 2008). Owing to significant pharmacological properties, overexploitation from forests due to an increasing market demand, this species has been categorized as a Critically Endangered (CR) species (Wani et al. 2021b) in the Northwestern Himalayas.

Aconitum heterophyllum is a perennial erect herb with about 30–90 cm height. The shoot is annual, while the root is biennial. The stems are simple or branched from base, glabrous and puberulous above, broad, ovate or orbicular or five-lobed and toothed, above three-fid or entire. The branches are absent or rarely one or two in number. Inflorescence is terminal but sometimes axial. The flowers are in racemes, 2.5 cm long, blue or greenish blue with purple veins and helmet shaped. The roots are tuberous, in pairs, whitish or light grey up to 3 cm long, 0.5–1.2 cm thick with conical ends, break easily and taste bitter. The mother and daughter tubers occur in pairs. The seeds are obpyramidal, 3–4 mm long, blackish brown, angles acute or more or less winged, faces smooth. Flowering occurs during July–August and fruiting during August–October. Seeds are collected during October–November (Kirtikar and Basu 1975). The tuberous roots (tradable part) of this species are harvested after the second year of growth during September–October (Peerzada et al. 2018). It contains a non-crystalline, non-toxic alkaloid called atisine. Other alkaloids found in this species are heteratisine, histine, heterophyllisine, heterophylline, heterophyllidine, atidine and histidine (Nagarajan et al. 2015; Balaramnavar et al. 2021). Aconitic acid, tannic acid, palmitic, stearic glycerides and vegetable mucilage are also present in addition to starch and sugars (Rajakrishnan et al. 2016). The roots are used as antipyretic, analgesic, antiperiodic, aphrodisiac, astringent, anti-venom, anti-inflammatory, anti-rheumatic and vermifuge. It is also used to treat piles and digestive and reproductive disorders (Sojitra et al. 2013; Peerzada and Sofi 2017). However, the aqueous extract of the root induces hypertension through action on the sympathetic nervous system, and in higher doses its use may become lethal (Kumar et al. 2016).

The per kg (dry weight) cost of *Aconitum heterophyllum* tuberous roots in Kashmir ranges between Rs. 4000 and 6000/– (Peerzada et al. 2018) with an annual demand of 200–500 MT at the national level (Goraya and Ved 2017). The ever-increasing demand for this species has led to its overexploitation from the wild. Owing to its increasing demand, decreasing natural populations from forests due to anthropogenic pressures including excessive harvesting, grazing and seed dormancy (Beigh et al. 2008; Srivastava et al. 2010), the National Medicinal Plants Board (NMPB) has enlisted *Aconitum heterophyllum* as a priority species for the promotion of cultivation outside forests with 75% subsidy (Peerzada et al. 2018). Apparently, the commercial cultivation of this species has not been taken up by any farmer in Kashmir due to its long gestation period (3–7 years), low seed germination, slow growth rate and meagre survivability. Moreover, the production within forests and natural habitats is usually preferred for high altitude medicinal plants like *Aconitum heterophyllum*, as there are serious apprehensions about variations in the biochemical compounds of wild and cultivated medicinal plant species (Ramatsobane and Anthony 2020). The climatic, edaphic, topographic and biotic factors, including the associated biodiversity, have a distinct role in the accumulation of biochemical compounds of plant species (Yeshi et al. 2022). Besides, various studies have established the potential impacts of climate change on the distribution of vegetation, particularly on grasslands and forest ecosystems (Rashid et al. 2015). The projected increase in average annual temperate of the Kashmir Himalayan region between

3.98 °C and 6.93 °C (Romshoo et al. 2020) by the end of this century will have devastating impacts on the vulnerable species. However, there is no specific information available on the effects of climate change on distribution of medicinal plants, as these plants have distinct spatial patterns and are influenced by diverse environmental factors (Dad and Rashid 2022). It is therefore crucial to identify the species-specific suitable habitats for restoration of Rare, Endangered and Threatened (RET) species. Such strategy would help in species restoration, rehabilitation, conservation and management of this prized natural wealth.

12.1.2 Habitat Suitability Modelling

Biodiversity in the fragile Himalayan region is highly vulnerable to the impacts of climate change. The abysmal consequences of climate change on habitat range shifts at the ecosystem and species level have threatening impacts on ecosystem functional traits (Shrestha et al. 2012; Wani et al. 2021a; Kumari et al. 2022). By the end of this century, the suitable habitats of various high mountain plant species are predicted to reduce significantly or disappear entirely due to increased earth temperature and fluctuating precipitation patterns aggravated by the global warming (Van de et al. 2007; Engler et al. 2011; Peerzada et al. 2016). The Rare, Endangered and Threatened (RET) and endemic species, which have narrow geographic distribution, undersized population and low reproductive capacity, are extremely susceptible to such variations and are at higher risk of extinction (Rew et al. 2020; Bobrowski et al. 2021; Megan 2021). Hence, the habitat suitability modelling of RET as well as endemic plant species is highly significant for monitoring, rehabilitation and conservation of their diminishing populations and natural habitats (Malhi et al. 2020). It also helps in predicting the distribution range and niche area (Pecl et al. 2017) with the help of field surveyed data in sync with the topographic and bioclimatic variables (Guisan and Thuiller 2005; Taleshi et al. 2019). Therefore, the species distribution models (SDMs) are an important ecological tool for forecasting the habitat suitability (Singh et al. 2017; Rew et al. 2020), restoring the degraded habitats (Guisan and Zimmermann 2000; Cengic et al. 2020) and re-introduction of indigenous species (Elith and Leathwick 2009; Fois et al. 2016; Condro et al. 2021) as well as predicting the climate change impact on species reproduction and future distributions (Rodríguez et al. 2007; Cengic et al. 2020). The SDMs combine species occurrence data with environmental variables (temperature, precipitation, elevation, geology and vegetation) to create a model that represents the distribution of species with respect to different ecological attributes (Condro et al. 2021; Kumari et al. 2022).

Although there are a number of statistical methods in practice to create a species distribution model, we have used MaxEnt for the present study. MaxEnt is a machine learning programme that uses a statistical technique called maximum entropy which makes predictions from incomplete information (Engler et al. 2011; Tovar et al. 2013; Kaky et al. 2020). MaxEnt models the geographic distribution of species using occurrence data and environmental variables on the principle of a uniform probability distribution (Mouquet et al. 2015; Dad and Rashid 2022).

Hence, modelling the suitable habitats and anticipated impacts of climate change can be an important restoration tool for *Aconitum heterophyllum*. In this study we attempted to study the role of different bioclimatic/environment variables on the habitat distribution of *Aconitum heterophyllum* and modelling of the current distribution range across the Kashmir Himalaya and evaluated the niche area using models to formulate management strategies.

12.2 Materials and Methods

12.2.1 Study Area

Kashmir, one of the two provinces of the Jammu and Kashmir UT, is situated on the northern periphery of India (Fig. 12.1). It extends from 33°.20′ and 34°.54′ N latitudes and from 73°.55′ and 75°.35′ E longitudes, covering an area of 15,948 km^2. About 64% area of the Kashmir is mountainous, and the valley is encircled by Pir Panjal Mountains in South and Southwest and the Great Himalayas in the North and East. Kashmir valley comprises of sedimentary, metamorphic and igneous rocks ranging from Precambrian to Recent. The region has temperate, sub-alpine and alpine climatic conditions, dominated by pine forests dispersed with diversity of high-value medicinal and aromatic plants.

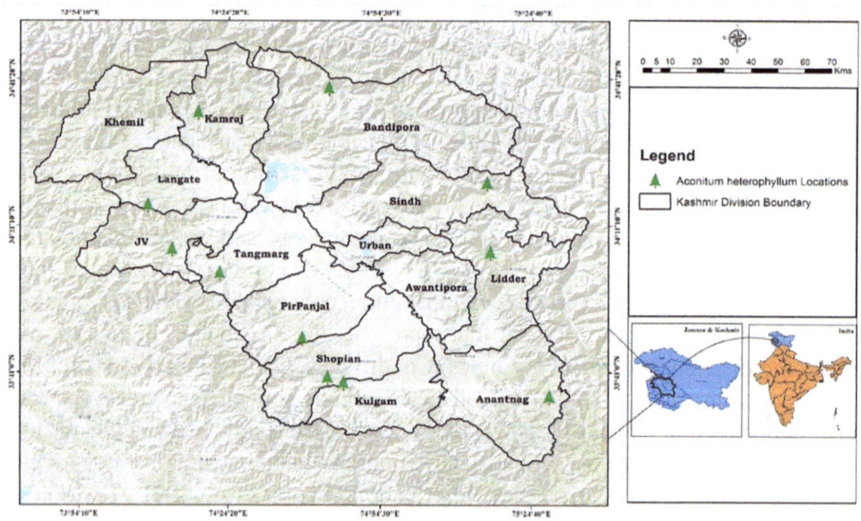

Fig. 12.1 Location map showing sample points across forest divisions of the study area J & K UT, India

12.2.2 Field Observation and Collection of Occurrence Data

Extensive field surveys were conducted across the 12 forest divisions of Kashmir (Table 12.1) falling in Northwestern Himalayan range, for field observation and collection species occurrence data. Around 12 GPS points of occurrences of *Aconitum heterophyllum* were collected based on direct/indirect evidence as well as through field observations assisted by the forest department. Accuracy of the each occurrence data was carefully checked before usage. The reference points were taken after spatial thinning and used for modelling the habitat distribution of *Aconitum heterophyllum*. Since it is a Critically Endangered species (Wani et al. 2021b), points were distributed far away from each other rather than clumped. In the general practice, more occurrence points tend to give more accurate results, but that was not possible in this case, as *Aconitum heterophyllum* is a rare and endangered species and is sparsely found in rough and steep terrains.

12.2.3 Environmental Layer Selection

Environmental layers are as much necessary as bioclimatic variables, but that is depending again on the study area as well as species habitat response over the time period. *Aconitum heterophyllum* is the species which is mostly affected by few environmental layers as per the ground observations like NDVI, landcover, slope, aspect and elevation, so these were included to run the MaxEnt model. Landuse-landcover (LULC) was derived from ESRI 2020 Global Landcover datasets of the classified landcover classes with a resolution of 30 m which was later recoded to 1 arc-second resolution to match it with the spatial resolution of other bioclimatic/ environmental parameters. The predominant classes include forest cover surrounding the Kashmir valley followed by agriculture land which is present around urban areas of Srinagar as well (Fig. 12.2). Area statistics was performed after the derived landcover from ESRI datasets.

Shuttle Radar Topography Mission (SRTM) digital elevation model (DEM) was utilized to derive an elevation map. Furthermore, slope and aspect were derived from the same DEM. Elevation range was found to be between 1031 m and 5354 m (Figs. 12.3, 12.4 and 12.5). Slope map was derived from elevation based on Horn's equation available in QGIS tool. Study area has diverse slopes ranging from 0° slopes in the valley to 76.95° in the mountain ridges. Predominantly the slope's direction (aspect) is found in the Eastern and North East sides especially in the southern part of the study area, whereas the upper region of the mountain ridges shows variation, however mostly in NW, SW and NE directions. Band 4 and Band 5 were used for generating NDVI (Fig. 12.6) as the ratio of these bands is used as per the NDVI formula. NDVI is a dimensionless quantity which provides the values between +1 and −1. From 0 to 1, it represents from sparse vegetation to dense vegetation, whereas values less 0 represent the complete absence of vegetation indicating water or ice. It is computed using Eq. (12.1) (Mehta et al. 2021):

Table 12.1 Sample points of the species across the different forest divisions of Kashmir

Record no.	District	Forest division	Location	Habitat	Altitude (m)	Vulnerability to	Common name
1	Bandipora	Bandipora Forest Division	Gurez, Kanzalwan	Sub-alpine/open slope	2637	Overexploitation/seeds eaten by birds	Mohand
2	Kupwara	Kamraj Forest Division	Machil	Sub-alpine/grassy slope	2278	Overexploitation/grazing	Patris
3	Kupwara	Langate Forest Division	Monabal	Alpine/ridges with moderate shade	3291	Overexploitation/grazing/seeds eaten by birds	Mohand/ Manglu/Patris
4	Baramulla	Jhelum Valley Forest Division	Boniyar	Sub-alpine/moist shady slope	2355	Overexploitation/forest degradation/grazing	Patris/Atis
5	Baramulla	Gulmarg Forest Division	Afharwat	Alpine meadow/open rocky slope	2944	Overexploitation/grazing	Pewakh
6	Budgam	Pir Panjal Forest Division	Yusmarg	Sub-alpine/steep slope	2710	Overexploitation/grazing	Patris
7	Srinagar	Dachigam National Park	Harwan Hills	Alpine/shady moist slope	2926	Grazing, soil erosion	Patris
8	Ganderbal	Sindh Forest Division	Sonmarg	Alpine/shady moist slope	3070	Forest degradation/ overexploitation	Pewakh
9	Kulgam	Kulgam Forest Division	Aharbal	Sub-alpine/grassy slope	2418	Overexploitation/grazing	Pewakh
10	Shopian	Shopian Forest Division	Hirpora	Sub-alpine/moist grassy slopes	2501	Overexploitation/grazing	Patris
11	Anantnag	Anantnag Forest Division	Daksun	Alpine/moist grassy slopes	2901	Overexploitation/grazing	Patris
12	Anantnag	Lidder Forest Division	Aru	Sub-alpine/open rocky slopes	2658	Overexploitation/grazing/soil erosion	Patris

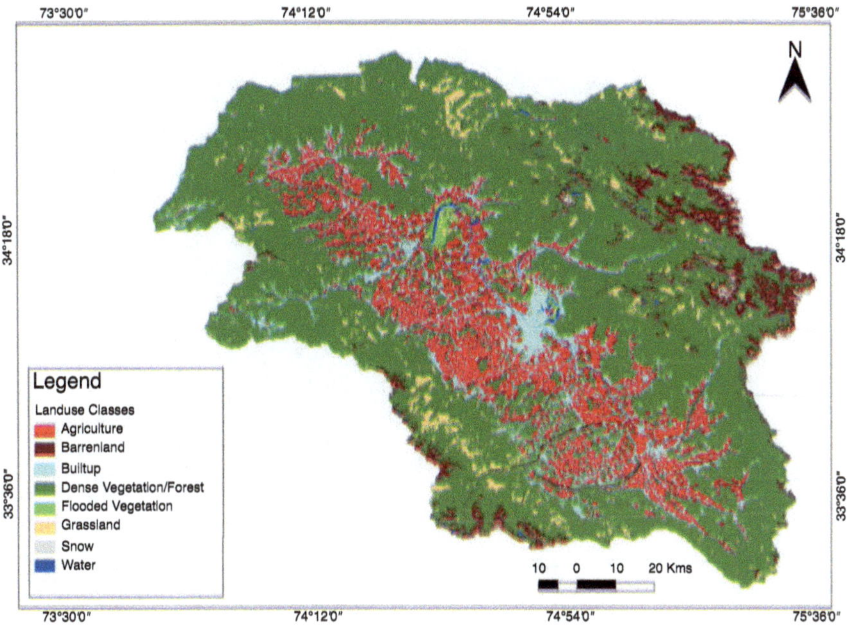

Fig. 12.2 Landuse-landcover (LULC) map of the study area

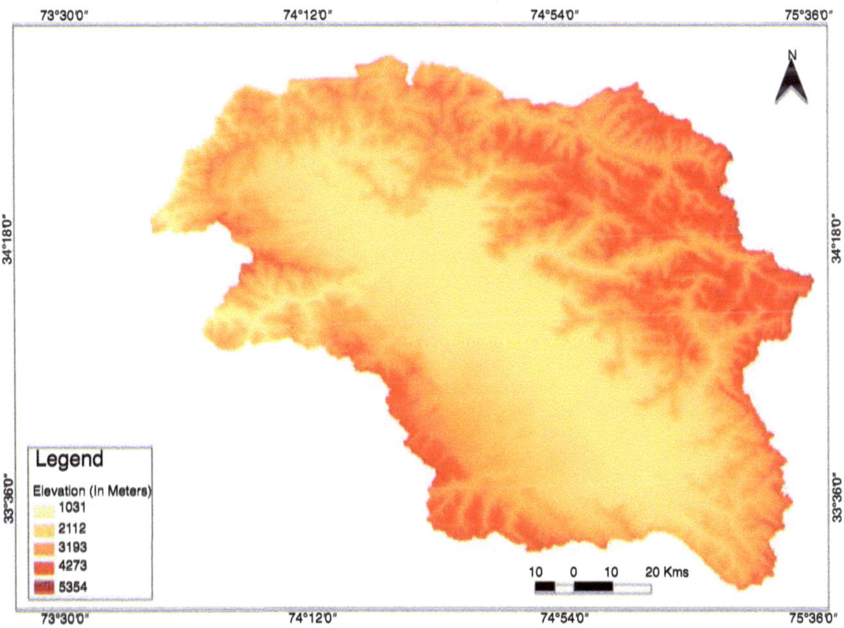

Fig. 12.3 Elevation map of the study area

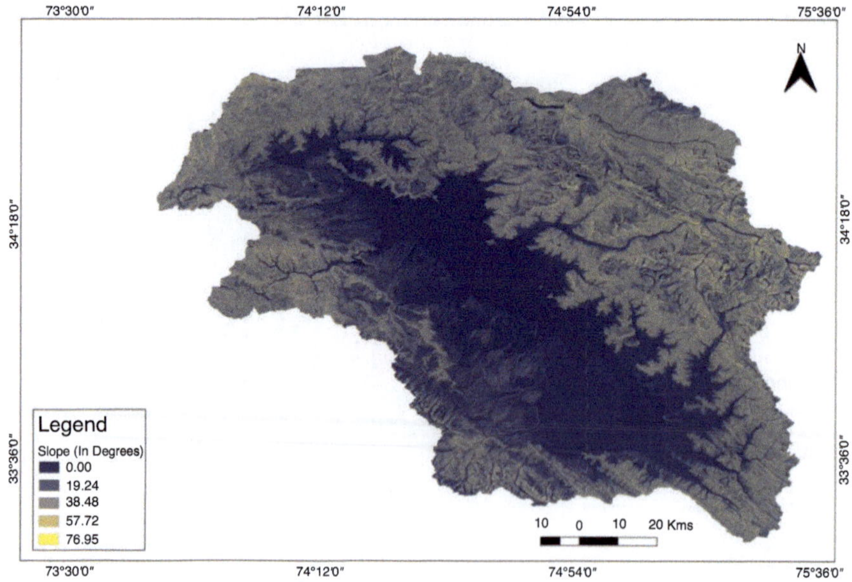

Fig. 12.4 Slope map of the study area

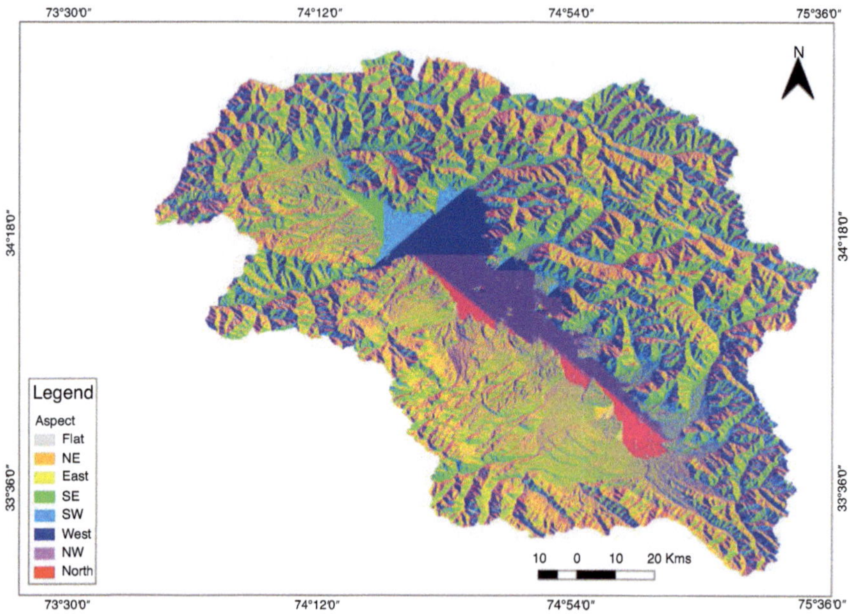

Fig. 12.5 Aspect map of the study area

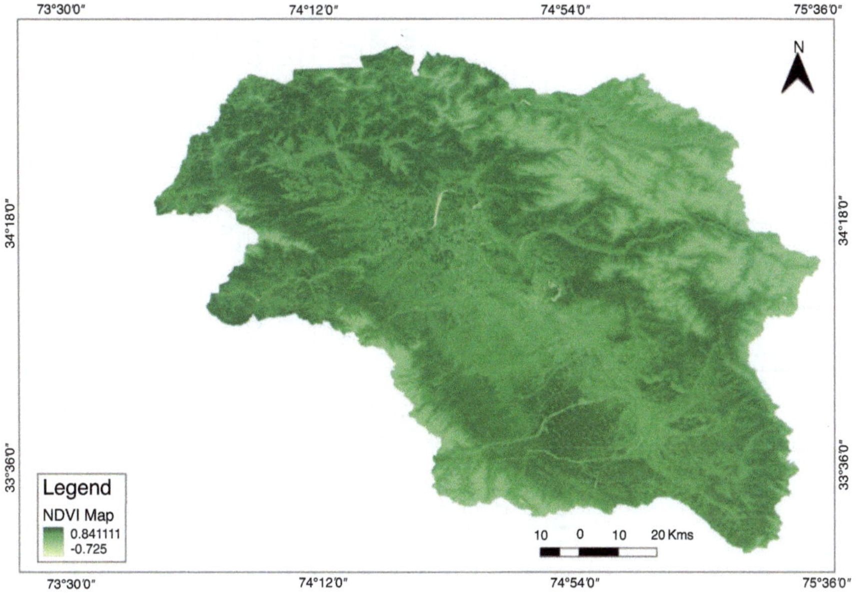

Fig. 12.6 NDVI (normalized difference vegetation index) map of the study area

$$\mathrm{NDVI} = (\mathrm{Band}\,5 - \mathrm{Band}\,4)/(\mathrm{Band}\,5 + \mathrm{Band}\,4) \tag{12.1}$$

12.2.4 Bioclimatic Variable Selection

Bioclimatic variables were selected from the WorldClim website, as many bioclimatic variables are correlated with each other, and we have performed Pearson's correlation among various bioclimatic and environmental variables using Microsoft Excel (Table 12.2). The most popular technique for analysing numerical variables is Pearson's correlation approach, which assigns a value between 0 and 1, with 1 denoting total positive correlation and 0 denoting total negative correlation (Boslaugh and Watters 2008). Variables that represent the coefficient of correlation (r) values greater than 0.7 were shortened to a single variable. Out of the 19 bioclimatic variables, 4 variables were applied in this study to run the MaxEnt model, viz. (a) bio1, annual mean temperature; (b) bio4, temperature variability; (c) bio12, annual precipitation; and (d) bio15, precipitation seasonality (coefficient of variation) obtained from WorldClim (https://www.worldclim.org), a database of global weather and climate data (Hijmans et al. 2005) and environmental layers such as NDVI, slope, aspect, elevation and landcover. Resampling was performed on all layers in order to have a uniform resolution of 1 km^2 to run the MaxEnt model using the bioclimatic as well as environmental variables.

Table 12.2 Pearson's correlation matrix

Bioclimatic variables	Aspect	bio1	bio2	bio3	bio4	bio5	bio6	bio7	bio8	bio9	bio10	bio11	bio12
Aspect	1.00												
bio1	0.02	1.00											
bio2	0.04	0.89	1.00										
bio3	0.17	0.97	0.83	1.00									
bio4	−0.24	−0.73	−0.47	−0.86	1.00								
bio5	−0.04	0.86	0.96	0.75	−0.32	1.00							
bio6	0.12	0.97	0.80	0.99	−0.87	0.73	1.00						
bio7	−0.22	−0.57	−0.22	−0.72	0.96	−0.09	−0.75	1.00					
bio8	0.14	0.89	0.79	0.89	−0.70	0.77	0.90	−0.56	1.00				
bio9	0.04	1.00	0.85	0.98	−0.79	0.81	0.99	−0.65	0.89	1.00			
bio10	−0.05	0.95	0.95	0.85	−0.48	0.98	0.84	−0.28	0.84	0.91	1.00		
bio11	0.10	0.98	0.82	0.99	−0.86	0.75	1.00	−0.73	0.89	0.99	0.86	1.00	
bio12	−0.22	0.25	−0.15	0.28	−0.56	−0.17	0.37	−0.70	0.27	0.32	0.01	0.35	1.00
bio13	−0.23	0.43	0.16	0.41	−0.52	0.17	0.49	−0.55	0.49	0.47	0.29	0.48	0.86
bio14	−0.32	0.40	0.03	0.36	−0.50	0.07	0.46	−0.59	0.44	0.44	0.23	0.44	0.95
bio15	0.13	−0.17	0.26	−0.21	0.48	0.26	−0.28	0.65	−0.06	−0.24	0.06	−0.26	−0.86
bio16	−0.35	0.49	0.24	0.41	−0.40	0.30	0.49	−0.43	0.53	0.51	0.41	0.49	0.81
bio17	−0.28	0.53	0.17	0.51	−0.63	0.18	0.59	−0.69	0.51	0.58	0.35	0.58	0.93
bio18	0.05	−0.03	−0.42	0.12	−0.57	−0.51	0.18	−0.76	0.06	0.06	−0.33	0.15	0.85
bio19	−0.45	0.32	−0.01	0.25	−0.35	0.04	0.33	−0.45	0.18	0.36	0.19	0.34	0.86
Elevation	0.14	−0.88	−0.75	−0.80	0.53	−0.80	−0.81	0.40	−0.77	−0.87	−0.87	−0.82	−0.32
Landcover	0.26	0.56	0.54	0.53	−0.24	0.60	0.49	−0.14	0.54	0.53	0.60	0.50	−0.24
NDVI	−0.12	0.62	0.57	0.64	−0.52	0.53	0.64	−0.42	0.60	0.63	0.58	0.64	0.21
Slope	0.36	0.02	0.20	0.07	0.07	0.13	−0.02	0.16	−0.18	−0.01	0.08	−0.01	−0.62

Bioclimatic variables	bio13	bio14	bio15	bio16	bio17	bio18	bio19	Elevation	Landcover	NDVI	Slope
Aspect											
bio1											
bio2											
bio3											
bio4											
bio5											
bio6											
bio7											
bio8											
bio9											
bio10											
bio11											
bio12											
bio13	1.00										
bio14	0.91	1.00									
bio15	−0.54	−0.76	1.00								
bio16	0.94	0.94	−0.54	1.00							
bio17	0.94	0.98	−0.74	0.93	1.00						
bio18	0.57	0.67	−0.77	0.42	0.65	1.00					
bio19	0.80	0.89	−0.81	0.83	0.89	0.52	1.00				
Elevation	−0.55	−0.50	0.22	−0.63	−0.60	0.06	−0.48	1.00			
Landcover	−0.17	−0.09	0.09	−0.08	−0.03	−0.32	−0.13	−0.40	1.00		
NDVI	0.34	0.29	−0.03	0.39	0.38	0.09	0.12	−0.60	−0.08	1.00	
Slope	−0.70	−0.70	0.26	−0.68	−0.63	−0.46	−0.48	0.02	0.27	−0.14	1.00

12.2.5 Species Distribution Modelling (SDM)

For the SDM, MaxEnt software was utilized which predicts the distribution based on the principle of maximum entropy to find species range with the highest probability uniformly combining environmental data based on occurrence data for as respective extent (Phillips et al. 2006). Firstly, occurrence points of species were added along with all the variables (bioclimatic and environment). Secondly, MaxEnt parameters were set as per the model requirement with the number of iterations and occurrence points. MaxEnt parameters with the basic, advanced and experimental mode as well as the number of output required as graphs, plots and output file formats were specified. The entire modelling process was performed with minimum 25 iterations while using 75% of the data for training and 25% for testing.

12.3 Results and Discussions

The model evaluation is performed by MaxEnt inbuilt based on receiver operating characteristic (ROC) curve which provided a mean AUC (area under curve) of 0.719 which signifies reasonable model performance. The response curves as depicted in (Figs. 12.7, 12.8, 12.9 and 12.10) give the probability of presence with respective values of variables given other environmental variables have average sample value. Based on these response curves, it can be inferred that the species will have higher presence in higher precipitation areas (mostly above 800 mm precipitation). Moreover, the probability of species presence will be higher where less precipitation

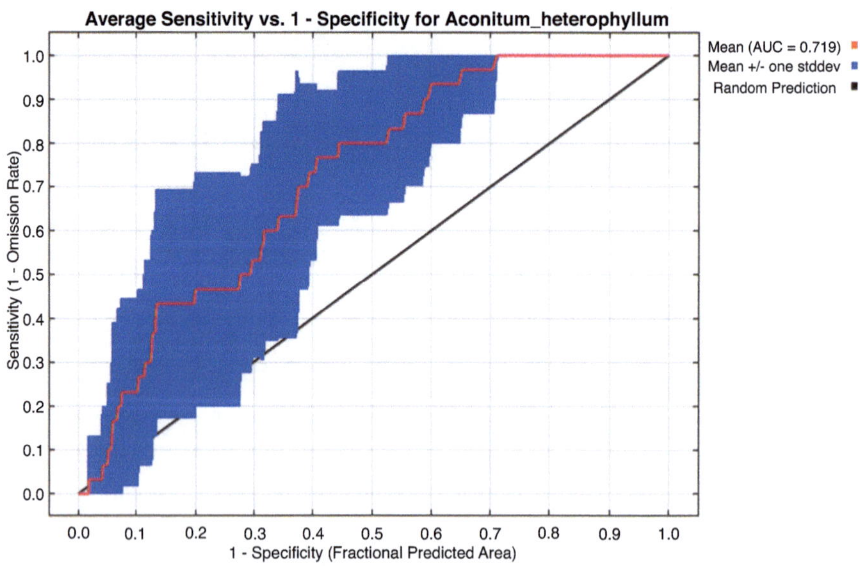

Fig. 12.7 ROC (receiver operating characteristic) curve showing specificity

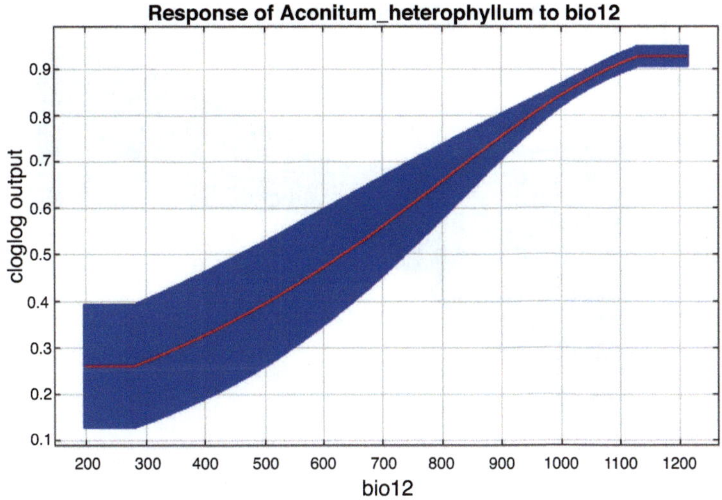

Fig. 12.8 Response curve showing predicted probability of species presence with respect to values of the variable bio12 (annual precipitation)

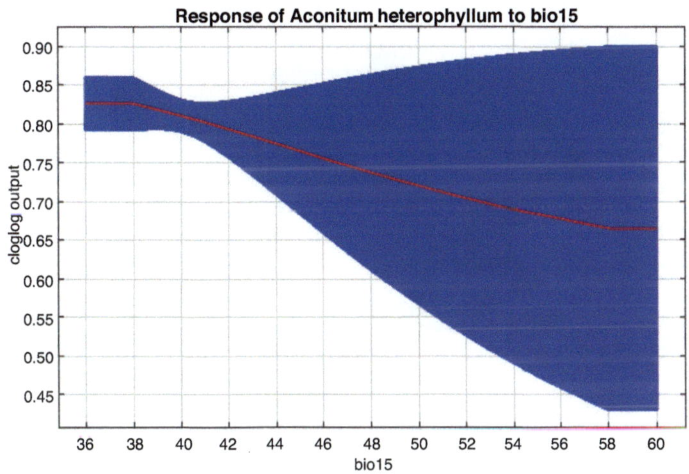

Fig. 12.9 Response curve showing predicted probability of species presence with respect to values of the variable bio15 (precipitation seasonality)

seasonality (<40) is found, i.e. standard deviation of monthly precipitation estimates as percents mean of the annual mean. Furthermore, as these species will have higher probability of occurrence in a higher degree of slope comparatively, this can be due to alpine environment. The habitat estimation of *Aconitum heterophyllum* in Kashmir Himalayas revealed that the highly suitable area is 210.08 km^2 (1.28%), moderately suitable area 6155.56 km^2 (37.62%) and not suitable area 9995.32 km^2 (61.09%) for the current scenario, respectively (Fig. 12.11).

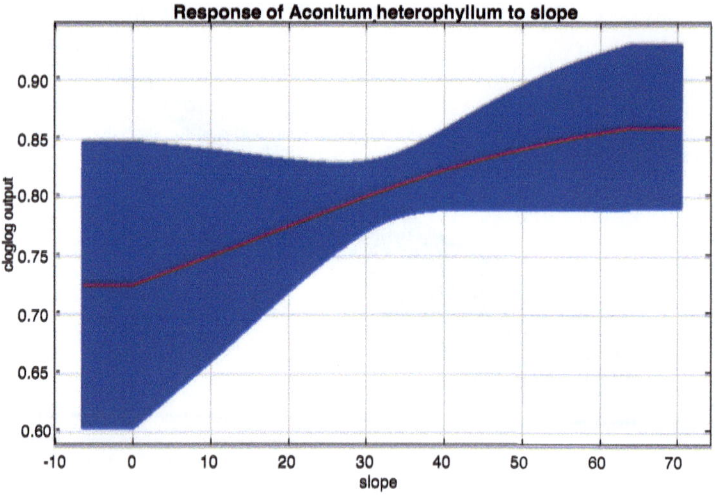

Fig. 12.10 Response curve showing predicted probability of species presence with respect to values of the variable slope

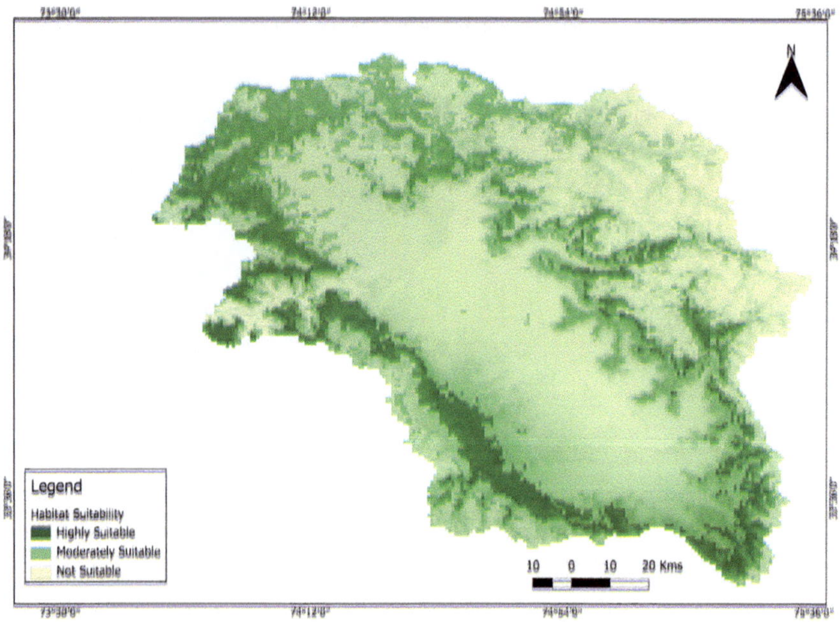

Fig. 12.11 Habitat suitability map of the *Aconitum heterophyllum* in Kashmir Himalayas

Table 12.3 Analysis of the different variable contributions

Variable	% Contribution	Permutation importance
Landcover	54	21.9
bio12	34.3	37.8
Slope	6.7	26
bio15	4.9	10.6
Elevation	0.1	2.5
Aspect	0.1	0.2
bio1	0	1
NDVI	0	0
bio4	0	0

The percent contributions of the variables are shown in Table 12.3. The landcover has the highest percent contribution (>50%) for the model followed by bio12 (annual precipitation) (>30%) and slope (>6%), whereas the annual precipitation has the highest permutation importance, i.e. the model performance when the values of the variables are randomly shuffled to evaluate. Furthermore, landcover and bio15 precipitation seasonality also had significant permutation importance. Response curves for the remaining variables vary through different algorithms, indicating that their influence in regulating the potential distribution of *Aconitum heterophyllum* varied to a greater extent in case of more occurrence points.

It is reported that some variables also affect the distribution range of *Aconitum heterophyllum*, surrounding the Kashmir valley due to change in the future climatic scenario; however, two additional bioclimatic variables were utilized, viz. bio2 (mean diurnal range) and bio8 (mean temperature of the wettest quarter). Nevertheless, that study utilized an ensemble model focusing more on the future climatic scenario and its effects on species distribution, and results show reduction in suitability area of *Aconitum heterophyllum* in all different scenarios except high altitude regions in Pir Panjal range which is similar to our results. However, the stark difference between their current scenario results and our model results shows that there is also higher suitability in forest divisions of Kamraj, Langate, Jhelum Valley, Tangmarg, etc. (Wani et al. 2022). Under the current climatic conditions, the habitat suitability model delineated and exhibited the predominance of the highest suitability habitat (HSH) for the assessed species in the highlands across area elevation ranging from 2200 m to 4200 m, whereas 4.6% of the upper study area was highly suitable for *Aconitum heterophyllum* (Dad and Rashid 2022). Results of this study finds support from Salam et al. (2020) who reported the high to very high habitat suitable areas for *Lagotis cashmeriana* (Royle) RUPR from the continuous alpine patches of Northwestern Himalayan region and medium to low habitat suitability areas as sub-alpine slopes among evergreen forests. The ranges of habitats including sub-alpine slopes, moist areas near ridges and exposed rocky slopes near streams as well as ground vegetation under the *Abies pindrow* (Fir) forests were found to be the most high probability areas for *Aconitum heterophyllum* in Kashmir. Such areas could be used for in situ conservation and restoration of the species after careful consideration of onsite conditions. The current study also demonstrated that biotic variables, especially human activities, have an impact on this species' habitats. Therefore, both in situ and ex situ conservations of this species are urgently needed.

12.4 Conclusion

It is concluded that the most suitable habitat for *Aconitum heterophyllum* is the periphery of the Kashmir valley. The highly suitable area accounted for 210.08 km^2 which was mostly higher elevation alpine forest areas surrounding the Kashmir valley, whereas moderately suitable areas were found to be 210.08 km^2, and the remaining majority of the study area which is not suitable mostly constitute agricultural land, urban build-up and permanent snow-covered patches, as indicated in the LULC comparison. It would be wise to have maximum occurrence points for any further investigation on this species, which can help in predicting the futuristic scenarios over a period of time. *Aconitum heterophyllum* is one of the most exploited medicinal plants from the Kashmir Himalayas due to its escalating demand in domestic and international markets. The anthropogenic pressure on its habitat and ruthless extraction from the wild has put the limited population of this species under extreme stress. It is therefore high time that this species is rehabilitated and conserved in the areas identified in this study. The species-specific targeted restoration efforts and campaign will help in conservation and sustainable extraction of the species.

Acknowledgement Authors are grateful to Mr. Shabir Ahmad (PI Division) and Mr. Wajahat Wasim (PP Division) of J & K Forest Department for providing the KMZ files of Forest Divisions of Jammu and Kashmir. Thanks are also due to Dr. Ashfaq Mir of FBT Division of the Faculty of Forestry, SKUAST Kashmir, for helping with the plagiarism check of this manuscript.

References

Balaramnavar VM, Kour M, Ahuja A (2021) *Aconitum heterophyllum*—an overview. J Pharma Res Int 33:242–247

Beigh SY, Nowchoo IA, Iqbal M (2008) Cultivation and conservation of *Aconitum heterophyllum*: a critically endangered medicinal herb of the Northwest Himalayas. Int J Geogr Inf Syst 11:47–56

Bobrowski M, Weidinger J, Schwab N, Schickhoff U (2021) Searching for ecology in species distribution models in the Himalayas. Ecol Model 458:109693

Boslaugh S, Watters PA (2008) Statistics in a nutshell: a desktop quick reference, Ch. 7. O'Reilly Media, Sebastopol, CA. ISBN-13: 978-0596510497

Cengic M, Rost J, Remenska D, Janse JH, Huijbregts MAJ, Schipper AM (2020) On the importance of predictor choice, modelling technique, and number of pseudo-absences for bioclimatic envelope model performance. Ecol Evol 10:12307–12317

Condro AA, Prasetyo LB, Rushayati SB, Santikayasa IP, Iskandar E (2021) Predicting hotspots and prioritizing protected areas for endangered primate species in Indonesia under changing climate. Biology 10:154

Dad JM, Rashid I (2022) Differential responses of Kashmir Himalayan threatened medicinal plants to anticipated climate change. Environ Conserv 49:33–41. https://doi.org/10.1017/S03768922000030

Elith J, Leathwick JR (2009) Species distribution models: ecological explanation and prediction across space and time. Annu Rev Ecol Evol Syst 40:677–697

Engler R, Randin CF, Thuiller W, Dullinger S, Zimmermann NE, Araujo MB, Guisan A (2011) 21st century climate change threatens mountain flora unequally across Europe. Glob Chang Biol 17:2330–2341

Fois M, Cuena-Lombrana A, Fenu G, Cogoni D, Bacchetta G (2016) The reliability of conservation status assessments at regional level: past, present and future perspectives on *Gentianalutea* L. ssp. lutea in Sardinia. J Nat Conserv 33:1–9

Goraya GS, Ved DK (2017) Medicinal plants in India: an assessment of their demand and supply. National Medicinal Plants Board, Ministry of AYUSH, Government of India, New Delhi and Indian Council of Forestry Research & Education, Dehradun. https://www.rcfceast.org/wp-content/uploads/2019/02/Medicinal_Plants_in_India_An_Assessment_of_their_Demand_and_Supply.pdf

Guisan A, Thuiller W (2005) Predicting species distribution: offering more than simple habitat models. Ecol Lett 8:993–1009

Guisan A, Zimmermann NE (2000) Predictive habitat distribution models in ecology. Ecol Model 135:147–186

Hexa-Research (2017) Herbal medicine market size and forecast, by product (tablets & capsules, powders, extracts), by indication (digestive disorders, respiratory disorders, blood disorders), and trend analysis, 2014–2024. https://www.hexaresearch.com/research-report/global-herbal-medicine-market

Hijmans RJ, Cameron SE, Parra JL, Jones PG, Jarvis A (2005) Very high resolution interpolated climate surfaces for global land areas. Int J Climatol 25:1965–1978

Kaky E, Nolan V, Alatawi A, Gilbert F (2020) A comparison between ensemble and Max Ent species distribution modelling approaches for conservation: a case study with Egyptian medicinal plants. Ecol Infor 60:101150

Kirtikar KR, Basu BD (1975) Indian medicinal plants, vol 1, 2nd edn. Lalit Mohan Basu, Allahabad

Kumar V, Malhotra N, Pal T, Chauhan RS (2016) Molecular dissection of pathway components unravels atisine biosynthesis in a non-toxic Aconitum species, Aconitum heterophyllum Wall. 3 Biotech 6:106

Kumari P, Wani IA, Khan S, Verma S, Mushtaq S, Gulnaz A, Paray BA (2022) Modelling of *Valerianawallichii* habitat suitability and niche dynamics in the Himalayan region under anticipated climate change. Biology 11:498. https://doi.org/10.3390/biology11040498

Malhi Y, Franklin J, Seddon N, Solan M, Turner MG, Field CB, Knowlton N (2020) Climate change and ecosystems: threats, opportunities and solutions. Philos Trans R Soc 375:20190104

Megan E (2021) Re-conceptualizing the role (s) of science in biodiversity conservation. Environ Conserv 48:151–160

Mehta A, Shukla S, Rakholia S (2021) Vegetation change analysis using normalized difference vegetation index and land surface temperature in greater Gir landscape. J Sci Res 65(3):1–6

Mouquet N, Lagadeuc Y, Devictor V, Doyen L, Duputié A, Eveillard D et al (2015) Predictive ecology in a changing world. J Appl Ecol 52:1293–1310

Muir G, Simona S (2018) Making NWFPS visible: disentangling definitions and refining methodologies. In: Proceedings of the IIQ. Agenda Item 10.2: overview of forestry and environment statistics in AP region, 27th session, Port Denarau, Fiji, 19–23 March 2018

Nagarajan M, Kuruvilla GR, Kumar KS, Venkatasubramanian P (2015) Pharmacology of *Ativisha, Musta* and their substitutes. J Ayurveda Integr Med 6(2):121–133

Pecl GT, Araújo MB, Bell JD, Blanchard J, Bonebrake TC, Chen IC, Williams SE (2017) Biodiversity redistribution under climate change: impacts on ecosystems and human well-being. Science 355:eaai9214

Peerzada IA, Sofi AH (2017) Traditional use of medicinal plants among tribal communities of Bangus Valley, Kashmir Himalaya, India. Stud Ethno Med 11(4):318–331

Peerzada IA, Reddy M, Wani I, Panse SS (2016) Impact of climate change Vis-à-Vis anthropogenic interventions on natural resources in Kashmir, India—an overview. J Appl Nat Sci 8(1): 489–493

Peerzada IA, Sofi AH, Singh O (2018) Tradable NTFPs vis-a-vis medicinal plants of Kashmir Himalayan forests. Expressions Print & Graphics Pvt. Ltd., Dehradun. ISBN-978-81-929285-8-6

Peerzada IA, Chamberlain J, Reddy M, Dhyani S, Saha S (2021) Policy and governance implications for transition to NTFP-based bioeconomy in Kashmir Himalayas. Sustainability 13:11811. https://doi.org/10.3390/su132111811

Peerzada IA, Islam MA, Chamberlain J, Dhyani S, Reddy M, Saha S (2022) Potential of NTFP based bioeconomy in livelihood security and income inequality mitigation in Kashmir Himalayas. Sustainability 14:2281. https://doi.org/10.3390/su14042281

Phillips SJ, Anderson RP, Schapire RE (2006) Maximum entropy modelling of species geographic distributions. Ecol Model 190(3-4):231–259

Pimm S, Russell G, Gittleman J, Brooks T (1995) The future of biodiversity. Science 1995(269): 347. https://doi.org/10.1126/science.269.5222.347

Rajakrishnan R, Lekshmi R, Samuel D (2016) Analytical standards for the root tubers of ativisha *Aconitum heterophyllum* Wall. ex Royle. Int J Sci Res Publ 6(5):531–534

Ramatsobane MM, Anthony JA (2020) Comparative phytochemical constituents and antioxidant activity of wild and cultivated *Alepidea amatymbica* Eckl & Zeyh. Bio Med Res Int 13: 5808624. https://doi.org/10.1155/2020/5808624

Rashid I, Romshoo SA, Chaturvedi RK, Ravindranath NH, Sukumar R, Jayaraman M et al (2015) Projected climate change impacts on vegetation distribution over Kashmir Himalayas. Clim Change 132:601–613

Rew J, Cho Y, Moon J, Hwang E (2020) Habitat suitability estimation using a two-stage ensemble approach. Remote Sens (Basel) 12:1475

Rodríguez J, Brotons L, Bustamante J, Seoane J (2007) The application of predictive modeling of species distribution to biodiversity conservation. Divers Distrib 13:243–251

Romshoo SA, Bashir J, Rashid I (2020) Twenty-first century-end climate scenario of Jammu and Kashmir Himalaya, India, using ensemble climate models. Clim Change 162:1473–1491

Salam N, Reshi ZA, Shah MA (2020) Habitat suitability modelling for *Lagotis cashmeriana* (ROYLE) RUPR., a threatened species endemic to Kashmir Himalayan alpines. Geol Ecol Landsc 6:1–11. https://doi.org/10.1080/24749508.2020.1816871

Shrestha UB, Gautam S, Bawa KS (2012) Widespread climate change in the Himalayas and associated changes in local ecosystems. PLoS One 7:e36741

Singh L, Tariq KM, Chandra S, Bhatt ID, Nandi SK (2017) Ecological niche modelling: an important tool for predicting suitable habitat and conservation of the Himalayan medicinal herbs. ENVIS Bull Himal Ecol 25:154

Sojitra NA, Patel RK, Dixit RB, Dixit BC (2013) Synthesis and antimicrobial study of pyrimidinone substituted 4(3H)-quinazolinone derivatives. J Org Chem Curr Res 2:2. https://doi.org/10.4172/2161-0401.1000118

Srivastava N, Sharma V, Barkha K, Dobriyal AK, Jadon VS (2010) Polyphenolics free DNA isolation from different types of tissues of *Aconitum heterophyllum* Wall-endangered medicinal species. J Plant Sci 5:414–419

Tahir M, Peerzada IA, Sabeena N, Amir FB, Dar M (2019) Potential of medicinal plants in Kashmir Himalayas: a review. J Pharmacogn Phytochem 9(1):1629–1631

Taleshi H, Jalali SG, Alavi SJ, Hosseini SM, Naimi B, Zimmermann NE (2019) Climate change impacts on the distribution and diversity of major tree species in the temperate forests of Northern Iran. Reg Environ Chang 19:2711–2728

Tovar C, Arnillas CA, Cuesta F, Buytaert W (2013) Diverging responses of tropical Andean biomes under future climate conditions. PLoS One 8:e63634

Van de VCM, Weiss SB, Ernst WG (2007) Plant species distributions under present conditions and forecasted for warmer climates in an arid mountain range. Earth Interact 11:1–33

Wani IA, Verma S, Kumari P, Charles B, Hashim MJ, El-Serehy HA (2021a) Ecological assessment and environmental niche modelling of Himalayan rhubarb (*Rheum webbianum* Royle) in northwest Himalaya. PLoS One 16:e0259345

Wani TA, Kaloo ZA, Dangroo NA (2021b) *Aconitum heterophyllum* Wall. ex Royle: a critically endangered medicinal herb with rich potential for use in medicine. J Integr Med 19(6):545–554

Wani ZA, Ridwan Q, Khan S, Pant S, Siddiqui S, Moustafa M, Ahmad AE, Yassin HM (2022) Changing climatic scenarios anticipate dwindling of suitable habitats for endemic species of Himalaya—predictions of ensemble modelling using *Aconitum heterophyllum* as a model plant. Sustainability 14:8491. https://doi.org/10.3390/su14148491

WHO (2002) World Health Organization. WHO/EDM/TRM/, Geneva. WHO Report; p 19, 21

Yeshi K, Crayn D, Ritmejeryte E, Wangchuk P (2022) Plant ' secondary metabolites produced in response to abiotic stresses has potential application in pharmaceutical product development. Molecules 27:313. https://doi.org/10.3390/molecules27010313

Application of Species Distribution Modeling for Conservation and Restoration of Forest Ecosystems

13

Shilky, B. S. P. C. Kishore, Gajendra Kumar, Purabi Saikia, and Amit Kumar

Abstract

Climate change and habitat fragmentation are responsible for creating unstable and isolated populations of various rare, endangered, and threatened plant species in their natural habitats. Such species face unprecedented extinction risks due to changing climatic conditions, anomalous population growth, and their high dependency on natural forests. The use of species distribution models (SDMs) for forest restoration and conservation has gained a growing trend due to its significant contribution to improving the success rates of forest restoration projects. A critical aspect of the planning and conservation of forests is the selection of suitable management strategies that match the needs of the RET plant species in both the present and future climates. SDMs can help the development of integrated conservation strategy through (1) concentrating future survey and research efforts on areas with a high likelihood of occurrence, (2) assisting in the selection of areas for conservation and restoration, and (3) designing future research questions, such as those required to forecast climate change reactions. The present chapter emphasizes the use of SDMs for the conservation and restoration of natural forests from further degradation and to fulfill the growing demands of forest goods and services for the sustainable economic development of forest-dependent communities and the nation.

Shilky · P. Saikia (✉)
Department of Environmental Sciences, Central University of Jharkhand, Ranchi, India
e-mail: shilky.20270201005@cuj.ac.in; purabi.saikia@cuj.ac.in

B. S. P. C. Kishore · G. Kumar · A. Kumar
Department of Geoinformatics, Central University of Jharkhand, Ranchi, India

S. Dhyani et al. (eds.), *Ecosystem and Species Habitat Modeling for Conservation and Restoration*, https://doi.org/10.1007/978-981-99-0131-9_13

Keywords

Species distribution models · Ecological niche modeling · Climate change · Conservation of biodiversity

13.1 Introduction

Globally, increasing habitat loss and degradation puts many of the world's species in jeopardy, resulting in population declines, genetic diversity loss, and even species extinction (Foley et al. 2005). Deforestation is the current and projected primary direct and indirect cause of species extinction. It is predicted that up to 21% of Southeast Asian forest species will disappear by the year 2100 due to previous and present deforestation (Sodhi et al. 2009). Furthermore, climate change can have a significant impact on ecological and biotic elements that influence the global distribution of habitats and species (Dar et al. 2019), such as a shift in the plant-pollinator relationship through phenological changes and regeneration failure due to stress on species or their native environments (Parmesan 2006; Reyer et al. 2013). Physiology, demography, dispersion, interspecific interactions, adaptation, and alteration of environmental factors are crucial processes influencing biodiversity to respond to global change (Urban et al. 2016). Some extinctions may happen as a direct result of habitat loss by removing all individuals, or they may happen indirectly as a result of aiding the spread of a disease or invading species, making it easier for humans to hunt them, or changing biophysical circumstances (Sodhi et al. 2009). Human-caused climate change, characterized by unprecedented variations in temperature and precipitation regimes, poses an additional threat to biodiversity and hastens climate fluctuations on ecosystems, which significantly impact conservation goals, resulting in the waste of vital and scarce conservation inputs (Millar et al. 2007). Global environmental change is causing species ranges to shift, fragment, and expand (Chen et al. 2011). Climate change is one of the primary factors influencing species dispersal potential, which has a significant impact on ecosystem structure, functioning, and shifts in species' geographical distribution all over the world because species always seek the best environment for survival (Peterson 2003; Feeley et al. 2012). It can cause a change in the mean of variables like temperature or precipitation and their variability (Seneviratne et al. 2012). Despite the significance of differences in mean values, there is evidence that plant distribution, survival, net primary productivity, and species diversity are influenced by extreme rather than average conditions (Knapp et al. 2002; Jentsch and Beierkuhnlein 2008). Rising temperatures may force species to migrate to higher elevations, hastening extinction in tropical regions (Angelo and Daehler 2013).

The degradation of the ecosystem is frequently caused by a failure to recognize its beneficial functions ranging from sustaining biodiversity; climate control; sediment storage; flood defense and storm buffering; soil, air, and water quality maintenance; support of food chains; and food and water to health and security, which our growing population requires today and in the future (Crooks and Turner 1999). Deforestation and forest degradation are probably the most well-known forms of ecosystem degradation because it changes the structure of the habitat so quickly and dramatically (Ehrlich et al. 2013). Ecosystem degradation has severe consequences for biodiversity and climate, prompting national, regional, and global goals for ecosystem restoration (Díaz et al. 2019). None of the negotiated global objectives for preserving life on Earth and halting land and ocean degradation has been fully met (UNEP 2021), and only 6 of the 20 Aichi Biodiversity Targets have been partially met (CBD 2020). Therefore, the United Nations has declared the years 2021–2030 the UN Decade on Ecosystem Restoration, with the Bonn Challenge and the New York Declaration on Forests aiming to restore 350 Mha of degraded forests globally by 2030 (Mansourian et al. 2021). Habitat restoration is a critical strategy for protecting and restoring ecosystems and preventing species extinction (Polak and Saltz 2011). People's prosperity and well-being depend on the process of preventing and reversing degradation, which leads to better ecosystem services and restored biodiversity through a range of strategies that are dependent on local conditions and community choices (UNEP 2021). On the other hand, conservation is a strategy for dealing with species extinctions, habitat loss, and ecosystem degradation caused by increased human population and activity (Marvier 2013). Species (re)introduction is one of the most effective ecological engineering strategies for restoring depleted species populations, degraded habitats, and ecosystems (Polak and Saltz 2011). Identification of high-priority conservation areas and prospective habitats for the reintroduction of RET plant species depends on thorough analyses of the variables that affect species rarity, proper land management, restoration techniques, and the development of more reliable prediction models (Maschinski and Haskins 2012).

The ecological niche is defined as a set of ecological conditions that allow species to persist and propagate (Grinnell 1917). In contrast, the fundamental niche that a species is compelled to fill due to interactions with other species is subdivided into the realized niche (Hutchinson 1957). Habitat distribution modeling, also known as ecological niche modeling (ENM), has provided new insights into the factors influencing species distribution over time and space (Kozak et al. 2008; Kumar et al. 2020). ENM is a strategy that uses computer algorithms to forecast species distribution in a geographical area based on a mathematical model of the species' ecological niche (Adhikari et al. 2012). It analyzes changes in habitat suitability over time in a particular scenario of environmental change and the relative appropriateness of habitat in regions where the species is not known to exist (Warren and Seifert 2011). When site occupancy data are insufficient and there aren't enough resources for more data collection, they explain habitat needs and help construct distribution estimates, both of which are essential to fulfilling endangered species conservation goals (Hirzel et al. 2006). The distribution of species in space and time is influenced by a variety of variables, including their environmental surroundings, and some of them may have a direct impact, while others may only have an indirect effect (Westgate et al. 2014).

13.2 Application of Species Distribution Models (SDMs) in the Conservation and Restoration of Forest Ecosystems

According to an ideal landscape composition for biodiversity conservation, more than 40% of the landscape should be covered in forest, with 10% of that being large tracts of forest and the remaining 30% being smaller fragments (Arroyo-Rodríguez et al. 2020). Existing forests must be conserved, degraded forests must be restored, and there must be a connection between the dispersed forest patches to improve the quantity of habitat (Watling et al. 2020). The consequences of habitat fragmentation on biodiversity will be less apparent if there are enough forests (Arroyo-Rodríguez et al. 2020). Additionally, forests are important for protecting water supplies, producing food, and providing additional ecosystem services, including carbon sequestration, mitigating climate change, restoring soil fertility, and maintaining air quality (Melo et al. 2021). In conservation and restoration research, SDMs are a valuable tool for predicting the habitat appropriateness of target species and are frequently used to evaluate relationships between species occurrence and environmental parameters (Lyon et al. 2019). The importance of SDMs in decision-making for the conservation and restoration of forest ecosystems is expanding quickly (Elith and Leathwick 2009) because SDM is a useful tool for forest management and may be used to foresee how climate change would affect forests (Guisan et al. 2017). Although open-access and worldwide species occurrence databases are limited and only available for a small number of surveyed locations (Rondinini et al. 2006), they offer the opportunity to analyze the data for developing conservation and restoration strategies (Jetz et al. 2012). Regarding conservation management decisions, information for unsurveyed or future colonized regions due to invasion and climate change is required, which is not available in any databases (Giljohann et al. 2011). This information may be obtained by anticipating species occurrences utilizing their environmental and climatic suitability along with spatial environmental data using various SDMs (Peterson et al. 2011). The prediction of habitat alterations using ENM/SDM approaches is essential for aligning human-assisted efforts utilized in banj oak recruitment and regeneration in degraded oak forest areas in the Central Himalayas to be coordinated with applicable silviculture practices (Dhyani et al. 2020). SDMs make it possible to evaluate the risk of targeted activities based on the possibility of under- and over-protected errors, which aids in prioritizing the implementation based on only a limited number of ideas and the cost-effectiveness of the approach (Schwartz 2012). It may evaluate various strategies and their capacity to preserve biodiversity under the expected climate of the present and the future (Hunt et al. 2020). Forest management strategies may fully use the potential of SDMs for the protection and restoration of forest ecosystems by establishing a structured decision-making process (Di Febbraro et al. 2018; Frans et al. 2021).

13.3 Application of SDMs in the Management of Threatened Species

Understanding the primary forces shaping species' geographic ranges is critical for SDMs (Pulliam 2000). Large-scale ecological models incorporating data have recently advanced conceptually and practically (Domisch et al. 2016; Merow et al. 2017). These models can provide the change of species distributions and abundances over time and space and are helpful in the study of eco-biogeography, conservation biology, evolution, dispersion and migration, species invasion, metapopulation, and climate change (Isaac et al. 2020). Ecologists can predict future species distributions based on the environment of existing places since the core principle behind SDMs believes that a species' existence in a particular area is strongly reliant on the environment (Austin 2007). The basic idea behind SDMs aims to connect species occurrence data with environmental factors to learn more about the target species' ecology and evolution (Elith and Franklin 2013). Data on species' presence and absence are statistically related to environmental predictor factors (Guisan and Thuiller 2005), which are crucial resources for concentrating conservation efforts on species with insufficient distribution data (Gogol-Prokurat 2011). It can predict the possible distribution of a species in different regions over time and space by combining observed occurrences of species with environmental factors (Requena-Mullor et al. 2019). It needs accurate occurrence data on the target species, which may be found in herbarium records, published scientific publications, or web sources, though there may be some issues with such data (Heberling et al. 2019). Numerous mathematical models can be used to fit, select, and evaluate correlative SDMs (Xu et al. 2015). The mathematical algorithms include (1) profile methods, which are simple statistical techniques that use, for example, environmental distance to known sites of occurrence, such as BIOCLIM (Nix 1986; Nix and Busby 1986) and DOMAIN (Carpenter et al. 1993); (2) regression methods, such as forms of generalized linear models (Nelder and Wedderburn 1972); and (3) machine learning methods, like maximum entropy or MaxEnt model (Jaynes 1957) (Table 13.1).

SDMs can provide maps of the probabilities of a species occurring in a given area under a set of environmental conditions. It helps in identifying suitable habitats for conserving various economically and ecologically important plant species whose natural populations are in decline (Jiménez-Valverde and Lobo 2007). It assesses the effects of land-use change on the habitats of threatened and various highly exploited economically important plant species (Rodríguez et al. 2007) and the impact of global climate change on their distributions (Thuiller et al. 2008). The human footprint includes a variety of human activities that have an indirect or direct influence on natural ecosystems, and they considerably impact the habitat distribution patterns of RET species (Sun et al. 2020). SDM heavily relies on the effect of various environmental variables, such as the human footprint (Beans et al. 2012), along with climatic factors to estimate the distribution of endangered species (Feng et al. 2020). Additionally, SDMs offer a robust predictive framework to identify populations that have not yet been discovered, and they may significantly increase the possibilities of discovering new populations by focusing their search on areas

Table 13.1 List of algorithms used for ecological niche modeling

Species distribution models (SDMs)		Sources
Correlative SDMs	Climate envelope models (CEMs)	Hijmans and Graham (2006)
	Bioclimatic models	Jeschke and Strayer (2008)
Mechanistic SDMs	Process-based models or biophysical models	Kearney and Porter (2009)
Mathematical algorithms of ecological niche modeling (ENM)		
Profile techniques	BIOCLIM model	Booth et al. (2014)
	DOMAIN model	Carpenter et al. (1993)
	Ecological niche factor analysis (ENFA)	Hirzel et al. (2002)
	Mahalanobis distance model	Mahalanobis (1936)
	Isodar analysis model	Morris (1987)
Regression-based techniques	Generalized linear model (GLM)	Nelder and Wedderburn (1972)
	Maxlike model	Royle et al. (2012)
	Favorability function (FF)	Real et al. (2006)
	Generalized additive model (GAM)	Murase et al. (2009)
	Multivariate adaptive regression splines (MARS)	Friedman (1991)
Machine learning techniques	Artificial neural networks (ANN)	Rosenblatt (1958)
	Support vector machines (SVM)	Cortes and Vapnik (1995)
	Boosted regression trees (BRT)/gradient boosting machines (GBM)	Elith et al. (2008)
	Random forest (RF) model	Breiman (2001)
	XGBoost (XGB) model	Chen and Guestrin (2016)
	Genetic algorithm for the rule set production (GARP)	Holland (1975))
	Maximum entropy (MaxEnt)	Phillips et al. (2006)

where there is a high probability of a species occurrence (McCune 2016). For instance, five new populations of *Gymnocladus assamicus* Kanjilal ex P.C. Kanjilal, a critically endangered tree species endemic to northeastern India, were located using SDM (Menon et al. 2010). Besides, the potential distribution of various economically significant tree species, such as *Bauhinia vahlii* (Wight & Arn.) Benth. (Thakur et al. 2022), *Butea monosperma* (Lam.) Kuntze (Tiwari et al. 2021), and *Boswellia serrata* Roxb. (Rajpoot et al. 2020), and ecologically significant tree species, such as *Pterocarpus marsupium* Roxb. (Kumar et al. 2020), was mapped in tropical forests of India using maximum entropy (MaxEnt) modeling. Moreover, habitat expansions result in species interactions between different species (Ancillotto et al. 2016). The bird *Psittacula krameri*, commonly known as rose-ringed parakeet, was previously adapted to habitats within their native ranges where humans predominated, but due to intraspecific niche differences and differential propagule pressure, their niche has been expanded across their original ranges where previously they were not populated (Cardador et al. 2016). SDMs can be used to predict the degree of habitat fragmentation for RET animal species, like the *Rhinoceros unicornis*, which can only be found in remote protected areas (Mukherjee et al. 2020). SDMs are also used in habitat connectivity mapping of carnivorous animals such as *Panthera onca* (Ramirez-Reyes et al. 2016), *P. uncia* (Holt et al. 2018), *P. tigris* (Suttidate et al. 2021), and *Melursus ursinus* (Puri et al. 2015).

13.4 Application of SDMs in the Management of Invasives

One of the primary causes of species extinction and biodiversity loss worldwide is biological invasion, which is typically caused by invasive alien species (Mooney and Drake 2012). The number of invasive species that successfully establish themselves in new habitats has been growing through time and habitat, and as a result, their effects are likely to be seen across all ecosystems, harming the native species (Chornesky and Randall 2003). For instance, *Rhinella marina*, the invasive cane toad, in northern Australia poses a serious threat to the *Acanthophis praelongus*, the northern death adder, belonging to the low vulnerability group (Phillips et al. 2010). Climate matching techniques offer the potential for identifying risk regions that are appropriate for the establishment of invasive species since invasive species can establish themselves outside of their natural range if the ecosystem there is remarkably similar to their original distributional area (Guisan et al. 2013). Early response strategies necessitate surveying and monitoring risk regions under invasion danger to detect infestations in the early phases of invasion (Peterson 2003). SDMs may be utilized to create invasion risk maps by identifying anticipated danger zones based on a species' climate adaptation (Srivastava et al. 2019). These maps are crucial for quick response and decision-making, early detection of invasive species, and identifying potential areas where invasive species could proliferate, spread, or cause damage (Jeschke and Strayer 2008). Additionally, they could assist with decisions relating to pest control, such as those involving regional quarantines, international trade laws, and survey design (Venette et al. 2010). Recent research on the use of SDMs for invasion management focused on identifying potential range shifts of species due to climate change (Padalia et al. 2015), estimating disease risk (He et al. 2019), invasion ecology (Taucare-Ríos et al. 2016), and the results of human activity and modifications to land use and cover (Wilson et al. 2013; Gallardo et al. 2015) on invasive plant species distribution. Invasion behaviors with change in elevation of *Chromolaena odorata* (L.) R.M. King & H. Rob., *Ageratum conyzoides* L., *Ageratina adenophora* (Spreng.) R.M. King & H. Rob., *Parthenium hysterophorus* L., and *Lantana camara* L. in the Himalayan region were studied using SDMs, and the suitable areas of *P. hysterophorus* and *A. conyzoides* will decrease by 2070, while other species will spread to newer regions (Lamsal et al. 2018). Vegetation indices had the most significant influence on the prediction of the possible range of invasion of *Prosopis juliflora* (Sw.) DC., followed by soil indices, biophysical variables, and water indices (Ahmed et al. 2021). Identification of invasion hotspots across nations or regions and conservation planning for protected areas are possible using the SDM techniques (Thapa et al. 2018).

13.5 Uses of Species Distribution Modeling Approaches in Policy Planning

SDMs are an important tool for the conservation of communities and ecosystems or species since they focus on ensuring redundancy and robustness for sustainable future populations (Redford et al. 2011). Accurate information about the habitat distribution of species is a vital requirement for their effective conservation planning and establishing appropriate, attainable conservation priorities (Liu et al. 2013). The findings of SDM are affected by degrees of uncertainty in the applied model and incomplete information about the species, which impacts decision-making and conservation planning (Srivastava et al. 2019). To enhance transparency and legitimacy in decision-making, models must be continually improved, and the criteria used to create and assess these models must be updated to reflect new capabilities and innovations (Sofaer et al. 2019). Decision-makers should identify the desired aim that is intended, and the uncertainties need to be addressed in model predictions so that the users may grasp the significance of the various forms of mistakes reducing the likelihood of uncertainty in SDMs (Ferraz et al. 2021). SDMs have discovered pathways that might simply access between protected areas (PAs) despite temperature changes (Nuñez et al. 2013), determining functional consistency in the development of PAs (Gallagher et al. 2013), key population variables that were connected to global change models and identified vital ecosystems (Bonnot et al. 2011), and enabled planning for extreme weather events to protect an endangered species (Bateman et al. 2012). The use of SDMs in policy planning is a challenging task (Addison et al. 2013) and, if not thoroughly evaluated, may impede its performance at numerous stages, from the imprecise formulation of the modeling aims to the improper use of the modeling outputs, via various unanticipated challenges across different stages of modeling (Guillera-Arroita et al. 2015). Model creation necessitates a thorough conception of decision aims and needs, as well as the incorporation of essential constraints into the modeling framework (Schmolke et al. 2010). Various challenges can also be overcome by the researchers working with stakeholders and incorporating their feedback in the model-making process (Boaz et al. 2018).

13.6 Limitations of SDMs

Multiple levels of uncertainty can exist in SDMs such as GARP or MaxEnt, including conceptual (scale and extrapolation), methodological (data resolution, sample design, and size, etc.), and algorithmic concerns (accuracy and model choice) (Engler et al. 2017). When employing complicated tools like SDMs, some uncertainty is unavoidable (Thuiller et al. 2019), and the use of SDMs has drawn criticism for a variety of reasons, including the specific algorithms used, the underlying principles, and sample biases of species occurrence data (Barbet-Massin et al. 2018). A well-known general impact of the SDMs is that geographic sample bias and geographic coordinate errors are both highly prevalent during data collection (Cayuela et al. 2009). When employing presence-only data, pseudo-absences are frequently used to create models that affect the outcomes of the model as a whole (Baxter and Possingham 2011). To create actual model predictions in the gradient between potential and realized distributions and to significantly condition the resultant model in the absence of trustworthy absence data, the pseudo-absence selection strategy is crucial (Chefaoui and Lobo 2008). There are often two sorts of mistakes in SDM prediction: (1) suitable habitat regions are incorrectly forecasted as unsuitable (false negatives), and (2) unsuitable habitat regions are incorrectly predicted as suitable (false positives). Both of these errors can have a detrimental impact on the decision-making process (Franklin 2010). False negatives in invasion management are more problematic because they increase the likelihood that the invasion extent will be underestimated and lead to poor decisions, whereas false positives can waste time and resources on surveillance in undesirable areas (Baxter and Possingham 2011). A viable alternative to threshold-based appropriate and unsuitable categorization would be the direct integration of uncertainty (Beale and Lennon 2012). Furthermore, researchers using SDMs must accept the restriction of poor forecasts, which denotes the need for further distribution data for the targeted taxon, as poor model outputs are generated by inadequate data rather than incorrectly designed algorithms (Cayuela et al. 2009). Reorienting SDMs toward the collective characteristics of biodiversity rather than specific entities, such as species assemblages or communities, might be a more advantageous course of action (Ferrier 2002). It may be used to identify locations where rare or vulnerable species are most likely to coexist with specific other species, though this does not resolve the problem (Golicher et al. 2008). The development of technology to enable the efficient integration of expert knowledge with SDM techniques is a significant problem for the future that warrants further study (Cayuela et al. 2009). A larger sample size and focus on sample prevalence could enhance the predictive accuracy of the model rather than the total number of background points because the number of background points varies depending on the modeling approach used, and in models like MaxEnt, the accuracy after a few hundred background points stabilizes (Liu et al. 2019). The geographic scope of the study should be acceptable, and sampling should take place in all possible convenient areas considering the historical distribution knowledge of the species (Cooper and Soberón 2018; Araújo et al. 2019).

13.7 Recommendation and Future Research Prospects

- Compared to a single model, the ensemble model's results reduce uncertainty and bias, enabling the generation of more justified and trustworthy forecast maps.

- To help end users with scenario planning, studies using SDMs should pay more attention to measuring, quantifying, and reducing uncertainty.
- To address the problems caused by the lack of data, emphasis should be placed on the inclusion of a citizen science database, and it must be cleaned and corrected before using this data for modeling.
- SDM studies on migratory species with high mobility should focus on seasonal niches.
- All major factors determining species range limits including both biotic, climatic, edaphic, and topographic must be considered in SDMs to acquire meaningful data from modeling approaches to use in conservation planning.

13.8 Conclusions

There is no disputing the importance of SDMs in maintaining and restoring forest ecosystems. SDMs are being used more frequently to predict the consequences of climate change or to determine the probable habitat range of threatened plant species. It may also be used to guide conservation decision-making for identifying the habitats of threatened plant species, creating protected areas, and managing species invasion, translocation, and reintroduction. Besides, it can help in conservation management by locating threatened plant species, detecting habitat alterations brought on by invasive species and climate change, pinpointing areas where threatened plant species should be relocated, or locating target areas for invasive species mitigation. It could be challenging to employ SDMs or other habitat suitability models when a landscape contains natural or artificial features that significantly differ from their current range. Therefore, researchers and management practitioners should innovate the methodology based on using SDMs for specific purposes. The stakeholders should focus more on communication, establishing a trustworthy and open connection, translating scientific findings into conservation goals, and sharing results and progress with all stakeholders.

References

Addison PF, Rumpff L, Bau SS et al (2013) Practical solutions for making models indispensable in conservation decision-making. Divers Distrib 19(5–6):490–502

Adhikari D, Barik SK, Upadhaya K (2012) Habitat distribution modelling for the reintroduction of *Ilex khasiana* Purk., a critically endangered tree species of northeastern India. Ecol Eng 40:37–43

Ahmed N, Atzberger C, Zewdie W (2021) The potential of modelling *Prosopis juliflora* invasion using Sentinel-2 satellite data and environmental variables in the dryland ecosystem of Ethiopia. Eco Inform 68:101545

Ancillotto L, Strubbe D, Menchetti M et al (2016) An overlooked invader? Ecological niche, invasion success and range dynamics of the Alexandrine parakeet in the invaded range. Biol Invasions 18(2):583–595

Angelo CL, Daehler CC (2013) Upward expansion of fire-adapted grasses along a warming tropical elevation gradient. Ecography 36(5):551–559

Araújo MB, Anderson RP, Márcia Barbosa A et al (2019) Standards for distribution models in biodiversity assessments. Sci Adv 5(1):1–12
Arroyo-Rodríguez V, Fahrig L, Tabarelli M et al (2020) Designing optimal human-modified landscapes for forest biodiversity conservation. Ecol Lett 23(9):1404–1420
Austin M (2007) Species distribution models and ecological theory: a critical assessment and some possible new approaches. Ecol Model 200(1–2):1–19
Barbet-Massin M, Rome Q, Villemant C et al (2018) Can species distribution models really predict the expansion of invasive species? PLoS One 13(3):e0193085
Bateman BL, VanDerWal J, Johnson CN (2012) Nice weather for bettongs: using weather events, not climate means, in species distribution models. Ecography 35(4):306–314
Baxter PW, Possingham HP (2011) Optimizing search strategies for invasive pests: learn before you leap. J Appl Ecol 48(1):86–95
Beale CM, Lennon JJ (2012) Incorporating uncertainty in predictive species distribution modelling. Philos Trans R Soc B Biol Sci 367(1586):247–258
Beans CM, Kilkenny FF, Galloway LF (2012) Climate suitability and human influences combined explain the range expansion of an invasive horticultural plant. Biol Invasions 14(10):2067–2078
Boaz A, Hanney S, Borst R et al (2018) How to engage stakeholders in research: design principles to support improvement. Health Res Policy Syst 16:60
Bonnot TW, Thompson FR III, Millspaugh JJ (2011) Extension of landscape-based population viability models to ecoregional scales for conservation planning. Biol Conserv 144(7): 2041–2053
Booth TH, Nix HA, Busby JR, Hutchinson MF (2014) BIOCLIM: the first species distribution modelling package, its early applications and relevance to most current MAXENT studies. Divers Distrib 20(1):1–9
Breiman L (2001) Random forests. Mach Learn 45(1):5–32
Cardador L, Carrete M, Gallardo B et al (2016) Combining trade data and niche modelling improves predictions of the origin and distribution of non-native European populations of a globally invasive species. J Biogeogr 43(5):967–978
Carpenter G, Gillison AN, Winter J (1993) DOMAIN: a flexible modelling procedure for mapping potential distributions of plants and animals. Biodivers Conserv 2(6):667–680
Cayuela L, Golicher DJ, Newton AC et al (2009) Species distribution modeling in the tropics: problems, potentialities, and the role of biological data for effective species conservation. Trop Conserv Sci 2(3):319–352
CBD (2020) Aichi Biodiversity Targets, September 2020. https://www.cbd.int/sp/targets/. Accessed 20 Aug 2022
Chefaoui RM, Lobo JM (2008) Assessing the effects of pseudo-absences on predictive distribution model performance. Ecol Model 210(4):478–486
Chen T, Guestrin C (2016) Xgboost: a scalable tree boosting system. In: Proceedings of the 22nd acm sigkdd international conference on knowledge discovery and data mining, pp 785–794
Chen IC, Hill JK, Ohlemüller R et al (2011) Rapid range shifts of species associated with high levels of climate warming. Science 333(6045):1024–1026
Chornesky EA, Randall JM (2003) The threat of invasive alien species to biological diversity: setting a future course. Ann Mo Bot Gard 90(1):67–76
Cooper JC, Soberón J (2018) Creating individual accessible area hypotheses improves stacked species distribution model performance. Glob Ecol Biogeogr 27(1):156–165
Cortes C, Vapnik V (1995) Support-vector networks. Mach Learn 20(3):273–297
Crooks S, Turner RK (1999) Integrated coastal management: sustaining estuarine natural resources. In: Nedwell DB, Raffaelli DG (eds) Advances in ecological research book series: Estuaries, vol 29. Elsevier, Amsterdam, pp 241–289
Dar JA, Subashree K, Raha D et al (2019) Tree diversity, biomass and carbon storage in sacred groves of Central India. Environ Sci Pollut Res 26(36):37212–37227

Dhyani S, Kadaverugu R, Pujari P (2020) Predicting impacts of climate variability on Banj oak (*Quercus leucotrichophora* A. Camus) forests: understanding future implications for Central Himalayas. Reg Environ Chang 20(4):1–13

Di Febbraro M, Sallustio L, Vizzarri M et al (2018) Expert-based and correlative models to map habitat quality: which gives better support to conservation planning? Glob Ecol Conserv 16: e00513

Díaz SM, Settele J, Brondízio E et al (2019) The global assessment report on biodiversity and ecosystem services: summary for policymakers. Intergovernmental Science-Policy Platform on Biodiversity and Ecosystem Services, p 60

Domisch S, Wilson AM, Jetz W (2016) Model-based integration of observed and expert-based information for assessing the geographic and environmental distribution of freshwater species. Ecography 39(11):1078–1088

Ehrlich PR, Kremen C, Ehrlich AH (2013) Human impacts on ecosystems: an overview. In: Levin SA (ed) Encyclopedia of biodiversity: reference work, 2nd edn. Academic Press, Elsevier, New York, pp 153–161

Elith J, Franklin J (2013) Species distribution Modeling. In: Levin SA (ed) Encyclopedia of biodiversity: reference work, 2nd edn. Academic Press, Elsevier, New York, pp 692–705. https://doi.org/10.1016/B978-0-12-384719-5.00318-X

Elith J, Leathwick JR (2009) Species distribution models: ecological explanation and prediction across space and time. Annu Rev Ecol Evol Syst 40:677–697

Elith J, Leathwick JR, Hastie T (2008) A working guide to boosted regression trees. J Anim Ecol 77(4):802–813

Engler JO, Stiels D, Schidelko K et al (2017) Avian SDMs: current state, challenges, and opportunities. J Avian Biol 48(12):1483–1504

Feeley KJ, Malhi Y, Zelazowski P et al (2012) The relative importance of deforestation, precipitation change, and temperature sensitivity in determining the future distributions and diversity of Amazonian plant species. Glob Chang Biol 18(8):2636–2647

Feng L, Sun J, Shi Y et al (2020) Predicting suitable habitats of Camptotheca acuminata considering both climatic and soil variables. Forests 11(8):891

Ferraz KMPM d B, Morato RG, Bovo AAA et al (2021) Bridging the gap between researchers, conservation planners, and decision makers to improve species conservation decision-making. Conserv Sci Pract 3(2):e330

Ferrier S (2002) Mapping spatial pattern in biodiversity for regional conservation planning: where to from here? Syst Biol 51(2):331–363

Foley JA, DeFries R, Asner GP et al (2005) Global consequences of land use. Science 309(5734): 570–574

Franklin J (2010) Mapping species distributions: spatial inference and prediction. Cambridge University Press, Cambridge

Frans VF, Augé AA, Fyfe J et al (2021) Integrated SDM database: enhancing the relevance and utility of species distribution models in conservation management. Methods Ecol Evol 13(1): 243–261

Friedman JH (1991) Multivariate adaptive regression splines. Ann Stat 19(1):1–67

Gallagher RV, Hughes L, Leishman MR (2013) Species loss and gain in communities under future climate change: consequences for functional diversity. Ecography 36(5):531–540

Gallardo B, Zieritz A, Aldridge DC (2015) The importance of the human footprint in shaping the global distribution of terrestrial, freshwater and marine invaders. PLoS One 10(5):e0125801

Giljohann KM, Hauser CE, Williams NS et al (2011) Optimizing invasive species control across space: willow invasion management in the Australian Alps. J Appl Ecol 48(5):1286–1294

Gogol-Prokurat M (2011) Predicting habitat suitability for rare plants at local spatial scales using a species distribution model. Ecol Appl 21(1):33–47

Golicher DJ, Cayuela L, Alkemade JRM et al (2008) Applying climatically associated species pools to the modelling of compositional change in tropical montane forests. Glob Ecol Biogeogr 17(2):262–273

Grinnell J (1917) Field tests of theories concerning distributional control. Am Nat 51:115–128
Guillera-Arroita G, Lahoz-Monfort JJ, Elith J et al (2015) Is my species distribution model fit for purpose? Matching data and models to applications. Glob Ecol Biogeogr 24(3):276–292
Guisan A, Thuiller W (2005) Predicting species distribution: offering more than simple habitat models. Ecol Lett 8(9):993–1009
Guisan A, Tingley R, Baumgartner JB et al (2013) Predicting species distributions for conservation decisions. Ecol Lett 16(12):1424–1435
Guisan A, Thuiller W, Zimmermann N (2017) Habitat suitability and distribution models with applications in R. In: Guisan A, Thuiller W, Zimmermann N (eds) Habitat suitability and distribution models: with applications in R ecology, biodiversity and conservation. Cambridge University Press, Cambridge, p I
He Y, Chen G, Potter C et al (2019) Integrating multi-sensor remote sensing and species distribution modeling to map the spread of emerging forest disease and tree mortality. Remote Sens Environ 231:111238
Heberling JM, Prather LA, Tonsor SJ (2019) The changing uses of herbarium data in an era of global change: an overview using automated content analysis. BioScience 69(10):812–822
Hijmans RJ, Graham CH (2006) The ability of climate envelope models to predict the effect of climate change on species distributions. Glob Chang Biol 12(12):2272–2281
Hirzel AH, Hausser J, Chessel D et al (2002) Ecological-niche factor analysis: how to compute habitat-suitability maps without absence data? Ecology 83(7):2027–2036
Hirzel AH, Le Lay G, Helfer V et al (2006) Evaluating the ability of habitat suitability models to predict species presences. Ecol Model 199:142–152
Holland J (1975) Adaptation in natural and artificial systems. The University of Michigan Press, Ann Arbor
Holt CDS, Nevin OT, Smith D et al (2018) Environmental niche overlap between snow leopard and four prey species in Kazakhstan. Eco Inform 48:97–103
Hunt TN, Allen SJ, Bejder L et al (2020) Identifying priority habitat for conservation and management of Australian humpback dolphins within a marine protected area. Sci Rep 10(1): 1–14
Hutchinson GE (1957) Concluding remarks. Population studies: animal ecology and demography. Cold Spring Harb Symp Quant Biol 22:415–427
Isaac NJ, Jarzyna MA, Keil P et al (2020) Data integration for large scale models of species distributions. Trends Ecol Evol 35(1):56–67
Jaynes ET (1957) Information theory and statistical mechanics. Phys Rev 106(4):620
Jentsch A, Beierkuhnlein C (2008) Research frontiers in climate change: effects of extreme meteorological events on ecosystems. C R Geosci 340(9–10):621–628
Jeschke JM, Strayer DL (2008) Usefulness of bioclimatic models for studying climate change and invasive species. Ann N Y Acad Sci 1134(1):1–24
Jetz W, McPherson JM, Guralnick RP (2012) Integrating biodiversity distribution knowledge: toward a global map of life. Trends Ecol Evol 27(3):151–159
Jiménez-Valverde A, Lobo JM (2007) Threshold criteria for conversion of probability of species presence to either-or presence-absence. Acta Oecol 31(3):361–369
Kearney M, Porter W (2009) Mechanistic niche modelling: combining physiological and spatial data to predict species' ranges. Ecol Lett 12(4):334–350
Knapp AK, Fay PA, Blair JM et al (2002) Rainfall variability, carbon cycling, and plant species diversity in a Mesic grassland. Science 298(5601):2202–2205
Kozak KH, Graham CH, Wiens JJ (2008) Integrating GIS-based environmental data into evolutionary biology. Trends Ecol Evol 23:141–148
Kumar A, Kumar A, Adhikari D et al (2020) Ecological niche modeling for assessing the potential distribution of *Pterocarpus marsupium* Roxb. in Ranchi, eastern India. Ecol Res 35(6): 1095–1105
Lamsal P, Kumar L, Aryal A et al (2018) Invasive alien plant species dynamics in the Himalayan region under climate change. Ambio 47(6):697–710

Liu C, White M, Newell G et al (2013) Species distribution modelling for conservation planning in Victoria, Australia. Ecol Model 249:68–74

Liu C, Newell G, White M (2019) The effect of sample size on the accuracy of species distribution models: considering both presences and pseudo-absences or background sites. Ecography 42(3): 535–548

Lyon NJ, Debinski DM, Rangwala I (2019) Evaluating the utility of species distribution models in informing climate change-resilient grassland restoration strategy. Front Ecol Evol 7:33

Mahalanobis PC (1936) A note on the statistical and biometric writings of Karl Pearson. Sankhyā 2(4):411–422

Mansourian S, Berrahmouni N, Blaser J et al (2021) Reflecting on twenty years of forest landscape restoration. Restor Ecol 29(7):e13441

Marvier M (2013) Conservation and people. In: Levin SA (ed) Encyclopedia of biodiversity: reference work, 2nd edn. Academic Press, Elsevier, New York, pp 221–229

Maschinski J, Haskins KE (2012) Plant reintroduction in a changing climate: promises and perils. Island Press, Washington, DC, p 432

McCune JL (2016) Species distribution models predict rare species occurrences despite significant effects of landscape context. J Appl Ecol 53(6):1871–1879

Melo FP, Parry L, Brancalion PH et al (2021) Adding forests to the water–energy–food nexus. Nat Sustain 4(2):85–92

Menon S, Choudhury BI, Khan ML et al (2010) Ecological niche modeling and local knowledge predict new populations of *Gymnocladus assamicus* a critically endangered tree species. Endanger Species Res 11(2):175–181

Merow C, Wilson AM, Jetz W (2017) Integrating occurrence data and expert maps for improved species range predictions. Glob Ecol Biogeogr 26(2):243–258

Millar CI, Stephenson NL, Stephens SL (2007) Climate change and forests of the future: managing in the face of uncertainty. Ecol Appl 17(8):2145–2151

Mooney HA, Drake JA (eds) (2012) Ecology of biological invasions of North America and Hawaii, vol 58. Springer Science & Business Media, Berlin

Morris DW (1987) Ecological scale and habitat use. Ecology 68(2):362–369

Mukherjee T, Sharma LK, Saha GK et al (2020) Past, present and future: combining habitat suitability and future landcover simulation for long-term conservation management of Indian rhino. Sci Rep 10(1):1–12

Murase H, Nagashima H, Yonezaki S et al (2009) Application of a generalized additive model (GAM) to reveal relationships between environmental factors and distributions of pelagic fish and krill: a case study in Sendai Bay, Japan. ICES J Mar Sci 66(6):1417–1424

Nelder JA, Wedderburn RW (1972) Generalized linear models. J R Stat Soc Ser A (General) 135(3):370–384

Nix HA (1986) A biogeographic analysis of Australian elapid snakes. Atlas Elapid Snakes Aust 7: 4–15

Nix H, Busby J (1986) BIOCLIM, a bioclimatic analysis and prediction system. Annual Report CSIRO, CSIRO Division of Water and Land Resources, Canberra, pp 59–60

Nuñez TA, Lawler JJ, McRae BH et al (2013) Connectivity planning to address climate change. Conserv Biol 27(2):407–416

Padalia H, Srivastava V, Kushwaha SPS (2015) How climate change might influence the potential distribution of weed, bushmint (*Hyptis suaveolens*)? Environ Monit Assess 187(4):1–14

Parmesan C (2006) Ecological and evolutionary responses to recent climate change. Annu Rev Ecol Evol Syst 37:637–669

Peterson AT (2003) Predicting the geography of species' invasions via ecological niche modeling. Q Rev Biol 78(4):419–433

Peterson AT, Soberón J, Pearson RG et al (2011) Ecological niches and geographic distributions (MPB-49). Princeton University Press, Princeton

Phillips SJ, Anderson RP, Schapire RE (2006) Maximum entropy modeling of species geographic distributions. Ecol Model 190(3–4):231–259

Phillips BL, Greenlees MJ, Brown GP et al (2010) Predator behaviour and morphology mediates the impact of an invasive species: cane toads and death adders in Australia. Anim Conserv 13(1):53–59

Polak T, Saltz D (2011) Reintroduction as ecosystem restoration. Conserv Biol 25(3):424–427

Pulliam HR (2000) On the relationship between niche and distribution. Ecol Lett 3:349–361

Puri M, Srivathsa A, Karanth KK et al (2015) Multiscale distribution models for conserving widespread species: the case of sloth bear *Melursus ursinus* in India. Divers Distrib 1(9): 1087–1100

Rajpoot R, Adhikari D, Verma S et al (2020) Climate models predict a divergent future for the medicinal tree *Boswellia serrata* Roxb. in India. Glob Ecol Conserv 23:e01040

Ramirez-Reyes C, Bateman BL, Radeloff VC (2016) Effects of habitat suitability and minimum patch size thresholds on the assessment of landscape connectivity for jaguars in the Sierra Gorda, Mexico. Biol Conserv 204:296–305

Real R, Barbosa AM, Vargas JM (2006) Obtaining environmental favourability functions from logistic regression. Environ Ecol Stat 13(2):237–245

Redford KH, Amato G, Baillie J et al (2011) What does it mean to successfully conserve a (vertebrate) species? Bioscience 61(1):39–48

Requena-Mullor JM, Maguire KC, Shinneman DJ et al (2019) Integrating anthropogenic factors into regional-scale species distribution models—a novel application in the imperiled sagebrush biome. Glob Chang Biol 25(11):3844–3858

Reyer CP, Leuzinger S, Rammig A et al (2013) A plant's perspective of extremes: terrestrial plant responses to changing climatic variability. Glob Chang Biol 19(1):75–89

Rodríguez JP, Brotons L, Bustamante J et al (2007) The application of predictive modelling of species distribution to biodiversity conservation. Divers Distrib 13(3):243–251

Rondinini C, Wilson KA, Boitani L et al (2006) Tradeoffs of different types of species occurrence data for use in systematic conservation planning. Ecol Lett 9(10):1136–1145

Rosenblatt F (1958) The perceptron: a probabilistic model for information storage and organization in the brain. Psychol Rev 65(6):386

Royle JA, Chandler RB, Yackulic C et al (2012) Likelihood analysis of species occurrence probability from presence-only data for modelling species distributions. Methods Ecol Evol 3(3):545–554

Schmolke A, Thorbek P, DeAngelis DL et al (2010) Ecological models supporting environmental decision making: a strategy for the future. Trends Ecol Evol 25(8):479–486

Schwartz MW (2012) Using niche models with climate projections to inform conservation management decisions. Biol Conserv 155:149–156

Seneviratne SI, Nichols N, Easterling D et al (2012) Changes in climate extremes and their impacts on the natural physical environment. In: Field CB, Barros V, Stocker TF et al (eds) Managing the risks of extreme events and disasters to advance climate change adaptation. A special report of working groups I and II of the Intergovernmental Panel on Climate Change. http://ipcc-wg2.gov/SREX/images/uploads/SREX-Chap3_FINAL.pdf. Accessed 20 Aug 2022

Sodhi NS, Brook BW, Bradshaw CJA (2009) Causes and consequences of species extinctions. Princeton University Press, Princeton, pp 514–520. https://assets.press.princeton.edu/chapters/s5_8879.pdf. Accessed 20 Aug 2022

Sofaer HR, Jarnevich CS, Pearse IS et al (2019) Development and delivery of species distribution models to inform decision-making. BioScience 69(7):544–557

Srivastava V, Lafond V, Griess VC (2019) Species distribution models (SDM): applications, benefits and challenges in invasive species management. CAB Rev 14(20):1–13

Sun J, Qiu H, Guo J et al (2020) Modeling the potential distribution of Zelkova schneideriana under different human activity intensities and climate change patterns in China. Glob Ecol Conserv 21: e00840

Suttidate N, Steinmetz R, Lynam AJ (2021) Habitat connectivity for endangered Indochinese tigers in Thailand. Glob Ecol Conserv 29:e01718

Taucare-Ríos A, Bizama G, Bustamante RO (2016) Using global and regional species distribution models (SDM) to infer the invasive stage of *Latrodectus geometricus* (Araneae: Theridiidae) in the Americas. Environ Entomol 45(6):1379–1385

Thakur KK, Bhat P, Kumar A et al (2022) Distribution mapping of Bauhinia vahlii Wight & Arn. In India using ecological niche modelling. Trop Ecol 63:286–299

Thapa S, Chitale V, Rijal SJ et al (2018) Understanding the dynamics in the distribution of invasive alien plant species under predicted climate change in Western Himalaya. PLoS One 13(4): e0195752

Thuiller W, Albert C, Araujo MB et al (2008) Predicting global change impacts on plant species distributions: future challenges. Perspect Plant Ecol Evol Syst 9(3–4):137–152

Thuiller W, Guéguen M, Renaud J et al (2019) Uncertainty in ensembles of global biodiversity scenarios. Nat Commun 10(1):1–9

Tiwari S, Ghosh B, Vaidya SN et al (2021) Modeling potentially suitable lac cultivation zones of *Butea monosperma* to promote livelihood security in rural India. Vegetos 34(3):630–637

United Nations Environment Programme (2021) Becoming #Generationrestoration: ecosystem restoration for people, nature and climate, Nairobi. https://wedocs.unep.org/bitstream/handle/20.500.11822/36251/ERPNC.pdf

Urban MC, Bocedi G, Hendry AP et al (2016) Improving the forecast for biodiversity under climate change. Science 353:aad8466

Venette RC, Kriticos DJ, Magarey RD et al (2010) Pest risk maps for invasive alien species: a roadmap for improvement. BioScience 60(5):349–362

Warren DL, Seifert SN (2011) Ecological niche modeling in Maxent: the importance of model complexity and the performance of model selection criteria. Ecol Appl 21(2):335–342

Watling JI, Arroyo-Rodríguez V, Pfeifer M et al (2020) Support for the habitat amount hypothesis from a global synthesis of species density studies. Ecol Lett 23(4):674–681

Westgate MJ, Barton PS, Lane PW et al (2014) Global meta-analysis reveals low consistency of biodiversity congruence relationships. Nat Commun 5(1):1–8

Wilson JW, Sexton JO, Jobe RT et al (2013) The relative contribution of terrain, land cover, and vegetation structure indices to species distribution models. Biol Conserv 164:170–176

Xu ZL, Peng HH, Peng SZ (2015) The development and evaluation of species distribution models. Acta Ecol Sin 35(2):557–567

Part III

Habitat Suitability Modeling for Protecting Animals and Their Habitat

Habitat Suitability Analysis of Asiatic Elephants (*Elephas maximus*) in the Tropical Moist Deciduous Forest of Assam Using Analytic Hierarchy Process (AHP)

14

Tanvi Hussain, Sarbeswar Kalita, and Arup Kumar Misra

Abstract

Asiatic elephant (*Elephas maximus*) listed as Endangered in the IUCN Red List is the only living species of genus *Elephas*, distributed in the Indian sub-continent and Southeast Asia. The decreasing number of Asiatic elephant is attributed to habitat destruction and human-elephant conflict (HEC) due to human encroachment in the forest areas. Therefore there is a need to understand the habitat suitability of Asiatic elephants and conserve them in situ. The objective of the study is to analyse the habitat suitability and map the corridor of Asiatic elephants in two reserve forests, namely, Barduar and Mayang Hill, situated on both sides of Chandubi Lake in Kamrup District. Remote sensing (RS) and geographic information system (GIS) have been playing an important role in decision-making and conservation of natural resources. Utilization of analytic hierarchy process (AHP) using spatial information and habitat suitability inputs, i.e. landcover, proximity to water, slope, aspect and elevation in GIS platform through weighted overlay, results in habitat suitability map of the species. The output raster is the resultant of untamed life mapping by providing weightage to the combination of factors contributing to their habitat suitability in the AHP model. The map generated will be useful for the conservation of Asiatic elephants in their natural habitat as well as synchronizing human activities to reduce HEC.

T. Hussain (✉) · S. Kalita
Department of Environmental Science, Gauhati University, Guwahati, India
e-mail: tanvihussain@gauhati.ac.in

A. K. Misra
Pollution Control Board, Assam, Guwahati, India

S. Dhyani et al. (eds.), *Ecosystem and Species Habitat Modeling for Conservation and Restoration*, https://doi.org/10.1007/978-981-99-0131-9_14

Keywords

Tropical forest · Asiatic elephant · Habitat suitability · Analytic hierarchy process (AHP) · Human-elephant conflict (HEC)

14.1 Introduction

Deforestation, forest degradation and rapid conversion of forests to agricultural land, human settlement and various other developmental activities have resulted in fragmentation and loss of wildlife habitat. During the past few decades, wildlife habitats and wilderness areas have shrunk due to unprecedented human growth and developmental activities (Sanare et al. 2015), especially in Northeast (NE) India. There is a growing need to conserve wildlife habitats and wild species in situ, for which a detailed understanding of the change of their habitat condition both in time and space is required. Although these fragmentation and habitat loss have affected all wildlife species like rhinoceros (*Rhinoceros unicornis*), tiger (*Panthera tigris*), leopard (*Panthera pardus*), etc., elephants (*Elephas maximus*) have become the focal point for conflict and conservation issues. Human-elephant conflicts (HEC) have an untoward impact on local communities like damage to crops and property and loss of human lives which incite fear, anger and hostility among the local communities resulting in retaliatory killings of elephants and undermining of conservation efforts. Forests of the Himalayan foothills and Northeast India accommodate one of the last remaining adequate elephant populations but are severely threatened at the same time (Choudhury 1999; Sukumar 2006). Hence, these areas are high-priority areas for elephant conservation with a distinct call for mitigation of HEC (Gureja et al. 2002). The wild elephant population in the country is estimated to be 29,964, with Karnataka reporting the highest population at 6049, followed by Assam at 5719 (MoEFCC 2017). Therefore, Assam is one of the important strongholds for the survival of Asian elephants.

Remote sensing (RS) and geographic information system (GIS) help understand the wildlife habitat in time and space, its past and present landuse/landcover characteristics, the rate at which the landuse is changing and the factors associated with the changes. Understanding the spatial characteristics of wildlife habitats and the requirements such as food availability, proximity to water, topography, etc. serves as the basis of habitat suitability analysis of the species or a specific species living in that area (Parihar et al. 1986; Davis and Goetz 1990; Roy et al. 1995; Kushwaha and Hazarika 2004; Imam and Kushwaha 2013). Analytic hierarchy process (AHP) is a decision-supporting tool which runs on the principle of pairwise comparisons of priorities and relies on the judgement of decision-makers to derive a priority scale (Saaty 1977). AHP helps decompose the factors on which a species relies for its survival into priorities, and pairwise comparison of priorities helps evaluate the combination of factors on a mathematical basis. Weightage obtained from AHP can serve as inputs in GIS platform for weightage overlay analysis to work out a habitat suitability map of wildlife for its conservation (Ying et al. 2007;

Imam and Tesfamichael 2013). Habitat suitability analysis will stand as a very important tool for study and conservation of wildlife in its natural domain and the reduction of man-animal conflict or specifically human-elephant conflict (HEC) in the areas where humans dwell in close proximity to wildlife habitats.

At present, there are only two living species of the biggest terrestrial animal belonging to two different genera, *Loxodonta africana* (African elephant) and *Elephas maximus* (Asian elephant). *L. africana* has two distinct sub-species found in the Savannas of Central and Western African, and *E. maximus* has at least three sub-species distributed in Sri Lanka (*E. maximus maximus*), Asia and Indian sub-continent (*E. maximus indica*) and Sumatran island (*E. maximus sumatranus*) (Hildebrandt et al. 2006). Attributed as the 'heritage animal of India', the species has found its place in the IUCN Red List as Endangered animal. There are ten (10) known elephant landscapes in India: East-Central landscape (South-West Bengal-Jharkhand-Orissa), Kameng-Sonitpur landscape (Arunachal-Assam), Eastern-South Bank landscape (Assam-Arunachal), Kaziranga-Karbi Anglong-Intaki landscape (Assam-Nagaland), North Bengal-Greater Manas landscape (Assam-West Bengal), Meghalaya landscape (Meghalaya), Brahmagiri-Nilgiri-Eastern Ghats landscape (Karnataka-Kerala-Tamil Nadu-Andhra), Anamalai-Nelliyampathy-High Range landscape (Tamil Nadu-Kerala), Periyar-Agasthyamalai landscape (Kerala-Tamil Nadu) and North-Western landscape (Uttarakhand-Uttar Pradesh). Of the ten elephant landscapes, four are in Northeast India of which three are in Assam itself. There is a growing need to conserve this gigantic terrestrial species. In order to conserve this species in situ, a habitat suitability analysis has been carried out in the tropical deciduous forest of Barduar and Mayang Hill Reserve Forests of Loharghat Range, Kamrup West Forest Division, Assam, India.

14.1.1 Human-Elephant Conflict in Assam and Northeast India

The gigantic mammals are listed under Schedule I of the Wild Life (Protection) Act, 1972 (GOI 2003), which is the highest level of protection granted to any species in India. In Northeast India only 25% of the elephant habitat is located within protected areas which accounts for the high likelihood of HEC in the region. The prime cause of HEC is the loss of habitat and shortage of food owing to a decrease in forest cover. Since 1950 more than half of the elephant habitat has been lost due to the conversion of forest-covered lands to cropland, human settlement areas and other developmental activities (Choudhury 2004). In NE India crops are cultivated on the hill slopes (shifting cultivation), in areas interspersed at the forest edges and the river plains. HEC occurs over crop raiding when elephants travel to the plains of Assam from the Himalayan foothills of Arunachal Pradesh and the hilly areas of Assam as well. Human settlement in the forest areas and at the forest fringes is another major cause of HEC, and most of these conflicts occur in the elephant movement routes, human-encroached elephant habitat areas and small forest patches.

The contiguous habitat in Kameng (Arunachal Pradesh)-Sonitpur (Assam) supports a large elephant population that witnessed HEC since the 1990s due to crop raiding and damage to property resulting in human deaths which in retaliation caused unexpected harming, and as many as 30 elephants were killed by poisoning during 2001–2002 (Choudhury 2002). Due to human encroachment in the area for various reasons including ethnic clashes in 2002, not less than 50% of the elephant habitat has lost in Sonitpur and nearby areas (Bist 2002; Choudhury 2002). The foothills of Barail range in Assam and Jaintia Hills in Meghalaya harbour as many as 100 elephants around the 1950s, and by the 1980s, the elephant population started declining due to habitat fragmentation in the adjoining hill ranges. The elephants wandered aimlessly in the region due to the loss of habitat in their range and killed 41 people from 1991 to 1997. Later the government sponsored the shooting of 'rogue' elephants, and by the end of the decade, the mammals extirpated from the area (Choudhury 2001a, 2004). Elephant habitats in Garo Hills of Meghalaya have been fragmented and destroyed over the increased practice of shifting (*jhum*) cultivation, logging, coal mining and poaching for ivory (Williams and Johnsingh 1996; Gurung and Lahiri Choudhury 2000; Datta-Roy et al. 2009). On the other hand, HEC has a different context in the hilly areas of Manipur, Nagaland and Mizoram wherein elephants are killed for their flesh, which is a native delicacy (Choudhury 2001b).

14.2 Materials and Methods

14.2.1 Study Area

The study was carried out in Mayang Hill (from 91°29′27.831″E and 25°50′34.412″N to 91°29′24.993″E and 25°51′48.923″N) and Barduar Reserve Forest (RF) (from 91°26′19.40″E and 26°0′30.856″N to 91°25′32.258″E and 25°52′30.95″N) lying adjacent to the outer hilly ranges of Khasi Hills with an area of 9239.35 ha located in Kamrup District, Assam, India (Fig. 14.1). A tropical climate prevails in the forests, and distinct seasons can be recognized; May to mid-October is the rainy season, when maximum monsoon rainfall occurs and average rainfall is about 1300 mm. The winter experiences only occasional showers. Both the reserve forests are drained by numerous streams. The Chandubi Lake lies in between the two reserve forests. The forests may be classified as tropical moist deciduous forest sub-categorized into Kamrup Sal forest and mixed moist deciduous forest (Champion and Seth 1968).

14.2.2 Data

In this study, Landsat-8 (OLI and TIRS) satellite imagery and ASTER Digital Elevation Model (DEM) were downloaded from the EarthExplorer user interface of the US Geological Survey (USGS) and Survey of India digital toposheets obtained from the Geological Survey of India (GSI), Government of India (Table 14.1).

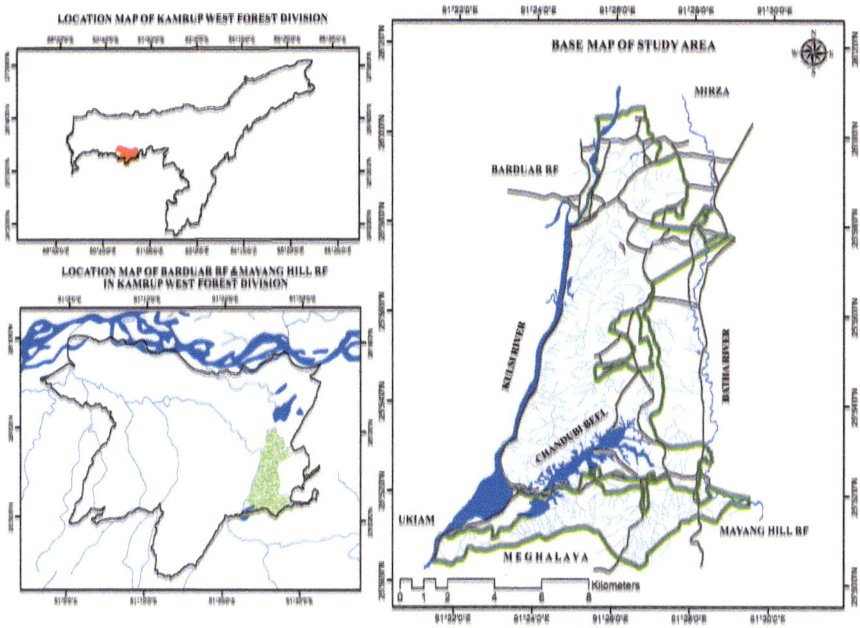

Fig. 14.1 Locational map of the study area

Table 14.1 Details of the data used for spatial mapping

Data	Path and row	Bands	Spatial resolution (in meter)	Period of acquisition
Landsat-8 OLI and TIRS[a] satellite imagery	137/42	11	30 (bands 1, 2, 3, 4, 5, 6, 7 and 9) 15 (band 8) 100 (bands 10 and 11)	March 2016 November 2017
ASTER[b] DEM	–	14	30	November 2011
SOI Toposheet[c]	78O/5	–		2005

[a]Landsat-8 Operational Land Imager (OLI) and Thermal Infrared Sensor (TIRS)
[b]Advanced Spaceborne Thermal Emission and Reflection Radiometer (ASTER)
[c]Survey of India (SOI) Toposheet

14.2.3 Generation of Spatial Database

The generation of a spatial database for the study has been carried out using ArcGIS version 10.1 software and Google Earth has also been used for identification of location and routes. The flowchart of database generation has been shown in Fig. 14.2.

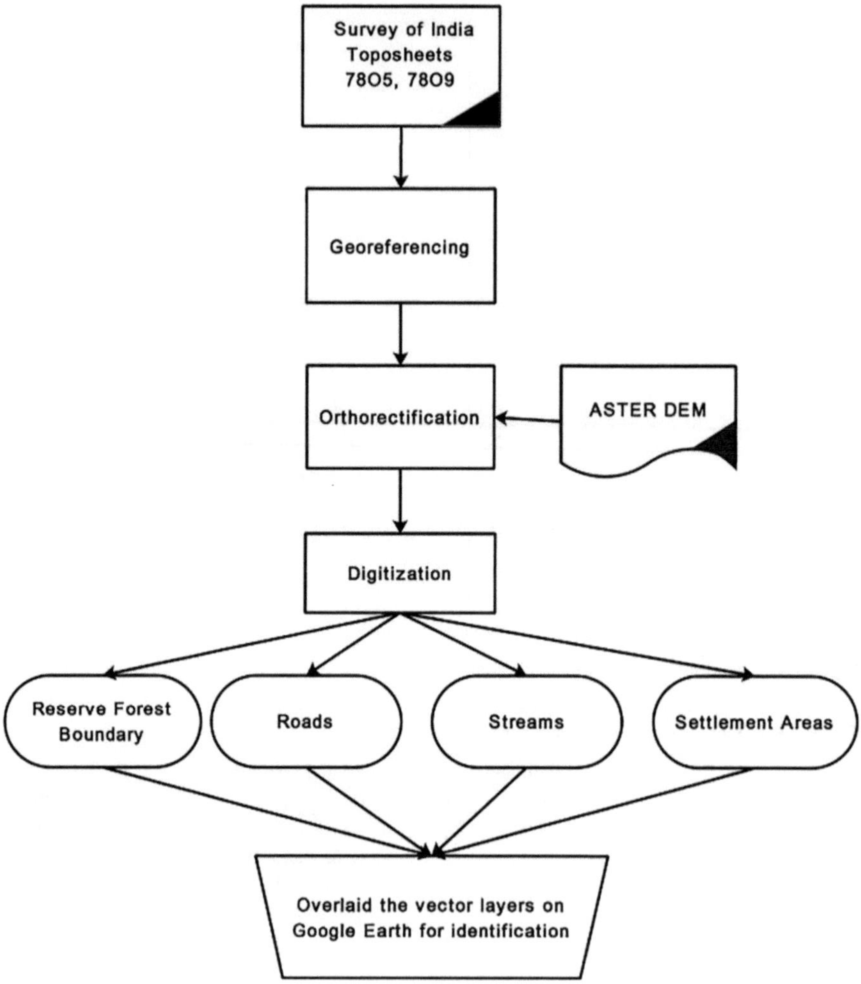

Fig. 14.2 Flowchart showing the generation of spatial database

14.2.4 Preparation of Forest Type, Forest Cover and Forest Strata Layers

Spatial analysis for the generation of forest type, forest cover and forest strata layers has been carried out using ERDAS Imagine version 2013 and ArcGIS version 10.1 software. The flowchart has been shown in Fig. 14.3.

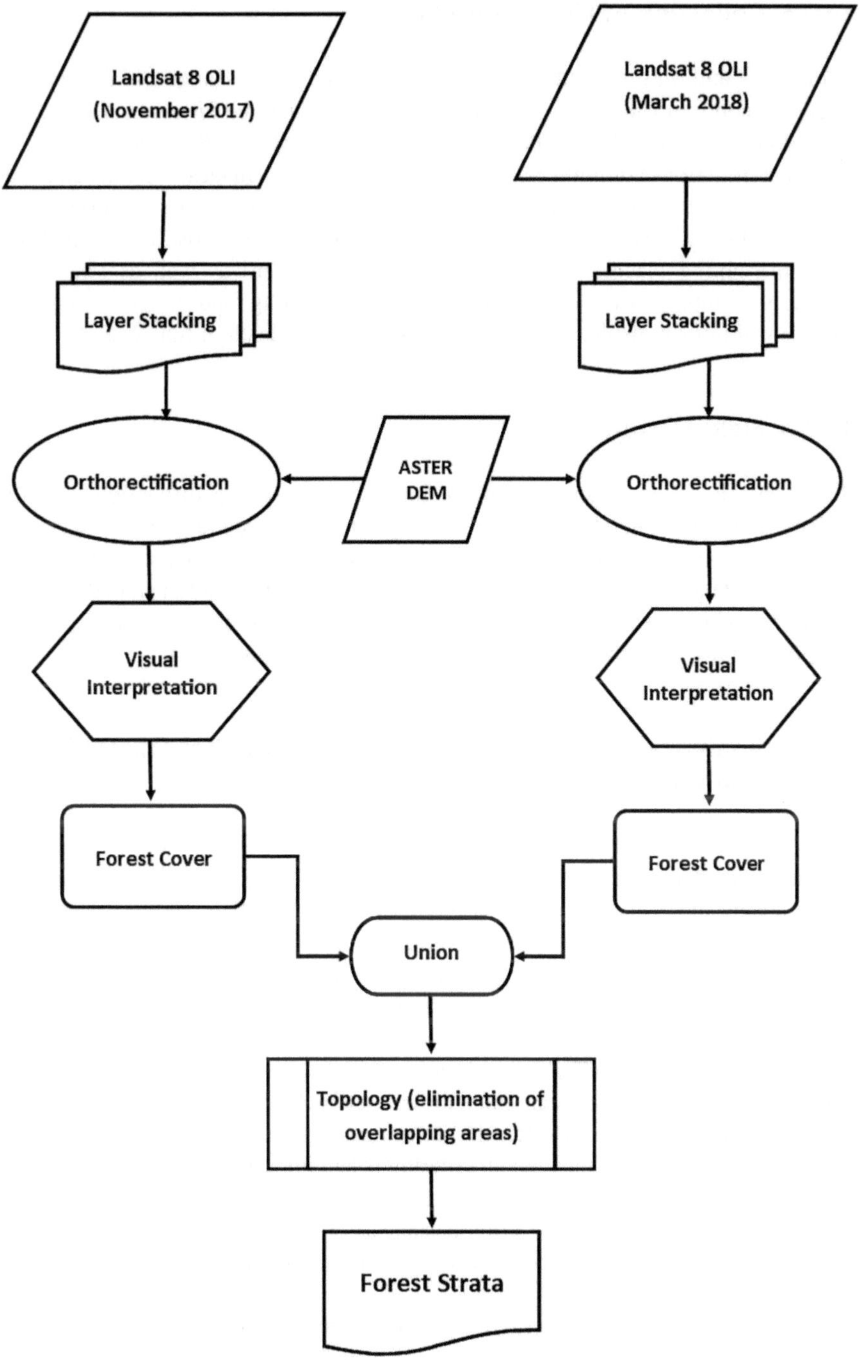

Fig. 14.3 Flowchart of the generation of forest cover, forest and forest strata spatial layers

14.2.5 Collection of Field Information

Barduar RF and Mayang Hill RF lie in the Loharghat Forest Range and the reserves consist of approximately 17 forest villages. A semi-structured interview and a group discussion were carried out with the village headman (*Gaon Bura*) along with the members of various committees, natives of all the forest villages, officials and staff of the Forest Department, Govt. of Assam, to gain insight into human-elephant conflict (HEC) in the two RFs. The information recorded were most frequently Asian elephant-sighting places, pathways of movement of herd as well as lone bulls, assembly points, season and locations of crop raiding and property damage, locations of human-elephant coincidence and injury or death of humans.

14.2.6 Field Survey

Based on the inputs received from the interview and group discussion with the natives and other literature, a field survey based on centre line transect was conducted. The Chandubi Lake where the herds of the gigantic mammal are mostly spotted was taken as the centre and line transects of $n = 15$ of 5 km and 10 km long and 100 m wide (based on terrain feasibility) were covered on foot to trace elephant footprints, dung and other signs of elephant movement (Fig. 14.4).

14.2.7 Analytical Hierarchy Process (AHP) for Habitat Suitability Analysis

Analytical hierarchy process (AHP) is a very flexible and efficient decision-making tool introduced by Thomas L. Saaty in 1980. AHP works on the principle of decomposing the decision-making into criteria and sub-criteria and alternatives on which the decision-making is based (Malczewski 2006). The criteria and sub-criteria are compared pairwise in a matrix assigning weights based on their relative importance. In AHP the weights for different criteria and sub-criteria are assigned in such a way that if there are n number of criteria or sub-criteria ($C_1 \dots .C_n$) to be compared then the real matrix will be $n \times n$ with $n(n - 1)/2$ number of evaluations. A square matrix A is computed. Let each entry to the matrix A be a_{ij} where i and j are two criteria and $i \neq j$. If $a_{ij} > 1$, then the ith criterion is more important than the jth criterion, while if $a_{ij} < 1$, then the ith criterion is less important than the jth criterion. The entries a_{ij} and a_{ji} should be such that $a_{ij} \times a_{ji} = 1$. In the pairwise comparison matrix, the entry for the same criteria such as $a_{ii} = 1$ or $a_{jj} = 1$. The relative importance between two criteria is measured on a numerical scale from 1 to 9, as shown in Table 14.2.

If it is assumed that the ith criterion has strong importance over the jth criterion, then the entry in the matrix will be $a_{ij} = 5$, and for the jth criterion over the ith criterion, it will be a reciprocal, i.e. $a_{ji} = 1/5$. Such a matrix is said to be a reciprocal matrix. The weights are consistent and transitive such that a_{ik} for all i, j and k is

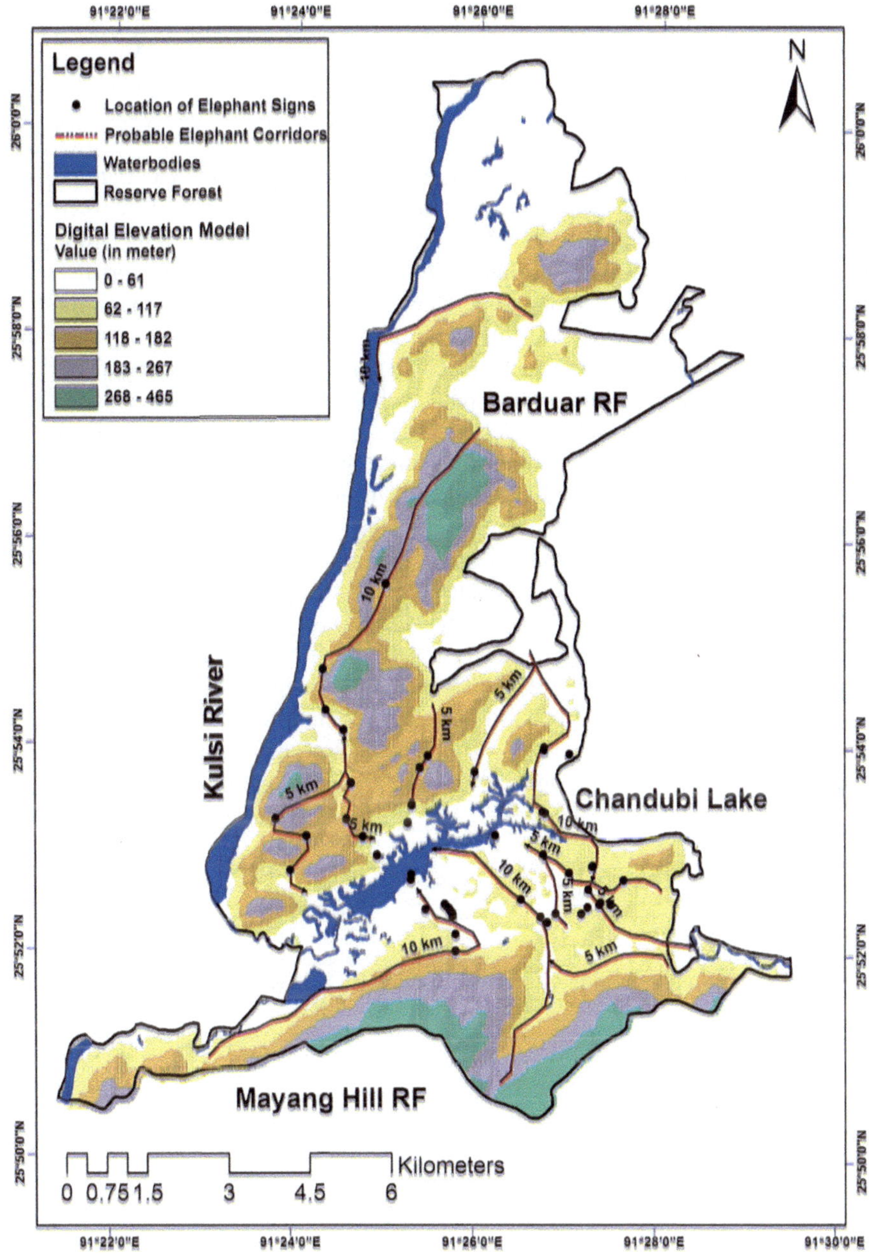

Fig. 14.4 Map showing line transects and points of elephant footprints and dung traced in the field study

Table 14.2 RCI values for different values of *n*

n	1	2	3	4	5	6	7	8	9
RCI	0	0	0.58	0.90	1.12	1.24	1.32	1.41	1.45

$a_{ik} = a_{ij} \times a_{jk}$. The values in Table 14.2 are only suggestive and translate the decision-maker's qualitative evaluations of the relative importance between two criteria into numbers. AHP uses a 9-point measurement scale ranging from 1 to 9 wherein the odd values are absolute weights like 1 denotes equal importance, 3 is moderate importance, 5 is strong importance, 7 is very strong importance and 9 is extreme importance (Sanare et al. 2015). It is also possible to assign the even values which are intermediate values and do not correspond to a precise interpretation. AHP involves direct participation to find out the final result, and it has gained wide applications in different fields of decision-making such as suitability analysis, site selection, product selection, marketing and business, etc. (Ayalew et al. 2005). Once the pairwise matrix is computed, the geometric mean and normalized weights for the criteria or sub-criteria are obtained by dividing the weights assigned to each criterion by the sum of the weights of that criteria or sub-criterion (Chowdhury et al. 2010). After the normalized matrix is computed, the average weight for each criteria or sub-criteria is calculated which is the eigenvector ω (of order n) or priority or score for respective criteria or sub-criteria and λ is the eigenvalue, a comparison matrix is consistent if $\lambda = n$. For matrices involving human judgement, the condition like $a_{ik} = a_{ij} \times a_{jk}$ does not hold good as human judgements are inconsistent to a greater or lesser extent; in such cases if $\lambda_{\max} = n$, then the judgements have turned out to be consistent. To calculate λ the eigenvector is multiplied with each of the entries in the judgement matrix and added:

$$\lambda = \sum_{j=1}^{n} a_{ij} \cdot \omega_j$$

The maximum eigenvalue ($\lambda_{\max}\omega$) is divided by n to obtain $\lambda_{\max}$:

$$\lambda_{\max} = \frac{\lambda}{n}$$

Finally, a consistency index (CI) can be calculated from:

$$\mathrm{CI} = \frac{(\lambda_{\max} - n)}{(n-1)}$$

It is necessary to calculate CI against judgements basically for random judgements. Saaty (1980) has calculated large samples of random matrices of increasing order and the CIs of those matrices. A true consistency ratio (CR) is calculated by dividing the CI for the set of judgements by the random consistency index (RCI) for the corresponding random matrix:

$$\mathrm{CR} = \frac{\mathrm{CI}}{\mathrm{RCI}}$$

Saaty (1980) suggests that if that ratio is less than 0.1, the judgement is consistent and if it exceeds 0.1, the set of judgements may be too inconsistent to be reliable. In practice, CRs of more than 0.1 have to be accepted sometimes. But if CR equals zero, the judgement is perfectly consistent.

14.2.8 Habitat Suitability Analysis

Decision-making in the first place involves gathering information, Asiatic elephants require a large range of habitats and provision for suitable habitat is the key requirement for the survival of the gigantic mammalian population. Asiatic elephants prefer areas with greater landcover (Sanare et al. 2015) and heterogeneity (Datye and Bhagwat 1995; Desai and Hedges 2010; Mandal and Das Chatterjee 2021) in close proximity to water (Sharma et al. 2020) such as mixed forest and riverine forest since these types of forests are potential sources of food (Sukumar 2003; Yamamoto-Ebina et al. 2016). These gigantic mammals have a tendency towards lower slopes (Sharma et al. 2020) and avoid monoculture forests since this type of forest does not have much food variability. Elephants also avoid humans while searching for food in the forest (Gaynor et al. 2018). Based on their requirements and habitat characteristics, the suitability analysis hierarchy has been decomposed into the following levels, shown in Fig. 14.5 (Table 14.3).

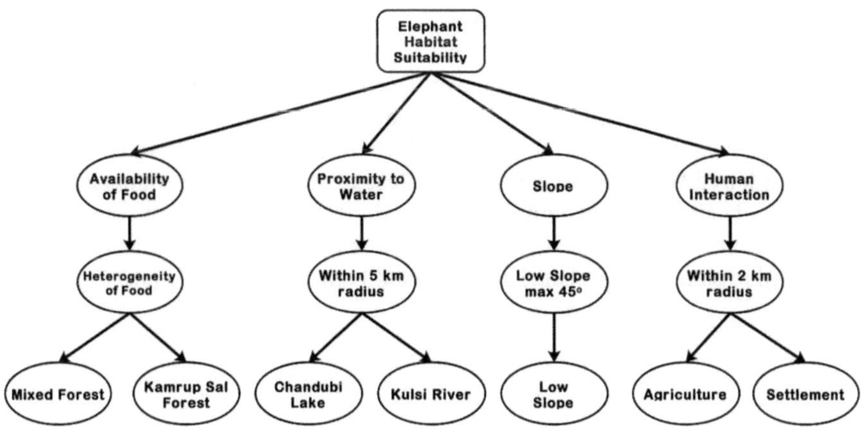

Fig. 14.5 Factors considered for elephant habitat suitability analysis

Table 14.3 Weighted factors considered for elephant habitat suitability analysis

Landcover	Proximity to water	Human interaction	Slope	Suitability
Dense mixed forest	Within 5 km	2 km away	≤45°	9 (highly suitable)
Open mixed forest	Within 5 km	2 km away	≤45°	7 (strongly suitable)
Dense Sal forest	Within 5 km	2 km away	≤45°	5 (moderately suitable)
Open Sal forest	Within 5 km	2 km away	≤45°	3 (suitable)

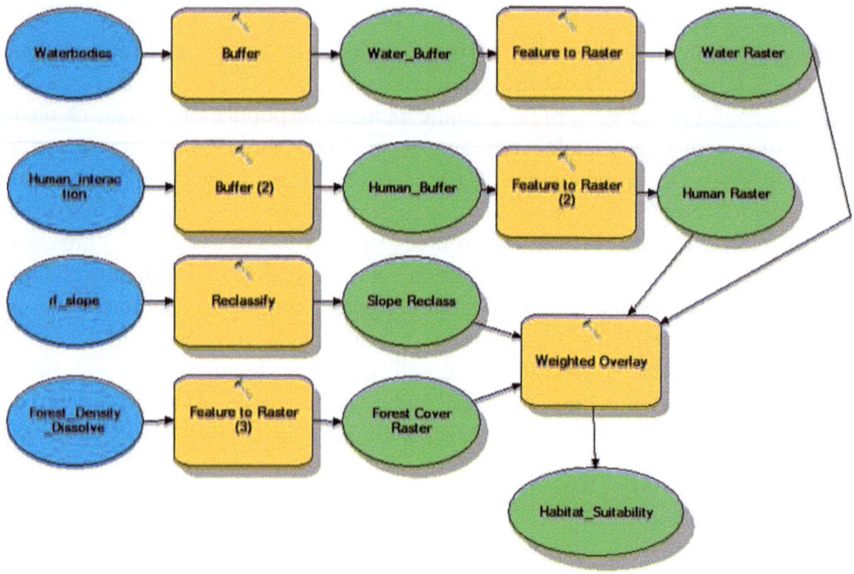

Fig. 14.6 Process flowchart of the weighted overlay model for habitat suitability

14.2.9 Weighted Overlay Model

Weighted overlay model is a general overlay analysis process processed in a single tool which reclassifies the input raster values into a common evaluation scale of preference or suitability, risk, etc. The weighted overlay model only accepts raster as inputs such as different forest classes or landuse classes with different input integers as values. The model was built in ArcView 10.2, the weightage derived for each criterion from AHP was incorporated in the input raster layers and the model was processed to obtain the elephant habitat suitability map for the study area (Fig. 14.6).

14.3 Results and Discussions

14.3.1 Forest Type

The forest-type map has been prepared based on different spectral signatures taken from seasonal (mid-October to mid-January and end of January to end of March) satellite imageries and classified as per Champion and Seth's forest-type classification (1968) shown in Fig. 14.7. The forest-type area statistics show that 41.83% of

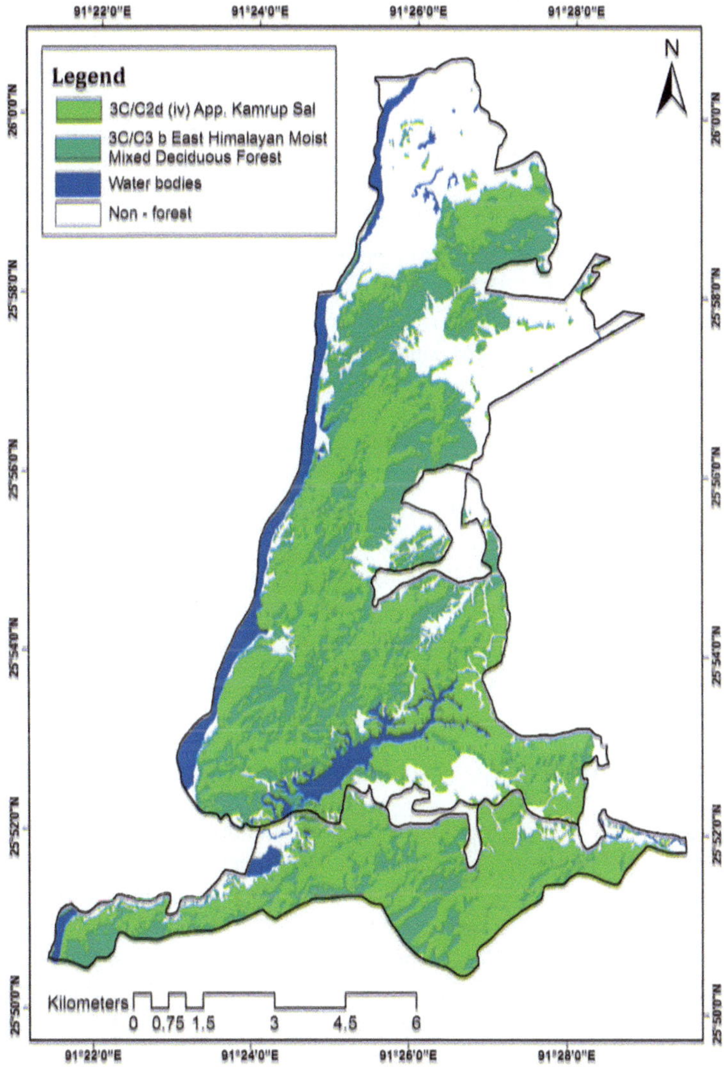

Fig. 14.7 Forest-type map of the study area

Table 14.4 Area statistics of forest types in the study area

Forest-type classes	Area (in ha)	Area (in %)
3C/C2d (iv) App. Kamrup Sal forest	3865.22	41.83
3C/C3b East Himalayan moist mixed deciduous forest	2432.07	26.32
Water bodies	560.33	6.06
Non-forest	2381.73	25.78
Total	9239.35	100

Table 14.5 Area statistics of forest cover and other landuse in the study area

Forest cover categories	Area (in ha)	Area (in %)
Dense forest	1840.60	19.92
Open forest	4456.69	48.24
Water bodies	560.33	6.06
Cropland	1563.63	16.92
Rural settlement	818.10	8.85
Total	9239.35	100

the area is dominated by 3C/C2d (iv) App. Kamrup Sal (Table 14.4) and 26.32% of the area is covered by 3C/C3b East Himalayan moist mixed deciduous forest. Water bodies cover 6.06% area, and 25.78% of the area is non-forest which attributes to other landuse classes such as cropland, settlement areas, etc.

14.3.2 Forest Cover, Water Bodies and Other Landuse

The Barduar and Mayang Hill RFs covering an area of 9239.35 ha have 19.92% of dense forest (Table 14.5) which exists mostly on the southern part of the forest stretch in the Mayang Hill RF (Fig. 14.8). Open forest dominates 48.24% of the area of which the majority lies in Barduar RF, and water bodies constitute 6.06% of the total area. Cropland and rural settlement occupy 16.92% and 8.85% of the area, respectively.

14.3.3 Forest Strata

In the present study, forest stratification refers to the horizontal stratification of the forest canopy. The forest strata map (Fig. 14.9) has been obtained by integrating the forest type and forest cover vector layers. The schema of the horizontal forest stratification has been shown in Table 14.6.

The forest strata area statistics of the two RFs (Table 14.7) show that Kamrup Sal open forest covers 24.52% and East Himalayan moist mixed deciduous open forest covers 23.72% of the total area. On the dense forest classification, Kamrup Sal dense forest covers 17.31% area, and East Himalayan moist mixed deciduous dense forest covers only 2.61% of the area.

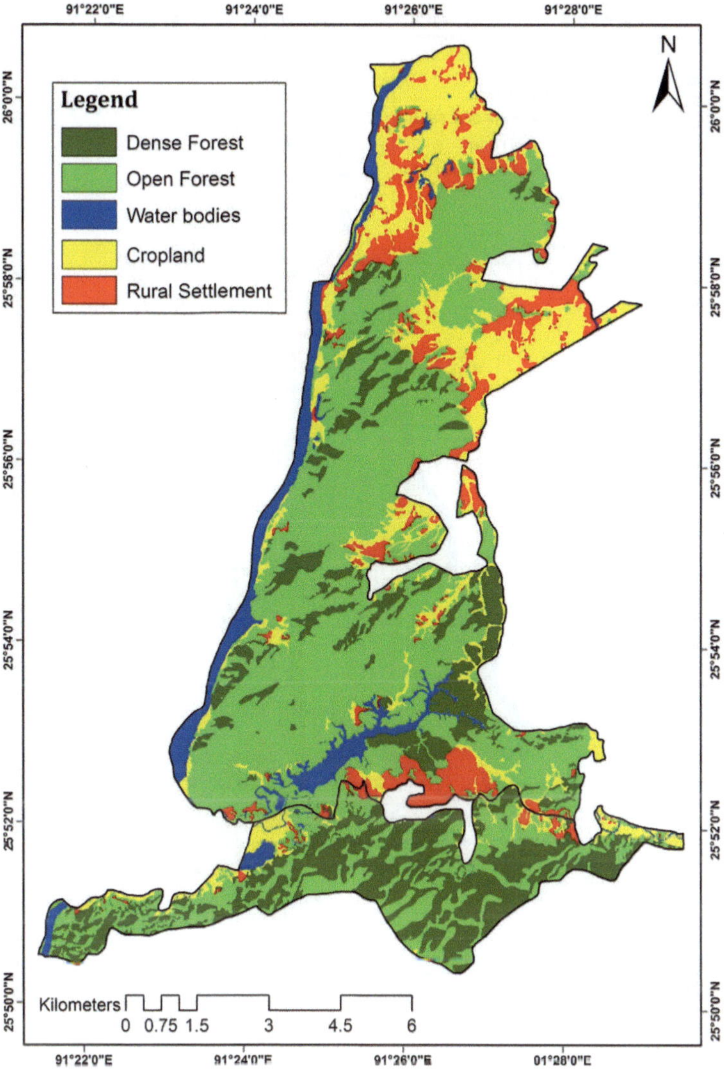

Fig. 14.8 Forest cover map of the study area

14.3.4 Identification of Suitable Habitat Zones

A pairwise comparison matrix was built based on the habitat suitability criteria for Asiatic elephants. The giant mammals prefer to dwell in dense vegetation covered in close proximity to water and away from human habitation. Table 14.8 shows the pairwise comparison matrix and the scores assigned to each pair of criterion.

After building the pairwise comparison matrix, a normalized matrix was computed, and the eigenvector for each criterion was derived. The priority

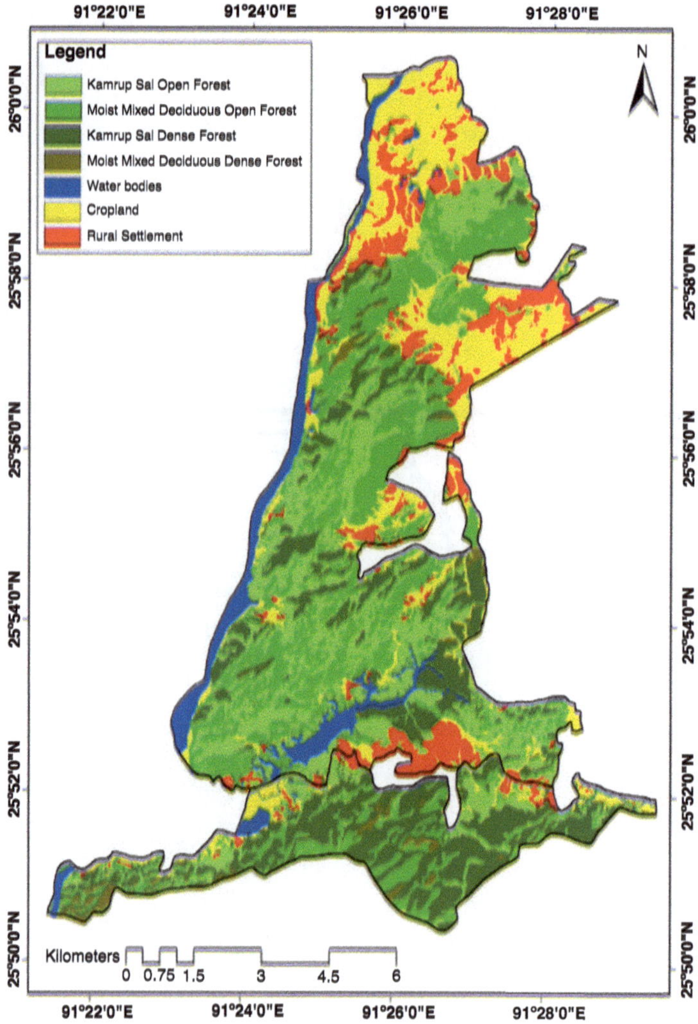

Fig. 14.9 Forest strata map of the study area

percentage shows (Graph 14.1) that for elephant habitat suitability the foremost factor is the availability of food in the forest which also shelters and sheds the gigantic mammals, followed by proximity to water, topography (slope) and disturbance or avoiding human interaction.

The eigenvalue (λ) for each criterion was as follows: forest (4.09), proximity to water (4.40), slope (4.11) and human interaction (4.25). The maximum eigenvalue (λ_{max}) was found to be 4.212, and the consistency index (CI) obtained was 0.070. The consistency ratio (CR) for this matrix $n = 4$ and $R = 0.90$ was 0.07. Since a CR for this matrix is less than 0.1, therefore the judgement is consistent.

Table 14.6 Forest strata classes based on the union of forest cover and forest type in the study area

Forest cover		Forest type	Forest strata
Forest	Dense	3C/C3b East Himalayan moist mixed deciduous forest	3C/C3b East Himalayan *moist mixed deciduous dense forest*
	Open		3C/C3b East Himalayan *moist mixed deciduous open forest*
	Dense	3C/C2d (iv) App. Kamrup Sal forest	3C/C2d (iv) App. *Kamrup Sal dense forest*
	Open		3C/C2d (iv) App. *Kamrup Sal open forest*
Water bodies	River	Water bodies	Water bodies
	Lake		
Other landuse	Cropland	Non-forest	Cropland
	Rural settlement		Rural settlement

Table 14.7 Area statistics of forest strata in the study area

Forest strata classes	Area (in ha)	Area (in %)
Kamrup Sal open forest	2265.55	24.52
Mixed deciduous open forest	2191.16	23.72
Kamrup Sal dense forest	1599.67	17.31
Mixed deciduous dense forest	240.91	2.61
Water bodies	560.33	6.06
Cropland	1563.63	16.92
Rural settlement	818.10	8.85
Total	9239.35	100

Table 14.8 Pairwise comparison matrix

j → *i* ↓	Human interaction	Proximity to water	Slope	Forest	Geometric mean
Human interaction	**1**	0.33 (1/3)	0.33 (1/3)	0.2 (1/5)	0.39
Proximity to water	3	**1**	4	0.5 (1/2)	1.57
Slope	3	0.25 (1/5)	**1**	0.25 (1/5)	0.66
Forest	5	2	4	**1**	2.51
Total	12	3.58	9.33	1.95	–

14.3.5 Asian Elephant Habitat Suitability

The respective weights derived for each criterion from AHP (Table 14.9) were fitted in the model, and the weight overlay model (Fig. 14.6) was executed to obtain the elephant habitat suitability map (Fig. 14.10) which has been classified into four categories, viz., low, moderate, moderately high and high.

The Asian elephant habitat suitability area statistics for the study (Table 14.10) show that 24% of the area has high habitat suitability for Asian elephants, 12.73% of

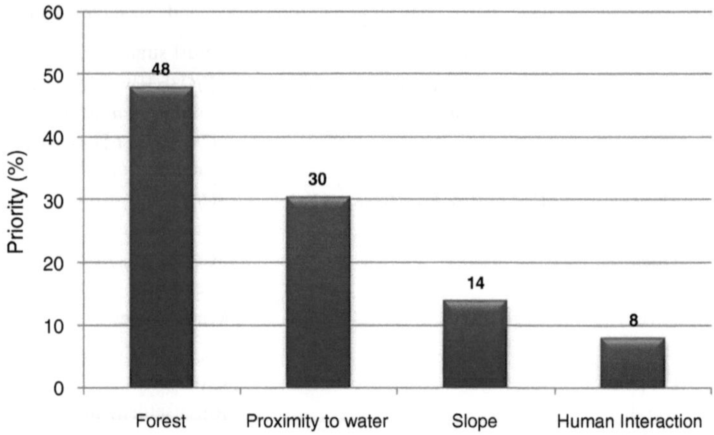

Graph 14.1 Priority (in %) of the habitat suitability factors

Table 14.9 Area statistics of the suitable habitat for Asian elephants in the study area

Habitat suitability categories	Area (in ha)	Study area (in ha)	Area under suitability categories (in %)
High	2217.15	9239.35 (i.e. 100% of the area)	24.00
Moderately high	1176.54		12.73
Moderate	3229.57		34.95
Low	1302.55		14.10
Total (low suitability has not been considered)	6632.26		71.68

the area is moderately suitable and 34.95% is moderately suitable in the study area. Only 14.10% of the area has low suitability. Of the total study area, 71.68% is suitable for the Asian elephants as a habitat. Forest cover is one of the important factors for Asian elephants' habitat suitability along with other factors like forest cover density, less fragmented forest, food variability and availability of water. Elephants have a tendency to rest and stay in moist dense forest areas of relatively low temperatures (Forman 1995; Mandal and Chatterjee 2018, 2019). Therefore, it is very crucial that the contiguous forest stretch of Mayang Hill RF on the Assam-Meghalaya state border and Barduar RF adjoining the Kulsi River and Chandubi Lake be given due emphasis for conservation and restoration of the areas under low suitability to intensify the habitat suitability in the area.

14.3.6 Pathways of Elephant Movement and Identification of Human: Elephant Conflict Areas

Based on the information received from the informants and field data of locations of elephant traces and HEC places, the pathways of elephant movement and areas of HEC have been generated in the GIS environment (Fig. 14.10). In Fig. 14.10 the orange-coloured dots represent the locations of elephant sighting, assembling points

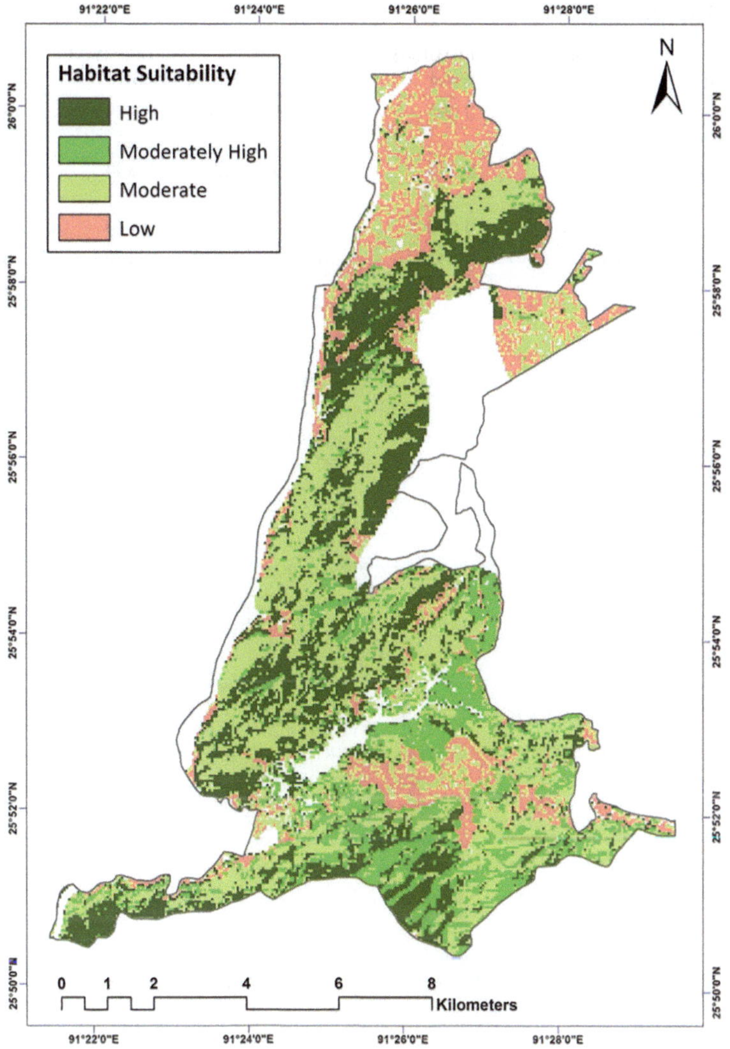

Fig. 14.10 Asian elephant habitat suitability map of the study area

and traces like footprints, faeces, etc., and the red-dotted lines represent the pathways of elephant movement in the area which have been traversed on the field and connected with the orange dots in GIS environment. Most of the human-elephant interaction occurs near the road passing along the Chandubi Lake between Barduar and Mayang Hill RF (Fig. 14.11). The point marked A, B and C in Fig. 14.11 are the locations where so far human casualties have been reported due HEC in the study area (Fig. 14.12).

Table 14.10 Normalized matrix

Criterion	Human interaction	Proximity to water	Slope	Forest	Eigenvector (ω)	Priority (%)
Human interaction	0.083	0.093	0.036	0.103	0.079	8
Proximity to water	0.250	0.279	0.429	0.256	0.304	30
Slope	0.250	0.070	0.107	0.128	0.139	14
Forest	0.417	0.558	0.429	0.513	0.479	48
Total	1	1	1	1	1	100

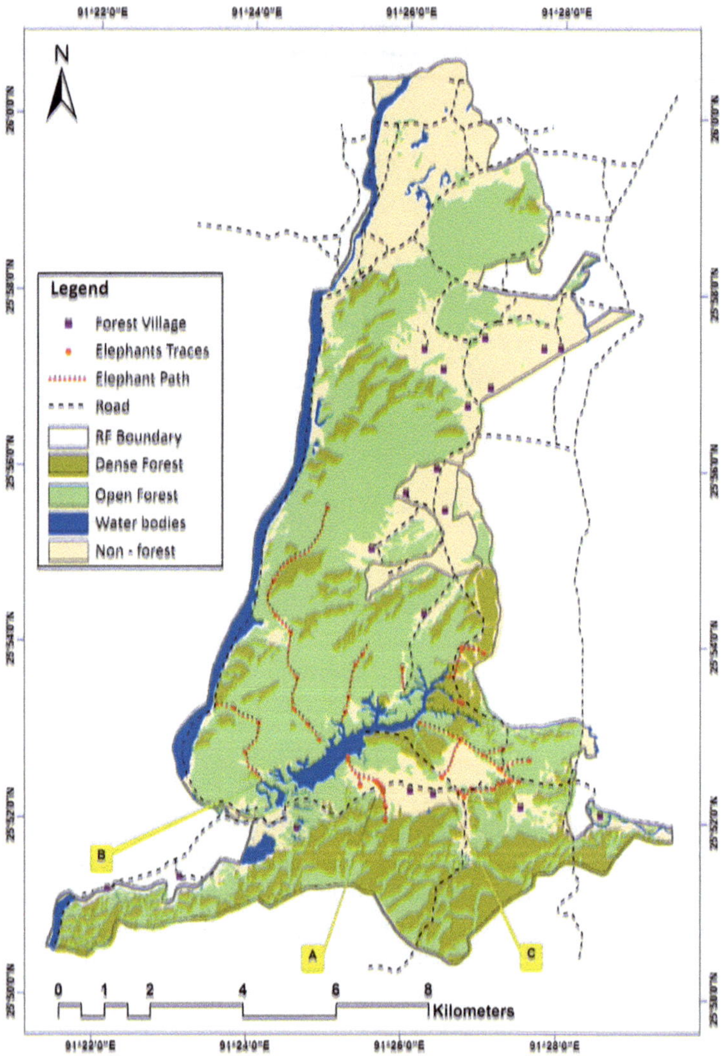

Fig. 14.11 Pathways of elephant movement of HEC in the study areas

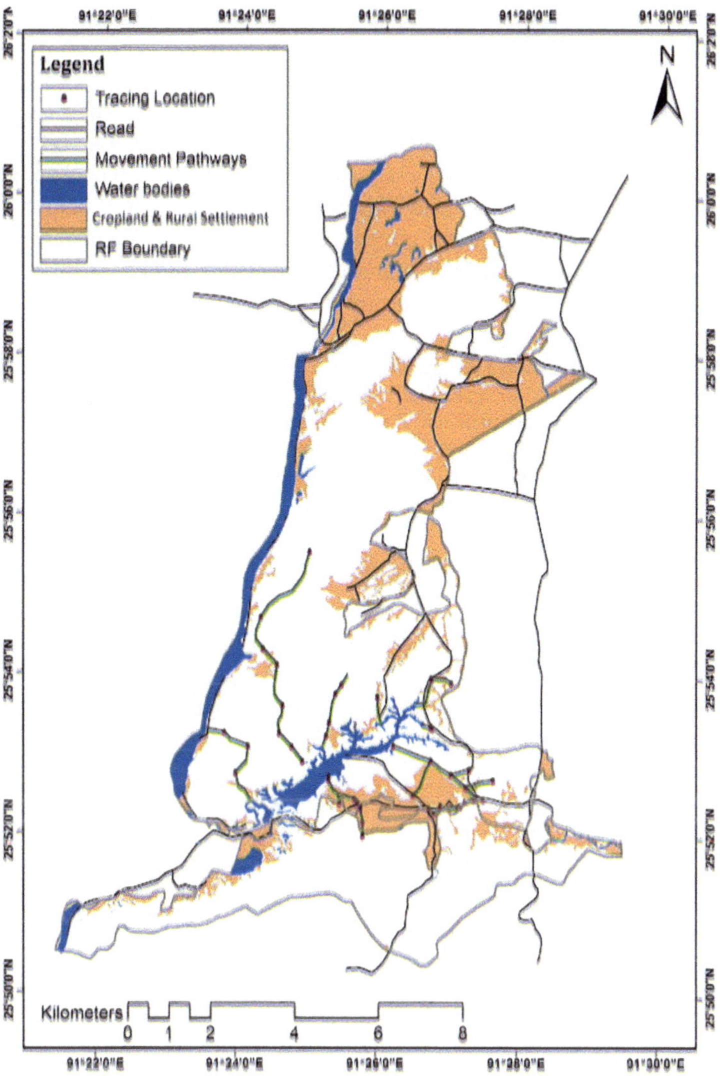

Fig. 14.12 Asian elephant movement pathways and coincidence with cropland and rural settlement

14.4 Conclusion

The study finds some of the ecological factors essential for the survival and habitat suitability of Asian elephants and demarcates the habitat suitability areas using geospatial tools. The gigantic mammals find areas suitable for the habitat with sufficient food, shade for resting and secured shelters away from disturbance.

Therefore, forest cover and availability of food are crucial for movement and habitat selection. Elephants are long-ranging animals which move in search of food and shelter from low suitability areas to suitable or highly suitable areas and given the presence of humans and road networks in the forest areas in many instances result in HEC (Koirala et al. 2016; Kumara et al. 2017). The study will be useful for improving the habitat quality and conservation of Asian elephant habitat in the present study area. The demarcated habitat suitability areas will help ecologists, conservationists, foresters and the government for managing elephant movement pathways, enhance forest and food sources for elephants and check on HEC in the present area. The study may also be useful for further research in this field.

Acknowledgements The authors are thankful to the natives and village headmen of the forest villages, the forest range officer and other officials of Loharghat Range, Kamrup West Forest Division, Department of Environment and Forest, Govt. of Assam. They also extend their special thanks to Shri Mozammil Hoque, forest guard, for helping us during field data collection and constantly being with us for safety and security purposes. The authors are also thankful the US Geological Survey for the availability of the data.

Declaration The authors declared that this is an original research work and no part of it has been produced or published anywhere.

Conflict of Interest The authors declare no conflict of interest.

References

Ayalew L, Yamagishi H, Marui H, Kanno T (2005) Landslides in Sado Island of Japan: Part II. GIS-based susceptibility mapping with comparisons of results from two methods and verifications. Eng Geol 81(4):432–445. https://doi.org/10.1016/j.enggeo.2005.08.004

Bist SS (2002) Conservation of elephants in NE India: past, present and future. Newslett Rhino Found Nat NE India 4:7–10. https://www.researchgate.net/publication/323294826_Conservation_Status_of_Asian_Elephants_in_Southern_Assam_India

Champion HG, Seth SK (1968) A revised survey of the forest types of India. Manager of Publications, New Delhi

Choudhury A (1999) Status and conservation of the Asian Elephant Elephas maximus in north-eastern India. Mammal Rev 29(3):141–174. https://doi.org/10.1046/j.1365-2907.1999.00045.x

Choudhury AU (2001a) Wild elephant extinct in Cachar. Newslett Rhino Found Nat NE India 3(7)

Choudhury AU (2001b) The wild elephant Elephas maximus in Mizoram. J Bombay Nat Hist Soc 98(3):439–441. https://archive.org/details/biostor-151530

Choudhury AU (2002) Massive habitat loss for primates in Assam's Sonitpur district. Asian Primates 8(1–2):18–20. https://www.asbb.gov.in//Downloads/HumanElephantConflictIndia04.pdf

Choudhury A (2004) Human–elephant conflicts in Northeast India. Hum Dimens Wildl 9(4): 261–270. https://asbb.assam.gov.in/sites/default/files/swf_utility_folder/departments/asbb_lipl_in_oid_7/portlet/level_1/files/HumanElephantConflictIndia04.pdf

Chowdhury A, Jha MK, Chowdary VM (2010) Delineation of groundwater recharge zones and identification of artificial recharge sites in West Medinipur district, West Bengal, using RS, GIS and MCDM techniques. Environ Earth Sci 59(6):1209–1222. https://doi.org/10.1007/s12665-009-0110-9

Datta-Roy A, Ved N, Williams AC (2009) Participatory elephant monitoring in South Garo Hills: efficacy and utility in a human-animal conflict scenario. Trop Ecol 50(1):163. https://citeseerx.ist.psu.edu/viewdoc/download?doi=10.1.1.580.1253&rep=rep1&type=pdf

Datye HS, Bhagwat AM (1995) Home range of elephants in fragmented habitats of Central India. J Bombay Nat Hist Soc 92(1):1–10. https://ia801806.us.archive.org/10/items/biostor-152409/biostor-152409.pdf

Davis FW, Goetz S (1990) Modelling vegetation pattern using digital terrain data. Landsc Ecol 4(1): 69–80. https://www.researchgate.net/profile/Scott-Goetz-2/publication/225622925_Modeling_vegetation_pattern_using_digital_terrain_data/links/0912f50b7ec0416b78000000/Modeling-vegetation-pattern-using-digital-terrain-data.pdf

Desai AA, Hedges S (2010) Notes from the co-chairs IUCN/SSC Asian Elephant specialist group. Gajah 3. https://www.researchgate.net/profile/Dr-Arumugam/publication/341110573_Estimating_Asian_Elephant_Population_in_Dindugul_Kodaikanal_and_Theni_Forest_Divisions_Western_Ghats_Tamil_Nadu/links/5fd8638f92851c13fe893630/Estimating-Asian-Elephant-Population-in-Dindugul-Kodaikanal-and-Theni-Forest-Divisions-Western-Ghats-Tamil-Nadu.pdf#page=9

Forman RTT (1995) Land mosaics: the ecology of landscapes and regions. Cambridge University Press, New York

Gaynor KM, Branco PS, Long RA, Gonçalves DD, Granli PK, Poole JH (2018) Effects of human settlement and roads on diel activity patterns of elephants (Loxodonta Africana). Afr J Ecol 56(4):872–881. https://doi.org/10.1111/aje.12552

GOI (2003) The Wild Life (Protection) Act Amendment 2002. Gazette notification. Government of India, New Delhi. https://parivesh.nic.in/writereaddata/MINISTRY%20OF%20LAW%20AND%20JUSTICE.pdf

Gureja N, Menon V, Sarkar P, Kyarong SS (2002) Ganesha to Bin Laden: Human-Elephant Conflict in Sonitpur District of Assam. Wildlife Trust of India, New Delhi. https://wti.org.in/wp-content/uploads/2017/03/pub_ganesha_bin-laden.pdf

Gurung S, Lahiri Choudhury DK (2000) Project: Elephant-human conflict in Asia state report on Meghalaya, India. Pt. I. (1992–1999). Asian Elephant Research and Conservation Centre, Bangalore

Hildebrandt TB, Göritz F, Hermes R, Reid C, Dehnhard M, Brown JL (2006) Aspects of the reproductive biology and breeding management of Asian and African elephants Elephas maximus and Loxodonta Africana. Int Zoo Yearb 40(1):20–40. https://doi.org/10.1111/j.1748-1090.2006.00020.x

Imam E, Kushwaha SPS (2013) Habitat suitability modelling for Gaur (Bos gaurus) using multiple logistic regression, remote sensing and GIS. J Appl Anim Res 41(2):189–199. https://doi.org/10.1080/09712119.2012.739089

Imam E, Tesfamichael GY (2013) Use of remote sensing, GIS and analytical hierarchy process (AHP) in wildlife habitat suitability analysis. J Mater Environ Sci 4(3):460–467

Koirala RK, Ji W, Aryal A, Rothman J, Raubenheimer D (2016) Dispersal and ranging patterns of the Asian Elephant (Elephas maximus) in relation to their interactions with humans in Nepal. Ethol Ecol Evol 28(2):221–231. https://doi.org/10.1080/03949370.2015.1066872

Kumara HN, Azeez PA, Singh A, Pal A, Nadu T (2017) Ecology of elephant (Elephas maximus) in South-West Bengal including population dynamics, migratory pattern, feeding habits and human-elephant conflict. Annual Report (April 2016—March 2017). Sálim Ali Centre for Ornithology and Natural History, Coimbatore, Tamil Nadu. http://project.wbfbcp.org/upload/annualreport21.pdf

Kushwaha SPS, Hazarika R (2004) Assessment of habitat loss in Kameng and Sonitpur Elephant Reserves. Curr Sci 87(10):1447–1453

Malczewski J (2006) GIS-based multicriteria decision analysis: a survey of the literature. Int J Geogr Inf Sci 20(7):703–726. https://doi.org/10.1080/13658810600661508

Mandal M, Chatterjee ND (2018) Quantification of habitat (forest) shape complexity through geo-spatial analysis: an ecological approach in Panchet forest division in Bankura, West Bengal.

Asian J Environ Ecol 6:1–8. https://doi.org/10.9734/AJEE/2018/40085. https://www.researchgate.net/profile/Mrinmay-Mandal-3/publication/323774029_Quantification_of_Habitat_Forest_Shape_Complexity_through_Geo-Spatial_Analysis_An_Ecological_Approach_in_Panchet_Forest_Division_in_Bankura_West_Bengal/links/5aae9ce80f7e9b4897c03427/Quantification-of-Habitat-Forest-Shape-Complexity-through-Geo-Spatial-Analysis-An-Ecological-Approach-in-Panchet-Forest-Division-in-Bankura-West-Bengal.pdf

Mandal M, Chatterjee ND (2019) Forest core demarcation using geo-spatial techniques: a habitat management approach in Panchet Forest division, Bankura, West Bengal, India. Asian J Geogr Res 2(2):1–8. https://www.researchgate.net/profile/Mrinmay-Mandal-3/publication/332352963_Forest_Core_Demarcation_Using_Geo-Spatial_Techniques_A_Habitat_Management_Approach_in_Panchet_Forest_Division_Bankura_West_Bengal_India/links/5cb0cc22a6fdcc1d498ff085/Forest-Core-Demarcation-Using-Geo-Spatial-Techniques-A-Habitat-Management-Approach-in-Panchet-Forest-Division-Bankura-West-Bengal-India.pdf

Mandal M, Das Chatterjee N (2021) Geospatial approach-based delineation of elephant habitat suitability zones and its consequence in Mayurjharna Elephant Reserve, India. Environ Dev Sustain 23(12):17788–17809. https://doi.org/10.1007/s10668-021-01412-1

MoEFCC (2017) Synchronised Elephant Population Estimation India-2017. Project Elephant Division. Ministry of Environment, Forests and Climate Change, Government of India, New Delhi

Parihar JS, Panigrahy S, Parihar JS (1986) Remote sensing based habitat assessment of Kaziranga National Park. Wildlife habitat evaluation using remote sensing techniques, Indian Institute of Remote Sensing/Wildlife Institute of India, Dehra Dun. pp 157–164

Roy PS, Ravan SA, Rajadnya N, Das KK, Jain A, Singh S (1995) Habitat suitability analysis of Nemorhaedus goral–a remote sensing and geographic information system approach. Curr Sci 68(8):685–691. https://d1wqtxts1xzle7.cloudfront.net/55688029/24097262-with-cover-page-v2.pdf?Expires=1666435942&Signature=QLU96Pp0YcRQw6OI76xGgv~saXEUeCn8g0~vMZ3rBC23KbhnWCD7LOH0vB5yxipYOmLCKZ2UcZBRAa~c4XEJ6MgzI~2~pY53JuQUts6xNBeYx1LkkyonrwFTKpTOi2jmncDQ1sDOnSkf0ZtrtXgDjzMoO0LvIktuV7AUgCIZbkN6mXVxrLymYzYCRC2Q5Avfcn3AYe6VJiKTDRvqgZ3oi5aPFljXCZYKjYqGp3IOI3T0 K5VVeuB5S~EphQCB9ybNJvqFClx8RncrFTlMCJT59axrc7buydf6q1VSFgcXyU2dmuCoOHlzWDYBvaVyhLkw3qFNcAp3Npfr7ClyBaBejw__&Key-Pair-Id=APKAJLOHF5GGSLRBV4ZA

Saaty TL (1977) A scaling method for priorities in hierarchical structures. J Math Psychol 15(3): 234–281. https://superdecisions.com/sd_resources/Paper_ScalingMethod.pdf

Saaty T (1980) The analytic hierarchy process (AHP) for decision making. In Kobe, Japan, pp 1–69. http://www.cashflow88.com/decisiones/saaty1.pdf

Sanare JE, Ganawa ES, Abdelrahim AMS (2015) Wildlife habitat suitability analysis at Serengeti National Park (SNP), Tanzania Case Study Loxodonta sp. J Ecosyst Ecograph 5:164. https://doi.org/10.4172/2157-7625.1000164

Sharma P, Panthi S, Yadav SK, Bhatta M, Karki A, Duncan T, Poudel M, Acharya KP (2020) Suitable habitat of wild Asian elephant in Western Terai of Nepal. Ecol Evol 10(12):6112–6119. https://doi.org/10.1002/ece3.6356. https://onlinelibrary.wiley.com/doi/pdf/10.1002/ece3.6356

Sukumar R (2003) The living elephants: evolutionary ecology, behavior, and conservation. Oxford University Press, New York. 478 pp. ISBN 0-19-510778-0

Sukumar R (2006) A brief review of the status, distribution and biology of wild Asian elephants Elephas maximus. Int Zoo Yearb 40(1):1–8. https://doi.org/10.1111/j.1748-1090.2006.00001.x

Williams AC, Johnsingh AJT (1996) Threatened elephant corridors in Garo Hills, north east India. Gajah 16:61–68. https://www.asesg.org/PDFfiles/Gajah/16-61-Williams.pdf

Yamamoto-Ebina S, Saaban S, Campos-Arceiz A, Takatsuki S (2016) Food habits of Asian elephants Elephas maximus in a rainforest of northern Peninsular Malaysia. Mammal Stud 41(3):155–161. https://doi.org/10.3106/041.041.0306

Ying X, Zeng GM, Chen GQ, Tang L, Wang KL, Huang DY (2007) Combining AHP with GIS in synthetic evaluation of eco-environment quality—a case study of Hunan Province, China. Ecol Modell 209(2–4):97–109. http://www.irantahgig.ir/wp-content/uploads/40018.pdf

Factors Affecting the Habitat Suitability of Eastern Swamp Deer (*Rucervus duvaucelii ranjitsinhi* Groves, 1982) in Manas National Park and Implication for Terai Grassland Restoration

15

Anukul Nath, Nazrul Islam, Shahid Ahmad Dar, Alolika Sinha, Bibhuti Prasad Lahkar, and Sonali Ghosh

Abstract

Conservation management to aid in the recovery of threatened species requires an understanding of their habitat availability and preference. Species distribution modelling can help delineate critical habitats to frame conservation decisions, particularly for a habitat-specialist species. Swamp deer is a grassland-obligate species, with three subspecies identified based on physical and geographic variations. Of these, the eastern swamp deer has restricted distribution and occurs only in two protected areas in the Brahmaputra valley of Assam, India. With assisted conservation efforts, the swamp deer population has revived in Manas National Park from an erstwhile heavily reduced remnant population. Through this paper an attempt has been made to analyse the patterns of swamp deer occurrence as determined by habitat variables using random forest algorithm models. The results indicate that the optimal habitats of swamp deer are the large grassland patches with wet climatic conditions, measured by the precipitation and evapotranspiration, within the broad grassland habitat of the park. The

A. Nath
Wildlife Institute of India, Dehradun, Uttarakhand, India

N. Islam (✉)
Gauhati University, Guwahati, Assam, India

S. A. Dar
University of Kashmir, Srinagar, Jammu and Kashmir, India

A. Sinha · B. P. Lahkar
Aaranyak, Guwahati, Assam, India

S. Ghosh
Department of Environment and Forest, Guwahati, Assam, India

S. Dhyani et al. (eds.), *Ecosystem and Species Habitat Modeling for Conservation and Restoration*, https://doi.org/10.1007/978-981-99-0131-9_15

findings have significant implications for the conservation of the threatened grassland habitat and its obligate species in the Terai grasslands of the region.

Keywords

Random forest · Species distribution modelling · World heritage site · Habitat management · Conservation

15.1 Introduction

Large herbivores play crucial ecological roles including seed dispersal and nutrient cycling, which have significant effects on forest structure and regeneration, and benefit other species in the environments they co-occur (Danell et al. 2006; Ripple et al. 2015). Habitat degradation, resource depletion, hunting and human-animal conflict have resulted in the rapid decline in the herbivore population with around 60% of them facing extinction (Ripple et al. 2015; Ceballos et al. 2017; Lindsey et al. 2017; Trouwborst 2019). The consequences of such human influences on extinction are magnified by the species' low population numbers and specialised habitat needs (Wallach et al. 2015; Ripple et al. 2016).

The Indian subcontinent has one of the most diverse large herbivore populations in southern and south-eastern Asia (Johnsingh et al. 2004; Ahrestani et al. 2011). The IUCN Red List of Threatened Species categorises 12 of the 15 terrestrial big herbivores as 'Threatened' indicating that the subcontinent's large herbivore species assemblages are now experiencing significant conservation challenges (Ripple et al. 2015). The subcontinent's habitat-specific large herbivores, swamp deer, greater one-horned rhinoceros and wild buffalo, are more vulnerable to extinction due to their inherent species biology and various anthropogenic activities and require accurate information on distribution, population size and various habitat parameters to ensure their survival in the future (Karanth et al. 2010). Due to their low density and cryptic behaviour, producing such information is difficult for many of these species, especially those residing in a mosaic of protected and non-protected environments near human settlements (Jathanna et al. 2003; Linkie et al. 2013; Marshal 2016).

The swamp deer, or barasingha (*Rucervus duvaucelii*), is an example of a South Asian habitat specialist and an endemic large herbivore (Qureshi et al. 2004; Tewari and Rawat 2013). Historically, the species was distributed throughout the Indo-Gangetic plains and the southern Himalayas covering India, Bangladesh, southern Nepal and Pakistan (Schaller 1967; Groves 1982; Sankaran 1989). Three subspecies have been defined based on physical and geographical variations: western swamp deer *Rucervus duvaucelii duvaucelii* (Cuvier 1823) restricted to the Terai grasslands of northern India and south-western Nepal; hard-ground Barasingha *Rucervus duvaucelii branderi* (Pocock 1943) restricted to Madhya Pradesh; and eastern swamp deer *Rucervus duvaucelii ranjitsinhi* (Groves 1982) restricted to the Brahmaputra valley (Groves 1982; Gopal 1992; Schaller 1967; WII 2017).

In the closing decades of the twentieth century, swamp deer saw a significant decline owing to widespread poaching and alteration of suitable habitats (Sankaran 1990; Singh 1970; Qureshi et al. 2004; Ahmed and Khan 2008; Saikia et al. 2012). The species is categorised as 'Vulnerable' on the IUCN Red List of Threatened Species (Duckworth et al. 2015) and included in Schedule I of the Indian Wildlife (Protection) Act, 1972. In Assam, the eastern swamp deer (ESD henceforth) was known to occur in the Brahmaputra valley's flat alluvial plains with tall grasses and in the Terai grasslands of flat to moderately hilly terrain, particularly in the Manas landscape near the southern foothills of Bhutan (Schaller 1967). By the late 1980s, only two remaining populations of this subspecies were reported from Kaziranga and Manas National Parks in Assam (Lahan and Sonowal 1973).

Before the socio-political upheaval in the Manas landscape, a robust population of ESD with about 500 individuals were reported in the Terai grassland of Manas National Park (Choudhury 1997). During the civil unrest, the species was almost extirpated from the protected area along with other grassland-dependent species such as greater one-horned rhinoceros (*Rhinoceros unicornis*) and pygmy hog (*Porcula salvania*) (Goswami and Ganesh 2014; Saikia et al. 2012). With the restoration of governance and renewed conservation efforts in the park, surveys were undertaken to confirm the species' presence. Indirect evidences such as pellets and antlers in the grassland habitat, along with the anecdotal sighting reports of the park rangers, indicated the presence of a highly diminished swamp deer population in the park (Das et al. 2009). In 2013, photographic evidence of the species further established the existence of swamp deer in Manas National Park (Borah et al. 2013). To supplement the existing population of the deer translocation of ESD in a phase-wise approach was planned. In 2014, 19 ESD were translocated from Kaziranga National Park followed by another 17 individuals in 2017 (Ahmed et al. 2016; Ghosh and Mathur 2020). Since its restocking, the population increased by twofold, and the recent population estimation recorded the presence of 121 swamp deer (Islam et al. 2022). However, the grassland habitat of which the species is obligate has been on the spatial decline (Sarma et al. 2008) and largely impacted by the invasion of invasive alien plants (Nath et al. 2019).

While it is confirmed that with greater protection ESD has rebounded from the brink of extinction in Manas National Park and has repopulated several areas of the park which are dominated by grasslands, there is potential for additional expansion, provided sound scientific and management interventions are in place to reduce the anthropogenic pressure and ensure the recovery of suitable habitats (Islam et al. 2022). To this end, the current study attempts to delineate suitable habitats within Manas National Park and suggests conservation strategies for restoring its habitat.

Delineating a species' distribution and optimal habitat is a necessary step in formulating conservation strategies for species management at the habitat or landscape level (Ortega-Huerta and Peterson 2004). In general, species distribution models (SDM) give a measure of a species' occurrence probability in a geographic region and help in identifying a habitat that is critical for target species management (Araujo and Williams 2000; Graham et al. 2004; McFarland et al. 2013). In Manas, an increase in the population of swamp deer (Islam et al. 2022) demands the implementation of specific habitat management strategies, as well as the identification of environmental, geographical, landscape and anthropogenic factors influencing the habitat suitability of the species. Here, we used the random forest algorithm model (Biau 2012) to find out a suitable habitat for swamp deer within the broad grassland habitat in the park. The findings of this study would help different stakeholders in developing long-term conservation strategies for swamp deer habitats which also support several other threatened habitat-specialist species.

15.2 Study Area

Manas National Park (MNP henceforth) is a strategic conservation area in the Jigme Dorji-Manas-Bumdeling conservation landscape in the eastern Himalayan eco-region, located (26°35′–26°50′N, 90°45′–91°15′E; Baksa and Chirang districts of Assam) at the junction of Indo-Gangetic, Indo-Malayan and Indo-Bhutan realms (Wikramanayake et al. 2002). MNP covering an area of 500 km^2 represents the core portion of the notified Manas Tiger Reserve that spans in an area of 2837 km^2 (Fig. 15.1). The park stretches on both banks of the Manas-Beki River and is bounded to the north by Bhutan's international boundary to the south by human habitations, west by the First Addition to Manas NP and in the east by Daodhara Reserve Forest (RF).

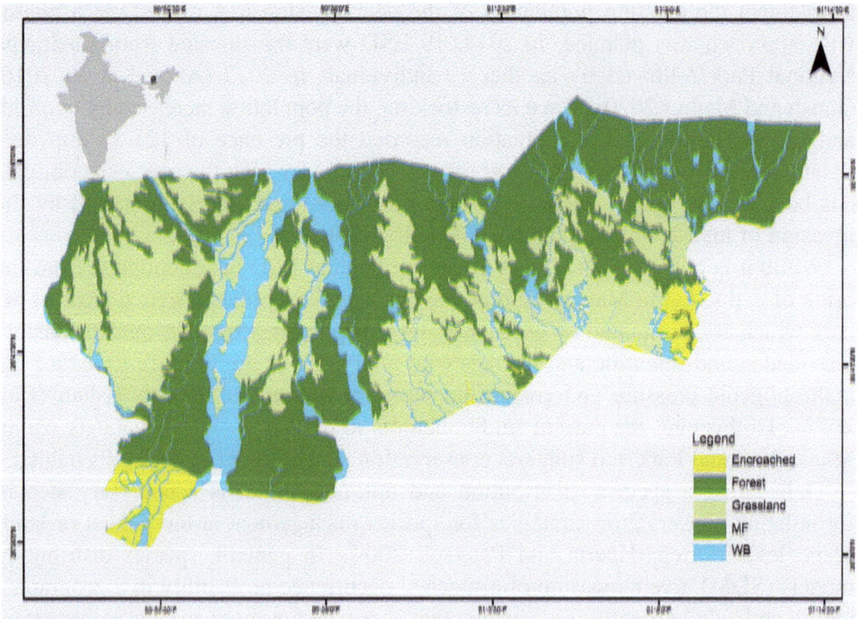

Fig. 15.1 Land use type (MF, mixed forest; WB, water body) of Manas National Park. (*Source:* Roy et al. 2016)

MNP is located in the *Terai-Duar* eco-region which is characterised by tall savannah-type grasslands interspersed with Sal (*Shorea robusta*)-dominated/moist mixed deciduous forests, along with tracts of swamps dominated by floodplain vegetation. The region once characterised the entire lowland region along the Himalayan foothills extending in India, Nepal and Bhutan into the Indo-Gangetic plains. Fire, floods and the edaphic climax resulted in a domination of natural grassland species which is now limited to protected areas along the Terai belt (Lehmkuhl 1994; Das et al. 2022).

The landcover type of MNP can broadly be categorised into woodland, which consists mostly of tree species from the semi-evergreen forest and moist mixed deciduous forest (Sarma et al. 2008). The semi-evergreen category covers 233.31 km^2 of the park, with the majority of it being in the north and extreme southwest. Savannah grasslands are dominated by tall grasses like *Narenga porphyrocoma*, *Imperata cylindrica*, *Phragmites karka*, *Arundo donax*, *Saccharum spontaneum*, *Themeda arundinacea*, *Saccharum procerum* and *Vetiveria zizanioides* with trees like *Bombax ceiba* and *Dillenia pentagyna* interspersed in the grasslands. This category has a total area of 161.98 km^2. Alluvial grasslands, which cover approximately 44.49 km^2 of the area, are distinguished by pristine patches of grasslands and the presence of waterlogging condition during the rainy season.

15.3 Methodology

15.3.1 Swamp Deer Occurrence Data

The presence records of swamp deer were compiled from various sources such as published literature (Das et al. 2009; Islam et al. 2022) and after consultation with subject experts and researchers working in the study zone. A few anecdotal reports on the species occurrence based on indirect evidence like fresh pellet groups were also included. The presence locations of the species were superimposed on 6.25 ha grid cells to remove multiple presence points, and only one presence point per grid cell (Brown et al. 2017) was retained. This resulted in using 82 locations for modelling the habitat suitability of the species (see Fig. 15.2).

15.3.2 Generating Pseudo-absences

Obtaining true absence data is especially challenging for a rare species with low density. However, these species typically have a high conservation value. When the actual absence data are unavailable, as in this study, pseudo-absences are frequently substituted (Brotons et al. 2004; Ferrier et al. 2002; Pearce and Boyce 2006). We generated 1000 pseudo-absences, which is more than ten times the number of presence records. The pseudo-absences were generated in a spatially random pattern within the MNP and by creating a buffer of up to 30 km on the southern side of the park.

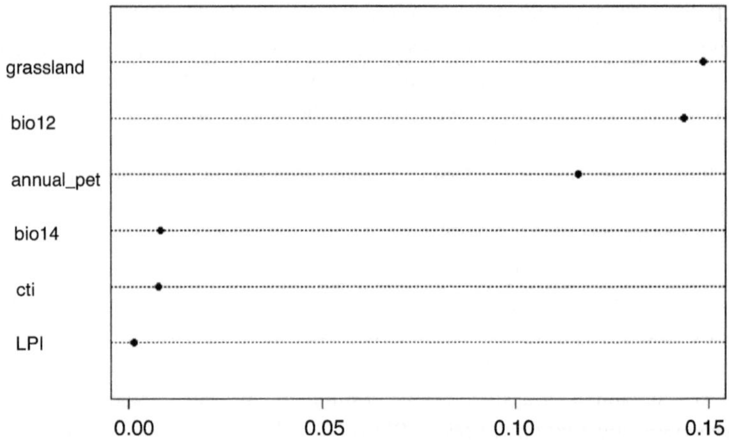

Fig. 15.2 Variable importance plot for the predictor variables based on model improvement ratio (MIR) from random forest classifications used for predicting occurrence of eastern swamp deer in Manas National Park

15.3.3 Habitat Variables

A total of 36 habitat variables that are potentially related to habitat selection of eastern swamp deer in northeast India based on published literature (Paul et al. 2020) were used as input variables for the suitability model. These variables represent climate, landscape composition, topography and anthropogenic influences in the region. All variables were projected to the 46 R UTM projection and resampled to a 6.25 ha spatial resolution in ArcGIS. Continuous variables were resampled using the bilinear interpolation method. The complete list of variables, their description and source are given in the Appendix.

The random forest algorithm (Breiman 2001) was used to develop multivariate models to predict the probability of swamp deer occurrences. To identify the most parsimonious model, the variables were filtered using a two-step procedure, e.g. step 1 (a multicollinear function under utilities in the R package (Ihaka and Gentleman 1996)) was applied to assess the potential correlation among all possible pairs of the above-mentioned variables and removed the variables ($n = 24$) that were highly correlated ($P < 0.05$), and step 2 (a model improvement ratio (MIR)) was used to retain the most important variables (Murphy et al. 2010). The MIR uses the permuted variable importance, represented by the mean decrease in out-of-bag (OOB) error, standardised from 0 to 1. The variables are subsets using 0.10 threshold increments, with all variables above the threshold retained for each model. This subset is always performed on the original model's variable importance to avoid over-fitting (Svetnik et al. 2004). Each subset model was compared, and the model that exhibited the lowest total OOB error and lowest maximum within-class error was selected. The variable importance and partial dependency plots for each variable selected in the final model (Dar et al. 2021) were generated. The partial dependency plots are useful in illustrating the relationship between the predicted probability of swamp deer occurrence and each habitat variable in the model. Prior to all random forest modelling, the minimum number of trees required by testing 10,000 boot-strap samples was determined and examined when the out-of-bag error ceased to improve. It was determined that the OOB error stabilised between 1000 and 1500 trees. The random forest algorithm was performed using the R package 'random forest' (Liaw and Wiener 2002).

An imbalance between presence and absence classes may cause prediction and model-fit bias (Chawla et al. 2003; Chen et al. 2004). To fix this, we iteratively down-sampled the majority class by randomly choosing 2 × [n of minority] and running a new random forest model with random subsets while preserving the minority sample size. We built a covariance matrix to capture sample distributions of independent habitat characteristics. The iterating models were stopped when the cumulative subset data covariance matrix was equivalent to the entire data covariance matrix ($P < 0.0001$). Our final model was an ensemble of randomly subsampled majority data models. The ESD probability distribution was generated using the proportion of majority votes across all branches. We also created variable significance and partial dependency plots for every habitat variable in the random forest model. Finally, our partial dependency graphs show how each predictor variable influences model predictions when all other predictor factors are controlled and describe the connection between ESD incidence and predictor variables.

15.3.4 Model Validation

Random permutations ($n = 99$) as well as cross-validation using the resampling approach was used to assess the model performance, and one-tenth of data was held as a validation set for each permutation (Evans and Murphy 2018). A number of performance matrices, including OOB error rate, model error variance and Kappa index of agreement, were produced by the cross-validation permutations. The OOB error rate measures the proportion of OOB samples that are incorrectly classified, and the Kappa index of agreement is a measure of agreement between predicted presences and absences with actual presences and absences corrected for an agreement that might be due to chance alone. The statistical range of the Kappa index of agreement ranges from 0 to 1: <0 values indicate no agreement, 0–0.20 as slight, 0.21–0.40 as fair, 0.41–0.60 as moderate, 0.61–0.80 as substantial and 0.81–1 as almost perfect agreement (Landis and Koch 1977).

15.4 Results

15.4.1 Variable Importance

The final model included six variables after variable selection with the model improvement ratio (Fig. 15.2). The most important variables were the proportion of grassland, average annual precipitation (bio12), annual potential evapotranspiration (PET), precipitation of the driest month (bio14), compound topography index (CTI) and largest patch index (LPI) for grassland.

15.4.2 Response of Swamp Deer to Habitat, Topography and Environmental Variables

Eastern swamp deer occurrence had a positive association with all the climatic variables (Fig. 15.3). As expected, swamp deer occurrence probability was the highest in the places where annual precipitation measured more than 3000 mm. Similarly, the probability of swamp deer increased sharply (<1400 mm) with the increase in annual potential evapotranspiration. Precipitation of the driest month (bio14) was also related to swamp deer occurrence. As expected, the proportion of grassland habitat has a positive association with swamp deer occurrence. Apart from that, the largest patch index of grassland has a linear response with the occurrence

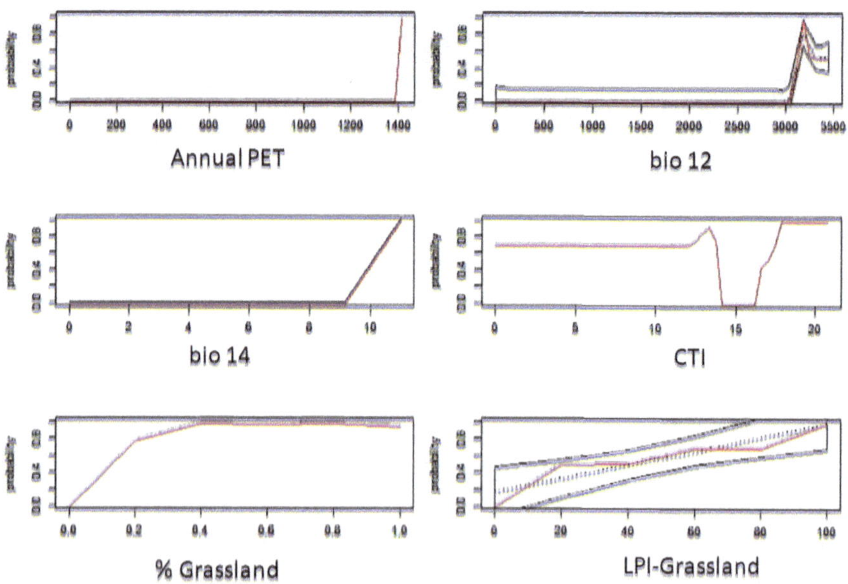

Fig. 15.3 Partial dependency plots representing the marginal effect of habitat variables on the predicted occurrence of eastern swamp deer

probability of the study species. The topographic variable (CTI) had non-linear relationships with swamp deer occurrences (Fig. 15.3). The compound topographic index showed a bimodal relationship with swamp deer occurrence, with the highest probability at >17 CTI. However, the species tend to avoid overlapping edges of woodland and grassland (CTI range, 14.5–16; Fig. 15.3).

15.4.3 Habitat Suitability Model

The habitat suitability map produced by the random forest model performed exceptionally well ($P < 0.001$, OOB error rate = 0.04; Table 15.1 and Fig. 15.4). The model had high accuracy (PCC 94.10%, 0.94) with high sensitivity and specificity (Table 15.1). The area under the ROC curve (AUC) was 0.94 (Table 15.1), indicating excellent model performance in predicting the occurrence of swamp deer in Manas National Park. The ESD was found to prefer the habitat of Kuribeel,

Table 15.1 Cross-validated performance of random forest habitat suitability model for the eastern swamp deer in Manas National Park, Assam, India

Performance matrix	Value
Accuracy (PCC)	94.1
Cohen's kappa	0.88
Area under the ROC curve	0.94
True skill statistics	0.88
Sensitivity	0.93
Specificity	0.95
Cross-validation OOB error	0.04
Cross-validation error variance	2.82E-05

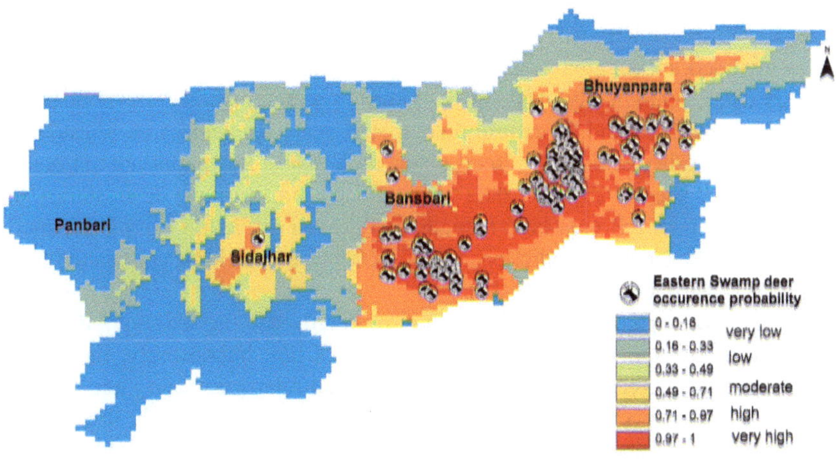

Fig. 15.4 The habitat suitability map showing the predicted occurrence probability of eastern swamp deer in Manas National Park. The map displayed areas of low to high suitability represented in a gradient from the lowest probability of swamp deer occurrence (blue) to the highest (red)

Bangale-haatdhowa, Uchila, Rupohi and Makhibaha sites (management units) within the park. The current model also showed preference in the Sidhajhar grassland in the western part of Manas.

15.5 Discussion

Our analysis provided insights into the patterns of habitat preference of eastern swamp deer in the Terai grassland of Manas National Park. Specifically, the model shows that optimal swamp deer habitat in Manas includes areas of relatively high grassland proportion along with large patches of grassland habitat and wet climatic conditions, as measured by precipitation and evapotranspiration. Annual precipitation and mean annual potential evapotranspiration are sufficient variables to capture water and energy aspects of species niches. Annual total precipitation (bio12) approximates the total water inputs and is therefore useful when ascertaining the importance of water availability to a species distribution. Additionally, the importance of precipitation during the driest month is an important factor as the species is dependent on the wet areas of grassland ecosystems. Subsequently, evapotranspiration is regarded as a strong predictor of net primary productivity of terrestrial ecosystems (Rosenzweig 1968) and is especially important to the swamp deer in parts of Terai because they predominantly forage on grasses (Nawaz 2008). It was found that the occurrence probability increased with an increase in precipitation in the driest month (bio14: the total precipitation that prevails during the driest month). The driest month is useful if extreme precipitation conditions during the year influence a species potential range.

We carefully constructed the swamp deer habitat suitability modelling for this investigation and incorporated important factors based on the species ecology. Although it is well known that species distribution modelling for habitat specialists produces better predictions (Connor et al. 2018; Rhoden et al. 2017), our study design included a focal area-based modelling approach, accurate presence locations (primary data), ecologically meaningful covariates with temporal correspondence to occurrence locations and an appropriate spatial scale for prediction (Araujo and Guisan 2006; Aubry et al. 2017; Beerkircher et al. 2009). It is important to note that we employed pseudo-absence data throughout studies since determining real absence from these tall grassland habitats had proven difficult. Overall, our model produced highly accurate predictions (AUC of 0.94), as would be anticipated for an obligate species (Connor et al. 2018).

The annual population estimation has revealed that the ESD population is increasing in Manas National Park (Islam et al. 2022). Interestingly, the ESD in Manas have been recorded from different wet-alluvial grasslands and swampy habitats of the park since its restocking and thus indicating that the translocated groups have suitably adapted in the wild, dispersed and occupied suitable habitats. Currently, the ESD population is confined to the central (Bansbari) and the eastern (Bhuyanpara) ranges of the park. The occurrence probability increased recently in another area—Sipajhar—of the central range which was earlier unoccupied by the species. There is a likelihood that the species might colonise the western range (Panbari), as the area has few large patches of swampy areas adjacent to the Manas River, in otherwise dry grassland habitat. Swamp deer recover from near-extinction status in Manas to a viable population and have dispersed within the park. There is a high potential, with proper scientific and management interventions to expand the population by addressing habitat enrichment and restoration. Ecologically, swamp deer plays an important role as a prey base for large predators (Lahkar et al. 2020) besides contributing significantly to the overall maintenance of the grassland ecosystem.

15.5.1 Conservation Strategies

From the above, it is well established that increasing the suitable habitat for eastern swamp deer would require specific management interventions that are proposed in the following pages:

Habitat management: A recent study (Nath et al. 2019) highlighted that approximately 30% of the existing grassland habitat in Manas is affected due to invasion by two invasive alien plants (IAPs)—*Chromolaena odorata* and *Mikania micrantha*. It is also predicted that in the absence of any management intervention, large tracts of prime grassland habitat are vulnerable to further collapse due to the onslaught of these invasive species. Further, proliferation of woody plants like *Bombax ceiba* and *Dillenia pentagyna* is supported by annual fires, and pure patches are being converted to a savannah type of habitat. Therefore, a systematic plan to manage the IAPs and supplement the invaded areas with native grass species using the technique of manual uprooting of the IAPs and creating a grass nursery will aid in grassland restoration (for details see Sinha et al. 2022). Furthermore, stakeholders associated with habitat and species management in Manas may take further help from Corporate Social Responsibility (CSR)-generated fund for the long-term sustainability of the eastern swamp deer in the region. The approach of CSR is to look at ecological, social and economic aspects when it comes to sharing responsibility (Ghosh and Mathur 2020). Therefore, Manas landscape provides a strong opportunity to get further support from CSR.

Management of grassland fires: Terai grasslands are largely a fire climax. Nonetheless, systematic controlled burning regime, along with the timing of inducing fire for managing the grasslands, is crucial. The grassland burning season (December to March) corresponds with the breeding season for the swamp deer. The fire not only destroys the cover and food, but it may also be fatal for escaping animals. In many cases fire is induced by graziers who seek to bring in their cattle, as well as opportunistic hunters. Active fire management (as a control burn) in the early winter (November–January) or a checkered burning regime with maintenance of fire lines is therefore prescribed.

Regulate livestock overgrazing: Instances of livestock grazing in the Bansbari range and parts of Bhuyanpara range of the park have been reported by the authors (pers comm). This damages the habitat and cover for wild animals, causes soil compaction and promotes invasive species. Cattle can also be a source of zoonotic infections that may be transmitted to wild animals. Therefore, stall-feeding of cattle can be promoted to restore swamp deer habitat.

Intensive study on vegetation dynamics: According to research, the park has lost more than 40% of its prime grassland habitat (short alluvial grasslands) owing to natural succession facilitated by yearly burning and floods (Ghosh et al. 2014). The spread of invasive species is another crucial aspect that must be examined on a spatiotemporal scale. The use of high-resolution satellite data (1 m) for detailed mapping of landcover categories will aid in comprehending landscape processes.

Undertake long-term research program: To develop effective management strategies, it is necessary to investigate and understand the impact of each disturbance factor on the grassland habitat of the species. Thus, studies on fire ecology, for example, the impacts of burning timing, frequency of burning and soil water content of grass species, need to conduct to come up with that appropriate management plans for different habitat types within the broad grassland habitat in the park.

15.6 Conclusion

Species distribution models have assisted biodiversity conservation by integrating research into policy and decision-making processes. In addition, conservation planning and execution need the identification of habitat regions where land preservation and management may increase the viability of a single or group of endangered species (McFarland et al. 2013). We give the first prediction map of suitable habitat for swamp deer in Manas National Park as well as the variables influencing their habitat preference. It is hoped that the results of the current research may encourage policy and management authorities to act on the management and restoration of swamp deer habitat.

Acknowledgements The authors would like to thank the Assam Forest Department and Bodoland Territorial Region for issuing necessary permits for various field-based studies that form part of this paper. The Field Directorate of Manas Tiger Project and the forest frontline staffs are thanked for the logistics support.

Appendix: List of 36 Habitat Variables Used for Habitat Suitability Modelling for Eastern Swamp Deer

Predictor variable	Source	Description	Units
Group 1: Climate			
Annual potential evapotranspiration (PET)	ENVIREM (http://envirem.github.io; Title and Bemmels 2018)	Mean monthly estimates	mm/month
PET seasonality		Monthly variability in potential evapotranspiration	mm/month

(continued)

Predictor variable	Source	Description	Units
PET of the coldest quarter		Mean monthly PET of the coldest quarter	mm/month
PET of the driest quarter		Mean monthly PET of the driest quarter	mm/month
PET of the warmest quarter		Mean monthly PET of the warmest quarter	mm/month
PET of the wettest quarter		Mean monthly PET of the wettest quarter	mm/month
Maximum temperature of the coldest month		Maximum temperature of the coldest month, i.e. January	°C
Minimum temperature of the warmest month		Minimum temperature of the warmest month, i.e. June	°C
Climatic moisture index		A metric of relative wetness and aridity	–
Bio1	WorldClim (https://www.worldclim.org/; Hijmans et al. 2005)	Mean estimates	°C
Bio2		Mean estimates	°C
Bio3		Mean estimates	°C
Bio4		Mean estimates	°C
Bio5		Mean estimates	°C
Bio6		Mean estimates	°C
Bio7		Mean estimates	°C
Bio8		Mean estimates	°C
Bio9		Mean estimates	°C
Bio10		Mean estimates	°C
Bio11		Mean estimates	°C
Bio12		Mean estimates	mm
Bio13		Mean estimates	mm
Bio14		Mean estimates	mm
Bio15		Mean estimates	mm
Bio16		Mean estimates	mm
Bio17		Mean estimates	mm
Bio18		Mean estimates	mm
Bio19		Mean estimates	mm
Group 2: Vegetation			
Land cover and vegetation indices			
Grasslands	Roy et al. 2016		%
Largest patch index		Calculated based on the grassland land cover type using the software FRAGSTATS	%
Group 3: Topographic			
Elevation	CGIAR-CSI	SRTM elevation data at 90 m resolution	m
		Calculated based on the elevation data using the Geomorphometry and	–

(continued)

Predictor variable	Source	Description	Units
Compound topographic index (CTI)		Gradient Metrix Toolbox in ArcGIS (Evans and Cushman 2009)	
Water bodies	Roy et al. 2016		%
Distance to rivers	HydroSHEDS database (http://hydrosheds.cr.usgs.gov)	River networks and was used to calculate the distance to river variable	m
Group 4: Disturbance			
Human footprint	Last of the Wild, v2 (http://sedac.ciesin.columbia.edu/wildareas/)	Anthropogenic impacts on the environment for the period 1995–2004, Last of the Wild Data Version 2, 2005	%
Distance to roads	DIVA-GIS (diva-gis.org/gdata)	Road networks and was used to calculate the road density variable	m

References

Ahmed K, Khan JA (2008) Status, population structure and conservation of swamp deer (*Cervus duvaucelii duvaucelii*) in Dudhwa Tiger Reserve, Uttar Pradesh, India. Int J Ecol Environ Sci 34(2):75–82

Ahmed A, Barman R, Choudhury B, Sarma R, Ashraf NVK, Kaul R, Menon V (2016) Supplementation of eastern swamp deer in Manas National Park, Assam, India. In: Soorae PS (ed) Global re-introduction perspectives: case-studies from around the globe. IUCN/SSC Reintroduction Specialist Group and Abu Dhabi, UAE, Environment Agency, Gland, Switzerland and Abu Dhabi, pp 178–181

Ahrestani FS, Heitkonig IMA, van Langevelde F, Vaidyanathan S, Madhusudan MD, Prins HHT (2011) Moisture and nutrients determine the distribution and richness of India's large herbivore species assemblage. Basic Appl Ecol 12(7):634–642. https://doi.org/10.1016/j.baae.2011.08.008

Araujo MB, Guisan A (2006) Five (or so) challenges for species distribution modelling. J Biogeogr 34(2):201–212. https://doi.org/10.1111/j.1365-2699.2006.01584.x

Araujo MB, Williams PH (2000) Selecting areas for species persistence using occurrence data. Biol Conserv 96:331–345. https://doi.org/10.1016/S0006-3207(00)00074-4

Aubry KB, Raley CM, McKelvey KS (2017) The importance of data quality for generating reliable distribution models for rare, elusive, and cryptic species. PLoS One 12(6):e0179152. https://doi.org/10.1371/journal.pone.0179152

Beerkircher L, Arocha F, Barse A, Prince E, Restrepo V, Serafy J, Shivji M (2009) Effects of species misidentification on population assessment of overfished white marlin tetrapturus albidus and roundscale spearfish T. georgii. Endanger Species Res 9(2):81–90. https://doi.org/10.3354/esr00234

Biau G (2012) Analysis of a random forests model. J Mach Learn Res 13(1):1063–1095

Borah J, Sharma T, Azad K et al (2013) Photographic evidence of the swamp deer in Manas. Oryx 47(4):481. https://doi.org/10.1017/S0030605313001063

Breiman L (2001) Random forests. Mach Learn 45(1):5–32

Brotons L, Thuiller W, Araujo MB, Hirzel AH (2004) Presence-absence versus presence-only modelling methods for predicting bird habitat suitability. Ecography 27(4):437–448

Brown JL, Bennett JR, French CM (2017) SDMtoolbox 2.0: the next generation Python-based GIS toolkit for landscape genetic, biogeographic and species distribution model analyses. Peer J 5: e4095. https://doi.org/10.7717/peerj.4095

Ceballos G, Ehrlich PR, Dirzo R (2017) Biological annihilation via the ongoing sixth mass extinction signaled by vertebrate population losses and declines. Proc Natl Acad Sci U S A 114:6089–6096. https://doi.org/10.1073/pnas.1704949114

Chawla NV, Lazarevic A, Hall LO, Bowyer KW (2003) SMOTEBoost: improving prediction of the minority class in boosting. In: Lavrac N, Gamberger D, Todorovski L, Blockeel H (eds) 7th European conference on principles and practice of knowledge discovery in databases. Knowledge discovery in databases: PKDD 2003. Lecture Notes in Computer Science, 2838. Springer, Heidelberg, pp 107–119. https://doi.org/10.1007/978-3-540-39804-2_12

Chen C, Liaw A, Breiman L (2004) Using random forest to learn imbalanced data. University of California, Berkeley, p 24. http://oz.berkeley.edu/users/chenchao/666.pdf

Choudhury A (1997) Checklist of the mammals of Assam. Gibbon Books with ASTEC, Guwahati

Connor T, Hull V, Vina A, Shortridge A, Tang Y, Zhang J, Wang F, Liu J (2018) Effects of grain size and niche breadth on species distribution modeling. Ecography 41:1270–1282. https://doi.org/10.1111/ecog.03416

Cuvier G (1823) Recherches sur les ossemens fossiles (Vol. 5, No. 1). chez G. Dufour et E. d'Ocagne

Danell K, Bergstrom R, Duncan P et al (eds) (2006) Large herbivore ecology, ecosystem dynamics & conservation. Cambridge University Press, Cambridge

Dar SA, Singh SK, Wan HY, Kumar V, Cushman SA, Sathyakumar S (2021) Projected climate change threatens Himalayan brown bear habitat more than human land use. Anim Conserv 24(4):659–676. https://doi.org/10.1111/acv.12671

Das JP, Sinha A, Talukdar BK (2009) Swamp deer in Manas: present status and feasibility of restocking. Preliminary report submitted to Manas Project Tiger Directorate, Assam Forest Department, p 14. htps://www.ruford.org

Das D, Banerjee S, Lehmkuhl J, Krishnaswamy J, John R (2022) The influence of abiotic and spatial variables on woody and herbaceous species abundances in a woodland–grassland system in the Eastern Terai of India. J Plant Ecol 15(1):155–167

Duckworth JW, Kumar NS, Pokharel CP et al (2015) Rucervus duvaucelii. The IUCN Red List of Threatened Species e.T4257A22167675. https://doi.org/10.2305/IUCN.UK.2015-4.RLTS.T4257A22167675.en

Evans JS, Cushman SA (2009) Gradient modeling of conifer species using random forests. Landsc Ecol 24(5):673–683. https://doi.org/10.1007/s10980-009-9341-0

Evans JS, Murphy MA (2018) rfUtilities. R package version. 2.1–3. https://cran.rproject.org/package=rfUtilities

Ferrier S, Watson G, Pearce J, Drielsma M (2002) Extended statistical approaches to modelling spatial pattern in biodiversity: the North-East New South Wales experience. I. Species-level modelling. Biodivers Conserv 11:2275–2307

Ghosh S, Mathur VB (2020) The deer that rode a car: role of CSR in Natural Resource Conservation Corporate Biodiversity Management for Sustainable Growth, p 59. ISBN:978-3-030-42702-3

Ghosh S, Lahkar BP, Deuti K, Nath A, Banerji-Bhattacharya G, Brahma N, Das JP, Sinha A, Lahkar D, Barukial B, Charles LP, Adhikari AS (2014) Survey for endangered Grassland Species in Manas National Park. Survey Report submitted to Field Directorate Tiger Project, Manas. Govt of Assam, p 21

Gopal R (1992) Resurrection of the Branderi Barasingha, Cheetal. J Wildl Preserv Soc India 31(1 & 2):1–5

Goswami R, Ganesh T (2014) Carnivore and herbivore densities in the immediate aftermath of ethno-political conflict: the case of Manas National Park, India. Trop Conserv Sci 7:475–487. https://doi.org/10.1177/194008291400700308

Graham AJ, Atkinson PM, Danson FM (2004) Spatial analysis for epidemiology. Acta Trop 91 (3):219–225

Groves CP (1982) Geographic variation in the Barasingha or swamp deer. J Bombay Nat Hist Soc 79:620–629. https://www.biodiversitylibrary.org/page/48745060
Hijmans RJ, Cameron SE, Parra JL, Jones PG, Jarvis A (2005) Very high resolution interpolated climate surfaces for global land areas. Int J Climatol 25(15):1965–1978
Ihaka R, Gentleman R (1996) R: a language for data analysis and graphics. J Comput Graph Stat 5(3):299–314
Islam N, Ahmed A, Barman R et al (2022) Revival of Eastern Swamp Deer *Rucervus duvaucelii ranjitsinhi* (Groves, 1982) in Manas National Park of Assam, India. J Threat Taxa 14(1): 20488–20493
Jathanna D, Karanth KU, Johnsingh AJT (2003) Estimation of large herbivore densities in the tropical forests of southern India using distance sampling. J Zool 261(3):285–290
Johnsingh AJT, Ramesh K, Qureshi Q, David A, Goyal SP, Rawat GS, Rajapandian K, Prasad S (2004) Conservation status of tiger and associated species in the Terai Arc Landscape. Wildlife Institute of India, Dheradun, pp viii+-110
Karanth KK, Nichols JD, Karanth KU, Hines JE, Christensen NL Jr (2010) The shrinking ark: patterns of large mammal extinctions in India. Proc R Soc B 277:1971–1979. https://doi.org/10.1098/rspb.2010.0171
Lahan P, Sonowal R (1973) Kaziranga wildlife sanctuary, Assam. J Bombay Nat Hist Soc 70(2): 245–278
Lahkar D, Ahmed F, Begum RH et al (2020) Inferring patterns of sympatry among large carnivores in Manas National Park—a prey-rich habitat influenced by anthropogenic disturbances. Anim Conserv 24(4):589–601. https://doi.org/10.1111/acv.12662
Landis JR, Koch GG (1977) An application of hierarchical kappa-type statistics in the assessment of majority agreement among multiple observers. Biometrics 33:363–374. https://doi.org/10.2307/2529786
Lehmkuhl JF (1994) A classification of subtropical riverine grassland and forest in Chitwan National Park, Nepal. Vegetatio 111(1):29–43. https://doi.org/10.1007/BF00045575
Liaw A, Wiener M (2002) Classification and regression by random Forest. R News 2(3):18–22
Lindsey PA, Chapron G, Petracca LS, Burnham D, Hayward MW, Henschel P, Hinks AE, Garnett ST, Macdonald DW, Macdonald EA, Ripple WJ, Zander K, Dickman A (2017) Relative efforts of countries to conserve world's megafauna. Glob Ecol Conserv 10:243–252. https://doi.org/10.1016/j.gecco.2017.03.003
Linkie M, Guillera-Arroita G, Smith J, Ario A, Bertagnolio G, Cheong F, Clements GR, Dinata Y, Duangchantrasiri S, Fredriksson G, Gumal MT, Horng LS, Kawanishi K, Khakim FR, Kinnaird MF, Kiswayadi D, Lubis AH, Lynam AJ, Maryati MM, Ngoprasert D, Novarino W, O'Brien TG, Parakkasi K, Peters H, Priatna D, Rayan DM, Seuaturien N, Shwe NM, Steinmetz R, Sugesti AM, Sunarto SME, Umponjan M, Wibisono HT, Wong CCT, Zulfahmi (2013) Cryptic mammals caught on camera: assessing the utility of range wide camera trap data for conserving the endangered Asian tapir. Biol Conserv 162:107–115. https://doi.org/10.1016/j.biocon.2013.03.028
Marshal JP (2016) Survival estimation of a cryptic antelope via photographic capture-recapture. Afr J Ecol 55:21–29. https://doi.org/10.1111/aje.12304
McFarland KP, Rimmer CC, Goetz JE, Aubry Y, Wunderle JM Jr, Sutton A, Townsend JM, Sosa AL, Kirkconnell A (2013) A Winter Distribution Model for Bicknell's Thrush (*Catharus bicknelli*), a conservation tool for a threatened migratory songbird. PLoS One 8(1):e53986. https://doi.org/10.1371/journal.pone.0053986
Murphy MA, Evans JS, Storfer A (2010) Quantifying Bufo boreas connectivity in Yellowstone National Park with landscape genetics. Ecology 91:252–261
Nath A, Sinha A, Lahkar B et al (2019) In search of Alien: factors influencing the distribution of *Chromolaena odorata* L. and *Mikania micrantha* Kunth in the Terai grasslands of Manas National Park, India. Ecol Eng 131:16–26
Nawaz M (2008) Ecology, genetics and conservation of Himalayan brown bears. PhD Thesis, Norwegian University of Life Sciences

Ortega-Huerta MA, Peterson AT (2004) Modelling spatial patterns of biodiversity for conservation prioritization in north-eastern Mexico. Divers Distrib 10(1):39–54

Paul S, Sarkar D, Patil A et al (2020) Assessment of endemic northern swamp deer (Rucervus duvaucelii duvaucelii) distribution and identification of priority conservation areas through modeling and field surveys across North India. Glob Ecol Conserv 24:15. https://doi.org/10.1016/j.gecco.2020.e01263

Pearce JL, Boyce MS (2006) Modelling distribution and abundance with presence-only data. J Appl Ecol 43:405–412

Pocock R (1943) The larger deer of British India. J Bombay Nat Hist Soc 43:553–572

Qureshi Q, Sawarkar VB, Rahmani AR, Mathur PK (2004) Swamp deer or barasingha (Cervus duvaucelii Cuvier, 1823). In: Sankar K, Goyal SP (eds) Ungulates of India, ENVIS bulletin: wildlife and protected areas, vol 7. Wildlife Institute of India, Dehradun, pp 181–192

Rhoden CM, Peterman WE, Taylor CA (2017) Maxent-directed field surveys identify new populations of narrowly endemic habitat specialists. PeerJ 5:e3632. https://doi.org/10.7717/peerj.3632

Ripple WJ, Newsome TM, Wolf C, Dirzo R, Everatt KT, Galetti M, Hayward MW, Kerley GIH, Levi T, Lindsey PA, Macdonald DW, Malhi Y, Painter LE, Sandom CJ, Terborgh, Van VB (2015) Collapse of the world's largest herbivores. Sci Adv 1(4):e1400103. https://doi.org/10.1126/sciadv.1400103

Ripple WJ, Chapron G, Lopez-Bao JV, Durant SM, Macdonald DW, Lindsey PA, Bennett EL, Beschta RL, Bruskotter JT, Campos-Arceiz A, Corlett RT, Darimont CT, Dickman AJ, Dirzo R, Dublin HT, Estes JA, Everatt KT, Galetti M, Goswami VR, Hayward MW, Hedges S, Hoffmann M, Hunter LTB, Kerley GIH, Letnic M, Levi T, Maisels F, Morrison JC, Nelson MP, Newsome TM, Painter L, Pringle RM, Sandom CJ, Terborgh J, Treves A, Van VB, Vucetich JA, Wirsing AJ, Wallach AD, Wolf C, Woodroffe R, Young H, Zhang L (2016) Saving the world's terrestrial megafauna. BioScience 66(10):807–812. https://doi.org/10.1093/biosci/biw092

Rosenzweig ML (1968) Net primary productivity of terrestrial communities: prediction from climatological data. Am Nat 102(923):67–74

Roy PS, Meiyappan P, Joshi PK, Kale MP, Srivastav VK, Srivasatava SK, Behera MD, Roy A, Sharma Y, Ramachandran RM, Bhavani P, Jain AK, Krishnamurthy YVN (2016) Decadal land use and land cover classifications across India, 1985, 1995, 2005. ORNL DAAC, Oak Ridge. https://doi.org/10.3334/ORNLDAAC/1336

Saikia BP, Rabha A, Saikia PK (2012) Manas: last destination of Swamp Deer (*Rucervus duvaucelii*) after Kaziranga in North East India. In: Rabha A, Saikia BP (eds) Manas: our good ol' darling, 1st edn. Bookland Publishers, Guwahati, pp 78–84

Sankaran R (1989) Status of the Swamp Deer in Dudhwa National Park (1988–1989). Technical report no. 14. BNHS, Bombay

Sankaran R (1990) Status of the Swamp Deer (Cervus duvauceli duvauceli) in the Dudwa National Park, Uttar Pradesh, India. J Bombay Nat Hist Soc 87(2):250–259

Sarma PK, Lahkar BP, Ghosh S et al (2008) Land-use and land-cover change and future implication analysis in Manas National Park, India using multi-temporal satellite data. Curr Sci 95:223–227

Schaller GB (1967) The deer & the tiger. University of Chicago Press, Chicago

Singh A (1970) The swamp deer of North Kheri. In: IUCN 11th Technical Meeting. Problems of threatened species. IUCN Pub New Series 18, vol 2, pp 52–54

Sinha A, Nath A, Lahkar BP et al (2022) Understanding the efficacy of different techniques to manage *Chromolaena odorata* L., an invasive Alien plant in the sub-Himalayan tall grasslands: toward grassland recovery. Ecol Eng 179:106618., ISSN 0925-8574. https://doi.org/10.1016/j.ecoleng.2022.106618

Svetnik V, Liaw A, Tong C, Wang T (2004) Application of Breiman's random forest to modeling structure-activity relationships of pharmaceutical molecules. In: Roli F, Kittler J, Windeatt T (eds) Multiple classifier systems, Lecture notes in computer science. Springer, Heidelberg, pp 334–343

Tewari R, Rawat GS (2013) Studies on the food and feeding habits of swamp deer (*Rucervus duvaucelii duvaucelii*) in Jhilmil Jheel Conservation Reserve, Haridwar, Uttarakhand, India. ISRN Zoology, 278213, p 6. https://doi.org/10.1155/2013/278213

Title PO, Bemmels JB (2018) ENVIREM: an expanded set of bioclimatic and topographic variables increases flexibility and improves performance of ecological niche modeling. Ecography 41 (2):291–307

Trouwborst A (2019) Global large herbivore conservation and international law. Biodivers Conserv 28:3891–3914. https://doi.org/10.1007/s10531-019-01856-y

Wallach AD, Izhaki I, Toms JD, Ripple WJ, Shanas U (2015) What is an apex predator? Oikos 124: 1453–1461. https://doi.org/10.1111/oik.01977

Wikramanayake E, Dinerstein E, Loucks C, Olson D, Morrison J, Lamoreux J, McKnight M, Hedao P (2002) Terrestrial ecoregions of the Indo-Pacific: a conservation assessment. Island Press, Washington, DC

Wildlife Institute of India (2017) National Studbook of Swamp Deer (*Rucervus duvaucelii*). Wildlife Institute of India, Dehradun and Central Zoo Authority, New Delhi. TR No 2017/011, 41 pp

Evaluating Potential Habitats of Chital, Sloth Bear and Jungle Cat in Selected Areas of Central Indian Landscape

16

G. Areendran, Aroma Caroline John, C. S. Abhijitha, Krishna Raj, and Kumar Ranjan

Abstract

Fragmentation has now emerged as a major global problem, with anthropogenic activities regarded as one of the main causes, primarily for affecting the habitat suitability. Habitat suitability is influenced by several elements, the most significant of which are the structural components of land use and topography. With the aid of remote sensing and GIS tools, habitats were assessed using a multi-criteria approach and the habitat suitability modelling. Land Use/Land Cover and topographic characteristics were used in MaxEnt distribution model to assess the habitat suitability in the heterogeneous landscape of Central India. Significant overlaps of potential habitats were observed between the species mostly within the protected areas.

Keywords

Habitat suitability · Fragmentation · MaxEnt · Species distribution · Protected area

16.1 Introduction

Habitat suitability assessments are recommended for conservation efforts on the basis that they will promote or preserve the survivability and diversity of wildlife species in the habitat units. The loss, fragmentation and alteration of habitats due to

G. Areendran (✉) · C. S. Abhijitha · K. Raj · K. Ranjan
IGCMC, WWF-India, New Delhi, India
e-mail: gareendran@wwfindia.net; abhijithacs@wwfindia.net; kraj@wwfindia.net; kranjan@wwfindia.net

A. C. John
Amity University, Noida, Uttar Pradesh, India

S. Dhyani et al. (eds.), *Ecosystem and Species Habitat Modeling for Conservation and Restoration*, https://doi.org/10.1007/978-981-99-0131-9_16

land use change and the introduction of alien pests endanger the health and diversity of the wildlife population.

Changes in land cover, which are often caused by anthropogenic land use alterations, affect the land-atmosphere interactions on which ecosystem services rely, which can further hamper the functioning of the ecosystem (Nduati et al. 2013). Observation of LULC (land use/land cover) changes is therefore critical (Nduati et al. 2013). There are numerous factors that can control the survival of certain species, and observing them can help in forecasting future trends and targeting conservation measures towards species in severe need. Developments in worldwide ecological regions, rare species and protected areas indicate that current and future urbanization has the potential to cause localized but considerable biodiversity damage (Mcdonald et al. 2008). Such studies illustrating LULC transformation may be beneficial in decision-making and policy direction and related administrative actions. They will also be important in determining the causes and effects of LULC transformation (Kumar and Singh 2021).

A tool for explaining or predicting key ecological factors for various species, such as species distribution and habitat quality, is habitat suitability modelling (Prajapati et al. 2015). The protection and management of vulnerable species depend heavily on the knowledge of regional distributions and habitat suitability; hence ecologically niche models (ENM) are useful tools in conservation biology because they may be used to map potentially suitable habitats (Jackson and Robertson 2011). Diverse techniques for ecological niche modelling are among the new software tools being used to address concerns of biodiversity. Niches may be recreated by associating the data on species occurrences with data sets that summarize climatic, topographic, edaphic and other "ecological" dimensions. Environmental modelling is an interdisciplinary field of study that calls for expertise in both biology and geographic information systems (GIS) to offer accurate geographical data for analysis, as both are necessary for the competent collecting of primary data and analysis of outcomes (Lissovsky and Dudov 2021). The combinations of environmental variables that are most closely associated with observed species presences may then be identified and projected onto landscapes to identify suitable regions (Soberon and Peterson 2005). The framework is related to the maximum entropy (MaxEnt) principle that forecasts the likelihood of species distribution under limitations based on environmental variables and species occurrence data (Teng et al. 2021). Since there are often insufficient reliable locations available for mapping the distribution of many species, MaxEnt has a relatively low location requirement for accurate model generation which is a very advantageous feature (Hernandez et al. 2006). Presence-only data and environmental data for the entire study region are needed. It can make use of both continuous and categorical data, as well as interactions between several factors (Phillips et al. 2006). One of the most reliable metrics for assessing the precision of model prediction is the AUC value of the area under the ROC curve (Wang et al. 2021). The distribution with the maximum entropy from every single environmental variable is chosen as the optimal distribution by MaxEnt using environmental variables and known sample information at a single point in time (Yang et al. 2021).

When using the jackknife method, the model is run while excluding one variable at a time. This provides information about the performance of every single variable in the model depending upon the crucial role played by each variable towards describing the distribution of the species and the amount of unique information offered by the variables. Since this can highlight strongly connected variables, as a result, the user can assess whether percent contribution values are likely to be influenced by these correlations (Baldwin 2009). The accuracy of MaxEnt is constrained by the approach and data used during modelling (Boral and Moktan 2021) despite the fact that it is an effective tool for modelling endangered species (Gebrewahid et al. 2020).

This paper is aimed at habitat suitability of chital, sloth bear and jungle cat, the forest dwelling species, in the highly fragmented Central Indian landscape.

16.2 Study Area

The study area, Central Indian landscape, extends from 21° 11′N–76°12′E to 22°59′ N–81°16′E which encompasses the wildlife reserves and corridors of Melghat tiger reserve, Satpura tiger reserve, Kanha tiger reserve and Pench tiger reserve along with their connecting corridors. Following the trend of the Satpura range, the area falls in the states of Maharashtra, Madhya Pradesh and Chhattisgarh within the districts of Amravati, Nagpur, Bhandara, Gondiya, Dewas, Sehore, Hoshangabad, Narsimhapur, Mandla, Dindori, Burhanpur, West Nimar, East Nimar, Betul, Chhindwara, Seoni, Balaghat, Kawardha and Rajnandgaon in Central India.

Hosting some of India's greatest dense forests, as well as diverse plant and animal species and indigenous people, Central India is widely regarded as the core of India's wildlife (Fig. 16.1).

16.3 Data and Methodology

See Figs. 16.2 and 16.3.

16.3.1 Data Sources

The images were procured from the US Geological Survey (USGS) EarthExplorer website. The data for the years 2020–2021 was taken from Landsat 8 to 9 OLI/TIRS. The OLI on Landsat 8 collects images spanning 11 spectral bands in different wavelengths of visible, near-infrared, shortwave infrared, panchromatic, cirrus and thermal infrared. It has a temporal resolution of 16 days and covers a swath of 185 kilometres. All of the bands are quantized to 8-bit data. The images were taken in the post-monsoon period ranging from the months of November to February for the years 2020–2021. Eight satellite images covered the target landscape area.

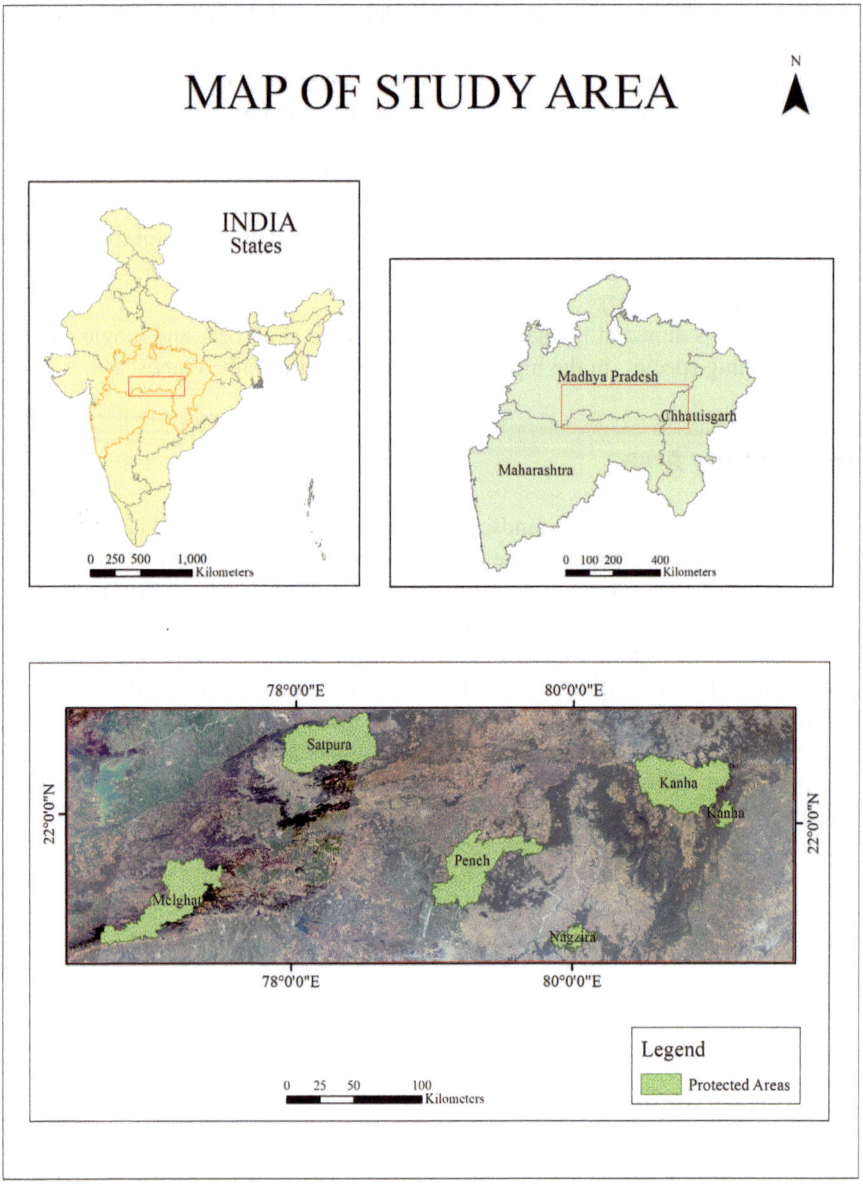

Fig. 16.1 Map of the study area

SRTM digital elevation model data was also acquired from the USGS EarthExplorer website. Shape files for railways were procured from Bhukosh GSI site. Bioclimatic variables were downloaded from Worldclim.org. Point location data was procured from Global Biodiversity Information Facility (GBIF) database for the period of 2015–2021.

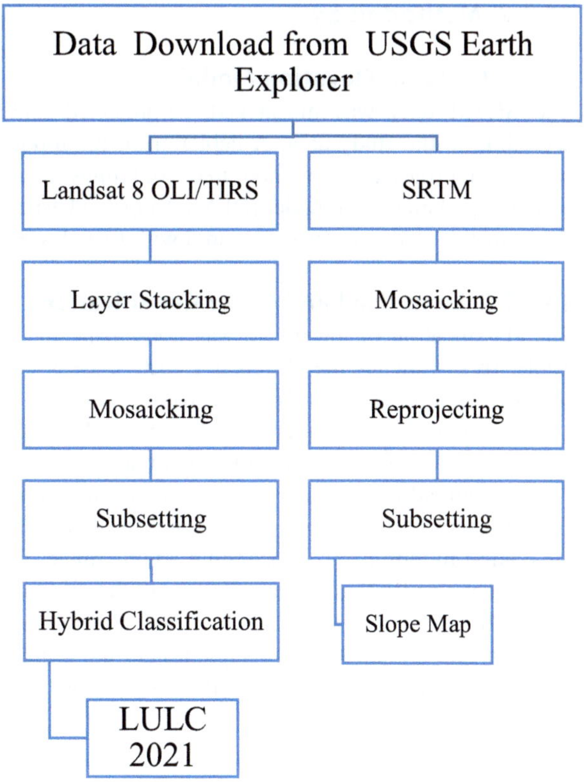

Fig. 16.2 Data processing flow chart 1

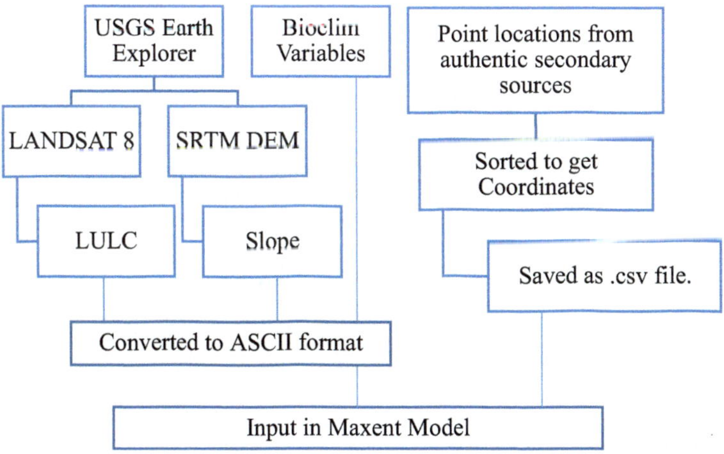

Fig. 16.3 Data processing flow chart 2

16.3.2 Methodology

16.3.2.1 Digital Elevation Model

The SRTM data was mosaicked, re-projected and then subsetting was done according to the study area. A depiction of the topographic surface of the earth's bare ground (bare earth) without trees, buildings or other surface items is provided by the digital elevation model (DEM). Slope and aspect elements were extracted. A slope map of study area was prepared with five classes of slope degrees (Fig. 16.4).

16.3.2.2 Land Use/Land Cover Classification

The classification of the image was done using the hybrid classification method. Table 16.1 showing the different LULC classes considered for the analysis is provided as follows (Figs. 16.5 and 16.6):

According to the methodological framework, factors used for the analysis were land use/land cover, slope and the bioclimatic variables which included:

Bio1, annual mean temperature; Bio2, mean diurnal range; Bio3, isothermality; Bio5, maximum temperature of the warmest month; Bio6, minimum temperature of the coldest month; Bio7, temperature annual range; Bio8, mean temperature of the wettest quarter; Bio9, mean temperature of the driest quarter; Bio10, mean temperature of the warmest quarter; Bio11, mean temperature of the coldest quarter; Bio12, annual precipitation; Bio13, precipitation of the wettest month; Bio14, precipitation of the driest month; Bio15, precipitation seasonality; Bio16, precipitation of the

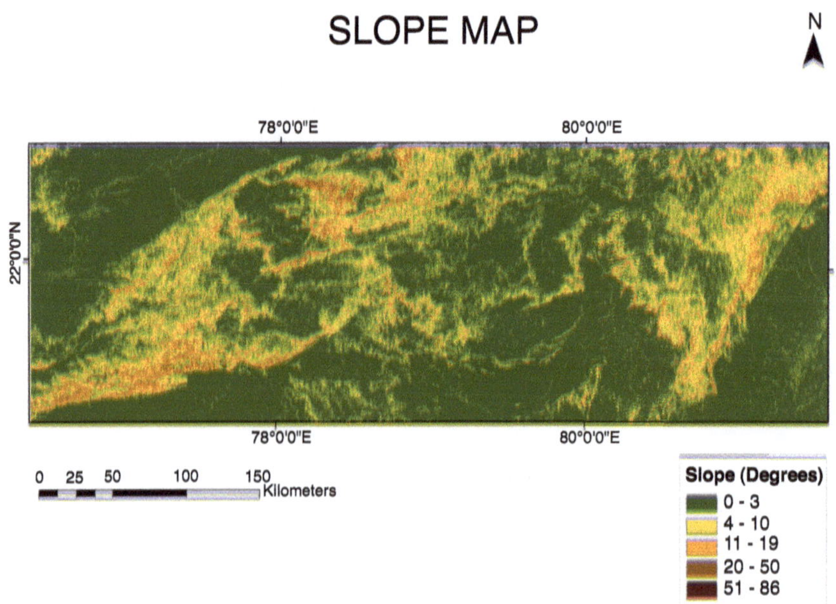

Fig. 16.4 Slope map

Table 16.1 Land use/land cover classes

S. no	Land use/land cover classes
1.	Dense forest
2.	Open forest
3.	Cultivation
4.	Open scrub
5.	Fallow land
6.	Barren land
7.	Built-up
8.	Water body
9.	River bed
10.	Natural vegetation

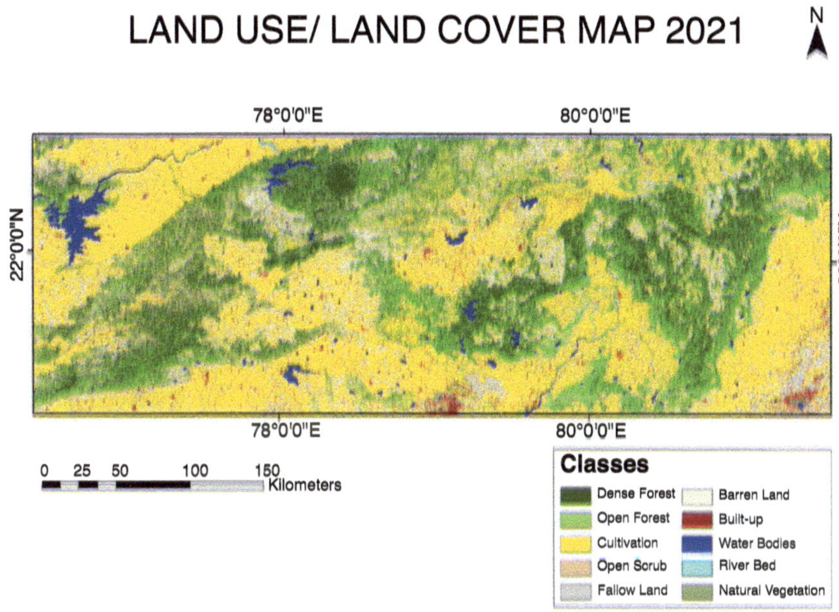

Fig. 16.5 Land use/land cover classification

wettest quarter; Bio17, precipitation of the driest quarter; Bio18, precipitation of the warmest quarter; and Bio19, precipitation of the coldest quarter (WorldClim n.d.).

The bioclimatic variables were clipped according to the study area and resampled to the layer of slope. A few layers had mismatch in the rows and columns of pixels and had to be corrected using the resample and extract by mask tools along with setting the processing extent and raster analysis to match the layers. All the layers were projected to the same projection system WGS 1984. The layers were then

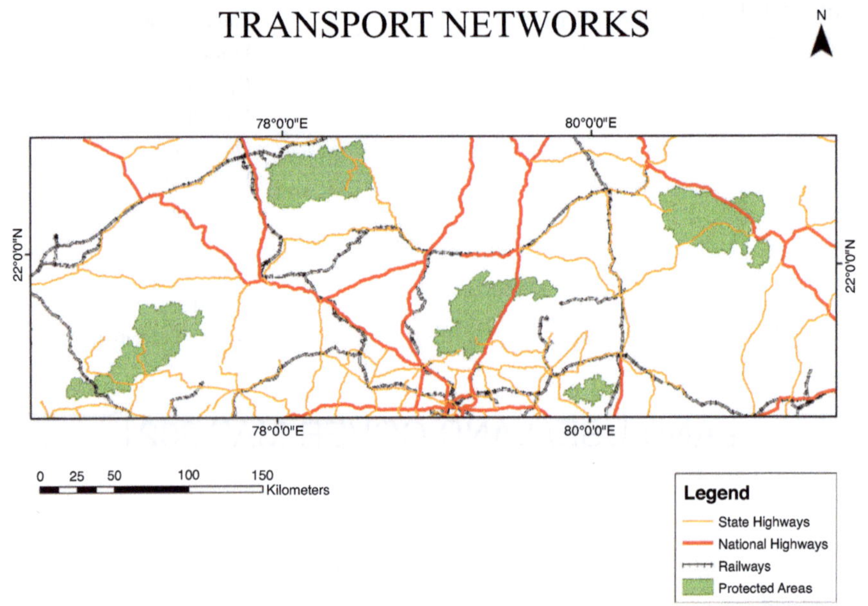

Fig. 16.6 Transport networks

converted to the ASCII format. The point location data of species were sorted to get the latitude and longitude of occurrence and saved in a .csv format. The layers were used in the MaxEnt model keeping LULC as categorical and the rest as continuous variables.

16.4 Results and Discussions

The majority of the protected areas of the Central Indian landscape such as Pench Tiger Reserve, Kanha Tiger Reserve and Satpura Tiger Reserve are observed to be very suitable zones (that are shown in blue- to green-coloured zones) which are mostly the core areas of these protected regions forming a high to moderately suitable zone. The zone in yellow suggests a marginally appropriate area. The final brown zone is an area that is completely unsuited for chital (Figs. 16.7 and 16.8).

Looking at the AUC jackknife for chital, it is possible to estimate the amount of influence each variable has. Histograms of each variable are calculated where relations are depicted with either just one variable, which is represented by a light blue-coloured histogram, the value without any variables is indicated by a dark blue colour and that with all variables is displayed by a red colour.

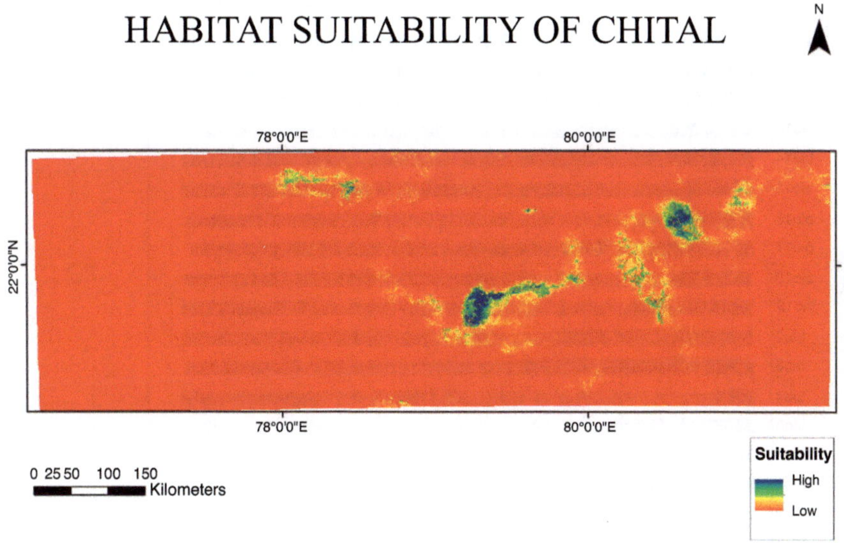

Fig. 16.7 Habitat suitability of chital

As seen from the jackknife, the model has taken into consideration almost each variable showing that it has an influence on the habitat suitability of chital (Fig. 16.9).

The suitable habit for the sloth bear as indicated is within the areas of Satpura Tiger Reserve and the denser forest regions highlighted in a shade of dark blue to green. The moderately suitable habit is shown in yellow, while the least suitable regions are shown in brown (Fig. 16.10).

The observed jackknife for sloth bear also shows a considerable influence of almost all variables with a few having higher influence (Fig. 16.11).

As it can be noted, the jungle cat has a suitable habitat spanning the regions of protected areas and dense forest cover as shown by dark blue shade. Moderately suitable habitats are shown in yellow tones, while the least suitable zones are shown in brown (Fig. 16.12).

The habitat for jungle cat is highly influenced by a handful of variables as can be observed from the jackknife plot (Fig. 16.13).

The potential habitats of the species under study have significant overlaps especially within the protected regions of Satpura Tiger Reserve for sloth bear and jungle cat, Pench Tiger Reserve for chital and jungle cat and Kanha Tiger Reserve for sloth bear and jungle cat again as well as within Nagzira Reserve.

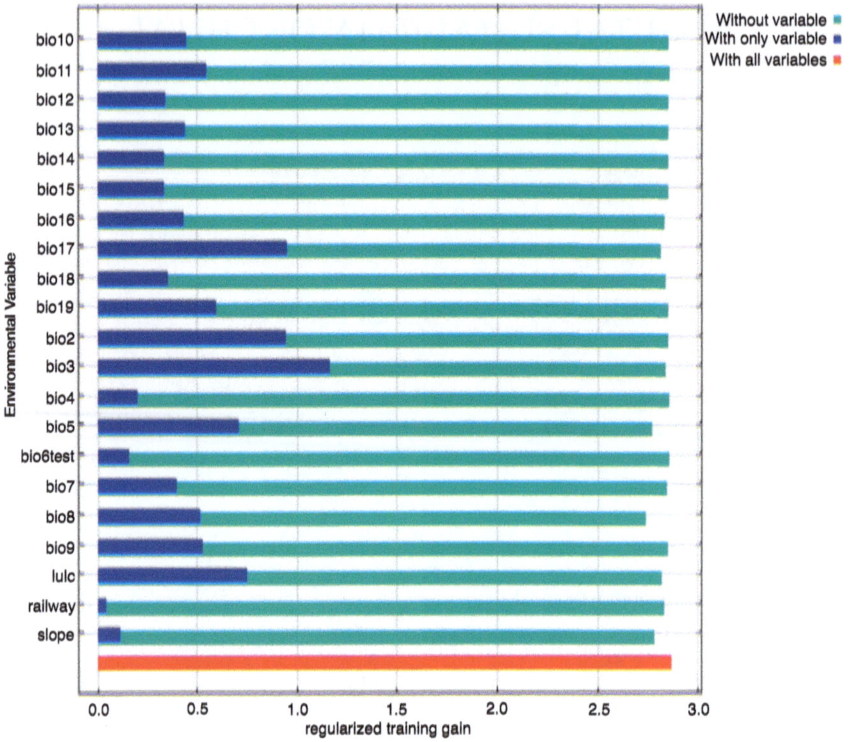

Fig. 16.8 Jackknife of regularized training grain of chital

16.5 Conclusion

With a large distribution range and multiple big populations, chital populations in India are believed to be stable. Habitat destruction, poaching and anthropogenic conflict may represent potential areas of conflict for India's vast mammalian diversity. Declared vulnerable by the IUCN, sloth bear in India still has few suitable habitats present, but as a result of habitat loss and poaching, the species is currently under constant threat. Problems with anthropogenic pressure are widespread, particularly for carnivores. The encroachment of agricultural regions has a significant negative impact on jungle cats as they can cause further human wildlife conflict. The primary risks to jungle cats are habitat destruction, fragmentation, fuel wood collecting and poaching.

This study aimed to draw attention to the Central Indian landscape as it has a vast and fragmented area that has both the potential of providing suitable habitats to many species and at the same time faces the threat of further fragmentation. This study can be used for further conservation and monitoring efforts within this region.

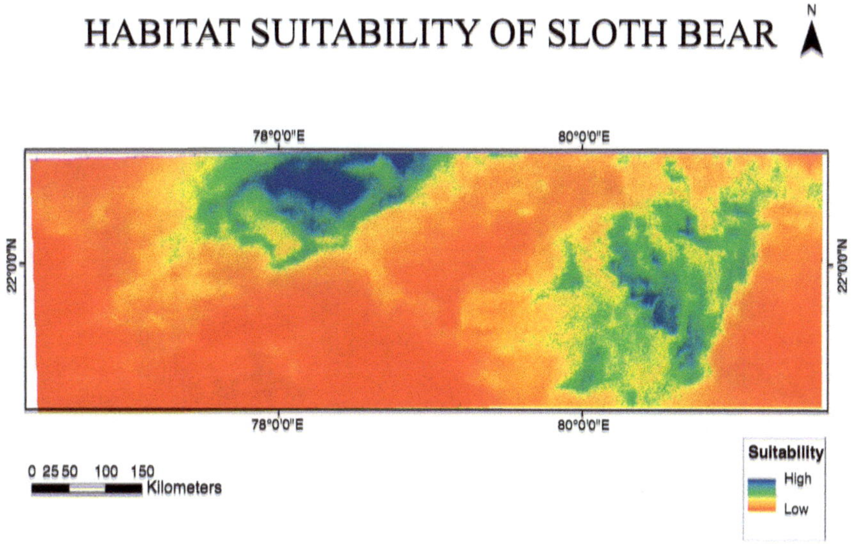

Fig. 16.9 Habitat suitability of sloth bear

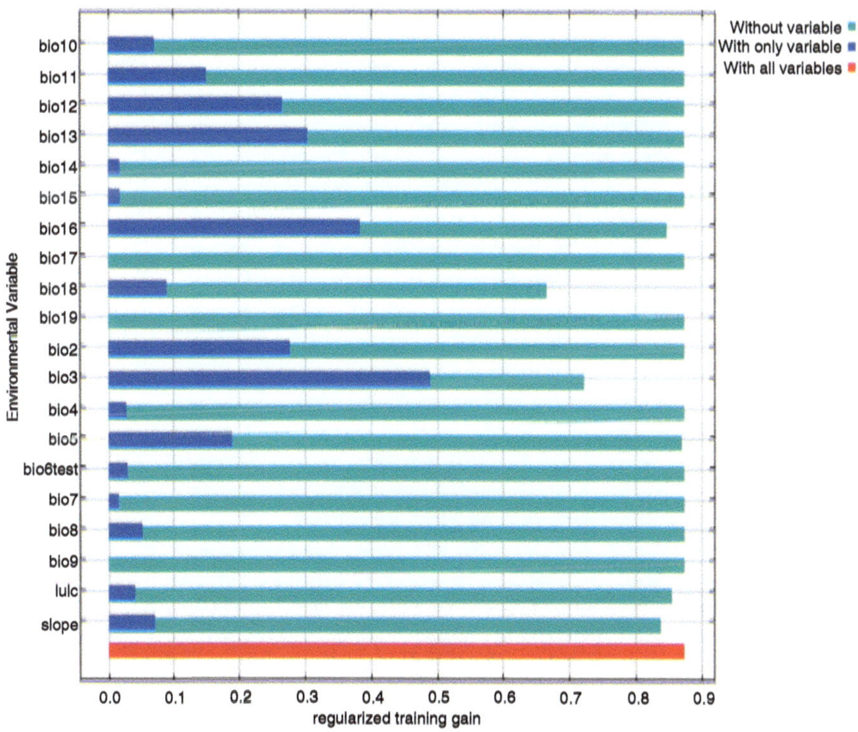

Fig. 16.10 Jackknife of regularized training grain of sloth bear

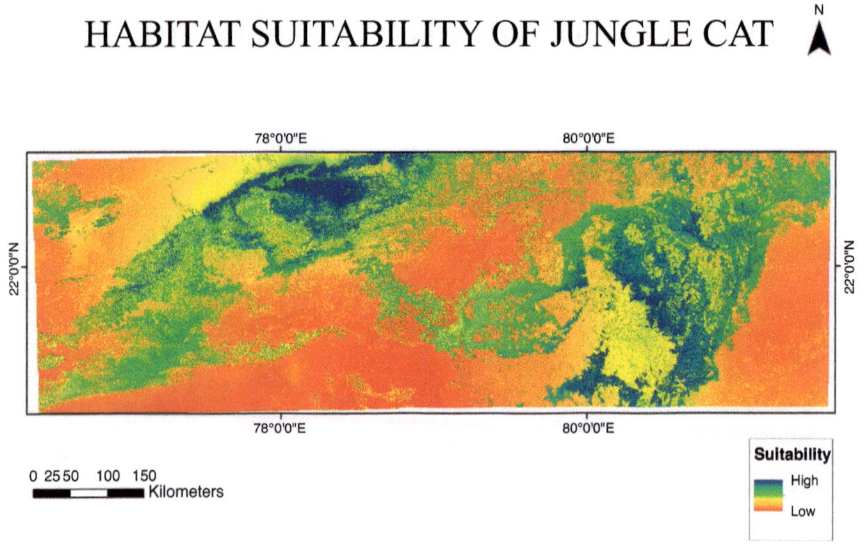

Fig. 16.11 Habitat suitability of jungle cat

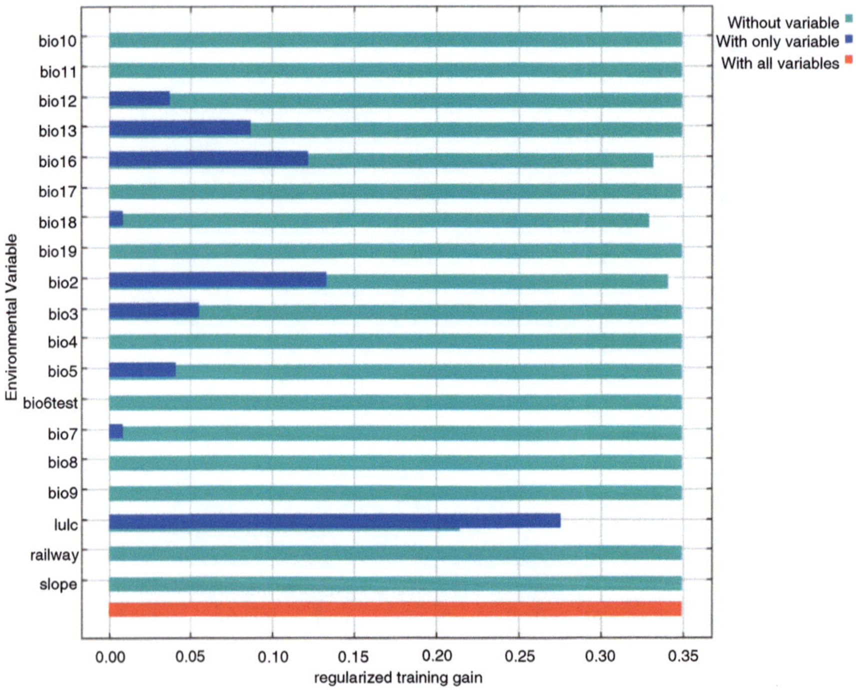

Fig. 16.12 Jackknife of regularized training grain of jungle cat

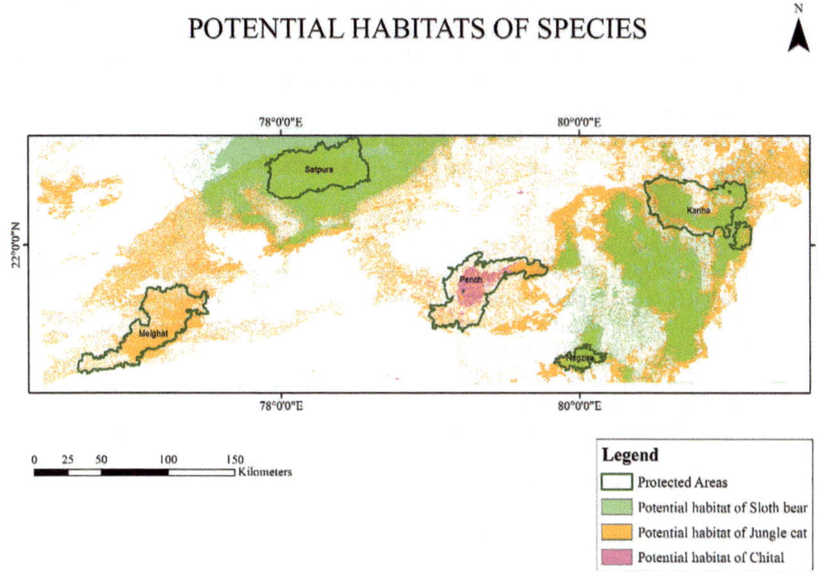

Fig. 16.13 Potential habitats of species

References

Baldwin RA (2009) Use of maximum entropy modeling in wildlife research. Entropy 11(4): 854–866. https://doi.org/10.3390/e11040854

Boral D, Moktan S (2021) Predictive distribution modeling of *Swertia bimaculata* in Darjeeling-Sikkim eastern Himalaya using MaxEnt: current and future scenarios. Ecol Process 10(1). https://doi.org/10.1186/s13717-021-00294-5

Gebrewahid Y, Abrehe S, Meresa E, Eyasu G, Abay K, Gebreab G, Kidanemariam K, Adissu G, Abreha G, Darcha G (2020) Current and future predicting potential areas of *Oxytenanthera abyssinica* (A. Richard) using MaxEnt model under climate change in Northern Ethiopia. Ecol Process 9(1). https://doi.org/10.1186/s13717-019-0210-8

Hernandez PA, Graham CH, Master LL, Albert DL (2006) The effect of sample size and species characteristics on performance of different species distribution modeling methods. Ecography 29(5).773–785. https://doi.org/10.1111/j.0906-7590.2006.04700.x

Jackson CR, Robertson MP (2011) Predicting the potential distribution of an endangered cryptic subterranean mammal from few occurrence records. J Nat Conserv 19(2):87–94. https://doi.org/10.1016/j.jnc.2010.06.006

Kumar S, Singh R (2021) Geospatial applications in land use/land cover change detection for sustainable regional development: the case of Central Haryana, India. Geo Environ Eng 15(3): 81–98. https://doi.org/10.7494/geom.2021.15.3.81

Lissovsky AA, Dudov SV (2021) Species-distribution Modeling: advantages and limitations of its application. 2. MaxEnt. Biol Bull Rev 11(3):265–275. https://doi.org/10.1134/s2079086421030087

Mcdonald RI, Kareiva P, Forman RTT (2008) The implications of current and future urbanization for global protected areas and biodiversity conservation. Biol Conserv 141(6):1695–1703. https://doi.org/10.1016/j.biocon.2008.04.025

Nduati EW, Mundia CN, Ngigi MM (2013) Effects of vegetation change and land use/land cover change on land surface temperature in the Mara ecosystem. Int J Sci Res 2(8):22–28

Phillips SB, Aneja VP, Kang D, Arya SP (2006) Modelling and analysis of the atmospheric nitrogen deposition in North Carolina. Int J Global Environ Issues 6(2–3):231–252. https://doi.org/10.1016/j.ecolmodel.2005.03.026

Prajapati RK, Triptathi S, Mishra RM (2015) Habitat suitability analysis for chital (*Axis axis*) using geo-spatial technology pf Panna National Park (M.P.) India. Int J Adv Res Sci Technol 4(6):427–434

Soberon J, Peterson AT (2005) Interpretation of models of fundamental ecological niches and species' distributional areas. Biodivers Inform 2(0):1–10. https://doi.org/10.17161/bi.v2i0.4

Teng SY, Su NJ, Lee MA, Lan KW, Chang Y, Weng JS, Wang YC, Sihombing RI, Vayghan AH (2021) Modeling the habitat distribution of *Acanthopagrus schlegelii* in the coastal waters of the eastern Taiwan strait using maxent with fishery and remote sensing data. JMSE 9(12):1442. https://doi.org/10.3390/jmse9121442

Wang R, Jiang C, Liu L, Shen Z, Yang J, Wang Y, Hu J, Wang M, Hu J, Lu X, Li Q (2021) Prediction of the potential distribution of the predatory mite *Neoseiulus californicus* McGregor in China using MaxEnt. Global Ecol Conserv 29(July):e01733. https://doi.org/10.1016/j.gecco.2021.e01733

WorldClim. (n.d.). Historical climate data. https://www.worldclim.org/

Yang X, Jin X, Zhou Y (2021) Wildfire risk assessment and zoning by integrating maxent and GIS in Hunan Province, China. Forests 12(10):1299. https://doi.org/10.3390/f12101299

Habitat Suitability Modeling of *Tor tor* (Hamilton, 1822) in the Indian Drainage Systems Using MaxEnt

17

Ranjit Mahato, Gibji Nimasow, Oyi Dai Nimasow, and Santoshkumar Abujam

Abstract

Freshwater ecosystems are severely affected by the alteration of habitat due to various anthropogenic influences including overexploitation of resources, construction of dams, soil and water pollution, land use/land cover change, etc. Habitat suitability modeling is an essential step toward the conservation of freshwater species including all types of mahseers. In this study, we predicted the habitat suitability of *Tor tor* in Indian river systems using the MaxEnt method. The performance of the model was good with the area under the curve (AUC) value of 0.86 and the true skill statistic (TSS) of 0.711. The model predicted a total river length of 9214.15 km (1.83%) and 17,155.93 km (3.42%) as highly suitable and suitable, respectively. The results show that Northeast India, the Himalayan region, and the Gangetic plain have highly suitable habitats for *T. tor*. Therefore, the model outcomes could help formulate policies for the conservation of *T. tor*.

Keywords

Habitat · Modeling · *Tor tor* · River system · Conservation

R. Mahato · G. Nimasow (✉)
Department of Geography, Rajiv Gandhi University, Rono Hills, Doimukh, Arunachal Pradesh, India
e-mail: ranjit.mahato@rgu.ac.in; gibji.nimasow@rgu.ac.in

O. D. Nimasow
Department of Botany, Rajiv Gandhi University, Rono Hills, Doimukh, Arunachal Pradesh, India
e-mail: oyidai.nimasow@rgu.ac.in

S. Abujam
Department of Zoology, Rajiv Gandhi University, Rono Hills, Doimukh, Arunachal Pradesh, India

S. Dhyani et al. (eds.), *Ecosystem and Species Habitat Modeling for Conservation and Restoration*, https://doi.org/10.1007/978-981-99-0131-9_17

17.1 Introduction

Freshwater species play an important role in the respective ecosystem in association with the surrounding environment. Fishes are an important source of animal protein in the world and are widely recognized as a good source for maintaining body health (Ganaie and Sharma 2021). The quantification of these relationships portrays the core of predictive geographical modeling in ecology to solve the crucial problem of how various environmental factors are controlling the distribution of species and communities (Guisan and Zimmermann 2000). The diversity of freshwater species is declining swiftly worldwide, and the population of freshwater wildlife has declined by 83% since 1970 (WWF 2018; Acreman et al. 2019). The huge concentration of population and urban centers around the freshwater ecosystems has threatened the freshwater species diversity including habitat loss and degradation, water pollution, overfishing, etc. (Arthington et al. 2016).

Species distribution models (SDMs) connect the species' occurrences with the prevailing environmental conditions and extrapolate the suitable habitat of a species (Domisch et al. 2015a). In other words, it estimates the distribution of a species within the geographic space and time with the help of environmental information as well as occurrence records (Mahato et al. 2022). Generally, SDMs can be characterized as a methodology that is based on core ecological and biogeographical principles about the association of species distributions with the physical environment (Elith and Franklin 2017). SDMs are frequently named correlative or statistical models, habitat models, or ecological niche models and are broadly separated into two categories, namely, correlative and process-based or mechanistic models (Srivastava et al. 2019). The outcomes of SDMs are also applied as input data for other investigations, especially in the field of conservation planning, impact assessment, and land use planning to identify suitable locations for habitat restoration and reintroduction of species (Elith and Franklin 2017).

Tor tor, commonly known as mahseer or Tor Barb, is the type of species of the genus Tor (Pinder et al. 2019). It was first described from the Mahananda, a tributary of the Ganges flowing through Northeast Bengal, India, by Hamilton (1822). It is a well-known game and food fish which inhabits the streams and rivers of the mountainous area and fast-flowing rivers of the plains and mostly prefers clear, fast-flowing waters having a stony, pebbly, or rocky surface (Shrestha 1997). The species has been described to reach 150 cm Total Length (TL) (Mishra 1959) and gain a maximum weight of 68 kg (Talwar and Jhingran 1991). *T. tor* is a native species of south Himalayan rivers extending from Pakistan to Myanmar and also toward the rivers of South India (Rayamajhi et al. 2018). The southernmost native distribution of *T. tor* was believed to be the Narmada River in Madhya Pradesh (Desai 2003). However, it was also discovered from the Godavari and Krishna River basins (Lal et al. 2013), which raises the question of whether it is native to tropical Peninsular India or expanded through artificial propagation (Pinder et al. 2019). *T. tor* has been earlier assessed as "Near Threatened" in the IUCN Red List due to swiftly decreasing populations, but recently it has been reassessed as "Data Deficient" (Rayamajhi et al. 2018). The present work is an attempt to predict the suitable habitats of *T. tor* in the Indian river systems by combining species occurrences, environmental variables, and geospatial technology.

17.2 Materials and Methods

17.2.1 Study Area

T. tor is native to the rivers of the Himalayas, Central India, Northeast India, and northern Peninsular India. Hence, the study area includes the rivers of the whole country. India is the seventh largest country in the world and the third largest in Asia. It is located between 8° 4′ N and 37° 6′ N latitudes to 68° 7′ E and 97° 25′ E longitudes. It shares an international border with Afghanistan and Pakistan in the northwest; China, Bhutan, and Nepal in the north; and Myanmar and Bangladesh in the east. The southern part of the country is also bounded by the Bay of Bengal to the east and the Arabian Sea to the west (Fig. 17.1). The biogeographic regions of India have been also considered to comprehend the habitat suitability of *T. tor* (Fig. 17.2).

17.2.2 Occurrence Records

The occurrence records of *T. tor* have been collected from both primary and secondary sources. The majority of the occurrence records were downloaded from the Global Biodiversity Information Facility (GBIF) using the *dismo* package of R (version 4.0.3) following (Hijmans et al. 2011). The primary data was collected from different rivers of Arunachal Pradesh using a handheld Global Positioning System (GPS). All the occurrence records have been filtered to remove records with no coordinate information. Initially, a total of 301 occurrence records were generated. We used a snapping tolerance of 3 km to move the occurrence points to the closest freshwater pixel, and the points falling beyond this range have been removed. Further, the *spThin* package in R was used to remove the biases from the records. The algorithm retained only one record within a 10-km spatial grid and returned a dataset of 95 occurrence records for the final model execution.

17.2.3 Selection of the Best Environmental Variables

Initially, a set of 51 environmental parameters (1-km spatial resolution) have been downloaded from the EarthEnv project (www.earthenv.org) using R (Table 17.1). The sources of the data include WorldClim, Consensus Land Cover, HydroSHEDS, and World Soil Information (ISRIC) as explained by Domisch et al. (2015b). Various statistical techniques face multicollinearity problem due to the presence of highly correlated information among the variables (Miller 2010), which negatively affects the model performance and poses difficulties in interpreting the relative performance of variables in the model predictions (Dormann et al. 2013; Manzoor

Fig. 17.1 Map of the study area

et al. 2018). The complex features created by MaxEnt are often highly correlated. So, it is recommended to minimize such highly correlated environmental variables (Merow et al. 2013). A threshold of $|r| > 0.7$ correlation coefficients between predictor variables is a proper indicator of collinearity (Dormann et al. 2013; Manzoor et al. 2018; Sony et al. 2018; Farrell et al. 2019; Feng et al. 2019). Similarly, the variance inflation factor (VIF) must be computed to check the effects of the multicollinearity of the variables. It is suspiciously high when VIF for an independent variable is greater than 5 or 10 (Tsagris and Pandis 2021). Initially, 17 (Table 17.2) out of 51 variables were selected based on the correlation

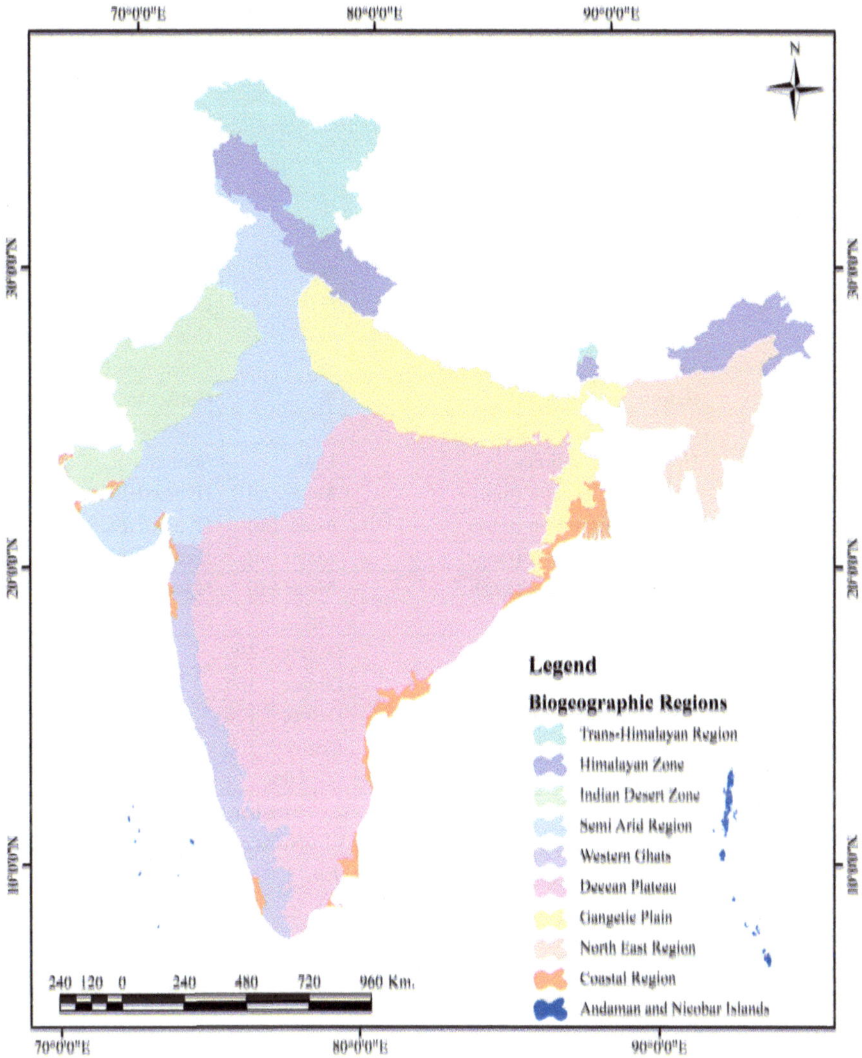

Fig. 17.2 Biogeographic regions of India

coefficients of |*r*| < 0.7 and VIF less than 5 using *usdm* package in *R* (Naimi, 2016). Out of these, ten variables with more than 1% contribution were selected (Table 17.3). The final model was performed with three replications to predict the suitable habitat of *T. tor* in the Indian river systems.

Table 17.1 Environmental data used for species distribution modeling of *Tor tor*

Variable code	Upstream average land cover	Unit	Source
Lc_avg_01	Evergreen/deciduous needleleaf trees	%	Consensus Land Cover
Lc_avg_02	Evergreen broadleaf trees	%	Consensus Land Cover
Lc_avg_03	Deciduous broadleaf trees	%	Consensus Land Cover
Lc_avg_04	Mixed/other trees	%	Consensus Land Cover
Lc_avg_05	Shrubs	%	Consensus Land Cover
Lc_avg_06	Herbaceous vegetation	%	Consensus Land Cover
Lc_avg_07	Cultivated and managed vegetation	%	Consensus Land Cover
Lc_avg_08	Regularly flooded vegetation	%	Consensus Land Cover
Lc_avg_09	Urban/built-up	%	Consensus Land Cover
Lc_avg_10	Snow/ice	%	Consensus Land Cover
Lc_avg_11	Barren	%	Consensus Land Cover
Lc_avg_12	Open water	%	Consensus Land Cover
Variable code	Upstream elevation	Unit	Source
Elv_01	Upstream elevation (min)	Meter [m]	HydroSHEDS
Elv_02	Upstream elevation (max)	Meter [m]	HydroSHEDS
Elv_03	Upstream elevation (range)	Meter [m]	HydroSHEDS
Elv_04	Upstream elevation (avg)	Meter [m]	HydroSHEDS
Variable code	Upstream slope	Unit	Source
Slope_01	Upstream slope (min)	Degree * 100	HydroSHEDS
Slope_02	Upstream slope (max)	Degree * 100	HydroSHEDS
Slope_03	Upstream slope (range)	Degree * 100	HydroSHEDS
Slope_04	Upstream slope (avg)	Degree * 100	HydroSHEDS
Variable code	Stream length and flow accumulation	Unit	Source
Flow_acc_01	Stream length	Number_cells	HydroSHEDS
Flow_acc_02	Flow accumulation	Number_cells	HydroSHEDS
Variable code	Upstream averaged hydroclimatic variables (bioclim)	Unit	Source
Hydroclim_01	Annual mean temperature	[°C] * 10	WorldClim
Hydroclim_02	Mean diurnal range (mean of monthly (max temp–min temp))	[°C] * 10	WorldClim
Hydroclim_03	Isothermality (BIO2/BIO7) (×100)	* 100	WorldClim
Hydroclim_04	Temperature seasonality (standard deviation ×100)	[°C] * 10	WorldClim
Hydroclim_05	Max temperature of the warmest month	[°C] * 10	WorldClim
Hydroclim_06	Min temperature of the coldest month	[°C] * 10	WorldClim
Hydroclim_07	Temperature annual range (BIO5–BIO6)	[°C] * 10	WorldClim
Hydroclim_08	Mean temperature of the wettest quarter	[°C] * 10	WorldClim
Hydroclim_09	Mean temperature of the driest quarter	[°C] * 10	WorldClim

(continued)

Table 17.1 (continued)

Variable code	Upstream averaged hydroclimatic variables (bioclim)	Unit	Source
Hydroclim_10	Mean temperature of the warmest quarter	[°C] * 10	WorldClim
Hydroclim_11	Mean temperature of the coldest quarter	[°C] * 10	WorldClim
Hydroclim_12	Annual precipitation	[mm]	WorldClim
Hydroclim_13	Precipitation of the wettest month	[mm]	WorldClim
Hydroclim_14	Precipitation of the driest month	[mm]	WorldClim
Hydroclim_15	Precipitation seasonality (coefficient of variation)	* 100	WorldClim
Hydroclim_16	Precipitation of the wettest quarter	[mm]	WorldClim
Hydroclim_17	Precipitation of the driest quarter	[mm]	WorldClim
Hydroclim_18	Precipitation of the warmest quarter	[mm]	WorldClim
Hydroclim_19	Precipitation of the coldest quarter	[mm]	WorldClim
Variable code	**Soil upstream average**	**Unit**	**Source**
Soil_avg_01	Soil organic carbon	[g/kg]	ISRIC
Soil_avg_02	Soil pH in H_2O	pH * 10	ISRIC
Soil_avg_03	Sand content mass fraction	%	ISRIC
Soil_avg_04	Silt content mass fraction	%	ISRIC
Soil_avg_05	Clay content mass fraction	%	ISRIC
Soil_avg_06	(Coarse fragments >2 mm fraction) volumetric	%	ISRIC
Soil_avg_07	Cation exchange capacity	[cmol/kg]	ISRIC
Soil_avg_08	Bulk density of the fine earth fraction	[kg / m3]	ISRIC
Soil_avg_09	Depth to bedrock (R horizon) up to maximum 240 cm	[cm]	ISRIC
Soil_avg_10	Predicted probability of occurrence (0–100%) of R horizon across sub-catchment	%	ISRIC

17.2.4 Model Setup and Evaluation

MaxEnt (version 3.4.4) was used to predict the habitat suitability of *T. tor*. MaxEnt is a machine learning algorithm that performs best with high prediction accuracy when used in small sample sizes and presence-only datasets (Ji et al. 2020; Li et al. 2022). It possesses several advantages over other methods such as good performance with incomplete datasets, short running time, simple operation, needs small sample size, and high simulation precision (Li et al. 2020). The settings of the model include logistic output, cross-validate replicate run type, response curves, jackknife measures of variable importance, ten percentile training presence, and rest kept as default settings (Mahato et al. 2022). The results ranging between 0 and 1 have been classified into five classes of suitability, viz., <0.1 unsuitable; 0.1–0.3, slightly suitable; 0.3–0.5, moderately suitable; 0.5–0.7, suitable; and >0.7, highly suitable.

Table 17.2 Variance inflation factor and contribution of selected variables

Sl. no.	Variable code	VIF	Initial contribution	Initial permutation importance
1.	Hydroclim_14	1.28	30.0	9.2
2.	Lc_avg_01	2.59	25.9	9.4
3.	Lc_avg_03	2.18	14.0	7.6
4.	Elv_01	1.53	5.3	29.8
5.	Lc_avg_02	2.23	5.1	7.4
6.	Hydroclim_03	1.38	4.8	3.7
7.	Lc_avg_04	2.18	4.6	5.5
8.	Soil_avg_07	2.51	3.4	8.4
9.	Hydroclim_09	4.47	2.5	3.8
10.	Slope_01	1.56	1.1	8.0
11.	Lc_avg_05	1.81	0.9	1.1
12.	Soil_avg_09	1.82	0.7	1.2
13.	Lc_avg_11	2.26	0.6	2.3
14.	Lc_avg_09	1.26	0.4	0.8
15.	Lc_avg_12	1.43	0.2	1.2
16.	Soil_avg_04	1.35	0.2	0.3
17.	Lc_avg_08	1.20	0.1	0.1

The stepwise methodology followed in this study is shown in Fig. 17.3. The threshold-independent, area under the curve (AUC) of the receiver operating characteristic (ROC) curve and true skill statistic (TSS) were used to evaluate the overall performance of the model.

17.3 Results

17.3.1 Model Performance

The model results show the distribution of *T. tor* is mostly influenced by Hydroclim_14 with 31.2% followed by Lc_avg_01 (26.9%) and Lc_avg_03 (14.7). Elevation, Lc_avg_02, and Hydroclim_03 were other important predictors with over 5% contributions. The rest of the variables such as Lc_avg_04, Soil_avg_07, and slope also slightly influenced the distribution of *T. tor* (Table 17.2). The performance of the model was reasonably consistent and good with a mean AUC of 0.860 and a TSS value of 0.711.

17.3.2 Habitat Suitability Modeling of *Tor tor*

The final results of habitat suitability were considered in terms of the total river length (km). Out of the total river length of 502,255.03 km, the model showed 9214.15 km (1.83%) as highly suitable and 17,155.93 km (3.42%) as suitable

Table 17.3 Contribution of selected variables

Sl. no.	Variable code	Contribution (%)	Permutation importance (%)
1.	Hydroclim_14	31.2	11.2
2.	Lc_avg_01	26.9	14.2
3.	Lc_avg_03	14.7	8.5
4.	Elv_01	5.7	31.9
5.	Lc_avg_02	5.2	8.1
6.	Hydroclim_03	5.2	5.4
7.	Lc_avg_04	4.4	3.5
8.	Soil_avg_07	3.0	8.7
9.	Hydroclim_09	2.6	4.0
10.	Slope_01	1.1	4.6

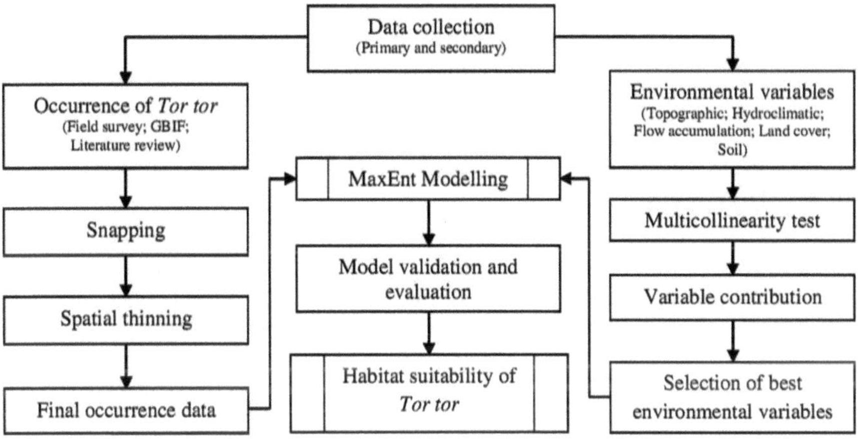

Fig. 17.3 Flowchart of the methodology

followed by 37,352.21 km (7.44%) as moderately suitable, and 159,173.13 km (31.69%) as slightly suitable. More than half of the study area, i.e., 279,359.61 km (55.62%), falls under the unsuitable category (Table 17.4 and Fig. 17.4). The results were also separately calculated for each biogeographic region of India. Out of the ten biogeographic regions, the highly suitable habitat has been predicted in only five regions. Among these regions, Northeast India showed the highest percentage (9.44%), followed by the Himalayan region (8.16%), the Gangetic plain (5.76%), the semi-arid region (0.57%), and the Deccan Plateau (0.15%) (Table 17.4 and Fig. 17.4). The model predicted Andaman and Nicobar Islands (100%), Trans-Himalayan region (99.35%), coastal region (92.00%), and Indian Desert region (85.61%) as unsuitable for *T. tor*.

A close examination of the model outputs reveals the rivers of Northeast India as suitable for the targeted species (Fig. 17.4). The north and south bank tributaries of the mighty river Brahmaputra appear to be the best home of *T. tor*. The northern

Table 17.4 Habitat suitable for *Tor tor* in the Indian river system

Categories	Suitability index	River length (km)	Percentage
Unsuitable	<0.1	279,359.61	55.62
Slightly suitable	0.1–0.3	159,173.13	31.69
Moderately suitable	0.3–0.5	37,352.21	7.44
Suitable	0.5–0.7	17,155.93	3.42
Highly suitable	>0.7	9214.15	1.83
Total		*502,255.03*	*100*

tributaries of the Brahmaputra such as Kameng, Dikrong, Subansiri, Siang, etc. have been predicted as suitable habitat for the species. Besides, a portion of the Barak River system also falls under suitable habitat. In the central and northern parts of the country, Ganga and its tributaries fall under the suitable category. Narmada, Tapi, and Godavari River systems have been also found suitable for *T. tor*. Some of the reservoirs, namely, Indra Sagar Reservoir of Madhya Pradesh, Gobind Sagar, and Maharana Pratap Sagar of Himachal Pradesh, also fall under suitable habitat for the species (Table 17.5).

17.4 Discussion

The distribution and occurrences of fish communities are highly influenced by the surrounding environment. The habitat of fish is largely influenced by various factors such as the meandering of streams, the gradient of river banks, riparian vegetation, and stream flow dynamics (Gebrekiros 2016). Based on the quality of water and the associated environmental factors especially rainfall and temperature, the species can travel to large rivers for breeding and feeding (Kaushik and Bordoloi 2016). The model finds a positive association of *T. tor* with precipitation of the driest month, land cover classes (evergreen/deciduous needleleaf trees, deciduous broadleaf trees, and evergreen broadleaf trees), upstream minimum elevation, and isothermality.

T. tor is widely distributed in Bangladesh, Bhutan, India, Myanmar, Nepal, and Pakistan. In India, the species is distributed over Arunachal Pradesh, Assam, Bihar, Haryana, Manipur, Meghalaya, Nagaland, Punjab, Sikkim, Uttar Pradesh, and Uttaranchal (Khajuria and Langer 2016). The model predicted the drainage system of the Himalayan foothill zone, Central India, Northeast India, and southern India (Narmada, Godavari, and Tapi) as the suitable habitats of *T. tor*. The results are in agreement with the earlier studies (Lal et al. 2013; Pinder et al. 2019). The results also show the highly suitable habitats over the hilly and mountainous rivers, which may be attributed to better growth in the rivers with a rocky surface (Brraich and Saini 2019) and clear and swiftly flowing water along with optimum water temperature and pH level (Patil and Saxena 2021).

The inland drainage systems provide fewer chances of migration to better environmental conditions and are frequently confined within the landlocked water bodies (Roy et al. 2021). Human populations are always settled along water bodies

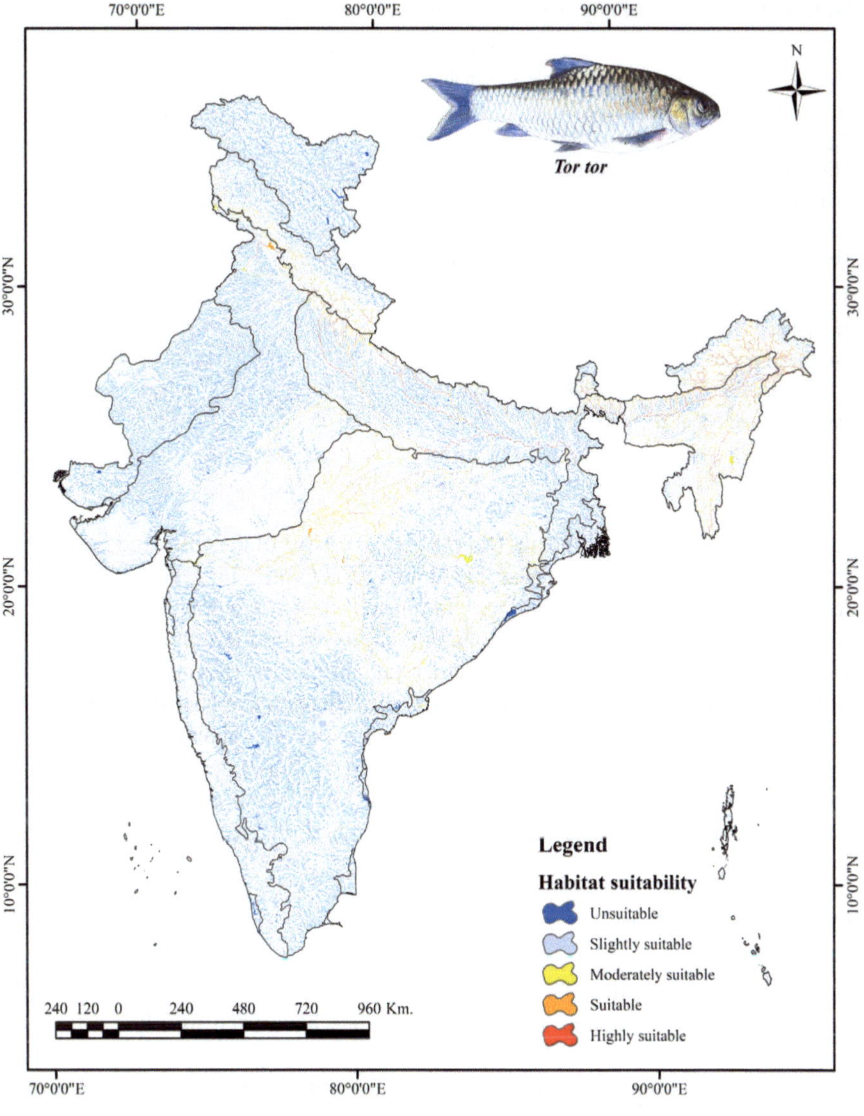

Fig. 17.4 Habitat suitability mapping of Indian drainage system

including rivers; hence, the rivers are severely affected by human influences (Zeng et al. 2022). The fish habitats have been altered or degraded resulting in the declining trend of fish assemblage, extinction of species, and replacement by other species (Aadland 1993; Gebrekiros 2016). In general, anthropogenic influences alter the natural environment through changing land use/land cover, modifying flow regimes, constructing river dams, and polluting soil and water as well as climate change which indirectly influences the extinction of various native species and/or

Table 17.5 Habitat suitability of *Tor tor* in different biogeographic regions (in %)

Zones	Unsuitable	Slightly suitable	Moderately suitable	Suitable	Highly suitable
Trans-Himalayan region	99.35	0.43	0.22	0.00	0.00
Himalayan zone	40.98	23.13	16.08	11.65	8.16
Indian desert zone	85.61	14.39	0.00	0.00	0.00
Semi-arid region	54.63	39.22	3.84	1.74	0.57
Western Ghats	57.53	40.38	2.07	0.02	0.00
Deccan Plateau	48.04	39.39	9.64	2.78	0.15
Gangetic plain	62.78	21.83	4.85	4.78	5.76
Northeast region	18.18	38.65	20.88	12.85	9.44
Coastal region	92.00	6.83	1.17	0.00	0.00
Andaman and Nicobar Islands	100.00	0.00	0.00	0.00	0.00

introduction of non-native species (Su et al. 2021). Most of the cities and large urban centers in India are located near the drainage network. In general, mahseer is decreasing very hastily in their population and size in Central India because of many reasons including overexploitation, water pollution, habitat destruction, domestic effluent, and use of insecticides and pesticides (Nautiyal and Dwivedi 2020), particularly the exploitation rate of *T. tor* (Dwivedi and Nautiyal 2012). *T. tor* has been reported to possess high nutrition and economic values (Dey et al. 2015). Therefore, there is an urgent need of evolving sound conservation strategies for the species in the country before it's too late. The results of the present study could help formulate plans and policies for the conservation and management of *T. tor*.

17.5 Conclusion

The study predicted the habitat suitability of *T. tor* in the Indian river systems through the MaxEnt model. The suitable habitats have been mostly predicted in the North Eastern and the Himalayan region. Although the species is widely distributed in the country, the model results show limited occupancy of the species in certain concentrated pockets. Keeping in view the reported declining population of the species due to various anthropogenic activities, there is a need of exploring the potential habitats of the species. The model predictions, particularly the moderately suitable areas, should be explored and confirmed about the occurrence of the species. Moreover, the predicted suitable and highly suitable areas need to be prioritized for the conservation and restoration of the species in the future. However, our study is constrained by the vastness of the study area. Therefore, we recommend further research at biogeographic/regional levels based on the results of the present study.

Acknowledgments The authors are thankful to the Soil and Limnological Laboratory, Department of Geography, Rajiv Gandhi University, Rono Hills, Doimukh, Arunachal Pradesh (India), for facilitating necessary infrastructural and laboratory facilities to accomplish this work. The authors also acknowledge Consensus Land Cover, WorldClim, HydroSHEDS, World Soil Information (ISRIC), and GBIF for making available various data sources used in this study.

References

Aadland LP (1993) Stream habitat types: their fish assemblages and relationship to flow. N Am J Fish Manag 13(4):790–806. https://doi.org/10.1577/1548-8675(1993)013%3C0790:SHTTFA%3E2.3.CO;2

Acreman M, Hughes KA, Arthington AH, Tickner D, Dueñas M (2019) Protected areas and freshwater biodiversity: a novel systematic review distils eight lessons for effective conservation. Conserv Lett 13(1):e12684. https://doi.org/10.1111/conl.12684

Arthington AH, Dulvy NK, Gladstone W, Winfield IJ (2016) Fish conservation in freshwater and marine realms: status, threats and management. Aquat Conserv Mar Freshwat Ecosyst 26(5): 838–857. https://doi.org/10.1002/aqc.2712

Brraich OS, Saini SK (2019) Ichthyofaunal diversity of Ranjit Sagar wetland situated in the northwestern Himalayas. Eur J Environ Sci 9(2):106–113. https://doi.org/10.14712/23361964.2019.14

Desai VR (2003) Synopsis of biological data on the tor mahseer *Tor tor* (Hamilton, 1822). FAO Fisheries Synopsis. No. 158. FAO, Rome, 36 pp

Dey A, Sarkar K, Barat S (2015) Evaluation of fish biodiversity in rivers of three districts of eastern Himalayan region for conservation and sustainability. Int J Appl Res 1(9):424–435

Domisch S, Amatulli G, Jetz W (2015a) Near-global freshwater-specific environmental variables for biodiversity analyses in 1 km resolution. Sci Data 2(1):1–13. https://www.nature.com/articles/sdata201573

Domisch S, Jähnig SC, Simaika JP, Kuemmerlen M, Stoll S (2015b) Application of species distribution models in stream ecosystems: the challenges of spatial and temporal scale, environmental predictors and species occurrence data. Fundam Appl Limnol 186(1–2):45–61. https://doi.org/10.1127/fal/2015/0627

Dormann CF, Elith J, Bacher S, Buchmann C, Carl G, Carré G, Marquéz JRG, Gruber B, Lafourcade B, Leitão PJ, Münkemüller T, McClean C, Osborne PE, Reineking B, Schröder B, Skidmore AK, Zurell D, Lautenbach S (2013) Collinearity: a review of methods to deal with it and a simulation study evaluating their performance. Ecography 36(1):27–46. https://doi.org/10.1111/j.1600-0587.2012.07348.x

Dwivedi AC, Nautiyal P (2012) Stock assessment of fish species *Labeo rohita*, *Tor tor* and *Labeo calbasu* in the rivers of Vindhyan region, India. J Environ Biol 33(2):261–264

Elith J, Franklin J (2017) Species distribution modeling. Reference Module in Life Sciences, Elsevier, In, pp 1–15. https://doi.org/10.1016/B978-0-12-809633-8.02390-6

Farrell A, Wang G, Rush SA, Martin JA, Belant JL, Butler AB, Godwin D (2019) Machine learning of large-scale spatial distributions of wild turkeys with high-dimensional environmental data. Ecol Evol 9(10):5938–5949. https://doi.org/10.1002/ece3.5177

Feng X, Park DS, Liang Y, Pandey R, Papeş M (2019) Collinearity in ecological niche modeling: confusions and challenges. Ecol Evol 9(18):10365–10376. https://www.ncbi.nlm.nih.gov/pmc/articles/PMC6787792/

Ganaie SA, Sharma GD (2021) Isolation and molecular identification of aspergillus species from tor mahseer (*Tor tor*) in river Narmada. Ann Rom Soc Cell Biol 25(4):19940–19946. https://annalsofrscb.ro/index.php/journal/article/view/8805

Gebrekiros ST (2016) Factors affecting stream fish community composition and habitat suitability. J Aquacult Mar Biol 4(2):00076. https://doi.org/10.15406/jamb.2016.04.00076

Guisan A, Zimmermann NE (2000) Predictive habitat distribution models in ecology. Ecol Model 135(2–3):147–186. https://doi.org/10.1016/s0304-3800(00)00354-9

Hamilton F (1822) An account of the fishes found in the river Ganges and its branches. Hurst, Robinson, and Co., Edinburgh, London

Hijmans RJ, Phillips S, Leathwick J, Elith J (2011) Dismo package for R (Version 4.0.3). https://cran.r-project.org/package=dismo

Ji W, Han K, Lu Y, Wei J (2020) Predicting the potential distribution of the vine mealybug, *Planococcus ficus* under climate change by MaxEnt. Crop Prot 137:105268. https://doi.org/10.1016/j.cropro.2020.105268

Kaushik G, Bordoloi S (2016) Ichthyofauna of Ranganadi River in Lakhimpur, Assam, India. Check List 12(2):1872–1872. https://doi.org/10.15560/12.2.1872

Khajuria B, Langer S (2016) Distribution record on abundance of *tor putitora* in Jammu waters. Int J Fish Aquat Stud 4(1):341–347

Lal KK, Singh RK, Pandey A, Gupta BK, Mohindra V, Punia P, Dhawan S, Verma J, Tyagi LK, Khare P, Jena JK (2013) Distributional records of tor mahseer *Tor tor* (Hamilton, 1822) from southern India. J Appl Ichthyol 29(5):1086–1090. https://doi.org/10.1111/jai.12017

Li J, Fan G, He Y (2020) Predicting the current and future distribution of three Coptis herbs in China under climate change conditions, using the MaxEnt model and chemical analysis. Sci Total Environ 698:134141. https://doi.org/10.1016/j.scitotenv.2019.134141

Li Z, Liu Y, Zeng H (2022) Application of the MaxEnt model in improving the accuracy of ecological red line identification: a case study of Zhanjiang, China. Ecol Indic 137:108767. https://doi.org/10.1016/j.ecolind.2022.108767

Mahato R, Abujam S, Bushi D, Nimasow OD, Nimasow G, Das DN (2022) Distribution modelling of *tor putitora* (Hamilton, 1822), an endangered cyprinid in the Himalayan river system using MaxEnt. Acta Ecol Sin. https://doi.org/10.1016/j.chnaes.2022.01.004

Manzoor SA, Griffiths G, Lukac M (2018) Species distribution model transferability and model grain size–finer may not always be better. Sci Rep 8(1):1–9. https://www.nature.com/articles/s41598-018-25437-1

Merow C, Smith MJ, Silander JA Jr (2013) A practical guide to MaxEnt for modeling species' distributions: what it does, and why inputs and settings matter. Ecography 36(10):1058–1069. https://doi.org/10.1111/j.1600-0587.2013.07872.x

Miller J (2010) Species distribution modeling. Geogr Compass 4(6):490–509. https://doi.org/10.1111/j.1749-8198.2010.00351.x

Mishra KS (1959) An aid to identification of common fishes of India and Pakistan. Rec Indian Museum 57:1–320

Naimi B, Araújo MB (2016) Sdm: a reproducible and extensible R platform for species distribution modeling. Ecography 39:368–375. https://doi.org/10.1111/ecog.01881

Nautiyal P, Dwivedi AC (2020) Growth rate determination of the endangered mahseer, *Tor tor* (Hamilton 1822) from the Bundelkhand region, Central India. J Fish Res 4(2):7–11

Patil N, Saxena S (2021) A preliminary study on habitat ecology of mahseer (*Tot tor* Hamilton) in western zone of Narmada river and its tributaries. Int J Innov Res Technol 8(4):629–637

Pinder AC, Britton JR, Harrison AJ, Nautiyal P, Bower SD, Cooke SJ, Lockett S, Everard M, Katwate U, Ranjeet K, Walton S, Danylchuk AJ, Dahanukar N, Raghavan R (2019) Mahseer (tor spp.) fishes of the world: status, challenges and opportunities for conservation. Rev Fish Biol Fish 29(2):417–452. https://doi.org/10.1007/s11160-019-09566-y

Rayamajhi A, Jha BR, Sharma CM, Pinder AC, Harrison A, Katwate U, Dahanukar N (2018) *Tor tor*. The IUCN Red List of threatened species, 2018–2. http://eprints.bournemouth.ac.uk/31561/1/Tor%20tor.pdf

Roy S, Ray S, Saikia SK (2021) Indicator environmental variables in regulating the distribution patterns of small freshwater fish *Amblypharyngodon mola* in India and Bangladesh. Ecol Indic 120:106906. https://doi.org/10.1016/j.ecolind.2020.106906

Shrestha TK (1997) The mahseer in the rivers of Nepal disrupted by dams and ranching strategies. R. K. Printers Teku, Kathmandu, Nepal, pp 1–259. https://www.cabdirect.org/cabdirect/abstract/20006782580

Sony RK, Sen S, Kumar S, Sen M, Jayahari KM (2018) Niche models inform the effects of climate change on the endangered Nilgiri Tahr (*Nilgiritragus hylocrius*) populations in the southern Western Ghats, India. Ecol Eng 120:355–363. https://doi.org/10.1016/j.ecoleng.2018.06.017

Srivastava V, Lafond V, Griess VC (2019) Species distribution models (SDM): applications, benefits and challenges in invasive species management. CABI Rev 2019:1–13. https://doi.org/10.1079/pavsnnr201914020

Su G, Logez M, Xu J, Tao S, Villéger S, Brosse S (2021) Human impacts on global freshwater fish biodiversity. Science 371(6531):835–838. https://doi.org/10.1126/science.abd3369

Talwar PK, Jhingran AG (1991) Inland fishes of India and adjacent countries, vol 1. Balkema, Rotterdam, 541 pp

Tsagrisa M, Nikolaos Pandisb N (2021) Multicollinearity. Stat Res Design 159(5):695–696. https://doi.org/10.1016/j.ajodo.2021.02.005

WWF (2018) Living planet report (2018) risk and resilience in a new era. WWF International, Gland, Switzerland. https://www.worldwildlife.org/pages/living-planet-report-2018

Zeng C, Wen Y, Liu X, Yu J, Jin B, Li D (2022) Impact of anthropogenic activities on changes of ichthyofauna in the middle and lower Xiang River. Aquacult Fish 7(6):693–702. https://doi.org/10.1016/j.aaf.2021.06.007

Part IV

Application of Modelling Tools and Approaches

Modelling the Influence of Marine Fishery Advisories on the Reduction of Carbon Dioxide Emissions for Odisha Under Varying Climate Change Scenarios Using CMIP Models: An Evidence-Based Approach for Policymaking

18

Sudip Kumar Kundu and Harini Santhanam

Abstract

Long-term emissions of carbon dioxide by mechanised boats can negatively impact marine environment through drastic reduction in target fish stocks, increase in by-catches and low profits for fishers. These in turn trigger a vicious cycle of overcapacity fishing to compensate for loss of profits and severely impact marine biodiversity conservation and sustainability. India's Marine Fishery Advisories (MFAs), developed and disseminated by INCOIS, is a valuable knowledge product to optimise the fishing expeditions to fish accumulation zones under the lowest disturbance to the marine environment, with significant co-benefits of lowering the emissions from industrial fishing activities. Simulating decadal, annual scale emissions for RCP 4.5 and 8.5 using a CMIP5-based modelling approach in the present study showed negligible effects under the former and significant rise in the latter scenario. Investigations also revealed that MFAs have the capacity to reduce the annual emissions by 220,667.5 tonnes/year for the Bay of Bengal region near Odisha. The present study provides an evidence-based policy approach for promoting and enabling the use of MFAs as knowledge products of geospatial technology, not just to meet the targets of Sustainable Development Goals 14b (SDG 14b) but also to contribute towards national-level conformance to international standards for prevention of maritime pollution such as 'MARPOL 73/78' and its subsequent amendments.

S. K. Kundu · H. Santhanam (✉)
Department of Public Policy (DPP), Manipal Academy of Higher Education (MAHE), Manipal, Bengaluru Campus, Bengaluru, India
e-mail: harini.santhanam@manipal.edu

S. Dhyani et al. (eds.), *Ecosystem and Species Habitat Modeling for Conservation and Restoration*, https://doi.org/10.1007/978-981-99-0131-9_18

Keywords

Carbon dioxide emissions · CMIP5 · Marine Fishery Advisories · Mechanised fishing crafts · Odisha · SDG 14 · MARPOL

18.1 Introduction

One-third of the total world population is dependent on the oceans for their livelihoods from fisheries, tourism and many other sectors (UNEP 2022; United Nations 2022). In the case of India, fishery and aquaculture sectors provide livelihoods to nearly 25 million fishers, while the number is double along the value chain (Department of Fisheries 2020b). Fishery sector in India also accounted for 1.24% of the total Gross Domestic Product (GDP) and 7.28% of the agriculture sector GDP for the years 2019–2020 (Department of Fisheries 2020a). The estimated total value of cumulative coastal and marine ecosystem services in India is to the tune of approximately 1.5 trillion Indian Rupees (INR) which contributed to 3.2% of the net national product during 2012–2013 (Kavi Kumar et al. 2016). One of the most significant ecosystem services rendered by marine ecosystems is carbon sequestration which helps in absorbing surplus carbon dioxide (CO_2) and keeps the global ecosystem cool through feedback mechanisms (United Nations 2022). However, the coastal regions are extremely vulnerable to extreme weather events (e.g. cyclones, storm surges, etc.), environmental degradation (e.g. coastal erosions, etc.), pollution and overfishing (UNEP 2022). Global warming due to the rising level of human-induced greenhouse gases (GHGs; dominated by CO_2) impacts the marine environment in the form of sea-level rise and many other extreme weather conditions (Zickfeld et al. 2017).

The global transportation sector has been treated as the fastest-growing one of the major contributors to climate change with one-fourth (almost 23%) of the CO_2 emissions added by the sector to the total emissions (Asian Development Bank 2020). Due to the enhancement in the fishing effort and efficiency in the marine fishery sector in India, the CO_2 emissions due to the operational diesel-based fishing crafts had also been increased subsequently from 0.30 million tonnes in the 1960s to 3.60 million tonnes in the 2010s (Vivekanandan et al. 2013). Mechanised fishing crafts contributed to approximately 88% of the total emissions from the marine fishery sector in India followed by motorised (~12%), while the emission was reported to be nil for the non-motorised sector (Vivekanandan et al. 2013).

Quantifications of the contributions of marine transportation sector to the emissions of Indian seas need close attention for Indian seas. The emission rate of carbon dioxide is also higher in the case of mechanised crafts (50.7 kg/h) compared to the motorised crafts (16.1 kg/h) in India. In this context, searching time for the fish accumulation zones in the open sea plays a crucial role in the consumption of diesel and, hence, CO_2 emissions. Therefore, the known and predicted location of the fish aggregation zones in the sea is very important in order to reduce the search time and fuel consumption. The ESSO-Indian National Centre for Ocean Information

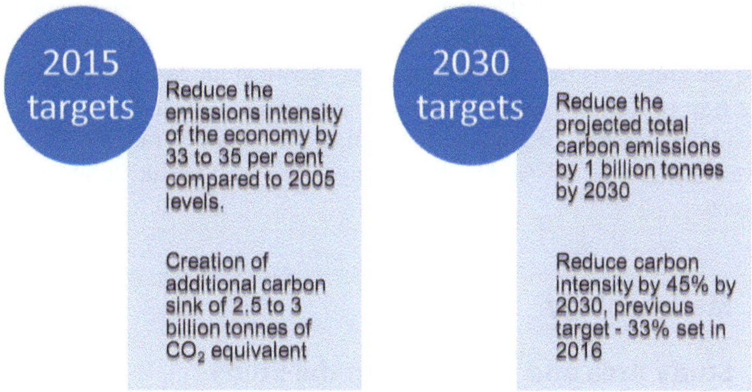

Fig. 18.1 Emission scenario of relevance to the marine fishing sector

Services (ESSO-INCOIS) has developed and disseminates Marine Fishery Advisories (MFAs) in the name of Potential Fishing Zone (PFZ) advisories to all the coastal communities on a daily basis subject to the availability of the cloud-free satellite data except for the fishing ban period imposed by the Government of India (ESSO-INCOIS 2020). PFZ advisories, short-term and reliable forecasts of the fish aggregation zones in the open sea, help fishers obtain the maximum catch with minimum effort by avoiding lengthy search time and hence reduce diesel consumption and CO_2 emissions from the fishing crafts in the Indian Ocean. In the case of Odisha, more than 500,000 fishers depend on marine fishing for their livelihoods across six coastal districts (CMFRI-DoF 2020).

India has aimed to reduce its CO_2 emissions (Fig. 18.1) by 33% to 35% by 2030 from the 2005 levels as a part of the 2015 Paris Climate Agreement (Reuters 2021).

The emissions levels of CO_2 for India were found to be at 2.63 billion tonnes in 2019 (before the emergence of the COVID-19 pandemic), contributing around 7.15% to the global annual emissions in the same year, when the per capita CO_2 emission in India was reported to be about 1.92 tonnes per person (Ritchie et al. 2020). While the global transport sector accounted for 24% of the total CO_2 emissions due to the fuel combustion, the emissions contributed by the Indian transport sector have been found to be about 13.5% to the total emission (Climate Action Tracker 2020). However, the share of transportation in marine and coastal ecosystems (i.e. mechanised and motorised fishing crafts) for India is not fully constrained. The emission intensity of CO_2 gas is higher for mechanised crafts compared to motorised crafts. For the capture of 1 tonne of marine fishes, on an average 1.18 tonne of CO_2 is emitted by mechanised crafts, compared to only 0.59 tonne for motorised crafts (Vivekanandan et al. 2013).

In this regard, the prediction of the emissions of anthropogenic CO_2 from the transportation sectors, especially in the marine fishery sector, is very crucial in the present scenario, especially with respect to emissions for the diesel-based marine fishing crafts (e.g. mechanised and motorised crafts). Global climate models (GCM) have been popularly used to simulate the anthropogenic CO_2 emissions worldwide

(Jones et al. 2013). GCM that was a part of the fifth phase of Coupled Model Intercomparison Project 5 (CMIP5) was utilised in the present study in order to compare the emissions of anthropogenic CO_2 between the last and current decades over the Bay of Bengal (BoB) under various warming scenarios. Considering the above, an effort has also been made to calculate the reduction rate of CO_2 emissions due to the usages of MFAs by the fishing crafts in Odisha during the fishing expeditions.

18.2 Methodology

18.2.1 Study Area and Data Used in the Study

The present study was carried out for the state of Odisha (Fig. 18.2), located on the north-eastern coast adjacent to the BoB. Odisha has six coastal districts, namely, Balasore, Bhadrak, Kendrapara, Jagatsinghpur, Puri, and Ganjam, from the north to south parts of the state. All along its 480-km-long coastline, Odisha has exhibited a high potential for marine fish production in India (Department of Fisheries 2020a). Approximately 83% of the total fish landings in Odisha are attributed to mechanised fishing from a total of 1748 mechanised crafts (CMFRI-FSI-DoF 2020; FRAD-CMFRI 2022). In general, mechanised crafts are associated with multi-day fishing trips lasting between 6 and 15 days in Odisha and with a diesel consumption of nearly 2000–3000 L per trip significantly adding to the emissions of CO_2.

In general, the PFZ advisories provide reliable and short-term predictions of the fish accumulation zones in the open sea (Subramanian et al. 2014). The basic inputs to generate this forecast are chlorophyll-a (Chl-a) concentration and sea surface temperature (SST). The satellite-derived Chl-a data are retrieved from the optical bands available in IRS-P4 OCM and MODIS-AQUA, while the thermal infrared channels of NOAA-AVHRR and ESA's Met-Op are used to generate SST data over the BoB and the Arabian Sea (ESSO-INCOIS 2020; Kundu et al. 2020).

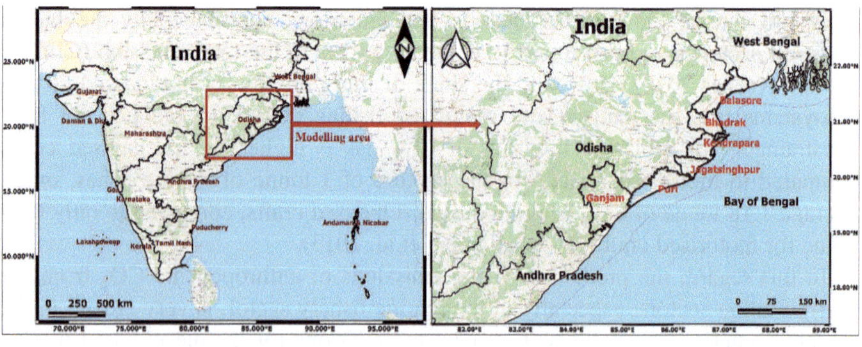

Fig. 18.2 Study area map of Odisha alongside the Bay of Bengal where the CMIP modelling approach simulated

The PFZ advisory developed by the ESSO-INCOIS is presently provided to different fish landing centres and fishing boat owners along the Indian coast daily subject to the availability of the cloud-free satellite data except during the fishing ban period imposed by the Government of India to preserve the juvenile stocks. The identification of PFZ using the remotely sensed data improves the catch size of fishes (by two- to fivefold) by minimising the effort (popularly known as catch per unit effort, CPUE) as well as reducing the diesel consumption as a result of the minimisation of the search time up to 30 to 70% per trip (Tummala et al. 2008). The PFZ advisories are also disseminated to Odisha state as in the case of the rest of the country, with uneven usage ratios across the different FLCs attributed to socio-technical constraints (Santhanam and Kundu 2022a).

18.2.2 Analyses of CMIP5 Scenarios

The fifth phase of the modelling framework of Coupled Model Intercomparison Project 5 (CMIP5) coordinated by the World Climate Research Programme (WCRP) was produced in order to improve the understanding of climate and also to provide the futuristic estimation of climate change which will be needful to those considering its possible consequences (Taylor et al. 2012). The CMIP5 model, namely, Model for Interdisciplinary Research on Climate Version 5 (MIROC5), has been used for the present comparison study in order to simulate the anthropogenic CO_2 emissions over the BoB under the Representative Concentration Pathways (RCP) 4.5 and RCP 8.5 during the 2010s and 2020s. The MIROC-ESM used in the present investigation was developed by the Japan Agency for Marine-Earth Science and Technology (JAMSTEC) in collaboration with institutions such as the University of Tokyo and the National Institute for Environmental Studies (NIES) and was selected for the ability to simulate the anthropogenic carbon dioxide emissions at the horizontal resolution (latitude × longitude) of around 1.4° × 1.4° (Watanabe et al. 2011).

RCPs used in the Fifth Assessment Report of the Intergovernmental Panel on Climate Change, considered as one of the latest generations of scenarios, are largely used to provide inputs to the climate models in the global context (Bjørnæs 2015). RCPs are time- and space-dependent trajectories of concentrations of various GHGs and pollutants resulting from anthropogenic activities, including changes in land use (Bjørnæs 2015). Of the different scenarios projected by the RCP, the RCP 4.5 scenario indicates a moderate emissions scenario and mitigation policies with the most probable baseline scenario where the global mean surface temperature will likely be restricted to 2.6°C by 2100 (van Vuuren et al. 2011). Under the "moderate" effect scenario represented by the RCP 4.5 scenario, it is understood that the marine species may not respond to the increase in emissions. On the other hand, RCP 8.5 denotes the scenario with high emissions and the absence of coherent management where the global mean surface temperature may arrive at 4.8°C by the end of the twenty-first century (van Vuuren et al. 2011; IPCC 2013). For the present analysis considering RCP 4.5 and RCP 8.5 for Odisha, two timescales are considered for, e.g. decadal changes (2010s versus 2020s) and year-wise changes (2020 versus

2021) in order to compare the anthropogenic carbon dioxide emission between RCP 4.5 and RCP 8.5 over the BoB adjacent to the Odisha coast. It has been assumed that the simulation of anthropogenic CO_2 will be higher under the RCP 8.5 scenario compared to RCP 4.5 over the BoB as it is expected to have higher emissions under the strongest warming scenario (IPCC 2013).

18.3 Results and Discussions

The present study compares the anthropogenic CO_2 emissions between two warming scenarios in decade-wise and year-wise over the Odisha coast adjacent to the BoB. The anthropogenic CO_2 concentration was simulated in ppmv from the MIROC5. Besides the comparison, the reduction in the emissions of CO_2 due to the usage of MFAs by mechanised and motorised crafts is also estimated for Odisha.

18.3.1 CMIP5 RCP Scenarios: Decadal

For the period 2011–2020 (Fig. 18.3), the simulated anthropogenic carbon dioxide emission ranged from 401.03 ppmv to 401.04 ppmv for RCP 4.5; on the other hand, CO_2 concentration varied between 410 ppmv and 414 ppmv under the RCP 8.5 scenario over Odisha region. However, the differences in the simulated versus actual anthropogenic CO_2 emissions recorded are of the order of 0.01 ppmv under RCP 4.5, while this was observed to be to the extent of 4 ppmv under the RCP 8.5 scenario.

For the investigation corresponding to the year 2020, the anthropogenic carbon dioxide under RCP 4.5 was simulated at around 411.04 ppmv, while the same ranged from 424 to 428 ppmv under RCP 8.5 (Fig. 18.4). Therefore, the differences in the emission are comparatively insignificant in the case of RCP 4.5 in 2020, while they are of the order of almost 4 ppmv change between RCP 4.5 and RCP 8.5.

In the post-pandemic year 2021, the carbon dioxide emission concentration was observed to be in the range of 413.290–413.291 ppmv under RCP 4.5, while the

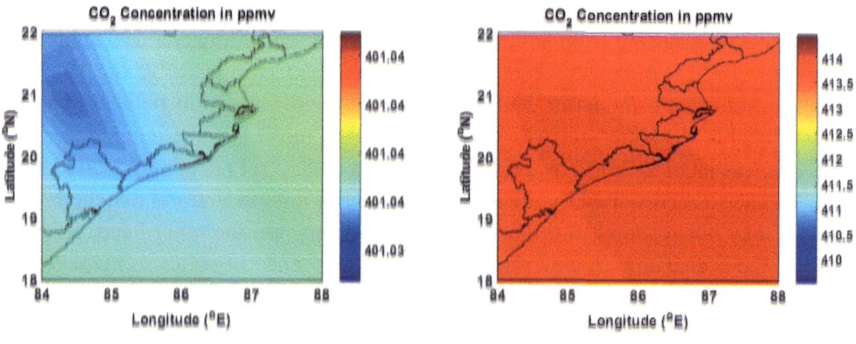

Fig. 18.3 Scenarios in the past decade, 2011–2020, simulated from CMIP5 corresponding to RCP 4.5 and RCP 8.5 for the Bay of Bengal region adjacent to Odisha

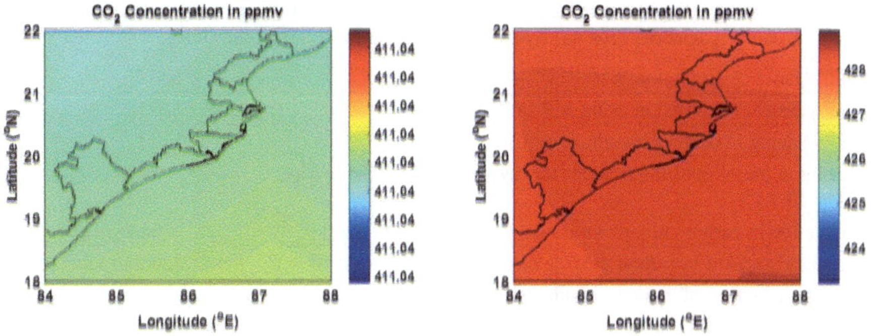

Fig. 18.4 CMIP5 models for RCP 4.5 and RCP 8.5 scenarios simulated for the pandemic year 2020 for Odisha

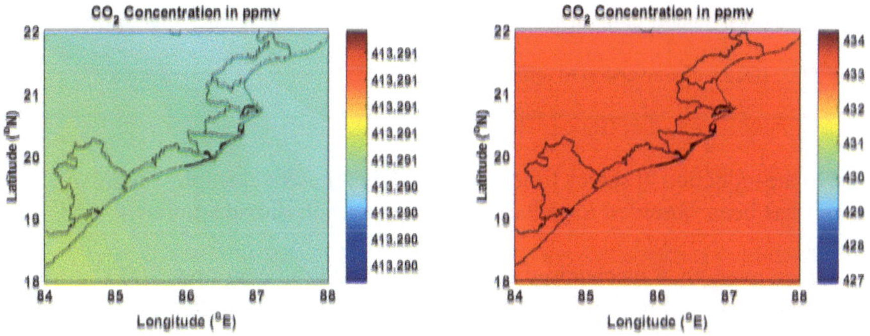

Fig. 18.5 CMIP5 models for RCP 4.5 and RCP 8.5 scenarios simulated for the post-pandemic year 2021 for Odisha

same ranged from 427 ppmv to 434 ppmv under RCP 8.5 in 2021 (Fig. 18.5). For this period as well, the differences under the RCP 4.5 scenario were observed to be very insignificant, while differences in the emissions recorded were of the order of around 7 ppmv under the RCP 8.5 scenario.

The above results indicate that while the decadal changes have been quite significant from 2011 to 2020 for both scenarios, RCP 4.5 as well as RCP 8.5, there is a marked difference between the simulated values for the two scenarios in both 2020 and 2021. The changes are quite insignificant for RCP 4.5, while they are high in the case of RCP 8.5.

18.3.2 Comparative Analyses of the Scenarios in Emission Reduction Versus the Use of MFAs

The search time for the fishing attributed to MFA usages has been reported to be reduced by 30%–70% per trip for mechanised fishing crafts (Tummala et al. 2008). Therefore, the respective consumption of diesel also expectedly reduces for fishing

Table 18.1 Estimation in the reduction of the carbon dioxide (CO_2) emissions due to the usages of Marine Fishery Advisories (MFAs) by the mechanised fishing sector for Odisha

Daily diesel consumption: (A)	300 L
Daily diesel consumption using MFAs: (B)	140 L
Daily saving in diesel consumption: (C = A–B)	160 L
Saving in diesel consumption per trip by assuming a 10-day trip: (D = 10 × C)	1600 L
Reduction in carbon dioxide emission per trip: (E = D × n_1)	4208 kg/trip
Annual reduction in carbon dioxide emission for one craft: (F = E × n_2)	126,240 kg/craft
Annual reduction of carbon dioxide emission for Odisha: (G = F × n_3)	220,667,520 kg/year
Annual reduction of carbon dioxide emission for Odisha: (H = G/n_4)	220,667.5 tonnes/year

$n_1 = 2.63$ kg/L (reduction in carbon emission per litre of diesel saved)
$n_2 = 30$ (number of annual trips undertaken as obtained from field survey)
$n_3 = 1748$ (total number of mechanised crafts operated in Odisha as of 2020 from CMFRI census report)
$n_4 = 1000$ (factor of conversion to tonnes)

due to the reduction of search time as reported therein. For example, saving of 1 L of diesel has been observed to reduce 2.63 kg of carbon dioxide emissions (NCAER 2015).

On an average, a mechanised fishing craft in Odisha consumes 300 L of diesel fuel per day in order to conduct fishing expeditions in the open sea while not utilising MFAs to reach PFZ. If the fishers can effectively make use of the MFAs for searching fish shoals, 160 L of diesel can be estimated to be saved per day. In this way, 1600 L of diesel can be saved per trip for a mechanised craft for a 10-day trip (Table 18.1). Therefore, a total volume of 48,000 L of diesel will be saved in a 1-year time span for one mechanised craft in Odisha assuming 30 trips per year (excluding the fishing ban period). Therefore, the reduction in the carbon dioxide emission is estimated in the present investigation to be 126,240 kg considering the reduction of 2.63 kg of carbon dioxide per litre of diesel saving due to the usage of MFAs. In a similar way (as detailed in Table 18.1), the reduction in carbon dioxide due to the usage of MFAs in Odisha will be 220,667,520 kg considering 1748 mechanised crafts are currently in operation in Odisha as reported CMFRI census (CMFRI-DoF 2020).

18.3.3 Futuristic Projections for 2030

The futuristic projection for the anthropogenic emissions of carbon dioxide has also been simulated from the MIROC5 model during 2021–2030 (Fig. 18.6). The simulated emissions under the RCP 4.5 scenario ranged from 423.917 ppmv to 423.918 ppmv; however, in the case of the RCP 8.5 scenario, the emission is

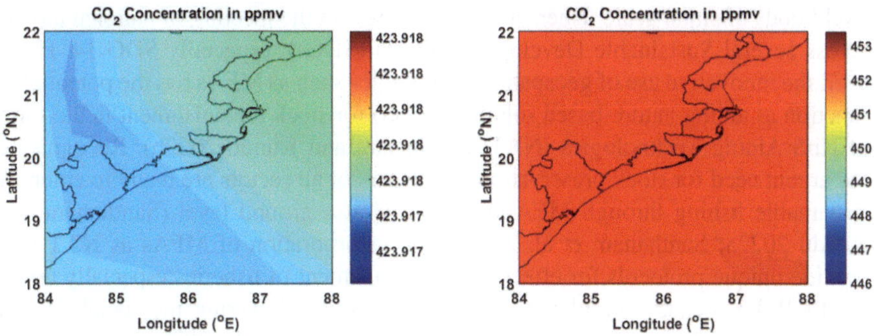

Fig. 18.6 Projected emissions from CMIP5 model for RCP 4.5 and RCP 8.5 scenarios for the period of 2021–2030

observed to be higher and ranged from 446 ppmv to 453 ppmv. Therefore, the differences in the RCP 4.5 scenarios within the ranges of emissions may not be significant for the present decade, while those for RCP 8.5 in the order of 7 ppmv are expected to show high increases in the emissions as shown in Fig. 18.6.

18.3.4 Discussions

It must be noted that though the differences in the ranges of anthropogenic carbon dioxide emission under RCP 4.5 are not significant in the case of the future projection too, the simulated emission is higher than that of the past decade under the same forcing scenario. On the other hand, the simulated emission under the warmest scenario was higher than that of RCP 4.5, and also the differences in the ranges varied significantly, from 4 ppmv to 7 ppmv for the Odisha region.

While it was evident during the pandemic year (2020) that the mechanised fishing crafts were not operated for a few months due to a nation-wide lockdown in India (Kundu and Santhanam 2021), in all probability, the CO_2 emissions may have been reduced attributed to the less operational crafts. However, it is difficult to quantify the extent to which the CMIP5 model may have simulated the reduction in the emissions at annual scales, although a slight enhancement in CO_2 emissions in the BoB in the post-pandemic years as projected in the model simulations appears to signify the rise in emissions from the periods of low emissions (2020, with successive pandemic lockdown periods) to elevated levels cumulatively (2021 post-pandemic lockdown year; Figs. 18.5 and 18.6).

It is essential to balance economic developmental needs with basic human needs such as food security and poverty alleviation; however a higher number of fleets of mechanised crafts with lengthier fishing trips being operated in the BoB will likely enhance the emissions of carbon dioxide in the coming decade and creating challenges for India to meet the net-zero emission commitment. On the other hand, the counter-productive impacts of the loss of fish stocks due to the disturbance of the marine environment as well as emissions can negatively impact the livelihoods of small-scale fishers as well and prevent the meeting of national targets across several Sustainable Development Goals (SDGs), especially SDG 14. However, the innovative use of geospatial technologies such as MFAs has the potential to function as aiding nature-based solutions (NbS) for stock replenishment in the form of NbS-Aiding Technologies (NAT; Santhanam and Kundu 2022b). Recognising the urgent need for stock preservation, the fishers of all sectors are ready to embrace sustainable fishing through MFA technologies at a ground level (Santhanam and Kundu 2022a; Santhanam et al. 2022). The incorporation of MFAs as NAT will provide unique protocols for effective co-management of fisheries especially under the RCP 8.5 scenario simulated in the present study, lowering the social cost of carbon for Odisha as well as India (Santhanam and Kundu 2022a, b). This can provide effective marine policy handles to formulate evidence-based policy frameworks at a national level to implement the use of critical, space-based technology products as MFAs with significant outcomes for fisher communities and marine ecosystem service regulation.

Further, the use of MFAs to lower emissions can be useful to implement measures related to emissions from vessel fleets in the ocean in conformance to the international legislations promulgated by 'The International Convention for the Prevention of Pollution from Ships, and subsequently amended by the protocol of 1978, the "MARPOL 73/78". The use of MFAs provides policy enablement for national conformance to the 2005 Amendment of MARPOL 73/78 which deals with the "Prevention of Air Pollution from Ships"' described in Annex VI, 2005 (Mantoju 2021).

18.4 Conclusions

In the present study, cumulative anthropogenic carbon dioxide emissions were modelled for the BoB coast near Odisha, under decadal (2011–2020; 2021–2030) and annual scales (2020 pandemic scenario and 2021 post-pandemic scenario). In general, the emissions are projected to be higher under the RCP 8.5 scenario than that of under RCP 4.5. Under the circumstances, the contribution of the large-scale mechanised fishing boats operating in the Odisha coast can add significant emissions to those of the BoB. The present study illustrated an evidence-based approach to assess the same, given that the total of 1748 mechanised crafts operated (approximately) in Odisha can potentially emit up to 413,752 tonnes of carbon dioxide every year if fishing expeditions are not optimised.

However, the proper usage of MFAs can help in reducing the search time for the identification of PFZ and minimising the diesel consumption for both large trawler and mechanised fishers. The carbon dioxide emission due to the operation of mechanised crafts in Odisha can also be decreased by approximately 53% under regular usage scenario; fishers for arriving at the PFZ aided by MFA have the potential to reduce carbon dioxide emissions by approximately 220,667.5 tonnes/year (220,667,520 kg/year) promoting sustainable fishing for Odisha alone.

In order to facilitate the regular usage of MFAs, incorporation of suitable policies for MFA-based sustainable fishing operations through public, private and institutional partnerships will be advantageous. The reduction in the carbon dioxide emissions will be co-beneficial to improve the sustainability of the marine ecosystem, species survival and stock preservation/diversification. The present investigation provided an example to illustrate that a robust targeted policy framework to make MFAs accessible and useable to all the marine fishers in India is critical to achieving lower emissions from industrialised fishing activities for India, at the same time with the opportunity to conform to international maritime regulations such as MARPOL 73/78 (Annex VI; 2005).

Acknowledgement The authors wish to thank *Padma Shree* Prof. S. Ayyappan, Chancellor, Central Agricultural University, Imphal, and Chairman, Karnataka Science and Technology Academy, Bengaluru, for his inspiration to study the current gaps in policy-based research for coastal and marine environments in India. SKK acknowledges the immense support of mechanised craft fishers in Odisha for providing valuable information on diesel consumption and other trip-related parameters for the research during the field survey for his doctoral research. The authors acknowledge the technical and infrastructural support provided to conduct research on technology-related policymaking at MAHE Bengaluru campus.

Conflict of Interest Authors declare no conflict of interest for this publication.

References

Asian Development Bank (2020) Reducing carbon emissions from transport projects. https://www.oecd.org/derec/adb/47170274.pdf

Bjørnæs C (2015) A guide to Representative Concentration Pathways. CICERO Center fior website: https://cicero.oslo.no/en/posts/news/a-guide-to-representative-concentration-pathways#:~:text=The new approach is built, including changes in land use. Accessed 27 Apr 2022

Climate Action Tracker (2020) Decarbonising the Indian transport sector: pathways and policies. https://climateactiontracker.org/documents/832/CAT_2020-12-09_Report_DecarbonisingIndianTransportSector_Dec2020.pdf

CMFRI-DoF (2020) Marine fisheries census 2016—Odisha, Kochi

CMFRI-FSI-DoF (2020) Marine fisheries census 2016—India, Kochi

Department of Fisheries (2020a) Handbook on fisheries statistics: 2020. http://dof.gov.in/sites/default/files/2021-02/Final_Book.pdf

Department of Fisheries (2020b) Pradhan Mantri Matsya Sampada Yojana (PMMSY). https://pmmsy.dof.gov.in/new-download

ESSO-INCOIS (2020) Potential Fishing Zone (PFZ) Advisory. Earth System Science Organization—Indian National Centre for Ocean Information Services website: https://incois.gov.in/MarineFisheries/PfzAdvisory. Accessed 27 Apr 2022

FRAD-CMFRI (2022) Marine Fish Landings in India 2020, CMFRI Booklet Series No. 25/2022. http://eprints.cmfri.org.in/15781/1/Marine Landings in India 2020.pdf

IPCC (2013) Summary for policymakers. In: Stocker TF, Qin D, Plattner GK, Tignor M, Allen SK, Boschung J, Midgley PM (eds) Climate hange 2013: the physical science basis. Contribution of working group I to the fifth assessment report of the intergovernmental panel on climate change, pp 3–29. https://www.ipcc.ch/site/assets/uploads/2018/02/WG1AR5_SPM_FINAL.pdf

Jones C, Robertson E, Arora V, Friedlingstein P, Shevliakova E, Bopp L, Tjiputra J et al (2013) Twenty-first-century compatible CO2 emissions and airborne fraction simulated by CMIP5 earth system models under four representative concentration pathways. J Climate 26(13): 4398–4413. https://doi.org/10.1175/JCLI-D-12-00554.1

Kavi Kumar KS, Anneboina LR, Bhatta R, Naren P, Nath M, Sharan A, Pednekar S et al (2016) Valuation of Coastal and Marine Ecosystem Services in India: Macro Assessment (No. 35). https://doi.org/10.13140/RG.2.2.22944.17926

Kundu SK, Santhanam H (2021) All pain and no gain: factors impacting local and regional sustainability due to COVID-19 pandemic with respect to the Indian marine fisheries. Curr Res Environ Sustain 3:100086. https://doi.org/10.1016/j.crsust.2021.100086

Kundu SK, Santhanam H, Srikanth R (2020) A technical assessment of the use of current geospatial technologies to derive marine fishery advisories in india and the way forward. In: Asian conference on remote sensing (ACRS2020), p 10. https://a-a-r-s.org/proceeding/ACRS2020/ifomqo.pdf

Mantoju CD (2021) Analysis of MARPOL implementation based on port state control statistics. J Int Maritime Saf Environ Affairs Shipping 5(3):132–145

NCAER (2015) Economic Benefits of Dynamic Weather and Ocean Information and Advisory Services in India and Cost and Pricing of Customized Products and Services of ESSO NCMRWF & ESSO-INCOIS. https://incois.gov.in/documents/ImpactAssessment-NCAER2015.pdf

Reuters (2021) India set to exceed emission cut targets. The Hindu: Business Line. https://www.thehindubusinessline.com/news/science/india-set-to-exceed-emission-cut-targets/article36027706.ece

Ritchie H, Roser M, Rosado P (2020) CO2 and greenhouse gas emissions. OurWorldInData.org website: https://ourworldindata.org/co2-and-other-greenhouse-gas-emissions. Accessed 27 Apr 2022

Santhanam H, Kundu SK (2022a) Assessment of socio-technical constraints of marine fishers in the utilisation of marine fishery advisories in southern Odisha, India. Anthropocene Sci. https://doi.org/10.1007/s44177-022-00014-4

Santhanam H, Kundu SK (2022b) Nature-based solutions (NbS) for sustainable development of the resource base and ecosystem services of marine and coastal ecosystems of India. In: Dhyani S, Basu M, Santhanam H, Dasgupta R (eds) Blue-green infrastructure across Asian countries. Springer, Singapore. https://doi.org/10.1007/978-981-16-7128-9_15

Santhanam H, Dhyani S, Benedict X (2022) Perspectives on reducing anthropogenic interferences and mainstreaming nature-based solutions for sustainable restoration of Pulicat lagoon, India: from research to policy and implementation. Mar Freshw Res. https://doi.org/10.1071/MF21242

Subramanian S, Manjulekshmi N, Narendra Pratap S, Janhavi K, Tejaswini P, Pastta MF (2014) A manual on the use of potential fishing zone (PFZ). In: ICAR Research Complex for Goa. https://doi.org/10.13140/2.1.5169.0565

Taylor KE, Stouffer RJ, Meehl G a (2012) An overview of CMIP5 and the experiment design. Bull Am Meteorol Soc 93(4):485–498. https://doi.org/10.1175/BAMS-D-11-00094.1

Tummala SK, Masuluri NK, Nayak S (2008) Benefits derived by the fisherman using potential fishing zone (PFZ) advisories. In: Frouin RJ, Andrefouet S, Kawamura H, Lynch MJ, Pan D, Platt T (eds) Remote sensing of inland, coastal, and oceanic waters, vol 7150, p 71500N. https://doi.org/10.1117/12.804766

UNEP (2022) GOAL 14: life below water. United Nations environment programme website: https://www.unep.org/explore-topics/sustainable-development-goals/why-do-sustainable-development-goals-matter/goal-14. Accessed 28 Apr 2022

United Nations (2022) Goal 14: conserve and sustainably use the oceans, seas and marine resources. United Nations website: https://www.un.org/sustainabledevelopment/oceans/. Accessed 28 Apr 2022 van Vuuren DP, Edmonds J, Kainuma M, Riahi K, Thomson A, Hibbard K, Rose SK et al (2011) The representative concentration pathways: an overview. Clim Change 109(1–2):5–31. https://doi.org/10.1007/s10584-011-0148-z

Vivekanandan E, Singh VV, Kizhakudan JK (2013) Carbon footprint by marine fishing boats of India. Curr Sci 105(3):361–366. http://eprints.cmfri.org.in/9900/#:~:text=In Indian marine fisheries%2C the,to 3.60 mt in 2010

Watanabe S, Hajima T, Sudo K, Nagashima T, Takemura T, Okajima H, Kawamiya M et al (2011) MIROC-ESM 2010: model description and basic results of CMIP5-20c3m experiments. Geosci Model Dev 4(4):845–872. https://doi.org/10.5194/gmd-4-845-2011

Zickfeld K, Solomon S, Gilford DM (2017) Centuries of thermal sea-level rise due to anthropogenic emissions of short-lived greenhouse gases. Proc Natl Acad Sci 114(4):657–662. https://doi.org/10.1073/pnas.1612066114

19 Impacts of Pollution on Tropical Montane and Temperate Forests of South Asia: Preliminary Studies by Postgraduate Students in India and Sri Lanka

K. Preeti, Malsha Tejhani, Vasundhara Pandey, Vedika Dutta, Piyali Das, Buddhika Weerakoon, Sudipto Chatterjee, Hemanthi Ranasinghe, and Sarath Nissanka

Abstract

Tropical montane and temperate forests of South Asia, specifically India and Sri Lanka, were studied for symptoms of pollution by postgraduate students from the respective countries. The study focused on multiple abiotic parameters to document deposition of pollutants on forests, centralizing on a particular pollution-sensitive species called lichens as the biotic component. Remotely sensed pollution data was extracted to estimate ground air pollution values, pH of bark, soil chemistry and lichen tissue nitrogen content which was collected over the identified sample sites in both countries. All tests together gave baselines for the forests' state of health. Independent tests on soil pH, conductivity, nitrate and others in both countries did not express any trend. Bark pH measured in Sri Lanka was higher in value than reported literature indicative of deviation from normal. Total nitrogen accumulation in lichen thallus from India was highest in most anthropogenically disturbed sites and least in lichens collected from interiors of the forest. In Sri Lanka, the lichen species, especially the pollution-sensitive ones, were highest in number and expressed growth forms farthest from the city, consequently having the lowest ambient pollution. These studies were compiled together as research findings conducted by postgraduate students through the funding from UNU—ProSPER.net—under the overarching support of SANH.

K. Preeti · V. Pandey · V. Dutta · S. Chatterjee (✉)
TERI School of Advanced Studies (TERI SAS), New Delhi, India
e-mail: s.chatterjee@terisas.ac.in

M. Tejhani · P. Das · H. Ranasinghe
University of Jayewardenepura, Colombo, Sri Lanka

B. Weerakoon · S. Nissanka
Peradeniya University, Kandy, Sri Lanka

S. Dhyani et al. (eds.), *Ecosystem and Species Habitat Modeling for Conservation and Restoration*, https://doi.org/10.1007/978-981-99-0131-9_19

Keywords

South Asia · Pollution · Forests · Nitrogen · Lichen

19.1 Introduction

In the times of climate change crisis, we have more problems than solutions. The concept of planetary boundaries (Rockström et al. 2009) offers a good metric to understand which aspects of our environment are out of balance. Nitrogen biogeochemical cycle is beyond the zone of uncertainty leading to over-enrichment of air, land and water with nutrients known to be limiting on Earth (de Vries 2021). Nitrogen has been the dominant gas but chemically intern as dinitrogen (N_2) makes up 78% of the atmosphere. Reactive nitrogen or Nr was exclusively formed through Biological Nitrogen Fixation (BNF) (Bobbink et al. 2010). This always kept the concentration of usable nitrogen in depleted amounts until Haber's process was discovered and popularized through fertilizers in the green revolution (Galloway et al. 2003). Now, the sources of Nr are beyond the Haber's process as the burning of fossil fuels also commonly releases SOx and NOx (Stevens et al. 2020).

Countries still in their industrial age, mostly developing countries, use fossil fuels such as coal and petroleum as their primary source of energy. This put developing countries at risk of degrading their biosphere integrity disproportionately than other parts of the world. Areas that are eco-sensitive are at a greater threat of harm. The Himalayas is one such site that offers a unique habitat for unique organisms. It is considered to have lower levels of pollution given its altitude, forest cover and lower urban pockets than flatlands owing to its difficult terrain. It is, therefore, home to many pollution-sensitive organisms. The temperate forests of the Himalayas have an abundance of an organism arising from a symbiotic association of fungi and algae called lichens. They have been historically used for biomonitoring air pollution. During the industrial revolution in the UK, it was found that lichens disappeared from locations close to sources of pollution. This led to the discovery of their sensitivity to SOx and NOx pollutants. Western Himalayas has been recently observed to show a sharp boom in economic growth and infrastructural development. It has led to increased tourism, urbanization, development of railways, industries and even improved farming supplemented with fertilizers. The expected impacts are disturbing the hydrological cycle, air pollution and land and water degradation leading to overall environmental deterioration. Its proximity to the Indo-Gangetic plains is also problematic since the entire region has been declared a global pollution hotspot. It is one of the most polluted sites in the world owing to the highly fertile soil that encourages heavy agriculture and industrialization. These factors made Uttarakhand, a state under the western Himalayas, a suitable study location. The tropical montane forests of Sri Lanka, however, are yet to reach such elevated levels of pollution. Therefore, they offer a good contrast also is a landscape unexplored for lichens and their capacity to indicate polluted sites. This research has opened that avenue by selecting three forests with varying proximity to a popular city called Kandy.

The comparative studies in the above-mentioned locations use lichens as the primary focus but measure several environmental pollution indicative parameters as an accessory to overall findings. The objective is to:

(a) Identify visible changes in lichen tissue nitrogen due to increased Nr pollution in and around the forests of Himalayas as well as Sri Lanka.
(b) Use accessory samples such as satellite imagery, bark and soil pH and lichen diversity to support the findings from lichen tissue composition.

19.2 Methodology

19.2.1 Study Area

Indian samples were collected from inside and outside forest at Chamoli district in the Garhwal division of Uttarakhand, 30.42°N 79.33°E, which belongs to the Lesser Himalayas region. Inside Chamoli, seven community forests were identified, namely, Jhinjhoni, Sankot, Bamiyala, Khanoli, Kimoli, Ghes and Koti. Samples from these sites were collected each month. A few sample gaps occurred due to COVID-19 restrictions as well as severe flooding and landslides in the study area but were consolidated to the best of scientific reasoning (Fig. 19.1).

In Sri Lanka, the forests selected were Udawatta Kele forest reserve, Gannoruwa forest reserve and Hanthana forest reserve close to Kandy city. Two transects in Udawatta Kele forest reserve and three transects in each Gannoruwa and Hanthana forest reserve were laid down. Transects were of the size 100 m × 5 m and were located at two places in Udawatta Kele forest reserve and three places of each Gannoruwa and Hanthana forest reserve so as to distribute from the most disturbed areas (close to the city) to far from the city. Each transect was created so as to have 2.5 m to either side, and the samples were collected within this range (Fig. 19.2).

19.2.2 Specific Analyses

The methodologies followed are described under several heads as follows:

The study considered four broad indicators of pollution to explore within the scope of the current study as follows:

(a) Identification of areas undergoing high and low pollution through remote sensing
(b) Delineation and exploration of N deposition on dominant forest trees of study locations
(c) Study of the soil chemistry with a focus on acidification due to pollutants

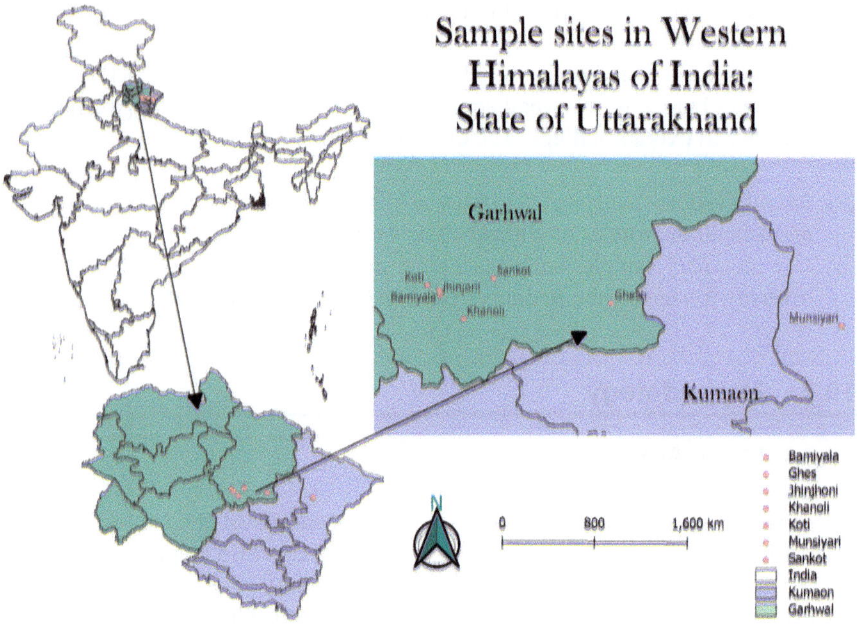

Fig. 19.1 Map showing study locations in India

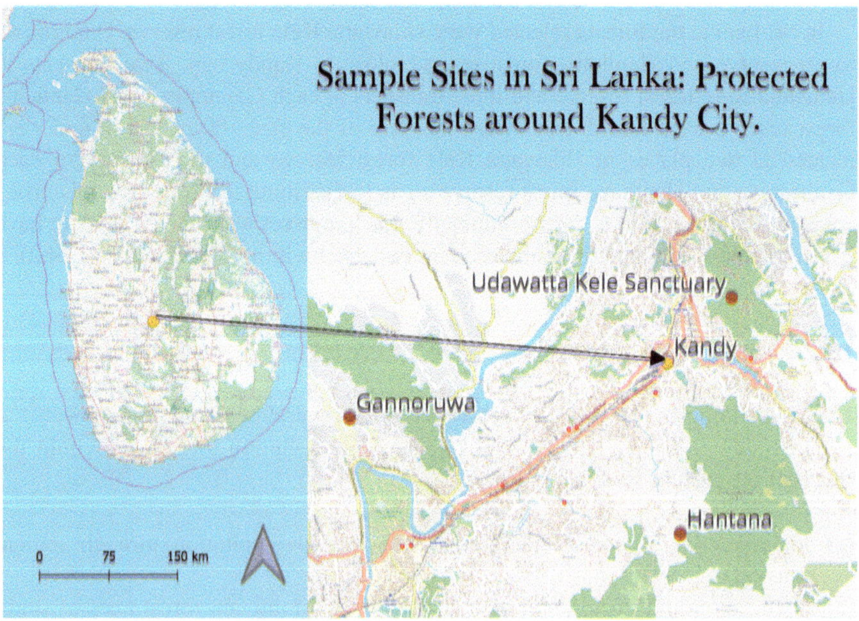

Fig. 19.2 Map showing study location in Sri Lanka

(d) Study of the bark pH and association of diversity of species reported to be sensitive to pollution

All these tests were conducted in seven community forests in Uttarakhand, India, and in three forests in Sri Lanka. The study had different methodologies depending on the level of restrictions on travel due to the COVID-19 pandemic. The sampling spanned across 12 months (February 2021–March 2022) in India with monthly sample collection of lichen, soil and twigs. In Sri Lanka transects were laid in all three forests, and one set of samples was collected from each transect.

Air pollution data—The air pollution data was acquired using Giovanni, a NASA-powered satellite product visualization tool. India acquired data for the state of Uttarakhand on the gaseous pollutants NO_2, SO_2 and PM2.5. Sri Lanka acquired satellite data on O_3, SO_2 and PM 2.5 for the forests studied. They also acquired data from government pollution monitoring department for the city of Kandy.

Lichen—In India, samples were collected from two villages: the village of Jhinjhoni and an eco-trek spot called Khaliya in Munsiyari village. From Jhinjhoni two landscapes were selected: inside the forest, called Inside Jhinjhoni, and outside the forest of Jhinjhoni called Outside Jhinjhoni. The reserved land for protection and eco-tourism in Munsiyari called Khaliya is referred to as Munsiyari everywhere in this paper hereafter. These locations offered three unique landscapes—inside forest for unpolluted air, outside forest for air with anthropogenic influence and eco-trek Munsiyari lying in between with little human influence due to treks but not a lot as it is also a protected site. Total nitrogen (TN) and dN15/dN14 were performed on lichens using the instrument IRMS (isotope-ratio mass spectrometer). This was to provide insight into the N deposition in the forests. The samples were sent to Birbal Sahni Institute of Palaeosciences (BSIP) Lucknow, India, as they had the facility for IRMS testing. In Sri Lanka, lichen specimens were observed through the light microscope (Nikon, SM5 621323). Characters such as thallus colour, shape, thallus structure, margins of lobes, presence/absence of isidia, soredia, external cephalodia, cyphellae, pseudocyphellae, cilia, rhizines, ascomata and conidiomata were observed under 40× magnification. The image of each lichen thallus was taken with clear details of the thallus and reproductive structures. Samples were then classified according to the microscopic observations. Species were identified using available field guides (Weerakoon 2015). Furthermore, lichens were classified according to indicator species of air pollution based on the available literature and findings of previous studies (Perlmutter 2010; Will-Wolf et al. 2015; Weerakoon 2013).

Soil quality—In India, convenient soil samples ($n = 53$) were collected both from inside and outside forest plots that were corresponding to the trees from which lichens, the primary pollutant indicator, were collected. The sample collection spanned from June to November 2021 despite restrictions associated with severe landslides and the pandemic, and the sampling effort throughout the study period was enhanced with utmost care. The collection process was facilitated by the members of community who worked with the Uttarakhand Youth and Rural

Development Centre (UYRDC), Chamoli. The soil samples were air-dried, sieved using a 2-mm sieve and mixed, and then 40 g of each was utilized for nitrate analysis via UV-Vis spectrophotometry at Food Analysis and Research (FARE) Labs Pvt. Ltd., Gurugram, Haryana, India. Twenty grams of each sample was used to form a 1: 2 mixture (Radojevic et al. 2007) for pH and conductivity analysis which was conducted at the Environmental Monitoring Lab, TERI SAS, using a pH meter "Thermo Scientific™ Orion™ 4-Star pH/ISE Benchtop Multiparameter Meter" and a standard conductivity meter. In Sri Lanka, four composite soil samples from each plot were collected using a soil auger from a depth of 0–30 cm at random places within the transect for the soil analysis. Samples were collected into sampling bags, and they were labelled with the name of the transect and the sample number and taken to the laboratory for soil sample analysis. In Sri Lanka, soil's electrical conductivity in 1:5 ratio, pH in 1:2.5 ratio, soil organic carbon (SOC) using ignition method and soil K content was measured as described by Warncke and Brown (1998) and Motsara and Roy (2008).

Twig/bark test—In India, unbranched twigs of approximately 1 cm diameter and 6 cm length were collected from trees in the study location. The trees were selected at random, and the twigs were collected from an accessible height. The twigs were cut to a length of 6 cm to ensure uniformity. The ends were then sealed with petroleum jelly to ensure that the extract is only from the outside of the bark. The pH of the bark was measured as the pH of the unbuffered aqueous solution in contact with the bark. The twigs were soaked in deionized water for 24 h, and the pH was recorded from the extract using a standard pH electrode. This methodology is adapted from Wolseley et al. (2009). In Sri Lanka bark pH analysis was done as mentioned by Kricke (2002). Pieces as thin as possible were removed from the surface of the bark on the tree using a knife or a chisel in order to measure only the outermost bark layer which has the closest association with the epiphytic vegetation. Then, 0.5 g of sampling materials was used for the analysis. Tiny bark samples were put into a 5-mL stoppered flask, and then bark samples obtained from different plants were soaked separately in distilled water with a constant volume of 5 mL for approximately 1 h at 80 °C. Finally, the pH value of each sample was measured using a standard pH meter (Consort C6010 3.3).

19.3 Results and Discussions

19.3.1 Study on Air Pollution Trends to Identify the Pollution Zones

From the maps it is evident that NO_2 is quite prominent as a pollutant, followed by PM2.5. The minimum and maximum values in these maps are the highest and lowest the state of Uttarakhand has seen in the decade 2010–2020. All the values in between are plotted as equal intervals in classes of 5. SO_2 has very low concentrations when visualized in the context of whole India but is shown to have high and low pollution areas when just Uttarakhand is brought into perspective. Polluted areas for SO_2 don't correspond with NO_2 and PM indicating that their source might be different as the

Table 19.1 Trend analysis for major pollutant in Uttarakhand

Pollutants	M-ktau	Two-sided p-value	Comments
NO_2	0.13	0.63839	Upward slope but statistically insignificant
PM2.5	0.127	0.64043	Upward slope but statistically insignificant
SO_2	−0.382	0.11947	Downward slope but statistically insignificant

high pollution areas are not near the valleys. PM 2.5 and NO_2 both have high pollution at low elevation which decreases as we move towards the north or high altitudinal areas. Man-Kendall test revealed that there exists a trend, but none of them are statistically significant. The atmospheric pollutants NO_2 and PM2.5 are actually moving upwards but have statistically insignificant p-value (Table 19.1).

The NO_2 and PM2.5 can be attributed partly to the high developmental activity, vehicular and other fuel-related emissions and partly the stubble burning that occurs every year in the Indo-Gangetic plains. The noticeable downward trend in all the graphs from 2019 is due to the COVID-19-linked lockdowns on all activities and movements.

In Sri Lanka, the air pollution is measured in two sets. The first is for the forests through remote sensing. It can be noticed here through the graph that the three locations have very similar pollution levels for NO_2 owing to the coarse resolution of satellite data. However, Hanthana forest shows consistently lower pollution level than the other two forests. Hanthana is also located farthest from Kandy city.

The data for Kandy city is however more variable.

The graphs are for NO_2, CO and PM10, respectively. It shows an evident dip in pollution levels for all pollutants during all quarters of 2021. It is attributable to the lockdown due to COVID-19.

19.3.2 Study on Lichen Tissue

The chlorolichen species, i.e. *Usnea* sp., *Everniastrum* sp., *Parmeloids* and *Ramalina* sp., are charted separately for location-wise analysis as they were found in all three sites, whereas cyanolichen was only found in Munsiyari.

Figure 19.6 shows total nitrogen accumulation in the four sampled chlorolichen species, and we find that Outside Jhinjhoni with most human interference, the value is highest and inside the forest it is lowest. Munsiyari has intermediate values. Figure 19.3 however does not give any conclusive inference for the dominant pollutant in our study location as the $\delta 15N$ values vary for location with every lichen species. This could be due to sampling since convenient sampling as per availability of lichen also means that distance from source of pollutant could not be standardized within the same location also for these lichen species. Figure 19.4 shows the levels of total nitrogen and corresponding $\delta 15N$ for them. The cyanolichens studied are *Peltigera* sp., *Loberia* sp. and *Leptogium* sp. They are consistently seen from Figs. 19.4 and 19.5 to have higher levels of total tissue nitrogen, owing to their nitrogen fixation ability due to the presence of heterocyst.

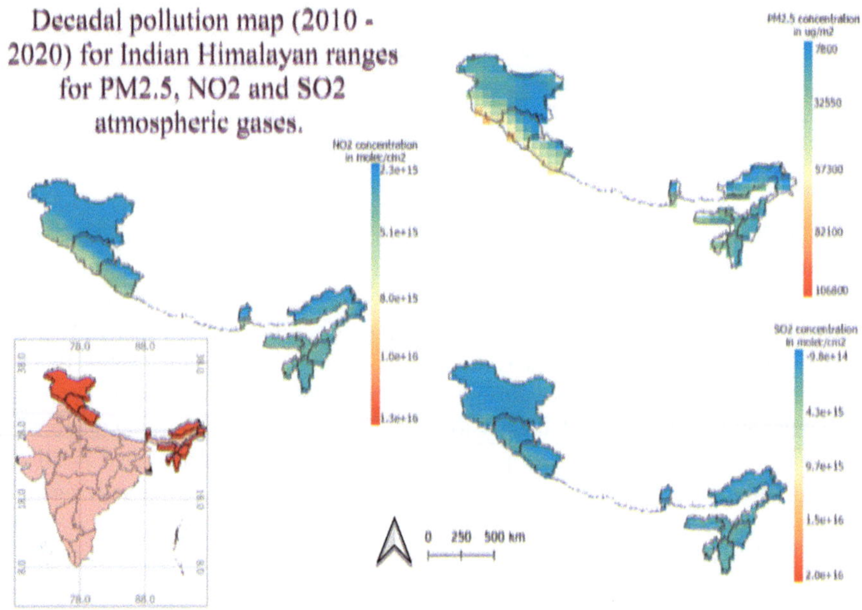

Fig. 19.3 Decadal pollution map of the region under study

These specialized cells could also be the reason for lower δ15N values as the N-fixation would have occurred from N_2 gas in the atmosphere which is 99.63% of δ14N (Diaz-Alvarez et al. 2018). Figure 19.6 is a dot plot showing a significant correlation of $R = 0.69$ between δ15N and total tissue nitrogen. It shows that the higher the total nitrogen in the tissue, the less negative will the δ15N value is expected to be, which says that more than emissions of NH_3 from agriculture and livestock and vehicular emission of NOx dominate the study locations. Higher negative values with high accumulation of Nr are seen when agriculture is a dominant activity, but here evidently fuel combustion, biomass burning and vehicular pollution are more dominant (Fig. 19.7).

19.3.3 Study on Lichen Diversity and Growth

The forests of Sri Lanka studied diversity of lichens under the pretense that location with higher pollution would be less habitable and would inherently have lower diversity than lesser polluted location. They also checked each lichen against pollution indicator list and found that the forest Hanthana, being the farthest from the city, also had the most diversity in forms as seen in Fig. 19.8 of lichen and the highest number of pollution-sensitive indicator lichen species as marked in Table 19.2.

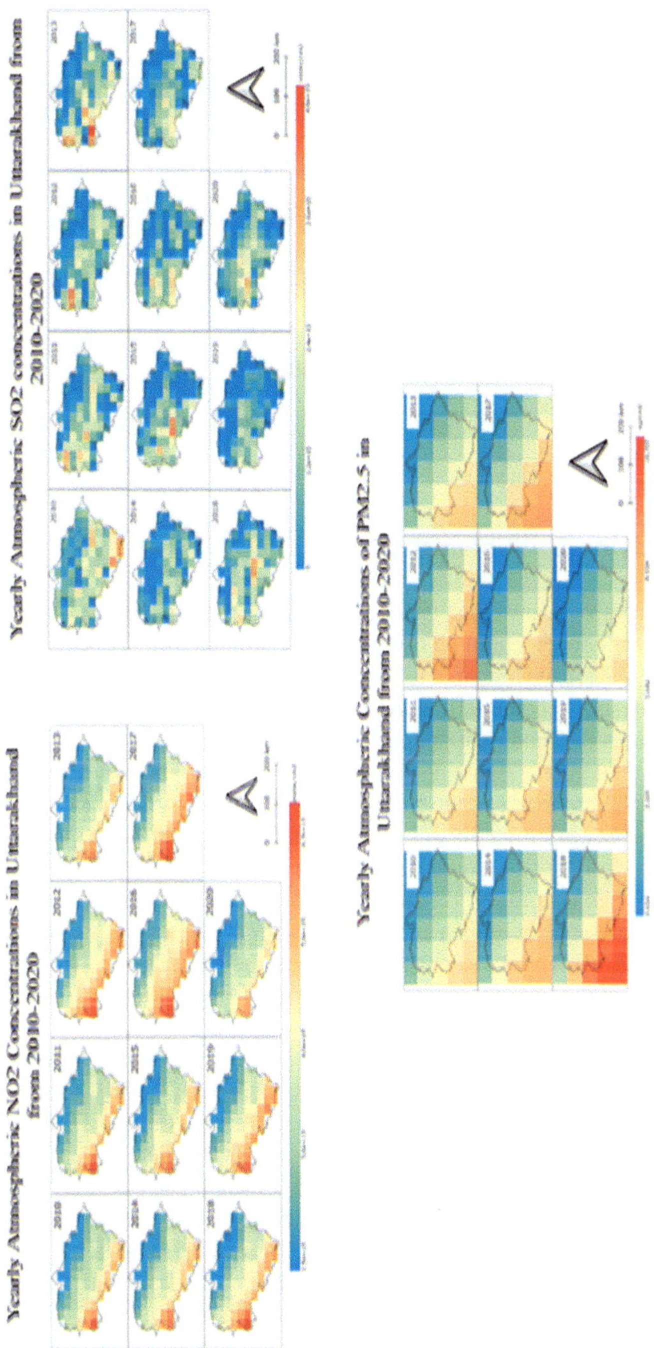

Fig. 19.4 Yearly pollutant concentration in the state of Uttarakhand, India

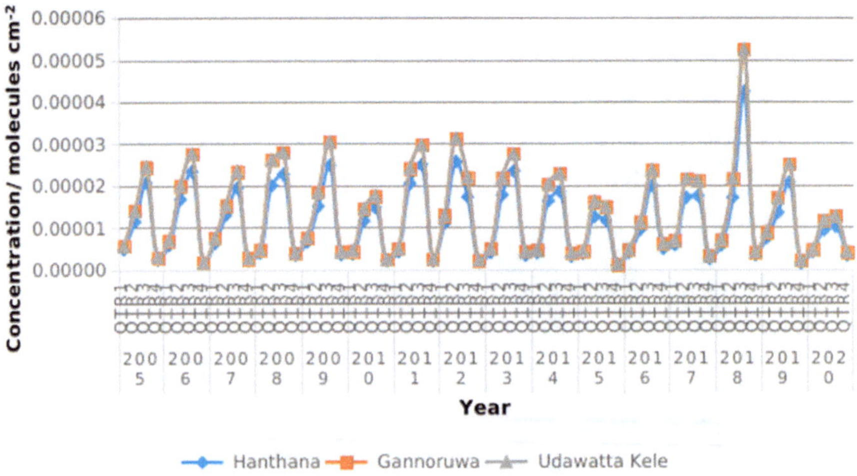

Fig. 19.5 Air pollution in study sites of Sri Lanka

19.3.4 Study on Soil Chemistry

From the line graph below (Fig. 19.9), a clear distinction in the nitrate level of soil inside and outside forest could be observed over the months. The level of nitrate is higher outside the forest compared to the soil nitrate of samples collected from inside the village forests. A potential explanation for the same could be that the forest canopy intercepts the incoming atmospheric nitrogen (Kennedy 2003); thus, inside the forest, vegetation may be acting as a buffer, but on the other hand, no such buffer is present outside the forest as the canopy cover is relatively less.

Higher amounts of nitrate in soil outside the forest could also be asserted to the fact that it is closer to the source of pollution such as fields that use Nitrogen, Phosphorus and Pottasium (NPK) fertilizers and vehicular pollution that is complementary to the presence of human and their activities in the region. A general decline in the nitrate level over the time period is also being reflected, yet it could be attributed to continuous improvement and increase in the sampling effort during the study.

A paired Wilcoxon or Mann-Whitney U test was performed to compare the medians of the nitrate level in both the regions using the "wilcox.test()" function in R studio as per which the median value of nitrate in soil was 20.5 mg/kg and 51.5 mg/kg inside and outside forest, respectively. The p-value for this comparative test was $p = 0.125$ making the difference non-significant, and thus the null hypothesis that the nitrate level inside and outside is statistically similar cannot be rejected.

Soil pH both inside and outside forest samples seemed to be in the same range, and hence, nearly overlapping with median value for the same was 4.83 and 5.09. The soil of the sites which sustain pristine oak forests naturally also lies in on the acidic side, thus corresponding to their general range. These values also correspond

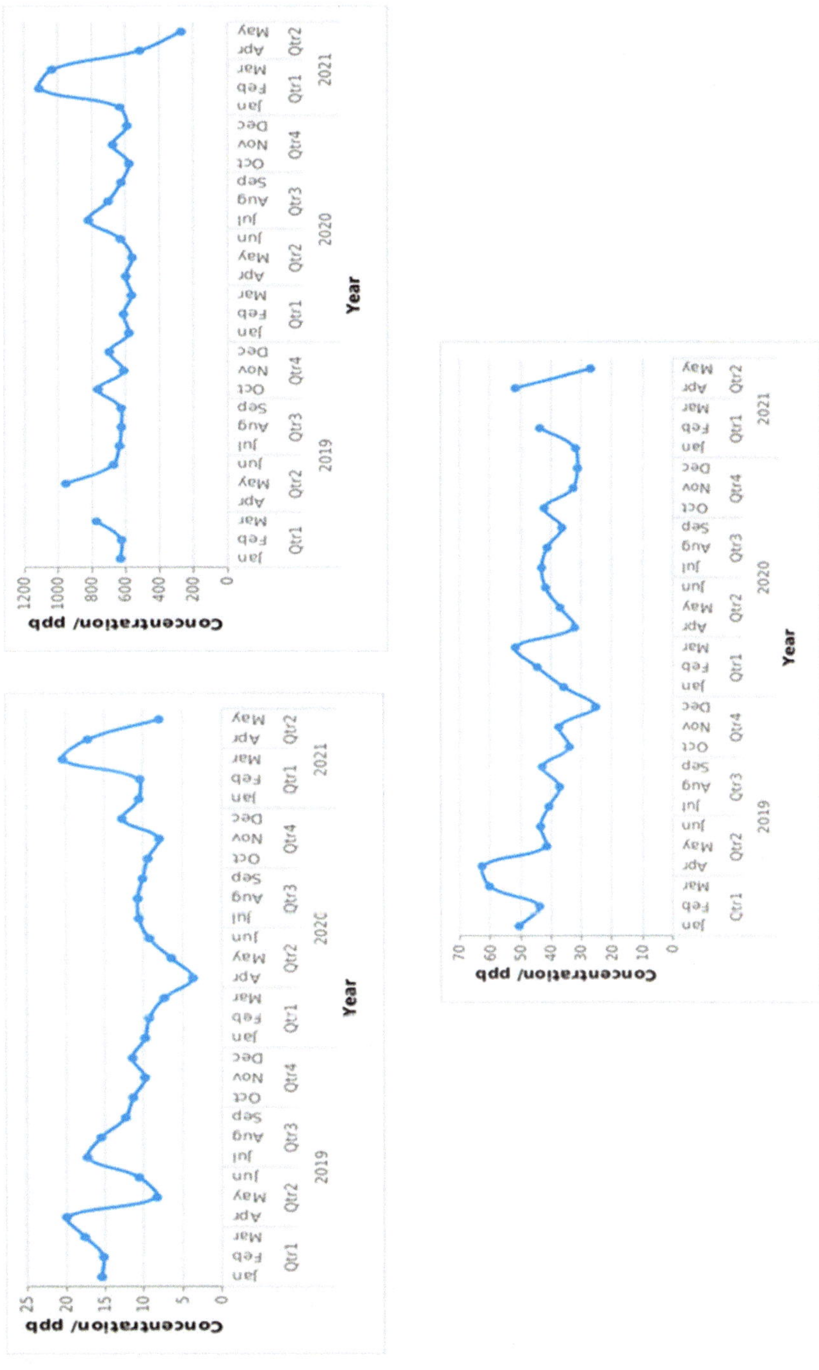

Fig. 19.6 Air pollutant concentrations over Kandy city in Sri Lanka

Fig. 19.7 Graphs showing N accumulation in lichen tissues

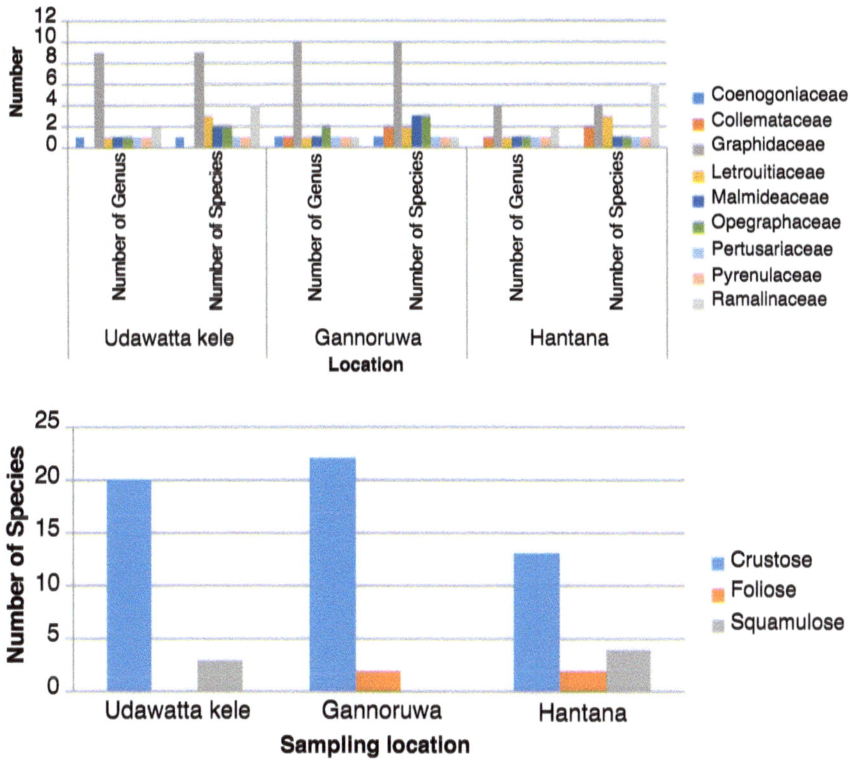

Fig. 19.8 Lichen species diversity in Sri Lanka

to the pH range of Garhwal region characterized by previous studies (Sheikh et al. 2010) (Fig. 19.10).

The soil electric conductivity is comparatively higher in the samples collected and analysed from outside forest sample locations which could be attributed to the higher nitrate levels. The incoming ammonia in the soil is converted into nitrate during which H+ ions are released which may be one of the contributing variables to the higher electric conductivity.

The median values for this parameter inside and outside forest are 0.261 and 0.226 S/m as determined by Wilcoxon test, yet it is statistically similar (p-value = 0.812) (Fig. 19.11).

In Sri Lanka, the soil pH and the electrical conductivity of each transect of Gannoruwa, Hanthana and Udawatta Kele forest reserves was tested. Out of this the average pH of transect one at Udawatta Kele sampling site was the lowest; however, the electrical conductivity was the highest of all. The average soil pH of the transects of the sampling sites ranged from G2 > G1 > G3 > U2 > H1 > H2 > H3 > U1 whereas the average electrical conductivity ranged from U1 > H1 > U2 > H2 > G2 > G1 > G3 > H3. This gave no clear trend however

Table 19.2 Lichen presence/absence survey in the forests of Sri Lanka

Family	Genus	Species	Udawatta Kele	Gannoruwa	Hanthana
Coenogoniaceae	*Coenogonium*	*Coenogonium* sp.	+	+	
Collemataceae	*Leptogium*	*Leptogium cyanescens*		+	+
Collemataceae	*Leptogium*	*Leptogium austroamericanum*		+	+
Malmideaceae	*Malmidea*	*Malmidea granifera*	+	+	+
Malmideaceae	*Malmidea*	*Malmidea barkeri*		+	
Ramalinaceae	*Phyllopsora*	*Phyllopsora* sp.	+		+
Ramalinaceae	*Phyllopsora*	*Phyllopsora* sp. *1*	+		
Ramalinaceae	*Phyllopsora*	*Phyllopsora* sp. 2	+		
Ramalinaceae	*Phyllopsora*	*Phyllopsora* sp. 3			+
Ramalinaceae	*Phyllopsora*	*Phyllopsora* sp. *4*			+
Ramalinaceae	*Phyllopsora*	*Phyllopsora* sp. 5			+
Ramalinaceae	*Bacidia*	*Bacidia millegrana*			+
Total		12	5	5	8

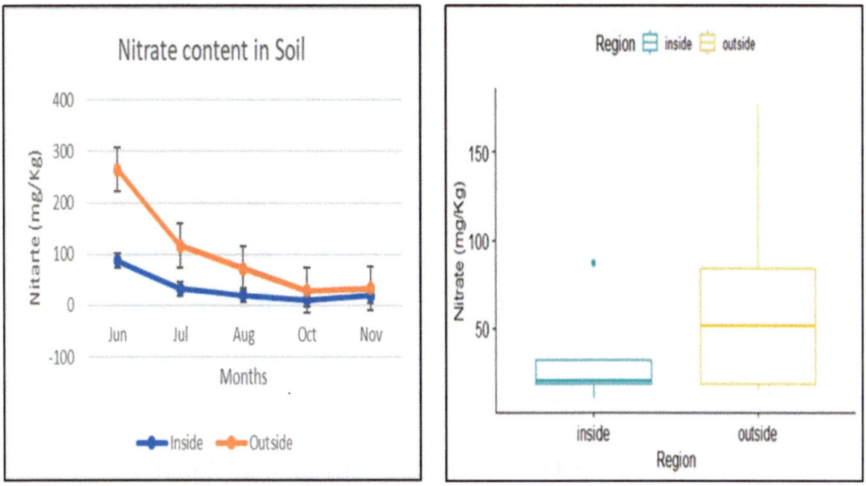

Fig. 19.9 Graphs showing soil nitrate content during June to November in Uttarakhand during the study period

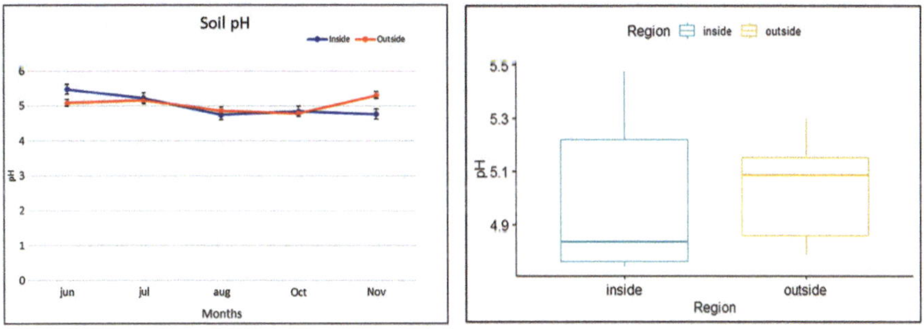

Fig. 19.10 Graphs showing soil pH range in Uttarakhand

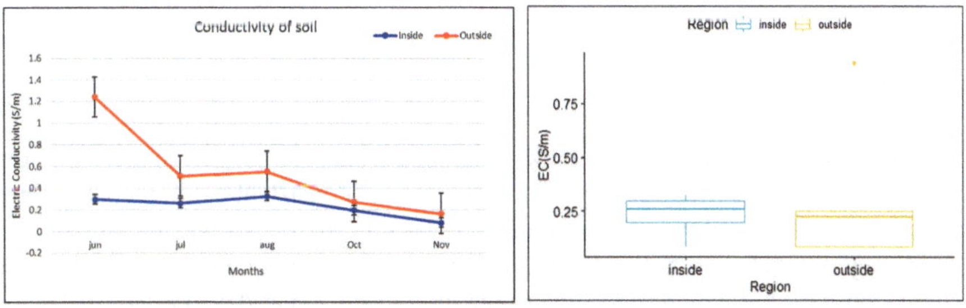

Fig. 19.11 Graphs showing soil conductivity in Uttarakhand during the study period

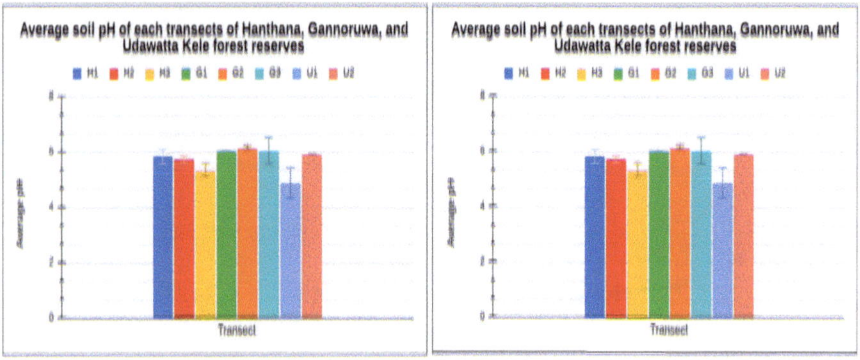

Fig. 19.12 Soil pH range in the sampling sites of Sri Lanka

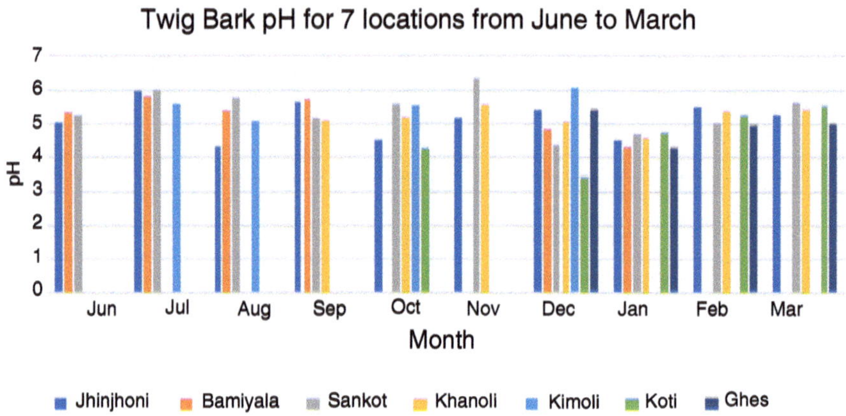

Fig. 19.13 Bark pH range for twig sampled from Uttarakhand

established a range of soil pH from 6 to 4 and conductivity ranging from 74 to 30 μS/cm (Fig. 19.12).

19.3.5 Study on Bark pH

Monthly test in India reveals the bark pH ranged from 3.38 to 6.32 in the 9 months from June 2021 to March 2022. This finding is consistent with the literature; therefore no abnormality was observed. The bark in Sri Lanka was measured, and their range lay between 5.4 and 6.13. This reading does not coincide with the past studies on bark pH which report the range at 2.2 to 4.7 (Grodzińska 1979). Therefore, the bark pH in Sri Lanka is more neutralized than expected. This can be attributed to many naturally occurring factors such as age of the tree, duration of storage of the bark, the level of insect infestation and the height of samples collected.

It can also be explained using ammonia pollution which is rising in both urban and natural landscapes. Given its basic nature, ammonia deposition can also be responsible since the existing literature is based on data recorded decades ago. But to definitively provide a reason, an extensive study would be required (Fig. 19.13).

19.4 Conclusion

Pollution studies are particularly important in not just urban areas to study the extent of pollution but also in locations that are remote and considered pristine. Indian Himalayas and tropical forests and Sri Lanka were observed to be clear examples that showed signs of pollution. In India high pollution levels were observed. Lichen thalli showed Nitrogen deposition. This is in conformity with existing literature. (Diaz-Alvarez et al. 2018). In Sri Lanka it was the bark pH that has turned basic as compared to the expected acidic value. The diversity and abundance of pollution lichens in forest Hanthana, farthest from city pollution, also indicate that pollution from neighbouring locations can stress even the pollution absorbing and purifying forests and the diversity within them. The study in both countries was conducted by university students to understand consequences of pollution through practical approach and benefit from cross-country knowledge exchange as well. Adding value to these datasets through ecosystem modelling is a vital step following sample data collection. Climate change and air pollution are highly interlinked, and effects of both biotic and abiotic components are evident. Mathematical modelling helps understand the importance of changes in pH, tissue N and soil chemistry can have and showcase broader implications they may have on ecosystems.

References

Bobbink R, Hicks K, Galloway J, Spranger T, Alkemade R, Ashmore M, Bustamante M, Cinderby S, Davidson E, Dentener F, Emmett B, Erisman J-W, Fenn M, Gilliam F, Nordin A, Pardo L, De Vries W (2010) Global assessment of nitrogen deposition effects on terrestrial plant diversity: a synthesis. Ecol Appl 20(1):30–59. https://esajournals.onlinelibrary.wiley.com/doi/abs/10.1890/08-1140.1

de Vries W (2021) Impacts of nitrogen emissions on ecosystems and human health: a mini review. Curr Opin Environ Sci Health 21:100249

Diaz-Alvarez EA, Cisneros RL, De E (2018) Biomonitors of atmospheric nitrogen deposition: potential uses and limitations. [online] ResearchGate. https://www.researchgate.net/publication/323727023_Biomonitors_of_atmospheric_nitrogen_deposition_Potential_uses_and_limitations. Accessed 17 Jan 2022

Díaz-Álvarez EA, Lindig-Cisneros R, de la Barrera E (2018) Biomonitors of atmospheric nitrogen deposition: potential uses and limitations. Conserv Physiol 6(1):coy011. https://academic.oup.com/conphys/article/6/1/coy011/4931295?login=true

Galloway JN, Aber JD, Erisman JW, Seitzinger SP, Howarth RW, Cowling EB, Cosby BJ (2003) The nitrogen cascade. BioScience 53(4):341. https://academic.oup.com/bioscience/article/53/4/341/250178

Grodzińska K (1979) Tree bark—sensitive biotest for environment acidification. Environ Int 2(3): 173–176

Kennedy F (2003) How extensive are the impacts of nitrogen pollution in Great Britain's forests? Forest Res Ann Rep Acc 2002–2003:66–75

Kricke R (2002) Measuring bark pH. In: Monitoring with lichens—monitoring lichens. Springer, Dordrecht, pp 333–336

Motsara M, Roy RN (2008) Guide to laboratory establishment for plant nutrient analysis. Food and Agriculture Organization of the United Nations, Rome

Perlmutter GB (2010) Bioassessing air pollution effects with epiphytic lichens in Raleigh, North Carolina, USA. Bryologist 113(1):39–50

Radojevic M, Bashkin V, Bashkin VN (2007) Practical environmental analysis. The Royal Society of Chemistry, Cambridge, pp 300–303

Rockström J et al (2009) A safe operating space for humanity. Nature 461(7263):472–475

Sheikh MA, Kumar M (2010) Nutrient status and economic analysis of soil in oak and pine forests in Garhwal Himalaya. J Am Sci 6(2):117–122

Stevens CJ (2020) The impact of air pollution on terrestrial managed and natural vegetation. Philos Trans A Math Phys Eng Sci 378:20190317. https://doi.org/10.1098/rsta.2019.0317

Warncke D, Brown JR (1998) Potassium and other basic cations. In: Brown JR (ed) Recommended chemical soil test procedures for the north central region, NCR publication no. 221. Missouri Agricultural Experiment Station, Columbia, MO, pp 31–33

Weerakoon G (2013) Some environmental factors influencing diversity of corticolous lichens in selected disturbed and undisturbed vegetation types in Knuckles Mountain range in Sri Lanka, A thesis submitted to the University of Sri Jayewardenepura for the award of the degree of doctor of philosophy in botany on lichenology. University of Sri Jayawardenapura, Sri Lanka

Weerakoon G (2015) Fascinating lichens of Sri Lanka, Dilmah conservation. Ceylon Tea Services PLC, Sri Lanka, 184 pp

Will-Wolf S, Jovan S, Neitlich P, Peck JE, Rosentreter R (2015) Lichen- based indices to quantify responses to climate and air pollution across northeastern USA. Bryologist 118(1):59–82

Wolseley PA, Leith ID, Dijk and Sutton MA (2009) Macrolichens on twigs and trunks as indicators of ammonia concentrations across the UK—a practical method. In: Sutton MA, Reis S, Baker SMH (eds) . Springer, Dordrecht, pp 101–108. https://doi.org/10.1007/978-1-4020-9121-69

Selection of Strategic Sampling Sites for River Quality Assessments Near Mined Areas as a Policy Handle for Low-Impact Development and Biodiversity Conservation: A Case Study of River Godavari

20

Jahnavi Sharma and Harini Santhanam

Abstract

Low-impact development (LID) synchronous with green infrastructure (GI) development is largely being perceived as a favoured approach for sustainable development and biodiversity conservation around mined out lands. While GI-based pathways can provide adequate insights for achieving the targets of land degradation neutrality in the case of mined lands, its crucial and apparent link to the water environment needs consideration to achieve LID in practical terms. The impacts of mining close to riverine environments are known to have spatially far-reaching effects. To detect the same, review the sampling strategy to arrive at the best-possible sampling strategy for effective restorative practices. Supporting this idea, the present study highlights the need of a geospatial modelling and visualization approach to review and design strategic sampling sites with case of River Godavari. The configuration facilitates the rapid investigation of adverse impacts of industrial pollution on land, in water or in air, and it directs the re-evaluation of distribution of the sampling locations to other land uses across the river. Such strategic sampling can provide data as policy handles to plan the nature of response (urgent, short or sustained) and spatial extent of eco-restoration of land and biodiversity.

J. Sharma
Wildlife Institute of India, Dehradun, India

H. Santhanam (✉)
Department of Public Policy (DPP), Manipal Academy of Higher Education (MAHE), Manipal, Bengaluru Campus, Bengaluru, India
e-mail: harini.santhanam@manipal.edu

S. Dhyani et al. (eds.), *Ecosystem and Species Habitat Modeling for Conservation and Restoration*, https://doi.org/10.1007/978-981-99-0131-9_20

Keywords

River Godavari · Water quality · Sampling distribution · GIS · Low-impact development

20.1 Introduction

The sixth sustainable development goal (SDG) focuses on clean water and sanitation with eight targets to be achieved by 2030. SDG targets 6.3, 6.4, 6.5 and 6.6 stresses on the specific need to protect and restore water systems, to implement integrated water resources management, to substantially increase water use efficiency across all sectors and to improve water quality by reducing pollution. Management of water for better environmental standards is crucial, and as a flowing body, its standards and frequent monitoring is essential to account for the dynamic changes (Hobbs 2008). To ensure the successful operation of the '*triple bottomline*' of societal, economic and environmental sustainability, it is essential to focus on not only the above-mentioned SDGs but also those integrated with the developmental aspects of India. Hence, the perspective of integrating development and eco-restoration is quite relevant in land-water resources management, especially with respect to the mining environment.

For river water quality sampling and monitoring, national mission on Clean Ganga in its five-tier approach is a mega project for river cleaning and rejuvenation project in India. To test water quality as per Central Pollution Control Board (CPCB), there are, at present, a total of 870 water monitoring stations in the country over 218 wells, 189 rivers, 53 lakes, 9 drains, 4 tanks, 3 creeks, 3 canals and 2 ponds. Water pollution changes in the physical characteristics of water such as colour, odour, faecal and organic matter contamination and toxic pollutants such as organic and heavy metals, river ageing, salinization, changes in river hydrology, contamination from agrochemicals and mining activities. The sources of contamination of rivers are classified as direct point sources, diffuse agricultural sources and diffuse urban sources as per the water testing and monitoring authorities in India. In addition to being a source of water, rivers are also sinks . The current norms for water quality set by the World Health Organization (WHO) is more stringent as compared to India, with 200 mg/L and a taste threshold of 500 mg/L (Rickwood and Carr 2017). The European Union[1] revised norms as per the tests conducted on 12 January 2021 for a total of 36 elements in the water such as Acrylamide, antimony, Bisphenol A, lead, nitrate, pesticides and seven biological parameters such as for *Clostridium perfringens* spores, coliform bacteria, enterococci, *Escherichia coli*, Heterotrophic

[1]Drinking water legislation in European Union

https://ec.europa.eu/environment/water/water-drink/legislation_en.html

https://www.consilium.europa.eu/en/documents-publications/public-register/public-register-search/results/?AllLanguagesSearch=False&OnlyPublicDocuments=False&DocumentNumber=5846%2F18&DocumentLanguage=EN

plate counts and somatic coliphages, considering the expected effects of seasonality and temporality of the source waters as well as the contributions of the natural run-offs, groundwater seepages into the rivers. Monitoring effort encapsulates the land-use changes using primary surveys, secondary data, or direct surveys. The CPCB stipulations require water samples to be collected both upstream and downstream of the established sampling points of a river in the jurisdiction of the respective state pollution control board by trained personnel from established monitoring sites. State Pollution Control Board take samples fortnightly (TSPCB, https://tspcb.cgg.gov.in/default.aspx). The collected samples are analysed at EMPRI laboratory (TSPCB, https://tspcb.cgg.gov.in/default.aspx). In 1978, CPCB took lead for water quality testing with the launch of Global Environmental Monitoring System (GEMS) in India. Initially, under GEMS, 24 surface water and 11 groundwater stations were launched. Apart from GEMS, the National Programme of Monitoring of Indian National Aquatic Resources (MINARS) for water quality monitoring started in 1984, with 113 stations spread over ten river basins. The monitoring stations are of three kinds with baseline, trend and impact or flux stations. All the major rivers of the country are included in the monitoring scheme starting from Brahmaputra, Ganga, Cauvery, Krishna, Narmada, Godavari, as well as their tributaries and distributaries. Of the total drainage basin in India, rivers form 82.4% of the drainage basin in the country. Currently, the list of parameters tested for water quality in India stands at 28. The water quality assessment tests water for physico-chemical parameters such as pH, turbidity, hardness of water, presence of nitrogen, phosphorus, potassium, lead, arsenic and heavy metals (TSPCB, https://tspcb.cgg.gov.in/default.aspx-yes). As per CPCB, the 28 parameters include 9 trace metals and 15 pesticides.

For a low-impact development in mining region, bio-retention and nutrient retention is sensitive to changes in pH and soluble salts of calcium and magnesium, whereas bacteria retention and water temperature attenuation could be potentially sensitive to pH, hardness and fertility changes (Dietz 2007). The hardness of water is an important parameter in the regulation of the biogeochemical cycle and nutrient cycling for redeveloping mined areas. Models such as MUSIC, SGWATER, L-THIA-LID evaluate bioretention, permeability pavement, differences in density development, impervious rain barrel, porous pavement and Swale[2] development (Eckart et al. 2017). The major objective of the work is to assess the strategic placement of observational cum sampling sites, which is necessary to plan LID (low-impact development) and LDN (land degradation neutrality) strategies for mined out lands. The study aims to re-evaluate the number of water quality sampling location and its distribution for River Godavari using geospatial technique. Additionally, application of geospatial techniques for monitoring changing conditions and statistical measures for stream network analysis could facilitate determination of sampling distribution in future for remaining rivers (Coraggio 2022; Paul et al. 2019; Brus and Knotters 2008; Sharp 1971).

[2]Swale is a low or hollow place, especially a marshy depression between ridges.

The deleterious impact of mining in a region is reflected by acid mine drainage, heavy metal residues, imbalances in nutrient cycling, poor water quality, loss of aquatic biodiversity, impact of rate of post mining ecological recovery and potential lung as well as other human health conditions (Srikanth and Nathan 2017; Garai and Narayana 2018; Sheoran et al. 2010; Singh 2017; Saini 2016; Ghose 1997). Inland water contamination in coal mine areas is a priority for research, as agricultural and industrial run-offs contaminate surface water (Garai and Narayana 2018). Sampling and monitoring the type, level, nature, persistent or residual activity, hydrological changes and spread of pollutant released in water sources, as well as accounting for the presence of mines, agricultural or other industries, is essential for developing low-impact recovery strategies for land uses in the aforementioned regions (Langereo 2017; Kay et al. 2006; Kilmartin 1989). The study asks a couple of questions: if the number of sampling location is enough for the length of the river; if the spatial spread of the sampling location is justified as per advanced or recent understanding of stream networks or sampling design. It is understood the distance between each sampling point and that of the river has to be ecologically optimum; synchronous sampling, spatial spread, frequency of monitoring sites and interval are essential for redesigning strategic water sampling sites for River Godavari (Brus and Knotters 2008; King and Hamel 2003). The assumption here is that the sampling plan requires revision.

20.2 Study Area

For the present investigation, we take the example of a study along a stretch of the River Godavari (Fig. 20.1) to highlight the scope for improvement in the present sampling network to enable short-term and long-term change detection synchronous to both the water environment (quality in terms of the hardness estimates) and the land environment (as changes in land use) with obvious repercussions to biodiversity conservation. The basin-wide study over an area of 200–400 sq. kilometre could provide closer estimation of sampling location requirement. It would make it easier to cover the length of the river as well. After the established water quality standard, a re-look at spatial spread of the sampling points using basin, watershed and stream order was facilitated to study representative sampling for water quality testing. The site of the study is the length of the River Godavari flowing though Telangana. It is a major water source and accounts for 73,201 km^2 basin for undivided state of Andhra Pradesh and Telangana. It traverses across the mineral-rich areas of Pranahita Godavari valley of the state and passes through the districts of Nizamabad, Adilabad, Karimnagar, Warangal and Khammam before flowing into the Bay of Bengal. In the study area of Godavari coalfields, there are six identified land use types. Overall, 60% of the land is agricultural which includes plantation, crop and fallow land. A series of coal mines are found on both sides of the river. Backed by the studies on impact of coal mines and prescribed by CPCB, environmental parameters are tested in and around coal mines fortnightly. Water quality data is collected and maintained by Telangana State Pollution Control Board (TSPCB) across the length of the river

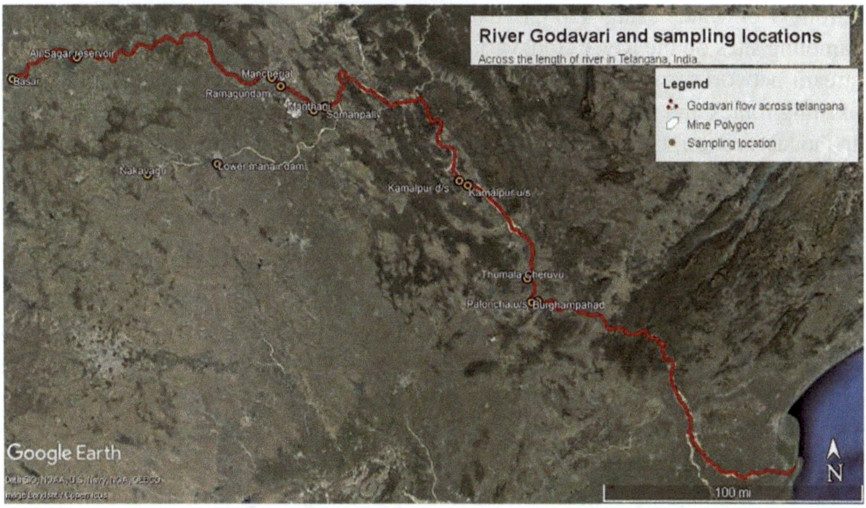

Fig. 20.1 The sampling stations distribution for River Godavari in Telangana where the field survey and studies on site selections were carried out

in Telangana. The industrial profile across the River Godavari in the state of Telangana has been attributed to causing water pollution in the river (Hussain et al. 2017). The predominant industries are coal mines and power plant, along with iron-based industries. The effluents from these industries lead to higher concentrations of heavy metals and other pollutants in the river (George et al. 2010). The study by Garai and Narayana (2018) in the Godavari coalfields has acknowledged increase in barren land as well as increased efforts of afforestation and plantation activity in the mining lease area in the decade ending in 2000.

20.3 Methodology

Freshwater ecosystem is essential for human as well as biodiversity of a region. It supports 7% of global biodiversity (Dudgeon et al. 2006). The mining, agricultural, industrial and day-to-day activities lead to soil, air and water pollution (Srikanth and Nathan 2017; Ghose 1997). To assess the ecological, agricultural, industrial impact on the freshwater ecosystem, a variety of tools are used such as the water quality test for physical, chemical and biological pollution using grab samples and hydrographs; geospatial analysis for the river flow using digital elevation model (DEM)-based stream order changes as well as water quality of riverine system; and indicator-based method to ensure suitability with the ecological health of the river (Davies 2014; WHO 2011; CPCB 2011; USGS 2006; Speight et al. 2004; King and Hamel 2003). The primary mode of water quality assessment for freshwater ecosystem is through on-site and lab-based sampling methods. The methodology adopted in the present study is presented under three sections. In the first section, geospatial mapping of hardness of water using TSPCB data is used to represent spatial distribution of sampling sites on the River Godavari. In the second section, DEM-based analysis of stream networks and stream order is used in the area for visual representation of existing routes and possibility for future strategic sampling for the river ecosystem. In the last section, the synchrony with biodiversity hotspots is tested as a measure to access the ecological suitability in the area.

20.3.1 Mapping of Hardness of Water as a Proxy

First, mapping of River Godavari (line string) across the state of Telangana was done on Google Earth. Then, the monitoring stations on the River Godavari were marked as locations along with data in description. Since TSPCB website provides sampling place names, and not latitude and longitude of sampling locations, the locations were marked as per closest access location to the river from those places. The latitude and longitude of sampling locations were received through Right to Information (RTI) filed on 6 September 2019 and the reply received on 23 September 2019. The sampling locations were verified using the latitude and longitude. The layer was saved as a Keyhole Markup Language (KML).[3] This layer was converted as a vector shape file and processed in Quantum GIS software (QGIS) and Arc GIS for further editing and map composition. In the vector layer obtained, representation of hardness of water was done in incremental size of spherical dots marked. The categorization of data was done according to World Health Organization (WHO) and United States Geological Survey (USGS) classification.

20.3.2 Mapping of Stream Networks

The DEM-based watershed delineation method is used for stream network design, calculating stream order and distances and proximity analysis for sampling sites (Luo 2011; Tarbotan 1991; Sharp 1971). In the second section for representation of stream and drainage in watershed delineation in Arc GIS, Cartosat-1 digital elevation model (DEM) of 30 m resolution at equator was downloaded in .tiff format (https://bhuvan.nrsc.gov.in/bhuvan_links.php). The Cartosat-1 satellite has a pair of Panchromatic cameras having an along-track stereoscopic capability with spatial resolution of 2.5 m in the horizontal plane with a swath of 27 km. A ready DEM of 30 m resolution is made available on Bhuvan (https://bhuvan.nrsc.gov.in/updates/bhuvan_jul2022.html), from which the DEM for present study area (a total of 12 tiles of 1 arc sec) was downloaded. The 12 tiles were stitched using ARC GIS Pro 1 released in 2001 by ESRI provided by UC Davis. For further watershed delineation and creation of drainage network, the following steps were followed.

[3] KML is a file format used to display geographic data in an Earth browser such as Google Earth (https://developers.google.com/kml/documentation/kml_tut).

The open layers plugin, the Google Satellite image, was used as the backdrop of the map for georeferencing and contextualization of the plotted map.

20.3.3 Suitability of Site for Synchrony with Biodiversity Hotspots

With regard to land degradation and freshwater ecosystem loss, assessing ecological and biodiversity health is essential for attaining sustainable development goals. The analysis of site suitability for biodiversity hotspots along the stream is conducted for characteristics such as suitability for quality change, land use, scale, water degradation, land degradation, magnitude of biodiversity change and magnitude of change. This change in suitability for biodiversity hotspot is crucial for green development, LID and LDN. This comparison provides a comparison of ecological health at the stream network.

20.4 Results and Discussions

20.4.1 Strategic Sampling Design to Arrive at Rational Location of Proposed Sampling Sites

Spatial distribution of water sampling is crucial for scale and assessment of water quality (Zhai 2022). While policies related to monitoring and evaluation of riverine ecosystems in proximity to mining environments from conservation perspectives in general indicate the need for a dense network, geospatial analyses of River Godavari indicated that the 100–200 km wide stretch of river did not possess a sampling point (Speight et al. 2004). Hence, through the present study, a rationale for the presence or absence of sampling points could be established with a basin-wide study as 35 sampling stations for a 73,000 km^2 basin river is not quite representative. Accordingly, the enablement of policies related to close-range monitoring of rivers as specified by CPCB, 35 sampling points—seven are clustered in first industrial area and six in the second cluster—were identified through field surveys and synoptic assessments. The areas where sampling points are stationed together happen to be industrial areas and could be studied as a basin for analysis. First area of cluster is dominated by coal mines and thermal power plants and the second area by medium and minor industry as per state level geographic and industrial presence data. A plausible explanation for this is the widely acknowledged and researched fact that industrial units are considered a major source of pollution. Surface mining impacts the physical, sequential and dynamic system of hydrological cycle over a large topography by altering hydraulic conductivity,[4] saturated

[4] *Hydraulic conductivity*: It is the rate of flow under a unit hydraulic gradient through a unit cross-sectional area of aquifer. In simpler terms, it is the material's (soil or plants) capacity to transmit water (http://www.aqtesolv.com/aquifer-tests/aquifer_properties.htm).

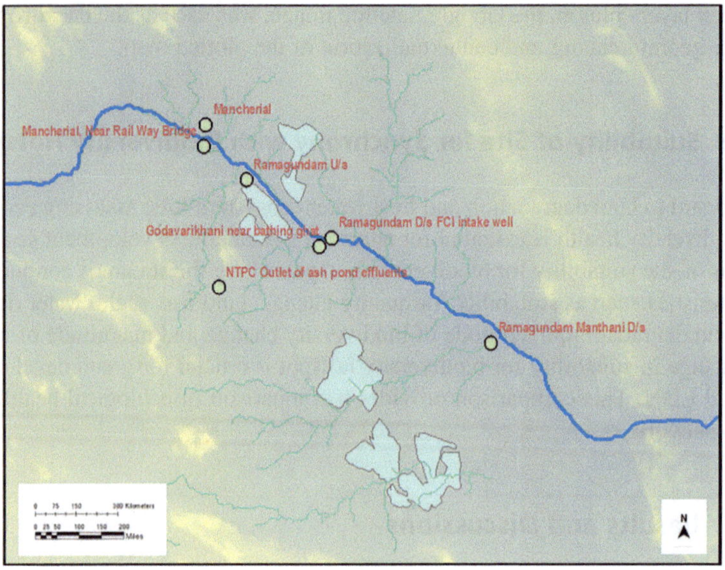

Fig. 20.2 Cluster of sampling station near coal mines (identified as first cluster in the text) where field investigations combined with geospatial analyses lead to identification of these monitoring sites

hydraulic conductivity,[5] macro-porosity and drainage rate at the mines and mine spoil, run-offs as well as biological contamination (Garai and Narayana 2018; Kay et al. 2006; King and Hamel 2003; Kilmartin 1989; Potter et al. 1988; Chong et al. 1985; Ward 1983). However, in order to conduct any comparative study for impact of activities or impact of land uses or precipitation at cotton plot in close proximity to the river or the selective impact of minor dyeing industries close to the river or how much stream and drainage network is apt to capture the difference between an assumed unpolluted stretch of river with that of the polluted one, further analysis could suggest if there is scope of improving the network design which could eventually provide more data points for river water quality analysis (Figs. 20.2 and 20.3).

20.4.2 Revise Number of Water Quality Parameters Sampling Location

The proposed revision for strategic water sampling could establish the number of strategic sampling in addition to water sampling distribution. One of the means to

[5] *Saturated hydraulic conductivity*: It is the quantitative measure of a saturated material's (soil) ability to transmit water when subjected to a hydraulic gradient (https://www.nrcs.usda.gov/wps/portal/nrcs/detail/soils/ref/?cid=nrcs142p2_053573).

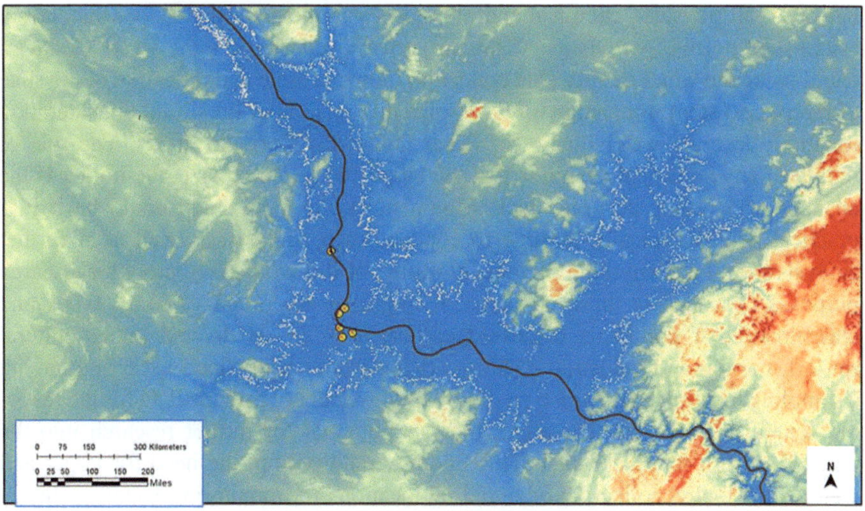

Fig. 20.3 The second cluster of sampling location at the River Godavari where field investigations combined with geospatial analyses lead to identification of the monitoring sites

achieve this is through geospatial representation to corroborate the sediment, soil characteristics, mine or agricultural run-off interaction with routes option (Coraggio 2022; Zhai 2022). In other words, the present geospatial analyses helped to explore whether the current network design resulted in the absence any relevant data. This does not imply sampling each and every metre or kilometre of the river length, rather a study on the network design could provide insights to improve sampling strategies in the country. Another unique aspect of finding this plausible gap is the method through which it was explored, i.e. through geospatial analysis, which provided the synoptic datasets crucial for conservation planning. The gap in network design could be visualized using geospatial tools, whereas in all probability these gaps would have gone unnoticed otherwise. This gap, if studied, could lead to a more comprehensive network design with the presence of more sampling locations for River Godavari and isolate the hotspots of biodiversity decline. In order to design a more efficient spatial distribution of sampling sites, the number of sampling points adds to improved ecosystem services, operational costs and data quality for a river ecosystem using geospatial or statistical techniques (Coraggio 2022; Zhai 2022; Dobriyal 2017).

20.4.3 Detect Water Quality Parameters for Possible Integrated Water Quality Assessment

The proposed revision for strategic water sampling could be extended to all the water quality parameters. The imprecise extent of spatial distribution for sampling water quality could lead to imprecise data collection. For effective data collection of river banks for all physical, chemical and biological parameters tested as per the current established norms. A revision of stream network redesign for the River Godavari could make the water quality data collection more accurate and reflective of the state of river ecosystem. Frequency of data collection is another important consideration for more accurate water quality of River Godavari (Corragio 2022). The rule of thumb of any statistical design is that the larger the sampling size, the higher is the chances of accruing accurate data. Arguments against the study could be if it is practical or economically feasible to conduct such analysis, as well as if further analysis would add anything of value or utility, or if it would be merely for research purposes. To address these concerns, it might be important to recall types and classification of pollutants discussed earlier; they are as diverse as agricultural run-off, mining and industrial residues arising from various land uses, and having a sampling location only for a type of pollutant would not provide data for the rest of the source of pollution. Therefore, there is potential for further research into the network design for river water. This finding could prompt the need for larger and well as distributed collection of sampling locations for monitoring water quality in India. Geospatial visualization is one of the techniques to determine the possible changes in existing water sampling sites. Using a geospatial multiparameter weighted approach could establish an integrated water quality assessment (Zhai 2022).

20.4.4 Synchrony with Biodiversity Hotspots

Table 20.1 shows the suitability of the sites for short-term change detection and long-term ecological monitoring. These sites were identified based on the field surveys in the area combined with the stream analysis.

It is evident from the above table that the NTPC outlet site is quite vulnerable to changes in all the three categories and indicates the need to plan better long-term in situ ash pond restoration or reuse policies. On the other hand, the predominantly high level of influence of the anthropogenic factors such as solid wastes and water pollution near the railway bridge deemed this site quite less suitable for change detection compared to the other, except for the case of land degradation (LD) from possible plastics and solid wastes pollution. It is therefore essential to note from the study that the biodiversity conservation is most essential around sites showing higher magnitudes of change detections. The present study indicates that though the sites vary between their characteristics to indicate LD or water quality changes, they remain central to plan adequate policies for biodiversity conservation as seen from the field-based assessments presented here.

It is important to note that Government of India has proposed sustainable landfill management as well as post-mining operations from land-use as well as water-use perspectives (PIB 2022). Further while, pit lake management has become a scientific methodology to innovate sustainable post-closure voids around the world (Younger and Wolkersdorfer 2004; Wolkersdorfer et al. 2020); recently, Santhanam and Srikanth (2019) discussed the applicability for Indian conditions. However, the

Table 20.1 Analyses of site suitability for long-term and short-term change detection synchronous to the water and land degradation as well as biodiversity changes

		Field-based interpretation of the nature and magnitude of change detection to plan sampling sites						
Proposed site	Cluster	Suitability as a site for water quality change detection *(yes/no)*	Suitability as a site for land-use change detection *(yes/no)*	Temporal scale of detection *(short* vs. *long)*	Expected magnitude of water degradation *(on a scale of 0–5, where 0 = nil and 5 = high)*	Expected magnitude of land degradation *(on a scale of 0–5, where 0 = nil and 5 = high)*	Expected magnitude of biodiversity loss *(on a scale of 0–5, where 0 = nil and 5 = high)*	Total magnitude of change detection *(normalized to a scale of 0–5)*
Mancherial	1	Yes	Yes	Long	3	4	4	3.7
Mancherial near railway bridge	1	Yes	Yes	Short	3	3	4	3.3
Ramagundam U/s	1	Yes	No	Short	4	4	4	4.0
Godavarikhani near bathing ghat	1	Yes	Yes	Short	4	4	5	4.3
Ramagundam D/s FCI intake well	1	Yes	Yes	Long	4	4	4	4.0
NTPC outlet of ash pond effluents	1	Yes	Yes	Long	5	5	5	5.0
Ramagundam Manthani D/s	1	Yes	Yes	Short	4	4	4	4.0

current investigation details that the use of the pit lake methodology alone may not suffice the integrated water resources management solutions in a mining environment in the vicinity of the riverine environments such as River Godavari. Further, despite the above discussions on the role of creating water assets for eco-restoration, the glaring lack of planning with respect to biodiversity conservation, which is critical for developing the natural feedbacks in the post-mining scenario, is apparent in many discourses. Given the circumstance, planning around well-established sampling nodes across a vast spatial scale provides the first-cut beginning towards long-term ecological monitoring of the restoration processes and pathways for sustainable natural resources management (e.g. Dhyani et al. 2022).

20.4.5 DEM-Based Integrated Strategic Water Sampling Design

The hydrological assessment for selecting site selection uses statistical measures, sampling methods, routing assessment and water assessment methods. A few of the contact and non-contact methods such as ultrasonic method, particle image velocimetry method, remote sensing method, electromagnetic method, flame method, weir method are used to establish the strategic site sampling points against operational, cost-effectiveness, accuracy, time-effectiveness, environmental impact (Dobriyal 2017). If the strategic sampling design is made using DEM-based study, it will require pre-requisite survey in the river belt. The DEM-based design for stream order and stream network could establish accurate stream hydrological processes (Tarbotan 1991; Sharp 1971). The DEM-based stream delineation for strategic water sampling design is effective for small scale (Luo 2011). The effectiveness of the DEM-based strategic sampling unit will have to be tested on the ground through the survey of stream network, water control project and routing processes in the River Godavari belt.

20.5 Conclusion

Overall, the present study provided useful discussions on the optimal distances and placement of sampling locations or the network design, which have not been mainstreamed among researchers and/or administrative bodies at a local or regional level with respect to riverine water monitoring in India. While majority of discussions is restricted to quality and standards, without a consensus on the optimal sample design and/or framework for deciding the representative sampling strategy, the objectives of LID and LDN remain unfulfilled. The strategic distribution of the sampling points proved that the notion of placement and distances may be very important considerations to aid rapid change detection across the land–water–air continuum in the case of mined regions. While it is assumed that the idea of sampling an entire river is cumbersome in practice, the use of geospatial methodology for deriving judicious placement of sampling sites as illustrated in the present study can facilitate effective assessments of the coupled land use, water quality and biodiversity of mined regions. The central argument presented here is to re-vision river sampling network design as part of the monitoring policy of mined areas through establishment of strategic sites. It is envisaged that the changes in the existing design could yield better data for water quality and thereby water-related policies in the country.

Acknowledgements We would like to acknowledge Dr. M. B. Rajani, Associate Professor, National Institute of Advanced Studies, IISc Campus, Bengaluru and Kuili Suganya, PhD Scholar, National Institute of Advanced Studies, IISc Campus, Bengaluru for their support and encouragement.

Conflict of Interest The authors declare no conflict of interest.

References

Brus DJ, Knotters M (2008) Sampling design for compliance monitoring of surface water quality: a case study in a polder area. Water Resour Res 44:W11410. https://doi.org/10.1029/2007WR006123

Central Pollution Control Board (2011) Impact of coal mine waste water discharge on surroundings with reference to heavy metals. CPCB, Bhopal. https://cpcb.nic.in/openpdffile.php?id=UmVwb3J0RmlsZXMvMjg0XzE0NzEzMzc5MDlfQ29hbE1pbmVXVzIwMTEucGRm

Chong SK, Becker MA, Hoare SH, Weaver GT (1985) Characterization of reclaimed mined land with and without topsoil. J Environ Qual 15(2):157–160

Corragio E, Han D, Gronow C, Tryfonas T (2022) Water quality sampling frequency analysis of surface freshwater: a case study on Bristol floating harbour. Front Sustain Cities 3:791595. https://doi.org/10.3389/frsc.2021.791595

Dhyani S, Bhaskar D, Santhanam H, Murthy IK (2022) Post-pandemic recovery through landscape restoration. Restor Ecol 30(5). https://doi.org/10.1111/rec.13617

Dudgeon D, Arthington AH, Gessner MO, Kawabata Z-I, Knowler DJ, Lévêque C, Naiman RJ, Prieur-Richard A-H, Soto D, Stiassny MLJ, Sullivan CA (2006) Freshwater biodiversity: importance threats status and conservation challenges. Biol Rev 81(02):163. https://doi.org/10.1017/S1464793105006950

Davies SP, Tsomides L (2014) Methods for biological sampling and analysis of Maine's rivers and streams. Department of Environment Protection, Augusta, ME, pp 1–31

Dietz ME (2007) Low impact development practices: a review of current research and recommendations for future directions. Water Air Soil Pollut 186:351–363. https://doi.org/10.1007/s11270-007-9484-z

Dobriyal P, Badola R, Tuboi C, Hussain SA (2017) A review of methods for monitoring streamflow for sustainable water resource management. Appl Water Sci 7:2617–2628

Eckart K, McPhee Z, Bolisetti T (2017) Performance and implementation of low impact development—a review. Sci Total Environ 607–608:413–443

Garai D, Narayana AC (2018) Land use/land cover changes in the mining area of Godavari coal fields of southern India. Egypt J Remote Sens Space Sci 21:375–381. https://doi.org/10.1016/j.ejrs.2018.01.002

George J, Thakur SK, Tripathi RC, Ram LC, Gupta A, Prasad S (2010) Impact of coal industries on the quality of Damodar river water. Toxicol Environ Chem 92(9):1649–1664. https://doi.org/10.1080/02772241003783737

Ghose MK, Kundu NK (1997) Shelf life of stock-piled topsoil of an opencast coal mine. Environ Conserv 24(1):24–30

Hobbs P, Suzan HH, Rascher J (2008) Management of environmental impacts from coal mining in the upper Olifants river catchment as a function of age and scale. Int J Water Res Dev 24(3): 417–431

Hussain J, Hussain I, Arif MD, Gupta N (2017) Studies on heavy metal contamination in Godavari river basin. Appl Water Sci 7:4539–4548

Kay D, Barbato J, Brassington G, Somer B (2006) Impacts of longwall mining to rivers and cliffs in the southern coalfield. In: Aziz N (ed) Coal: coal operators' conference. University of Wollongong & the Australasian Institute of Mining and Metallurgy, Wollongong, NSW, pp 327–336

Kilmartin MP (1989) Hydrology of reclaimed opencast coal-mined land: a review. Int J Surf Min Reclam Environ 3(2):71–82., Accessed 13 Sept 2019. https://doi.org/10.1080/09208118908944257

King KW, Hamel RD (2003) Considerations in selecting a water quality sampling strategy. Trans ASAE 46:63–73

Langereo AH, Tisseyre B, Roger JM, Scholasch T (2017) Test of sampling methods to optimize the calibration of vine water status spatial models. Preci Agric 19:365–378. https://doi.org/10.1007/s11119-017-9523-8

Luo Y, Su B, Yuan J, Li H, Zhang Q (2011) GIS techniques for watershed delineation of SWAT model in plain polders. 3rd international conference on environmental science and information application technology. Proc Environ Sci 10(2011):2050–2057

Paul R, Brindha K, Gowrisankar G, Tan ML, Singh MK (2019) Identification of hydrogeochemical processes controlling groundwater quality in Tripura, Northeast India using evaluation indices, GIS, and multivariate statistical methods. Environ Earth Sci 78:470. https://doi.org/10.1007/s12665-019-8479-6

Potter KN, Carter FS, Doll EC (1988) Physical properties of constructed and undisturbed soils. Am J Sci Soc 52:1435–1438

Press Information Bureau (PIB) (2022) Steps initiated for sustainable mining. PIB, Delhi. https://pib.gov.in/PressReleasePage.aspx?PRID=1813238. Accessed 4 Apr 2022

Rickwood C, Carr MG (2017) United Nations Environment Programme Global Environment Monitoring System (GEMS)/Water Programme. https://www.un.org/waterforlifedecade/pdf/global_drinking_water_quality_index.pdf. River basin share. Data on Godavari Basin

Saini V., Gupta, R.P., and Arora, M.K. (2016) Environmental impact studies in coalfields in India: a case study from Jharia coal-field; 53(January), 1222–1239

Santhanam H, Srikanth R (2019) Pit lakes as sustainable post-closure interventions for open-cast coal mines in the Indian context. In: Proceedings of World Aqua Congress 2019, 30th–31st October 2019, vol 1, pp 74–85

Sharp WE (1971) A topologically optimum water sampling plan for rivers and streams. Water Resour Res 7(6):1641–1646

Singh G (2017) Environmental issues with best management practice of coal mining in India; http://www.teriin.org/events/docs/gurdeep.pdf. Accessed 6 Mar 2018

Sheoran V, Sheoran AS, Poonia P (2010) Soil reclamation of abandoned mine land by revegetation: a review. Int J Soil Sediment Water 3(2):13

Speight VL, Kalsbeek WD, DiGiano FA (2004) Randomized stratified sampling methodology for water quality in distribution systems. J Water Resour Plan Manag 130:330–338. https://doi.org/10.1061/(ASCE)0733-9496(2004)130:4(330)

Srikanth R, Nathan HSK (2017) Towards sustainable development: planning surface coal mine closures in India. Contemp Soc Sci 13:30–43

Tarbotan DG, Bras RL, Iturbe IR (1991) On the extraction of channel networks from digital elevation data. Hydrol Process 5:81–100

USGS (2006) National field manual for the collection of water quality data. Collection of water samples. US Geological Survey. Techniques of Water Resources investigations, Book 9. Chapter A4. (Version 2.0, 9/2006)

Ward AD, Wells AG, Phillips RE (1983) Infiltration through reconstructed surface mined spoils and soils. Transactions 5:821–829

WHO (2011) Hardness in drinking-water. Background document for development of WHO guidelines for drinking-water quality. WHO/HSE/WSH/10.01/10/Rev/1, Switzerland. https://apps.who.int/iris/bitstream/handle/10665/70168/WHO_HSE_WSH_10.01_10_Rev1_eng.pdf?sequence=1&isAllowed=y. Accessed 13 Sept 2022

Wolkersdorfer C, Nordstrom DK, Beckie RD, Cicerone DS, Elliot T, Edraki M, Valente T, França SCA, Kumar P, Lucero RAO, Soler i Gil A (2020) Guidance for the integrated use of hydrological, geochemical, and isotopic tools in mining operations. Mine Water Environ 39(2):204–228

Younger PL, Wolkersdorfer C (2004) Mining impacts on the fresh water environment: technical and managerial guidelines for catchment scale management. Mine Water Environ 23:s2

Zhai M, Tao Z, Zhou X, Lv T, Wang J, Li R (2022) Water multi-parameter sampling design method based on adaptive sample points fusion in weighted space. Remote Sens (Basel) 14:2780. https://doi.org/10.3390/rs14122780

21 Ecological Niche Modeling Predicts the Potential Area for Cultivation of *Melia dubia* Cav. (Meliaceae): A Promising Tree Species for Agroforestry in India

Suresh Ramanan Sundaram, A. Arunachalam, Dibyendu Adhikari, U. K. Sahoo, and Kalidas Upadhyaya

Abstract

Ecological niche modeling (ENM) aids in delineating and demarcating the distributional range of a species, the potential distributional area of species in the future, and plans for targeted biodiversity surveys in new areas. We elucidate that ENM can also be applied to agroforestry and block plantations. Despite the technical difference between agroforestry and farm forestry, both these concepts encourage the plantation of woody plants in private and farmlands. Typically, the choice of species has a high influence on the economic success of agroforestry and farm forestry. There is a broad range of species that can be grown in these systems, but a diligent recommendation has to be made for the betterment of the farming community. In this study, we have delineated the niche of one of the most popularly promoted tree species, *Melia dubia* Cav. (Meliaceae), for agroforestry using the Maximum Entropy (MaxEnt) model. The model predicts that 3.92% of the total geographical area of India is highly suitable for *M. dubia*. Careful and dedicated tree management practices can be developed to ensure the success of the plantation in the less suitable areas.

S. R. Sundaram (✉) · A. Arunachalam
ICAR-Central Agroforestry Research Institute, Jhansi, Uttar Pradesh, India
e-mail: suresh.s@icar.gov.in

D. Adhikari
Plant Ecology and Climate Change Science, CSIR-National Botanical Research Institute, Lucknow, India

U. K. Sahoo · K. Upadhyaya
School of Earth Sciences and Natural Resource Management, Mizoram University, Aizwal, Mizoram, India

S. Dhyani et al. (eds.), *Ecosystem and Species Habitat Modeling for Conservation and Restoration*, https://doi.org/10.1007/978-981-99-0131-9_21

Keywords

MaxEnt · Niche · Meliaceae · Agroforestry · Extension

21.1 Introduction

India is a net importer of wood owing to the ban on green felling inside natural forests as well as the policy framed to protect the rich biodiversity for future generations. Thus, the country's wood and timber demand are mainly met by the trees outside forests (TOFs) along with the wood imported from other countries. As per the report released by the Centre for Science and Environment under the theme "*Wood is Good: But, is India doing enough to meet its present and future needs?*", TOFs provide the major portion of wood and timber demand. To quote the exact wordings, *In India, TOFs is defined as 'all trees growing outside recorded forest areas. TOF provides the meat of India's timber needs, and agroforestry and farm forestry are the backbones of TOF'* (Shrivastava and Saxena 2017). India became a leading example by adopting the National Agroforestry Policy in 2014, a first of its kind. One of the major goals of the policy was: *Meeting the ever-increasing demand of timber, food, fuel, fodder, fertilizer, fibre, and other agroforestry products; conserving the natural resources and forest; protecting the environment & providing environmental security, and increasing the forest/tree cover, there is a need to increase the availability of these from outside the natural forests* (GoI 2014).

Agroforestry as a land-use practice is very much relevant to United Nations' Sustainable Development Goals (SDGs) as it addresses 12 out of 17 SDGs (Arunachalam and Ramanan 2021). In most developing countries such as India where land is a scarce resource, a large proportion of land cannot be diverted for tree plantation and forestry practices. In this context, agroforestry is a viable option for meeting the increasing wood demand and also meeting the 33% green cover target of FAO (Joshi et al. 2011). Owing to its importance, India adopted the first agroforestry policy in 2014 (Ahmad et al. 2019).

Reviewing the tree plantations and agroforestry practices in the country, a few industrially important tree species such as *Eucalyptus*, *Casuarina*, and *Populus* got the major share owing to the demand from plywood and paper industries (Kulkarni 2013; Parthiban et al. 2019). Due efforts were made to find alternative species such as *Leucaena* and *Gmelina*. However, these efforts could not gain momentum in replacing the industrial tree species. Moreover, even the very fast-growing bamboo species could not replace the industrially valued species because of many reasons such as low pulp recovery and high silica content (Wang et al. 2021). Further, industrial species such as *Eucalyptus* had also gotten into controversial clout of high water usage and groundwater depletion (Morris et al. 2004; Reichert et al. 2021). This warranted a need to find an alternative tree species.

Melia dubia Cav. (Family: Meliaceae) is a native fast-growing tree species with clear bole and is recommended to be promoted as one of the viable choices for plantations and agroforestry (Chavan et al. 2022; Handa et al. 2020). Typically,

Meliaceae can be regarded as the timber family, given that important timber species such as *Azadirachta indica* A.Juss., *Cedrela odorata* L., *Toona ciliata* M. Roem., and *Swietenia macrophylla* King. belong to this family (Gupta et al. 2019). *M. dubia* as an agroforestry tree species can address various SDGs such as no poverty (SDG1) by increasing farmer's income, gender equity (SDG-5) by providing womenfolk with fuelwood for cooking, affordable and clean energy/sustainable energy solutions (SDG-7) wood fuel supply to poor farmers, responsible consumption and production, and ecological footprints (SDG-12) (Arunachalam and Ramanan 2021; van Noordwijk et al. 2018). Thus, this species has been promoted in different parts of the country (Kumar and Joshi 2021). In Kerala, this species has been cultivated on large scale compared to commercial crops such as rubber (Binu and Santhoshkumar 2019). Furthermore, there are clones/varieties released for this species at the country level. However, it has been reported that this species is native to the Western Ghats, so it will be significant to delimit the niche of this species, thereby delineating the areas where this species can be promoted for cultivation.

21.1.1 Ecological Niche Modeling for Agroforestry and Farm Forestry

Ecological niche modeling (ENM) aids in modeling and demarcating the distribution range of the species concerned while also predicting its stable habitats suited climatically under multiple projected climatic scenarios (Adhikari et al. 2018, 2019). Rajpoot et al. (2020) carried out niche modeling for *Boswellia serrata* Roxb. and reported that there is a potential chance for a decrease in the spread of the species in its natural distributions, thus recommending a long-term action plan for the conservation of this species. This sort of finding on the conservation of species based on niche modeling has been done for many floral and faunal species (Majumdar et al. 2019; Mipun et al. 2019; Pradhan et al. 2020).

After 1970s, agroforestry gained momentum as a scientific discipline (Nair et al. 2021). The definition of agroforestry clearly states it as a sustainable land-use practice where woody perennials/trees are grown along with crops and animals in spatial and temporal sequences (Arunachalam et al. 2021). Inherently, farm forestry and agroforestry promote the cultivation of trees in private/farmlands, i.e., trees outside forests. The success of agroforestry and farm forestry is highly dependent on the choice of species and thus ENM can play a significant role in pointing out the suitable area for a particular species. A study in Yunnan province applied ENM for identifying the climate space or niche of ten tree species and predicted the impact of climate change on the distribution of these species (Ranjitkar et al. 2016). Similarly, ENM based on the BiodiversityR package was used to delineate the "always-suitable" distributions of the *Xanthoceras sorbifolium* Bunge. in China (Wang et al. 2017). In Nepal, the potential zone for Himalayan alder species was delineated using Maximum Entropy (MaxEnt) and insisted that only 24% area of Nepal is suitable for *Alnus nepalensis*. All these studies attempted to provide practical solutions to one of the fundamental issues faced in agroforestry: selecting a tree/woody perennial species so that it is suitable for that region.

Thus, the objective of the present study is to model the niche of *M. dubia* and demarcate its present distributional ranges.

21.2 Materials and Methods

21.2.1 Data Collection

Eighty-four species occurrences records of *M. dubia* were compiled from field surveys carried out in the natural distribution range and published literature. The field surveys were carried out in the Western Ghats states such as Gujarat, Maharashtra, Karnataka, Kerala, and Tamil Nadu. A total of 19 bioclimatic variables of the current climatic conditions were used to delineate the niche of *M. dubia* (Table 21.1). All these 19 variables describe the annual trends in temperature and precipitation variability that may act as physiological constraints on the species and determine their geographic distribution (O'Donnell and Ignizio 2012). These variables have been derived from the average monthly climate data for minimum, mean, and maximum temperature and precipitation of the years 1970–2000 (Fick and Hijmans 2017). The dataset was downloaded as geoTiff files of 2.5 min spatial resolution from the WorldClim website (https://www.worldclim.org/).

Table 21.1 Description of 19 bioclimatic variables used (Source: O'Donnell and Ignizio 2012)

Variable code	Description	Unit
BIO1	Annual mean temperature	°C
BIO2	Mean diurnal range (mean of monthly (max temp – min temp))	°C
BIO3	Isothermality (BIO2/BIO7) (× 100)	%
BIO4	Temperature seasonality (standard deviation × 100)	%
BIO5	Max temperature of warmest month	°C
BIO6	Min temperature of coldest month	°C
BIO7	Annual temperature range (BIO5–BIO6)	°C
BIO8	Mean temperature of wettest quarter	°C
BIO9	Mean temperature of driest quarter	°C
BIO10	Mean temperature of warmest quarter	°C
BIO11	Mean temperature of coldest quarter	°C
BIO12	Annual precipitation	mm
BIO13	Precipitation of wettest month	mm
BIO14	Precipitation of driest month	mm
BIO15	Precipitation seasonality (coefficient of variation)	%
BIO16	Precipitation of wettest quarter	mm
BIO17	Precipitation of driest quarter	mm
BIO18	Precipitation of warmest quarter	mm
BIO19	Precipitation of coldest quarter	mm

21.2.2 Ecological Niche Modeling Process

Among the range of different algorithms (e.g., GLM, GBM, MaxEnt, SVM, XGBoost, NBM, RF) for species distribution modeling (Hao et al. 2012), the MaxEnt algorithm is preferred by the researchers owing to its simplicity and reliability of the results (Valencia-Rodríguez et al. 2021). MaxEnt software ver. 3.4.3 was used to model the niche of *M. dubia.* It uses occurrence records together with a summary of the environments in the accessible area, i.e., the background, to identify the environmental conditions and geographical area by the species of interest (Lantschner et al. 2019; Sillero and Barbosa 2021).

Model parameterization was done using a 10,000 background points: 500 iterations and a convergence threshold of 0.00001. Hinge, product, linear, and quadratic feature types were used to deal with model complexity, and overfitting was controlled using a regularization value of 1. Ten replicated model runs with cross-validations were implemented to assess model consistency. Performance was assessed based on the area under the curve (AUC) value. Analysis of variable contributions, jackknife procedure, and response curves were used to assess the importance of the predictor variables. The current distribution of *M. dubia* was predicted based on the *logistic* outputs, which were converted into binary maps to the climatic suitability and unsuitability with the condition of applying the threshold rule of 10 percentile training presence.

21.3 Result and Discussion

The average training and test AUCs were high (0.97 ± 0.002 SD and 0.958 ± 0.029 SD, respectively), which shows satisfactory model performance. The analysis of variable contribution showed that the temperature seasonality (49%) has the highest rank followed by the isothermality (19.9%), precipitation of the warmest quarter (9.2%), and precipitation of the coldest quarter (7.7%). These variables account for >85% of predicting the potential climatic niche of *M. dubia.* The jackknife analysis showed that the temperature seasonality contributed to the highest model gain (1.457) when used in isolation, while the precipitation of the coldest quarter decreased the gain predominately by 2.104 when omitted from the analysis (Table 21.2). This indicates the influence of these variables on the distribution of the species compared to other variables.

Response curves for the most important variables revealed that the climatic niche of the species is characterized by the temperature seasonality of 1.5°C (amount of temperature variation over a given year), the isothermality of 60–65%, precipitation of warmest quarter ~400 mm, and precipitation of coldest quarter ~1000–2000 mm (Fig. 21.1). Thus, the variables provide an estimate of the important climatic attributes of the species niche and potential distribution of *M. dubia* in India.

The MaxEnt model predicted that 1.66% of the total geographical area of India is highly suitable, moderate (2.26%), low (4.46%) respectively for *M. dubia.* Karnataka and Kerala have more suitability compared to Tamil Nadu, Maharashtra,

Table 21.2 Results of the analysis of variable contributions and Jackknife test of variable importance

	Analysis of variable contributions		Jackknife values of regularized training gain	
Bioclimatic variable codes	Percent contribution	Permutation importance	Without variable	With only variable
bio_1	0.3	0.3	2.1571	0.118
bio_2	0.6	0	2.1574	0.3785
bio_3	19.9	0.8	2.1603	1.3743
bio_4	49	58.1	2.1381	1.4465
bio_5	0.4	1.7	2.1495	0.6244
bio_6	0	0	2.1602	0.7318
bio_7	0.8	0.1	2.1602	1.0087
bio_8	5.5	15.3	2.1153	0.6437
bio_9	0.6	5.8	2.1387	0.3953
bio_10	0	0	2.159	0.6317
bio_11	1.7	0.5	2.1541	0.6367
bio_12	0.6	0.8	2.1542	0.3011
bio_13	2.3	0.8	2.1532	0.5
bio_14	0.9	1.3	2.147	0.0844
bio_15	0.3	0.2	2.1552	0.23
bio_16	0.1	0.2	2.1584	0.3465
bio_17	0.1	0	2.1583	0.0795
bio_18	9.2	6.2	2.0916	0.3433
bio_19	7.7	8.1	2.0604	0.5826

and Gujarat, along with certain areas in Andaman & Nicobar Islands and Lakshadweep.

The climate niche of *M. dubia* can be described via. climate parameters as the mean annual temperature is 22–24°C, mean temperature of the warmest quarter is 25–27°C, the mean temperature of the coldest quarter is 22–26°C, the maximum temperature of the warmest month is 33–34°C, a minimum temperature of coldest month is 20–22°C, temperature annual range is 10–13°C, mean diurnal range is 6–8° C, annual precipitation is 2200–2800 mm, precipitation in the wettest month is 800–1000 mm, precipitation in the driest month of the quarter is 90–100 mm, precipitation of wettest quarter is 1800–2000 mm, precipitation of driest quarter is 250–280 mm, precipitation of warmest quarter is 300–400 mm, and precipitation of coldest quarter is 1000–2000 mm.

Species of *Melia* are native to the Indo-China continent (Kumar and Joshi 2021), and owing to economic and medicinal utility, many species of this member have been introduced to different parts of the world. There is ambiguity regarding the interspecies differentiation among the individual *Melia* species. For instance, Gamble (1922) reported only *M. azedarach* L., *M. birmanica* Kurz, *M. composita* Willd., and *M. indica* (A.Juss.) Brandis from India. In the Flora of India (Hajra et al. 1997), *M. birmanica* and *M. composita* were synonymized under *M. dubia*, and

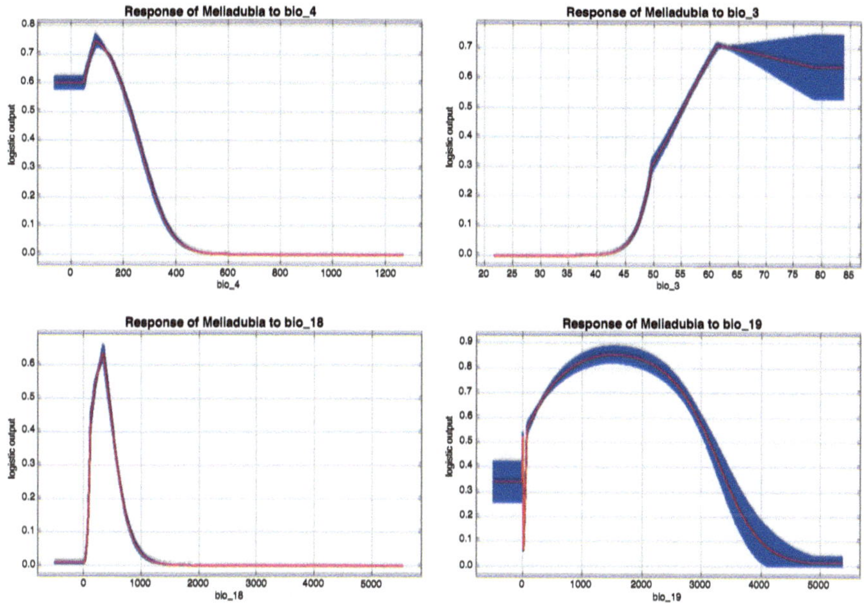

Fig. 21.1 Response curves elucidate the dependence of the predicted suitability on the selected variable as well as on the dependencies induced by correlations between the selected variable and other variables. The red curve shows the mean response of 20 replicated runs, while the blue shade represents ± 1 standard deviation

M. azedarach and *M. dubia* were recognized as two distinct species. There are research works that demarcated the difference between these species, yet there is no clear distinction in the present geographical distribution. It is reported that three different ecotypes exist in *M. dubia* and clear morphological segregation is reported as well (Kumar et al. 2022). In this context, this study defines the climatic niche of the *Melia dubia*o occurring in the Western Ghats. Champion and Seth (1968) reported that *M. dubia* belongs to the tropical moist and dry deciduous forests, which are characterized by the mean annual temperature of 24–27°C and 23.5–29° C and mean annual rainfall of 1200–3000 mm and 750–1900 mm, respectively.

The MaxEnt model indicates bioclimatic variable limits within the range of tropical moist and dry deciduous forests, thus corroborating the model predictions. Furthermore, a recent study has indicated the need for irrigation for *M. dubia* plantations established in the area having rainfall less than 1000 mm/year. The MaxEnt model indicated that there will be a declining trend in the suitability of the site having precipitation of less than 400 mm during the warmest quarter and precipitation of less than 1000 mm during the coldest quarter. Given that the sensitivity of this species, the planting materials sourced from the Western Ghats might not perform better in other areas than that indicated in Fig. 21.2, provided alternative irrigation and suitable silvicultural practices are adopted.

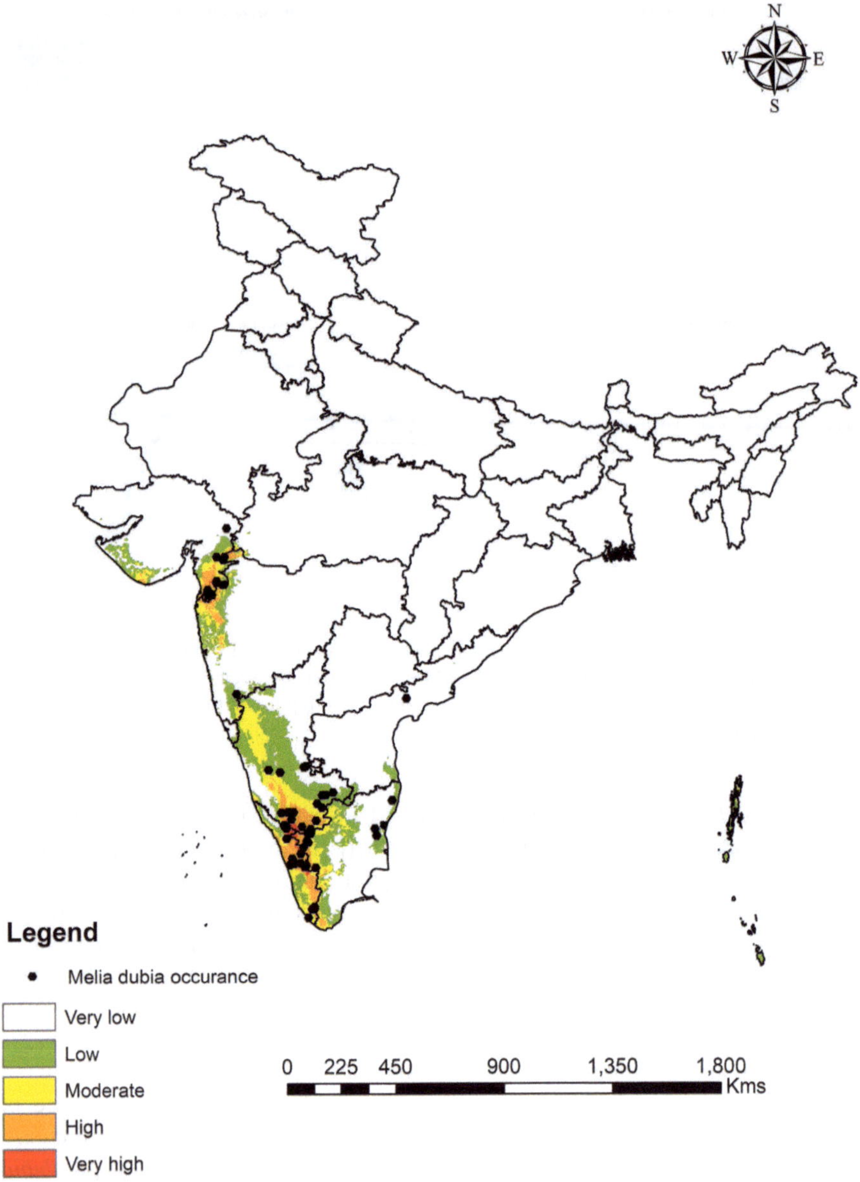

Fig. 21.2 Occurrence localities of *Melia dubia* overlaid on the modeled climatic suitability map of the species in India for current period. The different colored regions represent the climatically suitable area identified through applying 10 percentile threshold limit to the average probabilistic output

The model tentatively indicates that 3.92% of the geographical area of the country may be suited for *M. dubia* cultivation, especially for the planting material originating from the Western Ghats region. Given that there is ambiguity in the *Melia* species occurring in the North and North-eastern regions as separate species or ecotypes, there is a need for niche differentiation or niche overlap works to be carried out in the future. This sort of work demonstrates that niche-based analyses can help not only to identify suitable areas for the cultivation of a particular species but also to pertinently point out the suitable location of the germplasm/clones developed based on their native geographic distribution occurrence.

ENM can be used to collate more information with regard to agroforestry. There are instances where niche modeling has led to the development of suitability maps for different species as individuals as well as the combination of different woody perennials. For instance, an ensemble-based niche modeling has been used to create the "Suitability of key Central American agroforestry species under future climates: an atlas" which per se predicts the present area and projects the shift in the area for 54 important agroforestry tree species in the Central America countries of Belize, Guatemala, El Salvador, Honduras, Nicaragua, Costa Rica, and Panama (de Sousa et al. 2017). Agroforestry is inherent—the combination of different trees and crops together and therefore, collating the niche of different species together will aid in predicting suitable geographic areas for different agroforestry models. As there is due possibility for extrapolating the climate change impact, there is a way to predict the viability of different agroforestry models in the upcoming years (Ranjitkar et al. 2016). This sort of work can enable the policymakers to reframe the existing agroforestry policy and tree marketing guidelines to suit a particular species. From the agroforestry perspective in the farmer's field, it has always been the introduction of new tree species having better economic returns. For instance, the Indian Sandalwood tree (Santalum album L.) is native to southern India specifically, the Deccan Plateau. However, it is now cultivated in many parts of the country apart from its native distribution. There are a lot of speculations about the growth and yield from these new cultivation areas (Sandeep et al. 2020); in this regard ENM can tentatively provide a clear recommendation on species introduction. A similar sort of ENM work carried out in Nepal has provided inputs to the forest department to avoid Alnus nepalensis D.Don in the north-eastern part of India and to replace it with Alnus nitida (Spach) Endl. in combination with cardamom or tea as an intercrop (Rana et al. 2018). The results of the niche modeling of the trees and crops coupled with the fuzzy logic model have been used to determine the optimal tree crop combination too. This sort of result needs post-modeling data processing to make it more reliable and adaptable (Ranjitkar et al. 2021).

We believe that ENM can be applied in agroforestry in the following aspects: (1) to predict the suitability extent of newly developed clones/varieties, (2) to predict the present and future climatic suitability of a particular agroforestry model (tree + crop combination), (3) to prepare suitability maps/atlas for a particular region with the list of most suitable species, and (4) to facilitate policy formulation, thereby forewarning the promotion of new species in farmers' fields.

Acknowledgments The present study is part of Ph.D. work of the first author. We would like to acknowledge the support and guidance from the Indian Council of Agricultural Research, CSIR-NBRI, and Mizoram University. Financial support from the DST-NMSHE is thankfully acknowledged.

References

Adhikari D, Reshi Z, Datta BK, Samant SS, Chettri A, Upadhaya K, Shah MA, Singh PP, Tiwary R, Majumdar K, Pradhan A, Thakur ML, Salam N, Zahoor Z, Mir SH, Kaloo ZA, Barik SK (2018) Inventory and characterization of new populations through ecological niche modelling improve threat assessment. Curr Sci 114:519–531. https://doi.org/10.18520/cs/v114/i03/519-531

Adhikari D, Tiwary R, Singh PP, Upadhaya K, Singh B, Haridasan KE, Bhatt BB, Chettri A, Barik SK (2019) Ecological niche modeling as a cumulative environmental impact assessment tool for biodiversity assessment and conservation planning: a case study of critically endangered plant Lagerstroemia minuticarpa in the Indian eastern Himalaya. J Environ Manage 243:299–307. https://doi.org/10.1016/j.jenvman.2019.05.036

Ahmad F, Uddin MM, Goparaju L (2019) Agroforestry suitability mapping of India: geospatial approach based on FAO guidelines. Agr Syst 93:1319–1336. https://doi.org/10.1007/s10457-018-0233-7

Arunachalam A, Ramanan SS (2021) Agroforestry based land-use model for sustainable farming. Indian J Agron 66:101–110

Arunachalam A, Ramanan SS, Handa AK (2021) Administering agroforestry at the district level. Curr Sci 121:473–474

Binu NK, Santhoshkumar AV (2019) Genetic variability studies and plus tree selection of Melia dubia from Kerala part of Western Ghats. Green Farming 10(6):668–673

Champion HG, Seth SK (1968) A revised forest types of India. Manager of Publications, Government of India, Delhi, 511p

Chavan SB, Uthappa AR, Sridhar KB, Kakade V (2022) Scientific techniques for Melia dubia-based agroforestry systems: an emerging indigenous tree species for wood-based industries in India. Curr Sci 122:1451–1454

de Sousa K, van Zonneveld M, Imbach P, Casanoves F, Kindt R, Ordoñez JC, (2017) Suitability of key central American agroforestry species under future climates: an atlas. ICRAF Occasional Paper No. 26

Fick SE, Hijmans RJ (2017) Worldclim 2: new 1-km spatial resolution climate surfaces for global land areas. Int J Climatol. https://doi.org/10.1002/joc.5086

Gamble JS (1922) A manual of Indian timbers: an account of the growth, distribution, and uses of the trees and shrubs of India and Ceylon, with descriptions of their wood structure. S Low, Marston & Company Limited, pp xxvi + 868 + 20 plates

GoI (2014) National agroforestry policy. Department of Agriculture and Cooperation, New Delhi

Gupta S, Singh CP, Kishan-Kumar VS, Shukla S (2019) Machining properties of Melia dubia wood. Maderas Cienc y Tecnol 21:197–208. https://doi.org/10.4067/S0718-221X2019005000207

Hajra PK, Nair VJ, Daniel P (1997) Flora of India: Vol. IV: Botanical Survey of India, Calcutta, 494p

Handa AK, Chavan SB, Kumar V, Vishnu R, Ramnanan SS, Tewari RK, Arunachalam A, Bhaskar S, Chaudhari SK, Mohapatra T (2020) Agroforestry for income enhancement, climate resilience and ecosystem services. Indian Council of Agricultural Research, New Delhi

Hao C, yun Fan R, Ribeiro MC, Tan L, he Wu H, song, Yang J, feng, Zheng, quan W, Yu H (2012) Modeling the potential geographic distribution of black pepper (Piper nigrum) in Asia using GIS tools. J Integr Agric 11:593–599. https://doi.org/10.1016/S2095-3119(12)60046-X

Joshi AK, Pant P, Kumar P, Giriraj A, Joshi PK (2011) National Forest Policy in India: critique of targets and implementation. Small Scale For 10:83–96. https://doi.org/10.1007/s11842-010-9133-z

Kulkarni HD (2013) Industrial agroforestry: an Indian tobacco company (ITC) initiative. Indian J Agrofor 15:49–54

Kumar A, Joshi G (2021) Recent advances in *Melia dubia* Cav. ICFRE, Dehra Dun

Kumar R, Kumar A, Banyal R, Kumar M, Singh A, Yadav RK, Dobhal S, Sharma S (2022) Seed and seedling diversity delimitation and differentiation of Indian populations of Melia dubia cav. Saudi J Biol Sci 29:489–498. https://doi.org/10.1016/j.sjbs.2021.09.004

Lantschner MV, de la Vega G, Corley JC (2019) Predicting the distribution of harmful species and their natural enemies in agricultural, livestock and forestry systems: an overview. Int J Pest Manag 65:190–206. https://doi.org/10.1080/09670874.2018.1533664

Majumdar K, Adhikari D, Datta BK, Barik SK (2019) Identifying corridors for landscape connectivity using species distribution modeling of Hydnocarpus kurzii (King) Warb., a threatened species of the Indo-Burma Biodiversity Hotspot. Landsc Ecol Eng 15:13–23. https://doi.org/10.1007/s11355-018-0353-2

Mipun P, Adhikari D, Bora A, Bhat NA, Kumar Y (2019) Species distribution modelling of Brucea mollis Wall Ex Kurz in northeast India for its conservation. Plant Arch 19:3191–3196

Morris JIM, Ningnan Z, Zengjiang Y, Collopy J, Daping X (2004) Water use by fast-growing Eucalyptus urophylla plantations in southern China. Tree Physiol 24:1035–1044

Nair PKR, Mohan Kumar B, Nair VD (2021) An introduction to agroforestry: four decades of scientific developments, 2nd edn. Gewerbestrasse, Cham

O'Donnell MS, Ignizio DA (2012) Bioclimatic predictors for supporting ecological applications in the conterminous United States: U.S. Geol Surv Data Ser 691:10. https://doi.org/10.3133/ds691

Parthiban KT, Jude Sudhagar R, Cinthia Fernandaz C, Krishnakumar N (2019) Consortium of industrial agroforestry: an institutional mechanism for sustaining agroforestry in India. Curr Sci 117:30. https://doi.org/10.18520/cs/v117/i1/30-36

Pradhan A, Adhikari D, Chettri A (2020) Predicting the distribution of suitable habitats for Pandanus unguifer Hook. F.—a dwarf endemic species from Sikkim Himalayas, through ecological niche modeling. Int J Conserv Sci 11:145–152

Rajpoot R, Adhikari D, Verma S, Saikia P, Kumar A, Grant KR, Dayanandan A, Kumar A, Khare PK, Khan ML (2020) Climate models predict a divergent future for the medicinal tree Boswellia serrata Roxb in India. Glob Ecol Conserv 23:e01040. https://doi.org/10.1016/j.gecco.2020.e01040

Rana SK, Rana HK, Shrestha KK, Sujakhu S, Ranjitkar S (2018) Determining bioclimatic space of Himalayan alder for agroforestry systems in Nepal. Plant Divers 40:1–18. https://doi.org/10.1016/j.pld.2017.11.002

Ranjitkar S, Sujakhu NM, Lu Y, Wang Q, Wang M, He J, Mortimer PE, Xu J, Kindt R, Zomer RJ (2016) Climate modelling for agroforestry species selection in Yunnan Province, China. Environ Model Softw 75:263–272. https://doi.org/10.1016/j.envsoft.2015.10.027

Ranjitkar S, Bu D, Sujakhu NM, Gilbert M, Robinson TP, Kindt R, Xu J (2021) Mapping tree species distribution in support of China's integrated tree-livestock-crop system. Circ Agric Syst 1:1–11. https://doi.org/10.48130/CAS-2021-0002

Reichert JM, Prevedello J, Gubiani PI, Vogelmann ES, Reinert DJ, Consensa COB, Soares JCW, Srinivasan R (2021) Eucalyptus tree stockings effect on water balance and use efficiency in subtropical sandy soil. For Ecol Manage 497:119473. https://doi.org/10.1016/j.foreco.2021.119473

Sandeep C, Kumar A, Rodrigues V, Viswanath S, Shukla AK, Sundaresan V (2020) Morphogenetic divergence and population structure in Indian Santalum album L. Trees 34:1113–1129. https://doi.org/10.1007/s00468-020-01963-2

Shrivastava S, Saxena AK (2017) Wood is good: but, is India doing enough to meet its present and future needs? Centre for Science and Environment, New Delhi

Sillero N, Barbosa AM (2021) Common mistakes in ecological niche models. Int J Geogr Inf Sci 35:213–226. https://doi.org/10.1080/13658816.2020.1798968

Valencia-Rodríguez D, Jiménez-Segura L, Rogéliz CA, Parra JL (2021) Ecological niche modeling as an effective tool to predict the distribution of freshwater organisms: the case of the Sabaleta Brycon henni (Eigenmann, 1913). PLoS One 16:e0247876. https://doi.org/10.1371/journal.pone.0247876

van Noordwijk M, Duguma LA, Dewi S, Leimona B, Catacutan DC, Lusiana B, Öborn I, Hairiah K, Minang PA (2018) SDG synergy between agriculture and forestry in the food, energy, water and income nexus: reinventing agroforestry? Curr Opin Environ Sustain 34:33–42. https://doi.org/10.1016/j.cosust.2018.09.003

Wang Q, Yang L, Ranjitkar S, Wang JJ, Wang XR, Zhang DX, Wang ZY, Huang YZ, Zhou YM, Deng ZX, Yi L, Luan XF, El-Kassaby YA, Guan WB (2017) Distribution and in situ conservation of a relic Chinese oil woody species Xanthoceras sorbifolium (Yellowhorn). Can J For Res 47:1450–1456. https://doi.org/10.1139/cjfr-2017-0210

Wang T, Zhong Y, Wang C, Tong G (2021) A low capital method for silicon interference in bamboo Kraft pulping alkaline recovery system. J Clean Prod 315:128283. https://doi.org/10.1016/j.jclepro.2021.128283

Proportions of Change in the Airborne Particulate Matter (PM10) Concentrations Across Selected States in Peninsular India: A Study of Decadal, Pre-Pandemic Trends for Planning Restoration

22

Kiran Hungund, S. Varshini, and Harini Santhanam

Abstract

Air quality in India is being continuously assessed based on the data on air quality parameters, including PM10 and PM2.5 across different air monitoring stations, which are often unevenly distributed across the geographic and political boundaries within the country. Most of these stations form a part of networks operated by the Central and State Pollution Control Boards as well as in partnerships with research institutions such as the SAFAR and MAPAN. Further, the use of complex weather models and computational steps induce methodological complexities in deriving reliable patterns of air quality changes, which provide data for critical ecosystem and/or biodiversity assessments. The present study illustrates the use of a simple methodology to model the changes in the PM10 concentrations of Peninsular India in the form proportions of changes derived from a suite of geospatially derived datasets on land-use land cover, aerosol optical depths and the planetary boundary layers, as well as new metrics such as Blue to Built-up ratios, Green to Blue ratios and percentage of impervious surface area across Peninsular India during the period 2009–2019. The present study provides a methodological approach to assess the air quality changes as inputs to plan appropriate policy interventions.

K. Hungund · S. Varshini
Energy, Environment and Climate Change (EEC) Programme, National Institute of Advanced Studies (NIAS), Bangalore, India

H. Santhanam (✉)
Department of Public Policy (DPP), Manipal Academy of Higher Education (MAHE), Manipal, Bengaluru Campus, Bengaluru, India
e-mail: harini.santhanam@manipal.edu

S. Dhyani et al. (eds.), *Ecosystem and Species Habitat Modeling for Conservation and Restoration*, https://doi.org/10.1007/978-981-99-0131-9_22

Keywords

Air quality · PM10 · LULC · AOD · Peninsular India · Green to blue ratio

22.1 Introduction

Global studies on air quality and its deterioration demonstrated that elevated levels of particulate matter (PM) concentrations have a substantial effect on attributable mortality due to four causes, namely chronic obstructive pulmonary disorder (COPD), lung cancer, ischemic heart disease (IHD) and stroke (e.g. Dandona et al. 2020; Pope et al. 1995; MoHFW 2020). Despite the global and national recognitions of serious implications of the persistence of particulates PM10 in the air, country-wide assessments are often marred by factors such as the non-availability of adequate monitoring stations, lack of good quality datasets, the non-uniform coverage of the large area of interest that are typically investigated and the presence of numerous geographic barriers. In spite of the increasing impacts of the PM on the environmental and human health, air quality has taken centre stage in recent years; differential methodologies in the monitoring and assessment of the changes provide poor policy background to ensure the health and safety. Hence, developing better monitoring and evaluation frameworks for PM need innovative approaches to quickly detect changes synoptically and pay attention to areas where large-scale changes are observed. Under data-rich circumstances, the quality assurance and application of the environmental datasets can be expected to lead to successful interpretation of PM dynamics and thence the evolution of useful policy recommendations. In contrast, factors such as the availability of a poor quality of datasets that did not pass adequate quality assurance, absence of time series and/or absence of data on influential parameters (e.g. meteorological, climatological, environmental) and different levels of quality control of the monitoring instruments across different parts of the country can result in the use of sparse datasets to study the evolving patterns of air quality. These factors highly affect the formulation of best practices for air quality control, leading to adoption of rather 'quick-fixes' for resolving local-scale issues in the place of strong regulatory policies for summative improvements in the air quality as well as public health.

A typical investigation of the relationships of PM concentrations with land management as well as environmental characteristics demands the modelling of the parameters affecting the air quality characteristics. Such studies are reported over different resolutions: coarse or fine as per the scale of the investigations. However, both approaches have inherent advantages and disadvantages: for example, at a coarser resolution, it might not be possible to determine the nature of changes, impacts of local emissions, or specialised factors affecting air quality. At a finer scale, signals of the effects of the geography, impacts of the boundary layer and large-scale climatological characteristics may not provide accurate representations of the trends of air quality changes and/or the transboundary nature of the pollution. Integrating the coarser and finer resolution models may not always be possible, especially in the absence of reliable spatiotemporal time-series datasets. Thus, the concept of airsheds to study local scale changes becomes more important in place of investigating air quality changes across geopolitical borders such as state borders or national borders.

In India, the absence of adequate monitoring stations across the states has always been an issue of concern. For example, it has been reported that as per the WHO guidelines, an urban centre must possess at least 10–12 monitoring stations to adequately study the air pollutants trends at city level (Source: https://indianexpress.com/article/cities/mumbai/safar-framework-one-stop-solution-for-air-quality-management-7536259/). Successful projects such as SAFAR (Source: http://safar.tropmet.res.in/) and MAPAN (Source: https://www.ncess.gov.in/research-groups/atmospheric-science-group/laboratories/air-quality-monitoring-lab-aqml.html) have fulfilled these needs to an extent by providing a background for data collection in urban centres at higher resolutions. However, the need for greater number of air quality monitoring stations for sub-national, sub-regional, or peninsular-level assessment still remains unresolved, which is not easily and practically achievable owing to the cost factor, as well as the scientific manpower required to calibrate and collate the air quality datasets.

The Government of India (GoI) launched the National Clean Air Programme (NCAP) to tackle air pollution in 122 non-attainment cities. The NCAP aims to reduce the ambient air PM10 and PM2.5 concentrations to 20–30% by 2024 compared to the corresponding levels in 2017. To achieve this, the Central Pollution Control Board (CPCB) stated that there is a need for 800 Continuous Ambient Air Quality Monitoring Stations (CAAQMS) and 1250 manual monitoring stations compared to the current availability of 193 CAAQMS and 658 manual stations (NGT 2021). Moreover, most of the current AAQM stations are predominantly located in the National Capital Region (NCR) and the Indo-Gangetic Plain (IGP).

CAAQMS are expensive (Rs. 80 crores required to procure and install for only 25 CAAQMS) to install and maintain (MoEFCC 2020b). This strategy also suffers from other limitations (e.g. access to suitable locations for installation and maintenance). While geospatial techniques are used by ISRO and several agencies in India to perform a multitude of functions more efficiently and accurately than what is possible without them, the use of these versatile techniques for air pollution monitoring is not fully realised. A few models have been developed to relate PM pollution in India to geospatial parameters. However, these models suffer from limitations when they are applied to Peninsular India since they are largely based on the CAAQMS data from the highly polluted areas in the NCR and IGP. Hence, a robust mathematical model relating ambient air pollution in Peninsular India with geospatial and meteorological parameters is the need of the hour to monitor and forecast PM pollution levels in this region, which is less studied compared to the NCR and the IGP. Most of the existing air pollution models in India use air quality data extracted from the CAAQMS only. Since these stations are relatively few, the ground-truthing carried out in earlier studies is inadequate to model the PM pollution in this region in a reliable manner. The proposed investigation uses verified and validated data from multiple AAQM stations maintained by CPCB/SPCBs in

84 cities/towns in Peninsular India to identify the hotspots of changes in the air quality in the last decade in a pre-pandemic scenario. This enhanced level of ground-truthing will contribute to a more robust model that can be used for PM pollution control.

22.2 Methodology

22.2.1 Study Area

Air pollution levels in Peninsular India have different scenarios when compared to the Indo-Gangetic region. As most of the developed cities lie in this region, most of the emission comes from industrialisation and vehicular pollution. Apart from this emission, sea salt levels in the atmosphere increases whenever there is landfall of cyclones, which adds to the mixed concentration levels of PM in the atmosphere. Metropolitan cities such as Chennai, Mumbai and Hyderabad also have statistically negative trends as of Delhi from 2014 to 2019 (Singh et al. 2020). In 2020, 122 non-attainment cities within 11 states of southern cities under the NCAP had been identified for rehabilitation and restoration of ambient quality in India by 2024 (PIB 2020).

While approximately 127 towns/cities in Central and South India have at least one AAQ monitoring station operated by the concerned State Pollution Control Board (SPCB), the collection and analysis of data from these stations requires a higher level of effort and duty of care compared to data collected from CPCB's continuous monitoring stations limited to 12 towns/cities in Central and nine cities/towns in South India. This may be due to the following reasons:

1. The historical datasets from the AAQM stations maintained by the SPCB have not been digitised for various reasons, including the lack of adequate capacity in several SPCBs to complete this on time.
2. Since these datasets are derived largely from manual monitoring stations (compared to CPCB's continuous monitoring stations), they must be carefully verified and validated before using them in any model.

While the parameters and processes influencing PM pollution in Peninsular India need to be investigated in greater detail, the inadequate number of CPCB-maintained continuous AAQ monitoring stations in this region has posed a challenge to other researchers who rely only on CPCB stations for ground-truthing geospatial parameters affecting air pollution. Therefore, while the pollution trends in the northern, western and eastern parts of India have been studied separately by other researchers, lesser attention has been paid to the air pollution levels in Peninsular India. The diverse topography and the high level of variability in meteorological parameters in this region indicate the necessity to develop more granular PM pollution models with multiple geospatial and meteorological parameters using

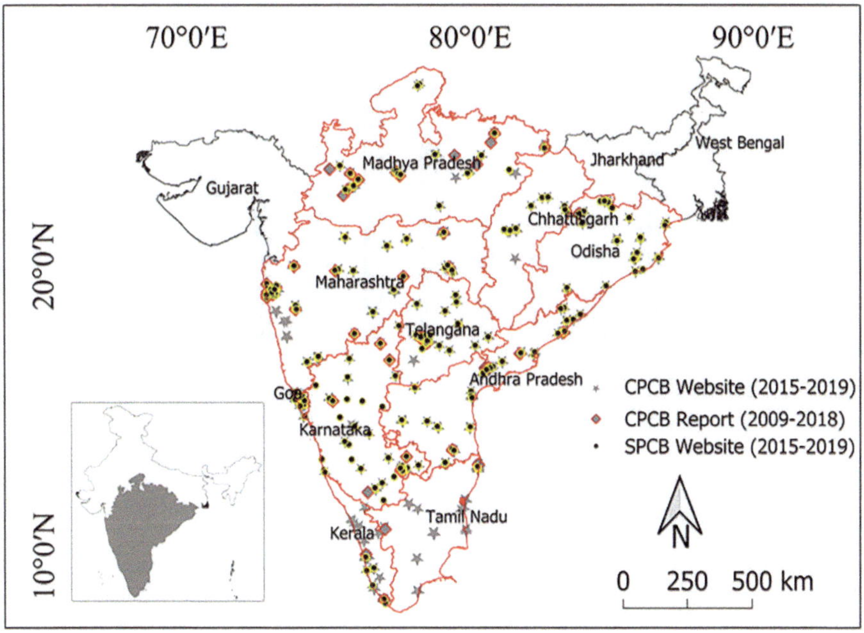

Fig. 22.1 Map of the study region and the state-wise distributions across India showing the distributions of the Central Ambient Air Quality Monitoring Stations (CAAQMS) and National Air Pollution Monitoring Programme (NAMP) stations of the CPCB and SPCB from where the spatio-temporal datasets were used for the present study

data from a larger number of AAQ monitoring stations spread across Peninsular India.

The National Clean Air Programme (NCAP) provided the scope to expand the effectiveness of the AAQ monitoring network across the country since it is critical to develop an action plan for the prevention, control and abatement of air pollution, as well as to enhance public awareness and capacity building (MoHFW 2020). There are currently 132 non-attainment cities throughout India under the NCAP located within the diverse geography of the Peninsular India region, over which the trends of air quality changes need to be reported. The present study provides a robust compilation of multisource datasets on air quality data in the public domain, including datasets contributed to public databases by CPCB and SPCB from multiple stations located in 84 cities/towns across Peninsular India. In the present study, the study area is divided into three parts: Central, Eastern and Western Peninsular India, consisting of ten states as shown in Fig. 22.1:

1. Madhya Pradesh and Chhattisgarh located in the Central Peninsular region of India.
2. Tamil Nadu, Pondicherry, Andhra Pradesh, Odisha and Telangana constituting the Eastern parts of Peninsular India.

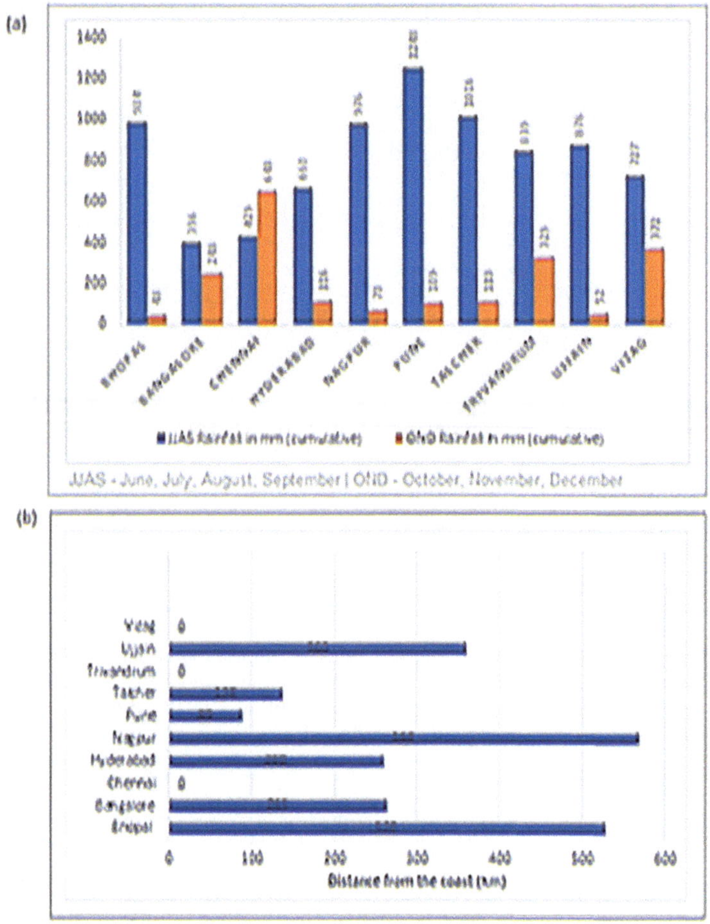

Fig. 22.2 The cumulative seasonal rainfall (in mm) and distances from the coast for selected locations within Peninsular India

3. Maharashtra, Kerala, Karnataka and Goa constituting the Western parts of Peninsular India.

The rainfall and distance from the coast for important sites within the area of interest in the Peninsular India are shown in Fig. 22.2a, b, illustrating the differential geographies and climatology of the three portions of the area studied. In general, the northeast monsoon is found to be less intense than the southwest monsoon in the study area across the different states, except for the Chennai metropolitan city, where the northeast monsoon was of higher magnitude during the period of investigation (Fig. 22.2a). No direct correlation was observed between the amount of rainfall, with the distances of the key cities from the coast as per the present study (Fig. 22.2b).

Table 22.1 List and descriptions of the datasets used in the present analyses as well as their sources

S. no.	Dataset	Description	Source for datasets
1	Basemap	Indian Boundary Shape file (Vector layer)	iGIS map
2	PM10	Point datasets of PM10 concentrations in different states and regions	CPCB, SPCB (annual datasets)—online datasets
3	LULC	Annual land use land cover (LULC) with scale of 250k	Downloaded from ISRO's Bhuvan Geoportal and remapped with georeferencing
4	AOD	INSAT 3D aerosol optical depth (AOD)	Downloaded from ISRO's VEDAS Geoportal remapped with georeferencing
5	PBL	Monthly planetary boundary layer (PBL) of ERA 5 reanalysis data	Downloaded from Copernicus Climate Data Store of European Centre for Medium-Range Weather Forecasts (ECMWF)

22.2.2 Data Collection and Analyses

22.2.2.1 Data Collection and Compilation

The annual data on the PM10 concentrations were obtained from the CPCB and state-wise SPCB repositories available in the public domain up to December 2021. The PM10 data gaps from 2009 to 2014 were also complemented using data referred to in the CPCB reports available in the public domain on the web, and the data from 2015 to 2019 were obtained from online repositories maintained by CPCB and SPCB. Aerosol optical depth (AOD) data from INSAT 3D was obtained from ISRO's VEDAS repository corresponding to the 15th day of every month at 12 p. m. in 2019 in order to capture the peak monthly and diurnal changes. Due to the non-availability of data for January and November alone, the respective data were downloaded for the 16th day and 19th day, respectively. The land use land cover (LULC) annual data of 25,000 scale were obtained from ISRO's Bhuvan Geoportal from 2009 to 2019. Monthly data on the planetary boundary layer (PBL) of ERA 5 reanalysis data were downloaded from Copernicus Climate Data Store. Table 22.1 lists the different datasets used in the study as well as their sources.

22.2.2.2 Data Analyses

A maximum of 625 station points were selected during the period 2009–2019 corresponding to the datasets on PM10. These were clustered for the state-wise distributions of station points, and the densities of the sampling sites were calculated by dividing the total number of stations by total area. It is found that the total station density of selected states in Peninsular India to be 0.000405/km^2. The station densities for the selected states in Peninsular India are shown in Fig. 22.3. The distributions of the AAQMS stations were observed to be quite uneven, with states with smaller areas (e.g. Goa) accounting for higher density of stations and hence

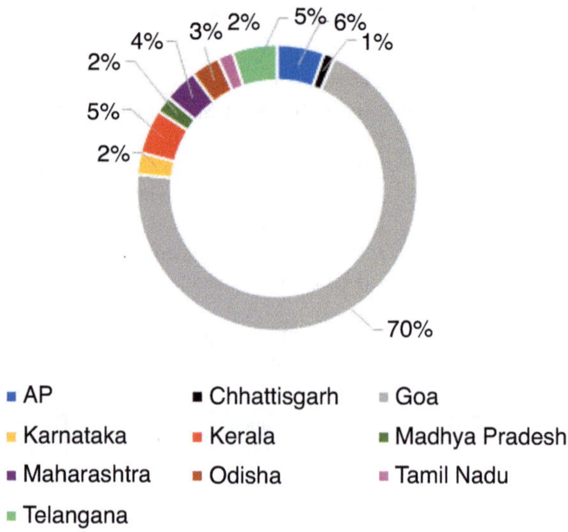

Fig. 22.3 State-wise densities (represented as percentages of the total number of locations) of the Ambient Air Quality Monitoring Stations (AAQMS) distributed over the total area of study obtained from the databases accessed online from the websites of Central and State Pollution Control Boards, for Southern, Central and Eastern India for the period 2009–2019

higher monitoring resolutions compared to other states encompassing greater geographic area but lesser station densities. However, it must be noted that the present study used only the data available in the public domain, while the possibility of the availability of data with a higher temporal resolution exists with the administrative units of the state or central boards.

The databases of the monthly, seasonal (summer, winter, monsoon), annual concentrations of PM10 were prepared for grids of sizes 10 km × 10 km (coarser model), which is usually adopted to survey population distributions at regional scales.

A geodatabase was created using the ground-level data in QGIS environment (QGIS 3.18.2 version), and a representation of the spatial distributions of the stations were created using the inverse distance weighting (IDW) interpolation for PM10 data selected areas of interest (AOI) using Eq. (22.1) as follows:

$$Z_j = \frac{\sum_i^n \frac{Z_i}{D_{ij}}}{\sum_i^n \frac{1}{D_{ij}}} \tag{22.1}$$

where

Z_i = value of known point,
D_{ij} = distance to known point,
n = user selected exponential and
Z_j = unknown point.

Further, the annual datasets on LULC, obtained from ISRO's Bhuvan Geoportal, were georeferenced and clipped to the study area (iGISmap). The root mean square error was calculated to be in the range between 0.05 and 0.09. Georeferencing of the INSAT 3D AOD data were performed, and the root mean square error was found to be in the range between 0.05 and 0.09. Monthly ERA 5 reanalysis data were spatially integrated across the study area and represented for 12 a.m. and 12 p.m. The PM10 data were also spatially interpolated and reclassified based on CPCB standard categories. The SCP plugin in QGIS and the land cover change function in post-processing were used to analyse the annual changes in PM10 concentrations.

During this process, the previous year's spatial datasets of PM10 were used for reference classification to compute the proportions of changes to generate the new PM10 values. Post-analyses reclassification was performed to categorise areas of no change, increase or decrease in annual PM10 concentration. In this way, the change detection maps were created from the reclassified outputs. The Raster layer unique value report generated for annual periods from the geospatial computation of the indices across the study area was used to delineate the reclassified image where the changes in the ratios were maximum or minimum across the years. Areas with an increase in the ratio corresponding to the increase in PM10 concentration area to total area were classified as 'red', whereas the areas showing a decrease in the ratios were classified as 'green'. The area statistics for 16 classes of LULC were obtained from ISRO's Bhuvan Geoportal, and the waterbody area was calculated by taking mean of minimum and maximum areas for the particular class. In the case of vegetation cover, the sum of areas under the classes of rabi crop, double/triple crop, plantation, deciduous forest, littoral swamp, shifting cultivation, kharif crop, Zaid crop, evergreen forest, degraded/scrub forest and grassland were computed. The Green to Blue ratio (GBR), Blue to Built-up area (BBA) and impervious surface area (ISA) were calculated as per the methodology described in Santhanam and Majumdar (2022) and Varshini et al. (2022).

22.3 Results and Discussions

22.3.1 Analyses of Land Use and Land Cover (LULC) and Aerosol Optical Densities (AOD) with Respect to PM10 Concentrations

Figure 22.4 shows the annual changes of LULC statistics of the study area during the period 2009–2015. An initial increase in the area for the LULC categories corresponding to built-up, waterbodies and vegetation areas followed by a decrease in the latter years were observed. In terms of the decadal change (2009–2019) area corresponding to built-up regions, extents of waterbodies and vegetation were observed to increase by 3.6%, 7.7% and 14.6%, respectively. In terms of annual changes, the percent increase in the areas of the built-up regions, waterbodies and vegetation were found to be in the range between 0.5% and 9%, 0.09% and 2%, and 0.6% and 14%, respectively. From these results, it is evident that the changes in

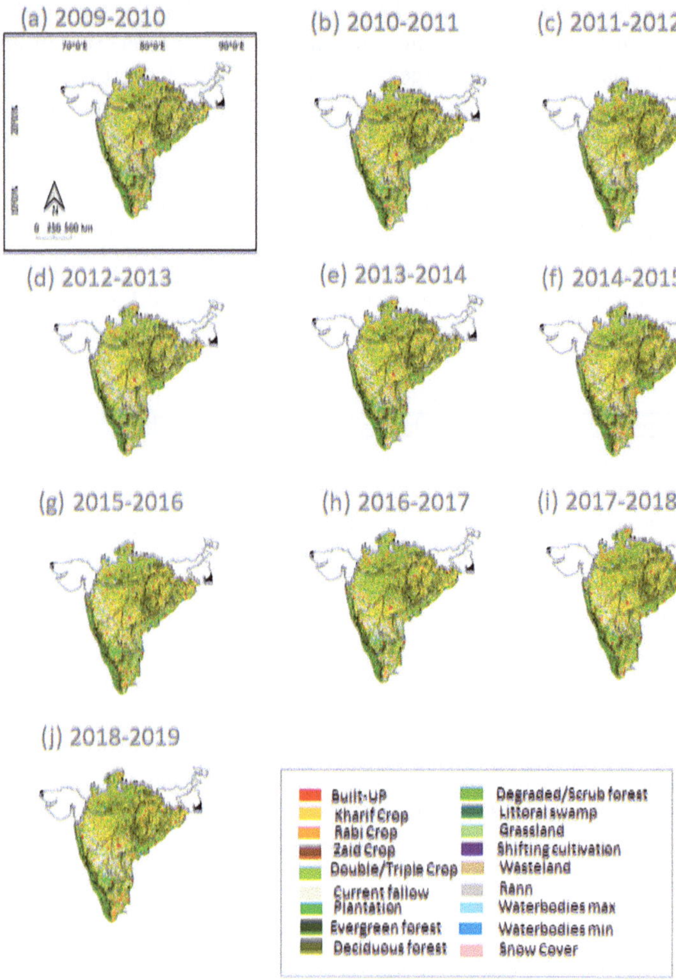

Fig. 22.4 Land use land cover (LULC) maps obtained from ISRO's Bhuvan Geoportal for the period 2009–2019 after georeferencing representing the pre-pandemic LULC scenario for Peninsular India

built-up area were observed to be very less when compared to the changes undergone by the areas of vegetation.

Gupta et al. (2020) from IIRS developed satellite-derived estimates of PM concentrations for the entire country by using a geographical weighted regression (GWR) model relating PM concentrations to INSAT 3D AOD and specific meteorological parameters and demonstrated a significant correlation ($r > 0.55$) over most parts of India except Central India and the Deccan Traps region ($r < 0.35$) where meteorological parameters are highly variable. Gupta et al. (2020) also reported that the spatial variability of air pollution in North India can be attributed to episodic

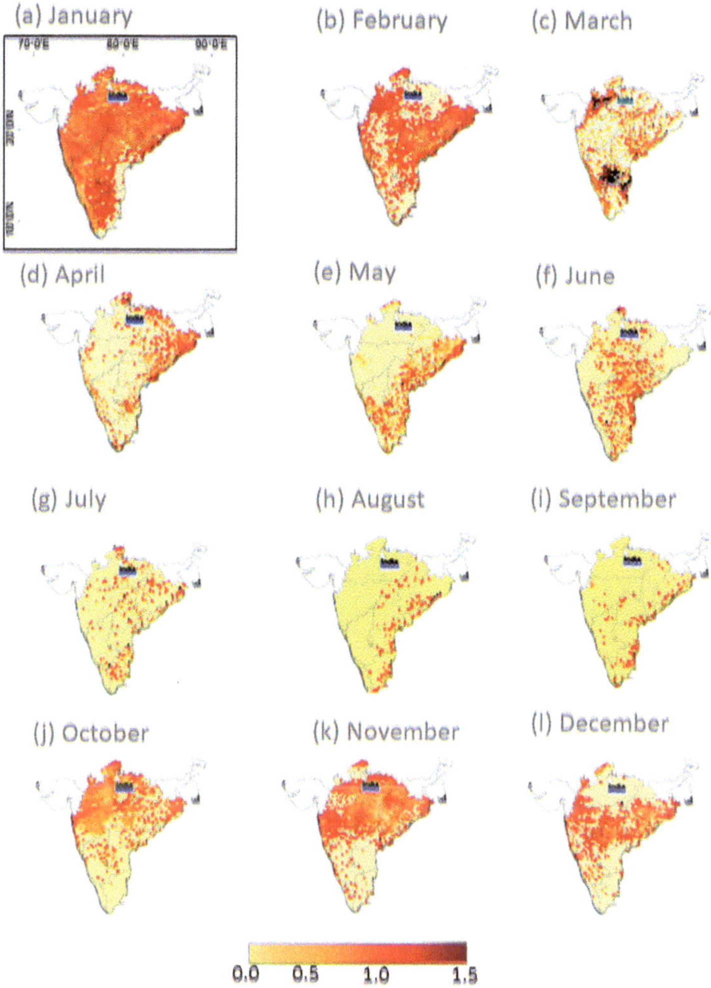

Fig. 22.5 Aerosol optical density (AOD) maps obtained from ISRO's Visualisation of Earth observation Data and Archival System (VEDAS) Geoportal (https://vedas.sac.gov.in/air quality monitoring/index.html) for the period January to December 2019 at 12 p.m. from INSAT data representing the pre-pandemic status of PM10 concentrations over ground-based monitoring stations of the Central Pollution Control Board (CPCB) in the study region

events such as dust and biomass burning in addition to the 'inversion' effect during winter. Further, the PM2.5/PM10 ratio was also found to be highest (0.7 ± 0.08) in South India, while it was least (0.59 ± 0.08) in West India. This indicates that anthropogenic emissions are the major source of air pollution in South India, while meteorology and background dust also contribute to the air pollution in North India. A similar trend is observed in Fig. 22.5, with generally low AOD in southern regions compared to the central, eastern and western regions.

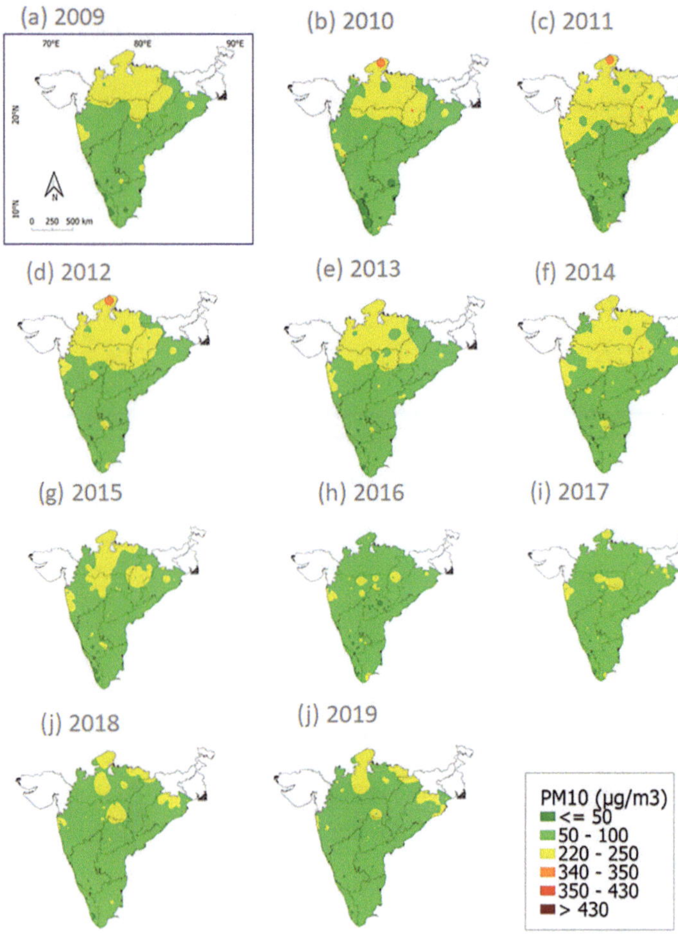

Fig. 22.6 Map of PM10 concentrations from ground data of CPCB and SPCB stations from the years 2009 to 2019

It is evident from Fig. 22.6 that the PM10 concentration levels in Madhya Pradesh and Chhattisgarh were found to be critical, whereas the concentration levels in Andhra Pradesh and Goa were higher; Tamil Nadu, Odisha, Maharashtra had medium to critical concentration levels; and Karnataka and Kerala had lower to critical and medium to higher concentration levels, respectively (CPCB 2009–2010). These concentration levels remain unchanged till 2012 except Kerala, where the concentration levels seem to be decrease from lower to medium range (CPCB 2011–12). Concentration levels in Andhra Pradesh shows critical levels when compared to previous years (CPCB 2015–16).

Further, it is observed that the concentration levels of PM10 shows decreasing trends over the years from 2009 to 2019. There are mainly four notable clusters evident from Fig. 22.6 between 2009 and 2019, which are as follows: dark green, which is less than 50 μg/m^3, green (50–100 μg/m^3), yellow (220–250 μg/m^3) and orange (340–350 μg/m^3). Although dark green and orange are not significant, it can be seen over southern and central parts of India, respectively. There has been a gradual decrease in concentration from 2009 to 2019. As observed, within the range of the concentration levels considered in the present study, most parts in Southern India does not fall above 50–100 μg/m^3, classified as indicated 'good air quality'. The central parts (i.e. Madhya Pradesh, Maharashtra and Chhattisgarh) are moderately polluted, which was observed from 2012 to 2015 and thereon small insignificant clusters are observed in the same locations from 2016 to 2019. The fluctuations in the concentration levels are greater in the central part of India, while the concentration levels are stagnant in other parts, specifically over the southern parts of India. As per the current study, the best air quality is observed spatially in almost all parts of Kerala over the years.

22.3.2 Deriving the Proportions of Change—Annual and Decadal Changes

Presently, PM10 measurements are reported only state-wise or city-wise (physical or geographic boundaries) and these do not correspond to actual airshed boundaries, which need to be established. Since we need to identify the probable airsheds over several years to understand the dynamicity of distributions, we looked at stable, ambient signals of air quality classification in pre-pandemic times to distinguish the possible changes in terms of proportions of changes. This is shown in Fig. 22.7 and discussed as follows.

Between 2009 and 2019, on a decadal scale, many areas studied show differential changes in the concentration levels. While looking at the year-to-year changes, it is observed that increased concentration levels are highlighted from 2010 to 2011, specifically over the central parts of India covering some parts of Madhya Pradesh, Maharashtra, Chhattisgarh and Odisha. No changes in the proportions between successive years were observed between 2012 and 2013, as well as from 2014 to 2015, except for some small parts of Maharashtra and Chhattisgarh, respectively. The decreased concentrations were also observed in some parts of Central India. Significantly, over some parts of Madhya Pradesh, Maharashtra and Chhattisgarh, the proportions of changes indicate a slight improvement in air quality. The current study also revealed that consistent good air quality was observed in the successive periods of 2014–2015 and 2015–2016.

Decadal changes from 2009 to 2019 are shown in Fig. 22.8, and the air quality has been observed to have progressively changed to 'good' during this decade (PM10 increased from 220–250 μg/m^3 to 50–100 μg/m^3). Although no significant changes in concentrations were observed over southern parts of India, enhanced air quality on a decadal scale is observed in central parts of India. Many parts of Madhya Pradesh and Chhattisgarh show good air quality in 2019 as compared to 2009 (Fig. 22.8a, b).

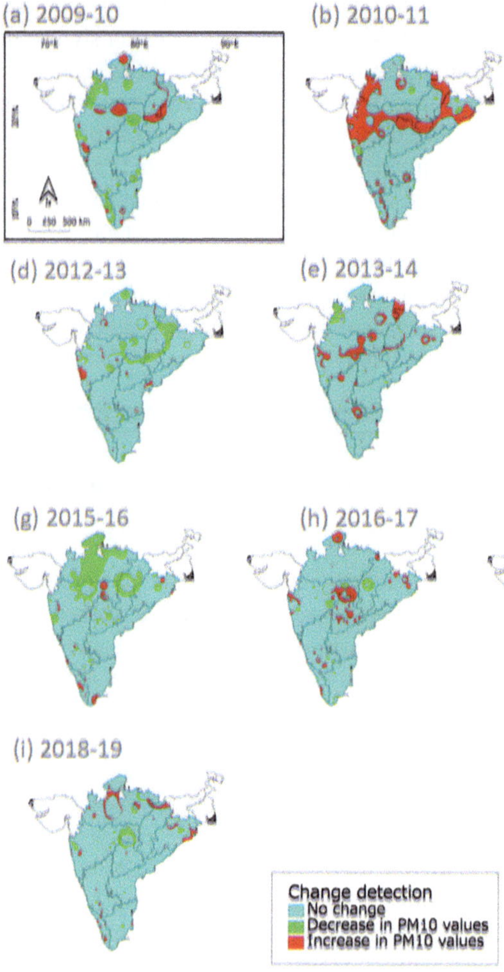

Fig. 22.7 Change detection map of PM10 concentrations for the period 2009–2019 in a pre-pandemic scenario. The changes here are represented as 'proportion of change' between successive years (e.g. 2009–2010, . . ., 2018–2019)

22.3.3 Estimation of BBA, GBA and ISA

The overall proportion of change in the green and red areas are 0.22 and 0.05, respectively, for the change between 2009 and 2019, indicating the slight increase in built-up areas during the decade as shown in Fig. 22.9. However, it is interesting to note the highest proportions of the red areas were mapped in 2010–11 and those of the green areas in 2015–16. Green area also showed a higher proportion of change to the total area in 2014–15 supporting lower PM10 and higher air quality.

As observed in Fig. 22.10, the proportion of Blue to Built-up area (BBA) varied between 40 and 41 throughout the decade (2009–2019), implying low level of

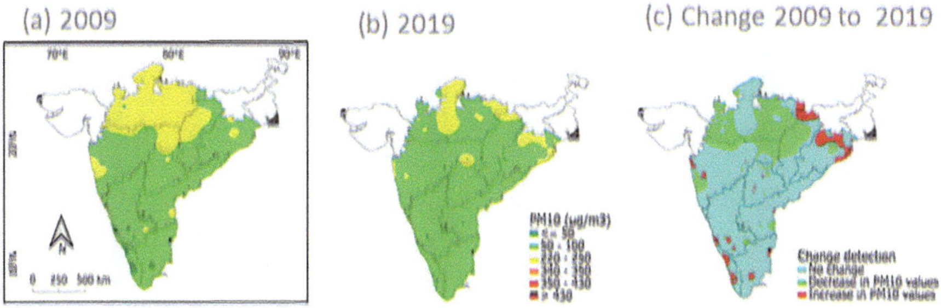

Fig. 22.8 Map of PM10 concentrations for the period 2009–2019, and change detection map of PM10 concentrations for the period 2009–2019

Fig. 22.9 Proportion of change in green area (decrease in PM10 concentrations) and red area (increase in PM10 concentrations) to the total area investigated between the years 2009 and 2019

changes to the water spread area. It was observed that ISA remained constant with less variations in the values. In the case of GBR, however, a sudden increase in the proportion (almost double) was observed from 2015 to 2019, indicating slight revegetation. This also coincides with the better air quality observed from Fig. 22.7f, g.

22.3.4 Changes in the PBL and Its Link to the Observed Proportions of Change

Figure 22.11 (i) and (ii) indicates the change in planetary boundary layers (PBL) at 12 a.m. and 12 p.m., respectively, during the years 2009–2019. Monthly changes have been derived and represented geospatially to understand any changes, which may be ascribed to diurnal changes and/or the seasonal changes. In terms of the diurnal changes from January to December, it is clear that there is a significant variation when compared to the height of PBL at 12 a.m. and 12 p.m. Higher PBL

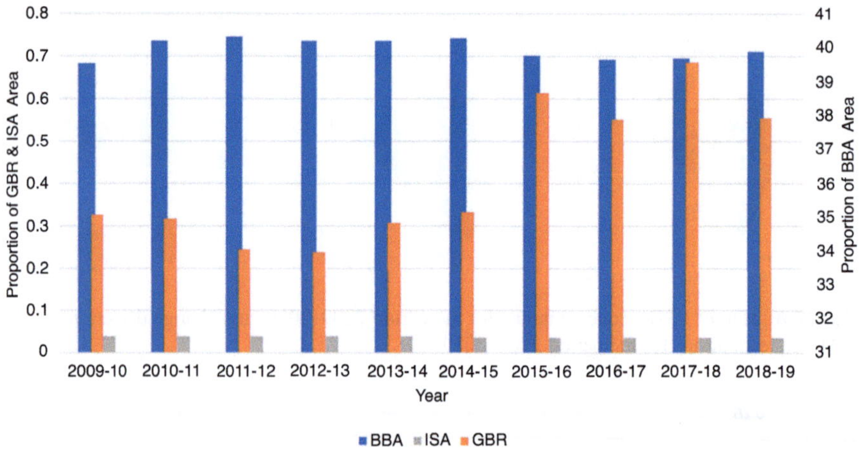

Fig. 22.10 Proportion of change in Blue to Built-up area (BBA), Green to Blue ratio (GBR) and impervious surface area (ISA) for the area of interest in the present study. The proportions of change for BBA, GBR and ISA were calculated using the data on LULC from Bhuvan Geoportal of India. (Source: https://bhuvan.nrsc.gov.in/home/index.php)

height is mainly attributed to the heating effect of the atmosphere, which is high during daytime and low during night-time. Consequently, high and low PBL height were observed at 12 p.m. and 12 a.m., respectively, which were selected as time windows for comparison across different seasons.

High PBL height is observed during summer season (March, April, May and June) due to intense heat during summer season as compared to other seasons. Black Carbon (PM10 and PM2.5) produced by most of the industries and vehicles during daytime is also one of the reasons for increased height in PBL levels due to its heat absorption characteristics. There is almost 80% reduction in PBL height from 3.5 to 0.8 km during 12 p.m. and 12 a.m., particularly in summer season. During monsoon and winter seasons, there is little heating effect produced, and hence low PBL level is observed.

22.4 Discussion and Conclusions

Air pollution is one of the major problems faced by India (especially in northern part of India) because of industrialisation and other anthropogenic activities. India has the ten most populated cities out of the 20 cities in the world and it is also the fifth most polluted country in the world, which was reported by World Health Organization (WHO) in 2019. Consistent increase in population has put undue demand for energy, which impacts the environment and Earth's atmosphere. For example, in Delhi the demand for energy increased on a decadal scale by 57.16% from 2001 to 2011 and further expected to increase to meet the greater energy demand (Kumar et al. 2015). Recent policies and regulations from the government have led to statistically

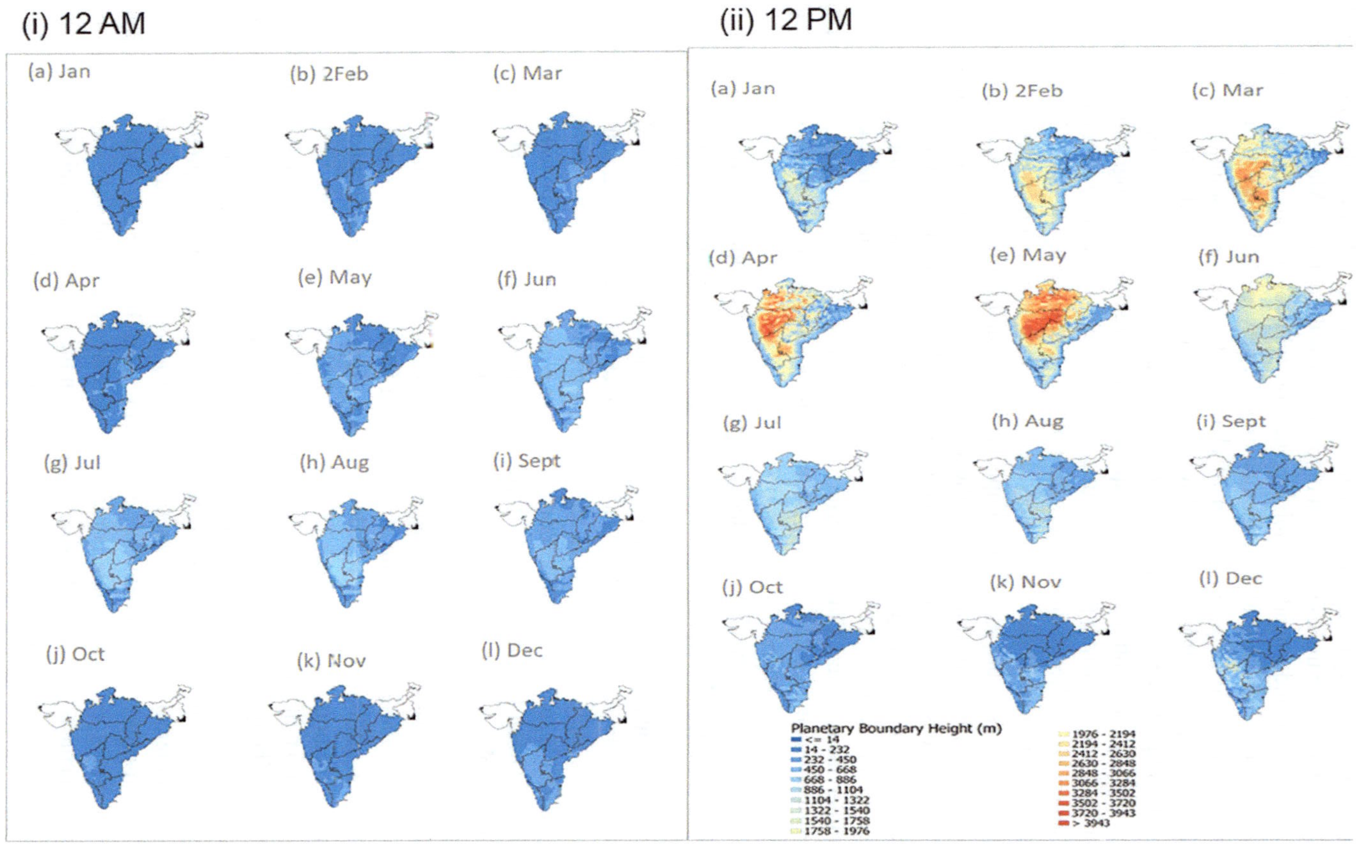

Fig. 22.11 Planetary boundary layer on 2019 at **(i)** 12 a.m. and **(ii)** 12 p.m. across the study area during the period 2009–2019 plotted on a monthly basis

significant decrease in particulate matter (PM) concentration (especially PM2.5) in Delhi, thus the negative trend is seen in 5 years from 2014 to 2019 (Singh et al. 2020). While such trends have been reported across the country (MoEFCC 2020a), it is difficult to understand the holistic scenario without a study of the trends and changes. The present study thus demonstrated a new methodology to monitor these changes reflected as a 'proportion of change' at a coarser resolution in order to identify the predominant 'airsheds' where the changes are mostly observed.

The importance of the present study is that it simplifies the methodology of deriving the sub-national level air quality across the whole geographic extent as opposed to the existing practice of integrating the city- or state-level measurements to reflect air quality changes across the political sate boundaries within India. While the movement of air parcels are very dynamic, such stratified monitoring cannot be helpful to study the associated impacts of air quality on the ecosystems or the bioclimatic divisions transcending the state or district boundaries alone, but require synoptic studies of airsheds to understand PM10 migrations. Although effective mathematical modelling can provide insights and also the delineation of airshed at higher resolutions (e.g. Guttikunda & Kopakka 2014; Guttikunda et al. 2019), simpler metrics save time and effort to delineate airsheds, which provide crucial information on air quality to support multi-proxy ecosystem analyses at sub-national or sub-regional scales.

The present study also provided an extensive and comprehensive pre-pandemic database of the PM10 concentrations for a decadal investigation from 2009 to 2019 compiled from multiple sources. The database was also expanded to include geospatial databases for meteorological parameters such as LULC and AOD. The overlay analyses of the critical land-use parameters, such as LULC with the airshed properties represented by AOD and PM10, were validated by the novel indices of urbanisation such as the proportions of change with respect to Blue to Built-up areas (BBA), Green to Built-up areas (GBR), percentage impervious surface area (ISA) and the planetary boundary layer (PBL). This extensive work resulting in the convergence of the datasets revealed the extents and magnitudes of changes, which is necessary to identify cross-over regions in Peninsular India region. Ambient data in stable pre-pandemic time provided for better constraint, and the data for reflection attributed to non-lockdown effects. Hence, the present study provides a unique methodology to assess proportions of changes according to the nature of stabilisation of the air environment and the land-use patterns in post-pandemic periods. The use of the indices as illustrated in the present study can be successfully used to study decadal-scale changes in the ecosystems at moderate resolutions with respect to land and water. Further, the present study can also provide a useful methodological approach to assess the proportional impacts of air quality as part of multi-proxy assessments of ecosystems and biodiversity. While the data distributions within the state borders vary largely across different states/divisions within a country, the data densities observed in the present study reinforce the need to identify the airsheds for the study of air quality unconfined to state/geopolitical borders.

Acknowledgements The authors wishes to thank Prof. R. Srikanth, Head, Energy, Environment and Climate Change Programme, NIAS, Bengaluru; Prof. P.G. Diwakar, ISRO Chair Professor, NIAS, Bengaluru; and Prof. Gufran Beig, Ashutosh Mukherjee Chair Professor, NIAS, Bengaluru for discussions that inspired them to pursue alternate perceptions of air pollution monitoring using the 'proportions of change' concept to understand the distribution of PM10 particles with relation to other environmental changes. The authors also wishes to thank ISRO, Bengaluru for the use of data products from Bhuvan and VEDAS geoportals for air quality studies. H.S. thanks Mr. Sameer Mishra (CSTEP, Bengaluru), Dr. Chanchal Chauhan (CSIR-NEERI), Mr. Aariz Ahmed, Mr. Sudip Kumar Kundu and Mr. Sujit Kumar Swain (NIAS, Bengaluru), who had contributed to the reconnaissance survey of air quality by helping with the compilation of PM10 datasets across Peninsular India in 2020–2021. The data used in the present study were all obtained from open data sources from the web portals of Central Pollution Control Board (CPCB), State Pollution Control Board (SPCB), Visualisation of Earth observation Data and Archival System (VEDAS) and Bhuvan Geoportal.

References

Dandona L et al (2020) Health and economic impact of air pollution in the states of India: the Global Burden of Disease Study 2019. https://doi.org/10.1016/S2542-5196(20)30298-9

Gupta A, Kant Y, Mitra D, Chauhan P (2020) Spatio-temporal distribution of INSAT-3D AOD derived particulate matter concentration over India. Atmospheric pollution research. https://www.sciencedirect.com/science/article/abs/pii/S1309104220302592

Guttikunda SK, Kopakka RV (2014) Source emissions and health impacts of urban air pollution in Hyderabad, India. Air Qual Atmos Health 7(2):195–207

Guttikunda SK, Nishadh KA, Gota S, Singh P, Chanda A, Jawahar P, Asundi J (2019) Air quality, emissions, and source contributions analysis for the greater Bengaluru region of India. Atmos Pollut Res 10(3):941–953

Kumar P, Khare M, Harrison RM, Bloss WJ, Lewis A, Coe H, Morawska L (2015) New directions: air pollution challenges for developing megacities like Delhi. Atmos Environ 122:657–661

MoEFCC (Ministry of Environment, Forest and Climate Change) (2020a) Continuous ambient air quality monitoring stations. Reply given by Hon Minister of Environment, Forest and Climate Change to question no 4514 in the Lok Sabha. http://164.100.24.220/loksabhaquestions/annex/173/AU4514.pdf. Accessed 20 Mar 2020

MoEFCC (Ministry of Environment, Forest and Climate Change) (2020b) Most polluted cities. Reply given by Hon Minister of Environment, Forest and Climate Change to question no. 973 in the Lok Sabha. http://164.100.24.220/loksabhaquestions/annex/174/AU973.pdf. Accessed 18 Sept 2020

MoHFW (Ministry of Health and Family Welfare) (2020) Deaths due to air pollution. Reply given by Hon Minister of Health and Family Welfare to question no 987 in the Lok Sabha. http://164.100.24.220/loksabhaquestions/annex/173/AU987.pdf. Accessed 7 Feb 2020

NGT (National Green Tribunal) (2021) Hearing on Original Application No. 681/2018, (With reports dated 05.04.2021 and 05.02.2021), Item No. 05. https://greentribunal.gov.in/gen_pdf_test.php?filepath=L25ndF9kb2N1bWVudHMvbmd0L2Nhc2Vkb2MvanVkZ2VtZW50cy9ERUxISS8yMDIxLTA0LTA4LzE2MTgzODgzNjkxNjI3OTgyMzU2MDc2YTU5MTdjNWQ1LnBkZg==

Pope CA, Thun M, Namboodiri M, Dockery D, Evans J, Speizer F, Heath C (1995) Particulate air pollution as a predictor of mortality in a prospective study of U.S. adults. Am J Respir Crit Care Med 151(669):674

Press Information Bureau (PIB), New Delhi (2020) Long-term, time-bound, National Level Strategy to Tackle Air Pollution-National Clean Air Programme (NCAP) To Achieve 20% to 30% Reduction in Particulate Matter Concentrations by 2024. Ministry of Housing & Urban Affairs. https://pib.gov.in/PressReleasePage.aspx?PRID=1655203. Accessed 16 Sept 2020

Santhanam H, Majumdar R (2022) Quantification of green-blue ratios, impervious surface area and pace of urbanisation for sustainable management of urban lake—land zones in India—a case study from Bengaluru city. J Urb Manag 11(3):310–320. https://doi.org/10.1016/j.jum.2022.03.001

Varshini S, Hungund K, Kundu SK, Santhanam H, Dhyani S (2022) Assessing ecological risks of urban air and water environment to analyse the scenarios for mainstreaming nature-based solutions: a case study of Bengaluru City, India. In: Dhyani S, Basu M, Santhanam H, Dasgupta R (eds) Blue-green infrastructure across Asian countries. Springer, Singapore. https://doi.org/10.1007/978-981-16-7128-9_8

Decomposition of Sunflower Cuttings and Its Impact on Soil Fertility of Rice Terraces (*Payoh*) in Banaue, Ifugao, Philippines

23

Damasa B. Magcale-Macandog, Milben A. Bragais, Marc Bryan Manlubatan, Jonson M. Javier, Marc Anthony F. Rabena, Jennifer D. Edrial, Kristina S. Mago, Teodorico L. Marquez Jr, Jerry Naayos, Randy Porciocula, and Sarena Grace L. Quiñones

Abstract

The study determined the nutrient content of sunflower stalks and leaves, the decomposition rate of varied amounts of sunflower cuttings applied to rice terrace—*payoh*, and the effect of sunflower cutting application on the soil fertility of *payoh* during the decomposition period. The nutrient content of sunflower stalks and leaves and other plant species was analyzed for N, P, and K contents using standard methodology. The experiment for sunflower decomposition was also set up in a *payoh* terrace. The statistical analysis used is two-way analysis of variance (ANOVA). The experiment was conducted in Barangay Poitan, Banaue, Ifugao, Philippines for a 12-week duration from September to December 2013. First, N, P, and K contents of sunflower stalks and leaves and other plant species from the study site were analyzed. Second, sixteen 1 m × 1 m plots treated with four replications of each four sunflower treatments (0, 500, 1000, and 2000 g sunflower cuttings per plot) were randomly set up following completely randomized design (CRD) to determine the bi-weekly decomposition of sunflower applied within the 12-week duration. Third, soil collected from the upper 10 cm of each of the 16 plots was analyzed for pH, organic matter, P, and K. Results revealed that N, P, and K contents of sunflower stalks and leaves are higher than other plant species used as fertilizer in the *payoh* plots, while all treatments with sunflower cuttings were fully decomposed 4 weeks after

D. B. Magcale-Macandog (✉) · M. A. Bragais · M. B. Manlubatan · J. M. Javier · M. A. F. Rabena · J. D. Edrial · K. S. Mago · T. L. Marquez Jr · J. Naayos · R. Porciocula · S. G. L. Quiñones
Institute of Biological Sciences, College of Arts and Sciences, University of the Philippines Los Baños, College, Laguna, Philippines
e-mail: dmmacandog@up.edu.ph

S. Dhyani et al. (eds.), *Ecosystem and Species Habitat Modeling for Conservation and Restoration*, https://doi.org/10.1007/978-981-99-0131-9_23

sunflower application. On the other hand, *payoh* plots treated with 2000 g of sunflower cuttings had significantly higher pH, organic matter, available P, and exchangeable K than other sunflower treatments. The application of 2000 g of sunflower cuttings to *payoh* plots results in significantly higher soil pH, soil organic matter, soil available P, and soil exchangeable K within 12-week duration.

Keywords

Tithonia diversifolia · Rice terrace · Decomposition · Soil fertility

Abbreviations

ANOVA	Analysis of variance
CRD	Completely randomized design
NPK	Nitrogen, phosphorus, potassium
OM	Organic matter

23.1 Introduction

The *muyong* forests (privately managed secondary forest or forest garden) and *payoh* (terraced paddies) are two of the five interconnected components of the agroecological zones in Ifugao that represent a hilly type or watershed model production system. In the production system, the *muyong* forest serves as the major component by providing water and stability to *payoh*. Given the location of the forests in the upper terrace, it also controls the overall physical soundness of the *payoh* by means of the water flowing from it (Butic and Ngidlo 2003). In addition, these forests also provide continuum supply of nutrients to the *payoh* (SITMo 2008). Hence, the long-term sustainability of the *payoh* is significantly dependent on the *muyong* forest (Butic and Ngidlo 2003; SITMo 2008). However, the *muyong* forests were increasingly under pressure from the need of cash income and growing upland population. Thus, some areas of the forests had been converted to coffee and banana for commercial plantings. Other areas were cleared to give way to residential development (Rondolo 2001). In effect, land conversions in the *muyong* forests might affect the release of nutrients from the slowly decomposing litters in the *muyong* forests to the terrace paddies, through the likely effects of changes in the structure of the forests to litter decomposition.

Hence, there is a need to supplement the probable reduction of nutrient release from the *muyong* forests to the *payoh* through biomass transfer of available organic materials. However, traditional organic materials used like crop *residues* and animal manures are typically unavailable in adequate quantities, labor intensive in processing and application, and low in nutrients (Palm and Rowland 1997). There is also the problem of competing uses of some organic materials such as fodder for livestock (Jama et al. 2000). Therefore, readily available and abundant organic materials with high nutrient content must be used. Various reports have shown the effective use of sunflower as soil amendment for rice (Nagarajah and Nizar 1982; Gusnidar et al. 2012), maize (Jama et al. 2000; Gachengco et al. 1999; ICRAF 1998; Gachengco 1996; Niang et al. 1996), and okra (Olabode et al. 2007). This is attributed to *Tithonia*'s high nutrient content and other beneficial properties. The use of readily available and abundant organic materials with high nutrient content is a good practice to supplement the nutrients required at the farm level. A good candidate for organic source of nutrients that satisfy these criteria is the *Tithonia diversifolia*, which is available in sufficient quantities in most farms in rice (*payoh*) terraces in Banaue, Ifugao.

T. diversifolia is commonly known as Mexican sunflower. It is a dicotyledonous shrub that belongs to the family Asteraceae. It is a succulent and soft shrub that grows up to a height of 1–3 m. The shrub produces seeds all year round, which are dispersed easily by wind, water, and animals (ICRAF 1997). It is also an aggressive weed that grows naturally everywhere due to its adaptability to most soils (Ojeniyi 2012; Ganunga et al. 1998). Hence, the plant is now widely spread in the humid and subhumid tropics in Central and South America, Asia, and Africa, aside from its country of origin in Mexico (Sonke 1997). In addition, *Tithonia* grows rapidly (Azeez 2020; Obatolu and Agboola 1993) and is capable of producing high biomass (Jama et al. 2000). It has fast decomposition due to its low lignin and polyphenol contents, C:N ratio, and high water content. It is an effective retriever of high amounts of nutrients from the soil (Jama et al. 2000). It is capable of attracting fungi and protecting the soil from soil erosion (Ojeniyi et al. 2012). Above all, the biomass of *Tithonia* has been found to have high nitrogen (N), phosphorus (P), and potassium (K) contents (Jama et al. 2000; Gachengco et al. 1999). Hence, *Tithonia* is an organic material that has a high potential for nutrient recycling, release, and supply.

Given these qualities of *Tithonia*, the use of plant's biomass in varying forms as soil amendment has been reported to effectively improve the growth and yield of various crops. The significant effect of incorporating *Tithonia* green biomass on maize yield was widely documented in Nigeria and Western Kenya (Gachengco et al. 1999; ICRAF 1998; Niang et al. 1996). The application of freshly crushed and dried ground *Tithonia* to soil pots was also shown to have significantly higher okra yield in Nigeria (i.e., number and weight of okra fruits per plant) than the application of *Tithonia* ash (Olabode et al. 2007). In addition, the use of *Tithonia* biomass was also reported to have significant effects on tomato (Liasu and Achakzai 2007) and yam (Adeniyan et al. 2008) yield. On the other hand, the combined effect of soil incorporation of tender green leaves and stems of *Tithonia* and second level of mineral fertilizer was reported to be comparable to the rice yield from the recommended mineral fertilizer application in Sri Lanka (Nagarajah and Nizar 1982). Likewise, *Tithonia* biomass was also proven to generate higher maize dry matter yield when incorporated or mulched *Tithonia* was combined with NPK fertilizer (Azeez 2010).

However, only the study of Nagarajah and Nizar (1982) is known to have documented the potential of *Tithonia* as a green manure for lowland rice and its combined effect with mineral fertilizer on rice yield. Recent studies have focused their attention in finding the appropriate application rate of *Tithonia* biomass with mineral fertilizer in rice paddies of Indonesia (Gusnidar et al. 2012; Hakim et al. 2012). There is no known study yet that explored the individual effect of *Tithonia* biomass application on the fertility of upland soil paddies, specifically terrace paddies. Hence, this study analyzed the nutrient content of wild sunflower stalks and leaves, determined the rate of decomposition of varied amounts of sunflower cuttings applied to *payoh*, and quantified the effect of sunflower cutting application on soil chemical properties of *payoh* during the decomposition period.

The results of this study will provide information on the nutrient content and speed of decomposition of sunflower stalks and leaves, and the application rate of sunflower cuttings that effectively improves the soil nutrients of the *payoh*. Generally, this study will provide relevant baseline information to rice farmers in Brgy. Poitan, Banaue, Ifugao on the potential use and benefits of sunflower cutting as soil fertilizer to terrace rice paddies.

23.2 Methodology

23.2.1 Nutrient Content of Sunflower Stalks and Leaves and Other Plant Species

Stalks and leaves of various plant species (i.e., sunflower, bolo, alangtin, rice, and azolla) commonly used as green manure/organic fertilizer in the *payoh* terraces were collected in areas surrounding the *payoh* terraces in Brgy. Poitan, Banaue, Ifugao. Nutrient content of sunflower stalks and leaves and other plant species was analyzed for N, P, and K contents using standard methodology. The nitrogen content was analyzed following the Kjeldahl method, total phosphorus was analyzed colorimetrically using phospo-molybdenum, and total potassium using flame photometer.

23.2.2 Sunflower Decomposition Experiment

The experiment for the sunflower decomposition was set up in a *payoh* terrace in Brgy. Poitan, Ifugao, Banaue. Sixteen 1 m × 1 m plots were set up by enclosing with polyethylene plastic and removing remaining rice stalks from the plots. Four sunflower treatments were applied in the experiment: 0, 500, 1000, and 2000 g sunflower cuttings per plot. Each treatment had four replicates that were arranged in completely randomized design (CRD). Soil samples collected from the upper 10 cm of each quadrat plot at bi-weekly intervals for 12 weeks were analyzed for pH, OM, P, and K contents.

23.2.3 Statistical Analysis of Data

Two-way analysis of variance (ANOVA) was used to determine the existence of significant difference in soil nutrients and soil properties of soil plots treated with varying amounts of sunflower cutting throughout the 12-week decomposition period. Specifically, the test determined if there was significant difference in soil nutrients and soil properties between sunflower treatments and between weeks.

23.3 Results and Discussion

23.3.1 Nutrient Content of Sunflower and Other Plant Species in *Muyong–Payoh* System

Sunflower stalks and leaves have higher N (2.61%), P (0.33%), and K (3.10%) contents (Table 23.1) compared with azolla, bolo, and alangtin plant commonly used as fertilizer in the *payoh* plots in Banaue, Ifugao. The P and K contents of sunflower stalk and leave samples fall within the range of P (0.2%–0.5%) and K (2.3%–5.5%) contents of 100 samples of wild sunflower leaves and stems from Kandy, Kegalle, Matale, and Kurunegala districts in Sri Lanka (Nagarajah and Nizar 1982), but falls below their result on N content range (3.2%–5.5%).

According to Gachengco et al. (1999), the average nitrogen (N), phosphorus (P), and potassium (K) concentrations of green leaf biomass of sunflower collected in East Africa is 3.5%, 0.37%, and 4.1%, respectively, on a dry weight basis. Comparably, samples collected within the top 50 cm of sunflower from several locations in West Sumatra had an average N, P, and K of 3.16%, 0.38%, and 3.45%, respectively (Hakim 2002). Nagarajah and Nizar (1982) also reported 3.2%–5.5% N, 0.2%–0.5% P, and 2.3%–5.5% K from 100 samples of sunflower leaves and tender stems from Sri Lanka. Different forms of 1 kg dried sunflower harvested in Western Kenya during the flowering stage also revealed N, P, and K concentrations of 1.76%, 0.82%, and 3.92%, respectively (Olabode et al. 2007).

In addition, green sunflower biomass was also found to have 1.8% calcium (Ca) and 0.4% magnesium (Mg) (Gachengco et al. 1999). On the contrary, a lower Ca (0.59%) and Mg (0.27%) contents of sunflower were reported by Rutunga et al. (1999).

Table 23.1 Nutrient contents of various plant species collected in the *muyong–payoh* system in Brgy. Poitan, Banaue, Ifugao, Philippines

Plant species	N (%)	P (%)	K (%)	Remarks
Bolo	1.63	0.19	1.22	Used as fertilizer
Rice (Oklan variety)	1.17	0.15	0.47	Rice variety
Azolla	1.45	0.24	0.45	Used as fertilizer
Alangtin	1.96	0.32	3.13	Used as fertilizer
Wild sunflower	2.61	0.33	3.10	Used as fertilizer

In comparison to other plant organic materials, average N concentrations (3.5%) of green leaf biomass are comparable to other N_2 fixing leguminous shrubs and trees (i.e., *Calliandra calothyrsus*, *Crotalaria grahamiana*, *Lantana camara*, *Leucaena leucocephala*, *Sesbania sesban*, and *Tephrosia vogelii*) (Gachengco et al. 1999). Dried samples of *Tithonia* were found to have significantly higher N, P, and K than dried *Chromolaena odorata* by 28%, 18%, and 17%, respectively. In addition, sunflower dried samples were significantly greater than dried *Panicum maximum* in terms of N, K, and Ca by 36%, 62%, and 54%, respectively (Olabode et al. 2007).

Compared to animal manure, sunflower's N concentration (1.76%) was not significantly different from N concentrations of poultry (1.78%) and swine manure (1.69%). The P content of sunflower (0.82%) was also not significantly different from the P content of swine manure (1.32%), but significantly higher than P concentrations of cattle manure (0.52%). In terms of K content, sunflower (3.92%) was significantly greater than those found in poultry (1.80%), cattle (0.95%), and swine manure (0.77%). However, sunflower had the lowest Mg concentration (0.0055%) among the organic materials compared (Olabode et al. 2007).

23.3.2 Sunflower Decomposition

Figure 23.1 shows the rapid transformation of sunflower cuttings applied (500, 1000, and 2000 g) to the soil plots 2 weeks after the sunflower application, which signals the rapid decomposition of sunflower cuttings. All soil plots treated with varying amounts of sunflower cuttings had fully decomposed by the 4th week, which is in accordance with the results reported by Nagarajah and Nizar (1982), who found that wild sunflower (green leaves and tender stems) applied to lowland rice paddies in the

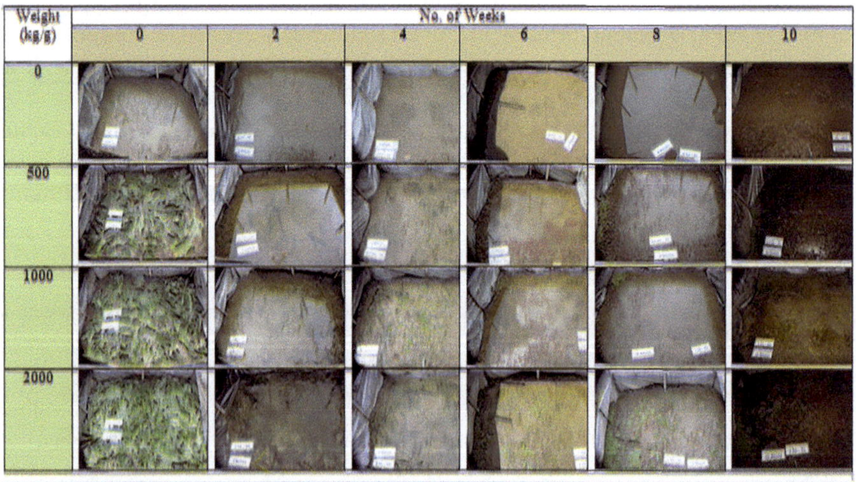

Fig. 23.1 Weekly decomposition of varied amounts of sunflower cuttings applied to each soil plots, Brgy. Poitan, Banaue, Ifugao, Philippines

mid-country wet zone of Sri Lanka had completely decomposed in about 3–4 weeks after planting.

In addition to the considerably high N, P and K contents of sunflower green biomass, it also has low lignin (6.5%) and polyphenol (1.6%) contents, which highlights its high potential as a soil amendment (Jama et al. 2000). Palm and Rowland (1997) also reported moderate lignin content and high soluble fraction as other strong points of the use of sunflower as organic material, which results in high biodegradability of the organic material. Dried samples of this weed were also shown to have a lower C/N ratio (8:1) than *Chromolaena odorata* (12:1) and *Panicum maximum* (30:1), which is an indication of a faster rate of decomposition (Olabode et al. 2007). Dried sunflower samples were also significantly better than dried *Chromolaena odorata* samples (25.72%) in terms of organic matter content (24.04%) (Olabode et al. 2007).

According to Gachengco et al. (1999), dry matter of sunflower takes a half-life of about 1 week to disappear during the rainy season in Western Kenya. A statistical comparison of species decomposition constants also showed that sunflower has the fastest decomposition rate, which was significantly different from leguminous crops at a P value of 0.05 (Partey 2010).

Fast decomposition of sunflower leaf biomass suggests a rapid release of nutrients, which is beneficial to short duration crops (i.e., vegetables and most annual crops) (Jama et al. 2000). However, the form of sunflower biomass to be applied must be a prime consideration as this affects the release of nutrients. According to Mafongoya and Nair (1997), drying of plant biomass leads to increase in polyphenol content and reduction in nutrient release. Hence, the application of sunflower green biomass is more beneficial to the soil than the application of dried sunflower biomass (Otuma et al. 1998).

23.3.3 Effects of Different Sunflower Application Rates on Soil Nutrients and Soil Properties of Closed *Payoh* System

23.3.3.1 Soil pH

At the start of the experiment, the pH of the soil was acidic, with values around 5 across all treatments of sunflower. Naturally, this is expected as the pH of soil under waterlogged conditions becomes acidic due to soil chemical reactions that take place under anaerobic conditions. The addition of sunflower at varying amounts resulted in an increase in soil pH 2–4 weeks after the application of sunflower. The plot with the highest treatment of sunflower (2000 g) showed the highest increase in soil pH (reaching up to pH of 6) on the 4th week after sunflower application.

Likewise, sunflower application of 500 and 1000 g resulted in high increase in soil pH for the two treatments 4 weeks after application (Fig. 23.2). These results conform the findings of Atayese and Liasu (2001), Hakim and Agustian (2003), and Hakim et al. (2009) that the use of sunflower as soil amendment results in higher soil pH.

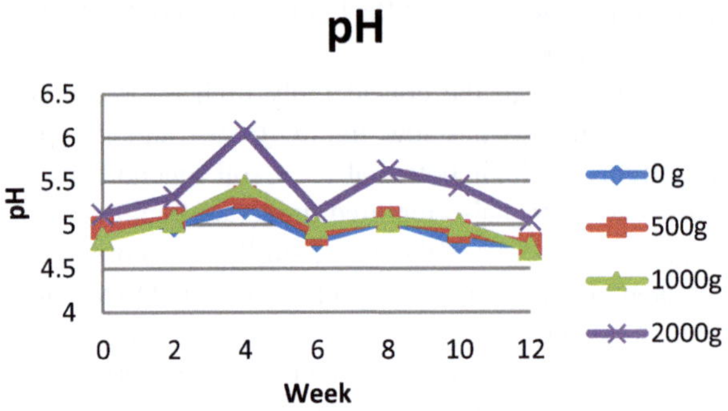

Fig. 23.2 Changes in soil pH of plots treated with varying levels (0, 500, 1000, and 2000 g) of sunflower during the 12-week decomposition period, Brgy. Poitan, Banaue, Ifugao, Philippines

Table 23.2 Analysis of variance (ANOVA) of soil pH between week, sunflower treatments, and interaction of week and sunflower treatments, Brgy. Poitan, Banaue, Ifugao, Philippines

Source of variation	SS	df	MS	*F*	*P* value	*F* crit
Sample (week)	4.590536	6	0.765089	16.71456	1.36E-12	2.208554
Columns (sunflower treatments)	3.591429	3	1.197143	26.15345	4.83E-12	2.713227
Interaction (week × sunflower treatments)	1.047321	18	0.058185	1.271131	0.228078	1.727955
Within	3.845	84	0.045774			
Total	13.07429	111				

After week 4, the soil pH across all treatments declined to pH levels around 5. Soil pH for all treatments rises again on the 8th week, but decreases back to around 5 pH levels on the 10th and 12th weeks (Fig. 23.2).

Based on the result of the two-way ANOVA test, the soil pH level across all treatments (*P* value = 1.36E-12) for each week (*P* value = 4.83E-12) was significantly different from each other (Table 23.2). That is, soil pH for 0, 500, 1000, and 2000 g of sunflower application was significantly different from each other for each week. Hence, this confirms that soil pH after the application of 2000 g of sunflower is significantly less acidic compared to other amounts of sunflower application.

Atayese and Liasu (2001) reported that soil under sunflower has higher pH. In addition to the improved soil fertility effect of sunflower by increasing soil pH, sunflower also reduces acid content and Al saturation of soil (Hakim and Agustian 2003; Hakim et al. 2009).

23.3.3.2 Soil Organic Matter (OM) Content

The soil organic matter (OM) content of the soil treated with sunflower increased 4 weeks after the application of sunflower. It was also shown that the soil organic matter increased in parallel with the rate of sunflower application (Fig. 23.3). Plots with the highest sunflower application (2000 g) had the highest soil organic matter content.

Fig. 23.3 Changes in soil organic matter (OM) content of plots treated with varying levels (0, 500, 1000, and 2000 g) of sunflower during the 12-week decomposition period, Brgy. Poitan, Banaue, Ifugao, Philippines

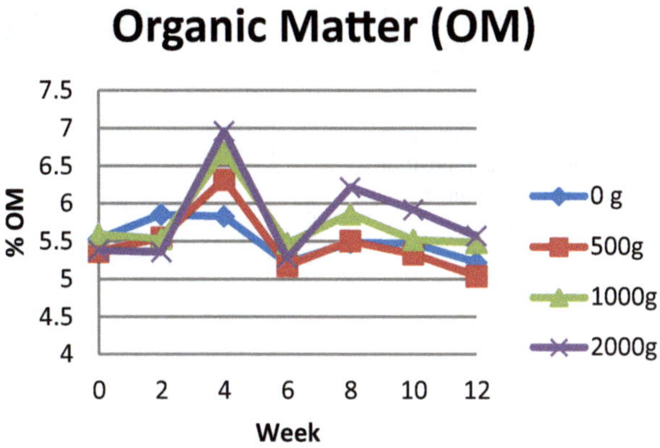

Table 23.3 Analysis of variance (ANOVA) of soil organic matter between week, sunflower treatments, and interaction of week and sunflower treatments, Brgy. Poitan, Banaue, Ifugao, Philippines

Source of variation	SS	df	MS	*F*	*P* value	*F* crit
Sample (week)	14.89465	6	2.482441	12.22524	7.71E-10	2.208554
Columns (sunflower treatments)	2.321496	3	0.773832	3.810879	0.012968	2.713227
Interaction (week × sunflower treatments)	4.293911	18	0.238551	1.174787	0.300548	1.727955
Within	17.05693	84	0.203059			
Total	38.56698	111				

This observation is confirmed by the result of the two-way ANOVA test, which revealed that organic matter for all treatments was significantly different from each other (*P* value = 7.71E-10) for each week (*P* value = 0.012968) (Table 23.3). In other words, the organic matter for the four applications of sunflower (0, 500, 1000, and 2000 g) was not the same for each week. Thus, soil organic matter content for 12 weeks is higher with higher applications of sunflower to the soil.

These findings are also in agreement with the results of Hakim and Agustian (2003) and Hakim et al. (2009) that soil application of sunflower increases soil organic matter.

23.3.3.3 Soil Available Phosphorus (P)

In general, soil available P for all treatments experienced large fluctuations within the 12-week decomposition period (Fig. 23.4), which might be related to factors affecting the availability of soil nutrients such as increase in soil microbial P biomass and decrease in soil P sorption (Jama et al. 2000).

Increase in soil available P for most of the 12-week decomposition period was only evident with the application of 2000 g of sunflower (Fig. 23.4). This might be attributable to large amount of sunflower applied to the soil, which is higher than the

Fig. 23.4 Changes in soil phosphorus (P) content of plots treated with varying levels (0, 500, 1000, and 2000 g) of sunflower during the 12-week decomposition period, Brgy. Poitan, Banaue, Ifugao, Philippines

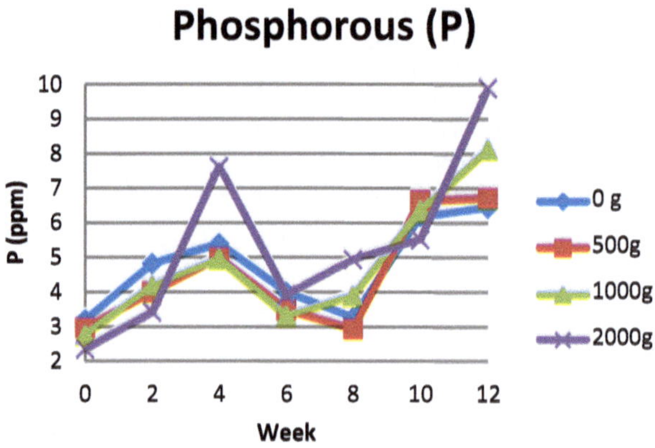

Table 23.4 Analysis of variance (ANOVA) of soil phosphorus between week, sunflower treatments, and interaction of week and sunflower treatments, Brgy. Poitan, Banaue, Ifugao, Philippines

Source of variation	SS	df	MS	*F*	*P* value	*F* crit
Sample (week)	592.7143	6	98.78571	8.618187	2.63E-07	2.208554
Columns (sunflower treatments)	183.8338	3	61.27795	5.345963	0.002031	2.713227
Interaction (week × sunflower treatments)	639.9143	18	35.55079	3.101495	0.000239	1.727955
Within	962.8475	84	11.46247			
Total	2379.31	111				

sunflower (5 t dry matter per ha) applied by Nziguheba et al. (1998) to maize crop that resulted in high increase in microbial P and larger reduction in sorbed P. On the other hand, the application of 500 and 1000 g of sunflower might be too low to induce a significant increase in soil available P in a flooded soil, as suggested by the results of Mutuo (2000) that a lower rate of application of sunflower biomass to maize crop does not result in an increase in soil microbial P. Results also show that the levels of soil available P for all treatments peak on the 12th week of decomposition (Fig. 23.4).

Furthermore, two-way ANOVA results revealed that soil available P for all sunflower treatments was significantly different from each other (*P* value = 0.002031) for each week (*P* value = 2.63E-07) (Table 23.4). Thus, this confirms that soil available P for 2000 g sunflower application was significantly higher for each week than other amounts of sunflower application.

23.3.3.4 Effect of Sunflower on Nutrient Release, Uptake, and Availability

During the decomposition process of sunflower, organic acids (i.e., acetic, propionic, salicylic, citric, succinic, and tartaric acids) were released, which potentially make the insoluble form of P and K to a more readily available form for plants to absorb. These organic acids also allow these insoluble P and K in the soil to potentially form chelates with Fe and Al (Gusnidar 2007).

Another strong point of sunflower as a soil amendment is its ability to attract fungi. According to Gusnidar (2007), soils under sunflower experienced increase in arbuscular mycorrhizal fungi spore, which enhances the nutrient absorption from the soil to the biomass.

Evidences also showed that soil incorporated with sunflower green biomass also induces rapid release of phosphorus. This is exhibited by the higher labile inorganic P in the acid soil of Western Kenya at 2 weeks incorporation of 15 kg P per ha from sunflower (8.1 mg P per kg) than those from the triple superphosphate (3.6 mg P per kg) (Nziguheba et al. 1998).

Studies also showed that sunflower application also increased the availability of nutrients by increasing P in soil microbial biomass and reducing P sorption by soil (Jama et al. 2000). Specifically, Nziguheba et al.'s (1998) findings revealed higher increase in microbial P (4.3 mg P/kg) and larger reduction in sorbed P (40 mg P/kg) from 15 kg P per ha as sunflower (equivalent to 5 t dry matter per ha) than as triple superphosphate (1.8 mg P/kg increase in microbial P and 19 mg P/kg reduction in sorbed P) after 2 weeks of application. However, lower application rate of sunflower (1.8 t dry matter per ha) to maize cropping did not result in an increase in microbial P (Mutuo 2000).

23.3.3.5 Soil Exchangeable Potassium (K)

Figure 23.5 shows that soil exchangeable potassium (K) increased with increasing amounts of sunflower cuttings applied. This might be attributed to the organic acids (i.e., acetic, propionic, salicylic, citric, succinic, and tartaric acids) produced by

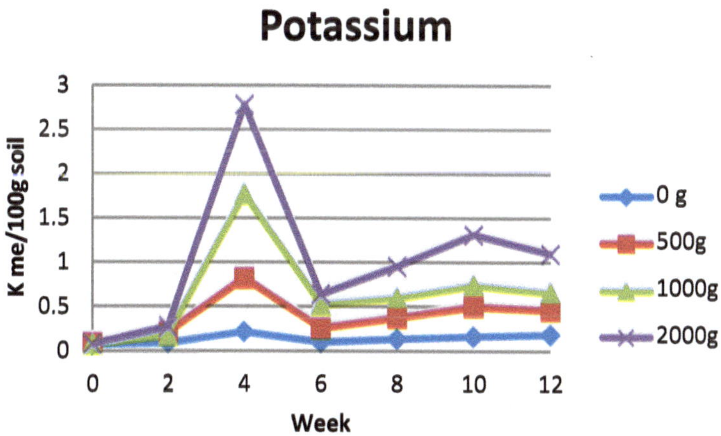

Fig. 23.5 Changes in soil exchangeable potassium (K) content of plots treated with varying levels (0, 500, 1000, and 2000 g) of sunflower during the 12-week decomposition period, Brgy. Poitan, Banaue, Ifugao, Philippines

Table 23.5 Analysis of variance (ANOVA) of soil potassium between week, sunflower treatments, and interaction of week and sunflower treatments, Brgy. Poitan, Banaue, Ifugao, Philippines

Source of variation	SS	df	MS	*F*	*P* value	*F* crit
Sample (week)	17.91227	6	2.985378	43.13314	1.22E-23	2.208554
Columns (sunflower treatments)	11.93265	3	3.97755	57.46817	2.69E-20	2.713227
Interaction (week × sunflower treatments)	9.986825	18	0.554824	8.016165	8.87E-12	1.727955
Within	5.8139	84	0.069213			
Total	45.64564	111				

sunflower during its decomposition that potentially transform the insoluble K to a form that are more readily absorbed by the crop (Gusnidar 2007).

Specifically, the peak increase in soil exchangeable K across three applications of sunflower cuttings (500, 1000, and 2000 g) was observed 4 weeks after sunflower application. Among the three treatments with sunflower cuttings, the highest soil exchangeable K was observed on soil plots treated with 2000 g of sunflower (Fig. 23.5).

This trend was further confirmed by the results of the two-way ANOVA test. The statistical test showed that soil exchangeable K across all treatments (P value = 1.22E-23) was different for each week (P value = 2.69E-20) (Table 23.5). Hence, a higher application of sunflower cuttings to the soil results in higher soil K.

23.3.3.6 Effect of Sunflower on Soil Properties

The findings of Atayese and Liasu (2001) revealed that soil under sunflower in the guinea savanna zone of Nigeria had higher N, P, K, Na, and Ca than bare soil. In comparison to soil cropped to cassava, soil fallowed to sunflower experienced an increase in soil N, available P, and exchangeable K by 206%, 41%, and 57%, respectively. Soil under sunflower and *Chromolaena* was insignificantly different from each other in terms of soil N and P, but the former was significantly greater than the latter's soil K by 18% (Onejiyi et al. 2012).

At different soil depths, soil chemical properties in terms of N (0.55%), P (8.3 mg/kg), K (0.47 cmol/kg), Mg (0.93 cmol/kg), Ca (3.8 cmol/kg), and organic matter (2.96%) at 0–5 and 5–10 cm depths were higher in soils fallowed to sunflower than the soils cropped to cassava. In addition, soils under sunflower had also higher N (0.29%), P (5.9%), Ca (2.9 cmol/kg), and organic matter (2.37%) values at 10–15 cm soil depth than the cropped soils. Results also showed that fallow plants like sunflower increase soil nutrients at a decreasing rate from 0 to 15 cm soil depth (Onejiyi et al. 2012).

In terms of soil physical properties, Atayese and Liasu (2001) also reported that soil under sunflower had higher pH, porosity, moisture content, arbuscular mycorrhizal fungi spores, and earthworm cast density and lower bulk density than bare soil. Likewise, Ojeniyi et al. (2012) found that soil fallowed to sunflower had significantly higher total porosity (62%) and significantly lower soil bulk density (1.0 g/cm^3) than soils cropped to cassava and fallowed to spear grass. Soil moisture content under sunflower (35.7%) was also insignificantly different from soil fallowed to spear grass (37.5%).

In addition to the improved soil fertility effect of sunflower by increasing soil pH, soil organic matter, and soil nutrients (i.e., N, P, K, Ca, and Mg levels), the weed also reduces acid content and Al saturation of soil (Hakim and Agustian 2003; Hakim et al. 2009).

23.4 Conclusion

Results revealed that nutrient content of sunflower stalks and leaves in terms of N, P, and K were relatively higher than other plant species used as soil fertilizer in the *payoh* plots of Banaue, Ifugao. In terms of rate of decomposition of sunflower, various amounts of sunflower cuttings applied were completely decomposed 4 weeks after the application of cuttings to the soil plots.

As to the effect of sunflower cuttings application on soil nutrients, an application of 2000 g of sunflower cuttings to 1 × 1 m soil plots resulted in significantly higher soil pH, soil organic matter, soil available P, and soil exchangeable K during the 12-week decomposition period than other amounts of sunflower cuttings applied.

In conclusion, the application of 2000 g of sunflower cuttings to soil plots in the *payoh* of Brgy. Poitan, Ifugao, Banaue results in significantly higher soil pH, soil organic matter, soil available P, and soil exchangeable K within 12-week duration.

Acknowledgments This study is part of the project entitled "Nutrient and Water Dynamics, Plant Biodiversity and Agricultural Productivity of *Muyong–Payoh* Systems in Banaue, Ifugao" funded by the National Research Council of the Philippines. We would like to thank our farmer collaborators in Brgy. Poitan and the Local Government Unit of Banaue, Ifugao for their support during the conduct of the study.

References

Adeniyan BO, Ojeniyi SO, Awodun MA (2008) Relative effect of weed mulch types on soil properties and yield of yam in Southwest Nigeria. J Soil Nat 2(3):1–5

Atayese MO, Liasu MO (2001) Arbuscular mycorrhizal fungi weeds and earthworm interactions in the restoration of soil fertility in the Guinea savanna region of Nigeria. Moor J Agric Res 3:104–109

Azeez JO (2010) Dynamics of carbon, nitrogen, phosphorus and potassium under different *Tithonia diversifolia* management systems in a tropical Alfisol: a greenhouse bioassay. J Agric Sci Environ 10(1):61–71

Butic M, Ngidlo R (2003) *Muyong* forest of Ifugao: assisted natural regeneration in traditional forest management. In: Dugan PC et al (eds) Advancing assisted natural regeneration (ANR) in Asia and the Pacific. Food and Agriculture Organization of the United Nations, Regional Office for Asia and the Pacific, Bangkok, Thailand, pp 23–27
Gachengo CN (1996) Phosphorus release and availability on addition of organic materials to phosphorus fixing soils. MSc thesis. Moi University, Eldoret, Kenya
Gachengo CN, Palm CA, Jama B, Othieno C (1999) Tithonia and senna green manures and inorganic fertilisers as phosphorus sources for maize in western Kenya. Agr Syst 44:21–36
Ganunga R, Yerokun O, Kumwenda JDY (1998) *Tithonia diversifolia*; an organic source of nitrogen and phosphorus for maize in Malawi. In: Waddington SR et al (eds) Soil fertility research for maize-based farming systems in Malawi and Zimbabwe, pp 191–194
Gusnidar (2007) Budidaya dan pemanfaatan Tithonia diversifolia untuk menghemat pemupukan N, P, dan K padi sawah intensifikasi. Disertasi Doktor Ilmu Pertanian Program Pascasarjana Unand, Padang, Indonesia
Gusnidar SY, Burbey Y, Saleh S, Andhika R (2012) Application of rice straw and Tithonia for increasing fertilizer use efficiency on Paddy soil. J Trop Soils 17(2):151–156
Hakim N (2002) Kemungkinan penggunaan *Tithonia diversifolia* sebagai sumber bahan organik dan unsur hara. J Andalas Bid Pert 38:80–89. Indonesian
Hakim N, Agustian (2003) Gulma Tithonia dan pemanfaatannya sebagai sumber bahan organik dan unsur hara untuk tanaman hortikultura. Laporan Penelitian Tahun I Hibah Bersaing XI/I. Proyek Peningkatan Penelitian Perguruan Tinggi DP3M Ditjen Dikti. Lembaga Penelitian Unand. Padang, Indonesian
Hakim N, Agustian MY (2012) Application of organic fertilizer Tithonia plus to control iron toxicity and reduce commercial fertilizer application on new Paddy field. J Trop Soils 17(2): 135–142
Hakim N, Mala Y, Agustian (2009) Pembuatan dan pemanfaatan pupuk organik Tithonia plus dalam penerapan metode SRI pada sawah bukaan baru. Laporan Hasil Penelitian KKP3T Tahun I. LP Unand dan Balitbang Pertanian Deptan. 61. Indonesian
ICRAF (1997) Using the wild sunflower, *T. diversifolia* in Kenya. International Centre for Research in Agroforestry, Nairobi, Kenya, p 5
ICRAF (1998) Annual report for 1997. International Centre for Research in Agroforestry, Nairobi, Kenya
Jama BCA, Buresh RJ, Niamg A, Gachenco CN, Nziguheba G, Amadalo B (2000) *Tithonia diversifolia* as green manure for soil fertility improvement in western Kenya. A review. Agr Syst 49:201–221
Liasu MO, Achakzai AK (2007) Influence of *Tithonia diversifolia* leaf mulch and fertilizer application in the growth and yield of potted tomato plants. Am Euras J Agric Environ 2: 335–340
Mafongoya PL, Nair PKR (1997) Multipurpose tree prunings as a source of nitrogen to maize under semiarid conditions in Zimbabwe. Nitrogen recovery in relation to pruning quality and method of application. Agr Syst 35:31–46
Mutuo PK (2000) Soil phosphorus pools following phosphorus fertilization and their relationship to maize yield in western Kenya. MSc thesis. Moi University, Eldoret, Kenya
Nagarajah S, Nizar BM (1982) Wild sunflower as a green manure for rice in the mid-country west zone. Trop Agric 138:69–78
Niang A, Amadalo B, Gathumbi S, Obonyo CO (1996) Maize yield response to green manure application from selected shrubs and tree species in western Kenya: a preliminary assessment. In: Mugah JO (ed) Proceedings of the first Kenya agroforestry conference on people and institutional participation in agroforestry for sustainable development. Kenya Forestry Research Institute (KEFRI), Muguga, Kenya, pp 350–358
Nziguheba G, Palm CA, Buresh RJ, Smithson CP (1998) Soil phosphorus fractions and absorption as affected by organic and inorganic sources. Plant and Soil 198:159–168

Obatolu CR, Agboola AA (1993) The potential of Siam weed (*Chromolaena odorata*) as a source of organic matter for soils in the humid tropics. In: Mulongoy K, Merxckx R (eds) Soil organic matter dynamics and sustainability in tropical agriculture. IITA/K.U. Leuven; Wiley, New York, pp 89–99

Ojeniyi SO, Odedina SA, Agbede TM (2012) Soil productivity improving attributes of Mexican sunflower (*Tithonia diversifolia*) and siam weed (*Chromolaena odorata*). Emir J Food Agric 24(3):243–247

Olabode OS, Sola O, Akanbi WB, Adesina GO, Babajide PA (2007) Evaluation of *Tithonia diversifolia* (Hemsl) a Gray for soil improvement. World J Agric Sci 3(4):503–507

Otuma P, Burundi C, Khabeleli A, Wasia E, Shikanga M, Mulogoli C, Carter SE (1998) Participatory research on soil fertility management in Kabras, Western Kenya: report of activities, 1996–1997. Tropical Soil Biology and Fertility Programme (TSBF), Nairobi, Kenya

Palm CA, Rowland AP (1997) Chemical characterization of plant quality for decomposition. In: Cadisch G, Giller KE (eds) Driven by nature, plant litter quality and decomposition, pp 379–392

Partey ST (2010) The agronomic qualities of the Mexican sunflower (*Tithonia diversifolia*) for soil fertility improvement in Ghana: an exploratory study. PhD Thesis. Kwame Nkrumah University of Science and Technology, Kumasi, Ghana

Rondolo MT (2001) Fellowship report. Tropical forest update. ITTO, Japan. 11(4)

Rutunga V, Karanja NK, Gachene CKK, Palm CA (1999) Biomass production and nutrient accumulation by *Tephrosia vogelli* and *Tithonia diversifolia* fallows during six-month growth at Maseno. Biotechnol Agron Soc Environ 3:237–246

SITMo (2008) Impact: the effects of tourism on culture and the environment in Asia and the Pacific: sustainable tourism and the preservation of the World heritage site of the Ifugao rice terraces, Philippines. Save the Ifugao terraces movement. UNESCO, Bangkok, Thailand, p 89

Sonke D (1997) Tithonia weed—a potential green manure crop. Echo Dev Notes 57:5–6

Opoku S, [illegible] [illegible]

[illegible] (2003) [illegible] Soil Science and [illegible]

[illegible]

[illegible]

[illegible]

Part V

Ecosystem and Species Modelling for Evidence-Based Decision Making

Forest Ecosystem Modeling for Policy Planning: A Review

24

Karun Jose, Aritra Bandopadhyay, A. Arya, and Rajiv Kumar Chaturvedi

Abstract

Vegetation modeling is an advanced tool that helps to understand the current forest ecosystem dynamics and provides a peek into future possibilities. In the era of climate change, projecting and monitoring different ecosystem elements and biodiversity are critical in supporting the management and conservation of forest ecosystems. Quantitative models are often used to understand and project the "impact of climate change" and the associated disturbances in forest ecology. Here we present a review of different ecosystem modeling approaches, exploring their potential applications to understand changing forest dynamics and climate change adaptation options in forest ecosystems. This comprehensive and comparative study helps us to get insights into the advantages and limitations of the various modeling-based approaches, providing a guideline for systematic execution of policy assessment according to a defined criteria (e.g., uncertainty management, data required, spatial and temporal dynamics, adaptation measures integration, and level of complexity). Further, we present an overview of ecosystem modeling and its usability for global policy planning in the forest sector. Finally, we suggest ways to use these advanced tools to help policy planning for conservation, restoration, and climate change adaptation in forest ecosystems.

Keywords

Vegetation modeling · Forest · Policy

K. Jose · A. Bandopadhyay · A. Arya · R. K. Chaturvedi (✉)
BITS Pilani, KK Birla Goa Campus, Sancoale, Goa, India
e-mail: rajivc@goa.bits-pilani.ac.in

S. Dhyani et al. (eds.), *Ecosystem and Species Habitat Modeling for Conservation and Restoration*, https://doi.org/10.1007/978-981-99-0131-9_24

24.1 Introduction

Greenhouse gas build-up in the atmosphere and rising temperatures have already caused widespread losses and damages to nature, ecosystems, and people (IPCC 2022). Observed climate change has caused substantial damages, and irreversible losses, to many of the terrestrial ecosystems across the world including the forest ecosystems. These changes include increase in burned area by wildfires, shifting of species poleward and to higher elevations, among other examples. Global temperatures have so far risen by only 1.1 °C, even this small change in global temperatures has already caused irreversible losses and damages in forest ecosystems across the world. Under different climate change scenarios, global temperatures are expected to rise to 2.5–4 °C range (IPCC 2021), even in India. Chaturvedi et al. (2012) suggest that under business-as-usual scenario, the temperatures are likely to rise to 3.3–4.8 °C by 2080s. It is important to understand as to how the projected climate change may affect forest ecosystems in different future warning scenarios. IPCC AR4, WG2 report concluded that one of the most advanced tools to assess the impact of climate change on vegetation dynamics/terrestrial ecosystems is dynamic global vegetation models (DGVMs) (Fischlin et al. 2007). Vegetation modeling, an emerging sophisticated tool, is being developed to understand ecosystem dynamics and predict future scenarios. The ecosystem model is defined as "a model that explains the interconnection between at least two ecosystem components, where the interactions are true ecological processes" (Tylianakis et al. 2008). Recently, unregulated anthropogenic emissions of warming gases and consequent climate change have been posing severe threats to the protected areas of the environment (IPCC 2022). Moreover, the rising population and demand for resources amplify agricultural expansion and extensive land-use changes, thereby destroying habitats and leading to species extinction (Newbold et al. 2014). Vegetation modeling will help the scientific community to monitor and understand complex environmental dynamics and develop long-term policy measures for effective management (Pasetto et al. 2018). Moreover, modeling biodiversity and ecology would further support the implementation of sustainable development whereby an understanding of resource utilization is obtained through this process (Niesenbaum 2019).

The concept of ecological modeling and its evolution began a century ago (Lotka 1925; Volterra 1926), but technological advancement, as seen within the past decade, has brought significant development in using these models (Chatzinikolaou 2013). The forest ecosystem is also uncertain due to the potential impacts of the changing climate (Keenan 2015; Nunes et al. 2021). Several forest simulation models predict that the forest composition and comprehensive coverage will cease in the future due to unpredictable consequences of climate change (Kirilenko and Sedjo 2007; d'Annunzio et al. 2015). Forest vegetation models have been coded to perform on various scales such as the leaf, stand, ecosystem, and regional and global levels incorporating various processes (such as photosynthesis, stomatal exchange, and evapotranspiration) (Hui et al. 2017). Farquhar's photosynthesis model estimates carbon budget and plant growth at leaf and canopy level, approximating the plant canopy to be a big leaf (Chen et al. 1999; Wang et al. 2017). On the regional and global scale, various ecosystem models have evolved; for instance, Schaefer et al. (2008) applied the Carnegie-Ames-Stanford Approach (CASA) model to estimate terrestrial biomass and carbon fluxes. They created a hybrid model by integrating the Simple Biosphere (SiB 2.5) model that provides biophysical and photosynthesis with the CASA model, which was able to project long-term carbon sources and singles that the individual models could not have. The terrestrial ecosystem carbon model (TECM) is another process-based model that explains the carbon dynamics of soils and plants within the terrestrial ecosystem (Wang et al. 2011). TECM mainly utilizes information on spatially explicit parameters in terrestrial ecosystems to calculate the estimates of carbon pool sizes and carbon fluxes. Schaphoff et al. (2018) provide an extensive overview of the latest version of LPJmL4, a process-based dynamic global vegetation model (DGVM) project, which is the consequence of climate and land use changes on the agriculture, terrestrial biosphere, and hydrological and carbon cycle. Joint U.K. Land Environment Simulator (JULES) is an improved model based on MOSES and TRIFFID DGVM, which includes a multitude of options for photosynthesis scaling from leaf to canopy, with the utmost intricate modeling of light interception profile through the vegetation (Clark et al. 2011).

Modeling helps policymakers anticipate the impacts of ecosystem degradation on human actions and projects future scenarios based on direct and indirect factors. Simulation of interaction between humans and the environment is essential for summiting the pathways to Sustainable Development Goals 2030. Despite all advancements in Vegetation modeling, it is evident that the research community has used a few models for management and decision-making processes, given the complexity of understanding mathematical models (DeAngelis et al. 2021). In this study, we attempt to review different Vegetation modeling approaches and explore their potential to understand forest dynamics and their applications in climate change adaptations.

24.2 Vegetation Modelling: From Correlative to Process-Based Approaches

Models are valuable tools for summarizing, arranging, and combining information or data into formats that enable the creation of probabilistic, quantitative, or Bayesian statements regarding the potential or future condition of the modeled entity (Duarte et al. 2003). Based on the complexity and degree of formalization, the Vegetation modeling can be sub-segmented into correlative, process-based, and expert-based models (Ferrier et al. 2016). Traditionally, the most common method of management was based on information provided by experts (Sutherland 2006). The term "expert" can be defined as one who attained a highly precise skill set in a specific field through learning experience (Kuhnert et al. 2010). An expert-based method generally comprises the following steps as described: deciding on how the information is to be used, what to bring out from it, designing the elicitation process, actual conducting the elicitation, and finally converting the output into quantitative statements that can be applied to a modeling approach (Martin et al. 2012). This approach has a time advantage over other models when the final decision is to be made exceptionally quickly with minimal data.

Correlative models use statistical techniques to develop the direct connection between biodiversity data (species abundance, richness, distribution) and environmental variables (Morin and Lechowicz 2008; Li et al. 2020). Based on actual observation data, correlative models generate information on biodiversity trends and their responses to the controlling factors, but they do not make an attempt to describe the mechanisms behind such patterns and reactions. They are usually used to forecast the future impact of environmental changes, the effects on biodiversity by human intervention, to help human production activities (increasing agricultural production), and to understand the ecological requirements for different species (Rahbek et al. 2007; Elith and Franklin 2013; Cobos et al. 2019). Since these models are designed based on data from the past state of the system, rapid decisions based on statistical relationship is feasible (Cuddington et al. 2013). However, under the current climate change conditions, models based on the previous data of a system are not suitable for future simulations (Williams et al. 2007). For example, many studies have predicted changes in species range based on climatic conditions in India. Models such as MaxEnt and SMCE use climatic data and species occurrence data of a particular location to develop a correlation and predict the future species range under climate change (Nimasow et al. 2016; Yadav et al. 2022). However, they deny including relevant ecological processes such as interspecific interactions and demographic relationships, which can also limit the species range, and their effect may not be included in future predictions.

Process-based models that work based on understanding critical ecological processes from a theoretical perspective give a suitable framework for including specific responses to changing environmental conditions (Cuddington et al. 2013). These are often more challenging to design than correlative models, because they need considerable information on factors that drive biodiversity patterns (Ferrier et al. 2016). There are many types of process models, for example, gap models, biogeochemical models, and DGVMs. Gap models are applied to investigate changes in vegetation and species interactions at significantly higher spatial resolution (plots the size of a single canopy gap or individual trees) across daily to yearly time steps. However, simulation of dynamics over several stands and cells is achievable. Biogeochemical models project carbon, water, and mineral (nutrient) cycles in terrestrial ecosystems such as forests. In climate change research, these models are widely applied to predict ecosystem net primary production, carbon flow, and storage. DGVMs project changes in vegetation attributes (such as leaf area and phenology) across annual to decadal time steps at vast geographical scales (Kerns and Peterson 2014) (more details on DGVMs are available in Sect. 24.3.2). However, Hybrid models are a combination of empirical and mechanistic components. There are two kinds of hybrid models: the first one integrates process-based empirical models by creating signal-transfer environment productivity functions, and the second one includes a causal structure with both empirical and mechanistic components (Luxmoore et al. 2002; Pretzsch 2009).

24.3 Vegetation Modeling at Leaf, Individual, Plot, Regional, and Global Levels

Vegetation models are designed at various scales, ranging from the leaf to the plant canopy and at the plot, regional, and global levels. These models mainly project phenomena such as photosynthesis and respiration, carbon distribution between plant organs, nitrogen uptake and mineralization, litter production, and Soil Organic Carbon (SOC), and these processes are used to understand the carbon fluxes between the atmosphere, soil, and plants (Hanson et al. 2004).

24.3.1 Leaf and Stand Models

At the leaf level, *Farquhar, von Caemmerer, and Berry* (FvCB) is the most commonly used model for projecting photosynthesis and leaf-level carbon and water fluxes (Rogers et al. 2017). The photosynthesis module predicts leaf-level carbon uptake based on biochemical or physiological characteristics, as well as the abiotic environment (intercellular CO_2 concentration and temperature). Similarly, stomatal modules connect the intercellular leaf space to the canopy air space and biophysically constrain carbon and water fluxes from the perspective of gas diffusion (Xu and Trugman 2021). Individual tree growth models such as *BWIN*, *Prognaus*, *Silva*, and *Moses* are widely used for predicting the influence of climate change on tree development, yield predictions, and ecosystem fluxes (Vospernik 2017). Most growth models are designed based on the mass balance method and consider organic matter decomposition, ecosystem fluxes (forest), and water balance. Hence, these models can evaluate above- and below-ground biomass production and assess carbon dynamics for a particular location (Hui et al. 2017). Table 24.1 represents some of the widely used individual and strand-level models (the table is classified based on type, spatial structure, and temporal structure).

Climate change affects specific physiological processes in plant species, such as photosynthesis, respiration, and growth, and can be investigated by different models. While certain models focus on the impact of elevated CO_2 concentration on the ecosystem, others, especially biogeochemical models, simulate the consequences of various climatic factors on the forest ecosystem carbon cycle. The physiological principles predicting growth (3-PG) model was developed to connect the traditional, mensuration-based growth and yield with process-based carbon balance models. Gross primary production (GPP) in forest ecosystems is mostly estimated using 3-PG process-based model at the stand level. By combining remote sensing and GIS techniques, the upgraded version of 3-PGS (physiological principles in predicting growth with satellite) estimates biophysical variables, including LAI (leaf area index), CWC (canopy water content), and FAPAR (fraction of absorbed

Table 24.1 The leaf and stand-level Vegetation models and classification based on temporal and spatial structure

Sl. no.	Model name	Type	Spatial structure	Temporal structure	Reference
1	3-PG	Process-based	Stand or cohort	Monthly	Almeida et al. (2004)
2	PNET (-C.N., -DAY)	Process-based	Stand	Monthly/daily	Aber et al. (1997)
3	BWIN PRO	Empirical model	Individual	5 year	Albrecht et al. (2011)
4	SIMWAL	Process-based	Individual	Hour	Balandier et al. (2000)
5	EMILIION	Process-based	Individual	1/50 day	Bosc (2000)
6	Hybrid	Process-based	Individual	Daily	Friend et al. (1997)
7	BALANCE	Process-based	Individual	Daily	Rötzer et al. (2010)
8	FORECAST	Hybrid model	Individual	Yearly	Kimmins et al. (1990)
9	TREE-BGC	Process-based	Individual	Daily	Korol et al. (1995)
10	FORGEM	Process-based	Individual	Daily	Kramer et al. (2008)
11	TREEMIG	Process-based	Cohort	Yearly	Lischke et al. (2006)
12	CO2FIX V.2	Empirical model	Cohort	Yearly	Masera et al. (2003)
13	WOODPAM	Process-based	Stand	Monthly	Peringer et al. (2013)
14	BIOME-BGC	Process-based	Stand	Daily	Pietsch et al. (2003)
15	SILVA	Hybrid model	Individual	5 year	Pretzsch et al. (2002)
16	YIELD-SAFE	Process-based	Individual	Daily	Van der Werf et al. (2007)

photosynthetically active radiation), which can be used to simulate forest biomass and productivity at regional level (Gupta and Sharma 2019).

Similarly, Yan et al. (2011) applied the PnET-CN model to describe the carbon sequestration potential using biogeochemical cycles of carbon (C) and nitrogen (N); they also validated the output using the data from coniferous forests in south China. EMILION model can be used to project the carbon budget of current branches based on their age and position within the crown, considering parameters such as distribution of light and interception, respiration, photosynthesis, transpiration, stomatal conductance, phenology, water transfer, and intra-annual growth by utilizing an object-oriented approach (Bosc 2000). FORECAST Climate model operates through a hybrid simulation approach, representing moisture and temperature availability on tree growth and survival and nutrient cycling, litter decomposition, and also representing the impact of growing CO_2 on water use efficiency (Seely et al. 2015).

24.3.2 Regional and Global Ecosystem Models

Understanding the ecosystem response to climate change on a global scale is essential both as a scientific question and for making policy decisions. The accuracy of regional models depends on how effectively the field data used for model development represents the region of interest (ROI), how accurate the environmental model driving variables (vegetation type, climate) represent the ROI, and the accuracy of the model prediction and observe data for the region (Olson et al. 2001). In this section, we will explain different DGVMs, which are mainly used globally and in India.

DGVM is a computational-based model that simulates terrestrial vegetation and the phenomenon and processes related to it; broadly speaking, the biogeochemical or hydrological cycles and the influence climatic parameters have on them. It is powerful enough to capture the transition in the forest ecosystem due to the influence of one or more input parameters from climatic variables to soil parameters (Kumar et al. 2018). Fischlin et al. (2007) suggested that one of the most advanced tools to assess the impact of climate change on vegetation dynamics/terrestrial ecosystems is dynamic global vegetation models (DGVMs). Prentice (1989) put forward the first outline for DGVMs (Fig. 24.1). In DGVMs, time series datasets are fed to replicate the ecological processes and the way they influence the establishment of dominant forest vegetation. DGVMs were needed because static vegetation was incapable of including the plant life cycle, and various cyclic processes such as carbon cycle and nitrogen cycle were not integrated, nor were considered the various anthropogenic and natural disturbances and climatic extremes (Quillet et al. 2010). The important processes represented in DGVMS are (1) terrestrial or surface processes, including energy flow and water budget; (2) carbon flux and plant growth as part of the carbon cycle; (3) plant establishment, completion, and mortality as vegetation dynamics; and (4) natural and anthropogenic disturbances such as a forest fire, overgrazing, land-use change, and storms (Korappath and Bilyaminu 2022). Table 24.2 represents a few DGVMs and required input parameters and outputs.

Although a PFT (plant functional types)-based approach is employed in most of the DGVMs rather than an individual species-based approach, much information about species type is suppressed on the regional scale rather than on the global scale, to the point where the dominant species may be excluded. The necessity to input high-resolution land use datasets for accurate energy and water cycle measures in coupled model systems such as RCM-DGVM improved model performance and accurate projections. It also requires modifying the parameters for their applicability at a regional scale (Myoung et al. 2011). In India, several studies are available where DGVMs have been applied to assess the impact of climate change on forest

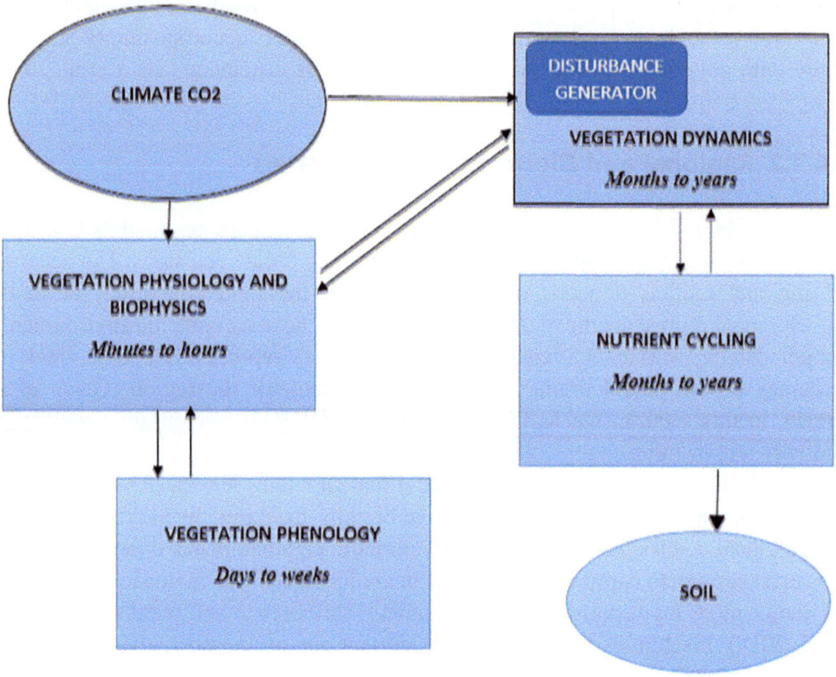

Fig. 24.1 The general framework and mechanisms of a DGVM and its time scale (adapted from Cramer et al. 2001)

ecosystems (Chaturvedi et al. 2011, 2012; Gopalakrishnan et al. 2011; Kumar et al. 2018).

24.4 Modeling and Policy-Making

The first National Forest Policy in India lead back to 1894, the British era. The policy was formulated to benefit the British Empire, restricting local people from utilizing forest resources and large-scale commercial deforestation by the East India Company. After independence, the National Forest Policy, 1952 was India's first forest policy; it was formulated with the concern about the need for efficient forest management and to prevent forest exploitation after the havoc of mindless deforestation during the colonial era. It incorporated every aspect that the world is concerned about today, such as protection measures, community interactions and administrative measures by the government, the scope for research, and annual budget allotment, which are mentioned and have evolved. It is also argued that to increase the forest cover to about one-third of the total land area today, we need even more robust and reliant policies to not only manage and protect the forest cover today but also the future and revive the already ailing forest regions. Making

Table 24.2 Major DGVMs that are broadly used in India and globally; we also represent the required inputs, outputs, and plant functional type (adapted from Aaheim et al. 2011, Kumar et al. 2018)

Model	Required input	PFTs	Output	Description
IBIS	1. Longitude and latitude (m) 2. Monthly mean temp. (°C) 3. Monthly mean temp. range (°C) 4. Minimum temp. ever recorded minus avg. temp. of the coldest month (°C) at that location 5. Mean "wet" days per month (days) 6. Monthly mean precipitation rate (mm/day) 7. Monthly mean relative humidity (%) 8. Monthly mean cloudiness (%) 9. Percentage of sand (%) 10. Percentage of clay (%)	Temperate broad-leaf evergreen Tropical broad-leaf evergreen Tropical broad-leaf drought-deciduous Temperate broad-leaf cold-deciduous Boreal conifer evergreen Boreal broad-leaf cold-deciduous Temperate conifer evergreen Boreal conifer evergreen Boreal broad-leaf cold-deciduous Boreal conifer cold-deciduous Evergreen shrub Cold-deciduous shrub	Average evapotranspiration Soil temperature Fractional cover of canopies Height of vegetation canopies Leaf area index NPP Total soil carbon and nitrogen Average sensible heat flux Vegetation types (IBIS classification) Total carbon from the exchange of CO_2	IBIS is recognized as the first model of its kind. It guides the researchers to develop improved global dynamic models with a better understanding to simulate the impacts of climate change on forests and their ecological processes. IBIS is a framework that combines land surface, vegetation dynamics, biogeochemical cycles, and hydrological processes. IBIS allows for the simulation of both short-term physiological processes and long-term ecosystem dynamics, which can be effectively included in atmospheric models. (Foley et al. 1996; Kucharik et al. 2000)
JULES	1. Longitude of the region 2. Temperature (°C) 3. Daily mean precipitation 4. Frequency of wet days 5. Incoming short- and long-wave	Broad-leaf trees Needle leaf trees C3 (temperate) grasses C4 (tropical)	Soil temperature Soil moisture Surface runoff Plant respiration Soil evaporation Gross primary productivity NPP Soil respiration	The Hadley Centre climate model includes the Joint U.K. Land Environment Simulator (JULES) to represent the land surface. It parameterizes the

(continued)

Table 24.2 (continued)

Model	Required input	PFTs	Output	Description
	radiation (W m^{-2}) 6. Diurnal temp. range (K) 7. Specific humidity 8. Wind speed	grasses Shrubs	Surface fluxes of heat Surface fluxes of carbon	hourly flows of energy, water, and CO_2 from the ground to the atmosphere. By developing seasonal stores of energy, water, and carbon budget, it can simulate changes in vegetation from decade to century (https://jules.jchmr.org/)
Biome-BGC	1. Altitude 2. Mean monthly values of precipitation (mm) 3. Temperature (°C) 4. Cloud cover (%) 5. Available water capacity of the topsoil 6. AWC of the subsoil	Tropical evergreen Temperate broad-leaved evergreen Summer green Tropical rain green Temperate evergreen conifer Boreal evergreen Temperate boreal deciduous Temperate grass Tropical/warm-temperate grass	1. Annual total precipitation (mm/yr) 2. Annual average air temperature (°C) 3. The annual maximum value of the projected leaf area index 4. Annual total evapotranspiration (mm/yr) 5. Annual total outflow (mm/yr) 6. Annual total NPP 7. Annual total net biome production	Biome-BGC is a model that estimates the fluxes and storage of energy, water, carbon, and nitrogen for the plant and soil components of terrestrial ecosystems. Because its algorithms depict physical and biological processes that influence energy and mass flows, it is a process model
LPJ	1. Daily air temperature (°C) 2. Precipitation (mm) 3. Solar radiation (W m^{-2}) 4. CO_2 concentration (ppm) 5. Soil texture (%) 6. Temperature (°C) 7. Soil water content	Tropical broad-leaved rain green Temperature needle-leaved evergreen Tropical broad-leaved evergreen Temperate broad-leaved evergreen Temperate	Vegetation structure PFTs Biomass carbon	Lund-Potsdam-Jena (LPJ) is a powerful model for studying the impacts of climate change on global vegetation (Sitch et al. 2003)

(continued)

Table 24.2 (continued)

Model	Required input	PFTs	Output	Description
		broad-leaved summer green Boreal needle-leaved evergreen Boreal needle-leaved summer green		
LPJmL	1. Temperature (°C) 2. Precipitation (mm) 3. Rainy days 4. Cloud cover (%) 5. Atmospheric CO_2 6. Soil texture (%) 7. Potential evapotranspiration 8. Soil temperature PFTs	Temperate needle-leaved evergreen Temperate broad-leaved evergreen Tropical broad-leaved evergreen Temperate broad-leaved summer green Tropical broad-leaved rain green Boreal summer green Boreal needle-leaved evergreen C3 herbaceous C4 herbaceous	GPP NPP Net ecosystem exchange (NEE) Autotrophic and heterotrophic respiration Vegetation carbon Soil carbon	LPJmL is a dynamic global vegetation, hydrology, and crop model that incorporates the carbon, water, and nitrogen cycles at the plant and soil levels. It is based on an extended Farquhar photosynthesis scheme, stomatal conductance mechanics, and functional and allometric principles, and it can represent managed and natural ecosystems and the biogeochemical fluxes between them

decisions that will have its impact, even after centuries, is not easy and needs scientific insights to formulate, thus compelling us to use the Vegetation model to get insights into the future.

Over the years, Vegetation models have become increasingly dynamic and are increasingly accepted to support computer-based forest policy-making by creating scenarios and projections representing the future of plant growth, forest productivity, carbon sink estimation, and other parameters. Ecology-based models are necessary for environmental arbitrament support and pro-environment policy formulation because they allow the effects of alternative management to be explored spatiotemporally and empirically. However, because environmental issues are so important, further evaluation of the model quality and applicability is essential, particularly if vegetation models are used to support decisions that impact the real world for the sustainability of the ecosystem. Modeling and policy-making interact in specific policy processes, but the relationship is less explored (Rykiel Jr 1996). We will try to discuss how Vegetation models support or might support the process of political decision-making processes. First, we go through the model evaluation process, which includes six steps, as identified by Jacqueline Augusiak and the team in 2014. The primary six elements of the evaluation process are (1) "data evaluation," scrutinizing the data used for model formulation and testing; (2) "conceptual model evaluation," understanding model complexity, design, and assumptions; (3) "implementation validation," testing the execution of equations used and the computer programs run; (4) "model output validation," comparisons of model output with the patterns that shape the model built and the calibrations made; (5) "model analysis" estimating model's sensitivity to parameter alteration; and (6) "model output corroboration," comparability of the model output with other datasets or different model output for the developmental purpose (Thacker et al. 2004). The multidimensional complexity of environmental concerns is addressed with the help of mathematical and statistical concepts and computer-based models; we need systematic checking of various building blocks of a model throughout its lifecycle and evolution to a guaranteed reduction in uncertainties and easy to use so that meaningful insights can be drawn, which will act as a basis for policy developmental plans.

The policy cycle can be summed up in four steps (Fig. 24.2): (1) "agenda or target setting," for achieving ecological sustainability; (2) "policy formulation and adaptation," by the governing bodies, guided by forest ecology experts; (3) "policy implementation," with the help of experts and computer-based modeling for predicting the future impacts of the agendas; and (4) "policy evaluation," analysis of the implemented policy and expanding the scope (Jordan 2001). The models act as an input for policymakers, or the policymakers' decision has to impact the modelers and sips into the models. For example, the t33% of forest cover India had been presenting as a goal to be met is a decision made by policymakers in 1952 and is still practically the basis of target fixing for all modelers working over the Indian region, thus influencing the model as well. So, it is essential to understand and realize how and when Vegetation models influence policy-making and how and when policymakers influence a model's built or structural design. The basic interaction between policy-making, society, forest ecosystem, and modeling is briefly described in Fig. 24.3.

24.4.1 Policy-Making: Ecological Sustainability and Conservation

The government of India has used outcomes of static and dynamic vegetation models to report to UNFCCC (United Nations Framework Convention on Climate Change) about the vulnerability of its forest ecosystem, as part of its various national

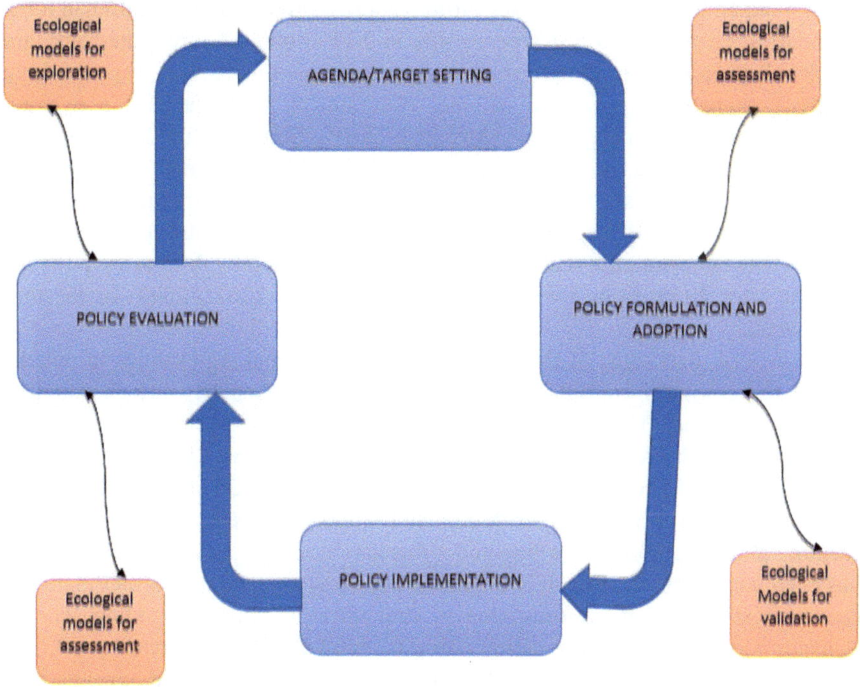

Fig. 24.2 The policy cycle and the possible use of models at various stages (adapted from Süsser et al. 2020)

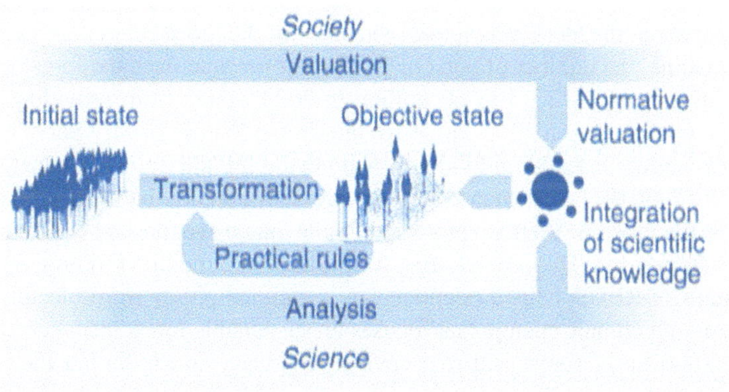

Fig. 24.3 Management concept for forest ecosystems. A system is converted from a starting state to a target state. Society's normative valuation and scientific knowledge contribute to the growth and accomplishment of the desired state (adapted from Süsser et al. 2021)

communications to the global body. For example, India's initial national communication to UNFCCC (MoEF 2004) used BIOME-3 vegetation response model to simulate the impact of climate change on Indian forests and to identify vulnerable grids in Indian forests. This analyses further reported projected shifts in Indian forest boundaries, changes in forest types, shifts in NPP, potential forest die-back, and possible loss or change in biodiversity under changing climate scenarios. Similarly, in 2012, as part of its second national communication, India used a dynamic vegetation model, namely "IBIS" (MoEFCC 2012). Similarly, the latest report to UNFCCC from China shows that according to the results of the multimodel ensemble analyses, the forest area exposed by decreasing NPP will reduce during low greenhouse gas (GHG) concentration scenarios. In contrast, it is also projected that at a high GHG concentration scenario, the forest area affected by decreasing NPP will increase after 2050, from 5.4% (2021–2050) to 27.6% (2071–2099) of the total forest area.

Let us look into some of the adaptive measures by making changes in policies related to ecological sustainability and conservation taken by various countries around the globe. The following discussed statistics of various countries are documented in the report "The Global Forest Goals Report, 2021" published by the Department of Economics and Social Affairs of the UN. Countries such as China and Liberia made clear guidelines to train and support research on tree breeding and seedling production for silviculture and afforestation. A forest carbon offset scheme has been initiated in the Republic of Korea, and New Zealand has further increased economic incentives for afforestation to strengthen its emission trading scheme. Ecuador formulated REDD+ action plans to reduce CO_2 emissions by 20% by 2025 through policy measures to reduce deforestation. Japan reported new financing methods such as forest environment tax, Nigeria launched green bonds, and Suriname raised the concession fee, and many other nations reported similar steps to promote sustainable forest management or forest growth. Canada, China, Serbia, Suriname, Lesotho, the Slovak Republic, and the United States of America have been vocal about the increasing interdependency of the forest ecosystem for employment. In China, the number of persons generating revenue from the forest increased from 52.47 million in 2015 to 60 million in 2020. Aside from providing roughly 196,000 employments in 2017 and 2018, the United States Forestry Service (USFS) employed about 955,400 individuals nationwide in the forest products sector. During 2017–2019, Uzbekistan restored more than 500,000 ha of an area prone to soil and water erosion. Vietnam protected fragile mangrove forests by getting shrimp farmers' help from UN-REDD and formulated an organic farming model. In Mongolia, UN-REDD helped people create a national policy for protecting forests and addressing climate change that focuses on sustainable forest management. India added 20,000 ha of forest and tree cover every year, and India led the world in official employment in the forest industry (6.23 million people employed).

24.5 Conclusion

In this review, we compare the various ecosystem modeling approaches that are being used to predict ecosystem dynamics to understand the forest change dynamics and climate change adaptation in forest ecosystems and assess their application in forest policy and planning. It is evident that different modeling approaches are undergoing fast evolution due to advancements in technology. These models are practical tools to evaluate various hypotheses and future climatic scenarios for effective decision-making and assess how policy decisions may impact the ecosystem. The future projections from these models can be used for formulating policy-making and sustainable environment plans. However, there is no model that can represent all the aspects of the ecosystem. Accepting the fact that "All the models have limitations, but they are useful," it is a big challenge for policymakers whose decisions may affect people's lives.

References

Aaheim A, Chaturvedi RK, Sagadevan AA (2011) Integrated modelling approaches to analysis of climate change impacts on forests and forest management. Mitig Adapt Strateg Glob Chang 16(2):247–266

Aber JD, Ollinger SV, Driscoll CT (1997) Modeling nitrogen saturation in forest ecosystems in response to land use and atmospheric deposition. Ecol Model 101(1):61–78

Albrecht AXEL, Kohnle ULRICH, Nagel JÜRGEN (2011) Übertragbarkeit empirischer statistischer Waldwachstumsmodelle: Prüf-und Anpassungsverfahren anhand des Beispiels BWinPro für Baden-Württemberg. AFJZ 182(1/2):11

Almeida AC, Landsberg JJ, Sands PJ (2004) Parameterisation of 3-PG model for fast-growing Eucalyptus grandis plantations. For Ecol Manag 193(1-2):179–195

Balandier P, Lacointe A, Le Roux X, Sinoquet H, Cruiziat P, Le Dizès S (2000) SIMWAL: a structural-functional model simulating single walnut tree growth in response to climate and pruning. Ann For Sci 57(5):571–585

Bosc A (2000) EMILION, a tree functional-structural model: presentation and first application to the analysis of branch carbon balance. Ann For Sci 57(5):555–569

Chaturvedi RK, Gopalakrishnan R, Jayaraman M, Bala G, Joshi NV, Sukumar R, Ravindranath NH (2011) Impact of climate change on Indian forests: a dynamic vegetation modeling approach. Mitig Adapt Strateg Glob Chang 16(2):119–142

Chaturvedi RK, Joshi J, Jayaraman M, Bala G, Ravindranath NH (2012) Multi-model climate change projections for India under representative concentration pathways. Curr Sci 103(7): 791–802

Chatzinikolaou E (2013) Use and limitations of ecological models. Transit Water Bull 6(2):34–41

Chen JM, Liu J, Cihlar J, Goulden ML (1999) Daily canopy photosynthesis model through temporal and spatial scaling for remote sensing applications. Ecol Model 124(2-3):99–119

Clark DB, Mercado LM, Sitch S, Jones CD, Gedney N, Best MJ, Pryor M, Rooney GG, Essery RLH, Blyth E, Boucher O, Harding RJ, Huntingford C, Cox PM (2011) The Joint U.K. Land Environment Simulator (JULES), model description – part 2: carbon fluxes and vegetation dynamics. Geosci Model Dev 4(3):701–722. https://doi.org/10.5194/gmd-4-701-2011

Cobos ME, Peterson AT, Osorio-Olvera L, Jiménez-García D (2019) An exhaustive analysis of heuristic methods for variable selection in ecological niche modeling and species distribution modeling. Eco Inform 53:100983

Cramer W et al (2001) Global response of terrestrial ecosystem structure and function to CO2 and climate change: results from six dynamic global vegetation models. Glob Chang Biol 7(4): 357–373

Cuddington K, Fortin MJ, Gerber LR, Hastings A, Liebhold A, O'Connor M, Ray C (2013) Process-based models are required to manage ecological systems in a changing world. Ecosphere 4(2):1–12

d'Annunzio R, Sandker M, Finegold Y, Min Z (2015) Projecting global forest area towards 2030. For Ecol Manag 352:124–133
DeAngelis DL, Franco D, Hastings A, Hilker FM, Lenhart S, Lutscher F, Tyson RC (2021) Towards building a sustainable future: positioning ecological modelling for impact in ecosystems management. Bull Math Biol 83(10):1–28
Duarte CM, Amthor JS, DeAngelis DL, Joyce LA, Maranger RJ, Pace ML, Running SW (2003) The limits to models in ecology. In: Models in ecosystem science, pp 437–451
Elith J, Franklin J (2013) Species distribution modeling. In: Encyclopedia of biodiversity, 2nd edn. Elsevier, pp 692–705
Ferrier S, Ninan KN, Leadley P, Alkemade R, Acosta LA, Akçakaya HR et al (2016) IPBES (2016): the methodological assessment report on scenarios and models of biodiversity and ecosystem services. Secretariat of the Intergovernmental Science-Policy Platform on Biodiversity and Ecosystem Services, Bonn
Fischlin A, Midgley GF, Price JT, Leemans R, Gopal B, Turley C, Rounsevell MDA, Dube OP, Tarazona J and Velichko AA (2007) Ecosystems, their properties, goods, and services. Climate change 2007: impacts, adaptation and vulnerability. Contribution of working group II to the fourth assessment report of the intergovernmental panel on climate change. (eds. Parry ML, Canziani OF, Palutikof JP, van der Linden PJ and Hanson CE, Cambridge University Press, Cambridge, 211–272
Foley JA, Prentice IC, Ramankutty N, Levis S, Pollard D, Sitch S, Haxeltine A (1996) An integrated biosphere model of land surface processes, terrestrial carbon balance, and vegetation dynamics. Glob Biogeochem Cycles 10(4):603–628
Friend AD, Stevens AK, Knox RG, Cannell MGR (1997) A process-based, terrestrial biosphere model of ecosystem dynamics (Hybrid v3. 0). Ecol Model 95(2-3):249–287
Gopalakrishnan R, Jayaraman M, Swarnim S, Chaturvedi RK, Bala G, Ravindranath NH (2011) Impact of climate change at species level: a case study of teak in India. Mitig Adapt Strateg Glob Chang 16(2):199–209
Gupta R, Sharma LK (2019) The process-based forest growth model 3-PG for use in forest management: a review. Ecol Model 397:55–73
Hanson PJ, Amthor JS, Wullschleger SD, Wilson KB, Grant RF, Hartley A, Cushman RM (2004) Oak forest carbon and water simulations: model intercomparisons and evaluations against independent data. Ecol Monogr 74(3):443–489
Hui D, Deng Q, Tian H, Luo Y (2017) Climate change and carbon sequestration in forest ecosystems. In: Handbook of climate change mitigation and adaptation, p 555, 594
IPCC (2021) Climate change 2021: the physical science basis. In: Masson-Delmotte V, Zhai P, Pirani A, Connors SL, Péan C, Berger S, Caud N, Chen Y, Goldfarb L, Gomis MI, Huang M, Leitzell K, Lonnoy E, Matthews JBR, Maycock TK, Waterfield T, Yelekçi O, Yu R, Zhou B (eds) Contribution of working group I to the sixth assessment report of the intergovernmental panel on climate change. Cambridge University Press, Cambridge, New York, NY. https://doi.org/10.1017/9781009157896
IPCC (2022) Climate change 2022: impacts, adaptation, and vulnerability. In: Pörtner H-O, Roberts DC, Tignor M, Poloczanska ES, Mintenbeck K, Alegría A, Craig M, Langsdorf S, Löschke S, Möller V, Okem A, Rama B (eds) Contribution of working group II to the sixth assessment report of the intergovernmental panel on climate change. Cambridge University Press, Cambridge, New York, NY, 3056 pp. https://doi.org/10.1017/9781009325844
Jordan A (2001) Environmental policy: protection and regulation. Int Encyclop Soc Behav Sci 7: 4644–4651
Keenan RJ (2015) Climate change impacts and adaptation in forest management: a review. Ann For Sci 72(2):145–167
Kerns B, Peterson DW (2014) An overview of vegetation models for climate change impacts. U.-S. Department of Agriculture, Forest Service, Climate Change Resource Center. www.fs.usda.gov/ccrc/topics/overview-vegetation-models

Kimmins JP, Scoullar KA, Apps MJ, Kurz WA (1990) The FORCYTE experience: a decade of model development. In: Proc Symp Forestry Canada Inf Rep, pp 60–67
Kirilenko AP, Sedjo RA (2007) Climate change impacts on forestry. Proc Natl Acad Sci 104(50): 19697–19702
Korappath S, Bilyaminu H (2022) Dynamic global vegetation models (DGVMs) and its applicability to climate simulations-a review. Ann Plant Sci 11:4587–4597. https://doi.org/10.21746/aps.2022.11.01.8
Korol RL, Running SW, Milner KS (1995) Incorporating intertree competition into an ecosystem model. Can J For Res 25(3):413–424
Kramer K, Buiteveld J, Forstreuter M, Geburek T, Leonardi S, Menozzi P, Van der Werf DC (2008) Bridging the gap between ecophysiological and genetic knowledge to assess the adaptive potential of European beech. Ecol Model 216(3-4):333–353
Kucharik CJ, Foley JA, Delire C, Fisher VA, Coe MT, Lenters JD, Gower ST (2000) Testing the performance of a dynamic global ecosystem model: water balance, carbon balance, and vegetation structure. Glob Biogeochem Cycles 14(3):795–825
Kuhnert PM, Martin TG, Griffiths SP (2010) A guide to eliciting and using expert knowledge in Bayesian ecological models. Ecol Lett 13(7):900–914
Kumar M, Rawat SPS, Singh H, Ravindranath NH, Kalra N (2018) Dynamic forest vegetation models for predicting impacts of climate change on forests: an Indian perspective. Indian J For 41(1):1–12
Li T, Xiong Q, Luo P, Zhang Y, Gu X, Lin B (2020) Direct and indirect effects of environmental factors, spatial constraints, and functional traits on shaping the plant diversity of montane forests. Ecol Evol 10(1):557–568
Lischke H, Zimmermann NE, Bolliger J, Rickebusch S, Löffler TJ (2006) TreeMig: a forest-landscape model for simulating spatio-temporal patterns from stand to landscape scale. Ecol Model 199(4):409–420
Lotka AJ (1925). Elements of physical biology. Williams and Wilkins, Baltimore, MD. Reprinted in 1956 as: elements of mathematical biology. Dover Publications, Mineola, NY
Luxmoore RJ, Hargrove WW, Tharp ML, Mac Post W, Berry MW, Minser KS, Peterson KD (2002) Addressing multi-use issues in sustainable forest management with signal-transfer modeling. For Ecol Manag 165(1-3):295–304
Martin TG, Burgman MA, Fidler F, Kuhnert PM, Low-Choy SAMANTHA, McBride M, Mengersen K (2012) Eliciting expert knowledge in conservation science. Conserv Biol 26(1): 29–38
Masera OR, Garza-Caligaris JF, Kanninen M, Karjalainen T, Liski J, Nabuurs GJ, Mohren GMJ (2003) Modeling carbon sequestration in afforestation, agroforestry and forest management projects: the CO2FIX V. 2 approach. Ecol Model 164(2-3):177–199
MoEF (2004) India's Initial National Communication to UNFCCC. Ministry of Environment and forests, Government of India. https://unfccc.int/documents/83420. Accessed 22 Sept 2022
MoEFCC (2012) India's second National Communication to UNFCCC. Ministry of Environment, Forest and Climate Change, Government of India. https://unfccc.int/documents/109438. Accessed 22 Sept 2022
Morin X, Lechowicz MJ (2008) Contemporary perspectives on the niche that can improve models of species range shifts under climate change. Biol Lett 4(5):573–576
Myoung B, Yong SC, Seon KP (2011) A review on vegetation models and applicability to climate simulations at regional scale. Asia-Pac J Atmos Sci 47(5):463–475
Newbold T, Hudson LN, Phillips HR, Hill SL, Contu S, Lysenko I, Purvis A (2014) A global model of the response of tropical and sub-tropical forest biodiversity to anthropogenic pressures. Proc R Soc B Biol Sci 281(1792):20141371
Niesenbaum RA (2019) The integration of conservation, biodiversity, and sustainability. Sustainability 11(17):4676. https://doi.org/10.3390/su11174676

Nimasow G, Nimasow OD, Rawat JS, Tsering G, Litin T (2016) Remote sensing and GIS-based suitability modeling of medicinal plant (Taxus baccata Linn.) in Tawang district, Arunachal Pradesh, India. Curr Sci 110(2):219–227

Nunes LJ, Meireles CI, Gomes CJP, Ribeiro NMA (2021) The impact of climate change on forest development: a sustainable approach to management models applied to Mediterranean-type climate regions. Plan Theory 11(1):69

Olson RJ, Johnson KR, Zheng DL, Scurlock JMO (2001) Global and regional ecosystem modeling: databases of model drivers and validation measurements. ORNL Technical Memorandum TM-2001/196. Oak Ridge National Laboratory, Oak Ridge, TN

Pasetto D, Arenas-Castro S, Bustamante J, Casagrandi R, Chrysoulakis N, Cord AF, Ziv G (2018) Integration of satellite remote sensing data in ecosystem modelling at local scales: practices and trends. Methods Ecol Evol 9(8):1810–1821

Peringer A, Siehoff S, Chételat J, Spiegelberger T, Buttler A, Gillet F (2013) Past and future landscape dynamics in pasture-woodlands of the Swiss Jura Mountains under climate change. Ecol Soc 18(3):11

Pietsch SA, Hasenauer H, Kučera J, Čermák J (2003) Modeling effects of hydrological changes on the carbon and nitrogen balance of oak in floodplains. Tree Physiol 23(11):735–746

Prentice IC (1989) Developing a global vegetation dynamics model: results of an IIASA summer workshop. https://pure.iiasa.ac.at/3223

Pretzsch H (2009) Forest dynamics, growth, and yield. In: Forest dynamics, growth and yield, pp 1–39

Pretzsch H, Biber P, Ďurský J (2002) The single tree-based stand simulator SILVA: construction, application and evaluation. For Ecol Manag 162(1):3–21

Quillet A, Peng C, Garneau M (2010) Toward dynamic global vegetation models for simulating vegetation–climate interactions and feedbacks: recent developments, limitations, and future challenges. Environ Rev 18:333–353

Rahbek C, Gotelli NJ, Colwell RK, Entsminger GL, Rangel TFLVB, Graves GR (2007) Predicting continental-scale patterns of bird species richness with spatially explicit models. Proc R Soc Lond Ser B Biol Sci 274(1607):165–174

Rogers A, Medlyn BE, Dukes JS, Bonan G, Von Caemmerer S, Dietze MC, Zaehle S (2017) A roadmap for improving the representation of photosynthesis in Earth system models. New Phytol 213(1):22–42

Rötzer T, Leuchner M, Nunn AJ (2010) Simulating stand climate, phenology, and photosynthesis of a forest stand with a process-based growth model. Int J Biometeorol 54(4):449–464

Rykiel EJ Jr (1996) Testing ecological models: the meaning of validation. Ecol Model 90(3): 229–244

Schaefer K, Collatz GJ, Tans P, Denning AS, Baker I, Berry J, Prihodko L, Suits N, Philpott A (2008) Combined simple biosphere/Carnegie-Ames-Stanford approach terrestrial carbon cycle model. J Geophys Res 113:G03034. https://doi.org/10.1029/2007JG000603

Schaphoff S, Von Bloh W, Rammig A, Thonicke K, Biemans H, Forkel M et al (2018) LPJmL4–a dynamic global vegetation model with managed land–Part 1: model description. Geosci Model Dev 11(4):1343–1375

Seely B, Welham C, Scoullar K (2015) Application of a hybrid forest growth model to evaluate climate change impacts on productivity, nutrient cycling and mortality in a montane forest ecosystem. PLoS One 10(8):e0135034

Sitch S, Smith B, Prentice IC, Arneth A, Bondeau A, Cramer W, Venevsky S (2003) Evaluation of ecosystem dynamics, plant geography and terrestrial carbon cycling in the LPJ dynamic global vegetation model. Glob Chang Biol 9(2):161–185

Süsser D, Ceglarz A, Gaschnig H, Stavrakas V, Giannakidis G, Flamos A, Lilliestam J (2020) The use of energy modelling results for policy-making in the E.U. Ther Deliv 1(1) Sustainable Energy Transitions Laboratory (SENTINEL) project

Süsser D, Ceglarz A, Gaschnig H, Stavrakas V, Flamos A, Giannakidis G, Lilliestam J (2021) Model-based policy-making or policy-based modelling? How energy models and energy policy interact. Energy Res Soc Sci 75:101984

Sutherland WJ (2006) Predicting the ecological consequences of environmental change: a review of the methods. J Appl Ecol:599–616

Thacker BH, Doebling SW, Hemez FM, Anderson MC, Pepin JE, Rodriguez EA (2004) Concepts of model verification and validation

Tylianakis JM, Didham RK, Bascompte J, Wardle DA (2008) Global change and species interactions in terrestrial ecosystems. Ecol Lett 11(12):1351–1363. https://doi.org/10.1111/j.1461-0248.2008.01250.x

van der Werf W, Keesman K, Burgess P, Graves A, Pilbeam D, Incoll LD, Dupraz C (2007) Yield-SAFE: a parameter-sparse, process-based dynamic model for predicting resource capture, growth, and production in agroforestry systems. Ecol Eng 29(4):419–433

Volterra V (1926) Variations and fluctuations of the numbers of individuals in animal species living together. Reprinted in 1931. In: Chapman RN (ed) Animal ecology. McGraw-Hill, New York, NY, pp 409–448

Vospernik S (2017) Possibilities and limitations of individual-tree growth models–a review on model evaluations. Die Bodenkultur J Land Manag Food Environ 68(2):103–112

Wang D, Ricciuto D, Post W, Berry MW (2011) Terrestrial ecosystem carbon modeling. In: Padua D (ed) Encyclopedia of parallel computing. Springer, p 2211. https://doi.org/10.1007/978-0-387-09766-4_395

Wang Q, Chun JA, Fleisher D, Reddy V, Timlin D, Resop J (2017) Parameter estimation of the Farquhar—von Caemmerer—Berry biochemical model from photosynthetic carbon dioxide response curves. Sustainability 9(7):1288

Williams JW, Jackson ST, Kutzbach JE (2007) Projected distributions of novel and disappearing climates by 2100 AD. Proc Natl Acad Sci 104(14):5738–5742

Xu X, Trugman AT (2021) Trait-based modeling of terrestrial ecosystems: advances and challenges under global change. Curr Clim Change Rep 7(1):1–13

Yadav S, Bhattacharya P, Areendran G, Sahana M, Raj K, Sajjad H (2022) Predicting impact of climate change on geographical distribution of major NTFP species in the Central India region. Model Earth Syst Environ 8(1):449–468

Yan Y, Wang S, Wang Y, Wu W, Wang J, Chen B, Yang F (2011) Assessing productivity and carbon sequestration capacity of subtropical coniferous plantations using the process model PnET-CN. J Geogr Sci 21:458–474

Ecological Carrying Capacity Modeling and Sustainability Assessment of the Seven Lakes of San Pablo City, Laguna, Philippines

25

Damasa B. Magcale-Macandog, John Vincent R. Pleto, Joseph G. Campang, Canesio D. Predo, Fatima A. Natuel, Ma. Grechelle Lyn D. Perez, Nethanel Jireh A. Larida, Yves Christian A. Cabillon, Sarena Grace L. Quiñones, and Jeoffrey M. Laruya

D. B. Magcale-Macandog (✉) · J. V. R. Pleto · J. G. Campang · Y. C. A. Cabillon · S. G. L. Quiñones
Institute of Biological Sciences, College of Arts and Sciences, University of the Philippines Los Baños, College, Laguna, Philippines
e-mail: dmmacandog@up.edu.ph

C. D. Predo
Institute of Renewable Natural Resources, College of Forestry and Natural Resources, University of the Philippines Los Baños, College, Laguna, Philippines

F. A. Natuel
College of Arts and Sciences, Laguna State Polytechnic University, San Pablo city, Laguna, Philippines

School of Environmental Science and Management, University of the Philippines Los Baños, College, Laguna, Philippines

M. G. L. D. Perez
School of Environmental Science and Management, University of the Philippines Los Baños, College, Laguna, Philippines

University of the Philippines Rural High School, College of Arts and Sciences, University of the Philippines Los Banos, College, Laguna, Philippines

N. J. A. Larida
Institute of Biological Sciences, College of Arts and Sciences, University of the Philippines Los Baños, College, Laguna, Philippines

University of the Philippines Visayas, Iloilo, Philippines

J. M. Laruya
College of Forestry and Natural Resources, University of the Philippines Los Baños, College, Laguna, Philippines

S. Dhyani et al. (eds.), *Ecosystem and Species Habitat Modeling for Conservation and Restoration*, https://doi.org/10.1007/978-981-99-0131-9_25

Abstract

Aquaculture was introduced in the seven lakes in the 1960s and has significantly contributed to the income of the fishing community, while ecotourism has boosted the local tourism industry in the city. However, these activities pose threats to the degradation of the environmental quality of the Seven Lakes. The sustainability of the aquaculture and tourism activities in the seven lakes was assessed using the ecological carrying capacity model. This model was composed of three criteria: biophysical, socioeconomic, and ecotourism. The biophysical criterion involved the collection of water quality indicators in the lake during the wet and dry seasons. Household survey and Key Informant Interviews were conducted to gather primary data on the socioeconomic and tourism indicators. The maximum value or limit of each indicator was based on water quality standards for the biophysical indicators, while the maximum values for the socioeconomic and tourism indicators were based on literature and secondary data from the local government unit (LGU) of San Pablo. Experts and key respondents ranked the various biophysical, socioeconomic, and ecotourism indicators following the Rank-Sum method to determine the weight of each indicator. Ecotourism lakes Yambo and Pandin had high (0.8549) and very high (1.2119) ecological carrying capacities (ECC), respectively, while Lake Mohicap had a medium sustainability index with an ECC value of 0.6500. Lake Sampaloc had a low sustainability index with only 0.0325 ECC value. Aquaculture lakes Calibato, Palakpakin, and Bunot had negative ECC values, indicating unsustainable ecosystems. Estimating and illustrating the recreational and aquaculture carrying capacities of the lakes can provide local policymakers a comprehensible overview of the potential consequences of the unrestricted proliferation of the various activities in the lakes.

Keywords

Ecological carrying capacity modeling · Seven lakes · Aquaculture · Ecotourism · Sustainable lake management

25.1 Introduction

Sitting 70 km away from Metropolitan Manila, San Pablo City is a chartered city in Laguna nestling the famous Seven Lakes (Fig. 25.1). These freshwater lakes were formed through phreatic eruption, a unique process where lava from Mt. San Cristobal intersected with the groundwater and created a steam-heated eruption forming crater-like depressions, which were later filled up with rainwater (Laguna Lake Development Authority [LLDA] 2006–2008).

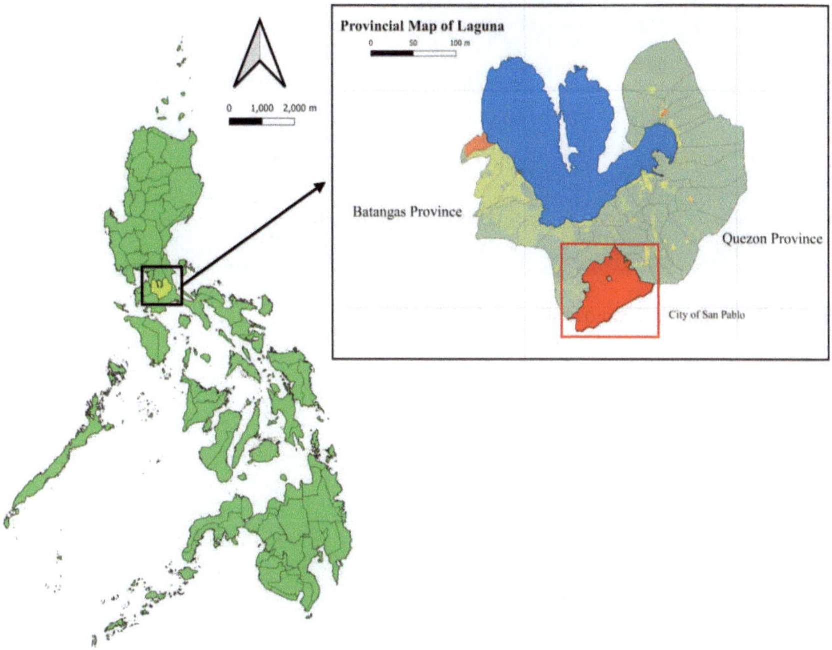

Fig. 25.1 Location map of San Pablo City

25.1.1 Seven Lakes of San Pablo City

The seven famous lakes in San Pablo are Lake Sampaloc, Lake Palakpakin, Lake Calibato, Lake Mohicap, Lake Pandin, Lake Yambo, and Lake Bunot. These freshwater lakes were formed through phreatic eruption, a unique process where lava from Mt. San Cristobal intersected with the groundwater and created a steam-heated eruption forming crater-like depressions, which were later filled up with rainwater. Since these lakes are all volcanic in origin, it shows the uniqueness of the city, its traditions, and culture.

Lake Sampaloc is the largest among San Pablo's Seven Crater Lakes. Located in the nearest proximity from the city proper, it is considered one of the prime tourist spots in the city (Fig. 25.2). The aquaculture industry in the lake includes freshwater fishes such as *tilapia*, *hito*, *dalag*, *ayungin*, bighead carp, and shrimps.

Lake Palakpakin is the shallowest among the seven lakes but second largest next to Lake Sampaloc. It is utilized mainly for fishing and aquaculture activities (Fig. 25.3). An increasing construction of fish cages resulted in limited open fishing ground for the fisherfolk.

Lake Calibato is the deepest of all the seven lakes (Fig. 25.4). It has the greatest volume of water in storage. Also, it is the highest lake in terms of elevation. Abundant fishes in the lake supply both the city of San Pablo and the municipality of Rizal.

Fig. 25.2 Fishing and ecotourism in Lake Sampaloc

Fig. 25.3 Fish cages and water hyacinth in Lake Palakpakin

Fig. 25.4 Fish pens in Lake Calibato

Lake Mohicap is the smallest lake. It is also the lowest lake in terms of elevation (Fig. 25.5). It is a major source of *tilapia* for Metro Manila and suburbs.

Lake Bunot is closest to Lake Sampaloc. It is used mainly for floating cages operation and aquaculture where most of the locals derive their source of income (Fig. 25.6).

Lake Pandin and Lake Yambo are known as "The Twin Lakes" (Fig. 25.7a, b). Both lakes are considered oligotrophic because of their deep clear lakes with low nutrient supplies, high dissolved oxygen level, and little organic matter. As San Pablo's best kept lakes, both are suitable for recreational activities such as swimming, picnics, and outings but not so much for aquaculture.

Fig. 25.5 Fish pens in Lake Mohicap

Fig. 25.6 Fish pens and nearby houses in Lake Bunot

Fig. 25.7 The twin lakes of San Pablo – Lake Pandin (**a**) and Lake Yambo (**b**)

The multiuses of these lakes, such as irrigation, domestic, navigation, sustenance fishing, tourism, and aquaculture, benefit the surrounding communities.

25.1.2 Aquaculture

Aquaculture is the breeding, raising, and harvesting of aquatic organisms in coastal or inland waters involving rearing interventions to enhance the production (National Oceanic and Atmospheric Administration [NOAA] 2021; Food and Agriculture

Organization [FAO] n.d.). It is an important food production industry and is one of the fastest developing sectors contributing about 50% of the world's fish used for food (FAO n.d.). Aquaculture in the Philippines has a long history and involves many species and farming practices in diverse ecosystems. Most of the production, which significantly contributes to the country's food security and employment, comes from the farming of seaweed, milkfish, tilapia, shrimp, carp, oyster, and mussel aquaculture in freshwater lakes in the Philippines (FAO n.d.). In the Seven Lakes of San Pablo City, Laguna, aquaculture first began in Bunot Lake in 1976 after the successful introduction of tilapia cage culture in Laguna de Bay by the Laguna Lake Development Authority (LLDA) in 1974. After some time, tilapia cage farming spread to other lakes such as Sampaloc, Palakpakin, Calibato, and Mohicap Lakes (Brillo 2015). By the late 1980s, fish pens and cages have become a common feature among the Seven Lakes. Tilapia farming extensively expanded and reached its peak in the late 1990s to the early 2000s, where the 10% area limit for aquaculture structures pursuant to the Fisheries Code of the Philippines (Republic Act [RA] 8550, section 51) was breached in most lakes (Brillo 2017). In the 1990s, illegal constructions of fish cages sprouted in the Sampaloc Lake covering approximately 70% of the total lake surface area during a certain period (Global Nature Fund [GNF] 2014). As a result, the lake has become extremely threatened. Removal of illegal constructions was already done but the increasing number of illegal fish cages became a problem again in 2012. Overcrowding of fish cages and overfeeding have resulted in eutrophication, occasional fish kills, and massive growth of green algae. Because of this, rehabilitation efforts were initiated by environmental organizations in order to conserve and protect the Sampaloc Lake along with six other crater lakes in San Pablo City, Laguna. At present, Lakes Sampaloc, Palakpakin, Calibato, and Bunot are principally exploited for aquaculture, particularly commercial tilapia and milkfish production via floating fish cages. While Lakes Pandin, Yambo, and Mohicap are principally used for ecotourism, few fish cages have been built in the lakes for the culture of the same species, *tilapia* and milkfish. Approximately 3500–5000 fingerlings of Nile *tilapia* usually sourced from Bureau of Fisheries and Aquatic Resources (BFAR) are reared in 10 m × 10 m fish cages and fed with commercial feeds until marketable size is reached (Guevarra et al. 2020).

Based on scientific findings, the greatest impact of aquaculture activities is perceived on water quality. It is important that farmers look into the water's chemical and physical properties affecting its quality since poor water quality affects the health and growth of the fish in the system (Towers 2015). Specific and different ranges of water quality variables such as salinity, temperature, pH, hardness, and oxygen concentration have varying effects among different fish species. Each species has its own optimum range of tolerance. Thus, it is really important for fisherfolks to make sure that the physical and chemical conditions of the water are within the optimum range of the fish under culture at all times as much as possible (The Fish Site 2015).

The physical impact of aquaculture practices like cage culture differs from one lake to another. In general, factors like accumulation of deteriorating materials such as nets, bamboos, and poles, buildup of feces, water hyacinth, and uneaten feeds

Fig. 25.8 Fish pens (**a**) and cages (**b**) in San Pablo

contribute to the gradual shallowing of lakes. Aquaculture provides a source of livelihood and protein for fish pen and cage operators as well as local fishing communities (Fig. 25.8).

25.1.3 Ecotourism

Ecotourism is a part of the tourism industry and market forces that influence the visitors' choice of destinations, transportations and accommodation, and food. By providing an income to the protected area via entrance fees, donations, and so on, ecotourism also provides income to service providers, including those in the community (Johnson and Thomas 1992).

Three of the seven crater lakes cater to ecotourism services. These are Lake Sampaloc and the twin lakes Pandin and Yambo. Lake Sampaloc is located at the city proper of San Pablo, and considered as the largest lake in terms of surface area. Tourists enjoy biking, jogging, fishing, and many aerobic activities such as Zumba and yoga in Lake Sampaloc. Among the three lakes, Sampaloc has had the most tourist arrivals since 2015 as stated by the Tourism Office of San Pablo. The twin lakes Yambo and Pandin cater to bamboo rafting services, fishing, picture-taking, and even swimming, since the waters of both lakes were classified as Class B waters, intended for recreational activities (Fig. 25.9). The ecotourism services serve as an income generating and source of their livelihood, assuring that the area is well-maintained and the resources around are protected. Lake Pandin management requires a fee of PhP 400 per person. This includes native food delicacies offered by the management and the services while staying in the lake for about 2 h. Meanwhile, Lake Yambo also offers the same amenities but with a fee of PhP 360 per person. In recent times, Lake Mohicap has also been opened to the public and is being conserved to be another ecotourism spot in San Pablo. Meanwhile, the other lakes in San Pablo are known to be sites of aquaculture of various fish species such as *carpa*, *hito*, *gurami*, *bangus*, *hipon*, *bitoo*, *kuhol*, *ayungin*, bighead carp, *tilapia*, and many others. Lake Sampaloc has the largest aquaculture among the

Fig. 25.9 Ecotourism in Lakes Pandin, Yambo, and Sampaloc

seven lakes, with Nile Tilapia (*Oreochromis niloticus*) as its main farmed fish (San Pablo City Comprehensive Land Use Plan [CLUP] 2015–2025; LLDA 2006–2008) (Fig. 25.9).

25.1.4 Ecological Carrying Capacity

Ecological carrying capacity (ECC) is the maximum service function a particular system can provide on the premise of maintaining its sustainability (Song et al. 2020). Lake's ecological carrying capacity includes self-maintenance, self-regulation, supply capacity of the lake, and human activities such as fishing, aquaculture, tourism, and living within the nearby areas. As expounded by Ross et al. (2013) and Lim (1995) in their articles, assessing the physical carrying capacity of the lake will define the total area suitable for aquaculture, as well as the threshold limit for tourism space, beyond which recreational facilities are saturated. The ecological carrying capacity will identify the magnitude of aquaculture production each lake can support without leading to significant changes to ecological processes, services, species biodiversity, populations, or communities in the environment, as well as the threshold limit for visitor use and consequent damage the lake can sustain without being degraded. ECC can be used as a reliable basis for environmental-economic decision-making. Estimating and illustrating the recreational and aquaculture carrying capacities of the lakes can provide the LGUs' (local government units) constituents and policymakers a comprehensible overview of the potential consequences of the unrestricted proliferation of the various activities the lakes are currently hosting.

In this study, biophysical, socioeconomic, and tourism data gathered and analyzed were used to develop ECC models of the seven lakes. The ECC models were used to assess the sustainability of the lakes' current environmental policies and management practices. Results of the sustainability assessment will be used as an instrument in the sustainable and effective policy planning of the local government of San Pablo City to address issues on food security and sustainability.

There are various ecological carrying capacity studies about lakes available, highlighting different parameters and factors. A study was prepared by the Lake Ripley Management District in 2003 to determine the recreational carrying capacity of Lake Ripley. In the report, the recreational carrying capacity refers to the number of watercrafts that can simultaneously operate on the lake without compromising user safety and causing environmental harm to the resource. Based on the findings, the average boating density exceeded the carrying capacity of the lake. The analysis suggested that there is a high probability of user conflict and environmental degradation on Lake Ripley due to overcrowding during mid-summer, weekends, and holidays. Another study by Reghunathan et al. (2016) highlighted the factors affecting the environmental carrying capacity of a freshwater tropical lake system in Vellayani Lake, India. Factor analysis was used to identify the factors controlling the carrying capacity of the lake, and hierarchical cluster analysis (HCA) was used to classify the lake. The results showed that the lake's carrying capacity to alkalinity is low due to ion deficiency. Acidity, mineralization, fertilizer, evaporation, and organic pollution factors are the controls of water quality during the pre-monsoon period. All factors except evaporation factor with additional runoff factor control the water quality during monsoon period. During post-monsoon, all factors as well as soil erosion factor influence the water quality. The result suggested that during pre- and post-monsoon, combating the acidic factor must be focused and runoff factor during monsoon season. Zeng et al. (2011) published a paper about an integrated approach for assessing aquatic ecological carrying capacity of Tai Lake Basin in China. According to the study, the aquatic ecological carrying capacity is an effective method for analyzing sustainable development in water management. The study used an indicator system that considers social and economic development as well as ecological resilience. To calculate the ecological index, normalized difference vegetation index (NDVI) was extracted from Moderate Resolution Imaging Spectroradiometer (MODIS) time series images, which was followed by spatial and temporal analysis of the vegetation cover. The study was conducted during the period 2000–2008 and it showed that there is a slight upward trend on the aquatic ecological carrying capacity, and the intensity of human activities exceeded the capacity in 2008.

25.1.4.1 Biophysical

In 2017, a study conducted by Paller et al. (2017) found that among the seven lakes, the most turbid ones were Mohicap (13.34 cm) and Palakpakin (17.37 cm), and can be classified as eutrophic to hypereutrophic using the Secchi disk visibility depth (SDVD) index. Other hydrological parameters of the Seven Lakes were also measured in the study and showed that Palakpakin lake, along with the other lakes, had mean surface dissolved oxygen level (6.80–9.63 mg/L), surface water temperature (22–32 °C), and mean pH levels (7.13–8.52) that were all within the standard levels. They also found out that Pandin lake was the least turbid lake (25 cm) using the SDVD index. As previously mentioned, other hydrological parameters of the seven lakes were measured in the study and showed that Pandin lake, along with the other lakes, had mean surface dissolved oxygen level (6.80–9.63 mg/L), surface water temperature (22–32 °C), and mean pH levels (7.13–8.52) that were all within the standard levels.

The water quality of Palakpakin lake was also assessed in another study in which Mendoza et al. (2019) observed the status and measured the water quality parameters of the Seven Lakes of San Pablo City to compare with published resources and literature from the 1930s to 2019 related to the Seven Lakes. One of the parameters measured in the study was the chlorophyll concentration to assess the trophic status of the Seven Lakes. It was found that Palakpakin lake, along with lakes Bunot and Calibato, were categorized to be eutrophic with 6.1–22 μg/L. In addition, it showed that the pH measurement of Palakpakin lake in August 2018 within 0–15 m depth is 5.16 ± 0.22, which is lower compared to the 6.5–8.5 ideal pH range for fish species according to the standard adopted for Class C fishery water. With pH levels less than the 6.5 limit, implications of low pH levels in lakes, specifically within 5.8–6.0, were reported to severely affect the growth and reproduction of aquatic organisms.

A more specific study by Navarrete et al. (2019) assessed the nutrient dynamics, phytoplankton diversity, sediment geochemistry, and water quality of Palakpakin lake during wet and dry seasons in four critical areas in the lake (inlet, center, sanctuary, and outlet) to comprehend its deteriorating ecological state. Results have shown that the lake has slightly alkaline water (pH 7.4–8.4) and the dissolved oxygen concentrations are 7.4 mg/L in the outlet, while the water in the inlet has 5.5 mg/L. Moreover, the turbidity (28 NTU) and Secchi depth (0.7–1.5 m) values obtained meant that the lake was eutrophic to hypereutrophic. The abundance of *Microcystis aeruginosa*, *Anabaena helicoidea*, and *Lyngbya* sp., which are indicator species of eutrophic to highly eutrophic waters, was also observed. The pH level of the lake supported the growth of *M. aeruginosa* as blue-green algae including the said species can grow at pH conditions of and greater than 6.5. Lastly, the concentration of available nutrients such as N and P in the center and sanctuary sediments was high, with phosphate concentrations of >2.0.5 mg/L, causing internal nutrient loading in the lake, indicating that the center and sanctuary are depositional areas of eroded sediments and aquaculture inputs such as excess feeds and fecal materials of fish. This nutrient loading was described to be a great factor in the increase in both density and diversity of the phytoplankton.

The study of Paller et al. (2021) on the review of the Seven Lakes of San Pablo stated that the most recent record of water quality of the Seven Lakes done by the LLDA in 2018 shows that the mean surface dissolved oxygen was still above the recommended level of 5.0 ppm for Class C water. The biochemical oxygen demand

(BOD) level of lake Bunot and Mohicap in 2018 exceeded the level of 7.0 ppm with a value of 7.25 ppm and 8 ppm, respectively. Ammonia levels of the Seven Lakes were too high and exceeded the level of 0.05 ppm except for Lake Yambo with only 0.03 ppm. The mean phosphate level of lake Sampaloc exceeded the recommended level of 0.5 ppm with a mean of 0.75 ppm.

25.1.4.2 Socioeconomic

San Pablo City, a first income class city known as the "City of Seven Lakes", has 80 barangays, of which 44 are classified as urban and 36 as rural. As of 2015, the city has a population of 266,068, and its population structure is dominated by younger age classes and is expanding rapidly (Philippine Statistics Authority [PSA] 2015). The land use is primarily agriculture with residential, commercial, industrial, special classes (educational, hospitals, parks and recreation, religious, and charitable areas), and government properties. Parks and recreational areas include the city's seven lakes, which offer different ecosystem services mainly ranging from provisioning to recreational services. These lakes are classified by the Laguna Lake Development Authority (LLDA) as Class C waters suitable for fishing and aquaculture. Since the 1980s, the LLDA, by virtue of Republic Act (RA) no. 4850, has the major role and jurisdiction in the monitoring and sampling of the seven lakes, same with the Laguna Lake and their tributaries, once every quarter to promote conservation and sustainable development (San Pablo CLUP 2015–2025; LLDA 2006–2008).

Each lake has its own operating local Fisheries and Aquatic Resource Management Councils (FARMC) and lake residents' organization. This is one great asset of San Pablo's ecotourism as community stakeholders actively involve themselves and participate in improving the lakes' sustainable development and management. Over the years, various community stakeholders have consistently participated and discussed problems and plans in dialogues and meetings conducted by the LLDA and the LGU of San Pablo. In fact, a multisectoral coalition composed of the Friends of the Seven Lakes Foundation, Inc. (FSLF), as well as civic and religious groups, launched the *Yakap sa Lawa* prayer rally and protest to prompt the administrative agencies to take notice of the conspicuous ecological problems that the lakes, especially Lake Sampaloc, were experiencing in the late 1990s. However, there had been some issues between proponents of ecotourism and fish farming of aquaculture lakes including Lakes Sampaloc, Bunot, Palakpakin, and Calibato. This is because the FARMC wants to have more space in these lakes for fish farming, while the Seven Crater Lakes and Watershed Management Council (SCLWMC) wants more area for ecotourism. This fragmentation is evident in the delayed construction of the Master Development Plan (MDP) of Lake Sampaloc. Distrust and divergence issues lasted for more than a decade until three of the seven lakes finally had their MDP in 2015. MDP is fundamental for each lake's effective management and is the basic enabler of ecotourism development. Since then, this plan has been serving as the guide that facilitates the regulation of fish cages or pens and designates specific areas for aquaculture and ecotourism, including their extent and arrangement (Brillo 2017).

Furthermore, it showed through the years that community stakeholders can be successful in developing the lakes surrounding them with support from a nongovernmental organization (NGO). For instance, the local FARMC and *Samahang Mangingisda ng Lawa ng Pandin* (SMLP) were assisted by the *Pundasyon ng Kalikasan* (PK), a local environmental group, in planning, organizing, and promoting the Pandin Lake Tour, a successful ecotourism enterprise initiated by a group of mostly women residents in Lake Pandin. At present, the lake women still continue to lead the operations of the lake's ecotourism enterprise. Indeed, the women residents of Lake Pandin play an important role in its sustainable development and management (Brillo 2017).

25.1.4.3 Tourism

Some of the seven crater lakes offer ecotourism, namely Sampaloc, Mohicap, and the twin lakes Yambo and Pandin. An essential consideration of transforming the approach of catering tourism into sustainable tourism involves social responsibility, a strong commitment to nature conservation and an integration of the local community in any tourist operation and development. As written in 2015 by the World Travel and Tourism Council (WTTC) and the Earth Council, the World Tourism Organization (WTO) defines sustainable tourism development as "meeting the needs of the present tourists, host region while protecting and enhancing the opportunities for the future". As stated in the framework of sustainable tourism development (Fig. 25.10) as cited by Alampay (2005), sustainable tourism development includes its basic elements to provide equity, ecological sustainability for the present and future generation, and economic stability, thus supporting the primary objectives of a sustainable tourism initiative.

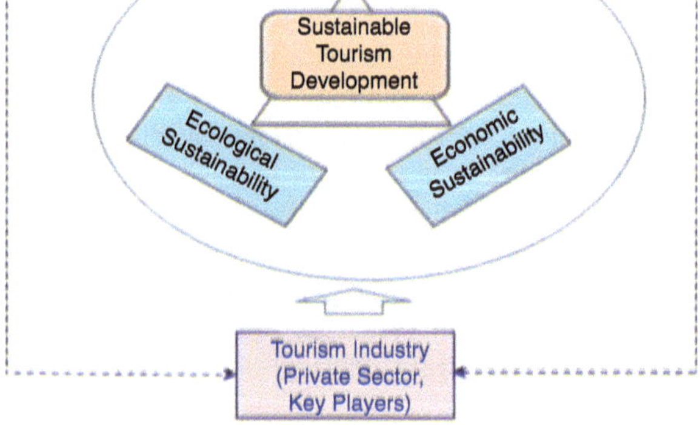

Fig. 25.10 Sustainable tourism development in the Philippines. Source: Alampay (2005)

While these lakes are utilized mainly for aquaculture and ecotourism, the integrity of the lakes is compromised. In fact, these lakes were proclaimed by the Global Nature Fund (GNF) as the "Threatened Lakes of the Year 2014" due to anthropogenic activities such as illegal squatting along the shores and its resulting water pollution, illegal fish pens, overcrowding fish cages, as well as the establishment of a number of commercial infrastructures nearby the lakes (LLDA 2006–2008). The aquaculture industry has significantly contributed to the income of the fishing community, while ecotourism has boosted the local tourism industry in the city. However, these activities pose threats to the degradation of the environmental quality of the Seven Lakes. Thus, it is necessary to examine their current condition and determine if it already exceeds the lakes' ecological carrying capacity (ECC). ECC is the maximum service function a particular system can provide on the premise of maintaining its sustainability (Song et al. 2020).

In this study, the sustainability of the aquaculture and tourism activities in the Seven Lakes was assessed using the ecological carrying capacity model. This model was composed of three criteria: biophysical, socioeconomic, and ecotourism. The biophysical criterion involved the collection of water quality indicators in the lake including temperature, dissolved oxygen, pH, BOD, total dissolved solids, conductivity, chlorophyll-*a*, nitrates, phosphates, and transparency during the wet and dry seasons. Household survey and Key Informant Interviews were conducted to gather primary data on the socioeconomic and tourism indicators. The socioeconomic indicators include the number and area of fish cages, stocking density, income from aquaculture, and nonfarming activities. Ecotourism indicators include number of tourists, facilities, manpower, cost and variety of activities, and number of hours spent in various tourism activities.

Estimating and illustrating the recreational and aquaculture carrying capacities of the lakes can provide the LGUs' constituents and policymakers a comprehensible overview of the potential consequences of the unrestricted proliferation of the various activities the lakes are currently hosting.

Generally, this study aims to harmonize the sustainable management of the seven lakes of San Pablo with the increasing density of cultivated areas, which has potential to offer business and employment opportunities. Specifically, the data gathered and analyzed will be used to estimate, model, and simulate the long-term dynamics and stability properties of each of the seven lakes, broken down into fundamental components of ecological, tourism, and socioeconomic carrying capacities, since these components act as the main approaches to estimating carrying capacity (Chougule 2011). The model to be generated by virtual applications will then be used as an instrument in the sustainable and effective policy planning of the local government of San Pablo City to address issues on food security and sustainability.

25.2 Methodology

25.2.1 Methodological Framework of the Ecological Carrying Capacity Study

This study aimed to derive a model of the biophysical, socioeconomic, and tourism carrying capacities of the Seven Lakes of San Pablo City (Fig. 25.11). The model will provide the stakeholders with a comprehensible framework of the overall condition of Seven Lakes' ecosystem, and an assessment of the lake's sustainability due to continued stress from aquaculture systems and recreational infrastructures and activities.

The results of the ecological carrying capacity modeling will be used to assess the sustainability of the lakes. The sustainability index ranking of each lake will be used in the crafting of management strategies toward the sustainable management of the lakes.

By integrating the framework and guidelines provided by related literature, this study will be able to derive a model of the estimated physical, ecological, social, and economic recreational and aquaculture carrying capacity of each lake belonging to the Seven Lakes of San Pablo City. This output will provide the stakeholders with a comprehensible framework of the overall condition of Seven Lakes' ecosystem, and a simulation of the predicted consequences of ecosystem degradation due to continued stress from aquaculture systems, and recreational infrastructures and activities.

This study would like to establish an evaluation index system to determine the ecological carrying capacity of the Seven Lakes of San Pablo, specifically on the socioeconomic, tourism, and water quality carrying capacity.

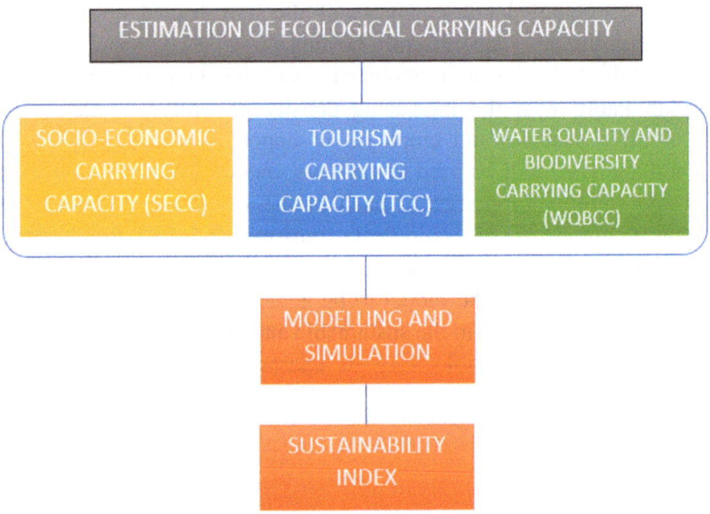

Fig. 25.11 Methodological framework for ecological carrying capacity modeling

25.2.2 Characteristics of the Seven Lakes of San Pablo

The seven lakes are generally small lakes with areas ranging from 20.5 to 99.2 ha (Table 25.1). Lake Sampaloc, located at the center of the city, is the largest among the seven lakes, while Lake Mohicap is the smallest lake. Lake Palakpakin is the shallowest among the seven lakes with a maximum depth of 7.7 m. Siltation is a problem in this lake. On the other hand, Lake Calibato is the deepest lake (156 m) and has the highest area occupied by fish pens (22.36%). The other two aquaculture lakes with high proportions of area occupied by fish pens are Lake Bunot (18.7%) and Lake Palakpakin (17.2%). Lake Sampaloc is the main ecotourism lake in San Pablo City with 1,967,261 visitors from 2015 to 2019 (Table 25.1). The twin lakes of Pandin and Yambo as well as Lake Bunot also attract local tourists.

25.2.3 Biophysical Assessment

25.2.3.1 Sampling Stations

The sampling design used in the study is purposive sampling to determine the impact of structures in and around the lakes. Four or five sampling stations were selected in each of the seven lakes (Figs. 25.12, 25.13, 25.14, 25.15, 25.16, 25.17, and 25.18). These stations were located near fish cage, near an outlet, littoral near houses, and littoral near vegetation and pelagic zone. Three replicate samples were collected at

Table 25.1 Characteristic features of the Seven Lakes of San Pablo

Lake	Location[a]	Area (ha)[a]	Maximum depth (in m)[b]	% Area occupied by fish pens (2018)[b]	Total tourist arrival (2015–2019)[c]
Bunot	Brgy. Concepcion	38.16	23	18.70	56,966
Calibato	Brgy. Sto. Angel	27.18	156	22.36	5237
Mohicap	Brgy. San Buenaventura	20.49	30.4	5.77	10,529
Palakpakin	Brgy. San Buenaventura, San Lorenzo and Dolores	54.49	7.7	17.25	3518
Pandin	Brgy. San Lorenzo, Sto. Angel	24	61.75	1.80	97,267
Sampaloc	Brgy. IV-A, IV-C, V-A, Concepcion, San Lucas	99.21	27.6	7.68	1,967,261
Yambo	Brgy. San Lorenzo	36	38	2.07	81,341

[a]San Pablo City Ecological Profile. City Planning and Development Office-City Government of San Pablo. San Pablo, City of Seven Lakes

[b]Mendoza, M.U., Briones, J.C.A., Itoh, M., Padilla, K.S.A.R., Aguilar, J.I., Okuda, N., & Papa, R.D.S. (2019). Small maar lakes of Luzon Island, Philippines: their limnological status and implications on the management of tropical lakes – a review. *Philipp J Sci, 148*(3): 565–578

[c]Data provided by the San Pablo Tourism Office (2019)

Fig. 25.12 Sampaloc lake sampling stations

Fig. 25.13 Mohicap lake sampling stations

each station. In situ parameters were measured using a YSI water quality multiparameter sensor. The parameters include dissolved oxygen, temperature, total dissolved solids, conductivity, salinity, and chlorophyll-*a*. Water quality parameters at the surface water and 5 m below the surface were measured for the in situ

Fig. 25.14 Bunot lake sampling stations

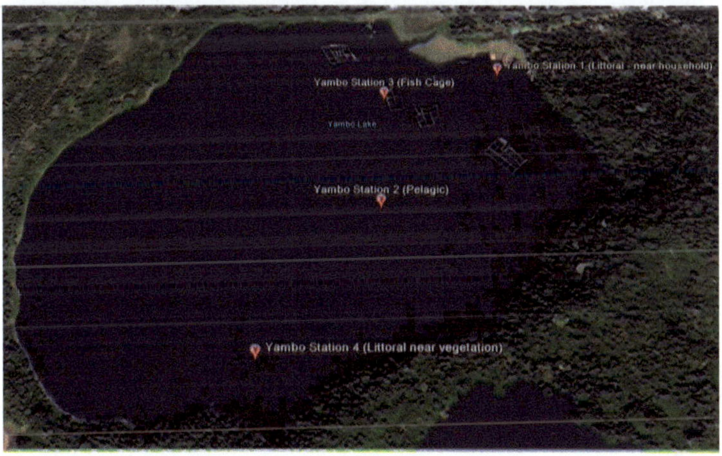

Fig. 25.15 Yambo lake sampling stations

parameters. A Secchi Disk was used to measure the transparency of water. Water samples were also collected for laboratory analysis. Three 6 L of water samples were used for BOD analysis. Three 500 mL of water samples were used for nitrogen and

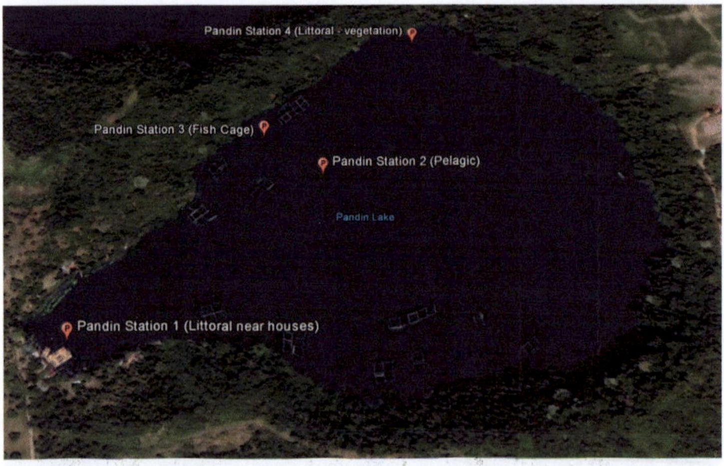

Fig. 25.16 Pandin lake sampling stations

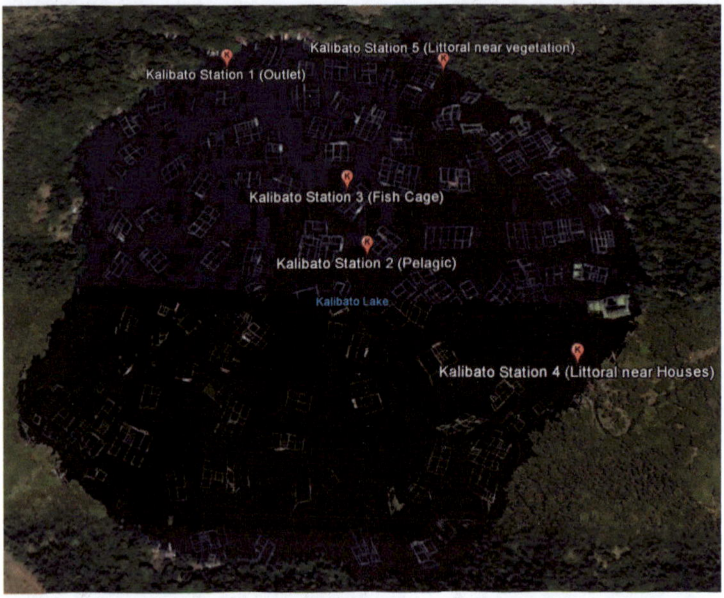

Fig. 25.17 Calibato lake sampling stations

phosphorus contents analysis. Water quality assessment was done on the following dates: August 2019, November 2019, and May 2021.

Gill netting, which is the most common type of fishing being employed by the locals, was used to determine the fish composition and abundance in the Seven Lakes. Samples of wild fish species per lake were collected with the assistance of the local fisherfolk. Horizontally oriented gillnets with different mesh sizes were set up

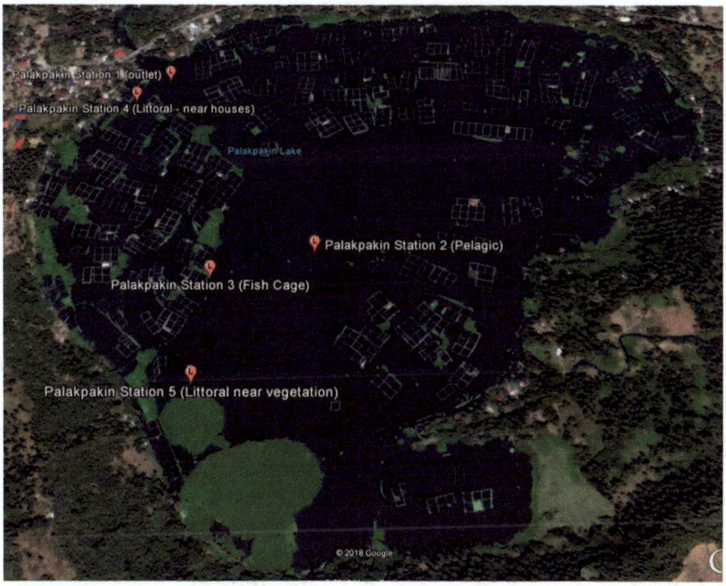

Fig. 25.18 Palakpakin lake sampling stations

in each lake at 5:00 PM and were retrieved at 7:00 AM. Fish samples were identified using several fish identification materials (Herre 1927, 1953; Vidthayanon 2007; Froese and Pauly 2019; Fitzgerald 1979; Behrends et al. 1982; Wu et al. 1983; Herder et al. 2012).

25.2.3.2 Floristic Survey Assessment

In Lakes Pandin and Yambo, a reconnaissance survey of macrophyte vegetation was conducted to identify sampling areas within the lakes where aquatic plants are present or located. The sampling points for both lakes were identified using a purposive sampling technique. The macrophyte vegetation is often characterized by its dispersed location within the lakes, which is affected by shifts in water current and wind direction (Dar et al. 2014).

The identification of aquatic macrophytes was done using the "A Field Guide of Aquatic Macrophyte Species Found in New York Lakes along with Potential Non-Native (Exotic) Invaders" (Johnson 2013) and "Aquatic Weed Identification" (UC ANR n.d.). A 10-m transect line was established over or through the macrophyte vegetation in each sampling location. Macrophytes species were identified and the extent of the macrophyte cover was determined in each transect. There was a total of five (5) sampling points for both lakes and each point had four replicates of a 10-m transect line.

A GPS (Garmin eTrex 10 Worldwide Handheld GPS Navigator) was used to record the coordinates of the sampling points, specifically the latitudinal and longitudinal locations of the transect lines. The identification of aquatic macrophytes along the transect line was done on-field and in the laboratory. During the field survey, identification was performed using the Plant ID application, as well as listing the names of macrophyte species and marking unidentified plant samples. Photographs of the unidentified macrophytes were taken for identification in the laboratory using identification materials including Plant List, StuartXchange, Aquagenixaquatics, Aquaplant, and Co's Digital. After the necessary data was gathered, the species count for each macrophyte was conducted using an offline application called Plant Population Calculator, and the frequency of macrophyte species in the area was calculated in the laboratory.

25.2.4 Water Quality and Biological Carrying Capacity (WQBCC)

The water quality and biological carrying Capacity (WQBCC) was obtained by summing the product of the following subindicators: water quality parameters, aquatic macrophytes, native fish species, introduced fish species, fish abundance, fish biomass, and volume of solid wastes, multiplied by its corresponding weight.

25.2.4.1 Socioeconomic Carrying Capacity (SECC)

Various stakeholders and experts were asked to rank the different socioeconomic indicators including the population living around each lake, annual income from aquaculture, income from tourism, area occupied by fish pens (ha), number of fish pens, and area for recreational activities based on importance. The resulting weights from all respondents were normalized and the average normalized weight for each sub-indicator was obtained. SECC was then calculated by obtaining the summation of the product of the aforementioned subindicators, multiplied by its calculated weight.

Aside from the collection of limited and actual values for each of the subindicators from literatures and secondary data, Key Informant Interviews (KII) and Focus Group Discussion (FGD) were also conducted to determine community participation in lake management and discuss the ecotourism plans in Lakes Pandin, Yambo, and Sampaloc as basis of the discussion. Participants of the KII and FGD included various stakeholders such as the Fisheries and Aquatic Resource Management Councils (FARMC) members, local government units (LGUs), tourism officers, *bantay-lawa*, fisherfolks, civic groups, women groups, and senior citizens.

Meanwhile, a Knowledge, Attitudes, and Practices (KAP) survey was constructed, pretested, and employed to the residents living near or around Lakes Sampaloc, Pandin, and Yambo in September 2019 to provide the researchers with insights regarding the disposal and management of aquaculture and tourism-related wastes in the area. These three lakes were chosen by the researchers to represent the Seven Lakes. Lake Sampaloc, as an urban lake, is also considered as the most populated of all the lakes since it is located in the city proper, and the lake with the highest number of fish cages and pens. On the other hand, the "twin lakes" Pandin and Yambo are the most popular ecotourism lakes in San Pablo, and the lakes with the least numbers of aquaculture structures (San Pablo CLUP 2015–2025).

Moreover, questions related to the residents' income from aquaculture and tourism-related activities were included in the KAP survey. KAP survey is a kind of household survey used primarily to collect data on what is known, believed, and done in relation to a particular topic (Zahedi et al. 2014).

In this study, three hundred seventy-six (376) of the total respondents are residents from the five barangays surrounding Lake Sampaloc, while ninety-four (94) are from merely Barangay Sulsuguin since it is considered the barangay closest to Lake Yambo. A total of sixty-four (64) respondents completed the household survey for Lake Pandin. The total number of respondents was computed using the Cochran's Formula and was equally divided per barangay through proportionate sampling.

25.2.4.2 Tourism Carrying Capacity (TCC)

The threshold limit of tourism activities and the number of individuals who can be supported within the natural resource limits of each lake were estimated using land-based surveys and management-by-objectives approaches. Parameters that were collected include population data, the number of tourists, and hotel daily reception number/day (Ding et al. 2015). Data on various variables that have noticeable effect on the ecotourism of the lake, such as the ecotourism awareness and the possibility of ecotourism disturbance on the ecology and education, seasonal or annual increase in crowd, modes of transportation, and the reason for visiting, were collected and incorporated in the management of carrying capacity.

The variables needed for computing carrying capacity were the size of the lake intended for tourism activities, points of activities, infrastructures and facilities available, and amenities for tourists. Following the manual of Calanog (2015) of the Department of Environment and Natural Resources—Ecosystem Research and Development Bureau (DENR-ERDB), the determination of the carrying capacity was based on the Boullon's (1985) carrying capacity mathematical model (BCCMM), which determined the standard requirement of the visitor, such as space, time, and other needs, while enjoying an activity, such as swimming.

In the BCCMM, the basic carrying capacity (BCC), potential carrying capacity (PCC), and real carrying capacity (RCC) were computed and then used in determining the total tourism carrying capacity (TCC) of an area.

25.2.4.2.1 Basic Carrying Capacity (BCC)

The parameters needed to identify BCC are the following: total size of the area used by the tourists and standard or average space requirement of visitors.

25.2.4.2.2 Potential Carrying Capacity (PCC)

PCC was calculated by computing the rotation coefficient (RC) of a specific tourism activity. RC was determined as the total number of hours the lake area was open for recreational activities (e.g., swimming) divided by the average number of hours the tourists enjoyed swimming. The average number of hours the tourists enjoyed recreational activities was based on responses obtained during the interviews.

25.2.4.2.3 Real Carrying Capacity (RCC)

RCC is the maximum permissible number of uses of an area once limiting (i.e., corrective) factors derived from the characteristics of the site (or standards/needs of the visitors) have been applied. RCC was computed by incorporating the limiting factors identified from the interviews and observations on-site. The following limiting variables were incorporated:

- Lf_1 (typhoons),
- Lf_2 (rainy days),
- Lf_3 (time available for swimming), and
- Lf_4 (quality of the lake as a tourism site, based on tourists' perceptions).

For the purpose of modeling, the tourism carrying capacity (TCC) was then obtained by adding the products of the following subindicators: number of tourists, hours tourists spent on the lake, facilities, manpower, cost and variety of activities, which were then multiplied by their corresponding weights (W_i).

25.2.5 Ecological Carrying Capacity (ECC) Modeling and Simulation

After the collection, analysis, and integration of the three components of carrying capacities (biophysical, socioeconomic, and tourism) of each lake, the water ecological carrying capacity (WECC) was developed following the approach of Ding et al. (2015). WECC models for each lake were developed to assess the effects of the increasing density of aquaculture systems and recreational activities and infrastructures to the lakes' sustainability. The model will help visualize the sustainability status of each lake related to the associated activities (Ding et al. 2015).

25.2.5.1 Model Framework

This study aimed to derive a model of the biophysical, socioeconomic, and tourism carrying capacities of the Seven Lakes of San Pablo City. The model will provide the stakeholders with a comprehensible framework of the overall condition of Seven Lakes' ecosystem, and an assessment of the lake's sustainability due to continued stress from aquaculture systems and recreational infrastructures and activities.

The results of the ecological carrying capacity modeling will be used to assess the sustainability of the lakes. The sustainability index ranking of each lake will be used in the crafting of management strategies toward the sustainable management of the lakes.

25.2.5.2 Indicators for Determining the Ecological Carrying Capacity

The socioeconomic criteria include the following indicators: population living around the lake, annual income from aquaculture, income from tourism, area occupied by fish pens, number of fish pens, and area allocated for recreational activities (Table 25.2). Tourism criteria include the following indicators: number of tourists, hours tourists spent on the lake, facilities, manpower, cost of activities,

Table 25.2 Indicators used to determine the ecological carrying capacity

Criteria	Indicator	Limited value	Actual value
Socioeconomic	Population living around the lake	Number of houses that can occupy the circumference of the lake multiplied by the average household size of San Pablo, which is 4	Number of houses around the lake observed on Google Earth and multiplied by the average household size of San Pablo, which is 4
	Income from aquaculture (annual)	Obtained from San Pablo City ecological profile	Obtained from San Pablo City ecological profile
	Income from tourism	Based on FGD and KII	Based on FGD and KII
	Area occupied by fish pens (ha)	Obtained from San Pablo City ecological profile	Obtained from San Pablo City ecological profile
	Number of fish pens	Obtained from San Pablo City ecological profile	Obtained from San Pablo City ecological profile
	Area for recreational activities (ha)	Obtained and measured during the field visit	Obtained and measured during the field visit
Tourism	Number of tourists	Based on the data given by the San Pablo Tourism Office	Based on the data given by the San Pablo Tourism Office
	Hours tourist spent on the lake	Based on FGD and KII obtained from the SKBMLP, and based on interview conducted with the tourist respondents	Based on FGD and KII obtained from the SKBMLP, and based on interview conducted with the tourist respondents
	Facilities	Based on FGD and KII obtained from the SKBMLP	Based on FGD and KII obtained from the SKBMLP
	Manpower	Based on FGD and KII obtained from the SKBMLP	Based on FGD and KII obtained from the SKBMLP
	Cost of activities	Based on FGD and KII obtained from the SKBMLP	Based on FGD and KII obtained from the SKBMLP
	Variety of water activities	Based on FGD and KII obtained from the SKBMLP	Based on FGD and KII obtained from the SKBMLP
Water quality and biological indicators	Water quality parameters (dissolved oxygen, temperature, pH, total dissolved solids, Chl-*a*, transparency, nitrate, phosphate, BOD, COD)	It is based on the recommended level set by the DENR AO 2016-08	Actual values from field data collection in 2019 and 2021.

(continued)

Table 25.2 (continued)

Criteria	Indicator	Limited value	Actual value
	Aquatic macrophytes	Based on 10% of the total surface area of each lake	Based on the surface area covered by macrophytes observed on Google Earth
	Native fish species	Based on historical data of fish biodiversity from the Participatory Rural Appraisal (PRA) and actual values from fish sampling in 2019 and 2021	Based on historical data of fish biodiversity from the Participatory Rural Appraisal (PRA) and actual values from fish diversity surveys in 2019 and 2021
	Introduced fish species	Based on historical data of fish biodiversity from the Participatory Rural Appraisal (PRA) and actual values from fish sampling in 2019 and 2021	Based on historical data of fish biodiversity from the Participatory Rural Appraisal (PRA) and actual values from fish diversity surveys in 2019 and 2021
	Fish abundance	Obtained from San Pablo City ecological profile	Obtained from San Pablo City ecological profile
	Fish biomass (kg)	Obtained from San Pablo City ecological profile	Obtained from San Pablo City ecological profile
	Volume of solid wastes (kg)	Obtained from FARMC members	Obtained from FARMC members

and variety of water activities. The water quality and biological criteria include the following indicators: water quality parameters (dissolved oxygen, temperature, pH, total dissolved solids, Chl-*a*, transparency, nitrate, phosphate, BOD, COD), aquatic macrophytes, native fish species, introduced fish species, fish abundance, fish biomass, and volume of solid wastes. The various sources of data for the limited and actual values were derived from San Pablo City ecological profile; San Pablo City Tourism Office; field measurements; DENR AO 2016-08; and conduct of PRA, FGD, and KII, with FARMC officers, *Samahan ng Kababaihan sa Pandin*, tourists, and households around the lakes. These actual and limited values for each indicator were entered into the carrying capacity models (Table 25.2).

The compiled data on the biophysical, socioeconomic, and tourism indicators affecting the lakes' ecosystem will then help in the generation of the threshold limit estimates for recreational infrastructures and visitor use of ecotourism lakes, the total area suitable for aquaculture and magnitude of aquaculture production without significant effect to the lake's ecosystem, and the value of stakeholder involvement and capacity of the lake to tolerate the presence of tourists.

25.2.5.3 Mathematical Model

The mathematical model used in the study was based on Ding et al. (2015). The calculation of the ECC is given by the following equations:

$$CS_i = CC_i / CC_{imax} \tag{25.1}$$

Equation (25.1) is used to represent the model of the carrying capacity level of each indicator. CC_i is the actual value of the indicator and CC_{imax} is the carrying capacity limit value of the indicator. CS_i, the capacity level of the indicator, was derived using Eq. (25.1).

After obtaining the carrying capacity level of each indicator, Eq. (25.2) was used to compute the carrying capacity (CC) of each criterion (SECC, TCC, and WQBCC):

$$CC = \sum CS_i \times W_i \tag{25.2}$$

W_i is the weight of each indicator, which was obtained from the results of the ranking survey conducted. Various stakeholders and experts were asked to rank the different indicators based on importance. The resulting weights from all respondents were normalized and the average normalized weight for each indicator was obtained.

Using Eq. (25.2), the socioeconomic carrying capacity (SECC), tourism carrying capacity (TCC), and water quality and biological carrying capacity (WQBCC) were computed.

Finally, the ecological carrying capacity (ECC) of each lake was derived by summing up the three component criteria (SECC, TCC, and WQBCC), given by the following Eq. (25.3):

$$ECC = SECC + TCC + WQBCC \tag{25.3}$$

The calculated ECC was compared with the sustainability index (Table 25.3) to assess the level of sustainability of each lake. Table 25.3 shows the sustainability index used in the study.

Table 25.3 Sustainability index used in determining the sustainability of the Seven Lakes

ECC value	Sustainability status
Below 0	Unsustainable
0–0.33	Low
0.34–0.67	Medium
0.68–1.00	High
Greater than 1	Very high

25.3 Results and Discussion

25.3.1 Biophysical Indicators

25.3.1.1 Water Quality Assessment and Monitoring

Water quality assessment of the seven lakes of San Pablo was done in between August and November 2019 and May 2021. The results of water quality monitoring showed that aquaculture lakes (Bunot, Calibato, Palakpakin, and Sampaloc) have months and stations with dissolved oxygen lower than 5.0 ppm (Fig. 25.19). A DO value lower than 5.0 ppm is already stressful and critical for fish survival (Pleto et al. 2018). These lakes are aquaculture lakes that experience fish kills periodically. Fishes that experience low levels of DO are more prone to disease and infection, as well as less efficient at converting food into energy and experience stunted growth (Kremer 2018). The low DO in water primarily results from excessive algal growth caused by nutrient enrichment, particularly nitrogen and phosphorus. When the algae die and decompose, the process consumes oxygen (Minnesota Pollution Control Agency [MPCA] 2009). The four aquaculture lakes have relatively high phytoplankton density. On the other hand, very high DO levels, especially during summer months, are due to the photosynthetic activities of aquatic plants and phytoplankton in water. Supersaturation occurs when oxygen is produced by algae more quickly than it can escape to the atmosphere. Another factor of supersaturated DO level is due to the rapid movement of water because of wind and fish movement that increases the rate of diffusion (Pleto et al. 2018).

The average temperature of the seven lakes is within the recommended range of 25–31 °C (Fig. 25.20). However, there are instances wherein it exceeded the upper limit especially during the summer months. Temperature is an important factor that

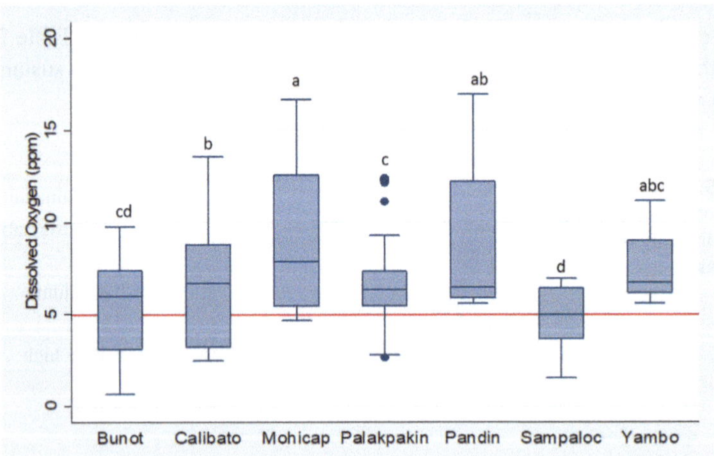

Fig. 25.19 Dissolved oxygen concentration (ppm) of the Seven Lakes of San Pablo (means with different letters are significantly different)

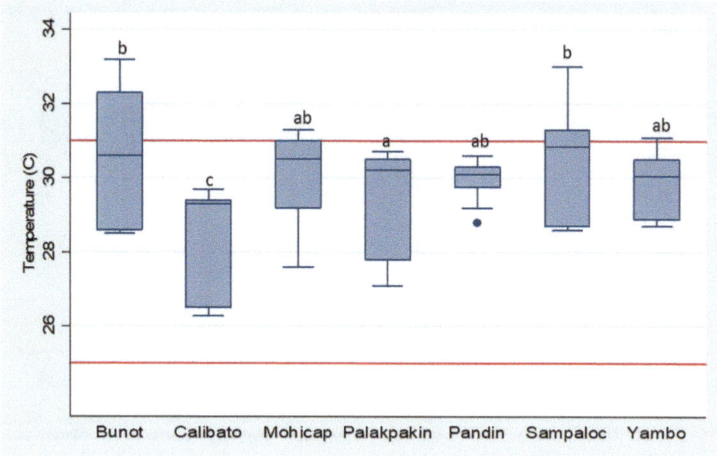

Fig. 25.20 Temperature (°C) level of the Seven Lakes of San Pablo (means with different letters are significantly different)

plays a major role in aquatic life and other water quality parameters. Its relationship with dissolved oxygen is that colder water or low temperature could hold more dissolved oxygen than warmer water. During the colder months of January and February, and toward the onset of the rainy season after a long warm dry season, the water temperature on the surface of the lake may become colder than the subsurface layers of the lake. This difference in water temperature may cause lake overturn that would cause a depletion of dissolved oxygen in the surface that can eventually lead to the occurrence of fish kill.

The water quality assessment and monitoring showed that the mean pH level of the seven lakes were within the recommended range of 6.5–9 set by the DENR (DAO 2016-08) (Fig. 25.21). Fishes can become stressed in water with pH levels ranging from 4.0 to 6.5 and 9.0 to 11.0 and death is certain at pH level less than 4.0 or greater than 11.0 (Ekubo and Abowei 2011). Changes in pH can cause an increase in solubility of phosphorus, making it available for algal growth that can cause eutrophication and eventually oxygen depletion (Munson et al. 2004).

Nutrients such as nitrogen and phosphorus play an important role in the growth of algae, which forms the base of the food web in the lake ecosystem. However, too much nitrogen and phosphorus could cause negative impacts on fish population. Excessive nutrients are known to cause eutrophication that may induce rapid algal growth. Although eutrophication is a natural process, anthropogenic activities hasten and make it worse. Excessive algal growth will eventually lead to algal die-off, and during the decomposition of dead algae, decomposers will take up oxygen in the water. This will cause a depletion of dissolved oxygen and eventually could lead to fish kill. The nitrate level of the seven lakes did not reach the recommended level of 7.0 ppm by the DENR (Fig. 25.22). However, there are stations in lakes Calibato and Pandin that exceed 7.0 ppm. Further, phosphate levels of the seven lakes exceeded

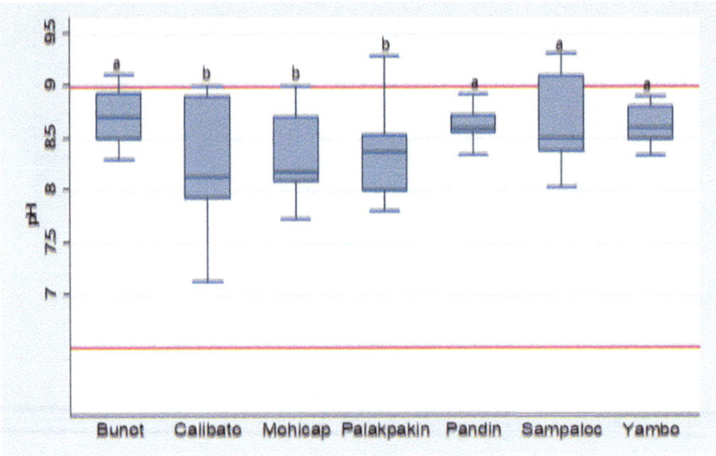

Fig. 25.21 pH level of the Seven Lakes of San Pablo (means with different letters are significantly different)

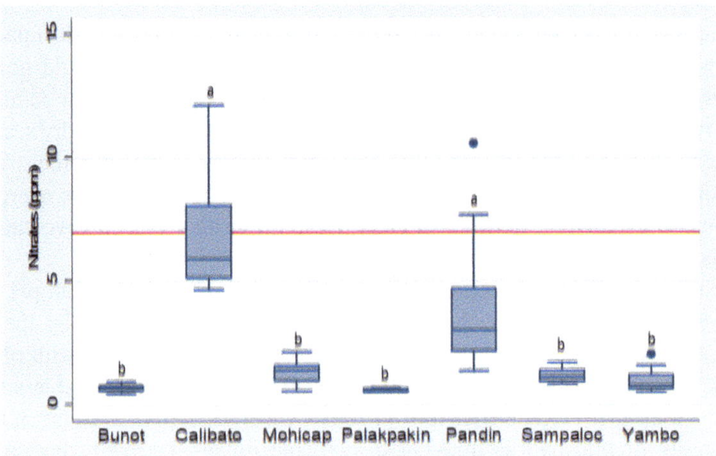

Fig. 25.22 Nitrates (ppm) level of the Seven Lakes of San Pablo (means with different letters are significantly different)

the recommended levels of the updated water quality guideline of the DENR for Class C water of 0.025 ppm (DAO 21-19). It also shows that aquaculture lakes (Bunot, Calibato, Palakpakin, and Sampaloc) have relatively high levels of phosphates compared to ecotourism lakes (Mohicap, Pandin, and Yambo) (Fig. 25.23).

Chlorophyll-*a* is the measure of the amount of algae that is growing in water. This can be used to classify the trophic condition of a lake. The Organization for Economic Co-operation and Development (OECD) suggested that a mean of

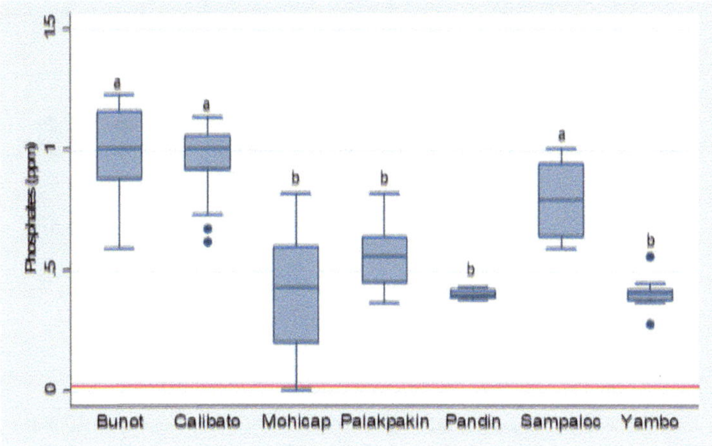

Fig. 25.23 Phosphates (ppm) level of the Seven Lakes of San Pablo (means with different letters are significantly different)

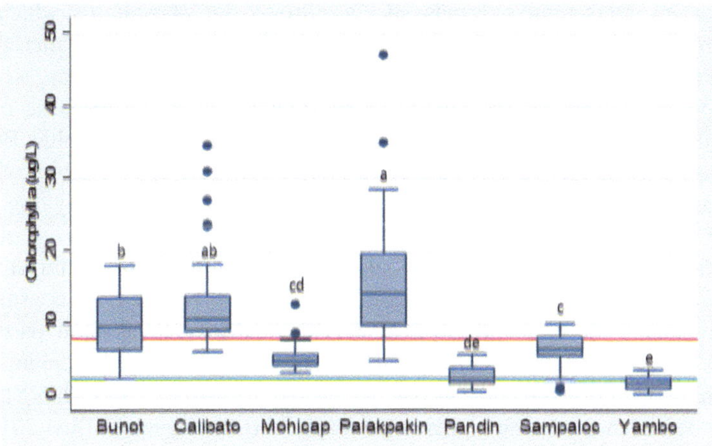

Fig. 25.24 Chlorophyll-*a* (µg/L) levels on the Seven Lakes of San Pablo (means with different letters are significantly different)

<2.5 µg/L indicates oligotrophic status or low level of nutrients in water; 2.5–8 µg/L is mesotrophic status, which means medium level of nutrients; and >8 µg/L is eutrophic wherein high nutrients are found in water, which can support dense phytoplankton and macrophyte population. This condition could deplete oxygen, which in turn could cause fish kill. Based on the assessment, it shows that Lakes Pandin and Yambo are considered oligotrophic (Fig. 25.24). Lakes Mohicap and Sampaloc are mesotrophic, while Lakes Bunot, Calibato, and Palakpakin are considered as eutrophic. The amount of chlorophyll-*a* in water is related to its temperature, nutrient content, light intensity, and wind. Anthropogenic activities such as

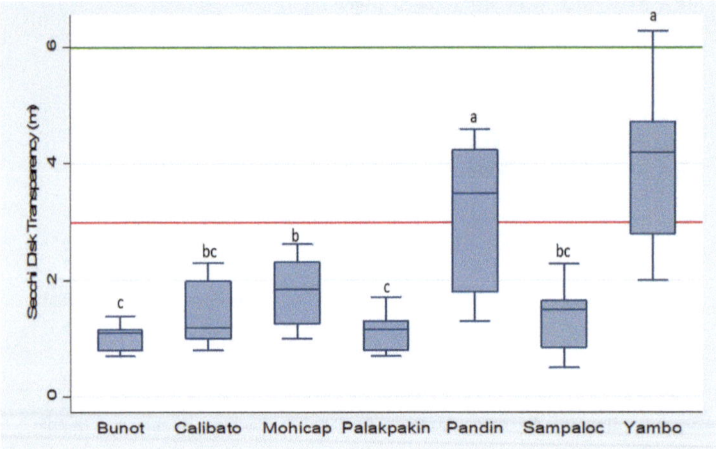

Fig. 25.25 Secchi Disk Transparency (m) on the Seven Lakes of San Pablo (means with different letters are significantly different)

improper sewage increase nutrients, such as nitrogen and phosphorus, which causes algal growth (Limno Loan Program Manual n.d.). Another parameter that can classify the trophic condition of a lake is the Secchi Disk Transparency. According to the OECD, a transparency level of greater than 6.0 m is considered as oligotrophic, 6.0–3.0 m is mesotrophic, and below 3.0 m is considered eutrophic. Based on the assessment, Lakes Pandin and Yambo are considered mesotrophic, while the other lakes are classified as eutrophic based on the range set by the OECD (Fig. 25.25).

Another important water quality indicator in lakes is the biochemical oxygen demand (BOD). It is the measurement of total dissolved oxygen consumed by microorganisms for biodegradation of organic matter such as food particles or sewage. The excess entry of cattle and domestic sewage from the nonpoint sources and increase in phosphate in the village ponds may be attributed to high organic load in these ponds, thus causing higher levels of BOD (Bhatnagar and Devi 2013). The DENR set the recommended level of BOD for Class C water to 7.0 ppm. Ekubo and Abowei (2011) set the aquatic system with BOD levels between 1.0 and 2.0 mg/L—considered clean; 3.0 mg/L—fairly clean; 5.0 mg/L—doubtful; and 10.0 mg/L—definitely bad and polluted. Based on the assessment, Lakes Bunot and Calibato exceeded the recommended level of 7.0 ppm, which is considered to be polluted according to Ekubo and Abowei (2011) (Fig. 25.26). The fish cage stations on the lakes have relatively high BOD levels and tend to exceed the limit.

25.3.1.2 Fish Biodiversity

Participatory Rural Appraisal (PRA) results show that fish composition in the Seven Lakes were very dynamic from 1940 to 2019, which can be attributed to species introduction (Fig. 25.27). Based on the local ecological knowledge of San Pablo City residents, introduced fish species are already present from the year 1940 to 1981,

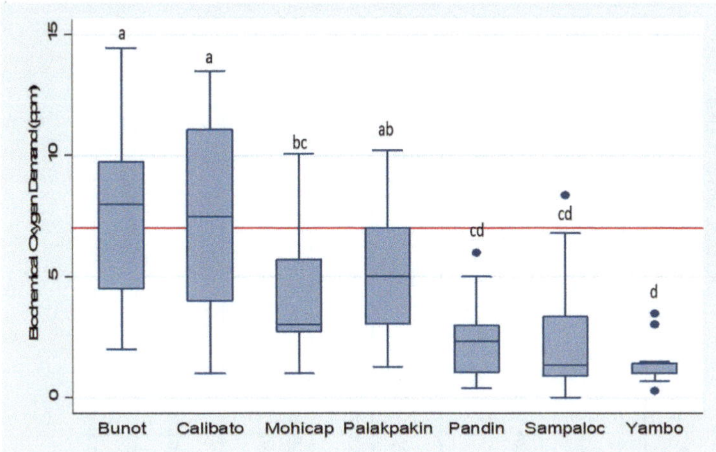

Fig. 25.26 Biochemical oxygen demand (BOD) (ppm) on the Seven Lakes of San Pablo (means with different letters are significantly different)

while some of the native species include *Glossogobius aureus*, *Giuris margaritacea*, and *Leiopotherapon plumbeus*. However, it can be noted that in the year 1981–2019, these native species became absent in some lakes. Aside from the loss of some native species in some lakes, the focus group discussion revealed that the abundance of fish caught during this period diminished.

Table 25.4 shows the fish species composition present in the Seven Lakes perceived by the community. The results indicate that *dalag* or mudfish is perceived to be present in most of the seven lakes, namely Sampaloc, Bunot, Yambo, Pandin, and Calibato. Consequently, there are various fish that they claimed are only present in a single lake, for instance, carp, tigreng tilapia, and black mask in Calibato; eel and starry goby in Sampaloc; tank goby in Bunot; and barangan in Yambo. Moreover, participants coming from Lake Calibato have given the highest number of fish species (11) present in their lake, followed by Lake Bunot (10) and Lakes Sampaloc and Yambo (8).

A total of 620 individuals were collected from the Seven Lakes from July to August 2019. The sampling was done during the onset of the wet season wherein the rainfall volume started to increase. Highest number of collected fish samples was recorded in Bunot Lake ($n = 379$), followed by Sampaloc Lake ($n = 68$), Mohicap Lake ($n = 62$), Pandin Lake ($n = 42$), Palakpakin Lake ($n = 30$), Yambo Lake ($n = 24$), and Calibato Lake ($n = 15$). The total number of fish collected was largely represented by *Amphilophus* sp. (red devil) from Bunot Lake and Sampaloc Lake and *Oreochromis niloticus* from Sampaloc Lake, Bunot Lake, Yambo Lake, Calibato Lake, and Palakpakin Lake (Fig. 25.28). The occurrence of *Vieja* sp., an aquarium fish, was first reported in Lakes Bunot, Calibato, Mohicap, and Palakpakin in 2017, but there was no record of mechanism of introduction. *O. niloticus* is commonly a cultured fish for aquaculture purposes. Results of Participatory Rural

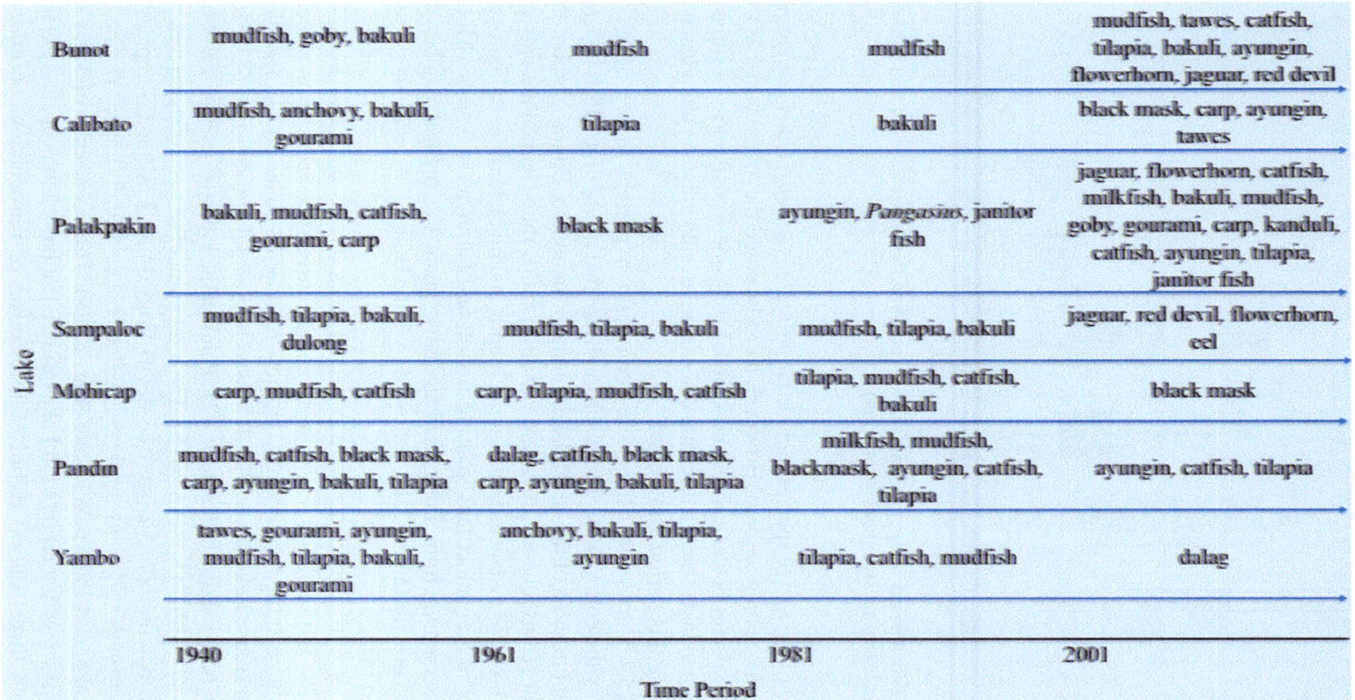

Fig. 25.27 Knowledge of San Pablo City residents on fish species present in the Seven Lakes

Table 25.4 Summary of fish species composition perceived by the community in the Seven Lakes

Fish species	Sampaloc	Bunot	Mohicap	Yambo	Pandin	Calibato	Palakpakin	*S*
Tawes (*Barbonymus gonionotus*)		+		+		+		3
Tigreng tilapia (*Tilapia* sp.)						+		1
Red devil (*Amphilophus* sp.)	+	+						2
Carp (*Cyprinus carpio*)						+		1
Bakuli (*Giuris margaritacea*)	+	+		+		+		4
Barangan (*Teuthis virgatus*)				+				1
Green flowerhorn (*Vieja* sp.)	+	+						2
Gourami				+		+		2
Igat or eel (*Anguilla* sp.)	+							1
Dugong or jaguar guapote (*Parachromis managuensis*)	+	+						2
Nile tilapia (*Oreochromis niloticus*)	+	+		+		+		4
Biya or tank goby (*Glossogobius giuris*)		+						1
Dulong or starry goby (*Mirogobius stellatus*)	+							1
Ayungin (*Leiopotherapon plumbeus*)		+		+		+		3
Dalag or mudfish (*Channa striata*)	+	+		+	+	+		5
Dilis or anchovies (Family *Engraulidae*)				+		+		2
Hito or catfish (*Clarias* sp.)		+	+		+	+		4
Black mask or largemouth bass (*Micropterus salmoides*)						+		1
S	8	10	1	8	2	11	0	

S number of species

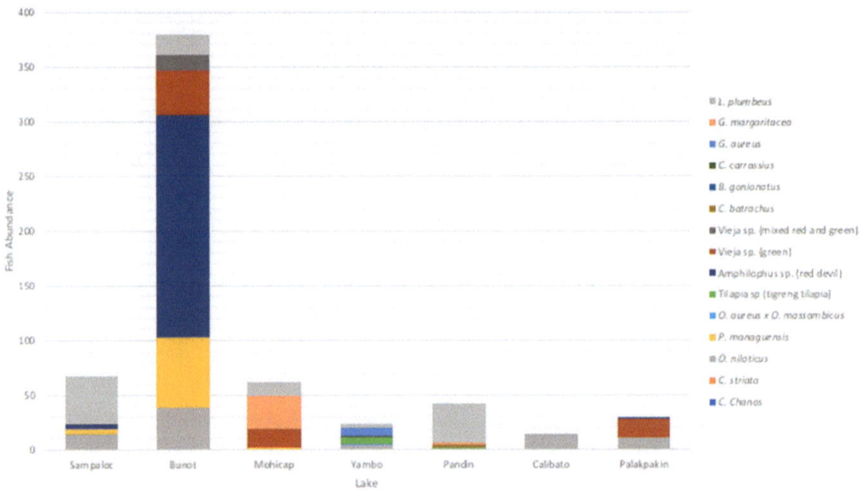

Fig. 25.28 Abundance of species collected from the Seven Lakes of San Pablo

Appraisal (PRA) revealed that they are being stocked both in fish cages and in the wild by the fish farmers. At present, it largely supports the local fisheries production in San Pablo City (Fig. 25.29).

25.3.1.3 Aquatic Macrophytes

25.3.1.3.1 Determining Aquatic Macrophyte Composition in Lake Pandin and Lake Yambo

The lakes are known for their ecotourism and aquaculture activities, and home to diverse aquatic plant species. *Hydrilla verticillata* was found in all transects around the lakes' periphery. In Lake Pandin, *E. crassipes* and *H. verticillata* were frequently identified together especially in points 1 and 2 (near households) and 5 (near vegetation). All the five aquatic macrophytes were recorded in sampling points 1 and 2, which were adjacent to the households. In Lake Yambo, *H. verticillata* was found in the lake's margin and was present in all sampling points, similarly with Lake Pandin. *Hydrilla verticillata*, *P. stratiotes*, and *I. aquatica* were found in sampling point 1, *P. stratiotes* and *E. crassipes* were found in sampling point 2, and *E. crassipes* was found in sampling point 3. Notably, only *P. stratiotes* was included in the sampling point.

As observed, *H. verticillata* was found to be the most abundant in Lake Pandin and Lake Yambo, possibly indicating the dominance of the species throughout the lakes' water. *Hydrilla verticillata* is native to Asia, in which it flourished in a wide range of freshwater environments, particularly lakes, resulting in its aggressiveness (Go Botany 2020). This aquatic plant is particularly aggressive in areas where water is regularly disturbed by anthropogenic activities, such as the twin lakes. *Hydrilla* produces the majority of its biomass near the water's surface, forming a dense mat

Fig. 25.29 Fish samples collected from seven lakes: (**a**) *Chanos chanos*, (**b**) *Channa striata*, (**c**) *Oreochromis niloticus*, (**d**) *Oreochromis* sp. (red tilapia), (**e**) *Parachromis managuensis*, (**f**) *Tilapia* sp., (**g**) *Vieja* sp. (red devil), (**h**) *Vieja* sp. (green flowerhorn), (**i**) *Vieja* sp. (mixed red and green flowerhorn), (**j**) *Clarias batrachus*, (**k**) *Barbonymus gonionotus*, (**l**) *Cyprinus carpio*, (**m**) *Glossogobius aureus*, (**n**) *Giuris margaritacea*, and (**o**) *Leiopotherapon plumbeus*

structure (Olson 2004). The ability of *Hydrilla* to thrive in a wide variety of environmental conditions allows it to exceed other aquatic plant species and dominate the water body, as it can be seen lining the lakes' water margin with a few other aquatic macrophytes.

According to a study by CABI (2016) on *H. verticillata*, it poses economic and ecological risks to water bodies and the aquatic ecosystem (CABI 2020). Along with its aggressive growth underwater, this macrophyte species may cause economic damage to the lake. *Hydrilla verticillata* was recorded in Lake Pandin and Lake Yambo, which are both ecotourism destinations that provided locals with livelihood and visitors with recreation. As part of the recreation activity for tourists, they may take a bamboo raft tour, known as a "balsa", to explore the area. However, due to the significant abundance of *H. verticillata*, it can hinder the movement of the bamboo raft, which can ultimately prevent tourists from visiting the lakes. With the decline in

Table 25.5 Summary of aquatic macrophytes recorded in Lake Pandin and Lake Yambo, San Pablo City, Laguna

No.	Family	Common name	Scientific name
1	Araceae	Water lettuce	*Pistia stratiotes* Linn.
2	Convolvulaceae	Water spinach	*Ipomeoa aquatica* Forsck.
3	Hydrocharitaceae	Water thyme	*Hydrilla verticillata* (L. f.) Royle
4	Nymphaeaceae	Waterlily	*Nymphaea rubra*
5	Pontederiaceae	Water hyacinth	*Eichhornia crassipes* (Mart.) Solms-Laub

tourist arrivals, the locals will experience a reduction in their daily to weekly income. Likewise, it may also disrupt the growth of other macrophyte species, resulting in a reduction in native aquatic plants. This suggests that the proliferation of *H. verticillata* in aquatic ecosystems, such as the Lake Pandin and Lake Yambo, may result in the reduction of other macrophytes and alteration in ecosystem structure and function. Furthermore, the assessment of the presence of aquatic macrophytes and their diversity reflects the status of the ecosystem of the lakes.

Lake Pandin and Lake Yambo are oligotrophic lakes that are known for their ecotourism and small-scale aquaculture activities. The lakes are considered as a freshwater ecosystem; thus, invasion of aquatic macrophytes is apparent in the area. Table 25.5 shows the list of aquatic macrophytes found in Lake Pandin and Lake Yambo, along with their common and scientific names.

A total of five species of aquatic macrophytes under five different families were recorded in Lake Pandin. These macrophytes were *H. verticillata*, *E. crassipes*, *P. stratiotes*, *I. aquatica*, and *N. rubra*. Notably, *N. rubra* is only present in Lake Pandin. It was stated by some locals that *N. rubra* was deliberately planted in the lake for aesthetic purposes. Meanwhile, only four species of aquatic macrophytes under four families were recorded in Lake Yambo. These macrophytes were *H. verticillata*, *E. crassipes*, *P. stratiotes*, and *I. aquatica*. These were the same species of macrophytes that were also found in Lake Pandin, but without the presence of *N. rubra*. In both lakes, *H. verticillata* was found to be covering both the lake's margin or periphery.

Table 25.5 shows the list of aquatic macrophytes recorded in both lakes such as *H. vertcillata*, *N. rubra*, *E. crassipes*, *I. aquatica*, and *P. stratiotes*. *Hydrilla verticillata* is a tropical and subtropical native plant to Asia and Indian subcontinents. It is considered invasive in foreign waters and aggressive in its native environment. It has been reported to infest a variety of freshwater habitats, including wetlands, rivers, and lakes, particularly Lake Pandin and Lake Yambo. It was observed along the margins of both lakes as a submerged macrophyte that is aggressively developing in the form of dense mats on shallow water (Valley and Bremigan 2002; Kissoon et al. 2013). *Hydrilla* can invade a deep-water environment where other native macrophytes struggle to survive and can effectively displace them. In the waters it invades, this macrophyte can be both beneficial and harmful. *Hydrilla* is known to tolerate a wide range of water conditions and disturbances, but its growth is restricted in areas with high salinity levels. Low-saline environments such as Lake Pandin and Lake Yambo provided an ambient habitat for a variety of aquatic macrophytes. According to studies, a dense population of *Hydrilla* may raise pH levels. Without any native predators to control its proliferation, the respondents control it by local harvesting that is led by the head of the community.

Table 25.6 shows each transect and cover the extent of aquatic macrophyte cover, and abundance in terms of count per species. In both lakes, *Hydrilla* has the highest cover and count. According to a study on *Hydrilla,* it has the potential to proliferate rapidly, posing a serious threat to the bodies of water it has infested (CABI 2020). As observed in Lake Pandin and Lake Yambo, *H. verticillata* was the most frequently recorded and abundant species of macrophyte. It is an Asian native that can thrive in any freshwater environment and has become aggressive (Go Botany 2020). It is particularly aggressive in areas where water is often disturbed, such as the twin lakes, because of ecotourism and aquaculture activities.

Other aquatic macrophytes such as *P. stratiotes*, *I. aquatica*, *N. rubra*, and *E. crassipes* were also abundant in Lake Pandin; however, they covered a smaller proportion of the lake surface than *H. verticillata.* In Lake Yambo, *P. stratiotes* and *I. aquatica* were both abundant but had less coverage, while *E. crassipes* had both less coverage and abundance than the other aquatic macrophytes (Fig. 25.30). This may be attributed to the ability of *H. verticillata* to thrive in a wide range of environmental conditions, allowing it an advantage over other aquatic macrophytes in terms of cover and abundance on the lake's periphery. Nevertheless, *H. verticillata* poses an ecological threat to the environment it has invaded. It forms dense mats underwater by growing its biomass near the surface and submerging the majority of its body in water (Olson 2004). It can clog canals and irrigation systems, as well as block the inlet or exit of any bodies of water, resulting in decreased water flow. The high coverage and abundance of *H. verticillata* can also cause difficulty in water access, resulting in fewer visitors and a reduction in aquatic species of flora and fauna.

Hydrilla verticillata, also known as water thyme, thrives in freshwater habitats such as streams, marshes, ponds, and lakes with a salinity of 7%. Similar to twin lakes, it can develop in oligotrophic (low nutrients) to eutrophic (high nutrients) conditions. As a result, it was the most frequently observed macrophyte in both lakes. *Hydrilla* has a higher dominance than other aquatic macrophytes, contributing to the high aquatic plant diversity of Lake Pandin with 435 individuals. The Jaccard index of similarity for Lake Pandin and Lake Yambo was 80%, indicating that macrophytes in the two lakes were identical (Fig. 25.30).

25.3.2 Socioeconomic Indicators

Table 25.7 shows the calculated carrying capacity (CC_i), which was derived by dividing the actual value to the carrying capacity limit value of each socioeconomic indicator. Meanwhile, Table 25.8 shows each indicator's actual and maximum values, which were used to compute the CC_i. Results showed that Lake Sampaloc has the highest calculated carrying capacity in terms of the population living around

Table 25.6 Length of cover and abundance of aquatic macrophytes recorded in Lake Pandin and Lake Yambo, San Pablo City, Laguna, using a 10-m transect line

		Cover (m)						Count					
Lake	Macrophytes	1	2	3	4	5	Total	1	2	3	4	5	Total
Pandin	*H. verticillata*	10	10	10	10	10	50	150	94	87	79	25	435
	P. stratiotes	10	5	0	0	0	15	21	14	0	0	0	35
	I. aquatica	10	10	0	0	0	20	17	11	0	0	0	28
	N. rubra	10	5	0	0	0	15	10	6	0	0	0	16
	E. crassipes	5	5	0	0	5	15	7	4	0	0	3	14
	Total	45	35	10	10	15	115	205	129	87	79	28	528
Yambo	*H. verticillata*	10	10	10	10	10	50	91	76	83	79	38	367
	P. stratiotes	11	4	0	0	0	15	38	30	0	0	0	68
	I. aquatica	20	0	0	0	0	20	20	0	0	0	0	20
	E. crassipes	0	10	3	0	0	13	0	7	6	0	0	13
	Total	41	24	13	10	10	98	149	113	89	79	38	468

Fig. 25.30 Aquatic macrophytes in Lakes Pandin and Yambo

Table 25.7 Calculated carrying capacity (CC_i) for each socioeconomic indicator for the seven lakes

	Lakes						
Indicators	Bunot	Calibato	Palakpakin	Sampaloc	Mohicap	Pandin	Yambo
Population living around the lake	0.059	0.005	0.019	0.078	0.002	0.004	0.001
Income from aquaculture (annual)	0.099	0.658	0.341	0.090	0.098	0.044	0.023
Income from tourism	–	–	–	−0.105	–	0.059	0.063
Area occupied by fish pens (ha)	0.067	−0.490	−0.272	−0.186	0.031	0.018	0.015
Number of fish pens	0.066	−0.476	−0.371	0.018	0.024	0.015	0.016
Area for recreational activities (ha)	–	–	–	0.020	0.005	0.017	0.004

the lake, followed by Lake Bunot. Lake Sampaloc is considered to be an urban lake as it is located in the city proper. In terms of annual income from aquaculture, Lake Calibato appeared to have the highest calculated carrying capacity, followed by Lake Palakpakin. Together with the other two aquaculture lakes (Bunot and Sampaloc), these two lakes have a high number of fish production yearly. Meanwhile, Lake Bunot was found to have the highest calculated carrying capacity both in terms of the number of fish pens and area occupied by fish pens (ha). According to a study conducted by Brillo (2015), Lake Bunot is in the worst condition among the seven lakes of San Pablo due to being oversaturated with fish pens and cages. This is also the lake with the highest number of flowerhorn found in this study.

Table 25.8 The actual and limited values of each criterion for the Seven Lakes of San Pablo

Criteria	Indicators	Bunot		Calibato		Palakpakin		Sampaloc	
		Limited value (CC_{imax})	Actual value (CC_i)	Limited value (CC_{imax})	Actual value (CC_i)	Limited value (CC_{imax})	Actual value (CC_i)	Limited value (CC_{imax})	Actual value (CC_i)
Socioeconomic	Population living around the lake	1024	208	884	20	1067	60	958	400
	Income from aquaculture (annual)	14,580,000	4,762,800	20,412,000	40,678,200	20,898,000	24,854,040	50,544,000	24,173,640
	Income from tourism	–						101,900	118,300
	Area occupied by fish pens (ha)	3	0.98	4.2	10	4.3	9	10.40	12.60
	Number of fish pens	300	98	420	837	430	511	1040	126.00
	Area for recreational activities (ha)	–						104	9.09
Tourism	Number of tourists	–	–	–	–	–	–	1019	1183.00
	Hours of tourist spent on the lakes	–	–	–	–	–	–	6	2.00
	Facilities (rafts, bathrooms, parking space, etc.)	–	–	–	–	–	–	12	6.00
	Manpower	–	–	–	–	–	–	30	13.00
	Cost of activities	–	–	–	–	–	–	100	50.00
	Variety of water activities	–	–	–	–	–	–	4	1.00

Water quality and biodiversity	Dissolved oxygen	5	4.66	5	5.03	5	5.32	5	5.43
	Temperature	31	30.00	31	28.11	31	29.07	31	30.00
	pH	9	8.65	9	8.09	9	8.23	9	8.66
	Total dissolved solids	250	177.25	250	436.87	250	0.00	250	215.92
	Chlorophyll-*a*	10	11.61	10	21.93	10	17.90	10	8.25
	Transparency	2	1.03	2	1.40	2	1.11	2	1.35
	Nitrate as nitrogen	7	0.67	7	5.84	7	0.53	7	1.25
	Phosphate as phosphorus	0.5	0.98	0.5	0.96	0.5	0.56	0.5	0.80
	Biochemical oxygen demand	7	7.61	7	7.47	7	5.48	7	2.34
	Chemical oxygen demand	100	17.83	100	8.50	100	13.25	100	15.50
	Aquatic macrophytes (sq m)	30,000	2243.41	42,000	3211.92	43,000	62,006.84	104,000	5112.04
	Native fish species	4	1.00	4	1.00	4	2.00	4	1.00
	Introduced fish species	1	6.00	1	2.00	1	5.00	1	4.00
	Fish abundance	1,231,662	294,000	1,231,662	144,353	1,231,662	1,534,200	1,231,662	1,492,200
	Fish biomass (kg)	221,699	52,920	221,699	25,984	221,699	276,156	221,699	268,596
	Volume of solid wastes	900	900	900	900	900	900	900	900

(continued)

Table 25.8 (continued)

Criteria	Indicators	Mohicap		Pandin		Yambo	
		Limited value (CC_{imax})	Actual value (CC_i)	Limited value (CC_{imax})	Actual value (CC_i)	Limited value (CC_{imax})	Actual value (CC_i)
Socioeconomic	Population living around the lake	607	12	725	32	582	12
	Income from aquaculture (annual)	6,804,000	4,082,400	9,720,000	3,013,200	9,720,000	1,020,600
	Income from tourism	0	0	73,600	16,200	38,520	8640
	Area occupied by fish pens (ha)	1.4	0.23	2	0.23	2	0.21
	Number of fish pens	140	23	200	23	200	21
	Area for recreational activities (ha)	14	0.5	20	1.67	20	0.50
Tourism	Number of tourists	0	0	296	81	214	48
	Hours of tourist spent on the lakes	6	2	6	2	6	2
	Facilities (rafts, bathrooms, parking space, etc.)	10	2	10	5	8	4
	Manpower	30	10	30	22	30	2
	Cost of activities	0	0	400	200	360	180
	Variety of water activities	4	2	4	3.00	4	3.00
Water quality and biodiversity	Dissolved oxygen	5	8.16	5	7.76	5	7.82
	Temperature	31	29.78	31	29.78	31	29.76
	pH	9	9.55	9	8.64	9	8.64
	Total dissolved solids	250	239.21	250	142.59	250	117.62
	Chlorophyll-*a*	10	12.77	10	3.16	10	1.93
	Transparency	2	1.80	2	3.14	2	3.95
	Nitrate as nitrogen	7	1.07	7	3.69	7	0.92
	Phosphate as phosphorus	0.5	0.40	0.5	0.40	0.5	0.41
	Biochemical oxygen demand	7	4.38	7	2.30	7	1.21
	Chemical oxygen demand	100	12.75	100	9.00	100	8.00
	Aquatic macrophytes (sq m)	14,000	1147.64	20,000	500.18	20,000	635.63

	Native fish species	4	2.00	4	3.00	4	4.00
	Introduced fish species	1	3.00	1	3.00	1	4.00
	Fish abundance	1,231,662	252,000	1,231,662	186,000	1,231,662	186,000
	Fish biomass (kg)	221,699	45,360	221,699	33,480	221,699	33,480
	Volume of solid wastes	900	200	900	350	900	350

In terms of income from tourism, Lake Yambo was found to have the highest calculated carrying capacity, followed by its twin lake, Pandin. Based on the water quality parameters evaluated by LLDA from 2002 to 2005, Lake Yambo has the best water quality of all the seven lakes. Apparently, Lake Sampaloc and the twin lakes are the only lakes found to have actual and carrying capacity limit values among the seven lakes as they are the only lakes being utilized for ecotourism purposes since the beginning of twentieth century. Lake Mohicap, however, has recently been included in the lakes, which are open for tourists. Furthermore, Lake Sampaloc was found to have the highest calculated carrying capacity in terms of the area for recreational activities, perhaps, because this lake is an open park with various entrance and exit ways, wherein both locals and tourists can enjoy its beautiful sceneries for free.

The seven lakes of San Pablo are considered important sources of livelihood for fishermen as well as recreational sites for locals and tourists. Results of the KAP survey implied that despite the high economic status and educational attainment of the people living near cities like San Pablo, where Lake Sampaloc is directly located, the residents can still have lower environmental awareness, attitudes, and practices. Perhaps, the main reason why residents of Lakes Yambo and Pandin as ecotourism lakes were observed to have higher KAP on SWM is their dedicated involvement in promoting sustainable tourism since these lakes are great source of livelihood for them, and they benefit from taking care of it as well. More so, the residents are being incentivized when they collect their waste and bring them to the nearest material recovery facility (MRF), particularly in Lake Yambo.

25.3.3 Tourism Indicators

The tourism carrying capacity (TCC) of Lakes Sampaloc, Yambo, and Pandin were evaluated. The variables needed for computing the tourism carrying capacity were the size of the lake intended for tourism activities, points of activities, infrastructures and facilities available, and amenities for tourists. Following the manual of Calanog (2015) of the Department of Environment and Natural Resources—Ecosystem Research and Development Bureau (DENR-ERDB), the determination of the carrying capacity was based on the Boullon's (1985) carrying capacity mathematical model (Fig. 25.31), which determined the standard requirement of the visitor, such as space, time, and other needs, while enjoying an activity, such as swimming.

Figure 25.32 showed the data on tourist arrivals during the last 5 years. It revealed that the most obvious peak of arrivals is at Lake Sampaloc. This is due to the accessibility in the area and is located within the city proper and has several entrances, while both Lakes Yambo and Pandin require an entrance fee and limited time (at most for 2 h). The year 2019 data was used to evaluate the capacity of the lake based on the number of tourists who visited the area.

In the BCCMM, the basic carrying capacity (BCC), potential carrying capacity (PCC), and real carrying capacity (RCC) were computed and then used in determining the total tourism carrying capacity (TCC) of an area (Fig. 25.33).

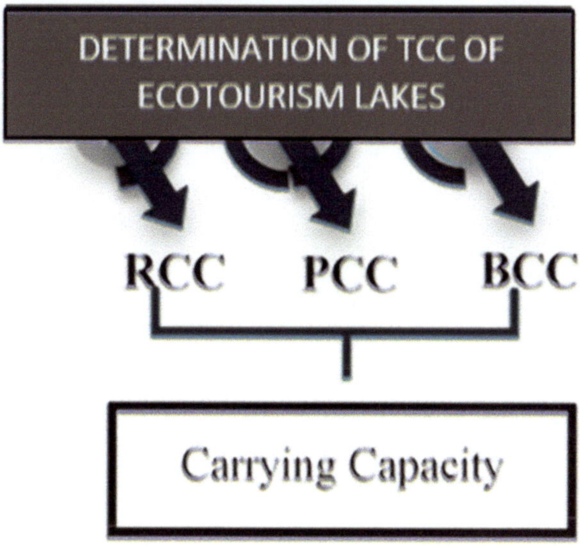

Fig. 25.31 Determination of the tourism carrying capacity of ecotourism lakes. *RCC* real carrying capacity, *PCC* potential carrying capacity, *BCC* basic carrying capacity. Source: Calanog (2015)

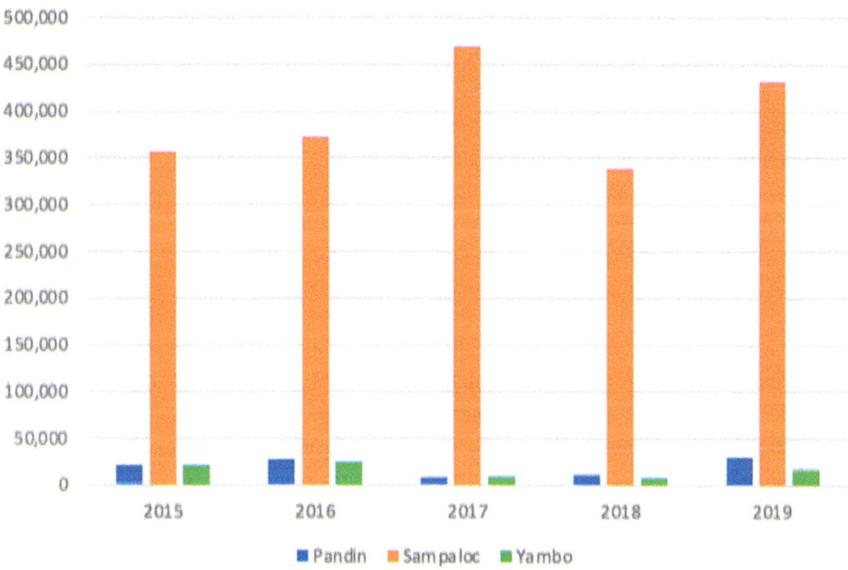

Fig. 25.32 Annual trend in the number of tourists visiting Lake Pandin, Lake Sampaloc, and Lake Yambo from 2015 to 2019. (Source: City Tourism Office of San Pablo)

The calculated RCCs for swimming, rafting, fishing, and sightseeing and photography in Lakes Pandin and Yambo were less at 296, 165, 106, and 284 persons per day and 214, 167, 108, and 168, respectively. When compared with tourist arrivals data per day in 2019, these RCCs have not been exceeded so far. However, Lake Sampaloc showed that the RCCs in all the present recreational activities such as

PANDIN		
Activities	RCC/day	Estimated Status
swimming	296	
rafting	165	
fishing	106	
sightseeing/ picture taking	284	
2019 ~tourist arrival per day: 81		

YAMBO		
Activities	RCC/day	Estimated Status
swimming	214	
rafting	167	
fishing	108	
sightseeing/ picture taking	168	
2019 ~tourist arrival per day: 49		

SAMPALOC		
Activities	RCC/day	Estimated Status
biking/walking/jogging/viewing/ picture taking	1019	
picnic	38	
zumba/aerobics	557	
2019 ~tourist arrival per day: 1183		

Fig. 25.33 2019 tourism carrying capacity estimation in Lakes Pandin, Yambo, and Sampaloc. *RCC* real carrying capacity. ● represents the number of tourists that does not exceed the RCC vs 2019 tourists' arrival. ● represents the number of tourists that exceeds the RCC vs 2019 tourists' arrival

picnicking, biking, jogging, and areas for exercise (Zumba and aerobics) have been exceeded with an estimate of 1183 tourists per day compared to the estimated computed carrying capacity of 1019, 38, and 557, respectively. The calculations made were also based on the set limiting factors—typhoons, rainy days, available time tourists may enjoy sightseeing and photography, and the rank of the activity based on the responses in the questionnaires.

This was based on the 250 tourists interviewed during the wet and dry season for Lake Pandin, and 24 and 75 tourists visited Lakes Yambo and Sampaloc, respectively, who responded to the Google forms distributed during the pandemic season. The TCC computation done was also based in consideration with the area intended for tourism activities, the recreational activities available for the tourists, as well as the declared limiting factors applicable in the site. Further recommendations in maintaining the lake's biophysical state as the tourists emphasize the preservation of the scenic view of Lake Pandin. However, the tourism association is recommended to enhance the tourist facilities in consideration with sustainable tourism.

25.3.4 Ecological Carrying Capacity (ECC) Modeling and Simulation

25.3.4.1 Carrying Capacity Level (CC_i) and Carrying Capacity Limit (CC_{imax})

The actual value of carrying capacity level (CC_i) and the carrying capacity limit value (CC_{imax}) of each indicator are presented in Table 25.8. These values were obtained from various sources as discussed in the Sect. 25.2. The carrying capacity level (CS_i) of each indicator was obtained using Eq. (25.1).

Lake Sampaloc has the highest actual value of people (400) living around the lake. The lake is located at the heart of the city and is surrounded by residential and commercial establishments. In all the seven lakes, the actual number of people living around the lake is way below the maximum values (Table 25.9). The income from aquaculture in Lakes Calibato and Palakpakin exceeded the maximum value of income based on the 10% of lake area covered by fish cages. Lake Sampaloc has the highest actual income from tourism compared with Lakes Pandin and Yambo. The areas occupied by fish pens in Lakes Calibato and Palakpakin exceeded the maximum area that can be covered by fish cages.

Lake Sampaloc has the highest actual value of tourists that visited the lake. In all the tourism lakes, the tourists spend about 2 h of recreation activities in the lake. The dissolved oxygen (DO) in Lakes Mohicap (8.16 ppm), Pandin (7.76 ppm), and Yambo (7.82 ppm) are well above the threshold of 5 ppm for Class C waters. Lake Bunot has DO level (4.66 ppm) a bit below the threshold level, while the other three lakes (Calibato, Palakpakin, and Sampaloc) have DO levels a bit above 5 ppm. The chlorophyll-*a* levels in aquaculture lakes including Calibato, Mohicap, Palakpakin, and Bunot were higher than the threshold level of 10. On the other hand, ecotourism lakes including Pandin, Yambo, and Sampaloc have chlorophyll-*a* levels well below the threshold level. Likewise, aquaculture lakes have phosphate levels above the threshold levels. Thus, aquaculture lakes are classified as eutrophic lakes, while ecotourism lakes are classified as oligotrophic lakes.

25.3.4.2 Carrying Capacity (CC)

For socioeconomic criteria, among the seven lakes, the carrying capacity for indicator "population living around the lake" is highest in Lake Sampaloc, followed by Lake Bunot. Lake Sampaloc is located at the center of San Pablo City surrounded by residential and commercial areas. Similarly, Lake Bunot is surrounded by local communities. The carrying capacity for the indicator "income from aquaculture" is highest in Lake Calibato, followed by Lake Palakpakin. Both of these lakes are aquaculture lakes with high fish production. For the indicator "income from tourism", ecotourism Lakes Pandin and Yambo have higher carrying capacity. For the indicator "area occupied by fish pens", Lakes Calibato, Palakpakin, and Sampaloc have negative carrying capacities due to the exceedance of the maximum 10% coverage of fish pens areas mandated by the Laguna Lake Development Authority. Lake Pandin has the highest carrying capacity for indicator "area for recreational activities".

For tourism criteria, the carrying capacity for the indicator "number of tourists" is highest for Lake Pandin. Lake Sampaloc has negative value for carrying capacity of the indicator "number of tourists" as the number of tourists that visit the lake exceed the maximum number that the lake can sustainably accommodate. Lake Yambo has the highest carrying capacity for the indicator "facilities" as the lake has a number of facilities that the tourists can use during their stay for recreation purposes in the lake. Lake Yambo also has the highest carrying capacities for several indicators including "manpower", "cost of activities", and "variety of water activities".

Table 25.9 Calculated carrying capacity (CC_i) for each indicator for the seven lakes

Criteria	Indicators	Bunot	Calibato	Palakpakin	Sampaloc	Mohicap	Pandin	Yambo
Socioeconomic	Population living around the lake	0.059	0.005	0.019	0.078	0.002	0.004	0.001
	Income from aquaculture (annual)	0.099	0.658	0.341	0.090	0.098	0.044	0.023
	Income from tourism	–	–	–	–0.105		0.059	0.063
	Area occupied by fish pens (ha)	0.067	–0.490	–0.272	–0.186	0.031	0.018	0.015
	Number of fish pens	0.066	–0.476	–0.301	0.018	0.024	0.015	0.016
	Area for recreational activities (ha)	–	–	–	0.020	0.005	0.017	0.004
Tourism	Number of tourists	–	–	–	–0.290	0.048	0.078	0.046
	Hours of tourist spent on the lakes	–	–	–	0.079	0.043	0.044	0.063
	Facilities (rafts, bathrooms, parking space, etc.)	–	–	–	0.113	0.032	0.113	0.063
	Manpower	–	–	–	0.057		0.070	0.178
	Cost of activities	–	–	–	0.054	0.048	0.071	0.079
	Variety of water activities	–	–	–	0.012	0.048	0.089	0.095
Water qualify and biodiversity	Dissolved oxygen	– 0.098	0.086	0.068	0.084	0.192	0.160	0.111
	Temperature	0.064	0.080	0.062	0.093	0.060	0.088	0.066
	pH	0.044	0.033	0.063	0.088	–0.062	0.079	0.061
	Total dissolved solids	0.046	–0.051	0.000	0.076	0.098	0.037	0.023
	Chlorophyll-*a*	– 0.073	–0.151	–0.132	0.068	–0.075	0.026	0.013
	Transparency	– 0.025	–0.057	0.042	0.044	0.053	0.064	0.058
	Nitrate as nitrogen	0.006	0.065	0.006	0.013	0.011	0.037	0.010
	Phosphate as phosphorus	– 0.123	–0.146	–0.080	–0.115	0.059	0.053	0.052
	Biochemical oxygen demand		–0.052	0.048	0.023	0.055	0.025	0.011

		–0.086						
	Chemical oxygen demand	0.013	0.004	0.008	0.010	0.009	0.006	0.005
	Aquatic macrophytes (sq m)	–0.002	0.005	–0.081	0.002	0.002	0.001	0.002
	Native fish species	0.016	0.012	0.027	0.009	0.026	0.044	0.088
	Introduced fish species	–0.154	–0.113	–0.257	–0.103	0.132	–0.066	–0.324
	Fish abundance	0.017	0.005	–0.064	–0.082	0.008	0.007	0.013
	Fish biomass (kg)	0.016	0.007	–0.055	–0.018	0.005	0.006	0.007
	Volume of solid wastes	–0.070	–0.074	–0.044	–0.035	0.011	0.022	0.010

Among the indicators for the water quality and biodiversity criteria, the indicator "dissolved oxygen" has the highest carrying capacity among all seven lakes. This is because the DO levels of all the lakes are well above the threshold level set for Class C waters. The indicator "Chlorophyll-*a*" is negative for all four aquaculture lakes including Bunot, Calibato, Palakpakin, and Mohicap. Lake Calibato, on the other hand, has the highest carrying capacity for nitrate-nitrogen. The carrying capacity for phosphates is negative for aquaculture lakes including Bunot, Calibato, Palakpakin, and Sampaloc. Likewise, the carrying capacity for the indicator "biological oxygen demand" is negative for lakes Bunot and Calibato. Lake Yambo has the highest carrying capacity for the indicator "native fish species" as it has the highest diversity of native fish species. The carrying capacity for the indicator "volume of solid wastes" is negative in aquaculture lakes including Bunot, Calibato, Palakpakin, and Sampaloc.

25.3.4.3 Indicator Weights

The calculated average weights of the various indicators for the three criteria (socioeconomic, tourism, and water quality and biological) are shown in Table 25.10. For each criterion (i.e., socioeconomic, tourism, and water quality and biological), the summation of weights of all indicators is equal to 1.00. The weights of the indicators vary among the seven lakes. The weight of the indicator "population living around the lake" is highest in Lake Palakpakin (0.329) and lowest in Lake Yambo (0.064). The weight of the indicator "income from aquaculture" is highest in Lake Calibato, followed by Lakes Bunot and Palakpakin. These three are among the four aquaculture lakes in San Pablo City. On the other hand, among the ecotourism lakes, Lake Yambo has the highest weight for the indicator "income from tourism", followed by Lake Pandin. The weights of the indicator "area occupied by fish pens" are highest in the aquaculture lakes Bunot and Calibato. The weight of the indicator "area used for recreational activities" is highest in Lake Pandin.

Comparing the values of the weights among the socioeconomic indicators, "income from aquaculture" is highest in aquaculture lakes Bunot and Calibato. In Lakes Palakpakin and Sampaloc, the indicator "population living around the lake" has the highest weights. In ecotourism lakes (Mohicap, Pandin, and Yambo), the weight of the indicator "income from tourism" is highest among the indicators.

Among the tourism lakes, the indicator "number of tourists" has the highest weight. Among the water quality and biological indicators, the indicator "dissolved oxygen" has the highest weight in Lakes Bunot, Calibato, Mohicap, and Pandin. The indicator "native fish species" has a high weight in Lake Yambo.

25.3.4.4 SECC, TCC, WQBCC, and ECC of the Seven Lakes

Aquaculture lakes Bunot and Mohicap have high socioeconomic carrying capacities (SECC) (Table 25.11). The negative SECC values of the other aquaculture lakes including Palakpakin, Calibato, and Sampaloc are attributed to the high exceedance of the number and area covered by fish cages over the maximum threshold values. Lake Yambo has the highest tourism carrying capacity (TCC) (0.5256), followed closely by Lake Pandin (0.4654). Aquaculture lakes have negative water quality and

Table 25.10 Calculated average weights of the different indicators used

Criteria	Indicators	Calculated average weights						
		Bunot	Calibato	Mohicap	Palakpakin	Pandin	Sampaloc	Yambo
Socioeconomic	Population living around the lake	0.292	0.225	0.119	0.329	0.095	0.188	0.064
	Income from aquaculture (annual)	0.302	0.330	0.164	0.287	0.142	0.187	0.218
	Income from tourism	–	–	0.238	–	0.266	0.091	0.282
	Area occupied by fish pens (ha)	0.204	0.206	0.188	0.130	0.154	0.153	0.140
	Number of fish pens	0.202	0.239	0.149	0.254	0.134	0.153	0.154
	Area for recreational activities (ha)	–	–	0.143	–	0.208	0.147	0.142
Tourism	Number of tourists	–	–	0.206	–	0.286	0.250	0.206
	Hours of tourist spent on the lake	–	–	0.143	–	0.131	0.238	0.190
	Facilities	–	–	0.214	–	0.226	0.226	0.127
	Manpower	–	–	0.095	–	0.095	0.131	0.190
	Cost of activities	–	–	0.167	–	0.143	0.107	0.159
	Variety of water activities	–	–	0.095	–	0.119	0.048	0.127

Criteria	Indicators	Calculated average weights						
		Bunot	Calibato	Mohicap	Palakpakin	Pandin	Sampaloc	Yambo
Water quality and biological indicators	DO	0.105	0.086	0.118	0.064	0.103	0.077	0.071
	Temperature	0.066	0.088	0.063	0.066	0.092	0.096	0.069
	pH	0.046	0.037	0.059	0.069	0.083	0.092	0.064
	TDS	0.064	0.029	0.103	0.081	0.064	0.088	0.049
	Chl-*a*	0.063	0.069	0.059	0.074	0.081	0.083	0.069
	Transparency	0.048	0.081	0.059	0.076	0.040	0.064	0.029
	NO_3	0.066	0.078	0.074	0.078	0.070	0.075	0.078
	PO_4	0.063	0.076	0.074	0.071	0.066	0.072	0.064
	BOD	0.079	0.049	0.088	0.061	0.077	0.068	0.061
	COD	0.072	0.051	0.070	0.059	0.070	0.064	0.064
	Aquatic macrophytes	0.031	0.071	0.029	0.056	0.029	0.040	0.054

(continued)

Table 25.10 (continued)

Criteria	Indicators	Calculated average weights						
		Bunot	Calibato	Mohicap	Palakpakin	Pandin	Sampaloc	Yambo
	Native fish species	0.063	0.049	0.051	0.054	0.059	0.037	0.088
	Introduced fish species	0.026	0.056	0.044	0.051	0.022	0.026	0.081
	Fish abundance	0.072	0.047	0.037	0.051	0.046	0.068	0.086
	Fish biomass (kg)	0.068	0.059	0.026	0.044	0.040	0.015	0.047
	Volume of solid wastes (kg)	0.070	0.074	0.048	0.044	0.057	0.035	0.027

Table 25.11 Socioeconomic carrying capacity (SECC), tourism Carrying capacity (TCC), water quality and biological carrying capacity (WQBCC), and ecological carrying capacity (ECC) of the Seven Lakes of San Pablo City

Lake	SECC	TCC	WQBCC	ECC
Bunot	0.2905	0	−0.4080	−0.1174
Calibato	−0.3038	0	−0.3437	−0.6476
Mohicap	0.1609	0.1698	0.3193	0.6500
Palakpakin	−0.2134	0	−0.3896	−0.6030
Pandin	0.1574	0.4654	0.5890	1.2119
Sampaloc	−0.0847	0.0244	0.1575	0.0325
Yambo	0.1219	0.5256	0.2072	0.8548

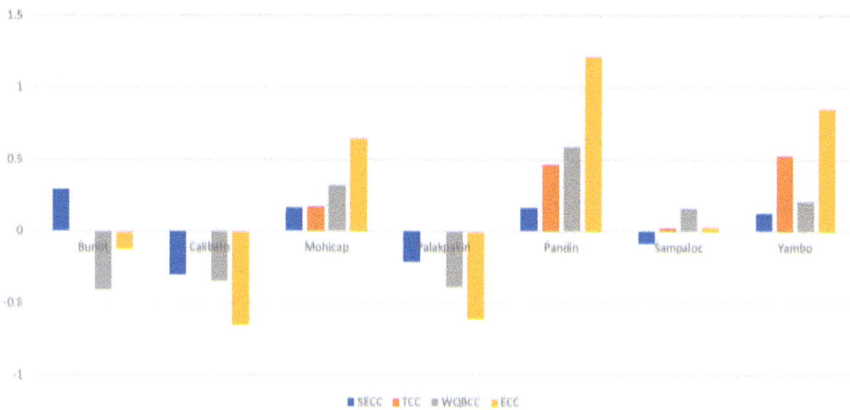

Fig. 25.34 Socioeconomic carrying capacity (SECC), tourism carrying capacity (TCC), water quality and biodiversity carrying capacity (WQBCC), and ecological carrying capacity (ECC) of the Seven Lakes of San Pablo City

biological carrying capacities (WQBCC), while lake Pandin has the highest WQBCC. Summing up all the three component carrying capacities yielded the ecological carrying capacity (ECC).

For Lake Pandin, the highest contributor to its high TCC is the presence of facilities in the lake. The tourists were satisfied with the available facilities such as the rafts, bathrooms, and parking space. For Lake Yambo, it is the manpower that significantly contributed to its high TCC (Fig. 25.34).

The water quality and biological carrying capacity (WQBCC) shows that Lakes Pandin and Yambo have very satisfactory water quality conditions, which is reflected by its WQBCC values of 0.5890 and 0.2072, respectively (Table 25.11). For Lakes Pandin and Yambo, the highest contribution to its WQBCC is the dissolved oxygen level. These lakes have a high dissolved oxygen concentration, which is suitable for aquatic organisms. On the other hand, Lakes Bunot, Calibato, and Palakpakin had negative WQBCC values of −0.4080, −0.3437, and −0.3896, respectively (Table 25.11). The contributory factors to the negative WQBCC values of these aquaculture lakes are the high concentration of phosphates, chlorophyll-*a*, and the presence of introduced species.

Overall, Lakes Yambo and Pandin had high (0.8549) and very high (1.2119) ecological carrying capacity (ECC) values, which translates to high and very high sustainability indices, respectively (Table 25.11). Lake Mohicap had a medium sustainability index with an ECC value of 0.6500, while Lake Sampaloc had a low sustainability index with an ECC value of only 0.0325. Lakes Calibato, Palakpakin, and Bunot had negative ECC values, which translates to unsustainable lake ecosystems. The water quality and biological carrying capacity (WQBCC) had significantly contributed to the negative ECC values of these aquaculture lakes (Fig. 25.34). The aquaculture management strategies such as excessive fish feeds and number of fish cages have contributed to the deterioration of water quality and biodiversity in aquaculture lakes.

Aquaculture lakes Bunot and Mohicap have high socioeconomic carrying capacities (SECC). The negative SECC values of the other aquaculture lakes including Palakpakin, Calibato, and Sampaloc are attributed to the high exceedance of the number and area covered by fish cages over the maximum threshold values. Lake Yambo has the highest tourism carrying capacity (TCC) (0.5256), followed closely by Lake Pandin (0.4654). Aquaculture lakes have negative water quality and biological carrying capacities (WQBCC), while Lake Pandin has the highest WQBCC. Summing up all the three component carrying capacities yielded the ecological carrying capacity (ECC).

The socioeconomic carrying capacity (SECC) shows that Lakes Calibato, Palakpakin, and Sampaloc are unsustainable with −0.3038, −0.2134 and −0.0847, respectively (Fig. 25.34). Lakes Bunot, Mohicap, Pandin, and Yambo have low sustainability index. Based on the analysis, it showed that the number and area occupied by fish pens affected the socioeconomic carrying capacity value for Lakes Calibato, Palakpakin, and Sampaloc, which makes it unsustainable.

For the tourism carrying capacity (TCC), the twin lakes Pandin and Yambo have medium sustainability index of 0.4654 and 0.5256, respectively. Lakes Mohicap and Sampaloc have low sustainability indexes. For Lake Pandin, the highest contributor to the TCC is the presence of facilities in the lake. The tourists were satisfied with the available facilities such as rafts, bathrooms, and parking space. For Lake Yambo, it is the manpower that significantly contributed to the TCC. This is due to the fact that the area is solely used for ecotourism and no built-up was seen around the lake. The civic organization welcomes tourists within Lake Yambo. In order to increase the sustainability index for Lakes Mohicap and Sampaloc, the LGU and community can focus on improving the facilities and the manpower to attract more tourists coming to the lake.

The water quality and biological carrying capacity (WQBCC) shows that Lakes Pandin and Yambo have very satisfactory water quality condition, which is reflected on its index score of 0.5890 and 0.2072, respectively (Table 25.11). For Lakes Pandin and Yambo, the highest contribution to its WQBCC is the dissolved oxygen level. These lakes have a high dissolved oxygen concentration, which is suitable for aquatic organisms. On the other hand, Lakes Bunot, Calibato, and Palakpakin had negative WQBCC of −0.4080, −0.3437, and −0.3896, respectively. The highest contribution to its negative WQBCC is the presence of introduced species.

Overall, Lakes Yambo and Pandin had high (0.8549) and very high (1.2119) ecological carrying capacity (ECC), which translates to high and very high sustainability index, respectively. Lake Mohicap had a medium sustainability index with an ECC value of 0.6500, while lake Sampaloc had a low sustainability index with an ECC value of only 0.0325. Lakes Calibato, Palakpakin, and Bunot had negative ECC values, which translates to an unsustainable ecosystem. The water quality and biological carrying capacity had significantly contributed to the negative outcomes of these lakes.

25.4 Summary and Conclusions

Anthropogenic activities such as aquaculture and ecotourism have largely contributed to the livelihood of local communities located around the Seven Lakes of San Pablo City. However, these activities pose threats to the degradation of the environmental quality of the Seven Lakes. Thus, it is necessary to examine the current ecological conditions and assess the sustainability of aquaculture and ecotourism activities in the seven lakes. An ecological carrying capacity modeling framework consisting of three criteria, namely socioeconomic (SECC), tourism (TCC), and water quality and biodiversity carrying capacities (WQBCC), was developed by the researchers to assess the sustainability of the lakes' current environmental policies and management practices.

In this study, a series of ecological and social methodologies were created and implemented to assess the sustainability of the aquaculture and tourism activities in the seven lakes. The biophysical criterion involved the collection of water quality indicators in the lake including temperature, dissolved oxygen, pH, BOD, total dissolved solids, conductivity, chlorophyll-*a*, nitrates, phosphates, and transparency during wet and dry seasons. Fish biodiversity and floristic surveys were also conducted in the lakes to provide further insights regarding the lakes' biodiversity carrying capacity and sustainability. Meanwhile, social surveys such as Key Informant Interviews, Participatory Rural Appraisal (PRA) activities, and Knowledge, Attitudes, and Practices (KAP) survey were conducted to gather primary data on the socioeconomic and tourism indicators. The socioeconomic indicators include the population living around the lake, annual income from aquaculture, income from tourism, area occupied by fish pens (ha), number of fish pens, and area for recreational activities. Ecotourism indicators include the number of tourists, facilities, manpower, cost and variety of activities, and number of hours spent in various tourism activities. Moreso, secondary data from the LLDA, LGU of San Pablo, and other stakeholders were used to supplement the primary data.

The values gathered from these sampling and surveys represented the actual value of the various indicators. The maximum value or limit of each indicator was based on water quality standards for the biophysical indicators, while the maximum values for the socioeconomic and tourism indicators were based on literature and secondary data from the LGU of San Pablo. Normalization of each indicator was done. Experts and key respondents ranked the various biophysical, socioeconomic, and ecotourism indicators following the Rank-Sum method to determine the weight of each indicator. The normalized value of each indicator was multiplied with its weight, and then summed for each criterion. The ECC is the summation for the three criteria.

Results showed that aquaculture lakes Bunot and Mohicap have high socioeconomic carrying capacities (SECC). The negative SECC values of the other aquaculture lakes including Palakpakin, Calibato, and Sampaloc are attributed to the high exceedance of the number and area covered by fish cages over the maximum threshold values. Lake Yambo has the highest tourism carrying capacity (TCC) (0.5256), followed closely by Lake Pandin (0.4654). Aquaculture lakes have negative water quality and biological carrying capacities (WQBCC), while Lake Pandin has the highest WQBCC. Summing up all the three component carrying capacities yielded the ecological carrying capacity (ECC).

Generally, it showed that ecotourism lakes Yambo and Pandin had high (0.8549) and very high (1.2119) ecological carrying capacities (ECC), which translates to high and very high sustainability indices, respectively. Lake Mohicap had a medium sustainability index with an ECC value of 0.6500, while Lake Sampaloc had a low sustainability index with an ECC value of only 0.0325. Aquaculture lakes Calibato, Palakpakin, and Bunot had negative ECC values, which translated to unsustainable ecosystems.

Estimating and illustrating the recreational and aquaculture carrying capacities of the lakes can provide the LGUs' constituents and policymakers a comprehensible overview of the potential consequences of the unrestricted proliferation of the various activities the lakes are currently hosting.

25.5 Recommendations

25.5.1 Future Research

This study further recommends for future researchers to study and analyze the economic carrying capacity of each lake as it will be beneficial for the local government unit to know their potential opportunities and concerns, which may be associated with the lakes' ecosystem services. Furthermore, assessment and monitoring models such as an integrative water quality index and Bayesian network could be developed to predict water quality and fish kill incidence in the aquaculture lakes of San Pablo. Monitoring of emerging contaminants such as microplastics and heavy metals as well as pathogenic organisms could be done in each lake of San Pablo.

25.5.2 Lakes Management and Conservation

In terms of aquaculture, this study further recommends for LLDA and LGU of San Pablo to continuously follow the 10% surface area limit for fish cages and pens in each lake. Closely monitoring and controlling of highly invasive fish species (e.g., Flowerhorns in Lake Bunot) is highly encouraged. Aquaculture lakes (Bunot, Calibato, and Palakpakin) are currently unsustainable and urgently need proper management strategies by the local government unit of San Pablo. Moreso, the government should make sure that the relocated illegal settlers from Lake Sampaloc and other lakes have alternative livelihoods already, so they will not likely go back near the lake to reside or catch fish for sustenance.

For the lakes' ecotourism, further studies on the basic carrying capacity of each lake can be done. The determination of the areas intended for an individual swimmer can be examined. The real carrying capacity may consider other factors that could affect the decision of the visitor in choosing the lake for recreational visit. Ecotourism facilities such as bathrooms and resting areas should also be improved in ecotourism lakes such as Pandin and Yambo.

For the local communities, organizations, and the local government unit (LGU) of San Pablo, the researchers recommend strengthening the strict implementation of rules and regulations on solid waste management, especially in the water bodies near public or urban places. Information, education, and communication activities such as open forums, free discussions, and seminars pertaining to proper disposal of fishing nets as well as tourism-related wastes should be done to protect and conserve the lakes even at the barangay level. A material recovery facility (MRF), similar to what Lake Yambo has, should also be constructed in each lake in San Pablo City to encourage people to reduce, reuse, and recycle (3Rs).

Acknowledgments This study would not be possible without the financial support and overall assistance provided by the Department of Science and Technology (DOST) and the National Research Council of the Philippines (NRCP) to the researchers throughout the study. The authors also wish to express their utmost gratitude for the guidance, support, secondary data provision, and cooperation from the Laguna Lake Development Authority (LLDA), Department of Environment and Natural Resources (DENR), Department of Tourism (DOT)—Region 4A, Laguna Provincial Office, San Pablo Mayor's Office, San Pablo City's Tourism and City Environment and Natural Resources Offices, Nagcarlan Mayor's Office, and Rizal Mayor's Office. They also wish to acknowledge the local residents nearby the seven lakes, including the FARMC (Fisheries and Aquatic Resource Management Councils) and *bantay-lawa* members, fisherfolks, civic groups, elderlies, women representatives, experts, and other stakeholders, who willingly spent their time participating in various discussions and interviews.

References

Alampay RB (ed) (2005) Sustainable tourism: challenges for the Philippines. Philippine APEC Study Center Network (PASCN) and the Philippine Institute for Development Studies (PIDS)

Behrends LL, Nelson RG, Smitherman RO, Stone NM (1982) Breeding and culture on the red-gold color phase of tilapia. J World Maricult Soc 13:210–220

Bhatnagar A, Devi P (2013) Water quality guidelines for the management of pond fish culture. Int J Environ Sci 3(6):1980–2009

Boullon RC (1985) Plantifacion del Espacio Touristico. Trillas (ed). Mexico. DE

Brillo BBC (2015) Development issues regarding Bunot Lake: the lesser Lake among the seven lakes of San Pablo City, Philippines. Lakes Reserv 20(3):155–165. https://doi.org/10.1111/lre.12096

Brillo BBC (2017) The governance of the seven crater lakes, San Pablo City, the Philippines. Asian J Water Environ Pollut 14(2):13–25. https://doi.org/10.3233/AJW-1700

CABI (2016) CABI compendium. CAB International, Wallingford. https://www.cabi.org

CABI (2020) Hydrilla verticillate. Invasive Species Compendium. https://www.cabi.org/isc/datasheet/28170

Calanog LA (2015) A Manual on Computing Carrying Capacity of Ecotourism Sites in Protected Areas. Ecosystems Research and Development Bureau, Department of Environment and Natural Resources, College, Laguna, Philippines

Chougule B (2011) Environmental carrying capacity and ecotourism development. Int J Econ Issues 4:45–54

Dar NA, Pandit AK, Ganai BA (2014) Factors affecting the distribution patterns of aquatic macrophytes. Limnol Rev 14(2):75–81

Ding L, Chen K, Cheng S, Wang X (2015) Water ecological carrying capacity of urban lakes in the context of rapid urbanization: a case of East Lake in Wuhan. Phys Chem Earth Parts A/B/C 89–90:104–113

Ekubo AA, Abowei JFN (2011) Review of some water quality management principles in culture fisheries. Res J Appl Sci Eng Technol 3(2):1342–1357

Fitzgerald WJ (1979) The red-orange tilapia - a hybrid that could become a world favourite. Fish Farm Int 6:26–27

Food and Agriculture Organization of the United Nations (n.d.) Aquaculture. https://www.fao.org/aquaculture/en/

Froese R, Pauly D (2019) Fishbase. http://www.fishbase.org.search

Global Nature Fund (2014) Intensive fish farming threatens Philippine crater lake - Lake Sampaloc is "Threatened Lake of the Year 2014". https://www.globalnature.org/bausteine.net/f/7999/PressReleaseofGNF_ThreatenedLakeoftheYear2014_Sampaloc.pdf?fd=2

Go Botany (2020) *Hydrilla verticillata* (L. f.) Royle water thyme. https://gobotany.nativeplanttrust.org/species/hydrilla/verticillata/

Guevarra RD, Paraso MGV, Lola MSEG (2020) Biomarker evaluation in Nile tilapia (*Oreochromis niloticus*) to assess the health status of aquaculture areas in the seven lakes of San Pablo. Philip J Sci 149(3):833–840

Herder F, Schliewen UK, Geiger MF, Hadiaty RK, Gray SM, Mckinnon JS, Walter RP, Pfaender J (2012) Alien invasion in Wallace's dreamponds: records of the hybridogenic "flowerhorn" cichlid in Lake Maranao, with an annotated checklist of fish species introduced to the Malili Lakes system in Sulawesi. Aquat Invasions 7:521–535

Herre AH (1927) Gobies of the Philippines and the China Sea. Monog Bureau Sci 23:1–352

Herre AH (1953) A checklist of Philippine fishes. Research Report Vol 20. Fish and Wildlife Service, United States Department of Interior, Government Publishing Office, Washington, DC, p 977

Johnson RL (2013) A field guide of aquatic macrophyte species found in New York lakes along with potential non-native (exotic) invaders. https://s3.amazonaws.com/assets.cce.cornell.edu/attachments/5879/A_Field_Guide_of_Aquatic_Macrophyte_Species...pdf?1421862332

Johnson P, Thomas B (eds) (1992) Perspectives of tourism policy. Mansell, London

Kissoon LT, Jacob DL, Hanson MA, Herwig BR, Bowe SE, Otte ML (2013) Macrophytes in shallow lakes: relationships with water, sediment and watershed characteristics. Aquat Bot 109:39–48

Kremer M (2018) The importance of dissolved oxygen in water for aquaculture. https://blog.jencoi.com/the-importance-of-dissolved-oxygen-in-water-for-aquaculture

Laguna Lake Development Authority (LLDA) (2006–2008) Water quality report of the seven crater lakes 2006–2008. LLDA-EQRD, Rizal

Lim LC (1995) The concepts and analysis of carrying capacity: a management tool for effective planning. Part I. Report produced under Project MY 0033. Malaysia, Petaling Jaya: WWF
Limno Loan Program Manual (n.d.). http://limnoloan.org/waterquality/chlorophyll_a/
Mendoza MU, Briones JCA, Itoh M, Padilla KSAR, Aguilar JI, Okuda N, Papa RDS (2019) Small maar lakes of Luzon Island, Philippines: their limnological status and implications on the management of tropical lakes – a review. Philipp J Sci 148(3):565–578
Minnesota Pollution Control Agency (2009) Low dissolved oxygen in water: causes, impact on aquatic life - an overview. www.pca.state.mn.us
Munson B, Axler R, Hagley C, Host G, Merrick G, Richards C (2004) Water on the web: understanding: water quality: parameters: Ph. https://waterontheweb.org/under/waterquality/ph.html
Navarrete IA, Dicen GP, Perez TR (2019) Towards integrated management of a shallow tropical lake: assessment of water quality, sediment geochemistry, and phytoplankton diversity in Lake Palakpakin, Philippines. Environ Monit Assess 191(485). https://doi.org/10.1007/s10661-019-7617-7
NOAA (2021) What is aquaculture? National Ocean Service website. https://oceanservice.noaa.gov/facts/aquaculture.html
Olson CM (2004) Introduced species summary project hydrilla (*Hydrilla verticillate*). http://www.columbia.edu/itc/cerc/danoff-burg/invasion_bio/inv_spp_summ/Hydrilla_verticillata.html
Paller VG, Corpuz MNC, Bandal MZ Jr (2017) Freshwater fish assemblages and water quality parameters in seven lakes of San Pablo, Laguna, Philippines. Asian J Biodivers. https://doi.org/10.7828/ajob.v8i1.995
Paller VG, Macandog D, De Chavez ER, Paraso MG, Tsuchiya MC, Campang J, Pleto JV, Bandal M, Cabillon YC, Elepano A, Macaraig JR, Mendoza S (2021) The seven lakes of San Pablo: assessment and monitoring toward a sustainable lake ecosystem. Philip Sci Lett 14(1): 157–178. https://scienggj.org/2021/PSL%202021-vol14-no01-p158-179-Paller%20et%20al.pdf
Philippine Statistics Authority (2015) Number of population in San Pablo City, Laguna, Philippines. https://psa.gov.ph/classification/psgc/?q=psgc/barangays/043424000
Pleto JVR, Arboleda MDM, Simbahan JF, Migo VP (2018) Assessment of the effect of remediation strategies on the environmental quality of aquaculture ponds in Marilao and Meycauayan, Bulacan, Philippines. J Health Pollut 8(20):181205. https://doi.org/10.5696/2156-9614-8.20.181205
Reghunathan V, Joseph S, Warrier U, Hameed S, Moses S (2016) Factors affecting the environmental carrying capacity of a freshwater tropical lake system. Environ Monit Assess 188(11): 615
Ross LG, Telfer TC, Falconer L, Soto D, Aguilar-Manjarrez J, Asmah R, Bermúdez J, Beveridge MCM, Byron CJ, Clément A, Corner R, Costa-Pierce BA, Cross S, De Wit M, Dong S, Ferreira JG, Kapetsky JM, Karakassis I, Leschen W, Little D, Lundebye AK, Murray FJ, Phillips M, Ramos L, Sadek S, Scott PC, Valle-levinson A, Waley D, White PG, Zhu C (2013) Carrying capacities and site selection within the ecosystem approach to aquaculture. In: Ross LG, Telfer TC, Falconer L, Soto D, Aguilar-Manjarrez J (eds) Site selection and carrying capacities for inland and coastal aquaculture, pp 19–46. FAO/Institute of Aquaculture, University of Stirling, Expert Workshop, 6–8 December 2010. Stirling, the United Kingdom of Great Britain and Northern Ireland. FAO Fisheries and Aquaculture Proceedings No. 21. Rome, FAO. 282 pp
San Pablo City Comprehensive Land Use Plan (2015–2025) City Planning and Development Office-City Government of San Pablo. San Pablo City, Laguna
Song M, Pan X, Pan X (2020) Assessment of China's marine ecological carrying capacity. In: Sustainable marine resource utilization in China: a comprehensive evaluation. https://doi.org/10.1016/B978-0-12-819911-4.00002-3
The Fish Site (2015) How to achieve good water quality management in aquaculture. https://thefishsite.com/articles/how-to-achieve-good-water-quality-management-in-aquaculture

Towers L (2015) Water quality: a priority for successful aquaculture. https://thefishsite.com/articles/water-quality-a-priority-for-successful-aquaculture

UC ANR (n.d.) Aquatic weed identification. https://ucanr.edu/sites/csnce/files/57535.pdf

Valley RD, Bremigan MT (2002) Effects of macrophyte bed architecture on largemouth bass foraging: implications of exotic macrophyte invasions. Trans Am Fish Soc 131:234–244

Vidthayanon C (2007) Overview on freshwater fishes of the Philippines. Lecture presented during national training course on freshwater fish identification, 18 October 2007, SEARCA. Zonal Center 2 PCMARD, IBS-UPLB, PIBCFI, Chester Zoo and World Fish Center, p 1–8

Wu JL, Hsu JC, Lou SK (1983) Esterase isozymes in *Oreochromis niloticus*, *O. aureus*, *O. mossambicus*, and Red tilapia. In: Fishelson L, Yaron Z (eds) Proc Int Symp on Tilapia in Aquaculture. Tel Aviv Univ Press, pp 281–290

Zahedi L, Sizemore E, Malcolm S, Grossniklaus E (2014) Knowledge, attitudes, and practices regarding cervical cancer and screening among Haitian health care workers. Int J Environ Res Public Health 11(11). https://doi.org/10.3390/ijerph111111541

Zeng C, Liu Y, Liu Y, Hu J, Bai X, Yang X (2011) An integrated approach for assessing aquatic ecological carrying capacity: a case study of Wujin District in the tai Lake Basin, China. Int J Environ Res Public Health 8(1):264–280

Assessment of the Contribution of Freshwater Ecosystem Services to the Hydropower Sector in the Kura–Araz Basin

26

Rovshan Abbasov and Marlon Flores

Abstract

This study focuses on the ecosystem services (ES) in the Kura–Araz basin. The study assesses the hydropower plant (HPP) dams' sector and reviews additional sectors including nature-based tourism, irrigated agriculture, and drinkable water supply. In addition, the study briefly discusses the role and value of ecosystem services that help to mitigate natural hazards related to poor ecosystem management. The study used a basic Targeted Scenario Analysis (TSA) approach. The TSA assesses current "business as usual (BAU)" ecosystems management practices and the current value of ecosystem services under BAU. It uses sector output indicators and compares them with potential "sustainable ecosystems management (SEM)" outputs to assess losses and potential gains (or losses) of shifting from BAU to SEM. The BAU approach is characterized by a focus on short-term gains (e.g., <10 years), externalization of impacts and their costs, and little or no recognition of the economic value of ES, which is typically depleted or degraded. Under SEM, the focus is on long-term gains (>10 years); also under SEM, the costs of impacts are internalized. BAU practices in freshwater ecosystem management have a high cost to the economy of Azerbaijan. Part of this high cost can be avoided by shifting to low-cost SEM practices. Despite the availability of several laws and regulations governing the administration and management of HPP and Dams in Azerbaijan, enforcement is weak. The legal framework is also incomplete, there is no means for law enforcement, and no measurable

R. Abbasov (✉)
Department of Geography and Environment, Khazar University, Baku, Azerbaijan
e-mail: rabbasov@khazar.org

M. Flores
United Nations Development Program, New York, NY, USA
e-mail: marlon.flores@undp.org

S. Dhyani et al. (eds.), *Ecosystem and Species Habitat Modeling for Conservation and Restoration*, https://doi.org/10.1007/978-981-99-0131-9_26

indicators or means to collect and evaluate it. Therefore, no results of the evaluation are fed into policy-making or to improve HPP/Dams management.

Keywords

Ecosystem services · Business as usual · Sustainable ecosystem management · Hydropower sector · Dam development · Floods · Droughts

26.1 Introduction

Freshwater bodies provide drinking and irrigation water, as well as important ecosystem services (Postel and Carpenter 1997). Water is a source of life, one of the most important conditions for providing food and recreation needs, and it is a habitat for a wide range of animals (Aylward et al. 2005). Freshwater sources play a significant role in everyday life and the economy of Azerbaijan, which is primarily located in semidesert and temperate climates (Scandizzo and Abbasov 2022a). The use and nonuse values of Azerbaijan's freshwater ecosystems have traditionally contributed to the well-being of water consumers. The use of water and related environmental commodities or services in the consumption and production process results in useful values (Abbasov and Smakhtin 2009; Abbasov and de Blois 2021).

Due to human-based factors, including unsustainable urban water consumption, industry and infrastructure development projects, agriculture, and increased expansion of the hydropower sector, freshwater ecosystems are among the most endangered habitats in Azerbaijan. In the upper watershed, sectors such as agriculture and forestry contribute by encouraging unsustainable forestry, farming, and husbandry (extensive/overgrassing) practices. These unsustainable practices have an adverse effect on freshwater ecosystems, which in turn has an adverse effect on basin management.

This study focuses on the ecosystem services (ES) in the KARB (Kura–Araz basin) part of Azerbaijan. The purpose of this study is not to assess the impact of hydropower plant (HPP)/Dams development on the environment. It is recognized, however, that HPP/Dam development can be significantly damaging to ecosystems below the HPP/Dam if not managed following rigorous environmental standards (BAU practices). Along with reviewing other industries such as nature-based tourism, irrigated agriculture, and drinkable water supply, the study evaluates the HPP/Dams sector. The research also briefly explores the value and role of ES in reducing natural risks brought on by ineffective ecosystem management.

26.2 Methodology

The study adopted a Targeted Scenario Analysis (TSA) methodology. The TSA evaluates the present value of ecosystem services under "business as usual (BAU)" ecosystem management practices. To evaluate the costs and possible benefits (or losses) of switching from BAU to SEM (sustainable ecosystems management), it compares sector output indicators with potential SEM outputs. The BAU approach is distinguished by a concentration on short-term advantages (e.g., 10 years), externalization of consequences and costs, and little to no awareness of the economic worth of ES, which is often depleted or deteriorated. Long-term gains (over 10 years) are the main emphasis of SEM, and the costs of consequences are internalized as well. Maintaining ecosystem services creates the possibility of a long-term flow of ecosystem products and services that can be considered when making decisions. As a realistic and economical means of realizing long-term earnings, SEM methods frequently assist ecosystem sustainability.

The TSA approach serves multiple purposes such as:

- Analyze the HPP/Dam sector and compare "poor" and "good" environmental management approaches to ascertain the possible economic gains or losses of engaging in productive activities.
- Provide policymakers and industry with information on the advantages and disadvantages of engaging in profitable activities that have an influence on ecosystem services.
- Assist government representatives and the business community in incorporating ecosystem management strategies into sectoral economic planning, company business plans, and investment strategies.
- Present financial (and social) justifications to boost political will and bolster financial backing for better freshwater and forestry ecosystem management.
- Based on data availability, the following indicators were selected to assess BAU and SEM impact (Table 26.1).

The TSA study includes the following five steps:

- Definition of the scope of analysis: Freshwater ecosystems/HPP and reservoirs.
- Definition of sectors in agreement with stakeholders, and assessment of data availability vis-à-vis potential indicators to be used.
- Based on data that is already accessible and first-hand research, the selection of indicators to establish the BAU baseline and prospective SEM intervention.
- Create values/scenarios for BAU and SEM approaches.
- Formulation of recommendations for policy- and decision-makers.

26.3 Ecosystem Services and Problems of Freshwater Ecosystems in Azerbaijan

The term biodiversity refers to "the variety of life on Earth at all its levels, from genes to ecosystems, and the ecological and evolutionary processes that sustain it" (Heywood and Watson 1995; Pimm et al. 1995; Abbasov et al. 2022; Dhyani et al. 2020, 2021; Scandizzo and Abbasov 2022b). Biodiversity includes not only species we consider rare, threatened, or endangered, but every living thing as well—even

Table 26.1 Sample indicators used to construct BAU/SEM scenarios

Sector indicators (5–10-year trends)	Applied in the study
Employment increase (no. of jobs) by subsector (direct, indirect, and induced)	
Income, average annual increase by subsector	√
Annual revenue from green taxes	√
Foreign exchange earnings (annual, from exports)	√
Sector investment (government)	√
Sector investment (private sector)	√
Damage costs (because of BAU practices)	√
Avoided damages costs (as a result of SEM practices)	√
Production trend (volume and value)	√
Sector production trend (as a percentage of GDP)	√

organisms we still know little about, such as microbes, fungi, and invertebrates. Biodiversity is important everywhere; species and habitats in your area as well as those in distant lands all play a role in maintaining healthy ecosystems. We need biodiversity to satisfy basic needs such as food, drinking water, fuel, shelter, and medicine (Maclaurin and Sterelny 2008).

An ecosystem is an interdependent system of living and nonliving organisms that share a common habitat (Daily and Matson 2008; Kumar et al. 2022; Weißhuhn et al. 2017; Abbasov et al. 2022, b). An ecosystem is also a completely independent system of living and nonliving organisms. The term "ecosystem services" describe how ecosystems both directly and indirectly benefit people (UNEP 2014; Carpenter et al. 2006). Numerous definitions of ecological services exist (Bennett et al. 2009; Daily 2003). Abbasov et al. (2022) defined ecosystem services (ES) as the process through which natural resources, such as trees, snow cover, and fertile soil, are transformed into beneficial outcomes, such as wood products, winter tourism, and arable land. ES might be characterized as a "service" offered to individuals by the natural environment (Scandizzo and Abbasov 2022a).

Water supply, pollination, seed distribution, climate regulation, water purification, nutrient cycling, and agricultural pest management are just a few of the services that ecosystems offer (Daily 1997; Dhyani et al. 2018). Thirty percent of human crops rely on the free pollination services provided by animals for many flowering plants (Costanza et al. 1997; Pearce and Atkinson 1993, Flores and Adeishvili 2012). ES are classified as provisioning, habitat, cultural, and regulating services (TEEB Foundations 2010):

- Provisioning services are ecological functions that explain how ecosystems produce materials or energy. They consist of items such as food, water, building supplies, and other things.
- Regulating services are the services that ecosystems offer by serving as regulators, such as controlling floods and disease outbreaks or regulating the

Table 26.2 Ecosystem services provided by freshwater ecosystems

Provisioning services	Regulatory services	Cultural services	Supporting services
Water (quantity and quality) for consumptive use (drinking, domestic use, and agriculture and industrial use)	Maintenance of water quality (natural filtration and water treatment)	Recreation and tourism (river-rafting, kayaking, hiking, and fishing as a sport, river viewing)	Role in nutrient cycling (role in the maintenance of floodplain fertility), primary production
Water for nonconsumptive use (for generating power and transport/ navigation)	Buffering of floods, erosion control through water and land interactions, and flood control infrastructure	Existence values (personal satisfaction and free-flowing rivers) Option values	Predator/prey relationships and ecosystem resilience
Aquatic organisms for food and medicines			

quality of the air and soil. Habitat/support services are directly linked to the habitats that support species and they have an indirect influence on human well-being and other ecosystem services.

- Cultural services are nonmaterial benefits including recreation and tourism, specifically eco-tourism.

The ecosystem services (ES) linked to hydropower and dam development are classified under the category of "provisioning." This refers to the human use of fresh water for domestic use, irrigation, power generation, and transportation (Millennium Ecosystem Assessment 2005; Postel and Carpenter 1997; Aylward et al. 2005; Green et al. 2015). Table 26.2 shows the classification of these ecosystem services.

In the last several decades the protection and management of freshwater sources became one of the major issues of modern societies. Threats to freshwater ecosystems have reached global scales and require urgent actions from water managers and policymakers (Gleick et al. 2001). These threats include climate changes, contamination of surface and groundwater sources, degradation of freshwater ecosystems, and deforestation (Table 26.3). The impact of these threats at the upper watershed level, where the catchment point is located, can severely affect HPP/Dam productivity, as well as other sectors depending on freshwater ecosystems such as irrigated agriculture.

Deforestation and unsuitable agricultural practices (extensive/overgrassing) in Azerbaijan are considered to be one of the most important factors that threaten HPP/Dams development. These unsustainable practices caused by poorly planned agriculture and land use result in increased erosion and change in water flows. In addition, there are other threats such as contamination of surface and groundwater. Table 26.3 provides an overview of threats to freshwater/forest ecosystems and its economic impact.

Table 26.3 Overall threats to freshwater/forest ecosystems

Threats to freshwater ecosystems	Caused by	Environmental consequences	Economic impact
Climate changes	Industrial and urban air pollution	Increased evaporation from water surfaces, reduced stream flows, and reduced quantity and quality of water	Reduced production of hydropower and agriculture Reduced revenue Loss of jobs Power shortages Reduced foreign exchange gains from exports Less revenue from taxes to government Reduction of pro-poor investments and poverty increase
Contamination of freshwater ecosystems	Industrial, agricultural, and urban effluents	Habitat pollution, reduced quality of water, and eutrophication	
Degradation of freshwater sources	Agricultural, industrial, and municipal water withdrawals	Reduced flows, narrowing and extinction of migration routes for fish, and habitat degradation	
Deforestation	Urbanization, agricultural development, and mass removal of forests	Erosion, landslides, riverbed sedimentation, increased turbidity, increased temperature, reduced oxygen, and increased BOD levels	

Currently, dam development faces rather serious problems in Azerbaijan. Large areas, forests, and irrigated lands were inundated during the development of dams. Most of the rivers in the Kura basin are the preferred spawning grounds for valuable sturgeon fish and serve as migration routes.

Ecosystems that sustain the hydropower industry may see a slow but long-lasting influence as a result of climate change. Warmer temperatures brought on by climate change can particularly disrupt the water cycle, throwing off the balance between precipitation and evaporation. As a result, excessive evaporation and low precipitation can cause drought in some locations, while excessive precipitation might cause drought in other areas. In addition, warm winter temperatures result in early melting and more rain than snow, which affects river water flows. Natural disasters such as floods, droughts, and storms have a direct impact on the water supplies used by various industries, including the hydropower industry (Abbasov 2018). Because water from HPP reservoirs can be used in a variety of industries, such as fishing, recreation, and agriculture, water scarcity often results in conflict across sectors.

26.3.1 Overview of the Hydropower and Dams Sector in the Kura–Araz Basin

Kura and Araz form the largest transboundary river system of the South Caucasus region. The origins of the Kura can be found in east Turkey, and it flows across the Ardahan plateau through Georgia and enters Azerbaijan. In Azerbaijan, the Kura crosses the Kura–Araks plain, where it joins with the Araks and finally flows into the

Caspian Sea (Map 1). The length of the river is 1364 km and the basin is 188,000 km.

The Kura River plays a vital role in both local and regional economies and has been used to generate energy, irrigation, and water supply in Turkey, Azerbaijan, and Georgia. Recently, there are 63 water reservoirs in the Azerbaijan Republic, 46 of which are located in the Kura–Araz basin (Fig. 26.1).

The hydropower sector plays an important role in the energy sector, contributing a considerable amount of electricity produced. The installed capacity of all power generations, in 2018, was 5842 MW, out of which 1147 MW came from the hydropower sector (or 18.2%).

Dams are utilized in Azerbaijan for a variety of purposes, including irrigation, hydropower generation, fishing, and recreation, but their primary function is energy production. Freshwater ecosystems are essential to the production of electricity. HPPs account for just 10% of all electricity output. A total of 3100 million kW/h of electricity was produced in 2010.

There are nine HPPs in the Kura basin with various power capacities. In Table 26.4, the information regarding the capacity and hydropower capacity of these reservoirs is given.

As noted, to assess the current characteristics of HPP management vis-à-vis economic impact, the study uses two scenarios: BAU and SEM. Below we give characteristics of BAU and options for the SEM interventions for the HPP in the Kura basin (Table 26.5). This table demonstrates the potential benefits of SEM

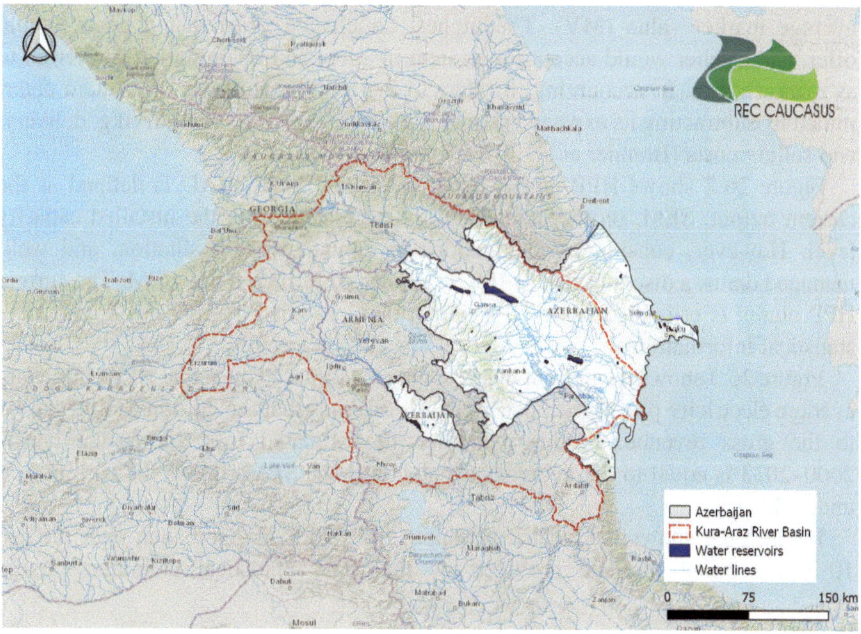

Fig. 26.1 Kura–Araz river basin (map of REC Caucasus)

Table 26.4 Characteristics of HPP reservoirs

No.	Water reservoir	Area (km^2)	Capacity of reservoir (km^3)	Installed capacity of HPP (MW)	Area of irrigated lands (ha)	Production in 2012 (MW/h)
1.	Mingechevir	605	15.73	402	970,000	1,400,000
2.	Shamkir	116	2.68	380	46,000	1,200,000
3.	Yenikend	23.2	1.58	150	6000	447,000
4.	Varvara	22.5	0.06	16	–	75,720
5.	Sarsang	14.2	0.565	50	120,000	–
6.	Araz	145	1.254	22	400,000	55,690
7.	Bilav		0.1	22	–	75,230
8.	Vaykhir		0.1	5	16,800	19,460
9.	Sugovushan	5.8	0.59	7.8	23,000	120,000

approaches. In order to define BAU baselines and possible SEM interventions, several indicators were used; for example, silting reservoirs, power generation, fishing, and recreation.

26.4 Economic Benefits from ES to HPP/Dams Development

This section includes information on existing and planned, large and small HPPs; the annual trends and forecast of electricity production of existing HPPs; and their average market value (MV). The highest anticipated price that a buyer would offer, and a seller would accept for an item in a free and open market is referred to as market value. In accounting, it refers to the replacement cost of an item determined by subtracting its expected selling price from the estimated carrying, delivery, and selling costs (Brenner et al. 2010; Krieger 2001).

Figure 26.2 shows HPP output under BAU and SEM. BAU is defined as the current output. SEM, on the other hand, is calculated using the installed capacity level. However, considering fluctuations in annual rainfall, siltation, and well-managed dams, a discount rate of 10% is applied to further define SEM. The current HPP output is estimated using HPP output data provided by HPPs or government statistical information.

Figure 26.3 shows the current market value (MV) based on actual HPP output and average electricity price (price per kW/h). It assumes that the current MV is similar to the gross revenue. In this example, the estimated loss during the period 2000–2013 is equal to the aggregated potential MV (SEM) minus the current MV under BAU.

Figure 26.4 shows the electricity output from HPP in Azerbaijan over the last 10 years, based on data provided by the State Statistical Comittee.[1] Large

[1] www.stat.gov.az.

Table 26.5 Characteristics of BAU and SEM practices in the hydropower sector

BAU	SEM
• Lack of an HPP sustainable development strategy, including watershed management plans and funding • Poor/absence of spatial planning policy, limited capacity • Deficient monitoring system • Outdated system to assess the availability of water resources ("sanitary flow"—10% of average annual water flow) • Development of HPPs in pristine ecosystems, including ecosystems in protected areas • Absence of freshwater ecosystems management plans • Deforestation and erosion in riverbeds in the upper and mid basins; weak law enforcement, absence of dam safety standards • Overgrazing causes erosion in upper and lower watersheds • Deteriorated or obsolete infrastructure (reservoirs, intake points and water canal network, pumping stations, silting control) • Lack of metering for domestic and industrial users, and water fees for irrigation	• Sustainable HPP development strategy, including watershed management plans and funding • Updated system for water availability criteria for sustainable development of HPPs Inventory of and monitoring of hydro-resources • Introduce spatial planning policy Freshwater and forest ecosystems management is an integrated part of the sector policy, and funding to implement is available • Deforestation and erosion control programs are available, and regulations are enforced • Grassland use and agriculture is planned and managed in river basins • Measures to eliminate illegal logging are enforced; erosion and sediments in water are decreasing • Adequate zoning and land use policies and enforcement • Infrastructure is maintained (reservoirs, intake points and water canal network, pumping stations, silting control) • Metering for domestic and industrial users, and water fees for irrigation are applied and collected • Adequate investment in maintenance and renovation • Increasing institutional capacity • Strong law enforcement

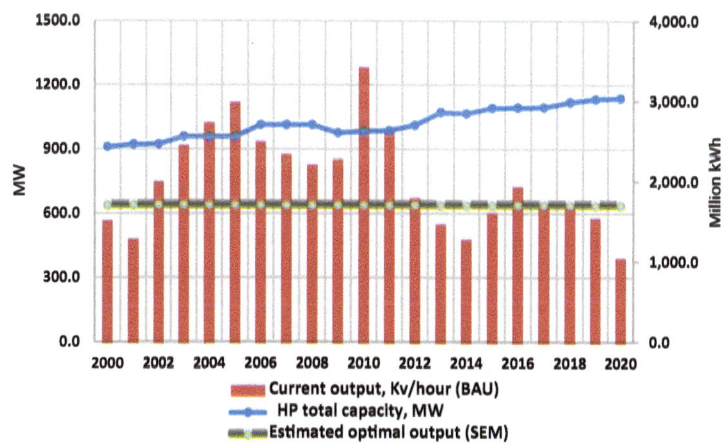

Fig. 26.2 HPPs output under BAU and SEM

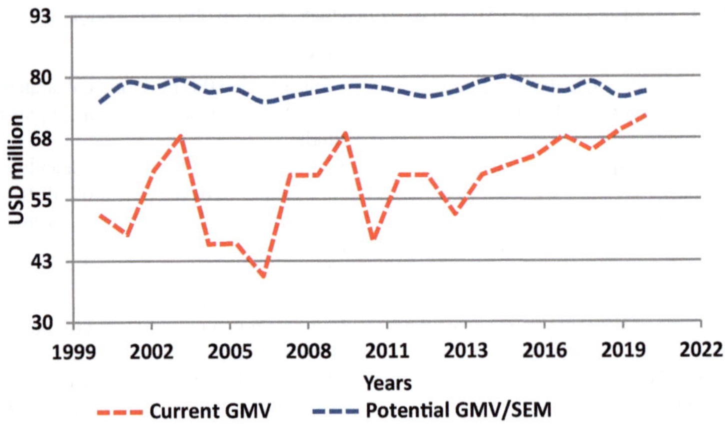

Fig. 26.3 Estimated loss and potential gain in gross market value (GMV) under BAU and SEM

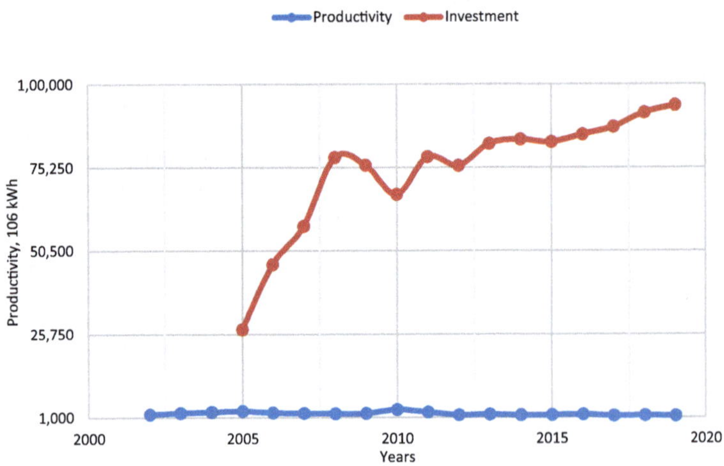

Fig. 26.4 Estimated productivity and infrastructure investment under BAU

investments were made in the HPP industry between 2005 and 2009, including the installation of new, cutting-edge generators in various HPP. The addition of these new generators helped to quickly increase the amount of electricity produced, but during the past 2 years, there has been a dramatic decrease in the amount of electricity produced. Little to no money was spent on watershed management during this time (the water factory). This is a typical BAU scenario; it can involve dam management issues, heavy silting, and deforestation. Figure 26.4 describes the level of investment in HPP/Dams infrastructure over the last 10 years. Despite the increasing trend for this period, the total amount of investments is rather low. Under the BAU scenario, investment in infrastructure and equipment is high; however, productivity is not sustained as illustrated in Fig. 26.4.

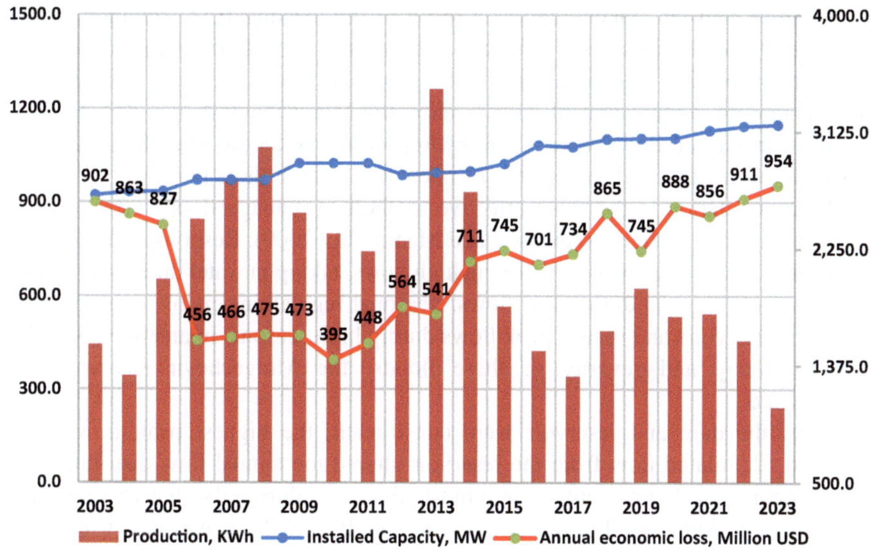

Fig. 26.5 Comparison of HPP installed capacity, total actual productions, and economic loss due to the sector's lower HP generation from 2003 to 2020 under BAU

Figure 26.5 illustrates economic losses in electricity production for the period of 2003–2012. It demonstrates that Azerbaijan's real HPP production is substantially lower than the combined installed capabilities of all HPP. For instance, although the Mingechevir HPP has a 402 MW installed capacity, its actual production in 2012 was only 159 MW. The effects of many causes could account for this disparity. The efficient management of dams is one straightforward explanation. It is thought that Azerbaijan's HPP dam management is BAU based on the significant discrepancy between installed capacity and actual production. To estimate economic losses in electricity production, we used the following formula:

$$EL = MP(IC - AP)$$

where EL is the economic losses for 1 year, MP is market prices for the electricity in 2012, IC is total installed capacity of all HPP in Kura basin, and AP is average price.

A BAU-based economic loss is shown in Fig. 26.5. This is an estimate of the discrepancy between the HPP's installed capacity and its actual output of electricity. The total economic loss under BAU from 2003 to 2012 is close to 4.5 billion USD (from 2000 to 2012, it is 6.4 billion USD), which is significantly more than the market value of the power produced during that time. Under SEM, an annual level of productivity that is close to 2000 kWh is expected, whereas under BAU, we see dramatic fluctuations in output.

Conflict between stakeholders is a result of the existing BAU situation, which includes decreased electricity production, decreased agricultural output due to less water available for irrigation, and flooding in downstream regions as a result of insufficient flood management. For instance, the Mingechevir dam and reservoir is used for irrigation, flood control, and hydropower production. Therefore, the management of the dam and reservoir is of interest to at least three parties.

Flow regulation is the most effective method to manage floods effectively in downstream part of rivers (Abbasov and Mahmudov 2009). Reservoirs can be used to balance the flow in rivers with spring high flows, taking in water during high flows and releasing it again during low flows. Seasonal regulations enable to accumulate water in reservoirs and reduce peak flows during high seasons. In an effort to reduce the frequency and severity of these floods, the Mingechevir reservoir was constructed.

After the dam and reservoir were constructed, the highest peak flows were reduced. Regulated flow from the reservoir altered the annual flow distribution of downstream, and flood events were almost eliminated during the first 15 years after construction. Although this was considered a shift to SEM, it was not sustainable, and mismanagement of the upper KARB, combined with other determining factors, resulted in increased floods and economic loss during the later years.

Reservoirs that are properly maintained should be able to store water during strong flows. Although little money is spent on dam repair, state-owned HPP and dam operators are interested in sustaining energy flow. For instance, Mingechevir reservoir acts as a flood prevention depository during the high-flow seasons, lowering the risk of flooding. To avoid a decrease in energy production, the Mingechevir reservoir was not emptied in 2010, nonetheless, before the peak flow season. Because the reservoir did not serve as a depository during the high flow, there were floods, inundations of 50,000 ha of irrigated land, and the loss of dwellings. Hydroelectric power stations in Azerbaijan produced less electricity by the end of 2013—nearly 75% less. This is a compelling argument in favor of switching from BAU to SEM.

Likewise, the SCS claimed that the Azerbaijani hydropower plant issue began at the end of 2012 and persisted throughout 2013. According to the data, electricity production at HPPs for January–October 2013 only totaled 1,209,106 kW/h, which is 24.5% less than for the same period in 2012. Estimates show that this will result in extra economic losses totaling USD 184,292,000 alone in 2011–2012. Over the years 2002–2012, the hydropower sector is estimated to have suffered a total economic loss of close to USD 4.5 billion.

26.5 Other Benefits and Risks of Hydropower Development

26.5.1 Nature-Based Tourism

Nature-based tourism is an important part of the world tourism industry. Growing interest and diminishing areas of open spaces make reservoirs very attractive in terms of nature-based tourism. They may be very important for tourism both in mountain and lowland regions. Reservoirs can be used for all types of recreational activities including rowing, surfing, swimming, and recreational fishing. Reservoirs of the Kura basin that support hydropower generation may be used for all the purposes.

Man-made natural attractions such as reservoirs that feed water into HPPs could enhance the tourism. Following safety standards, HPP reservoirs are used for outdoor watersports such as kayaking, canoeing, rowing, sport fishing, water skiing.

The assessment of nature-based tourism in any sector requires detailed analyses of all resources, including location, natural peculiarities, quality and quantity. Unfortunately, at the time of this study, there was no information available on tourism in the targeted reservoirs. Therefore, to evaluate the current recreational potential, a survey with five tourism experts was conducted. The survey was based in a very simple methodology that reflects the subjective opinions of these experts regarding real conditions around the reservoirs. Their opinions are included in Table 26.6.

For example, the recreational potential of the Mingechevir dam/reservoir is important. Mingechevir reservoir used to be one of the biggest Olympic rowing centers in the former Soviet Union.

Recently, a new rowing center with modern standards has been built. This center will likely increase rowing importance in the downstream part of the Kura river. In 2010, the new rowing center opened. The total area of the rowing center is 7.2 ha. The center's hotel may host 250 people simultaneously and it can accommodate 500 people to watch rowing games simultaneously. However, the hotel is only directed to serve sportsmen, and prices are considered too high for ordinary tourist. Local experts suggest that this center could host over 200,000 tourists every year. In Azerbaijan, the average tourist daily spending is higher than USD 150. This number can be used to produce a rough estimate of potential income from tourism. However, given the fact that these reservoirs are not used for tourism purposes, the estimated annual loss of revenue is roughly USD 180 million.[2]

Official statistics confirm that during the past 7 years, there has been significant growth in the number of tourists visiting Azerbaijan. In 2006, there were 218,982 person-days of foreign visitors served; by 2012, this number had increased to 674,435.[3] Investments in the tourism sector have been growing as well. This growth was accompanied by a gradual reduction of investments in the tourism sector.

[2] www.amaf.az.

[3] www.stat.gov.az.

Table 26.6 Tourism potential and challenges related to HPP/reservoirs

Reservoir	Potential	Key challenges
Mingechevir	The rowing center near the reservoir has a great potential. The reservoirs can be used for rowing, fishing, and surfing. Suitable climate conditions prevent freezing of water throughout the year, which makes reservoir very attractive for tourists and sportsmen. The potential annual number of tourists is 150,000 with an average of 2-day stay	High prices for lodging. Limited number of budget hotels. The absence of general services such as restaurants, car rental, trains from other cities, and so on
Shamkir and Yenikend	Attractive for fishing, surfing, and boating. Mountains and natural extremes are rather close. The proximity of such amenities makes the greatest contribution. The potential annual number of tourists is 50,000 with an average 2-day stay	Slight remoteness from the residential areas. No lodging opportunities. Not easily accessible
Araz	Attractive for fishing, surfing, and boating. Mountain areas and many types of natural springs are very close. There are opportunities for extreme tourism. Could be accessed from mountain regions of Turkey. Potential number of tourists is 150,000 with 2 days stay	Located directly on the border. Not easily accessible from the Baku
Sarsang	Attractive for fishing, surfing, and boating. Mountain areas and many types of natural springs are very close. Potential number of tourists is 150,000 with 2 days stay	High risk war zone. Not accessible

Moreover, investments in the tourism sector include mainly government expenditures in large infrastructure; nothing on freshwater ecosystems management. This is a typical unsustainable BAU practice that undermines the potential long-term development of the tourism sector. This trend is illustrated in Fig. 26.6.

It is worth noting that current tourism investments cover only central city of Baku and are made by international hotel companies. So far, there were no considerable government investments in regions, while number of tourists continues to increase. This is a typical BAU approach that results in additional pressure on ecosystems threatening potential long-term economic gains.

Nature-based tourism has great potential in the region. For instance, if 30% of these people are interested in nature-based tourism in the target region, and the average amount of daily spending per tourist is USD 110, including accommodation ($70), food ($20), transport ($10), and other expenses ($10), the total income to touristic enterprises could reach USD 30,000,000. Attractions such as the new Kura Olympic Rowing Center in Mingechevir, combined with other local natural

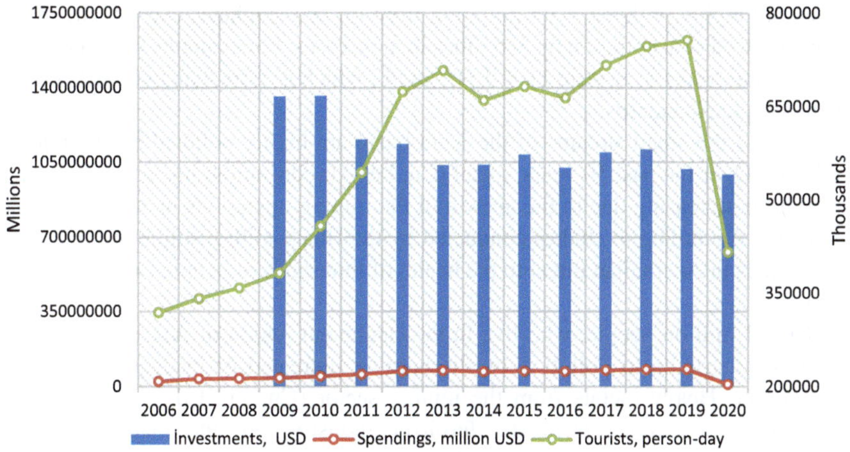

Fig. 26.6 The trend of investment and person/days served in the tourism sector in the KARB in Azerbaijan, BAU scenario (Source: State Committee of Statistics)

attractions, are essential to sustain the future economic benefits of nature-based in the region. However, in addition to poor investment in tourism, the Mingechevir reservoir tourism potential is at risk because of water pollution and sedimentation.

These estimated values show only part of the existing total spending and do not reflect the full potential of this sector. Generally, tourism sector in Azerbaijan is still weak and has poor incentives to develop. A shift to SEM, in addition to additional investment, includes changes in visa policy, making new regulations concerning tourism and creation market-driven mechanisms.

26.5.2 Drinkable Water

The increase in demand of drinkable water, in response to the growing population, indicates the importance of this sector in Azerbaijan. Water is indispensable for economic growth and poverty reduction (Scandizzo and Abbasov 2022a).

The largest city that is completely supplied from the reservoirs is Mingechevir, which is the fourth-biggest city in Azerbaijan, with a population of about 100,000. The water for Mingechevir is taken directly from the Mingechevir reservoir and then distributed to residential users in urban areas with no treatment.

Shirvan and Karabakh canals are not only major sources of irrigation water but also main sources of drinking water in most of places of Aran Economic District they cross.

The main water sources in Azerbaijan are the transboundary Kura and Araz rivers, which are affected by permanent pollution in the territory of neighboring Turkey, Iran, Georgia, and Armenia (Abbasov and Smakhtin 2009; Suleymanov et al. 2010). The quality of the drinking water is poor both in source and distribution points. The rivers of Kura and Araz, which are the main sources of water supply for

Aran, are highly polluted, with pollution from oil and sulphates exceeds the maximum allowed concentration by 4–5 times in most cases. For example, concentrations of As in the Araks river were 11.8–151.3 mg/L, which is more than two times higher than accepted standards.

Most of the small streams of the Kura basin are highly polluted by the mining industry. Over the past 50 years, metal (Cu, Fe, Al) concentrations in some streams have been increasing due to the growth of the mining operations in Azerbaijan and Armenia. According to studies conducted by the Blacksmith Institute in 2012, new gold mines in Azerbaijan threat to the health of thousands of people (www.az.dbisa.org).

The 11 million people, who live in the catchment region of Azerbaijan's water sources, contribute to pollution by discharging untreated or inadequately treated wastewater. Heavy metals from the mining sector (Cu, Zn, Cd, As), as well as ammonia and nitrates from the fertilizer industry, are the main pollutants. Concentrations might go up to nine times above average. Mineral oil and phenols are six to three times higher than average, respectively.

The Araz River is said to be among the most turbid in the world, and excessive turbidity raises the expense of treating water for drinking. Due to the noticeable sediment flows in these rivers, conventional treatment and huge facilities are needed to improve the water quality and minimize the burden of silt near the withdrawal point. The Kura withdrawal sites were constructed immediately following the Kura and Araz river confluence. Due to heavy pollution in upstream regions of the Kura basin, waterborne diseases in the downstream regions of the Kura basin ravage the health of thousands of rural people and result in huge economic losses (Scandizzo and Abbasov 2012).

The City of Baku is the second major user of the regulated Kura water. Nearly 25% of the Greater Baku area that has more than 4 million of residents are supplied by water withdrawal facilities located in a downstream part of the Mingechevir reservoir. The whole system has a total capacity of nearly 13.5 m^3/c (Scandizzo and Abbasov 2022a).

Water losses are a major issue in developing countries, seriously undermining efforts to develop sustainable water supply systems. Current estimates show that average water losses in Asian cities are around 50–60% of total water released to the networks, while for European countries these losses range between 10% and 40% of the total water supply.

Several estimates agree on an average consumption in Baku of 400 liters per capita per day. However, a World Bank survey (Scandizzo and Abbasov 2012) confirms that real consumption is nearly 170 L/day. Nevertheless, the apparently high individual water consumption rate is the result of several factors, mostly related to the poor condition of the transmission and distribution pipe network, domestic pipes and taps, as well as the absence of metering. As noted, the current system provides little or no incentive for consumers to conserve water; this in turn reduces water available in other parts of the network, and imposes higher operational costs on the systems. Since water leaks are very common, nearly 60% of the total water

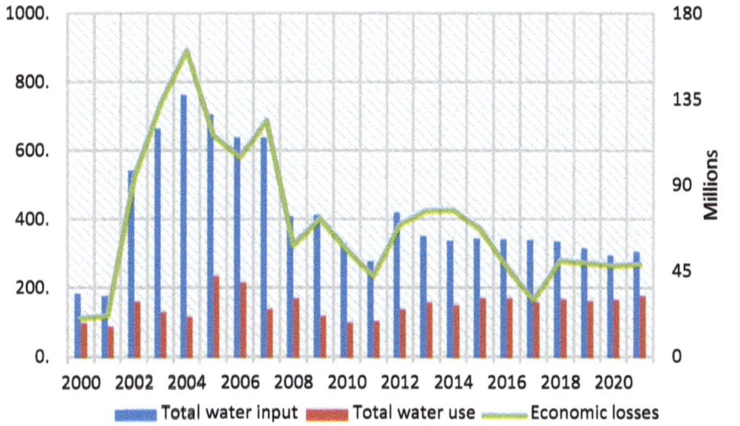

Fig. 26.7 Difference between total water input (million m^3), water use (million m^3), and economic losses (million USD) under BAU 2000–2021

input is not used and directly being mixed with wastewater. This is a typical BAU scenario.

A water balance can be built to calculate how much water is lost from a system based on measurements or estimates of the amount of water generated, consumed, and lost (together with any water imported or exported). The simplest water balance equation is:

$$TL = SI - C$$

where TL is the total loss of water, SI is the distribution network system input, and C is the consumption.

By estimating the difference between the amounts of water distributed and invoiced, it is possible to estimate total losses at the distribution network (Fig. 26.7). This information is presented by state-owned Azersu JSC. Using the aforementioned information and average cost of water (0.25 USD/m^3), economic losses can be estimated. Total economic loss over the period of 2003–2012 reaches USD 1 billion 68 million. These losses include maintenance and operational costs related to water transport, including treatment costs as well.

These losses are the manifestation of BAU scenario and may be partially avoided by SEM interventions. The implementation of a SEM strategy could considerably reduce these losses, particularly to control pollution sources and decrease water treatment costs.

26.6 Incidence of Natural Hazards

The most common hazard in Azerbaijan that could be linked to poor dam management is floods. For example, downstream part of the Mingechevir dam is often suffered from floods caused as a result of poor dam maintenance. As we have noted, one of the goals of the construction of Mingechevir dam was to reduce the frequency and severity of the floods. After the completion of the dam and reservoir, the highest peak flows were reduced. Regulated flow from the reservoir altered the annual flow distribution of downstream, and flood events were almost entirely eliminated during the first 16 years after construction (Abbasov and Mahmudov 2009).

However, according to studies related to investigation of the watershed erosion and channel silting confirm that the riverbed and reservoir silting was the main driver of the last year floods. (e.g., Abbasov 2011; Abbasov and Mahmudov 2009). Intensive deforestation and unsustainable agricultural practices in upper watersheds increase turbidity of water in rivers and streams. This increased volume of suspended sediments entering to the reservoirs from the upstream watersheds causes the reduction of the capacity of the reservoir, and during high water seasons, floods affect downstream areas.

Large floods have been a result of improper dam and watershed management since 1993. Recent floods in the target area have an average annual impact of 200,000–250,000 lives. For instance, in May 2010, 50,000 ha of farmland were flooded, tens of thousands of dwellings were demolished, and more than 240,000 people were affected. The estimated cost of the damage was USD 591 million. Poor upper basin management and dam management were the primary causes of this flood devastation (flow regulation).

As a result of flooding, the Government of Azerbaijan (GoA) boosted its state budget in 2010 by up to USD 425 million. In 2013, USD 180 million were invested to lessen the effects of floods. The estimated costs for 2014 come to over USD 185 million. The expenses over the previous 4 years has barely above USD 1 billion. The annual expenditures for preventing floods are displayed in Fig. 26.8. BAU is related to the high expense of the 2010 flood. By switching to SEM management, this expenditure might be decreased, for example, to just USD 20 million each year (Fig. 26.8).

In addition to comparing BAU and SEM scenarios, a comprehensive and yet simple cost–benefit analysis (CBA) can be used to guide management interventions. The first requirements of CBA are data on costs and benefits of an integrated SEM management program. Costs may include forest management, erosion prevention, dam management, canal cleaning, and construction of dykes along the canal. However, the GoA does not investment on upper watershed management. As a reference, the total investment on nature protection, in 2012, slightly exceeded USD 4 million and mainly covered recurrent costs of central and local offices of the environmental departments.

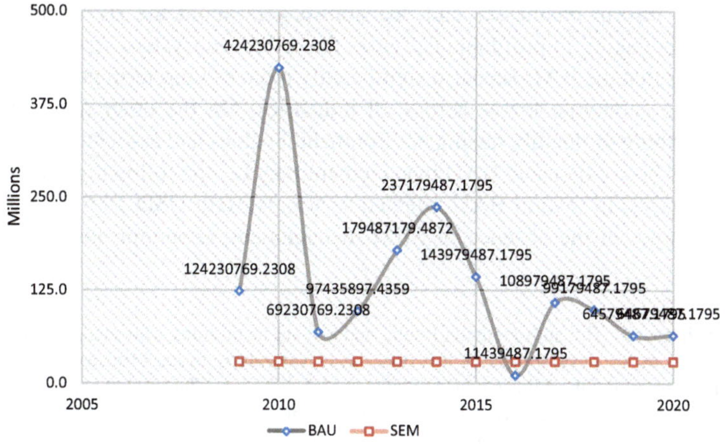

Fig. 26.8 Annual costs of floods under BAU and SEM and SEM 2008–2015. (Source: Author's estimates based on official data)

26.7 Conclusions

The economy of Azerbaijan is heavily impacted by BAU practices in managing freshwater ecosystems. By switching to low-cost SEM strategies, a portion of this high cost can be avoided. Although there are numerous laws and rules governing the management and administration of HPP and Dams in Azerbaijan, enforcement is inadequate. Additionally, there are gaps in the legislative framework, no mechanisms for enforcing the legislation, and no tools for gathering and analyzing data. As a result, the evaluation's findings are not used to develop policies or to enhance HPP/Dams management.

The potential effects of present ecosystem management techniques in the upper river basin are not taken into account in the current environmental impact assessments of HPP/Dam projects (small and large). The performance of HPP/Dams will suffer as a result, which could have further negative externalities affecting other industries such as irrigated agriculture, tourism, fisheries, and drinkable water supplies. The total cost of these negative externalities frequently exceeds the advantages now received by the HPP/Dams business.

A comprehensive package of interrelated policy reform measures is required at both the national and regional levels because improving ecosystem management in the upper watershed necessitates the involvement of numerous sectors, including HPP/Dams, agriculture, forestry, fisheries, tourism, and water supply. The introduction of sustainable HPP/Dams development in the Southern Caucasus is described as necessitating this "policy mix" package.

Lack of information and data limited the scope of this study; hence, additional research is needed, some of which would involve the establishment of primary data baselines. Basic scenarios (BAU/SEM), however, were developed wherever it was practical to inform businesses and decision-makers about the risks and financial rewards of engaging in profitable activities that affect ecosystem services.

It is clear that the BAU scenario significantly reduces long-term gains by causing considerable economic losses across all industries. On the other hand, the SEM might contribute to steadily raising environmental values and associated benefits. To demonstrate how expensive BAU management may be, the following rough total of economic losses in several BAU-affected sectors is used: USD 18.6 billion. It also illustrates how, in the absence of SEM management, economic losses may keep rising.

Acknowledgments The authors thank public and private institutions who made contributions at various stages of the development of this document, particularly to the Ministry of Ecology and Natural Resources and the Ministry of Agriculture of Azerbaijan. The authors are also thankful to Maka Bitsadze at WWF Caucasus Program Office who went to great lengths to facilitate the procurement process, provide research material, review drafts, and advised throughout the drafting process. The authors also thank Konul Ahmadova and Elshad Askerov from the WWF Baku office for their assistance and support.

References

Abbasov R (2018) Community-based disaster risk management in Azerbaijan. Springer

Abbasov RK, de Blois CLC (2021) Nature-based management scenarios for the Khojasan Lake. Sustain Water Res Manag 7(6):1–12

Abbasov RK, Mahmudov RN (2009) Analysis of non climatic origins of floods in the downstream part of the Kura River, Azerbaijan. Nat Hazards 50(2):235–248

Abbasov RK, Smakhtin VU (2009) Introducing environmental thresholds into water withdrawal management of mountain streams in the Kura River basin, Azerbaijan. Hydrol Sci J 54(6): 1068–1078

Abbasov R, Karimov R, Jafarova N (2022) Ecosystem and socioeconomic values of clean water. In: Ecosystem services in Azerbaijan. Springer, Cham, pp 71–121

Aylward B, Bandyopadhyay J, Belausteguigotia JC, Borkey P, Cassar AZ, Meadors L, Bauer C (2005) Freshwater ecosystem services. In: Ecosystems and human Well-being: policy responses, vol 3, pp 213–256

Bennett EM, Peterson GD, Gordon LJ (2009) Understanding relationships among multiple ecosystem services. Ecol Lett 12(12):1394–1404

Brenner J, Jimenez JA, Sarda R, Garola A (2010) An assessment of the non-market value of the ecosystem services provided by the Catalan coastal zone, Spain. Ocean Coast Manag 53(1): 27–38

Carpenter SR, Bennett EM, Peterson GD (2006) Scenarios for ecosystem services: an overview. Ecol Soc 11(1):29

Costanza R, d'Arge R, De Groot RS, Farber S, Grasso M, Hannon B, Limburg K, Naeem S, O'Neill RV, Paruel J, Raskin RG, Sutton P, Van den Belt M (1997) The value of the world's ecosystem service and natural capital. Nature 387:253–260

Daily G (ed) (1997) Nature's services: societal dependence on natural ecosystems. Island, Washington DC

Daily G (2003) What are ecosystem services. In: Global environmental challenges for the twenty-first century: resources, consumption and sustainable solutions, pp 227–231

Daily GC, Matson PA (2008) Ecosystem services: from theory to implementation. Proc Natl Acad Sci 105(28):9455–9456

Dhyani S, Lahoti S, Khare S, Pujari P, Verma P (2018) Ecosystem-based Disaster Risk Reduction approaches (EbDRR) as a prerequisite for inclusive urban transformation of Nagpur City, India. Int J Disaster Risk Reduct 32:95–105

Dhyani S, Karki M, Gupta AK (2020) Opportunities and advances to mainstream nature-based solutions in disaster risk management and climate strategy. In: Nature-based solutions for resilient ecosystems and societies. Springer, Singapore, pp 1–26

Dhyani S, Majumdar R, Santhanam H (2021) Scaling-up nature-based solutions for mainstreaming resilience in Indian cities. In: Ecosystem-based disaster and climate resilience. Springer, Singapore, pp 279–306

Flores M, Adeishvili M (2012) Economic valuation of the contribution of ecosystems in protected areas to economic growth and human well-being in Georgia. Prepared by ECFDC/GCCW/AMECO, UNDP/GEF project Catalyzing Financial Sustainability of Georgia's Protected Areas System

Gleick PH, Singh A, Shi H (2001) Emerging threats to the world's freshwater resources. A Report of the Pacific Institute for Studies in Development, Environment, and Security, Oakland, CA

Green PA, Vörösmarty CJ, Harrison I, Farrell T, Sáenz L, Fekete BM (2015) Freshwater ecosystem services supporting humans: pivoting from water crisis to water solutions. Glob Environ Chang 34:108–118

Heywood VH, Watson RT (1995) Global biodiversity assessment, vol 1140. Cambridge University Press, Cambridge

Krieger DJ (2001) Economic value of forest ecosystem services: a review. The Wilderness Society Publications 32p

Kumar R, Singh A, Pandey U, Srivastava P, Mehra S (2022) Mapping the extent of invasive species: an assessment based on high-resolution data for selected species in parts of Eastern Himalaya in Sikkim. In: Forest dynamics and conservation. Springer, Singapore, pp 249–259

Maclaurin J, Sterelny K (2008) What is biodiversity? In: What is biodiversity? University of Chicago Press

Millennium Ecosystem Assessment (MA) (2005) Ecosystems and human well-being: synthesis. Island Press, Washington DC

Pearce DW, Atkinson G (1993) Capital theory and measurement of sustainable development. Ecol Econ 8:103–108

Pimm SL, Russell GJ, Gittleman JL, Brooks TM (1995) The future of biodiversity. Science 269(5222):347–350

Postel S, Carpenter S (1997) Freshwater ecosystem services. In: Nature's services: societal dependence on natural ecosystems, p 195

Scandizzo PL, Abbasov R (2022a) Do people appreciate the economic value of water in Baku city of Azerbaijan? In: Blue-green infrastructure across Asian countries. Springer, Singapore, pp 193–220

Scandizzo PL, Abbasov R (2022b) Public interests and private incentives in designing an ecological payment systems. In: Forest dynamics and conservation. Springer, Singapore, pp 439–467

Small hydropower potential in Azerbaijan (2009) United Nations development program final report, Baku, Azerbaijan

Suleymanov B, Ahmedov M, Safarvoa K, Steinnes E (2010) Metals in main rivers of Azerbaijan: influence of transboundary pollution. Water Air Soil Pollut 213:301–310

TEEB Foundations (2010) In: Kumar P(ed) The economics of Ecosystems and Biodiversity: ecological and economic foundations. Earthscan, London, Washington

Weißhuhn P, Reckling M, Stachow U, Wiggering H (2017) Supporting agricultural ecosystem services through the integration of perennial polycultures into crop rotations. Sustainability 9(12):2267

Eutrophication Modeling of Chilika Lagoon Using an Artificial Neural Network Approach

27

Prasannajit Acharya, Pradipta R. Muduli, and Mira Das

Abstract

Chilika lagoon is the first Ramsar site in India located along the East Coast. Prediction of the eutrophication of such ecosystems is a key approach for a sustainable management perspective as it helps to formulate a management action plan. In the present study, a data-driven modeling approach, an artificial neural network (ANN), was used to predict eutrophication in the Chilika lagoon. Back-propagation neural network model was used to relate the major parameters that influence eutrophication indicators such as total nitrogen (TN), total phosphorus (TP), Secchi disc depth (SD), dissolved oxygen (DO), biological oxygen demand (BOD), pH, water temperature (WT), and Turbidity (TURB). The model evidenced an acceptable level of prediction when compared with the results of the field observations. This model's most important determinant variables were those with a high Random Forest (RF) model permutation relevance ranking, which reduced the network's structure and led to a more accurate and effective process. It demonstrated a high agreement between BOD and Turbidity. As per the TLI (trophic level index) estimation, the Chilika lagoon was observed to maintain an oligotrophic condition. However, there was a trophic switchover between the seasons and sectors. The study evidenced that the ANN was able to predict the indicators with reasonable accuracy, which could be proved as a valuable tool for the Chilika lagoon. This approach can be considered when

P. Acharya · M. Das
Department of Chemistry, Institute of Technical Education and Research, Siksha 'O' Anusandhan (Deemed to be University), Bhubaneswar, India
e-mail: prasannajitacharya9@gmail.com; miradas@soa.ac.in

P. R. Muduli (✉)
Wetland Research and Training Centre, Chilika Development Authority, Balugaon, India
e-mail: prmuduli.wrtc@gmail.com

S. Dhyani et al. (eds.), *Ecosystem and Species Habitat Modeling for Conservation and Restoration*, https://doi.org/10.1007/978-981-99-0131-9_27

developing a sustainable management and conservation action plan for Chilika and other similar aquatic ecosystems around the world.

Keywords

Annoyance · Artificial neural network · Chilika lagoon · Eutrophication · Mean squared error · Multi-layer perceptron · Root mean squared error

27.1 Introduction

In the area of management of water resources, data-driven models are commonly implemented since, in contrast to scientific and physical-based models (hydrological). They require significantly less effort in terms of the quantity and quality of the data required (Adnan et al. 2021). Artificial neural networks (ANNs) are computation models that find widespread use in the resources and environmental research areas (Oyebode and Stretch 2019). This computational model (ANN) is intended to function in a manner analogous to that of the human brain and nervous system. Specifically, ANNs are utilized for the prediction of numerous different subfields of aquatic ecology, such as benthic, planktonic communities, fishery assemblage, and bio-manipulation evaluation (Kim et al. 2019). Numerous ANN water quality modeling studies have been carried out over the course of the past two decades, and the findings of these studies have been extremely advantageous (Goethals et al. 2007). One of their primary advantages is the capacity to represent complicated and nonlinear processes, and the other is that they do not require hypotheses to be made about the distribution of data or the linkages among dependent and independent variables (Oyebode and Stretch 2019). In contrast to this, it gives users the ability to properly simulate external linkages despite having a limited understanding of the issue at concern (Bennett et al. 2013). Therefore, ANNs are considered to be the optimal choice for predicting marine habitats, which are identified by their intricate dynamics and nonlinear analyses (Kim et al. 2019).

Programmers that specialize in water quality monitoring, for instance, generate a great deal of data with complicated structures (Muduli et al. 2017). However, there are also many cases in which inadequate information and integrity may be a concern in ecological quality evaluation and management (Panigrahi et al. 2007; Nayak et al. 2004). The lack of data might be due to different causes, including malfunctioning sensors, insufficient financing, and unfavorable weather conditions during monitoring. In the research done by Sahu et al. (2014), data scarcity of water resources was brought to light as a problem. The authors successfully addressed this problem by incorporating a k-fold test set of data incorporated into the ANN's learning input set. In general, when predicting with ANNs, the phenomenon of information scarcity or small datasets is a concern in scientific domains, and the method of k-fold validation set is widely used (Cigizoglu and Kişi 2005).

The relationships between variables of response and prediction are assumed to be linear, and normal distribution is made by a significant number of statistical-based models on the quality of water (Kuo et al. 2007). However, ANNs can accurately show the nonlinear relationship among ecosystem variables (Maier et al. 2004). Additionally, ANNs are able to use known data as input without making any prior assumptions (Maier et al. 2004). A mapping of input and output variables is created by ANN, which is used for predictions of chosen outcomes with respect to appropriate inputs (Wei et al. 2001). By using an appropriate combination of interconnecting variables and model parameters, a neural network with multiple layers can generate an approximation of any smooth, measurable function that exists between the vectors of input and output (Agwu et al. 2020). ANN can model patterns in the dynamics of algal development and predict algal blooms using environmental variables (Smith et al. 2008). For instance, ANN models were required to predict the quality of river waters (Zhang and Stanley 1997; Singh et al. 2009), shallow lakes (Kuo et al. 2007), coastal areas (Lee et al. 2003a), and reservoirs (Young et al. 2011).

The primary objective of predictive modeling is to maximize accuracy; hence, machine learning (ML) methods are often carried out with input variables (predictor variables) and one or more output variables (target variables) (Phillips et al. 2008). To predict a water quality metric, all possible variables or a subset of them can be employed. Due to this, the model may contain either very high or very low numbers of inputs, and both of them are unacceptable (Maier et al. 2010). To solve this problem, a predictor variable selection stage was taken into account in this study to get rid of unnecessary data. To accelerate machine learning, the number of predictor variables was reduced. The goal of lowering the variable numbers (predictor) in ML was to increase the speediness of the algorithm's process of learning, improve predicted accuracy, and increase the interpretability of outputs (Motoda and Liu 2002; Mulia et al. 2013). Many factors influence the Chl-a level. A Random Forest (RF) approach was used to identify the utmost important predictor variables for the Chl-a level. RF modeling technique employs decision trees and is trained on a list of source variables (predictor) to accurately predict the variables of the output (Strobl et al. 2007). The RF method has a lot of possibilities. First off, there is no assumption made regarding predictor variable probability distribution. Second, it can manage a lot of factors and choose the most helpful ones from them (Park et al. 2015). RF forecasts sidestep the overfitting issue that plagues different types of regression techniques, which are nonlinear because they are derived from the average set of numerous basic models (Were et al. 2015). Since each tree is constructed using an arbitrary subgroup of the input data, no additional independent dataset is needed to assess the model's ability to predict outcomes (Motoda and Liu 2002). The RF approach is also suitable for natural ecosystems with high physicochemical diversity, leading to the eutrophication process.

Eutrophication is the primary factor responsible for the decline in water quality that has been observed in several wetlands, estuarine, and oceanic environments all over the world, which is one of the most critical problems that we face in the modern era (Smith et al. 2006). The eutrophication of freshwater lakes has serious socioeconomic ramifications that undermine human well-being. Experts are working on the design of modern tools for more effective comprehensive monitoring of the quality of water. A lake's ecosystem is so dynamic and diverse that even a minuscule fraction of external stress such as tourism activities with a biologically active component input can cause eutrophication. Because of this, advanced modeling techniques with different algorithms are employed for eutrophication forecast scheme preparation (Dynowski et al. 2019). ANNs have already been utilized in the modeling of eutrophication processes in lakes (Hadjisolomou et al. 2017). The modeling of eutrophication can also make use of methods such as linear regression and decision trees, as well as other methodologies. These approaches have the benefit of requiring adjustments to a limited number of parameters. However, as per expectation, performance might not be observed when the sample size (data) is inadequate or when a set of hypotheses regarding distributions (linear or Gaussian) are violated (Brown et al. 2020). Contrarily, the effect of nonlinearity, which is frequently recorded among relevant water quality parameters, does not affect artificial neural networks (ANNs) (Hadjisolomou et al. 2017).

As a Ramsar site, the conservation of Chilika lagoon's ecology and biodiversity is very crucial. Eutrophication and toxic bloom could be vital factors for the deterioration of the quality of water and health of the Chilika lagoon. This can cause an overabundance of the growth of aquatic plants and algae, which can throw off the natural equilibrium of aquatic life (Peetabas and Panda 2015). Indicators of water quality are commonly utilized to predict eutrophication levels in lake waters (Hadjisolomou et al. 2017; Hagan et al. 2014). However, for the Chilika lagoon, accurate estimation could be challenging for two main causes. Initially, the geographical and seasonal patterns are influenced by altering the environmental, geographical, and climatic factors. The interdependence and interrelationship of the variables could be the second factor that could contribute to the challenge of making reliable predictions (Hagan et al. 2014). Consequently, the investigation of methods that may quantitatively forecast eutrophication indicators is an essential endeavor to undertake. Techniques that aid in the development of efficient measures to avoid eutrophication is the need of the hour. Hence, considering the values of the lagoon and its conservation measures, the prediction of the Chilika lagoon, Odisha, using artificial neural networks (ANNs) is the goal of this proposed study.

27.2 Methodology

27.2.1 Study Area

Chilika lagoon, located in the state of Odisha, is the first Ramsar wetland site of India and has provided support to 200,000 people through tourism (birds and dolphins) and fishery as a source of income. The location of the lagoon is between the longitudes 85°05′ and 85°35′ East and the latitudes 19°28′ and 19°54′ North (Fig. 27.1). During the dry and wet seasons, a total water spread area of 704 km^2 and 1020 km^2, respectively, maintained in the lagoon (Acharya et al. 2022). The Chilika lagoon is inhabited by a variety of ecosystems, including saltwater, freshwater, and marine water environments, ranging from shallow to deeper waters (Barik et al. 2017). These ecosystems can be considered as four distinct geographical

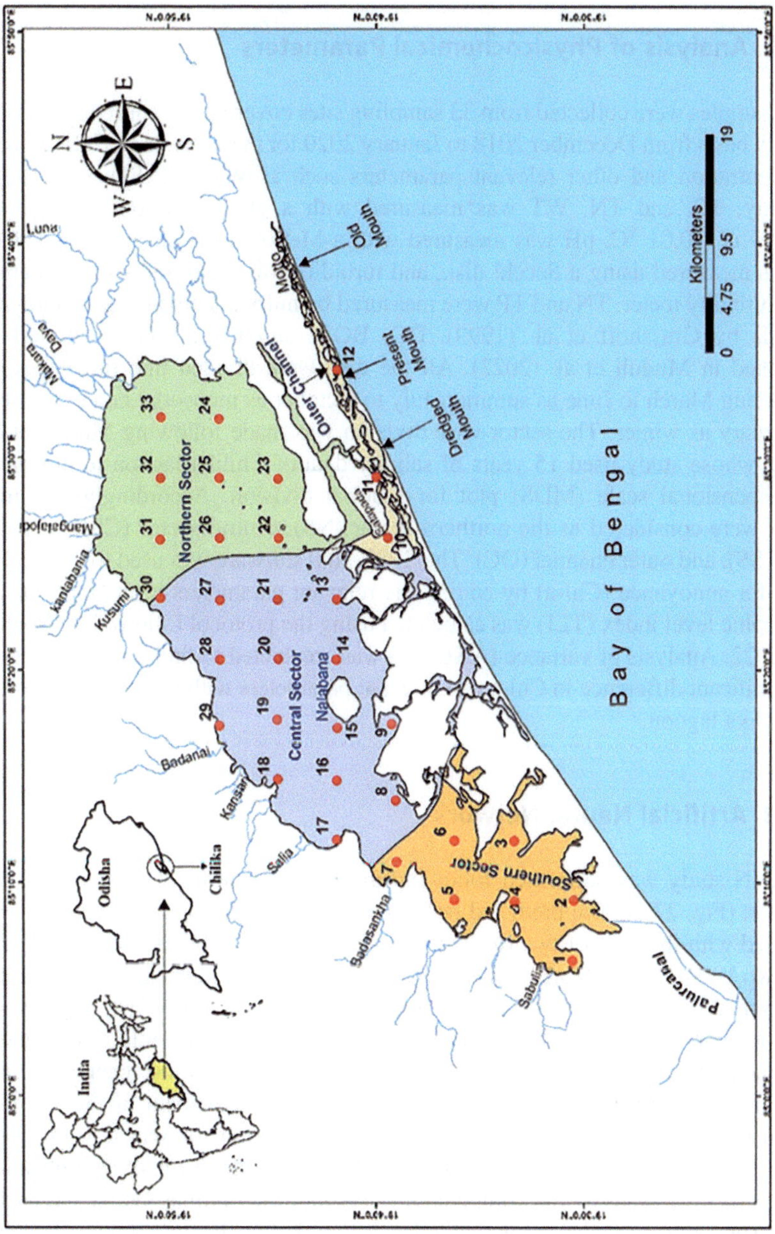

Fig. 27.1 Thirty-three sampling locations of the Chilika lagoon

sectors: outer channel, central channel, northern channel, and southern channel (Muduli et al. 2017).

27.2.2 Analysis of Physicochemical Parameters

Water samples were collected from 33 sampling sites covering the entire lagoon on a monthly basis from December 2018 to January 2020 for the analysis of Chlorophyll-a concentration and other relevant parameters such as WT, pH, SD, DO, BOD, Turbidity, TP, and TN. WT was measured with a glass thermometer with an accuracy of ±0.01 °C, pH was measured with a Metrohm pH electrode (±0.001), SD was measured using a Secchi disk, and turbidity was measured with a Thermo Orion turbidity meter. TN and TP were measured by nutrient autoanalyzer following methods by Grasshoff et al. (1999). DO, BOD, and Chl-a were measured as mentioned in Muduli et al. (2022). All the data were divided into three seasons considering March to June as summer, July to October as monsoon, and November to February as winter. The sector-wise division was made following Muduli et al. (2017), whose study used 15 years of salinity data of Chilika lagoon to prepare a multidimensional scale (MDS) plot for sectoral division. Accordingly, the four sectors were considered as the northern sector (NS), central sector (CS), southern sector (SS), and outer channel (OC). The SPSS 20.0 software was used to predict the model for annoyance (Chl-a) by adding the relevant parameters in the input layer. The trophic level index (TLI) was calculated using the protocol followed by Muduli et al. 2022. Analysis of variance (ANOVA) was conducted to find out if there were any significant difference in Chl-a and relevant parameters with respect to sectors of the Chilika lagoon.

27.2.3 Artificial Neural Network

The ANN study was performed following the stringent protocol as shown in the flowchart (Fig. 27.2). The predicted performance efficiencies of each network were evaluated with the help of two distinct varieties of neural network models. Results from two neural network techniques were compared and evaluated. This modeling study aimed to accomplish two different goals at the same time. First, an artificial neural network was trained by employing a method known as k-fold cross-validation. Then, the outcomes of the models were compared with those generated with various k values to the modeling results. In contrast, the element of the required amount of computational time is considered as a criterion for selecting the ideal model, with regard to the performance of ANN and the level of complication (amount of time required for computation). Second, the optimal model's capacity for an explanation was investigated in order to identify whether or not it was suitable for use as a tool for the management of water quality. In this study, the evidence was presented to demonstrate how well the ANN was able to forecast the Chlorophyll-a

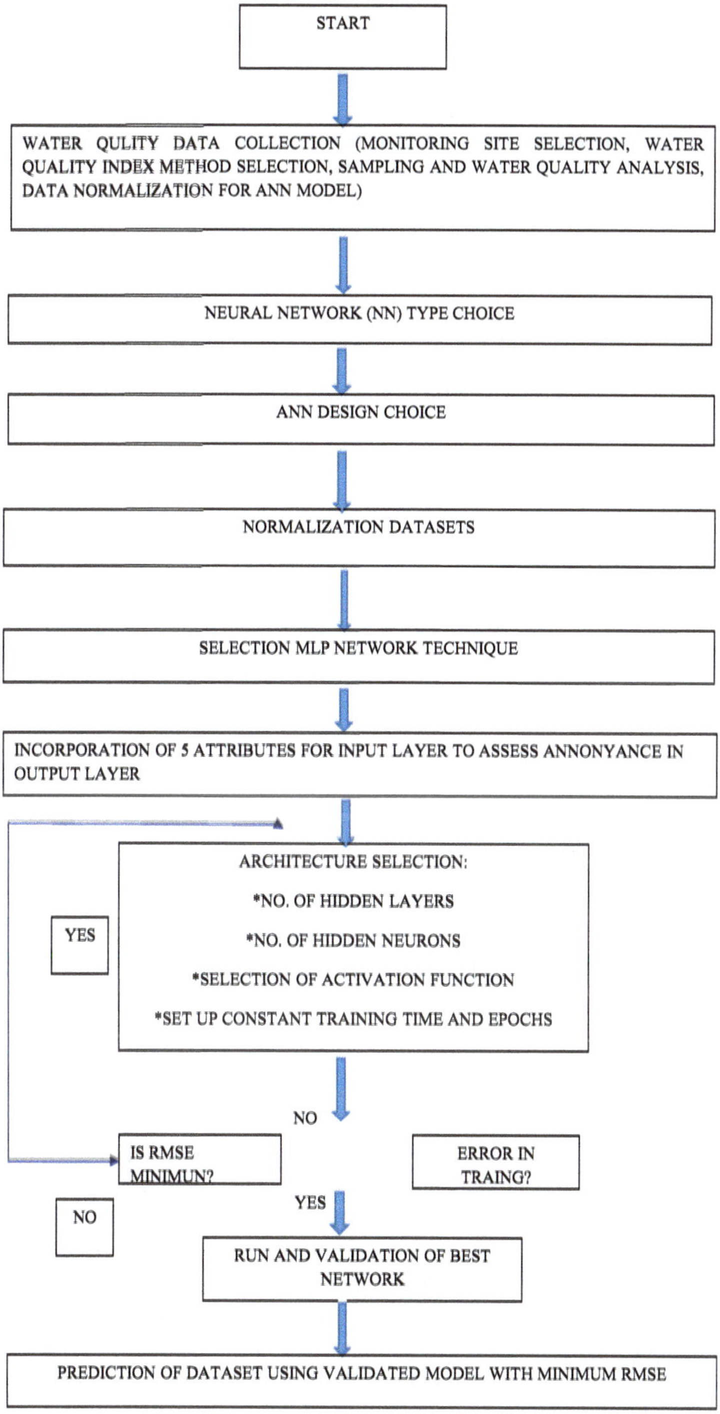

Fig. 27.2 Flow chart of artificial neuron network (ANN) applied for Chl-a prediction in Chilika lagoon

levels in Chilika lagoon. A sensitivity analysis algorithm was utilized, which was based on this model, to understand the impact of each environmental variable.

The objective of the study was to model the annoyance on the basis of input variables such as pH, DO, BOD, Turbidity, WT, SD, TP, and TN. The multilayer perceptron (MLP) network was selected for this study due to its ability to unfold complex relationships among the datasets. The best model was picked up, which has minimum MSE and RMSE value out of 2500 architectures (500 architectures for each neuron, i.e., 1–5).

27.2.4 Min-Max Normalization

Analyzing datasets without normalizing them reduces the efficacy of a dataset with a lower scale. So, data normalization is crucial when the attributes are of varying scales, and it aids in keeping all datasets within the same range (0–1) (Negnevitsky 2011). This method was extremely beneficial to the multilayer perceptron (MLP) technique. In the present study, the datasets were normalized as per the following equation:

$$x_\mu = (x - x_{\min})/(x_{\max} - x_{\min}) \tag{27.1}$$

where x_μ is normalized score, $x_{\max}$ is maximum value, $x_{\min}$ is minimum value, and x is value of each entry of dataset.

27.2.5 Structure of ANN

In the present study, SPSS 20.0 is used to find annoyance using a feed-forward architecture. For annoyance prediction, an MLP network with a sigmoid activation function is used. In general, the activation function transfers the generated values between 0 and 1 or −1 and +1, and so on (Hew and Kadir 2017). The sigmoid activation function, often known as the logistic function, is a popular ANN activation function that exists between 0 and 1 (Tian et al. 2015). A feed-forward neural network is a classification algorithm that is biologically inspired. In general, an ANN contains a number of basic neurons that act as processing units and are grouped in layers. Each unit is linked to the previous layer's factors. Each connection bears varying weights or strengths. The network's knowledge is encoded by the weights of the connections. The units of a neural network are also referred to as nodes.

In general, data enters the input layers, passes through hidden layers made up of multiple neurons, and finally reaches the output layer (Tian et al. 2015). When it is used as a classifier, there is no feedback between the layers during normal operation. Hence, it is known as feed-forward neural network. ANN models are created by using data to depict the association between variables of input (x) and output (y). In order to use the data for adjustment of weights to reduce error, a data-driven model is developed using the neuron shown in Eq. (27.2)

$$y = f(x) + \text{error} \tag{27.2}$$

where y is desired output variable and x is input variables.

When linear activation function is used in ANN, then neuron is considered as linear model with weights corresponding to slopes shown in Eq. (27.3)

$$f(x_1, x_{2\ldots}\ x_m) = a + w_1 x_1 + w_2 x_2 + \ldots w_m x_m \tag{27.3}$$

where x_1, x_2, ..., x_m are the nodes in input layer, a is standard error, w_1, w_2, ..., w_m are weighted coefficients to the corresponding nodes.

The eight input nodes (pH, DO, BOD, Turbidity, WT, SD, TP, and TN) were utilized to predict annoyance level in the output layer. The datasets with one constant hidden layer and a fixed ratio for testing (70%), training (20%), and hold out (10%) were run 500 times for each of the neurons (1–5) as illustrated in Fig. 27.3, following Kuo et al. (2007). The MSE and RMSE were calculated in each run to predict the best model out of 500 runs in each neuron. The effective model for irritation in each neuron was chosen and established on the lowest values of MSE and RMSE. Accordingly, the best architectural model was selected from among the top five models chosen from various neurons (picked from 500 runs in each neuron that is 1 neuron to 7 neurons). The weights and important bar graph were also used to assess the importance of the input factors.

The MSE and RMSE were calculated using Eqs. (27.4) and (27.5), respectively (Neill and Hashemi 2018)

$$\text{MSE} = \frac{1}{N} \sum_{i=1}^{n} (y_i - \widehat{y}_i)^2 \tag{27.4}$$

where N is number of elements of data from the whole test, y_i is real value e, and $\widehat{y}_i$ is predicted value.

$$\text{RMSE} = \sqrt{\text{MSE}} \tag{27.5}$$

27.2.6 Training and Testing Parameters

An input random number seed was used to determine the initial values of the weights, which were then assigned. The starting values of the weights were established by using a random starting method, and they were arbitrarily set anywhere between −0.1 and 0.1. This range was used to test all the developed models. All of the neural network models that were used for this research consisted of three layers, with nodes in adjacent levels being fully connected to one another. This meant that there was only one hidden layer, and it contained 20–60 hidden nodes. The learning rate ought to be selectable in order to expedite the process of training and arriving at a converged value for the weights. It was essential to establish a rate

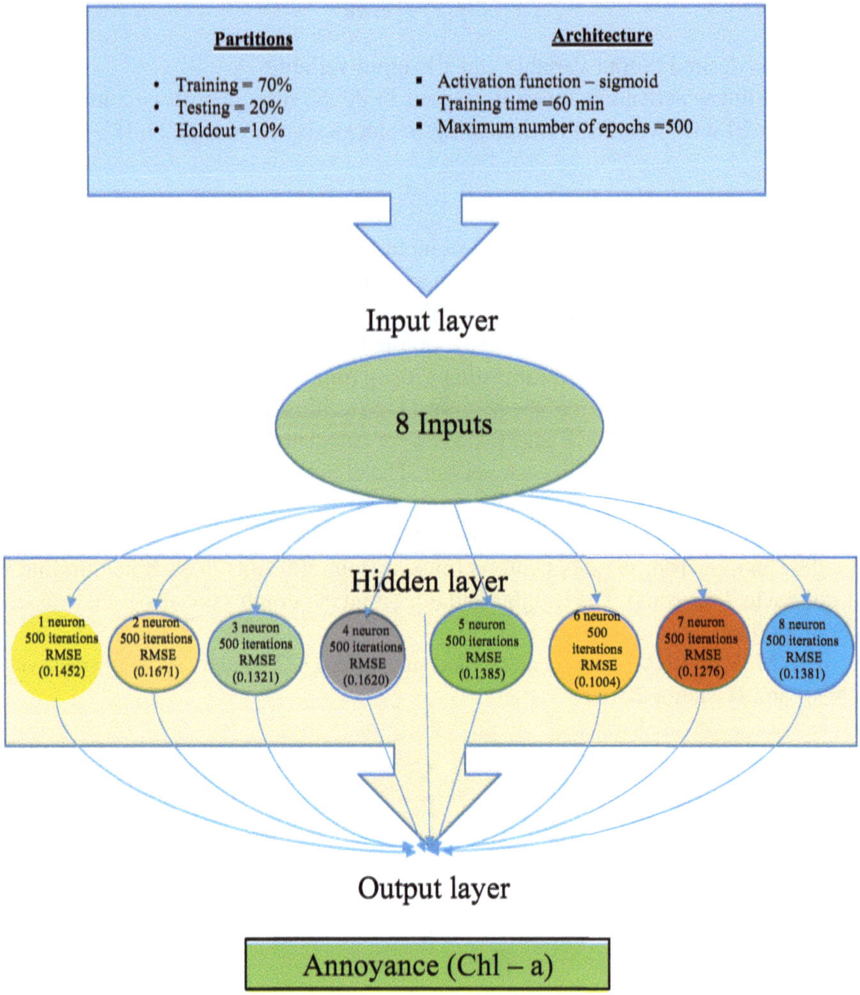

Fig. 27.3 Schematic representation of ANN model for prediction of Chl-a from relevant water quality parameters of Chilika lagoon

of learning that was sufficiently low to converge and at the same time also high to keep the amount of time needed for computing to a manageable level. Therefore, a learning rate of 0.05 was decided upon and the value of the momentum was decided to be 0.9.

27.2.7 Assessment of the ANN Model Performance

The accuracy of the model was determined by calculating four different metrics: the mean squared error (MSE), the root mean square error (RMSE), the sum of square errors (SSE), and the percentage error. The RMSE was used as a metric of goodness-of-fit to rate the effectiveness of the ANN models. Since RMSE efficiently characterizes an average measure of the inaccuracy in predicting changes in the eutrophication indicators, it was chosen as the parameter. Table 27.1 displays the models' RMSE and correlation coefficient, respectively. To determine the impact of the input variables and their contribution to the network output, more analysis would be required (Lee et al. 2003a, b).

27.2.8 Random Forest (RF)

The potential of the RF model to quantify the relative contribution of the numerous predictor variables to the outcome is an intriguing feature that can be helpful in selecting the most significant ones. The out-of-bag (OOB) methodology via permutation is a technique that assesses how effectiveness of the predictor variables toward response variable prediction, and it was used to rank the most significant variables (predictor). With increasing value of this measure, the impact of the predictor variable grows (Huang et al. 2015). The model error was expected to change depending on the permutation of predictor variable's values due to its influences on the prediction. The process entailed computing the gain in the MSE by OOB data permutation. The prediction error on the OOB portion of the data was recorded for each tree. The disparities between the two OOB errors were then normalized by the difference's standard deviation and averaged across all trees (Mitchell 2011).

Two simulations of the RF model were performed in the present study. Initially, all the seven physicochemical indicators for Chl-a concentrations prediction were evaluated. Second, whether the Chl-a predictors had any discernible geographical or seasonal dependence was examined. To put it another way, investigation was also made to know whether adding sample sites and seasons to the mix of observable factors would help with predictions of Chl-a concentrations. Since RF models can handle both quantitative and qualitative predictor variables, this was skilled by inserting two additional categorical predictor variables, one for each of the 33 stations and the three seasons as well.

27.3 Results and Discussion

27.3.1 Variability of Eutrophication Indicator Parameters

The average values of the eutrophication indicators such as pH, DO, BOD, Turbidity, WT, SD, TP, and TN recorded in the Chilika lagoon found to be significantly lower than the threshold values specified for good health of surface waters (Smith

Table 27.1 Descriptive statistics of relevant eutrophication indicator parameters used for Chl-a prediction for Chilika lagoon

	N	Minimum	Maximum	Mean	Std. deviation	Variance	Threshold	Reference
TN (μM)	429	0.29	135.90	39.27	29.42	865.82	46–86	Smith et al. (1999)
TP (μM)	429	0.17	48.51	1.33	2.74	7.53	0.97–3	Smith et al. (1999)
DO (ppm)	462	4.11	16.00	8.42	1.75	3.07	4	CPCB (1986)
BOD (ppm)	452	0.06	12.28	2.85	1.78	3.17	3	CPCB (1986)
pH	461	6.36	10.18	8.26	0.53	0.28	6.5–8.5	CPCB (1986)
Turb (NTU)	462	0.56	574.00	43.49	79.09	6256.19	30	CPCB (1986)
SD (m)	462	0.04	2.08	0.61	0.42	0.18	<2	Smith et al. (1999)
WT (°C)	462	11.84	35.50	27.89	3.68	13.55	–	–
Chl-a (μg L^{-1})	429	0.01	25.71	0.67	1.62	2.62	9	Smith et al. (1999)

et al. 1999; CPCB, 1986) (Table 27.1). These indicators found to be significantly varied with season and sector ($p < 0.05$) (Fig. 27.4a, b). As reported by several studies, these water quality parameters can be considered as indicators of health of Chilika lagoon (Barik et al. 2017; Muduli and Pattnaik 2020; Muduli et al. 2022). For instance, according to Smith et al. (1999), low SD, high TN, and TP reduce the light penetration in water, which in effect reduces the amount of photosynthesis and supports dominant respiration, which may decline DO levels. A lower DO may also be the indication of pollutants (total organic nitrogen and total organic phosphorous) in the ecosystem (Stefanidis and Papastergiadou 2019). As recommended by CPCB (1986), a minimum DO level of 4 mg L^{-1} required for fisheries and propagation of wildlife. Smith et al. (1999) recommended that a lake can be considered as oligotrophic, mesotrophic, eutrophic, and hyper-eutrophic if the Chl-a level of <3.5, 3.5–9, 9–25, and >25 maintained in the lake ecosystem. Accordingly, the present study results indicated that the Chilika lagoon remained in good health (oligotrophic condition) during the study period. However, if Chl-a data is not available, it will be inappropriate to predict the trophic state of the aquatic ecosystems. The ANN model in this case would be much helpful as the model is trained by earlier available data and predicts for the required period (Srisuksomwong and Pekkoh 2020).

27.3.2 Trophic Level Index

To find out whether the Chilika lagoon maintains a healthy trophic status, TLI was calculated using the filed observed data of TN, TP, SD, and Chl-a. The average TLI (<3) results of the entire lagoon for the complete study period showed that the lagoon maintained oligotrophic condition. However, the lagoon found to have trophic status switch over, i.e., from mesotrophic to eutrophic condition in some regions of the northern sector, particularly in the summer season. The SD was the main controlling factor to change the trophic status. The SD decline was due to the mixing of water with high turbidity due to sediment churning by wind action in low-depth regions (Barik et al. 2017; Jally et al. 2020; Muduli and Pattnaik 2020). Although all sectors recorded to be within oligotrophic range, the highest was found in the NS (2.4) followed by OC > CS > SS (1.4).

27.3.3 Correlation of Chl-a with Water Quality Parameters

Chl-a maintained a significant positive correlation with DO, BOD, Turbidity, and a negative correlation with SD (Table 27.2). Similar result was also recorded by Gebler et al. (2017) and Singh et al. (2009) for the Polish lowlands, Poland and Gomti River, Uttarakhand, respectively. No significant associations were identified with TN, TP, and WT. According to Huo et al. (2013), SD is a major component for lake eutrophication. This finding was mirrored in this investigation as Chl-a, which is considered as a proxy for eutrophication in aquatic ecosystems, (Smith et al. 1999) was negatively correlated with SD. This was also further supported by a positive

(a)

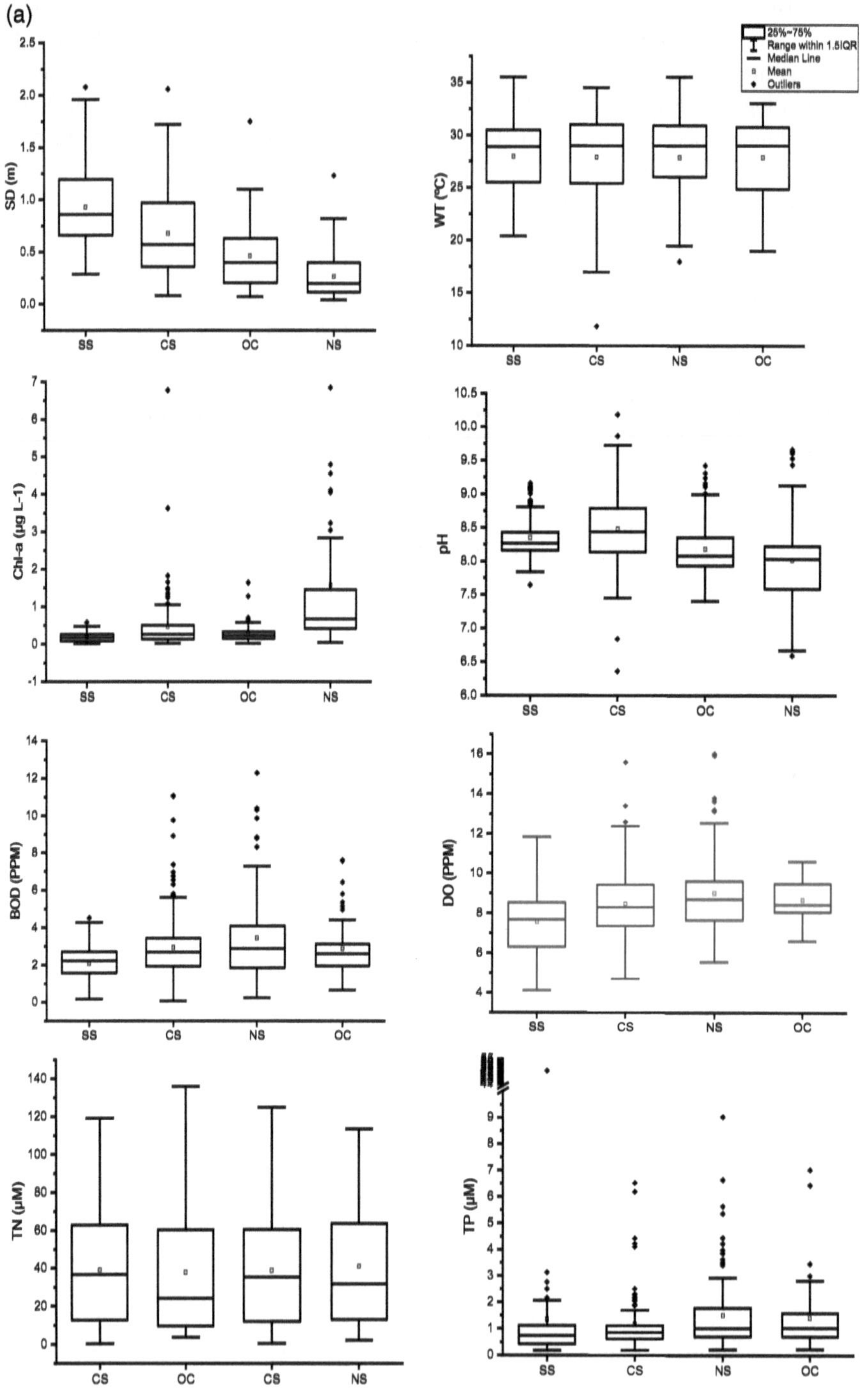

Fig. 27.4 (**a**) Sectoral variation of eutrophication indicator parameters of Chilika lagoon. (**b**) Seasonal variation of eutrophication indicator parameters of Chilika lagoon

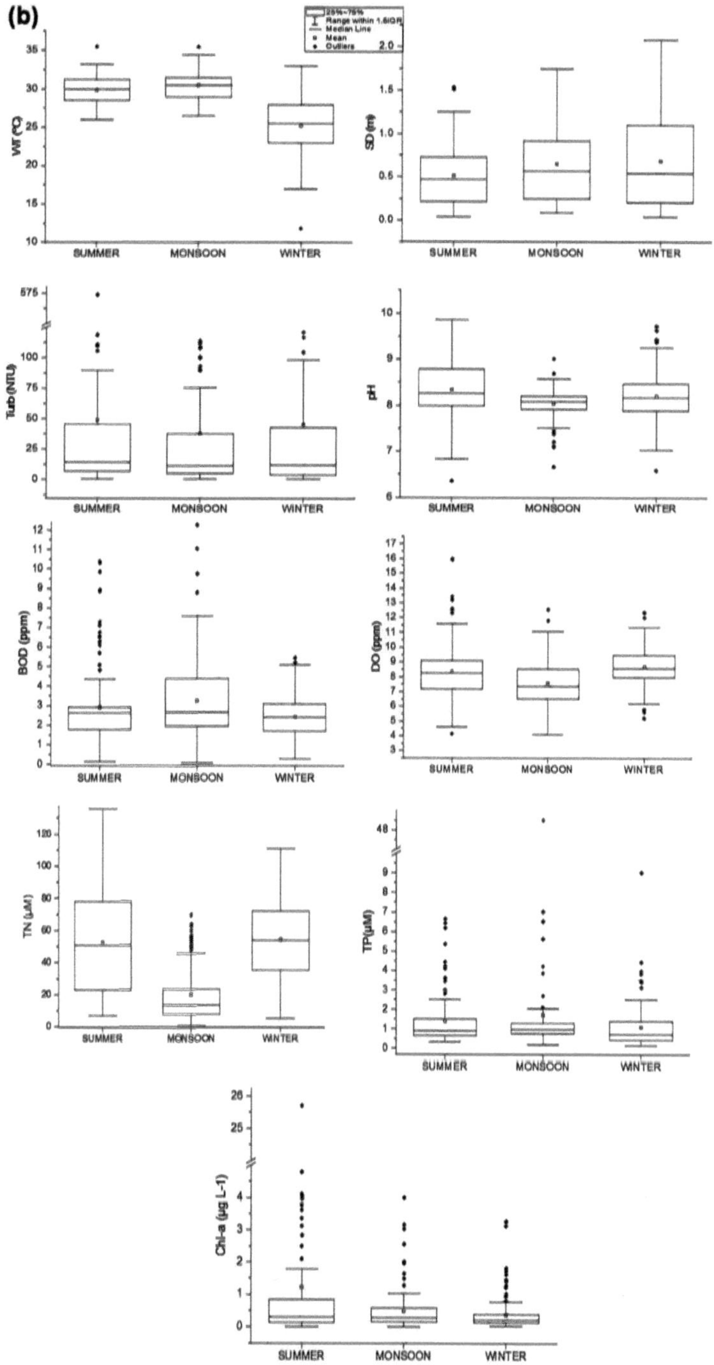

Fig. 27.4 (continued)

Table 27.2 Correlation of Chl-a with relevant environmental variables of Chilika lagoon

		TN	TP	DO	BOD	pH	Turb.	SD	WT
Chl-a	Pearson correlation	0.065	0.051	0.137**	0.285**	0.044	0.218**	−0.268**	0.012
	Sig. (two-tailed)	0.199	0.316	0.005	0.000	0.361	0.000	0.000	0.806
	N	396	396	429	419	428	429	429	429

*Correlation is significant at the 0.05 level (two-tailed)
**Correlation is significant at the 0.01 level (two-tailed)

correlation with turbidity. This indicated that the contribution of Chl-a is significant for the turbidity of the water column of the Chilika lagoon unlike other studies where total suspended solid (TSS) contributed the most (Patra et al. 2016). Correlation with BOD could be due to the fact that higher Chl-a was recorded in the NS, where comparatively higher organic load (BOD) sourced from riverine discharge has been recorded, and lower Chl-a was observed in the SS, with least observed BOD (Fig. 27.4a; Muduli and Pattnaik 2020). Significant correlation with DO indicates that the photosynthesis process was the dominating phenomena in Chilika lagoon, which controlled the DO level rather than physical phenomena such as wind action and surfing (Muduli et al. 2022).

27.3.4 Prediction Model for Chl-a Using ANN

Several studies predicted the major indicators for eutrophication, which are listed in Table 27.3. Most of the studies used long-term data, which could be taken in the present study due to certain data scarcity. However, an attempt was made using the available data of 1 year similar to other studies by Bhagowati et al. (2022) and Strobl et al. (2007). The selected effective models from 500 runs for various neurons with a constant hidden layer, input and output layer, and their MSE and RMSE values in this work has been listed in Table 27.4. The objective of 500 rounds of learning for multiple neurons was to minimize error and improve model accuracy (EL Idrissi et al. 2019). The MSE and RMSE values for the architecture "8 inputs → 1 hidden layer (1 neuron) → 1 output," "8 inputs → 1 hidden layer (2 neurons) → 1 output," "8 inputs → 1 hidden layer (3 neurons) → 1 output," "8 inputs → 1 hidden layer (4 neurons) → 1 output," "8 inputs → 1 hidden layer (5 neurons) → 1 output," "8 inputs → 1 hidden layer (6 neurons) → 1 output," "8 inputs → 1 hidden layer (7 neurons) → 1 output," and "8 inputs → 1 hidden layer (8 neurons) → 1 output" were 0.01817 and 0.134801, 0.017976 and 0.134073, 0.017465 and 0.132156, 0.019813 and 0.140757, 0.014658 and 0.12107, 0.01279 and 0.11054, 0.01543 and 0.12053, and 0.014720 and 0.13491, respectively. From Table 27.4, the best prediction model to measure annoyance was found for the architecture "8 inputs → 1 hidden layer (6 neurons) → 1 output". The MSE (0.014658) and RMSE (0.11054) values were the lowest of all eight best models of distinct neurons, and the network structure explaining the dynamics of Chl-a annoyance (Fig. 27.5). In order to eliminate the chances of overfitting the model, RMSE values were taken into account (Ooi and Tan 2016). Figures 27.5 and 27.6 depict a comparison of the model's actual and predicted values and the residuals plot, respectively. A similar approach was also made by Lee et al. (2003a) to establish a model for Chl-a prediction by incorporating the water quality parameters (TN, TP, BOD, WT) as input layer by observing the minimum MSE values of Chl-a.

The correlation coefficient for the model prediction generated during training and testing for each neural network, as well as the RMSE for the Chilika lagoon are shown in Table 27.4. It was clear that scenario 6 (Turbidity) indicated a score that was comparable to both throughout training and testing. Additionally, the ANNs

Table 27.3 Prediction of eutrophication through ANN in aquatic ecosystems around the world

Sl. no.	Place of study	Input parameter for ANN	Computed parameter by ANN	Period of data used	Reference
1	Lake Fuxian, Southwest China	TP, DO, SD, Chl-a	DO, TN, Chl-a, SD	5 year	Huo et al. (2013)
2	Keban Dam reservoir, Turkey	Chl-a	Chl-a	6 year	Karul et al. (2000)
3	Assam, India	Water quality, Chl-a	Chl-a	1 year	Bhagowati et al. (2022)
4	Amirkabir Reservoir, Iran	Water quality, Chl-a	Chl-a	12 year	Aria et al. (2019)
5	Imha Dam, South Korea	Chl-a	Chl-a	17 year	Mamun et al. (2020)
6	Maekuang River, Thailand	NH_3, NO_3, and PO_4; Secchi depth, BOD, WT, pH	Chl-a and *M. aeruginosa* cells	2 year	Srisuksomwong and Pekkoh (2020)
7	Florida and the Southern Blue Ridge, USA	Water quality, Chl-a	Chl-a	1 year	Strobl et al. (2007)
8	Menor lagoon, Spain	Water quality, Chl-a	Chl-a	4 year	Jimeno-Sáez et al. (2020)
9	Te-chi Dam, Taiwan	Water quality, Chl-a	DO, TP, Chl-a, SD	15 year	Kuo et al. (2007)
10	Tolo, Hong Kong	Water quality, Chl-a	Chl-a	3 year	Lee et al. (2003a)
11	Gomti River, India	pH, TA, total hardness, TSS, COD, NH_3, NO_3, chloride, PO_4	DO, BOD	10 year	Singh et al. (2009)
12	Chilika lagoon, India	TP, TN, Chl-a, SD	Chl-a	1 year	Present study

presented in Table 27.4 that produced the scenario turbidity were appropriate for the use in the prediction of Chl-a. Analysis of the primary parameters that influence the reproduction of algae can be accomplished through the process of photosynthesis. These factors include the TN and TP content of the water body, as well as WT and a variety of other chemical and physical elements. It was found that the amounts of TN and TP, in addition to the WT, had a positive correlation with the eutrophication (Chl-a) (Wang et al. 2017; Bui et al. 2017; Luo et al. 2014). As a consequence, these environmental parameters can be utilized to make accurate forecasts regarding the development of eutrophication and Chl-a of lagoon (Mowe et al. 2007).

Table 27.4 Tested ANNs for prediction of Chl-a in Chilika lagoon, Odisha

	Training							Testing		
HN	IL	HL	*N*	SSE	MSE	RMSE	*N*	SSE	SME	RMSE
TN	8	1	341	0.636	0.018171	0.134801	44	0.853	0.050176	0.224001
TP	8	1	341	0.737	0.017976	0.134073	44	0.504	0.038769	0.196899
DO	8	1	338	0.751	0.017465	0.132156	47	0.292	0.024333	0.155991
BOD	8	1	351	0.951	0.019813	0.140757	34	0.27	0.03375	0.183712
pH	8	1	344	0.557	0.014658	0.12107	41	0.456	0.032571	0.180476
Turb	8	1	321	0.521	0.01279	0.19054	44	0.621	0.03021	0.17537
SD	8	1	320	0.541	0.01543	0.12053	46	0.43	0.04214	0.14521
WT	8	1	340	0.761	0.1472	0.13491	41	0.328	0.042301	0.17423

IL input layers, *HL* hidden layers, *HN* hidden network, *N* numbers of datasets, *SSE* sum of squares error, *MSE* mean square error, *RMSE* root mean square error

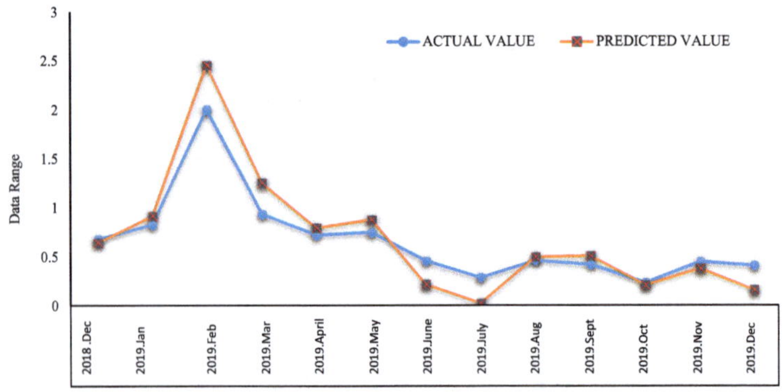

Fig. 27.5 Actual vs predicted values of Chl-a concentration in Chilika lagoon

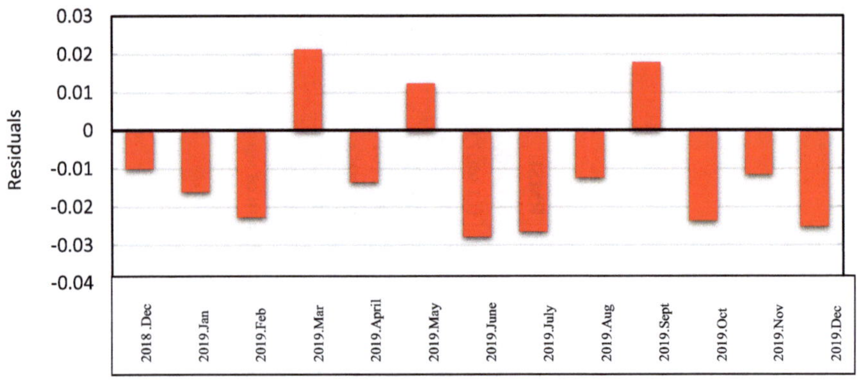

Fig. 27.6 Plot of the residuals model computed values of Chl-a in lagoon waters

Table 27.5 Sensitivity analysis showing the importance of eight environmental variables used for prediction of Chl-a annoyance (relative error: 0.198)

Variables	Importance	Normalized importance
BOD	0.385	100
pH	0.276	71.7
TP	0.116	30.2
DO	0.084	21.9
SD	0.06	15.5
Turbidity	0.056	14.6
WT	0.013	3.4
TN	0.01	2.6

27.3.5 Importance of Variables

The importance of the factors such as BOD, pH, TP, DO, SD, Turbidity, WT, and TN is demonstrated in Table 27.5. The model's good performance has been demonstrated in Fig. 27.5 by observing the relative error of 0.198. Lee et al.

(2003a, b) explained the model's performance for sensitivity and annoyance by taking the relative error into account. For the prediction of Chl-a annoyance, BOD of exposure was found to be most important, whereas TN was of least importance.

There was no distinct pattern produced by the ANN for other parameters, i.e., SD and WT. According to Hadjisolomou et al. (2018), modeling limnological parameters is a very specific instance activity, and the exact processes regulating limnological parameters are complicated and typically their relationships are not connected or studied easily. Since ANNs do not involve any hypotheses regarding the model or the distribution of the data, their data-driven nature makes it difficult to understand the linkages and interactions between the associated parameters (Teles et al. 2006). Second, despite the fact that developed ANN is an effective predictor of the Chl-a parameter, there are a variety of additional variables that could be crucial in understanding the patterns of Chl-a that have been observed. According to Napiórkowska-Krzebietke et al. (2020), cyanobacterial bloom dynamics and accumulation in lakes are influenced by a variety of variables, including wind speed, even though sediment resuspension by wind action is a typical phenomenon in shallow lakes. The sensitivity analysis findings in relation to the WT were consistent with a relevant modeling study of Chilika lagoon (Hadjisolomou et al. 2018) that looked at the interactions between environmental factors with the use of an unsupervised ANN. It was determined that the Chilika lagoon statistics were mostly related to the WT. Since the Chilika lagoon is a shallow lake, the seasonality effect has an impact on how well the temperature affects its operation.

A relatively rise in Chl-a that was found for a reverse change in the water temperature and relates to months with low temperature was another intriguing finding from the results of the sensitivity analysis. This increase might be due to additional meteorological conditions or factors that exist during the summer and monsoon. As an illustration, wind action is very heavy in these seasons and may cause the release of phosphate and nitrogen from the sediments, which used to support eutrophication (Liu et al. 2019). Likewise, based on the findings of the sensitivity analysis, it was determined that the SD parameter's role was difficult to comprehend because it was linked to complicated processes like action of wind and influx of nutrients from catchment during monsoon (Borowiak et al. 2020). Nevertheless, the ANN was able to link higher SD levels to higher algal production, which might indicate that higher nutrient levels had an impact on the productivity of the algae.

The mechanism of meteorological factors and eutrophication indicators are critical and hence the relationships are frequently nonlinear (Paerl 2006). The changes in the concentration of Chl-a levels in relation to the accompanying nutritional variations that were seen during the sensitivity analysis can be rationalized on the basis of this claim. The sensitivity analysis of ANN showed that the TP had a bigger effect on algal output than the TN. When TP concentration raised, the difference between TP's impact and that of TN became more apparent. The ANN simulation scenarios for the TN clearly demonstrated that the lagoon was encouraging algal production and that a decrease in TN level was associated with a decrease in Chl-a level. As a result, it was advised that nitrogen as well as phosphorus should be of primary emphasis for the lake's nutrient management. Additionally, it was found that reducing both nutrients at the same time caused Chl-a levels to drop more dramatically and was connected to the additive effects of DIN and TP parameters (Hadjisolomou et al. 2017). For instance, according to a recent study, Prespa lagoon is particularly dynamic in terms of increasing nutrient concentration, and even minor contributions of nutrients from water fowl are linked to cyanobacterial blooms (Verstijnen et al. 2021). This relationship between nutrients (TN and TP) and how the Chilika lagoon relates to eutrophication was clearly recorded by the ANN. The results of the ANN showed that a decrease in TP was connected to a decrease in Chl-a levels, and vice versa. The strong links between eutrophication and high influx of phosphorous recorded in freshwater ecosystems give credence to this modeling scenario that was related to TP disturbances (Heisler et al. 2008). The second scenario, which dealt with TN disturbances, likewise demonstrated a drop in Chl-a levels with increased TN. Because the TN and TP parameters behaved similarly, every rise in nitrogen levels was linked to an increase in algal productivity (Akagha et al. 2020).

27.3.6 Random Forest

Chl-a is a very crucial indicator of the level of eutrophication in water bodies (Lu et al. 2016). The Chl-a concentrations in the Chilika lagoon were ranged from 0.1 to 25.71 μg L^{-1}. The Chl-a concentrations in coastal ecosystems can have intricate interactions with TN and TP and water quality parameters (temperature, pH, salinity, dissolved oxygen, and Turbidity) (Jimeno-Sáez et al. 2020). RF is a useful method from ML algorithms to deal with complex relation among these variables. The Chilika lagoon data samples of the seven predictor variables (temperature, pH, salinity, dissolved oxygen, Turbidity, TN, and TP), and one target variable (Chl-a), served as the basis for training the RF model. The approach obtained an R^2 score of around 0.62 and an MSE of 0.28. Figure 27.7 showed the OOB technique using permutation's ranking of response factor in accordance with its validity. Only a few characteristics, such as Turbidity, Secchi depth, dissolved oxygen, and pH, made a discernible difference in the estimation of the Chl-a content. Chl-a pigment is vital for aerobic photosynthetic process. According to Frolov et al. (2012), the turbidity affects the intensity of sunlight in the water column, which affects how efficiently most algae can photosynthesize. This explains the high association between turbidity and Chl-a.

Oxygen is produced by algae throughout the day, and they consume during the night period. According to Bějaoui et al. (2016), oxygen is also acquired during algae decomposition. In agreement with that, the present study also demonstrated a considerable relation between dissolved oxygen levels and Chl-a concentrations. Additionally, numerous research has shown the close relationship between pH and Chl-a (Menendez et al. 2001). The additional predictor variables in the RF model were TN, TP, salinity, and temperature in descending order of significance.

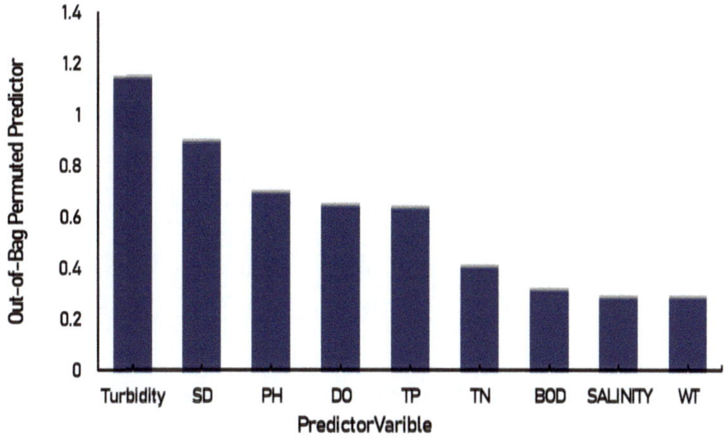

Fig. 27.7 Importance ranking the predictors for the "first" RF model for the prediction of Chl-a in the Chilika lagoon

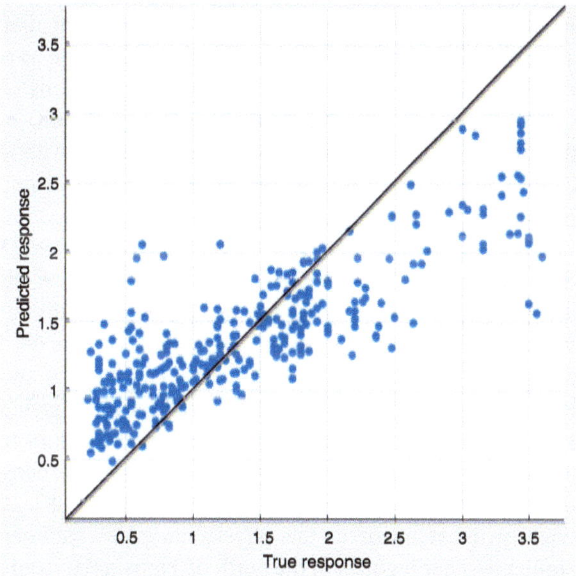

Fig. 27.8 Random Forest prediction of Chl-a concentrations in the Chilika lagoon considering physicochemical predictor factors. The difference between predicted and true responses is represented by the expected and observed values of Chl-a

A scatter plot that directly compares the forecasted and observed Chl-a levels has been illustrated in Fig. 27.8. The fitted RF model performed substantially better than the one Béjaoui et al. (2016) reported for the Bizerte lagoon ($R^2 = 0.51$), and similarly to the one reported for the Ghar el Melh lagoon ($R^2 = 0.64$) (Béjaoui et al. 2018). As a basis, the observed Chl-a concentrations for the Chilika lagoon were predicted with more accuracy. It is well known that the number of observed data is crucial to the predictive model's accuracy (Béjaoui et al. 2016). In the Chilika

Table 27.6 The regression coefficients between Chl-a levels and physicochemical factors in the Chilika lagoon were estimated using a statistical analysis (coefficients shown with (*) are significantly at *p*-value <0.05)

	Estimate	Std. error	*t*-value	*p*-value
Intercept	3.028	2.154	1.754	0.00757
TN	−0.947	0.19	−0.304	0.00932
TP	0.247	0.065	−0.612	0.00687
DO	0.849	0.662	−3.18	0.00905
BOD	−0.947	0.744	0.174	0.00401*
pH	0.372	−0.233	0.079	0.00421
Turb	0.329	0.72	−1.799	0.00358*
SD	0.34	0.349	0.33	0.00603
WT	−0.771	0.491	−1.794	0.00816

lagoon, we employed monthly observations that lasted for around 1 year. In addition to the RF model, a multivariate linear regression (MVLR) model was fitted for evaluation. The predictor variables and Chl-a concentrations linear model parameters (Estimate) were almost in agreement with the associations seen using the RF model described in Table.27.6. The two most significant predictors of Chl-a concentration levels were Turbidity and BOD (Table 27.6). The results of the RF model were quantitatively validated by the linear model. The R^2 for MVLR was approximately 0.21. It was clear that the RF model, as opposed to the MVLR, better represented the dependence of Chl-a concentrations on other variables. Using the OOB technique by permutation, the quality of results was ensured. Thus, given that the RF model has significant advantages over conventional correlative analyses, we can assert that it might be used to better comprehend more complicated interdependencies between variables.

This observation can support the assertion made by Maier et al. (2010) that developing ANN models does not require a linear method when the input variables are linked to the output of the model. Prior to doing any modeling analyses, all variables were gone through treatments to standardize their distribution. The MVLR's performance was not enhanced by the modifications, though. It was expected that there was a strictly nonlinear relationship among all of the variables. Due to anthropogenic contribution and the effects of hydroclimatic factors such as evaporation, temperature, precipitation, and so on, ecosystems are ruled by a multitude of complicated phenomena (Viaroli et al. 2008). Chl-a levels were shown to be linearly independent to the physicochemical parameters in recent investigations similar to other lagoons in the north of Tunisia (Béjaoui et al. 2016). ML approaches, which are well-known for their ability to process the complex nonlinear time series processes, were used to execute all of our modeling directly on the original data. It is useful for forecasting to fit a model directly without transformations.

To investigate the relationships between physical, chemical, and biological factors in coastal ecosystems, RF is a useful forecasting tool. According to Béjaoui et al. (2016), TN and DO are the main sources of Chl-a in the Bizerte lagoon. These two variables with the strongest correlations to the plankton dynamics in the Ghar Melh lagoon, according to Béjaoui et al. (2018), are temperature and turbidity. Despite the fact that the influence of the Chl-a predictor variables varied with respect to different studies, dissolved oxygen and Turbidity were typically among the key factors. Using a genetic algorithm-optimized back-propagation neural network, Li et al. (2017) determined that the temperature, turbidity, dissolved oxygen, total phosphorus and nitrogen concentrations, and total nitrogen concentration were the most important input variables for Chl-a. It is noteworthy to mention that the Chilika lagoon's unique ecosystem characteristics, such as the size of its water masses, varying eutrophic states, water depth, and its connectivity to the sea, can be used to explain the differences in RF results between the preceding ecosystems. Additionally, diverse modeling methodologies, as well as distinct field studies and laboratory analytical procedures, could have contributed to these disparities.

27.4 Conclusion

In the present study, a relatively small number of observations ($n = 384$) was used to develop an ANN and model the Chilika lagoon trophic status considering RMSE and MSE validation. The results evidenced that the created ANN can be used as an efficient predictor of Chl-a and can serve as a management tool for similar aquatic ecosystems. The performance of the model was verified on the basis of relative error of 0.198 and indicated that BOD and Turbidity played a crucial role for prediction of annoyance, whereas TN and WT played least important factor. The integrated approach of ANN model and TLI could be considered as an effective tool to understand the trophic level of any aquatic ecosystems. The present study performed with limited available dataset of TN and TP for 1 year. However, a long-term dataset would provide a better prediction and hence it is suggested for monitoring of these vital water quality parameters on a monthly basis on a long run and recalibrate the model on availability of more data. This would enable better machine learning/ training and better prediction of trophic status or aquatic health status parameters. The results from such studies could be a vital scientific input for formulation of management action plan and conservation of similar aquatic ecosystems around the globe. Hence, we propose the ANN as an effective tool for the computation of indicator parameters of Chilika lagoon trophic status and also it can be used in similar ecosystems to improve the understanding of trend of pollution.

Acknowledgments The authors are grateful to the CE, Chilika Development Authority (CDA), for providing required facilities at WRTC. This work was carried out with the funding support from the World Bank (ICZMP, Odisha) (CREDIT No. 4765-IN) and NPCA (National Plan for Conservation of Aquatic Ecosystems). Thanks are also due to Mr. Bibhuti B. Dora for preparing the Chilika GIS map. The supporting staff and researchers of CDA are also acknowledged for their generous help during the study.

Conflict of Interest There is no conflict of interest.

References

Acharya P, Muduli PR, Mishra DR, Kumar A, Kanuri VV, Das M (2022) Imprints of COVID-19 lockdowns on total petroleum hydrocarbon levels in Asia's largest brackish water lagoon. Mar Pollut Bull 174:113137. https://doi.org/10.1016/j.marpolbul.2021.113137

Adnan RM, Zounemat-Kermani M, Kuriqi A, Kisi O (2021) Machine learning method in prediction streamflow considering periodicity component. In: Deo R, Samui P, Kisi O, Yaseen Z (eds) Intelligent data analytics for decision-support systems in hazard mitigation. Springer, Singapore. https://doi.org/10.1007/978-981-15-5772-9_18

Agwu OE, Akpabio JU, Dosunmu A (2020) Artificial neural network model for predicting drill cuttings settling velocity. Petroleum 6(4):340–352. https://doi.org/10.1016/j.petlm.2019.12.003

Akagha SC, Nwankwo DI, Yin K (2020) Dynamics of nutrient and phytoplankton in Epe Lagoon, Nigeria: possible causes and consequences of reoccurring cyanobacterial blooms. Appl Water Sci 10(5):1–16. https://doi.org/10.1007/s13201-020-01190-7

Aria S, Asadollahfardi G, Heidarzadeh N (2019) Eutrophication modelling of Amirkabir Reservoir (Iran) using an artificial neural network approach. Lakes Reserv Res Manag 24:48–58. https://doi.org/10.1111/lre.12254

Barik SK, Muduli PR, Mohanty B, Behera AT, Mallick S, Das A, Samal RN, Rastogi G, Pattnaik AK (2017) Spatio-temporal variability and the impact of Phailin on water quality of Chilika lagoon. Cont Shelf Res 136:39–56. https://doi.org/10.1016/j.csr.2017.01.019

Béjaoui B, Armi Z, Ottaviani E et al (2016) Random Forest model and TRIX used in combination to assess and diagnose the trophic status of Bizerte Lagoon, southern Mediterranean. Ecol Indic 71: 293–301. https://doi.org/10.1016/j.ecolind.2016.07.010

Béjaoui B, Ottaviani E, Barelli E et al (2018) Machine learning predictions of trophic status indicators and plankton dynamic in coastal lagoons. Ecol Indic 95:765–774. https://doi.org/10.1016/j.ecolind.2018.08.041

Bennett C, Stewart RA, Beal CD (2013) ANN-based residential water end-use demand forecasting model. Expert Syst Appl 40(4):1014–1023. https://doi.org/10.1016/j.eswa.2012.08.012

Bhagowati B, Talukdar B, Narzary B, Ahamad K (2022) Prediction of lake eutrophication using ANN and ANFIS by artificial simulation of lake ecosystem. Model Earth Syst Environ 8:5289–5304. https://doi.org/10.1007/s40808-022-01377-8

Borowiak M, Borowiak D, Nowinski K (2020) Spatial differentiation and multiannual dynamics of water conductivity in lakes of the Suwalki landscape park. Water (Switzerland) 12(5):1277. https://doi.org/10.3390/W12051277

Brown MGL, Skakun S, He T, Liang S (2020) Intercomparison of machine-learning methods for estimating surface shortwave and photosynthetically active radiation. Remote Sens 12(3):1–13. https://doi.org/10.3390/rs12030372

Bui MH, Pham TL, Dao TS (2017) Prediction of cyanobacterial blooms in the Dau Tieng Reservoir using an artificial neural network. Mar Freshw Res 68(11):2070–2080. https://doi.org/10.1071/MF16327

Cigizoglu HK, Kişi Ö (2005) Flow prediction by three back propagation techniques using k-fold partitioning of neural network training data. Nord Hydrol 36(1):49–64. https://doi.org/10.2166/nh.2005.0005

CPCB (1986) Central Pollution Control Board. Primary water quality criteria for class SW-1 waters. In: The Environment (Protection) Rules 1986

Dynowski P, Senetra A, Źróbek-Sokolnik A, Kozłowski J (2019) The impact of recreational activities on aquatic vegetation in alpine lakes. Water (Switzerland) 11(1):173. https://doi.org/10.3390/w11010173

EL Idrissi T, Idri A, Bakkoury Z (2019) Systematic map and review of predictive techniques in diabetes self-management. Int J Inf Manag 46:263–277. https://doi.org/10.1016/j.ijinfomgt.2018.09.011

Frolov S, Ryan JP, Chavez FP (2012) Predicting euphotic-depth-integrated chlorophyll-a from discrete-depth and satellite-observable chlorophyll-a off Central California. J Geophys Res Ocean 117:1–7. https://doi.org/10.1029/2011JC007322

Gebler D, Szoszkiewicz K, Pietruczuk K (2017) Modeling of the river ecological status with macrophytes using artificial neural networks. Limnologica 65:46–54. https://doi.org/10.1016/j.limno.2017.07.004

Goethals PLM, Dedecker AP, Gabriels W, Lek S, De Pauw N (2007) Applications of artificial neural networks predicting macroinvertebrates in freshwaters. Aquat Ecol 41(3):491–508. https://doi.org/10.1007/s10452-007-9093-3

Grasshoff K, Kremling K, Ehrhardt M (eds) (1999) Methods of seawater analysis, 3rd edn. Wiley-VCH, Weinheim, 632 pages

Hadjisolomou E, Stefanidis K, Papatheodorou G, Papastergiadou E (2018) Assessment of the eutrophication-related environmental parameters in two Mediterranean lakes by integrating statistical techniques and self-organizing maps. Int J Environ Res Public Health 15(3):547. https://doi.org/10.3390/ijerph15030547

Hadjisolomou E, Stefanidis K, Papatheodorou G, Papastergiadou E (2017) Evaluating the contributing environmental parameters associated with eutrophication in a shallow lake by applying artificial neural networks techniques. Fresenius Environ Bull 26:3200–3208

Hagan R, Manktelow R, Taylor BJ, Mallett J (2014) Reducing loneliness amongst older people: a systematic search and narrative review. Aging Ment Health 18(6):683–693

Heisler J, Glibert PM, Burkholder JM, Anderson DM, Cochlan W, Dennison WC, Dortch Q, Gobler CJ, Heil CA, Humphries E, Lewitus A, Magnien R, Marshall HG, Sellner K, Stockwell DA, Stoecker DK, Suddleson M (2008) Eutrophication and harmful algal blooms: a scientific consensus. Harmful Algae 8(1):3–13. https://doi.org/10.1016/j.hal.2008.08.006

Hew TS, Kadir SLSA (2017) Applying channel expansion and self-determination theory in predicting use behaviour of cloud-based VLE. Behav Inform Technol 36(9):875–896. https://doi.org/10.1080/0144929X.2017.1307450

Huang J, Gao J, Zhang Y (2015) Combination of artificial neural network and clustering techniques for predicting phytoplankton biomass of Lake Poyang, China. Limnology 16:179–191. https://doi.org/10.1007/s10201-015-0454-7

Huo S, He Z, Su J, Xi B, Zhu C (2013) Using artificial neural network models for eutrophication prediction. Procedia Environ Sci 18:310–316. https://doi.org/10.1016/j.proenv.2013.04.040

Jally KS, Kumar Mishra A, Balabantaray S (2020) Estimation of trophic state index of Chilika Lake using Landsat-8 OLI and LISS-III satellite data. Geocarto Int 35:759–780. https://doi.org/10.1080/10106049.2018.1533593

Jimeno-Sáez P, Senent-Aparicio J, Cecilia JM, Pérez-Sánchez J (2020) Using machine-learning algorithms for eutrophication modeling: case study of Mar Menor lagoon (Spain). Int J Environ Res Public Health 17(4):1189. https://doi.org/10.3390/ijerph17041189

Karul C, Soyupak S, Çilesiz AF, Akbay N, Germen E (2000) Case studies on the use of neural networks in eutrophication modeling. Ecol Model 134(2–3):145–152. https://doi.org/10.1016/S0304-3800(00)00360-4

Kim DK, Park K, Jo H, Kwak IS (2019) Comparison of water sampling between environmental DNA metabarcoding and conventional microscopic identification: a case study in Gwangyang Bay, South Korea. Appl Sci (Switzerland) 9(16):3272. https://doi.org/10.3390/app9163272

Kuo JT, Hsieh MH, Lung WS, She N (2007) Using artificial neural network for reservoir eutrophication prediction. Ecol Model 200(1–2):171–177. https://doi.org/10.1016/j.ecolmodel.2006.06.018

Lee JHW, Huang Y, Dickman M, Jayawardena AW (2003a) Neural network modelling of coastal algal blooms. Ecol Model 159(2–3):179–201. https://doi.org/10.1016/S0304-3800(02)00281-8

Lee S, Ryu JH, Min K, Won JS (2003b) Landslide susceptibility analysis using GIS and artificial neural network. Earth Surf Process Landform 28(12):1361–1376. https://doi.org/10.1002/esp.593

Liu X, Zhang G, Sun G, Wu Y, Chen Y (2019) Assessment of Lake water quality and eutrophication risk in an agricultural irrigation area: a case study of the Chagan Lake in Northeast China. Water (Switzerland) 11(11). https://doi.org/10.3390/w11112380

Li X, Sha J, Wang ZL (2017) Chlorophyll-A prediction of lakes with different water quality patterns in China based on hybrid neural networks. Water (Switzerland) 9:524. https://doi.org/10.3390/w9070524

Luo W, Chen H, Lei A, Lu J, Hu Z (2014) Estimating cyanobacteria community dynamics and its relationship with environmental factors. Int J Environ Res Public Health 11(1):1141–1160. https://doi.org/10.3390/ijerph110101141

Lu F, Chen Z, Liu W, Shao H (2016) Modeling chlorophyll-a concentrations using an artificial neural network for precisely eco-restoring lake basin. Ecol Eng 95:422–429. https://doi.org/10.1016/j.ecoleng.2016.06.072

Maier HR, Morgan N, Chow CWK (2004) Use of artificial neural networks for predicting optimal alum doses and treated water quality parameters. Environ Model Softw 19(5):485–494. https://doi.org/10.1016/S1364-8152(03)00163-4

Maier HR, Jain A, Dandy GC, Sudheer KP (2010) Methods used for the development of neural networks for the prediction of water resource variables in river systems: current status and future directions. Environ Model Softw 25:891–909. https://doi.org/10.1016/j.envsoft.2010.02.003

Mamun M, Kim JJ, Alam MA, An KG (2020) Prediction of algal chlorophyll-a and water clarity in monsoon-region reservoir using machine learning approaches. Water (Switzerland) 12(1):30. https://doi.org/10.3390/w12010030

Menendez RGDP, Andino SG, Lantz G, Michel CM, Landis T (2001) Noninvasive localization of electromagnetic epileptic activity. I. Method descriptions and simulations. Brain Topogr 14:131–137

Mitchell MW (2011) Bias of the random forest out-of-bag (OOB) error for certain input parameters. Open J Stat 01:205–211. https://doi.org/10.4236/ojs.2011.13024

Mowe MAD, Mitrovic SM, Lim RP, Furey A, Yeo DCJ (2007) Tropical cyanobacteria blooms: a review. https://doi.org/10.4081/jlimnol.2014

Muduli PR, Pattnaik AK (2020) Spatio-temporal variation in physicochemical parameters of water in the Chilika lagoon. In: Finlayson C, Rastogi G, Mishra D, Pattnaik A (eds) Ecology, conservation, and restoration of Chilika lagoon, India. Wetlands: ecology, conservation and management, vol 6. Springer, Cham. https://doi.org/10.1007/978-3-030-33424-6_9

Muduli PR, Barik M, Acharya P, Sahoo I (2022) Variability of nutrient and their stoichiometry in Chilika lagoon India. In: Coastal ecosystems, pp 139–173. https://doi.org/10.1007/978-3-030-84255-0_7

Muduli PR, Barik SK, Mahapatro D, Samal R, Rastogi G, Tripathy M, Bhatt K, Pattnaik A (2017) The impact of tropical cyclone 'Phailin' on the hydrology of Chilika lagoon, India. Int J Environ Sci Nat Res 4(2). https://doi.org/10.19080/IJESNR.2017.04.555632

Mulia IE, Tay H, Roopsekhar K, Tkalich P (2013) Hybrid ANN-GA model for predicting turbidity and chlorophyll-a concentrations. J Hydro-Environ Res 7:279–299. https://doi.org/10.1016/j.jher.2013.04.003

Motoda H, Liu H (2002) Feature selection, extraction and construction. Commun IICM 5:67–72

Napiórkowska-Krzebietke A, Kalinowska K, Bogacka-Kapusta E, Stawecki K, Traczuk P (2020) Cyanobacterial blooms and zooplankton structure in lake ecosystem under limited human impact. Water (Switzerland) 12(5). https://doi.org/10.3390/W12051252

Nayak BK, Acharya BC, Panda UC, Nayak BB, Acharya SK (2004) Variation of water quality in Chilika lake, Orissa. Indian J Mar Sci 33(2):164–169

Negnevitsky M (2011) Artificial intelligence: a guide to intelligent systems, 4th edn. Addison-Wesley

Neill SP, Hashemi MR (2018) Fundamentals of ocean renewable energy: generating electricity from the sea. Academic Press

Ooi K-B, Tan G (2016) Mobile technology acceptance model: an investigation using mobile users to explore smartphone credit card. Expert Syst Appl 59. https://doi.org/10.1016/j.eswa.2016.04.015

Oyebode O, Stretch D (2019) Neural network modeling of hydrological systems: a review of implementation techniques. Nat Resour Model 32(1):e12189. https://doi.org/10.1111/nrm.12189

Paerl HW (2006) Assessing and managing nutrient-enhanced eutrophication in estuarine and coastal waters: interactive effects of human and climatic perturbations. Ecol Eng 26(1):40–54. https://doi.org/10.1016/j.ecoleng.2005.09.006

Park Y, Cho KH, Park J et al (2015) Development of early-warning protocol for predicting chlorophyll-a concentration using machine learning models in freshwater and estuarine reservoirs, Korea. Sci Total Environ 502:31–41. https://doi.org/10.1016/j.scitotenv.2014.09.005

Panigrahi S, Acharya BC, Panigrahy RC, Nayak BK, Banarjee K, Sarkar SK (2007) Anthropogenic impact on water quality of Chilika lagoon RAMSAR site: a statistical approach. Wetl Ecol Manag 15(2):113–126. https://doi.org/10.1007/s11273-006-9017-3

Patra JK, Das G, Paramithiotis S, Shin HS (2016) Kimchi and other widely consumed traditional fermented foods of Korea: a review. Front Microbiol 7:1493

Peetabas N, Panda RP (2015) Conservation and management of bioresources of Chilika Lake, Odisha, India. Int J Sci Res Publ 5(7):1–4

Phillips G, Pietiläinen OP, Carvalho L et al (2008) Chlorophyll-nutrient relationships of different lake types using a large European dataset. Aquat Ecol 42:213–226. https://doi.org/10.1007/s10452-008-9180-0

Sahu BK, Pati P, Panigrahy RC (2014) Environmental conditions of Chilika Lake during pre and post hydrological intervention: an overview. J Coast Conserv 18(3):285–297. https://doi.org/10.1007/s11852-014-0318-z

Singh KP, Basant A, Malik A, Jain G (2009) Artificial neural network modeling of the river water quality—a case study. Ecol Model 220(6):888–895. https://doi.org/10.1016/j.ecolmodel.2009.01.004

Smith JL, Boyer GL, Zimba PV (2008) A review of cyanobacterial odorous and bioactive metabolites: impacts and management alternatives in aquaculture. Aquaculture 280(1–4):5–20. https://doi.org/10.1016/j.aquaculture.2008.05.007

Smith JA, Jarman M, Osborn M (1999) Doing interpretative phenomenological analysis. In: Murray M, Chamberlain K (eds) Qualitative health psychology: theories and methods, Sage, London, pp 218–241. https://doi.org/10.4135/9781446217870.n14

Smith VH, Joye SB, Howarth RW (2006) Eutrophication of freshwater and marine ecosystems. Limnol Oceanogr 51(1 Part 2):351–355

Srisuksomwong P, Pekkoh J (2020) Artificial neural network model to prediction of eutrophication and Microcystis aeruginosa bloom. Emerg Sci J 4(2):129–135. https://doi.org/10.28991/esj-2020-01217

Stefanidis K, Papastergiadou E (2019) Linkages between macrophyte functional traits and water quality: insights from a study in freshwater lakes of Greece. Water (Switzerland) 11(5):1047. https://doi.org/10.3390/w11051047

Strobl RO, Forte F, Pennetta L (2007) Application of artificial neural networks for classifying lake eutrophication status. Lakes Reser Res Manag 12(1):15–25. https://doi.org/10.1111/j.1440-1770.2007.00317.x

Teles LO, Vasconcelos V, Pereira E, Saker M, Vasconcelos V (2006) Time series forecasting of cyanobacteria blooms in the Crestuma Reservoir (Douro River, Portugal) using artificial neural networks. Environ Manag 38(2):227–237. https://doi.org/10.1007/s00267-005-0074-9

Tian Z, Gu B, Yang L, Lu Y (2015) Hybrid ANN-PLS approach to scroll compressor thermodynamic performance prediction. Appl Therm Eng 77:113–120. https://doi.org/10.1016/j.applthermaleng.2014.12.023

Verstijnen YJM, Maliaka V, Catsadorakis G, Lürling M, Smolders AJP (2021) Colonial nesting waterbirds as vectors of nutrients to Lake Lesser Prespa (Greece). Inland Waters 11(2):191–207. https://doi.org/10.1080/20442041.2020.1869491
Viaroli P, Bartoli M, Giordani G, Naldi M, Orfanidis S, Zaldivar JM (2008) Community shifts, alternative stable states, biogeochemical controls and feedbacks in eutrophic coastal lagoons: a brief overview. Aquat Conserv Mar Freshwat Ecosyst 18(S1):S105–S117
Wang L, Wang X, Jin X, Xu J, Zhang H, Yu J, Sun Q, Gao C, Wang L (2017) Analysis of algae growth mechanism and water bloom prediction under the effect of multi-affecting factor. Saudi J Biol Sci 24(3):556–562. https://doi.org/10.1016/j.sjbs.2017.01.026
Wei B, Sugiura N, Maekawa T (2001) Use of artificial neural network in the prediction of algal blooms. Water Res 35(8):2022–2028. https://doi.org/10.1016/S0043-1354(00)00464-4
Were K, Bui DT, Dick ØB, Singh BR (2015) A comparative assessment of support vector regression, artificial neural networks, and random forests for predicting and mapping soil organic carbon stocks across an Afromontane landscape. Ecol Indic 52:394–403. https://doi.org/10.1016/j.ecolind.2014.12.028
Young WA, Millie DF, Weckman GR, Anderson JS, Klarer DM, Fahnenstiel GL (2011) Modeling net ecosystem metabolism with an artificial neural network and Bayesian belief network. Environ Model Softw 26(10):1199–1210. https://doi.org/10.1016/j.envsoft.2011.04.004
Zhang Q, Stanley SJ (1997) Forecasting raw-water quality parameters for the north Saskatchewan river by neural network modeling. Water Res 31(9):2340–2350. https://doi.org/10.1016/S0043-1354(97)00072-9

MIX
Papier aus verantwortungsvollen Quellen
Paper from responsible sources
FSC® C105338

If you have any concerns about our products,
you can contact us on
ProductSafety@springernature.com

In case Publisher is established outside the EU,
the EU authorized representative is:
Springer Nature Customer Service Center GmbH
Europaplatz 3, 69115 Heidelberg, Germany

Printed by Libri Plureos GmbH
in Hamburg, Germany